Nature-Based Wastewater Treatment Systems

Giving an account of successfully applied and recently developed green remediation technologies for water pollution control, this book describes the scope and applications of nature-based wastewater treatment technologies for environmental sustainability. The major focus is on associated eco-environmental concerns, recent technological developments, field studies, lessons learned, sustainability concerns, and future challenges. It also deals with the development of valuable bioresources together with wastewater treatment for the circular economy.

This book:

- Covers nature-based wastewater treatment systems for the efficient management of wastewater for the protection of precious water resources.
- Includes development and utilization of useful bioresources, bioenergy, and value-added products together with wastewater treatment for the circular economy.
- Discusses technological aspects such as design, operation, and maintenance, eco-friendliness, effectiveness, and sustainability concerns.
- Highlights technological advancements, field experiences, research gaps, recent developments, challenges, and future directions for further improvements.
- Reviews field studies and challenges between pollution sources, exposure pathways, and impacts on environmental quality and human health.

This book is aimed at graduate students and researchers in environmental engineering and sciences, environmental microbiology, and biotechnology.

Nature-Based Wastewater Treatment Systems

Emerging Approaches with Potential Resource Recovery Options

Edited by
Adarsh Kumar, Saroj Kumar, and Sheel Ratna

CRC Press is an imprint of the
Taylor & Francis Group, an **informa** business

Designed cover image: Shutterstock

First edition published 2025
by CRC Press
2385 NW Executive Center Drive, Suite 320, Boca Raton FL 33431

and by CRC Press
4 Park Square, Milton Park, Abingdon, Oxon, OX14 4RN

CRC Press is an imprint of Taylor & Francis Group, LLC

Library of Congress Cataloging-in-Publication Data
Names: Kumar, Adarsh, (Environmental microbiologist), editor. |
Kumar, Saroj, editor. | Ratna, Sheel, editor.
Title: Nature-based wastewater treatment systems : emerging approaches with
potential resource recovery options / edited by Adarsh Kumar, Saroj
Kumar and Sheel Ratna.
Description: First edition. | Boca Raton, FL : CRC Press, 2025. |
Includes bibliographical references and index. |
Identifiers: LCCN 2024018227 (print) | LCCN 2024018228 (ebook) |
ISBN 9781032450216 (hbk) | ISBN 9781032578163 (pbk) | ISBN 9781003441144 (ebk)
Subjects: LCSH: Phytoremediation. | Sewage sludge precipants—Recycling. |
Sewage—Purification—Technological innovations. | Sustainable
engineering—Technological innovations.
Classification: LCC TD755 .N44 2025 (print) | LCC TD755 (ebook) |
DDC 628.4—dc23/eng/20240701
LC record available at https://lccn.loc.gov/2024018227
LC ebook record available at https://lccn.loc.gov/2024018228

ISBN: 9781032450216 (hbk)
ISBN: 9781032578163 (pbk)
ISBN: 9781003441144 (ebk)

DOI: 10.1201/9781003441144

Typeset in Times
by codeMantra

This book is truly dedicated to our families for their selflessness, unconditional love, support, patience, forgiveness, understanding, educating us to date and being a hope to move forward in life. Without them, this book would not have been possible.

Contents

Preface

As explained in SDG-6 (Sustainable Development Goal-6), providing access to safe and adequate water to everyone, and protecting and restoring aquatic ecosystems for future needs is one of the prime responsibilities. However, about 80% of wastewater is discharged back without or after partial treatment worldwide, with serious environmental and public health consequences. Freshwater pollution is a serious concern globally nowadays. There is a public outcry against the continually rising problems of pollution to ensure a safe and healthy environment for all living creatures. A variety of organic and inorganic contaminants causes serious water pollution and toxicity to aquatic ecosystems. Several external pressures, such as rapid population growth, climate change, urbanization, and industrial establishment, are imposing significant concerns on sanitation amenities. As a result, we need an environmentally friendly sanitation approach that allows the treatment of wastewater while sustaining ecosystem services. There is an increasing trend toward the removal of these contaminants from wastewater through biological processes, particularly via nature-based treatment systems (NBTS) to protect freshwater ecosystems and public health. Being an eco-friendly technology, NBTS can solve the problem of freshwater pollution. It is a low-cost and self-driven process that utilizes microorganisms and plants or their enzymes to degrade and detoxify wastewater contaminants, thus promoting sustainable development. This includes connecting state-of-the-art technologies, particularly nature-based technologies. NBTS has long been utilized to treat several types of wastewater, extending back to the use of natural wetlands and ponds for wastewater discharge in ancient times. NBTS for wastewater treatment comprises natural remediation technologies, phytoremediation using native plants, ponds and soil infiltration, hydroponics and aquaponics, etc. Recently, there has been rising concern and the importance of NBTS as an alternative to traditional wastewater treatment systems. For example, treatment wetlands and stabilization ponds are NBS often used in decentralized wastewater treatment systems. They are probably a viable option for rural areas as well as urban and peri-urban areas that do not have access to combined centralized systems. This publication *Nature-Based Wastewater Treatment System* was developed as a response to the need for consolidated evidence based on the use of nature-based technologies for improved sanitation, with an emphasis on the co-benefits that these technologies can provide to both people and ecosystems. Additional benefits of NBTS use as part of wastewater systems include temperature regulation, carbon sequestration, groundwater recharge, flood control, production of biomass, providing habitat for plants and animals, and adding aesthetic values and recreation areas. Therefore, developing an understanding of these crucial benefits associated with NBTS through this book, *Nature-Based Wastewater Treatment Systems: Emerging Approaches with Potential Resource Recovery Options* can help municipalities, students, researchers, scientists, and professionals working in the field of biological wastewater treatment, water resource management and protecting the environment, public health sectors, microbiology, biotechnology, environmental sciences, and eco-toxicology, who aspire to work on the clean bioremediation technologies to protect the environment and public health.

This book describes the scope and applications of nature-based wastewater treatment technologies for environmental sustainability. The major focus of this book is on the associated eco-environmental concerns, recent technological developments, field studies, lessons learned, sustainability concerns, and future challenges. This book also deals with the development of valuable bioresources together with wastewater treatment for the circular economy. To address the problems of previous books in the proposed title, this book "provides readers with (i) all-inclusive knowledge on the pollution and toxicity profile of organic and inorganic contaminants; (ii) timely subject knowledge on the latest topics in the field of biological wastewater treatment; (iii) authoritative contributions on various aspects of eco-friendly wastewater treatment technologies from leading experts from around the world; (iv) up-to-date assessment of the field applications of NBWTS technologies;

(v) comprehensive review of current developments and future insights into wastewater treatment technologies; (vi) all-inclusive knowledge on the merits, demerits, sustainability concerns and opportunities in NBWTS technologies; (vii) review the knowledge gaps and future directions for the effective implementation of NBWTS on the field scale; (viii) technological advancements and lessons learned from the past field applications.

Overall, this book offers inclusive knowledge on emerging clean and green wastewater treatment technologies to lessen the dependency on freshwater resources and environmental resilience. At the individual level, this book will be helpful for laymen to protect our natural environment and health hazards by managing such pollutants through various eco-friendly strategies in a sustainable and effective manner.

Adarsh Kumar, Saroj Kumar, and Sheel Ratna
Editors

About the Editors

Adarsh Kumar is engaged in understanding the role and mechanism of bacterial ligninolytic enzymes (Laccase, Lignin peroxidase, and Manganese peroxidase) in the degradation and detoxification of residual organic and inorganic pollutants from industrial wastewater and value-added product recovery. He has qualified National Eligibility Test (NET) for lectureship in 2016. Currently, he is working with the District Environment Committee, Pilibhit, UP, framed by the Department of Environment, Forest and Climate Change, Government of Uttar Pradesh, India, and awarded as a certified Miyawaki Plantation Professional. He has published many scientific original research papers, book chapters, and proceedings magazines, as popular science articles in several peer-reviewed international and national journals. He has presented some papers at national and international conferences in his field. He has also served as a potential reviewer for various scientific journals in his research areas. He is a life member of the Association of Microbiologists of India (AMI), Indian Science Congress Association (ISCA), and International Society of Environmental Relationship and Sustainability (ISERS), India.

Saroj Kumar is currently working on nature-based wastewater treatment technologies for sustainable wastewater management and resource recovery for the circular economy. The major research focus of Dr. Kumar is on the management of water resources, especially through constructed wetlands technology. He has qualified National Eligibility Test (NET) for lectureship and Junior Research Fellowship in the years 2015 and 2016 in Environmental Science, conducted by the Indian Council of Agricultural Research (ICAR, India) and University Grants Commission (UGC, India), respectively. He has published several articles in high-impact, peer-reviewed international journals, chapters with international publishers, and popular science articles in magazines. He has also presented several papers at national and international conferences, events, and training programs in India. He is an editorial member of the *Journal of Plant* published by Science Publishing Group, USA and acts as a reviewer in several international journals. He is an active member of the Indian Science Congress Association (ISCA) and the Institute of Scholars (InSc), India. Dr. Kumar also received the Research Excellence Award from the Institute of Scholars India. Currently, he is working with the District Ganga Committee, Lakhimpur Kheri, under the administrative supervision of State Mission for Clean Ganga, Uttar Pradesh and National Mission for Clean Ganga, Ministry of Jal Shakti, Government of India. He is recognized as a certified Miyawaki Plantation Professional in India.

Sheel Ratna is currently working on wastewater treatment through bacterial-assisted constructed wetlands, adsorbent columns, and the production of value-added products utilizing wastewater as a substrate. He has qualified for several national-level exams such as ICAR-NET (2016) in Environmental Science, ICAR-NET (2017) in Agricultural Microbiology, UGC-NET/JRF (2017) in Environmental Science, and GATE-(2019) in Life Science. He has published several research articles in peer-reviewed international journals with high repute, and chapters with international publishers. He has also presented several papers at national and international conferences, and events in India. He is an active member of the Indian Science Congress Association (ISCA) and the Association of Microbiologists (AMI) of India.

Contributors

Kumar Abhishek
Department of Environment Forest and Climate Change
Government of Bihar
Patna, India

Priya Agarwal
Department of Civil Engineering
Sharda University
Greater Noida, India

K. Angappan
Department of Plant Pathology
Tamil Nadu Agricultural University
Coimbatore, India

Anusha Atmakuri
INRS Eau, Terre et Environnement
Québec, Canada

A. Bharani
Department of Environmental Sciences
Tamil Nadu Agricultural University
Coimbatore, India

Pawan Kumar Bhargawa
Rhizosphere Biology Laboratory, Department of Environmental Microbiology, School of Earth and Environmental Sciences
Babasaheb Bhimrao Ambedkar University (A Central University)
Lucknow, India

Sayan Bhattacharya
School of Ecology and Environment Studies
Nalanda University
Rajgir, India

Naveen Chand
Environment and Biofuel Research Laboratory, Department of Hydro and Renewable Energy (HRED)
Indian Institute of Technology Roorkee
Uttarakhand, India

Subhash Chandra
Department of Chemistry
B.S.N.V. P.G. College (University of Lucknow), Near Charbagh Station Road
Lucknow, India

Shraddha Chavan
INRS Eau, Terre et Environnement
Québec, Canada

Yasmin Cherni
Carthage University, Laboratory of Wastewater Treatment and Valorization
Water Research and Technologies Center (CERTE), Technopark Tourist Route of Soliman Nabeul
Soliman, Tunisia

Suchismita Das
Department of Life Science & Bioinformatics
Assam University
Silchar, India

V. Davamani
Tamil Nadu Agricultural University
Coimbatore, India

Nabarupa Dhar
Department of Commerce
Cachar College
Silchar, India

J. Dharani
Tamil Nadu Agricultural University
Coimbatore, India

Periyasamy Dhevagi
Department of Environmental Sciences
Tamil Nadu Agricultural University, Coimbatore, India

Hema Diwan
Indian Institute of Management Mumbai
Powai, India

Patrick Drogui
INRS Eau, Terre et Environnement
Québec, Canada

Priya Dubey
Department of Biosciences, Faculty of Science
Integral University
Lucknow, India

Venkatesh Dutta
Department of Environmental Science (DES), School of Earth and Environmental Sciences (SEES)
Babasaheb Bhimrao Ambedkar (A Central) University
Lucknow, India

S.K. Dwivedi
Department of Environmental Science, School of Earth and Environmental Sciences (SEES)
Babasaheb Bhimrao Ambedkar University (A Central University)
Uttar Pradesh, India

J. Ezra John
Department of Environmental Sciences
AC&RI, Tamil Nadu Agricultural University
Coimbatore, India

Supratim Ghosh
Department of Biological Sciences
BITS Pilani Hyderabad Campus
Hyderabad, India

Thangaraj Gokul Kannan
Department of Environmental Sciences
Tamil Nadu Agricultural University
Coimbatore, India

Humberto Raymundo Gonzalez Moreno
Wetlands and Environmental Sustainability Laboratory, Division of Graduate Studies and Research, Tecnológico Nacional de México/Instituto Tecnológico Superior de Misantla
Veracruz, México

Gareth Griffiths
Energy and Bioproducts Research Institute (EBRI)
Aston University
Birmingham, United Kingdom

Ekta Gupta
Institute of Environment and Sustainable Development
Banaras Hindu University
Varanasi, India

Namita Gupta
Department of Environmental Science
Babasaheb Bhimrao Ambedkar University
Uttar Pradesh, India

Vartika Gupta
Department of Botany (Environmental Sciences)
University of Lucknow
Lucknow, India
and
Environment Technologies Division
CSIR-National Botanical Research Institute
Lucknow, India

Lidia Elena Requena Hernández
International Trade Operations Area Tariff Classification and Customs Clearance, Universidad Tecnológica de Matamoros
Tamaulipas, México

Athar Hussain
Professor, Civil Engineering Department
Netaji Subhas University of Technology, West Campus
New Delhi, India

R. Jayashree
Department of Environmental Sciences
Tamil Nadu Agricultural University
Coimbatore, India

Nuno Jorge
Centro de Química de Vila Real (CQVR), Departamento de Química
Universidade de Trás-os-Montes e Alto Douro (UTAD)
Vila Real, Portugal (M.S.)

P. Kalaiselvi
Tamil Nadu Agricultural University
Coimbatore, India

Ganesan Karthikeyan
Department of Environmental Sciences
Tamil Nadu Agricultural University
Coimbatore, India

Raveendra Gnana Keerthi Sahasa
Department of Environmental Sciences
Tamil Nadu Agricultural University
Coimbatore, India

Adarsh Kumar
District Environment Committee, Department of Environmental, Forest and Climate Change, Pilibhit
Uttar Pradesh, India

Binilkumar Amarayil Sreeraman
Indian Institute of Management Mumbai
Powai, India

Manoj Kumar
Environment and Biofuel Research Laboratory, Department of Hydro and Renewable Energy (HRED)
Indian Institute of Technology Roorkee
Uttarakhand, India

Pavan Kumar
Department of Biotechnology
Bundelkhand University
Jhansi, India
and
ICAR-Indian Institute of Sugarcane (ICAR-IISR)
Lucknow, India

Rajesh Kumar
Rhizosphere Biology Laboratory, Department of Microbiology, School of Earth & Environmental Sciences
Babasaheb Bhimrao Ambedkar University (A Central University)
Lucknow, India

Saroj Kumar
District Ganga Committee, National Mission For Clean Ganga, Ministry of Jal Shakti
Lakhimpur Kheri, India

Ratna Dewi Kusumaningtyas
Chemical Engineering Department, Faculty of Engineering
Universitas Negeri Semarang
Semarang, Indonesia

T. Ilakiya
Tamil Nadu Agricultural University
Coimbatore, India

Marco S. Lucas
Centro de Química de Vila Real (CQVR), Departamento de Química, Universidade de Trás-os-Montes e Alto Douro (UTAD)
Vila Real, Portugal (M.S.)

Danielle Maass
Instituto de Ciência e Tecnologia (ICT)
Universidade Federal de São Paulo(UNIFESP)
São José dos Campos, SP, Brazil

Richa Madan
PhD Research Scholar, Department of Environmental Science
Gurukula Kangri (Deemed to be University)
Haridwar, India

M. Margunani
Faculty of Economy, Universitas Negeri Semarang
Semarang, Indonesia

Georgina Martínez-Reséndiz
Wetlands and Environmental Sustainability Laboratory, Division of Graduate Studies and Research, Tecnológico Nacional de México/ Instituto Tecnológico Superior de Misantla
Veracruz, México
and
Postdoctoral position by CONACYT (Consejo Nacional de Ciencia y Tecnología) in Tecnológico Nacional de México Campus Misantla
Veracruz, Mexico

Nidhi Mourya
School of Ecology and Environment Studies
Nalanda University
Bihar, India

Kelvin Mutugi Kithaka
School of Ecology and Environment Studies
Nalanda University
Bihar, India

Sampurna Nand
Environment Technologies Division
CSIR-National Botanical Research Institute
Lucknow, India
and
Department of Environmental Sciences
Dr. Ram Manohar Lohia Avadh University
Faizabad, India

S. Nikil
Department of Physics
PSG College of Arts and Science
Tamil Nadu, India

Subramanian Nithiyanantham
PG and Research Department of Physics
Thiru. Vi. Kalyanansundaram Govt Arts and Science College
Thiruvarur, India

A. Nivetha
Department of Chemistry
Bharathiar University
Tamil Nadu, India

E. Parameswari
Tamil Nadu Agricultural University
Coimbatore, India

Anju Patel
Environment Technologies Division
CSIR-National Botanical Research Institute
Lucknow, India
and
Plant Ecology and Environment Technologies
CSIR- National Botanical Research Institute
Lucknow, India

José A. Peres
Centro de Química de Vila Real (CQVR), Departamento de Química
Universidade de Trás-os-Montes e Alto Douro (UTAD)
Quinta de Prados, Portugal (M.S.)

C. Poornachandhra
Department of Environmental Sciences
Tamil Nadu Agricultural University, Coimbatore, India

Ramesh Poornima
Department of Environmental Sciences
Tamil Nadu Agricultural University
Coimbatore, India

Sanjeev Kumar Prajapati
Environment and Biofuel Research Laboratory, Department of Hydro and Renewable Energy (HRED)
Indian Institute of Technology Roorkee
Roorkee, India

M. Prasanthrajan
Department of Environmental Sciences
AC&RI, Tamil Nadu Agricultural University
Coimbatore, India.

Haniif Prasetiawan
Chemical Engineering Department, Faculty of Engineering
Universitas Negeri Semarang
Semarang, Indonesia

Prashant
Department of Environmental Science
Central University of South Bihar
Gaya, India

Priyanka
Department of Environmental Science, School of Earth and Environmental Sciences (SEES)
Babasaheb Bhimrao Ambedkar University (A Central University)
Lucknow, India

S.S. Rakesh
Department of Environmental Sciences
AC&RI, Tamil Nadu Agricultural University
Coimbatore, India

Sundarajayanthan Ramakrishnan
Department of Environmental Sciences
Tamil Nadu Agricultural University
Coimbatore, India

Sheel Ratna
Rhizosphere Biology Laboratory, Department of Microbiology, School of Earth & Environmental Sciences
Babasaheb Bhimrao Ambedkar University (A Central University) Vidya Vihar Raebareli Road
Lucknow, India
and
Department of Environmental, Forest and Climate Change
Ballia, India

Amin Retnoningsih
Biology Department, Faculty of Mathematics and Natural Sciences
Universitas Negeri Semarang
Semarang, Indonesia

Alan Antonio Rico Barragan
Department of Environmental Engineering, Tecnológico Nacional de México
Veracruz, México

Jesus Castellanos Rivera
Wetland and Environmental Sustainability Laboratory, Division of Postgraduate Studies and Research, Tecnológico Nacional de México/Instituto Tecnológico de Misantla
Veracruz, Mexico

Aditi Roy
Department of Botany (Environmental Science)
University of Lucknow
Lucknow, India

Gaurav Saini
Department of Civil Engineering
Netaji Subhas University of Technology
Delhi, India

C. Sakthivel
Department of Chemistry
Bharathiar University
Coimbatore, India

Luis Carlos Sandoval Herazo
Wetland and Environmental Sustainability Laboratory, Division of Postgraduate Studies and Research, Tecnológico Nacional de México/Instituto Tecnológico de Misantla
Veracruz, Mexico

M. Saratha
Department of Sericulture
Tamil Nadu Agricultural University
Coimbatore, India

S. Paul Sebastian
Tamil Nadu Agricultural University
Coimbatore, India

Jayaraman Sethuraman Sudarsan
School of Energy and Environment
NICMAR University (National Institute of Construction Management and Research)
Maharashtra, India

P.A. Shahidha
Department of Environmental Sciences
Tamil Nadu Agricultural University
Coimbatore, India

Parag Shil
Department of Commerce
Assam University
Assam, India

Siddharth Shukla
Department of Environmental Sciences
Dr. Ram Manohar Lohia Avadh University
Uttar Pradesh, India

Monika Singh
Institute of Environment & Sustainable Development
Banaras Hindu University
Uttar Pradesh, India

Pankaj Kumar Srivastava
Environment Technologies Division
CSIR-National Botanical Research Institute
Uttar Pradesh, India
Plant Ecology and Environment Technologies
CSIR-National Botanical Research Institute
Uttar Pradesh, India

Suchi Srivastava
Plant Ecology and Environment Technologies, CSIR- National Botanical Research Institute
Lucknow, India

R. Sunitha
Assistant Professor, Controllerate of Examinations
Tamil Nadu Agricultural University
Coimbatore, India

Ana R. Teixeira
Centro de Química de Vila Real (CQVR), Departamento de Química
Universidade de Trás-os-Montes e Alto Douro (UTAD)
Vila Real, Portugal (M.S.)

R.D. Tyagi
BOSK-Bioproducts
Québec (QC), Canada
and
Institute of Bioresource and Agriculture, Sino-Forest Applied Research Centre for Pearl River Delta Environment and Department of Biology
Hong Kong Baptist University
Hong Kong
and
School of Technology
Huzhou University
Huzhou, China

Anupriya Verma
Department of Civil Engineering
Netaji Subhas University of Technology
Delhi, India

Mohamed Ali Wahab
Carthage University, Laboratory of Wastewater Treatment and Valorization
Water Research and Technologies Center (CERTE)
Technopark Tourist Route of Soliman, Soliman, Tunisia

Dwi Widjanarko
Mechanical Engineering Department, Faculty of Engineering
Universitas Negeri Semarang
Semarang, Indonesia

Sucihatiningsih Dian Wisika Prajanti
Faculty of Economy
Universitas Negeri Semarang
Semarang, Indonesia

Bhoomika Yadav
INRS Eau, Terre et Environnement
Québec, Canada

Gautam Yadav
Indian Institute of Management Mumbai
Powai, India

Conrado Planas Zanutto
IPC - Institute for Polymers and Composites
University of Minho
Braga, Portugal

Talita Corrêa Nazareth Zanutto
Department of Chemical Engineering and Food Engineering (EQA)
Federal University of Santa Catarina (UFSC)
Florianópolis, SC, Brazil

Florentina Zurita
Environmental Quality Research Center, Centro Universitario de la Ciénega
University of Guadalajara
Jalisco, Mexico

Acknowledgments

The editors would like to express their profound feeling of reverence and extend heartfelt gratitude to the Divisional Forest Offices, Social Forestry Division, Pilibhit and South Kheri Forest Division, Lakhimpur Kheri, Department of Environment Forest and Climate Change, Government of Uttar Pradesh, India and Department of Environmental Microbiology, Babasaheb Bhimrao Ambedkar (A Central), University, Lucknow UP, India for providing the necessary infrastructure throughout the completion of this book.

We also gratefully acknowledge and extend our sincere thank you to all the contributors for their valuable contributions.

1 Water Present and Future
Prospects for Overcoming Scarcity

R. Jayashree, S. Nikil, A. Bharani, P.A. Shahidha, and R. Sunitha

1.1 INTRODUCTION

The driest areas of both China and India are where they cultivate their most water-intensive crops. Most of the globe won't have enough water to cover basic needs year-round by 2040 if this trend continues. The blue planet Earth has an abundance of water. The amount we have is more than 300 million trillion gallons. Although water can escape as vapour into the air or turn into ice, it never leaves the earth. But since 2% of that water is frozen at the poles and 97% of it is salty, all of humanity depends on just 1% of that water to thrive.

The sustainability of the natural resource base is threatened by water shortage, which has an impact on all social and economic sectors. When managing water resources, it is necessary to use an intersectional and interdisciplinary approach for the sake of optimising economic and public welfare in an unbiased manner without jeopardising the sustainability of crucial environs. Here arises the necessity to integrate several sectors. This integration must focus on people, their way of life, and the ecosystems that support them while taking development, supply, usage, and demand into consideration. For programmes to alleviate water shortages to be successful on the demand side, water production must be increased across all sectors. In addition, maintaining and restoring resources like rivers, wetlands, forests, and soils for purification and acquiring water from them is essential for boosting the supply of excellent-quality water. The globe may or may not be experiencing a water crisis right now, according to experts. The collection of reports from throughout the world can be used as evidence by those who hold this belief. The idea that a worldwide crisis need not result from a point source with broad effects but rather that a calamity can be composed of several identical episodes throughout the world, even if the incidents are unrelated to one another, is indisputable.

1.2 PERPETUAL CYCLE OF WATER SCARCITY AND POVERTY – A THREAT TO LIFE

Water shortage is mostly a muddle of poverty. The fate of the global poor is unclean water and inadequate sanitation. Poor children and families are the first to suffer from lack of cleanliness, while the rest of the world's residents gain from easy ingress to the water they require for household purpose. While the typical water usage in Europe and the US is about 500 litres per day, one in five people in the developing world lacks access to enough clean water. Furthermore, the poor pay more. According to a recent UNDP research, persons who have access to piped water often spend five to ten times as much per unit of water as those who live in developing country slums (UNDP, 2006). Water shortage does not just refer to dry spells or dried-up rivers for the poor. Primarily, it is about securing that people have the reliable and assured approach they require to safeguard their livelihoods. For the underprivileged, crisis is about how conservatories work and how equality and openness are ensured in choices that have an impact on their life. It concerns decisions on infrastructure development and how those decisions are handled. Organisations struggle to allocate resources fairly all around the globe. Water is essential for each and every creatures for the existence of life. While the Millennium Development Goals has designated an approach to safe water and sanitation

DOI: 10.1201/9781003441144-1

as important priorities, it is becoming increasingly clear that these measures alone are insufficient. Water is indispensable to the daily income or food production of millions of people. Water is essential for everyone who depends on it for a living, including farmers, small rural businesses, herders, and fishermen. However, as resources become more limited, a growing percentage of individuals experience the loss of their sources of income. People are most often impacted by water scarcity in rural regions. Irrigation still serves as the foundation of the rural economy in a significant portion of the developing globe. However, smallholder farmers, who frequently reside on marginal ground and rely mostly on rainfall for production, trump up the bulk of the world's poor. They are quite susceptible to many changes, including changes in market pricing as well as droughts and floods. However, plans for managing water resources seldom include precipitation since they typically only pay attention to surface and groundwater. To deal with water constraints, nations must properly incorporate rainwater into their policies.

1.3 DIVERSE ASPECTS OF WATER SCARCITY

Water is necessary for maintaining healthy ecosystems and for any socioeconomic growth. The demand for water resources increases as inhabitants grow and infrastructure asks for larger provisions of water for the household, agricultural, and business sectors. This leads to tensions, disputes between users, and undue pressure on the environment. It is extremely concerning how much pressure the world's freshwater supplies are under due to increased demand, wasteful use, and pollution. There are divergent ways of explaining the water crisis. Water stress is generally understood to be the condition in which the demands of all sectors, including the environs, cannot be fully met due to the successive impact of all users on the availability or quality of water.

The complexity of water poverty makes evaluation, challenging. It depends not just on the availability of water sources but also on how well-accessible clean, unpolluted water is to populations. It also depends on the resources that individuals who use the water need. To assess each home's subjective perception of welfare as a result of water poverty; we apply a Water Poverty Index technique employing Principal Component Analysis to build an index at the household level in ten villages within one big farming community (Liu et al., 2017). While the communities with greater access to fresh water had comparatively higher subjective well-being, the ones closest to contaminate water sources were unhappier. To reduce water poverty and improve subjective well-being in local communities, it is recommended that regional policies and environmental protection measures be implemented vigorously.

1.4 IS SHARING WATER A SOURCE OF DISPUTE OR A PROSPECT FOR COOPERATION?

The world is still consuming more water. The world's water supplies are always under stress as a result of causes including population expansion, industrialisation, increasing demand, pollution, and the effects of global warming: A rising number of locations have limited water resources. Every nation's economy depends on water to function and expand, raising everyone's level of life in the process. The social quality of life may be greatly impacted, leading to major intra-societal conflicts, if the supply of water is limited because of problems like misuse, pollution, or politics. Conflicts over allocation, such as those between various ethnic groups, industry and agriculture, or the urban and rural population, are one way these tensions express themselves. It depends on the degree of stress, the political structure in existence in a state, and any unique climatic or hydrological factors that may exist whether such water conflicts would result in violence (Fröhlich, 2012).

The Farakka salvo is designed to channel water from the Ganges into the Bhagirathi branch during the dry season in order to clear debris and improve Calcutta's navigability. Seventeen kilometres are spent upstream from Bangladesh's border. According to Bangladesh, this Indian diversion upstream would result in less water flowing into the Padma branch, which enters Bangladesh during the dry season. Bangladesh opposes this barrage because they believe that the decrease in flow during the dry season would lead the country's environment to be damaged (Crow and Singh, 2000). While India, the more powerful country, claimed that there wasn't enough information to finish the exercise, Bangladesh persisted in coming to a water-sharing arrangement. In spite of Bangladesh's constant protests, India started building the Farakka Barrage in 1960, sparking a violent conflict between the two countries that has persisted for decades. The second factor that fuels tension between the two nations is the disparity in approaches used to improve Ganges flow during its dry season so that both nations have access to adequate water during this period. India plans to build a major connecting canal to convey the excess water, in contrast to Bangladesh's proposal, which called for building storage facilities in Nepal. In Bangladesh, the Brahmaputra River meets the Ganga River. Each side disagreed with the other's suggestion. India vehemently opposes Bangladesh's proposal for augmentation because it doesn't want to involve other parties in the war and because it believes Nepal is unable to store the massive volumes of water that both India and Bangladesh want, contrary to Bangladesh's claims. However, Bangladesh rejected India's idea for a connecting canal because they believed it would split Bangladesh in half and subjugate one of its halves, which would be detrimental to Bangladesh in the long run (Sood and Mathukumalli, 2011). It is evident that due to their adamant disagreement on the concept of augmentation, India and Bangladesh have hit a deadlock in their discussion over how evenly the two riparian countries should split the Ganges River's benefits (Pham, 2018).

1.5 VULNERABILITY AND SUSTAINABILITY ISSUES PERTAINING TO WATER IN FRONTIERS

One of the most efficient ways to fight hunger and poverty is to increase rural areas' extent for production via agricultural prosperity since the majority of the world's impoverished live there. There are promising opportunities for increased output and diversification, albeit the future directions are challenging and unique in each place. In rural Africa, 80%–90% of households depend on raising and eating their own food. They more than anybody else experience the very real problem of water scarcity, which is made worse by persistent problems with desertification and soil erosion as well as difficulties obtaining inputs like as fertiliser and seed. Unforeseeable recurring rains and brief dry seasons of just a couple of days may regularly destroy agricultural output at critical times in the cropping cycle, leading to food shortages.

Due to the food, employment possibilities, and other alternate benefits it produces, irrigation offers a direct source of livelihood for thousands of millions of poor people in developing nations. Politicians hope that increased agricultural production, especially in Africa, would lift a huge number of people out of poverty by employing them as wage workers, producers of marketable items, or small-business owners away from farms. Rural farmers may produce their own food with the use of conservation agriculture or local methods for conserving land and water. These farming techniques make the most of the local farming conditions, account for the fact that the majority of poor farmers only possess a hand hoe, and provide some protection from the frequent water shortages. The ability of farmers to manage the little money produced inside of their own houses is crucial. It is important that farmers be given the chance to go past the first rung of the agricultural ladder and into the more commercialised subsistence farming types (Nikil and Bharani, 2022).

1.6 BUYING WATER, COLLECTING, STORING AND SECURING WATER THROUGH NONFINANCIAL MEANS

Water is a crucial component for all living creatures on the planet. Water scarcity is now a widespread issue across the world, impacting not just humans but also other animals. The fundamental reason for the rise in recent wild animal sightings in residential areas is that these creatures need water to thrive. However, there may be other factors as well. The main cause of this water shortage is humans. Therefore, it only makes sense that we begin conserving water right now and find efficient ways to do so in the future. Scientists are still working around the clock to find practical and economical ways to preserve and store water for the future. However, rainwater collection is an important method that is widely used.

The geography, temperature, geology, and seasonal rainfall brought on by the monsoons are the main factors affecting the accessibility of water supplies in India. The timing and quantity of rainfall, the storage and release of surface water, and the accessibility of groundwater all exhibit notable spatial differences as a result. Important cultural, social, political, and (increasingly) economic factors also affect how people obtain and use water for the environment, industry, agriculture, home use, and consumption. Rural residents' openness to secure portable water has historically been considered a social concern. Among rural India and other developing nations, the primary issue with access to drinking water has been identified as continuing to be women's labour. Some consequences of this labour were the loss of opportunities for employment and education, gender inequity, and women's empowerment. Due to this, access has come to be seen as the primary objective for rural drinking water during the last several decades. Even while the access gap is closing, the poorest people continue to suffer, which increases support for social programmes that prioritise universal access. When rural drinking water projects are undertaken in distant areas that are often sparsely populated by the poor, the economic viability gap widens; nevertheless, there are a few other considerations that further complicate things. First off, the amount of freshwater that is available for residential and drinking requirements is drastically decreasing due to environmental deterioration and rising abstraction for competing purposes. The lack of clarity on who is in charge of utilising and maintaining the water resources of the planet is another factor contributing to their dwindling availability.

1.7 RESOURCES FOR BIZARRE WATER AND CHANCES FOR WATER AUGMENTATION TO ATTAIN FOOD SECURITY IN REGIONS WITH LIMITED WATER RESOURCES

It is clear from the unique water cycle, which facilitates an annual rechargeable water reserve of around 7,000 m cube per capita (Shiklomanov, 2000), that there is always enough pure water to encounter the demands of the planet's current inhabitants. However, the yearly renewable water supply per capita in certain areas and nations is less than 500 m^3. Additionally, these regions' erratic water availability over time leads to severe occurrences. For instance, both floods and droughts commonly happen, sometimes in the same location or in nearby areas. In locations with greater than 40% of the world's residents, water demands already outstrip availability due to the uneven dissemination of freshwater resources and population density globally (Beitinger et al., 2000).

Water scarcity may affect up to 60% of the world's inhabitants by 2025, based on present statistics and anticipated future trends. When employing conventional resources, there have been improvements to the techniques for increasing water usage effectiveness. However, water-risk nations will need to depend more on the exploitation of bizarre water storages in order to slightly ease water lacunas. When used for irrigation, non-conventional water resources either come from specialist procedures like desalination or require the proper pre-treatment and soil-water or water-crop management practices. Desalination of saltwater and very brackish groundwater, rainfall collection, and irrigation using low-quality water sources are some ways that water-scarce regions might access these resources. Wastewater, agricultural runoff water and water-table aquifer with various salts are the sorts of low-quality fluids utilised for irrigation. In many underdeveloped nations, a sizable portion of the wastewater produced by the residential, commercial, and business sectors is utilised, either in its untreated or partially treated form, for crop cultivation. The primary issues with unregulated wastewater irrigation are the preservation of the ecosystem and public health. It is anticipated that groundwater and salty or sodic drainage water will be used more frequently in agriculture. This calls for adjustments to the current crop management, irrigation, and soil techniques for dealing with the upcoming escalation in salinity. Another alternative that could assist nations in arid regions to attain food security is the "physical" movement of food and water between basins, nations, and regions. A definition of "physical" transportation excludes pressure movement networks like rivers and canals. Examples of "physical" transportation include drawing pure water from surface water networks and transporting it through vast tubes or pipes by the ocean in enormous containers (Ariyoruk, 2003). Almost all water transit ideas are currently in the development stage. Virtual water is a term that has been used to emphasise the significance of water in the transfer of food among countries that have an abundance of water and those that do not (Allan, 2003).

1.7.1 Water Availability – Worldwide

Water makes up around 70% of the earth's surface, or 1,400 million cubic kilometres. This water is salty, though, as 97.5% of it is seawater. Only 1% of the entire water can be used by human beings since all the clear water cannot be extracted.

Water has long been seen as a free resource since it is so readily available. A reliable supply of high-quality water is an increasing worry, nevertheless, as a consequence of rising water deficit and water shortages. Because of the improper distribution of water resources throughout the continents, some nations have an abundance of water while many others are experiencing a water shortage. Similar to this, population increase on different continents is uneven, creating a significant mismatch between the current population and water supplies. With more than 60% of creatures living in areas where water is scarce, Asia comprises 36% of the global fresh water reservoirs (Table 1.1).

Despite having a reasonable distribution of rainfall, the country currently struggles to effectively utilise rainwater due to a lack of understanding and inadequate infrastructure for building dams and reservoirs. By 2050, there will be 27 million ha using canal irrigation, up from 17 million ha in 2000. By connecting the reservoirs and utilising 36 billion m^3 of manmade groundwater recharge, there is still room to raise the potential by 35 million ha (Government of India, 2009). Due to the following factors, India's need for water is rapidly growing (Amarasinghe & Williams, 2007):

- The urbanisation process will also have a significant effect on development. Urban regions were home to 28.2% of Indians in 2007, and by 2050, that percentage is projected to rise to 55.2%.
- Indians will have a per capita income of $6,735 in 2050, up from $468 in 2007. Water use will grow due to increased industrialisation as its share of the GDP rises from 29.1% in 2000 to 40% by 2050.
- By 2050, there will be an 80% rise in water crisis as a consequence of the expanding agriculture.

TABLE 1.1
Per Capita Water Accessibility in Indian Scenario

Year	Population (Million)	Per Capita Water Availability (m^3/year)
1951	361	5177
1955	395	4732
1991	846	2209
2001	1027	1820
2025	1394	1341
2050	1640	1140

Source: Government of India (2009).

1.8 FRAMEWORK FOR RISK MANAGEMENT OF WATER SCARCITY – DROUGHT MANAGEMENT, STRATEGIC RESOURCES AND TECHNOLOGY

Although it may be mistakenly believed to be an infrequent and random occurrence, drought is a natural component of climate. Though it may be found in almost every climatic zone, there are major regional differences in its properties. The distinction between aridity, which is limited to areas with little rainfall and is a permanent aspect of climate, and drought, is that the former is a transient aberration. A lack of precipitation for an extended length of time, often a season or more, that causes a water deficit for particular types of activities, populations, or environmental sectors is a wide definition of a drought. Droughts are often categorised as meteorological, agricultural, hydrological, and socioeconomic events in terms of typologies. The exposure to the natural hazard and the susceptibility of the society to the event determine the risk of drought for every region or population. The sensitivity of people to drought is complicated. Both affluent and poor nations are significantly impacted by drought, yet these consequences have quite different features. Additionally, there are significant differences in how well-equipped each nation, area, neighbourhood and organisation is to deal with drought. A vulnerability profile, which includes a study of vulnerability variables, is therefore a crucial tool for evaluating local risk. A crucial component of planning for drought risk reduction is the vulnerability profile.

Four primary areas of activity that take into account the goals of disaster management publications may be used to describe the components of a drought risk reduction framework (Figure 1.1):

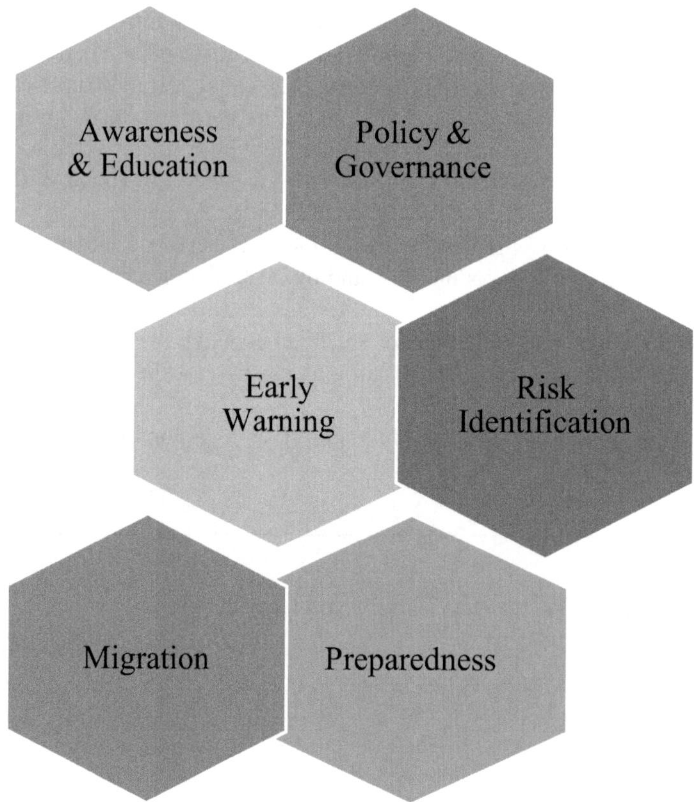

FIGURE 1.1 Components of disaster management.

Source: ISDR secretariat and National Drought Mitigation Centre, University of Nebraska-Lincoln, USA.

1. Policy and governance as a key component of political commitment and risk management for drought.
2. Drought risk spotting, recognition of its effects, and pioneer warning, including problem monitoring and assessment, evaluations of potential effects and the modelling of advance detection and communications networks.
3. To provide the groundwork for a culture of drought risk mitigation and susceptible communities, drought awareness and knowledge management are necessary.
4. To lessen the potential negative consequences of drought, effective drought mitigation and preparation measures must be implemented at the practice level rather than only as policy.

The coordination of efforts, information transfer, project support, and facilitation of efficient and inexpensive practices are all key functions performed in part by the regional and international communities.

The following guidelines should be followed when creating national and local drought risk reduction plans and when putting them into action:

1. Drought risk problems must be included in a process for sustainable development and catastrophe risk reduction, and this requires political commitment, strong institutions, and effective governance.
2. For policies to become practices, a bottom-up strategy with community involvement in decision-making and implementation is necessary.
3. Developing capacity and knowledge is often necessary to foster governmental commitment, capable organisations, and an informed law.
4. In order to manage drought and its effects, policies should provide a defined set of operational principles. They should also encourage the creation of preparation plans that detail how to accomplish these goals.
5. Rather than focusing simply on drought assistance, drought policies and programmes should incorporate mitigation and preparation.
6. The main elements of a drought strategy and plan are drought monitoring, risk assessment, and the development of suitable risk reduction strategies.
7. It is important to design and enforce policy procedures to guarantee that drought risk reduction policies are implemented.
8. To lessen the consequences of drought, sound long-term investment in mitigation and preparation strategies is crucial.

1.8.1 Risk Analysis Approach

1.8.1.1 Menace Estimation

The occurrence of droughts at different stages of severity and length affects the drought danger for countries and areas that are susceptible to drought. For the sake of clarity on these risks and how they differ over time and space, countries must develop holistic and unsegregated drought early warning methodologies that take into account seasonal variations and water provision factors.

1.8.1.2 Drought Impact Assessment

Similar to how knowing historical patterns in drought-related effects is essential for predicting future effects and comprehending shifting vulnerabilities, every drought has a different set of effects, based on the intensity, length, and geographic scope of the drought as well as the dynamic social environment. The effects of the drought may be broadly categorised as economic, environmental, or social for practical reasons, even if certain effects may genuinely cut across many sectors. These effects are signs of underlying weaknesses. Impact analyses are therefore an excellent place to start when identifying the root causes of drought-related response gaps. A drought impact assessment identifies industries, groups of people, or pursuits that are susceptible.

1.8.1.3 Threat Estimation

Threat estimation offers a paradigm for determining different causative agents of drought consequences. It emphasises focus on the various factors that contribute to susceptibility rather than the negative outcomes that follow aggravating events like drought. For instance, poorer agricultural productivity during a drought may be a direct effect of an absence of rainfall. Concerned state governments have the authority to take urgent action to alleviate the crisis brought on by the drought. When a natural disaster, such as a drought, strikes, the State Disaster Response Fund (SDRF), which is easily accessible to them, is used to launch the required relief efforts. For severe natural catastrophes, further financial support is considered from the National Disaster Response Fund (NDRF) and allowed based on a memo received from the state government in accordance with established process, bearing in mind the items and standards in fashion for aid.

1.9 CLIMATE CHANGE – MORE TROUBLED WATERS AHEAD

The consequences of climatic variations are a major threat to economic and social progress. The rate and direction of growth may have an effect on the rate and severity of climate change, as well as on the susceptibility of civilisations to its effects. As a result, it's abundantly evident that climate change and its effects need to be integrated into conventional economic policy, development initiatives, and foreign assistance programmes. However, more than just sitting ideal, the climate change and development groups need to dedicate fully to come up with solutions. This is due to the fact that the two groups do not have the same goals, often function on distinct time and spatial scales, and do not always communicate well with one another. There is a dire need for data-driven evaluation of the consequence of climate variation on development efforts, as well as practical assistance on how to appropriately include it in the broader context of other urgent societal issues. The OECD's Environment and Development Co-operation directorates worked together to produce Linking seasonal variation and Development, a draft on incorporating climate change solutions into development planning and assistance.

It is estimated that seasonal variation is accountable for 20% of the worldwide rise in water shortage. The effects of a warming planet are projected to exacerbate both drought and flood extremes, which will have an effect on bulk water supply. Without action to slow global warming, the UK Meteorological Office found in 2006 that by 2100, extreme droughts will occur every other year.

Increasing population and per capita usage are not the sole causes of fresh water scarcity. In addition to meteorological influences, cultural, technological, political, and institutional ones all play a role in the intricate dynamics of water supply vs water requirements. Because of cultural differences in traditions, income levels, gender roles, and other aspects of daily life, the average amount of water used by a single person in the world varies widely.

Politics plays a role in how and how much water is distributed to industrial, municipal, and agricultural users. The shortage issue will be resolved by increased efficiency and effectiveness brought on by more reasonable pricing and more equal water distribution. Institutional in the sense that governmental agency and other bureaucratic entities now control the majority of the administration of water reservoirs at the local, regional, and global levels. Separate entities that may or may not coordinate or integrate their policies and investments manage and purify water resources, plan for distribution, preserve the environment, set prices, and perform administrative duties. In addition to the repercussions of fresh water crisis in certain parts of the globe, the quality of existing fresh water should be a relevant concern.

1.10 MAKING WATER SCARCITY EVERYBODY'S BUSINESS

Water scarcity needs to be brought to public attention. At different levels, actions are required to deal with water scarcity:

- On a global scale, nations must work together more closely to address nation to nation water handling challenges within the network of laws, putting an emphasis on rules and regulations and communication as well as the drive to maximise the entire socioeconomic benefits of water.
- To address competing uses fairly and equitably and to better account for rising scarcity, policies and governance at the national level need to be adjusted. The creation of efficient conflict-resolution mechanisms will become increasingly crucial, and organisationally bound water policies are essential.
- Better management techniques are required at the local level in every industry, which will promote productivity, sustainability in water usage and sectorial water supply.

1.11 DESALINATED OCEANIC WATER AND HIGHLY MURKY GROUNDWATER

Seawater contains significant amounts of salt. More than 450 mmol/L of sodium ($Na+$) and roughly 55 dS/m of electrical conductivity (EC) are present in it. Because doing so would have an adverse impact on their health, seawater cannot be used directly for food by either people or animals without being treated to reduce its salt level (Qadir et al., 2003). The same holds true for freshwater which is incredibly brackish and has high levels of various salts. Since more than 50 years ago, seawater or severely brackish groundwater has been converted into high-quality freshwater via the desalination process. The application of this strategy in dry and semi-arid locations, as well as in nations bordering oceans or salt lakes, has been motivated by a lack of fresh water. The Middle Eastern nations are the biggest producers of freshwater from seawater. However, desalinating ocean water and extremely brackish groundwater to create pure water is an urgent necessity in a number of other nations. Desalination facilities are currently operational in more than 120 nations worldwide.

To maximise brine dilution, this necessitates a study into the best methods of brine disposal. Additionally, regular monitoring is necessary to determine any potential effects of brine on different marine environments. In several nations, research is being done to create environmentally friendly and effective brine disposal solutions (Latorre, 2002). To avoid desalination having a detrimental influence on nearby places and the larger ecosystem, it is critical that the appropriate legislation, policies, and procedures be in place.

1.12 RUNOFF WATER COLLECTED VIA WATER HARVESTING

In drought-prone and semi-arid areas, precipitation is infrequent and highly unpredictable both within and between seasons. In addition, because there is little vegetation cover and thin, crusting soils, a significant portion of the precipitation is lost during runoff and evaporation. These points significantly boost tactics that make the best use of runoff and rainfall. In areas where rainfall alone would be inadequate for crop growth, water harvesting is routinely practised. The different categories that water collection techniques might fall into are numerous. Water harvesting can be broadly categorised into two main groups: (i) macro-catchment and storm water harvesting and dividing systems, and (ii) micro-catchments techniques. Catchment areas for micro-catchment systems are no more than 100 m long. In concern to catch rainwater behind them and promote higher percolation, contour bunds are made of soil, stone, or rubbish embankments that are erected. Fruit trees and shrubs are typically grown using these systems. The meskat-type system is another kind of micro-catchment-based water harvesting device. In this approach, the farm is separated into a discrete catchment region that is situated directly above the cropped area rather than alternating catchment and cultivated sections. To increase drainage, the catchment region is frequently cleared of vegetation. A bund in the shape of a "U" surrounds the cultivable land to catch runoff. These techniques are used in Tunisia to grow tree species like the olive (*Olea europaea* L.) (Oweis et al., 2004). Balochistan, Pakistan, uses a similar technique known as Khushkaba for cultivating field crops (Oweis & Hachum, 2001). The benefits of rainwater harvesting systems is influenced by:

- The slope of the landscape,
- The soil's texture and structure,
- The availability of soil nutrients, and
- The depth of the farmed area.

1.13 CURRENT ADAPTATION STRATEGIES AND LOCAL STRATEGIES TO OVERCOME WATER SCARCITY

- **Create flood barriers to safeguard infrastructures –** Important infrastructure can be protected using flood barriers.
- **Construct the required aquifer storage and recovery infrastructure –** More inland water storage facilitates recharge when surface water flows exceed demand. This increases climate resilience for small or protracted droughts and makes use of seasonal variations in surface water runoff.
- **Install a low-head dam to isolate the freshwater pool from the saltwater wedge –** The salt water and freshwater boundary will fall further upstream in tidal estuaries as a result of rising sea levels and decreased freshwater flow brought on by drought. The water purity of surface water may be negatively impacted by boundary alterations upstream. This upstream movement can be stopped by building low-land dams over tidal estuaries.

1.13.1 Introduce Substitute or on-Site Electricity

Localised energy shortages could happen given that the challenge for electricity is expected to increase in the future. It may be wise to create "off-grid" sources as a suitable hedging tactic against electricity shortages. Furthermore, a redundant power source might offer resilience in cases where a natural disaster results in power shortages. On-the-spot facilities can come from biogas, solar, wind, and so forth. Electric utilities should be installed over anticipated flood levels, including backup equipment.

1.13.2 Management of Inflow in the Sewer System

More severe storms will result in greater infiltration and inflow of wet weather into sanitary and combined sewers. The capacity and performance of wastewater collecting systems and treatment facilities can be estimated using the sewer model. Modifications to the system that could be made to lessen those effects include steps to prevent infiltration, more receiver system capacity, offsite storage, etc.

1.13.3 Practice Water Conservation and Scarcity Management

Implementing water conservation strategies that reduce waste and inefficiency is an efficient and affordable way to satisfy the growing demand for water supplies. Public outreach is a crucial part of any campaign to conserve water. The most effective schedule of day to engage in water-intensive activities, as well as data on how to access household utilities like front-loading washers, low-flow toilets, and showerheads, are all part of outreach messaging. Additionally, other sectors can be the focus of outreach and education.

1.14 IMPROVING HEALTH THROUGH BETTER ACCESS TO WATER

1.1 billion people worldwide don't have access to an upgraded water supply, and 2.6 billion do not have an improved sanitation system. Although the majority of these people reside in rural regions, the number of urban residents who lack approach to sufficient water and sanitation facilities is quickly rising. The majority of people who lack access to appropriate sanitation facilities reside in Asia, but the majority of those who lack access to water do so in Africa. Water-related illnesses are widespread in many areas and are correlated with poor outreach to healthy water and fundamental sanitation. Ninety-three per cent of rural Indian households had access to upgraded water supplies in 2015. However, the Sustainable Development Goals have replaced the Millennium Development Goals, and the new baseline indicates that less than 49% of the rural population uses drinking water that is securely managed.

1.15 OPPORTUNITIES FOR PHYSICAL TRANSPORTATION OF PURE WATER, FOOD IMPORTS, AND THE "VIRTUAL WATER" NETWORK TO LEAD TO FOOD SECURITY IN UNDERDEVELOPED COUNTRIES

The 'physical' drift of water from water-surplus areas, where freshwater can be tapped and then transported, as well as the importation of food items from nations with sufficient water reservoirs and affordable food production costs are opportunities that can be exploited to help countries with water scarcity improve their food security. In the latter scenario, countries with limited water supplies import food instead of water. These nations conserve a significant amount of water that would otherwise need to be used for domestic food production due to water scarcity.

1.15.1 Physical Transportation of Freshwater

An approach that could be utilised to supplement freshwater needs in water-scarce locations is long-distance, large-scale freshwater transportation at the inter- and intra-country scales. The government of Libya built a 3,500-km pipeline to transport water to the coastal regions, where immense development and rising populations have severely strained the water supply available for homes, agriculture, and industry. More than $5 \times 106\,m^3$ of water per day is anticipated to be transported by the pipeline from the desert to the coastal districts. Only with strong regional collaboration and confidence between water exporting and importing nations, as well as those situated between them, can initiatives requiring cross-border water flow be brought to fruition.

1.15.2 Food Imports and the 'Virtual Water' Array

Since the start of global trade, countries have exchanged food. By bringing food in this way, nations having water crisis can limit using their own water to domestically produce the same amount of food or utility while facing a water shortage. They 'preserve' their water in this way. The concept of "virtual water" has been used to highlight the crucial section that water plays in the food trade between nations with water surpluses and those with water deficits, the latter of which must partially rely on food imports to achieve food security. The concept of "virtual water" compares the quantity of water that is incorporated in a crop that can be purchased internationally with the amount of water that would be needed to produce that crop domestically since it relates water, food, and trade (Allan, 2003). For instance, 1 kg of wheat requires roughly $1.1\,m^3$ of water to produce. Thus, if such sources are accessible, the water-scarce country also benefits from around $1.1\,m^3$ of virtual water for every kilogramme of imported wheat, frequently at a substantially cheaper cost than the price for the same amount of water generated from local water resources in the country itself (Chapagain and Hoekstra, 2003). Similar accounts are there for other foods, although it can be challenging to measure and compare the precise amounts of water required to grow foods in various agro-climatic zones (Rosegrant et al., 2002).

Agronomic, economic and political factors are all present in the trading of food commodities and "virtual water." The amount of water used to grow crops is a part of agronomy. The "virtual water" aspect is in line with the idea of integrated water management, which takes into account factors such as water supply and demand when choosing how to use scarce water resources most effectively (Bouwer, 2002).

It is predicted that most cereal prices will fall globally during the next 20 years, albeit more slowly than in the past (Rosegrant et al., 2002). This implies that the economies of some nations with limited access to water that import grains may benefit from a subsidised deal, especially when the global rates of agricultural items are lower than the price of production in these nations. Many nations with limited access to water will continue to rely partially on food imports while also producing some of their own food needs. Therefore, the export of goods using less water will give a chance for those nations with limited water resources to maximise the value of their finite water reservoirs. However, in the case of those nations where these resources are scarce, it is also necessary to take into account the significance of land and capital in agricultural imports and exports. Additionally, in nations where lowering unemployment is a key policy objective, labour-intensive agricultural cultivation and processing may offer possibilities (Wichelns, 2001). The phrase was then used by Professor Allan to highlight the idea that severe local water shortages might be very efficiently alleviated by global economic forces. To address critical water issues in an economic and political manner, he proposed the idea of "virtual water." Since 1995, the phrase has become more frequently used, and by the turn of the millennium, it had taken centre stage in numerous conversations about security and the scarcity of water (Allan, 2003). The phrase has faced criticism since then. Merrett (2003) suggested that the term "virtual water" be changed to "the crop water requirements of food," since he believed it to be nothing more or less than the amount of water required to produce agricultural goods.

1.16 MARGINAL QUALITY WATER RESERVOIRS

Wastewater produced for residential and commercial purposes, drainage water from irrigated land and surface runoff that has entered the drainage system; and groundwater from a variety of sources make up the category of waters with marginal quality. The important residues of cosmetics or microbes are present in marginal quality waters at levels that are excess than in pure water. These impurities include salts, metals, drop of medicines and other organic chemicals substances. These components have adverse impacts on vegetation, animals and people.

1.16.1 Wastewater from Household, Municipal, and Industrial Activities

Population development and the availability of necessities that raise living excellence have led to an increase in the urge for high grade water to meet the demands of the residential, municipal, and industrial sections in nations with limited water resources. There is a consequent increase in wastewater production. This could be utilised again after therapy in conjunction with effective management techniques. Urban wastewater is made up of several types of urban runoff, including few or all of the household wastewater produced as well as wastewater produced by hospitals, commercial organisations, industrial facilities, and other institutions.

Improving public awareness, using proper irrigation techniques, reducing human exposure, limiting the kinds of vegetation that can be employed with wastewater, restricting their use, disinfecting entity, ensuring institutional cooperation, lengthening land tenure, and increasing funding are some of the management options that should be used. The World Health Organization (WHO) is taking into account the circumstances encountered by these countries when amending measures for wastewater use in agriculture because it is not practical to outright avoidance of wastewater usage in many developing countries.

1.16.2 Agricultural Drainage Water

The salt content in the drainage water that results from flowing through these deposits can significantly surpass those expected to happen as a result of irrigation alone. Some small but potentially harmful substances may dissolve and be displaced by drainage fluids in certain geological conditions. Reusing agricultural water has become a significant point of irrigation as freshwater supplies become scarcer. Agricultural drainage water can be used for a variety of crop production systems depending on the amounts and types of salts present as well as the usage of suitable irrigation and soil management tactics. The best suitable vegetation for use as bio-drainage gadgets are profound tree species. Depending on the specifics of the location, a combo of bio-drainage and ordinary drainage systems may also be taken into consideration.

1.16.3 Salty Groundwater

Water bodies of poor standards, such as those that contain salt water, are prevalent in many water-scarce places. While sodic water has higher quantities of Na^+ than other cations, saline water has an excess of salts. Reactions between the water and the strata of the earth through which it flows to become hydrological water, as well as chemistry within the earth formations in which the groundwater is located, are the causes of such waters of mediocre quality. Saline aquifers can also appear close to saltwater sources. Drainage water from agricultural areas has an impact on the quality of numerous groundwater sources.

1.17 INTENSE PRODUCTION WITH MINIMAL WATER

The solution to the difficulties of future water scarcity lies in increasing water productivity. By 2050, the share of water used for farming, industry, and domestic purposes will have increased by 60%–90%, based on inhabitants, income, and presumptions about the environment's water needs. This increase will occur in the absence of further increase in the supply of water or significant changes in production models. Two-thirds of the world's energy demand, which will increase by 50% by 2030, will come from developing nations. Water productivity might, however, be increased more quickly. A combination of institutional measures that guarantee justice in access to water while promoting increased efficiency can go a long way toward reducing water shortages and ensuring environmental sustainability.

1.18 THE WAY FORWARD – WE CAN PERFORM STILL BETTER

Actions must be taken at the municipal, governmental, and river basin levels to address the water shortage. Additionally, it urges national governments to work more closely together to share in the advantages of managing water resources at the global and international levels. Due to the intersectoral character of the problem, managing water scarcity involves cooperation, the sharing of common goals and guiding principles, and coordinated action. The institutional segregation of models in the water development section is a significant obstacle to properly managing water scarcity in countries.

1.18.1 Uprooting Uncertainty for the Benefit of the Suffering Class

The many rural poor are those who experience daily struggles with water scarcity the most. This is especially true for people who live in remote places where food production is subject to climatic fluctuations and where poor water and sanitation service delivery has an unfavourable effect on their quality of life. Urban residents are also impacted by this pervasive and sneaky issue, particularly in Asian cities. In most regions of the world, more economic growth is necessary for reducing poverty, but growth quality and, in particular, how much it expands possibilities for the poor, are equally important. For such expansion, better water management can act as a catalyst. Water is a crucial input for many productive processes, opening chances for local business owners to supply equipment, build facilities, and offer services. High returns are produced by local investments, which are crucial since they also maintain positive effects on the local economy and have substantial multiplier effects. Under circumstances of greater water security and sustainability, there are numerous options to enhance impoverished people's capacity to escape poverty. Targeting the poor and promoting a variety of livelihoods can be accomplished through funding household water systems that make water available for household-based day-to-day activities including fruit production, pottery making and laundry. The best chance for promoting economic growth in many nations is through broad-based agricultural expansion.

1.18.2 Giving Priority to Social and Environmental Services

How to effectively assist shareholders in controlling their water crisis in an atmosphere of rising demand and dependency is a key topic today as water becomes scarce. This issue is particularly important for farming because it is the greatest worldwide user of water and struggles to secure the required amount of water reservoirs to face the demands of an expanding global population and control the effects of its operations on the base of natural resources. Collaborators in managing their water assets entail assisting them in decision-making and coming to an agreement on the essential mechanisms for sharing and assigning goods and services related to water. There is a need for valuation frameworks that take into account these three factors and put stakeholders at the centre of the process. The worth of water-related products and services can now be expressed in a variety of ways, using monetary quantities. These procedures are difficult and demanding in terms of the knowledge, time, and data needed for their use even though they have a great deal of potential value. This makes it difficult for them to be widely used, particularly in underdeveloped nations.

1.19 CONCLUSIONS

Droughts are also predicted to occur more frequently as a consequence of climate variations. As a result of these pressures, the institutions in the region are changing to enable better sustainable oversight of the region's water resources. In many of these nations, progress is still being made, which is encouraged by improved monitoring and management systems, growing regional cooperation, and most importantly, governmental awareness. For effective water management, it is essential to incorporate new technologies. These technologies can help you use conventional or freshwater resources more efficiently. Measures to combat drought must be incorporated into long-term plans for the use of water, land, and development as a whole. When managing water resources at the primary level, there is a chance to directly address policy choices as well as the requirements and issues of the natural water cycle. Systems for observing and early admonition are getting better all the time and are already being used in planning procedures. Finally, improved regional collaboration, deeper comprehension of the dynamics and social significance of resources and more effective monitoring systems offer hope for easing the current stresses on the water reservoirs in the ensuing decades. When there is a water shortage, people come up with a variety of ways to get enough water, which frequently entails getting it from shaky and awkward methods. Education and employment consequences particularly affect women and children because they are typically in charge of collecting water. Although development organisations largely acknowledge this truth, it nevertheless needs more scholarly investigation. More consideration must be given to the work done by kids to collect water in particular. As a whole, women, children and youth all participate in collecting water and preserving it.

REFERENCES

Allan, J. A., (2003). Virtual water-the water, food, and trade nexus: Useful concept or misleading metaphor? *Water International*, 28, 4–11.

Amarasinghe, B. M. W. P. K., & Williams, R. A. (2007). Tea waste as a low cost adsorbent for the removal of Cu and Pb from wastewater. *Chemical Engineering Journal*, 132(1–3), 299–309.

Ariyoruk, A. (2003). Turkish water to Israel? Policy Watch No. 782 (August 14, 2003). Washington, DC: The Washington Institute for Near East Policy.

Beitinger, T. L., Bennett, W. A., & McCauley, R. W. (2000). Temperature tolerances of North American freshwater fishes exposed to dynamic changes in temperature. *Environmental Biology of Fishes*, 58(3), 237–275.

Bouwer, H. (2002). Integrated water management for the 21st century: Problems and solutions. *Journal of Irri gation and Drainage Engineering*, 128, 193–202.

Chapagain, A. K., Hoekstra, A. Y. (2003). *Virtual water flows between nations in relation to trade in livestock and livestock products* (59 pp). Value of Water Research Report Series No. 13. Delft: UNESCO-IHE Institute for Water Education.

Crow, B., & Singh, N. (2000). Impediments and innovation in international rivers: The waters of South Asia. *World Development*, 28(11), 1907–1925.

Fröhlich, C. J. (2012). Water: Reason for conflict or catalyst for peace? The case of the Middle East. https://www.cairn.info/revue-l-europe-en-formation-2012-3-page-139.htm?contenu=article

Government of India. (2009). *Inter basin water transfer proposals. Task force on interlinking of rivers* (27 pp). New Delhi: Ministry of Water Resources.

Latorre, M. (2002). Design optimisation of SWRO plants for irrigation. In Proceedings of the international desalination association's world congress on desalination and water reuse, March 8–13, 2002, Manama, Bahrain.

Liu, J., Yang, H., Gosling, S. N., Kummu, M., Flörke, M., Pfister, S., & Oki, T. (2017). Water scarcity assessments in the past, present, and future. *Earth's Future*, 5(6), 545–559.

Merrett, S. (2003). Virtual water and Occam's razor. *Water International*, 28, 103–105.

Nikil, S., & Bharani, A. (2022). People's participation in restoration of water ecosystem under climate change scenario. In Suborna Roy & Chandan Kumar Panda (Eds.), *Climate change dimensions and mitigation strategies for agricultural sustainability* (Vol.II, pp. 1–24).

Oweis, T. Y., & Hachum, A. (2001). Improving water productivity in the dry areas of West Asia and North Africa. In J. W. Kijne, R. Barker, & D. Molden (Eds.), *Water productivity in agriculture: Limits opportunities for improvement* (pp. 179–198). Wallingford: CABI Publishing.

Oweis, T. Y., Hachum, A., Bruggeman, A. (2004). Indigenous water-harvesting systems in West Asia and North Africa. In *International center for agricultural research in the dry areas (ICARDA),* Aleppo, Syria, 173 pp.

Pham, T. T. T. (2018). Water-sharing conflict: A case study in the Ganga waters dispute between India and Bangladesh. SSRN. https://papers.ssrn.com/sol3/papers.cfm?abstract_id=3228964

Qadir, M., Boers, Th. M., Schubert, S., Ghafoor, A., & Murtaza, M. (2003). Agricultural water management in water-starved countries: Challenges and opportunities. *Agricultural Water Management*, 62, 165–185.

Rosegrant, M. W., Cai, X., & Cline, S. A. (2002). *World water and food to 2025* (322 pp). Washington, DC: International Food Policy Research Institute.

Shiklomanov, I. A. (2000). Appraisal and assessment of world water resources. *Water International*, 25(1), 11–32.

Sood, A., & Mathukumalli, B. K. P. (2011). Managing international river basins: Reviewing India-Bangladesh transboundary water issues. *International Journal of River Basin Management*, 9(1), 43–52.

UNDP. (2006). *Human development report. Beyond scarcity: Power, poverty and the global water crisis.* New York: United Nations Development Programme.

Wichelns, D. (2001). The role of 'virtual water' in efforts to achieve food security and other national goals, with examples from Egypt. *Agricultural Water Management*, 49, 131–151.

2 Nature-based Wastewater Treatment Technologies

Concept, Understanding, Opportunities and Future Prospectus

Saroj Kumar, Priyanka, Pawan Kumar Bhargawa, Subhash Chandra, Venkatesh Dutta, and Prashant

2.1 INTRODUCTION

Water resources are seriously affected by numerous external pressures, such as contamination, over abstractions, floods and droughts, as well as physical alterations including changes in land use, soil erosion, drainage, channelization and other barriers, etc. (Souliotis & Voulvoulis, 2022). Countries are facing wide-ranging challenges, from the reduction of natural resources such as water, ecosystem amenities, and other abiotic assets together with its degradation which poses serious pressure on human health and water security across the globe (Souliotis & Voulvoulis, 2021; Kumar et al., 2022a). Additionally, climate change further intensifies these problems, as the rise in temperature affects the hydrological cycles which enhances the episodes of extreme events (Tabari, 2020). As nature and humans establish the socioenvironmental linkage that forms dynamic and complex networks, any alterations in economic and societal variables may impact the form, quality and quantity of natural resources. Several drivers e.g., intensive agriculture, rapid population growth and land use change significantly influence the biogeochemical cycles, water quality, and biodiversity (Teixeira et al., 2014; Kumar et al., 2022b). The urbanization progression is continuously convoyed by growing impermeable surfaces, such as the extension of buildings and roads (Ercolani et al., 2018). These alterations decrease the rate of rainfall infiltration and rise the surface runoff, thus extending pressure on the city's/town's drainage system, particularly for extreme events, which probably lead the flooding situations (Grebel et al., 2013). Further, the goals of wastewater treatment technologies are shifting from linear to hybrid models such as together with the protection of aquatic ecosystems and human health to comprise lessening the waste generation, loss of rare resources, use of energy and water and development of various bioresources, enhancing resource use efficiency and nutrients recycling for the circular economy (Lundin et al., 2000; Kumar et al., 2022c). Recently, nature-based solutions (NBS) have been measured as an innovative approach that has gained extensive attention to lessen such expected incidences of catastrophes. The European Commission (2015) established the NBS system as an action encouraged by nature, and concurrently delivers several social, economic, and environmental benefits and promotes resilience. They are usually measured to design a range of small green infrastructures such as rain gardens, green roofs, porous pavements, and bio-retentions in catchments to bring the altered regimes back to previous hydrologic states (Eckart et al., 2017). Albert et al. (2021) have illustrated three application principles of NBS: (i) deliver constant benefits to nature, society, and the economy; (ii) act as an interdisciplinary approach that includes experience from current methods; and (iii) improve the application of NBS in actual real-life sceneries.

DOI: 10.1201/9781003441144-2

NBS is a comparatively novel concept that emerged from values linked to sustainability, resilience, and environmental management and services (Lafortezza et al., 2018). The major aim of developing NBS systems is to address complex socioenvironmental problems, promote the linkage of water, land, and biotic sources and merge the sectoral strategies of dissimilar scales (Faivre et al., 2017; Artmann et al., 2019). The key component of NBS is the operationalization of ecological amenities, the intangible and tangible goods and services created over the working of ecosystems by the provision of the economy for the welfare of human beings (Lafortezza & Chen, 2016; Maes et al., 2016). NBS emphasized making use of interferences to the manmade and natural setting that mimics the natural ecosystem processes and can satisfy a range of policy objectives such as increasing human well-being, ecosystem restoration, climate change adaptation, improving water resources, and their appropriate use and conservation of biodiversity, etc. Consequently, the proposal of such systems integrates ecological as well as social factors, such as the well-being of humans that affect the inclusive performance and health of the system. These actions are generally related to concepts such as natural resolutions, ecological engineering, ecosystem approaches, and green infrastructure as interdisciplinary interferences which can foster the effective implementation of strategies at a comparatively lower cost and create benefits for the society and environment (Dudley et al., 2010; Benedict & McMahon, 2012; EEA, 2017). Various studies advise that NBS are principally cost-effective and eco-friendly alternatives to tackle wastewater contamination (McGonigle et al., 2012; Souliotis & Voulvoulis, 2022). They can offer multiple benefits such as advancing water management, climate change mitigation and adaptation, protection and enrichment of biodiversity, coastal resilience, enhanced air quality, improved involvement of stakeholders, urban rejuvenation, recreation, social cohesion, and enhancements in public health through the restoration of ecosystem functions (Wise et al., 2010; Raymond et al., 2017; Kumar & Dutta, 2019). The application of NBS also has the potential to provide noteworthy economic outcomes through nutrient recovery and the development of valuable bioresources and job opportunities. For example, in Germany, the restoration of Emscher Landscape Park creates around 86,000 jobs in the previous 20 years (Lieuw-Kie-Song & Perez-Cirera, 2020).

2.2 NBS CONCEPT AND UNDERSTANDING

The term NBS emerged from what is known as biomimicry, which enables us to learn from, motivate by, and try to copy natural processes. The essential prerequisite for NBS to work efficiently is to gain an understanding of natural environmental processes and nature as vital infrastructure to offer crucial services to help communities and the ecosystem and turn social, ecological, and economic tasks into novel prospects (Stefanakis et al., 2021). They use the complex natural ecosystem processes, to attain the probable and desired results i.e., regulation of water flow and the capability to store carbon. Consequently, NBS can also be a vital component in enhancing life quality, prosperity, well-being, social and economic sustenance, biodiversity, green growth, and ecological sustainability (Kolokotsa et al., 2020; Spano et al., 2021). The International Union for Conservation of Nature (IUCN) employed a resolution (WCC-2016-Res-069-EN) in the World Conservation Congress and Members' meeting 2016, which defined the usage of NBS as "activities to guard, manage and reestablish altered ecosystems sustainably that brings ecological, economic and social benefits and support to build resilience (Stefanakis et al., 2021). NBS brings more and more diverse natural aspects and developments into the landscapes of cities through resource-efficient, locally accepted, and systemic interferences (Bulkeley, 2020). To differentiate NBS from conventional water resource management methods, the IUCN suggests eight basic principles upon which NBS is based as exhibited in Figure 2.1 (Cohen-Shacham et al., 2016). Approximately four billion people throughout the world face critical water shortages for at least one month of the year (Thapa et al., 2021). NBS is considered an umbrella concept that covers a wide range of activities based on the ecosystem processes for mitigation and adaptation in the

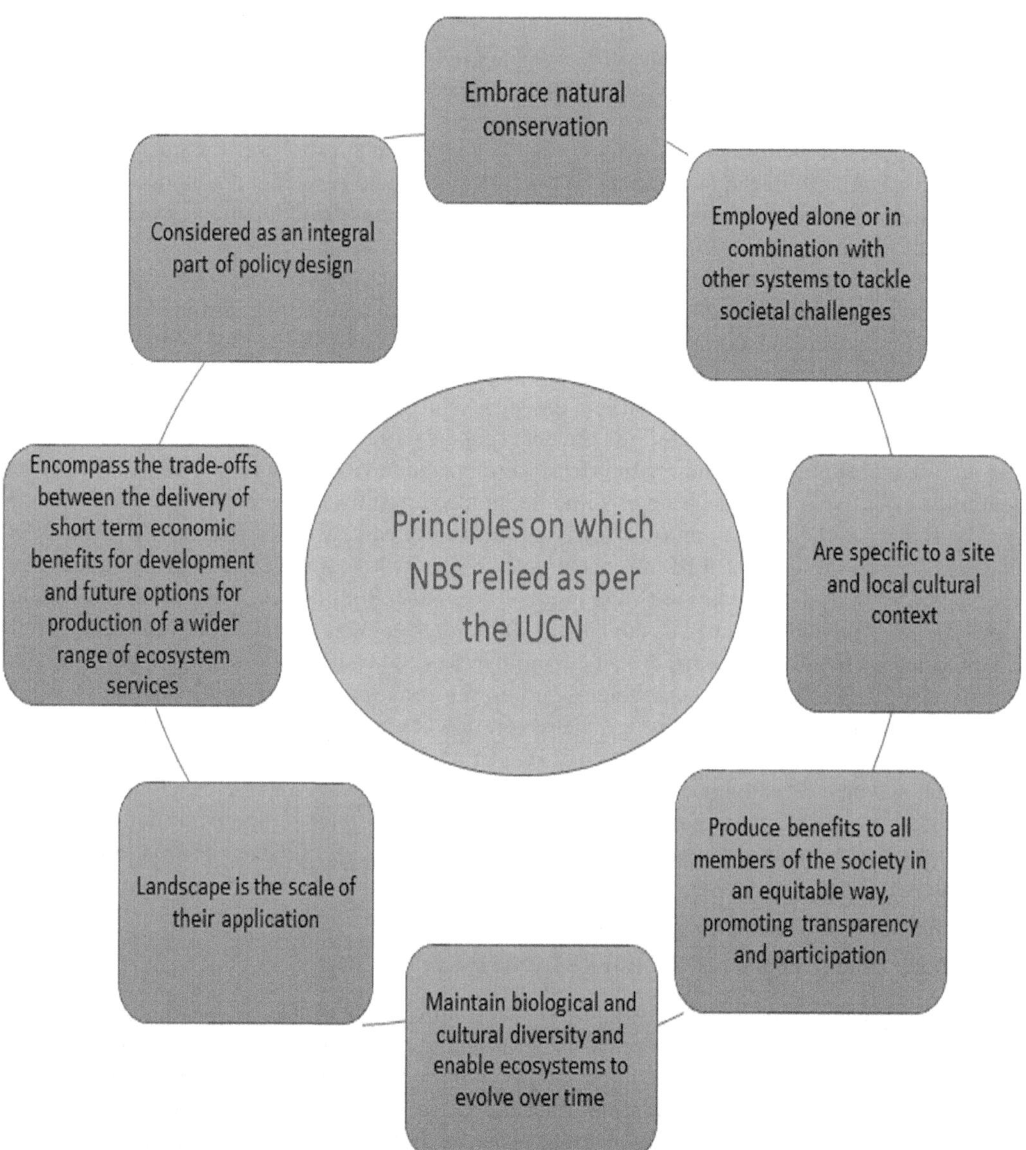

FIGURE 2.1 Eight principles on which NBS are based as explained by IUCN.

form of green infrastructure (Seddon et al., 2020). They can be viewed as the correspondent of "grey infrastructure", comprising various engineering actions and services utilized for accompanying or relieving purposes accomplished by ecosystems, like water collection, storage and purification, etc. Eggermont et al. (2015) elucidated the distinction between engineering actions and NBS and proposed a methodology to describe NBS whether engineering interferences are required, the sum of ecological facilities and groups of stakeholders targeted through their design and application. They define three dissimilar categories of NBS which include activities requiring negligible or no interferences in ecosystem functioning, strategies that impact the working of ecosystems and environment and indiscreet management practices or development of new ecosystems targeting to intensify the ecosystem services.

However, inland water resource management practices comprise the usage of buffer strips to alleviate diffuse contamination, guard biodiversity, minimize river bank erosion, and enhance the aesthetics of the surrounding. River restoration targeting to bring back connected waterbodies to their previous phase that delivers maximum ecosystem amenities, water retention actions to lessen the flood risk and artificial watercourses among others (Collentine & Futter, 2018; Blau et al., 2018; Cole et al., 2020).

Recently, NBS has shown the potential to manage stormwater runoff to some extent (Ahiablame & Shakya, 2016). Currently, various researcher throughout the globe have progressively shifted their attention from investigational fields to modeling at the urban catchment scale by examining the hydrological influences of NBS (Versini et al., 2016; Qiu et al., 2021). Consequently, different hydrological models are accepted for the forecasting of hydrological responses of NBS in urban settings (Her et al., 2017). Most of these modeling studies are based on the semi-distributed Storm Water Management Model (SWMM), which signifies the NBS as the fraction in every sub-catchment (Guo et al., 2019; Luan et al., 2019). However, certain current studies have explained entirely dispersed hydrological models to predict the hydrological influences of spatial distributions (Qiu et al., 2021). For example, a study conducted by Fry and Maxwell (2017) specified that the spatial location of NBS together with the primary flow direction is crucial as compared to the number of NBS employed within the watershed. A comparative assessment of conventional wastewater treatment systems and nature-based solutions for sustainable water resource management is provided in Table 2.1.

TABLE 2.1
Existing Wastewater Treatment Systems Vs Future Nature-Based Solutions for Sustainable and Efficient Water Resource Management

Existing Wastewater Treatment Systems	Future Nature-Based Solutions
Based primarily on linear approaches such as the treatment of wastewater, water supply and their evacuation	Working on alternative or multifunctional approaches such as together with wastewater treatment, development of various value-added products, energy, nutrient recovery, fertilizer (N, P, K), soil conditioner, etc.
Remote, out of sight	Can be designed anywhere including in buildings
Requires technical instruments and expertise	Based on ongoing ecosystem processes
Vulnerable	Resilient, robust
Static	Flexible
Repulsive	Attractive
Centralized	Distributed and optimized among sites of generation and use of products
Depending on a city or town-wide piping network	All possibilities from no piping network to several smaller and larger networks depending upon the existing situation
Largest possible, requesting economy of scale	All sizes, depending on the requirement
Leaving stormwater	Harvesting rainwater
All-in-one solution	Separation of sources wherever useful for product quality
Ineffective in the removal of emerging or new contaminants	Effective in the removal of new emerging contaminants by source control and systemic adaptation
Dealing with various types of water in urban settings in unconnected and distinct ways: supply water, wastewater, natural waters and stormwater, etc. or connect them in a way that makes their future use difficult, e.g., combined sewer	Integrate all types of waters in a unique system, a network system having several nodes enabling the use of all probable sources and keeping separate flows were required for future use

2.3 NBS AND THEIR SIGNIFICANCE TO WATER RESOURCE MANAGEMENT AND OTHER ECOSYSTEM SERVICES

Current studies on NBS evaluated their ability in climate resilience, cooling cities and reducing the effect of heat islands, restoration of waterbodies, biodiversity conservation and enhancing human health and well-being and altering the places and their sense in cities (Yang et al., 2018; Kondo et al., 2018; Frantzeskaki, 2019). They are promoted and demonstrated in offering numerous ecosystem amenities and as such, being illustrious from grey infrastructure explanations by being multifunctional results (Raymond et al., 2017). Furthermore, multifunctionality helps as an access point for encouraging and supporting NBS relevance in cities and towns, where space is inadequate, and green cover is under pressure from growth. The multifunctionality of NBS is usually understood on two different levels. Primary, the functional efficiency and study of ecosystem services, such as nutrient cycling, material decompositions, or plant biomass. The secondary values of nature to humans in the form of various social, economic, and ecological benefits and responding to biodiversity, ecological and climate plans in cities as exhibited in Figure 2.2 (Raymond et al., 2017; Hansen et al., 2019). It is supposed that NBS inevitably leads to socially just outcomes (Haase, 2017). Various researcher claims that NBS and their scaling may lead to displacement, social segregation, and inadequate dispersal of

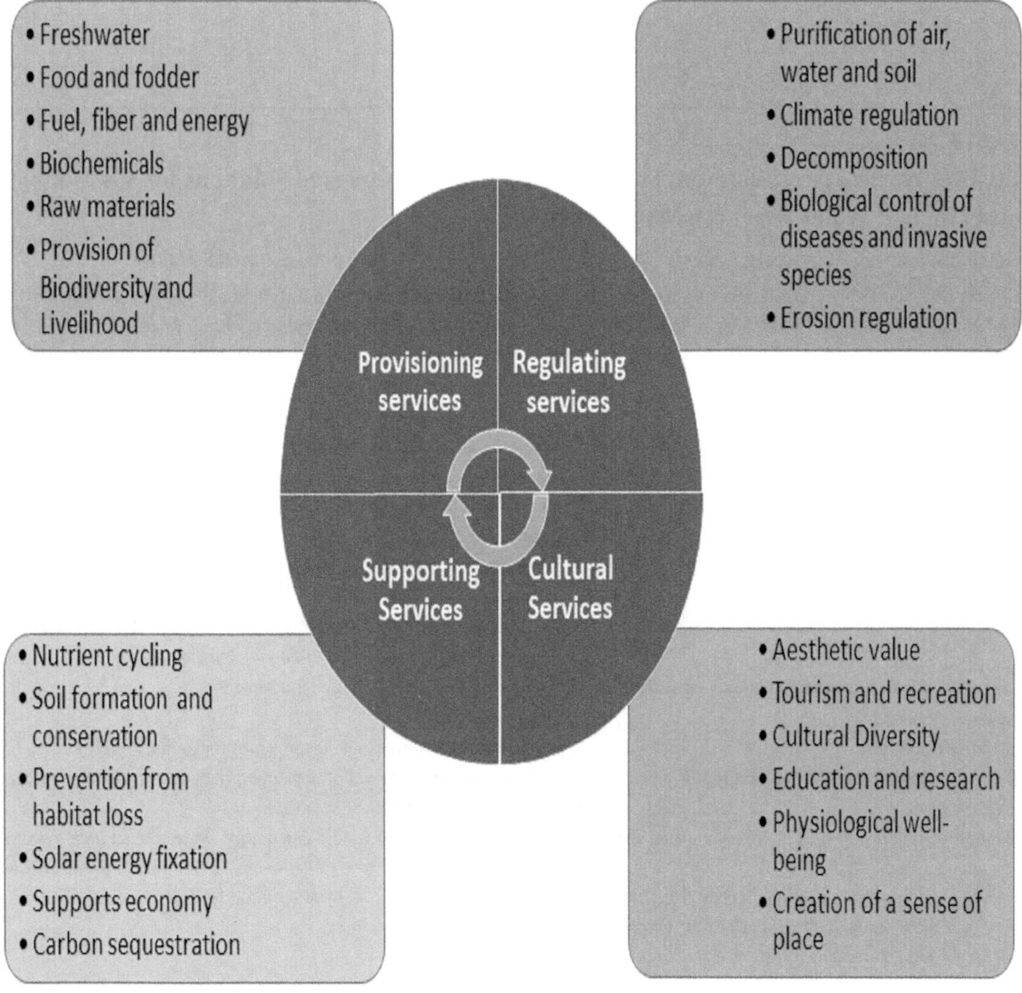

FIGURE 2.2 Major environmental services offered by NBS. (Modified from Jamion et al., 2022.)

benefits for marginalized and underprivileged communities (Anguelovski et al., 2018). NBS when providing benefits for human beings is also assisting nonhuman living organisms and ecosystems (Washington et al., 2018). However, such presumptions, are not accurate and need improved consideration. The planning of cities based on natural solutions, with its mainly human-centered lens, compromised with more combined strategies in their design and application processes that help and allow inclusivity and representation of nonhuman, multispecies (Pineda Pinto, 2020). Additionally, planning improvement and application is an integrally challenged process where multilevel political financial prudence shapes the transformation from justice-based thoughts to practical application. In this situation, multifunctionality planning becomes a crucial development goal that can link and be directed by ecosystem and environmental justice perspectives for NBS to bring social-environmental just results, it requests to be measured firstly, through a better understanding of the interactions and trade-offs of the various roles and benefits (Hansen et al., 2019), and secondly, by seeing dissimilar ways of appreciating their contributions to all living and non-living creatures including humans (Stephens et al., 2019; Yaka, 2019). It is critical to understand how NBS can be scaled up from proposals to apply for their mainstreaming in cities and towns (Hansen et al., 2019). To address this, we agreed on the assumptions of previous studies conducted by Engström et al. (2018) and Wang and Banzhaf (2018). They have suggested that a combined view of multifunctionality and design of NBS can advance our knowledge based on the interrelation of social-environmental functions and their benefits. The literature on NBS exhibited a partial emphasis on ecological justice, consequently, we may focus on such a perception as an epistemological redesign for conceptualizing and supervising the planning and design for their successful implementation. An environmental justice view encompasses the community of justice for everyone. It extends the magnitudes of justice to social–environmental processes: acknowledgment of their interactions, involvement of nature in sociopolitical matters, understanding of ecosystems' abilities, as well as confirming a fair and unbiased dispersal of benefits (Washington et al., 2018).

2.3.1 Constructed Wetlands (CWs) as a Promising Nature-Based Solution

Wetlands have been established to be the most effective, significant, and cheaper alternatives to foster various ecological services including carbon sequestering (Harenda et al., 2018; Derakhshan-Nejad et al., 2019; Kumar et al., 2021). They own a complex ecosystem such as marshes, lakes, and floodplains having water-saturated soil and offering several services to communities and the environment (Jamion et al., 2022). Wetlands have been depicted as 'the kidney of the landscape' including excessive subsidiary amenities as bioresources, sinks, and alterations in various chemical and biological elements (Mitsch, 2017). They sequester excessive CO_2 from the atmosphere despite improvements in water quality, flood mitigation, and wildlife protection as exhibited in Table 2.2 (Lorenz & Lal, 2018; Graves et al., 2020). Wetlands encompass the maximum carbon storage as the longest carbon pool and consequently take part in universal carbon cycling (Lorenz & Lal, 2018). However, the guidelines of the central mitigation bank do not consider the function of wetlands as carbon storage, it may be recognized as a crucial ecological service as stated by Means et al. (2016).

TABLE 2.2
Carbon Sequestration Potential of Different Waterbodies

Waterbody	tCO_2 Equivalent /hectare/year	Reference
Lakes	7.1	110
Ponds	16.12	
Floodplain	3.36	
CWs	13.35	113
Wetland meadow	0.169	114

Source: Modified from Souliotis and Voulvoulis (2022).

CWs integrate physicochemical and biological methods, delivering ecological amenities resulting from the interactions among plants, soil, water, and aquatic organisms, e.g., water, nutrient, carbon cycling, sedimentation and filtration, etc. Furthermore, different plants enhance the site's aesthetic by developing important wildlife habitations for numerous animals, such as fishes, insects, amphibians, songbirds, etc. (Rajpar & Zakaria, 2014; Kumar et al., 2020). They pose the great potential to exert CO_2 by sequestration and retain it as biomass and organic matter to adsorb contaminants and improve the water quality (Ward et al., 2017; Wong et al., 2018). Further, Ramsar Convention on wetlands highlights that mitigation and adaptation of climate change are all about carbon and water respectively (Sherren & Verstraten, 2013). Nevertheless, in previous years, pollution control, reduction in carbon, water supply, management of agricultural resources, and energy production have all been seen as distinct problems and to be solved individually. It is clear now that interconnections among nutrients, water, and carbon cycles can be utilized in a system that accomplishes several roles with improvement in the health of the ecosystem (Avellan et al., 2017).

CWs are previously utilized only for water reuse plans like gardening, nutrient and contaminants elimination in domestic locations, and for irrigation of crops, golf courses, public parks, aquaculture, groundwater recharge or to reinstate natural wetlands (Rossa et al., 2019; Nan et al., 2020; Kumar et al., 2022c). However, upcoming studies exhibited the opportunity of leveraging CWs as a probable mitigation strategy in dealing with the carbon–water nexus, concerning those engineered and natural physicochemical developments at the interface of the environment among water and carbon cycles (Masi et al., 2018; Were et al., 2019). Therefore, CWs can tackle a minimum of two crucial sustainable development goals (SDG 6 and 13) dealing with clean water and sanitation and climate action respectively through the carbon–water nexus. SDG 6 sets the basic guidelines for watershed management to safeguard human and ecosystem health by sustaining the optimum and appropriate water availability (UN-Water, 2013). Consequently, it is important to measure the safety of such sources by studying the roles and values of CWs. Further, CWs also help to lessen the urban heat effect, enhance the living standards of communities, and contribute to the combination of roles and values for the public (Masi et al., 2018; Kumar et al., 2021). Sometimes, the values and functions of CWs are confused and measured as identical, this leads to their inappropriate management and impacts the ecosystem's health seriously. Therefore, it is necessary to understand the concept of CWs based on their natural services and values through combined approaches to support the conservation, adaptation, and mitigation strategies to decrease climate change susceptibility and improve ecosystem health and their functions to achieve a sustainable development agenda in 2030.

2.3.2 Development of Green Infrastructure for Environmental Resilience

Green infrastructure (GI) is a current strategy to deal with matters that primarily originate in urban settings. It is regarded as a substitute for conventional infrastructure that is typically dependent on concrete and more precisely termed grey infrastructure. The name GI was originally stated as an open green space in the urban setting and generally linked with each other to offer numerous ecosystem services and/or to develop a new ecosystem (Mell, 2008; Stefanakis, 2019). The primary roles of GI were considered as the management of urban runoff and stormwater, lessening the urban heat islands effects, advancing air quality, etc. Though, in recent years the application of GI has been expanded further to embrace other benefits, such as various ecosystem services, the formation of new habitats for wildlife, and greenhouse gas emissions reduction, among others (Haines-Young & Potschin, 2012). Consequently, it is assumed that GI currently signifies an extensive range of solutions intended to grow urban resilience.

The shifting concern toward this route is due to the growing risks that modern cities face in severe and more regular appearance and an increasing degree of life-threatening events like floods and prolonged dry periods. In current scenarios, it can be understood that ecosystem services, i.e., all amenities and goods provided by nature to humans, own a tremendous value, which would not be ignored (Potschin & Haines-Young, 2011). Thus, it is evident that, under these circumstances, GI is strongly connected to sustainability.

The acceptance of sustainable development directs that economic development should complement environmental protection. Wise usage of natural capital is a requirement for both economic development and the constraint of the degradation of the environment. The sustainable strategies integrate the idea of intergenerational solidarity, which explains that the present needs must be met without compromising the capability to meet those of upcoming generations. Hence, the application of GI can better the present situations and contribute more efficiently to this direction as compared to existing grey infrastructure (Stefanakis, 2019).

The local administrations and governments experience growing challenges connected to stormwater urban and runoff management due to rapid urbanization and expansion of cities across the globe. Recent problems have to do with the presence of aging infrastructure, alterations in precipitation patterns, deforestation of the watershed, wetlands degradation, and wide-ranging use of impermeable surfaces such as roadways and parking lots that result in contamination of water resources urban and floods. This is expected that climate change further increases the incidence of these phenomena, and acceptance of GI can make a more valuable and targeted use of ecosystem services to alleviate the effects of climate change (Lennon, 2015). Furthermore, as cities are expanding continuously and becoming dense, there is a correspondingly rising mandate for enhanced sanitary and environmental conditions and a more intellectual way to use urban spaces.

The application of GI is further rising from the continuous global evolution toward a circular economy that trails the 5Rs rules, i.e., decrease water losses and improve the efficiency, reuse and recycle of water and wastewater, restore water of a definite quality to where it was withdrawal from, and recover bioresources. The ultimate aim of such initiatives is to improve the resilience of urban locations to the effects of climate change together with the lessening of greenhouse gases associated with water resources management (Masi et al., 2018). Therefore, it is apparent that GI is at the epicenter of the innovative revelation to attain superior circularity inside the urban systems.

2.3.2.1 Role of GI in the Management of Urban Water

GI performs a vital role in the management of urban water, particularly stormwater and its treatment. In modern cities, the management of stormwater runoff poses a major issue. This contributes to combined sewer overflow (CSO) specifically in older cities where municipal wastewater, stormwater, and urban runoff are collected and transported in the same pipe network to a centralized treatment plant (Barco et al., 2008; Barbosa et al., 2012). Most of the time the volume of receiving overflow is found higher than the treatment capacity of the plants, which indicates that the additional CSO and/or wastewater is discharged into nearby freshwater streams to avoid the burden on treatment plants, which sometimes results in flood events in urban settings (Rizzo et al., 2020) Further the pollution of surface waters and the corresponding damage to the recipient ecosystems, as during flood events water from the sewer washout, roads, and sanitary effluent carry a huge amount of nutrients, organics, solids, and infectious pathogens (Gasperi et al., 2010; Stefanakis, 2019). Consequently, the effective treatment of high volumes of water carrying high contaminant loads is required in several locations to attain the desired effluent quality. Management of such water volumes is also vital to avoid harm to civic and private infrastructure and public and ecosystem health. GI can also play a crucial role in wastewater treatment in peri-urban and urban settings. The recognized solution nowadays is the centralized methods, for example, the application of large end-of-the-pipe treatment plants using conventional technologies. These amenities are characteristically not observed as an eco-friendly and viable solution since they require heavy conventional materials for installations such as steel and concrete, are energy-intensive, need chemicals, and require greater working and maintenance costs (Stefanakis et al., 2014). Further, the installation of a traditional wastewater treatment plant in urban areas habitually degrades the neighboring area in terms of aesthetic and market value. From these, it is vibrant that there is an urgent need for new infrastructure that fulfills the necessities of wastewater treatment and cleanliness along with developing green spaces to moderately compensate for the deficiency of adequate green areas. Consequently, NBS and, particularly, environmental-friendly technologies such as CWs emerge as an ideal alternative that is efficient to deliver the desired values to the ecosystem and encourage water circularity (Table 2.3) (Masi et al., 2018).

TABLE 2.3
Different Types of Nature-Based Solutions along with their Potential Benefits

Type of NBS	Description	Potential to Alleviate Pressure	Benefits	References
Wetlands	Shallow ponds have different types of vegetation cover	Reduction in pollution load, habitat loss, and physical modification	Regulates water supply; water purification; flood mitigation; controls water temperature; enhances biodiversity, recreational and aesthetic value, the livelihood of neighbors, environmental resilience and provides educational opportunities and carbon sequestration, etc.	Karjalainen and Heikkinen, (2005); Harrington et al. (2013)
Riparian buffers	Having various vegetation between terrestrial ecosystems and water streams	Reduction in contamination load and physical changes	Water purification; flood and erosion reduction; controls water temperature due to shade; enhance biodiversity, recreational and aesthetic values	Volk et al. (2009); Stutter et al. (2012)
Reforestation	Trees and several other plants in the catchment	Can reduce the pollution load of different sources including sediments and prevent habitat loss and physical alteration	Water purification, regulates water supply, flood risk and erosion reduction, enhances biodiversity, tourism, carbon sequestration, and various livelihood prospects	Nisbet et al. (2011); Perni and Martínez-Paz (2013)
Relinking rivers to floodplains	Eradicating barriers along the path of the river	Potential to reduce pollution load of different sources, physical changes, and water withdrawal	Regulates water supply, water purification, flood risk and erosion reduction, nutrient replenishment and enhances biodiversity, and ecological resilience, provides recreational, livelihood, and educational opportunities	Schindler et al. (2016); Funk et al. (2019)
River and wetlands restoration	Restoring the natural functions of water bodies	Potential to reduce pollution load including sediment, habitat loss and fragmentation and can minimize manmade flow pressures	Regulates water supply, flood risk reduction, provision of biodiversity, recreation, aesthetics, and tourism	Gilvear et al. (2013); Darwiche-Criado et al. (2017)
Sustainable drainage systems (SuDS)	Potential to manage rainfall onsite	Reduces pollution load, physical changes; water abstractions, etc.	Groundwater recharge, water supply; reduction in flood risk, habitat loss, and water quality enhancement	Jones and Macdonald, (2007)
Soils and vegetated land	Sustaining good soil structure and foliage cover	Can reduce diffuse and point source contamination with sediment and physical alteration	Enhances soil quality and improves drainage, advances water quality, crop yields, and environmental resilience	Soana et al. (2021)

Source: Modified from Souliotis and Voulvoulis (2022).

2.4 FUTURE CONCERNS AND CHALLENGES

While the benefits of encouraging NBS for wastewater treatment and other environmental values are well recognized, as defined previously, there is still a lack of a fundamental understanding of the role of NBS in circular economy and emerging NBS concepts. As defined, a circular economy will alter the meaning we use natural capital, encourage the usage of renewable resources, plan out wastage and adverse effects, and restore natural arrangements. NBS is an important share of the strategies to attain such goals since their viewpoint is closely built on these values. They trail the circularity idea and are observed as enablers to the conversion from a linear to a circular community, as they carry novel strategies such as ecosystem services and biodiversity conservation that bring the additional provision of invention and job creation and move in the direction of sustainable growth (EC, 2015). Sustainability is depended on the utilization of environmental resources to minimize waste, energy, and emission, by reducing energy and material consumption for development in an NBS-based circular economy approach. Currently, most countries are not completely employing NBS for adaptation, despite the apparent capability of nature to deliver multiple economic, environmental, and climate resilience supports, as specified by the World Conservation and Monitoring Centre of the UN Environment Programme (de Lamo et al., 2020). A total of 70 out of the 167 countries, mostly low-income, include NBS actions under the Paris Agreement (Kapos et al., 2019). This may be due to the lack of understanding and/or awareness about NBS, policy and governance issues, and limited accessibility of information, evidence, and funding to acquire and scale up NBS. Together with these, the technical challenges are the key factors that conquer the widespread application of NBS (Kapos et al., 2019). Thus, numerous steps have been projected to encourage circular NBS approaches to overcome these challenges (GCA, 2019; Stefanakis et al., 2021).

2.4.1 Awareness Campaign on Environmental Values

Information should be circulated through the partnership and knowledge and experience exchange to various sectors with the help of civil societies, governments, NGOs, and other private agencies. Governments, NGOs, financial organizations, research bodies, corporate sectors, civil societies, and other participants may encourage the extensive application of NBS through monitoring, assessing, and sharing knowledge with different sectors. The worth of natural assets such as different waterbodies and ecosystems functioning together with the assistance they deliver must be understood particularly by legislators. For example, restoration of coastal wetlands can be two to five times more economical as compared to building non-natural barriers for controlling shores from sea wave erosion (Lawrence & Vandecar, 2015). The average cost required to restore mangroves is around $0.01/square foot, more economical than the generally developed grey structure. However, specific understandings are required for each province to recognize the ideal chances and approaches to employ NBS. Further, the opinions and knowledge of neighboring residents should also be considered in related policies (Stefanakis et al., 2021).

2.4.2 Promote NBS into Climate Change Mitigation and Adaptation Policies

Reduction in the emissions of greenhouse gas is one of the crucial goals as many countries are progressively adopting circular economy strategies, NBS is encouraged to perform a dynamic role in minimizing carbon footprint, particularly when employed at field scale, such as across cities and ecosystems (Stefanakis et al., 2021). NBS for climate adaptation is best employed at extensive scales to deliberate the connections established within and between the ecosystems and the dispersal of probable recipients and influences. The impact of climate change and susceptibility valuations should contain the study of influences on ecosystems and the consequences on people's vulnerability. Due to this, preparation, drafting, decision, and execution on climate adaptation would trail a system viewpoint together with NBS as an integral part since starting. For example, In Mexico, the administration declared one-third of the country's river basins as protected water reserves (around 124 million acres area) and safeguarding the supply of water for approximately 45 million people (GCA, 2019). Parallel policies can be efficiently implemented in different countries. A study conducted by McDonald, and Shemie. (2014) revealed that the restoration and conservation of forests and water in the 534 global major cities can efficiently control water flows and save up to $890 million every year.

2.4.3 Enhance Investment in NBS

As the innovative government preference for climate neutrality is in progress or already in place throughout the world, climate policies and the circular economy establishment are required in the upcoming days. The circular economy will need a technological force and noteworthy investment in novel technologies together with climate change mitigation and adaptation plans. Thus, in this novel circular strategy, financial organizations should advance new funding models that can sustain longstanding investment in NBS. It is well known that getting funding is one of the major barriers to the implementation of NBS and taking action for climate adaptation. Therefore, administrations can fascinate such investments by altering and re-organizing their guidelines, and grants, and offering incentives to private investors for investment in adaptation plans (Kapos et al., 2019). Further, the additional hindrance is the inappropriate understanding of the operation and capabilities of NBS. For example, Canada provides a $1.6 billion fund for the Mitigation and Adaptation of Public Disasters such as preventing societies from wildfires, droughts, flood risks, etc., by investing in nature-based infrastructure (Stefanakis et al., 2021). Currently, this fund employed a $20 million plan to restore salt marsh and improve levees in Nova Scotia (along the Bay of Fundy), which decreases the coastal floods that affect several thousand humans' lives with hectares of agricultural land (GCA, 2019).

2.4.4 Linking Nature-Based Strategies in Financial Settings

NBS are included and assessed in various financial conditions, procurement and industry standards with other policies among the different adaptation resolutions and their profits are evaluated for all possibilities. The solution is to integrate the present encounters with the possible solutions and the prevailing knowledge. However, a very slow change is evident toward the integration of NBS, while some global monetary establishments such as Asian Development Bank (ADB) and World Bank are progressively considering nature-based strategies in their financial settings. From the literature, it is evident that the application of NBS needs attention on three key aspects, viz., science, practice, and policy (EEA, 2015). For example, science is desired in evidence documentation to inspire policy development and to advise about proof-based interferences. Furthermore, to confirm the deployment of communities, individual involvement, and acceptance of NBS intervention, nearby people must be kept engaged and involved in project proposals, applications, and monitoring. Currently, NBS has gained significant political concern mainly relying on its role as a carbon sink. Nevertheless, there is a great and increasing body of suggestions that reinstated, or managed ecosystems are vital to support people and the economy and reduce the adverse climate change effects. Consequently, a strengthened and up-to-date evidence base is needed for climate change adaptation and to build the resilience of societies. This is required for new technologies, models, and methodologies that are presently established given the circular strategy that can verify the concept to validate the scalability of the resolution. Further, most studies on NBS are focused on the Global North, though the societies in the Global South are more susceptible to climate and can gain various profits from NBS use (Seddon et al., 2020). Several major global corporations such as Microsoft, Coca-Cola, Amazon, Mercedes-Benz, Henkel, Siemens, etc. have taken "The Climate Pledge." through a recent initiative considering the need to reduce climate change and understanding the value of NBS. The signatories call for a robust promise by corporations to achieve the net-zero carbon emission goal by 2040. They understood the functions of NBS in the decarbonization of corporate actions and developed an idea onward depending on four key principles for NBS. These are cutting emissions by improving carbon capture, protecting and conserving current ecosystems, being socially accountable and involving local groups and ecologically answerable and conserving biodiversity. Notably, these initiatives originate from the global corporate sector meanwhile such corporations can create awareness and can also establish examples for the international business areas. It is known that Amazon is the major company buying renewable energy, to cover their processes with 100% renewables by 2025. The corporation has also procured 100,000 electric vehicles with visualization to make 50% of their deliveries on net-zero carbon emissions by 2030 (Stefanakis et al., 2021).

2.5 CONCLUSION

Globally, countries are facing wide-ranging encounters, from reduction or exhaustion of natural wealth such as water, ecosystem amenities, and other biotic assets together with its degradation which poses serious pressure on humans and water security. Approximately four billion people throughout the world face critical water shortages for at least one month of the year. NBS is a comparatively novel idea that emerged from values linked to ecosystem resilience, sustainability, and amenities offered by nature. It addresses complex socioenvironmental problems, promotes the integration of water, land, and biotic components, and merges the sectoral strategies of different scales. The key components of NBS are the operationalization of ecological facilities, and the intangible and tangible goods and services created over the working of ecosystems for the welfare of communities. Further, they exhibited their ability in climate resilience, cooling cities, reducing the effect of heat islands, restoration of waterbodies, biodiversity conservation, and enhancing human health and well-being, etc. Several global companies such as Microsoft, Coca-Cola, Amazon, Mercedes-Benz, Henkel, Siemens, etc. have taken "The Climate Pledge." through a recent initiative considering the need to reduce climate change and understanding the value of NBS. The signatories call for a robust promise by corporations to achieve the net-zero carbon emission goal by 2040. Some global monetary institutions such as ADB and World Bank are progressively considering nature-based strategies in their financial settings. In addition to these, there is still a lack of an important understanding of the role of NBS in circular economy and emerging NBS concepts on a wide scale. Thus, information should be circulated through partnerships and knowledge and experience sharing across the various sectors with the help of civil societies, governments, NGOs, and other private agencies. Governments, NGOs, financial institutions, research bodies, civil societies, corporate sectors, and several other participants may encourage the extensive application of NBS through monitoring, assessing, and sharing knowledge with different sectors.

ACKNOWLEDGMENT

The authors are grateful to the Divisional Forest Office, South Kheri Forest Division, Lakhimpur Kheri and the Department of Environment Forest and Climate Change, Government of Uttar Pradesh, India for providing the necessary infrastructure throughout this study.

REFERENCES

Ahiablame, L., & Shakya, R. (2016). Modeling flood reduction effects of low impact development at a watershed scale. *Journal of Environmental Management*, 171, 81–91.

Albert, C., Brillinger, M., Guerrero, P., Gottwald, S., Henze, J., Schmidt, S., & Schröter, B. (2021). Planning nature-based solutions: principles, steps, and insights. *Ambio*, 50(8), 1446–1461.

Anguelovski, I., Connolly, J., & Brand, A. L. (2018). From landscapes of utopia to the margins of the green urban life: for whom is the new green city? *City*, 22(3), 417–436.

Artmann, M., Kohler, M., Meinel, G., Gan, J., & Ioja, I. C. (2019). How smart growth and green infrastructure can mutually support each other – A conceptual framework for compact and green cities. *Ecological Indicators*, 96, 10–22.

Avellan, C. T., Ardakanian, R., & Gremillion, P. (2017). The role of constructed wetlands for biomass production within the water-soil-waste nexus. *Water Science and Technology*, 75(10), 2237–2245.

Barbosa, A. E., Fernandes, J. N., & David, L. M. (2012). Key issues for sustainable urban stormwater management. *Water Research*, 46(20), 6787–6798.

Barco, J., Papiri, S., & Stenstrom, M. K. (2008). First flush in a combined sewer system. *Chemosphere*, 71(5), 827–833.

Benedict, M. A., & McMahon, E. T. (2012). *Green infrastructure: linking landscapes and communities*. Island Press,Washington, D.C., United States.

Blau, M. L., Luz, F., & Panagopoulos, T. (2018). Urban river recovery inspired by nature-based solutions and biophilic design in Albufeira, Portugal. *Land*, 7(4), 141.

Bulkeley, H. (2020). Nature-based solutions towards sustainable communities. Analysis of EU-funded projects. Directorate-General for Research and Innovation, European Commission, Publications Office of the European Union, Luxembourg.

Cohen-Shacham, E., Walters, G., Janzen, C., & Maginnis, S. (2016). Nature-based solutions to address global societal challenges. IUCN: Gland, Switzerland, 97, 2016–2036.

Cole, L. J., Stockan, J., & Helliwell, R. (2020). Managing riparian buffer strips to optimise ecosystem services: a review. *Agriculture, Ecosystems & Environment*, 296, 106891.

Collentine, D., & Futter, M. N. (2018). Realising the potential of natural water retention measures in catchment flood management: trade-offs and matching interests. *Journal of Flood Risk Management*, 11(1), 76–84.

Darwiche-Criado, N., Comín, F. A., Masip, A., García, M., Eismann, S. G., & Sorando, R. (2017). Effects of wetland restoration on nitrate removal in an irrigated agricultural area: the role of in-stream and off-stream wetlands. *Ecological Engineering*, 103, 426–435.

de Lamo, X., Jung, M., Visconti, P., Schmidt-Traub, G., Miles, L., & Kapos, V. (2020). Strengthening synergies: how action to achieve post-2020 global biodiversity conservation targets can contribute to mitigating climate change, UNEP-WCMC, (pp.1–13), 219 Huntingdon Road Cambridge CB3 0DL, United Kingdom.

Derakhshan-Nejad, Z., Sun, J., Yun, S. T., & Lee, G. (2019). Potential CO2 intrusion in near-surface environments: a review of current research approaches to geochemical processes. *Environmental Geochemistry and Health*, 41(5), 2339–2364.

Dudley, N., Stolton, S., Belokurov, A., Krueger, L., Lopoukhine, N., MacKinnon, K., & Sekhran, N. (2010). Natural solutions: protected areas helping people cope with climate change, 20103347633, IUCN/WCPA "Parks for Life" Coordination Office, Slovenia.

EC. (2015). The EU and nature-based solutions. https://research-and-innovation.ec.europa.eu/research-area/environment/nature-based-solutions_en. Accessed 13 January 2021.

Eckart, K., McPhee, Z., & Bolisetti, T. (2017). Performance and implementation of low impact development-A review. *Science of the Total Environment*, 607, 413–432.

EEA. (2015). Exploring nature-based solutions – The role of green infrastructure in mitigating the impacts of weather- and climate change-related natural hazards. Publications Office of the European Union, Luxembourg.

EEA. (2017). Green infrastructure and flood management: promoting cost-efficient flood risk reduction via green infrastructure solutions. Copenhagen, Denmark. https://www.eea.europa.eu/publications/green-infrastructure-and-flood-management.

Eggermont, H., Balian, E., Azevedo, J. M. N., Beumer, V., Brodin, T., Claudet, J., & Le Roux, X. (2015). Nature-based solutions: new influence for environmental management and research in Europe. *GAIA-Ecological Perspectives for Science and Society*, 24(4), 243–248.

Engström, R., Howells, M., Mörtberg, U., & Destouni, G. (2018). Multi-functionality of nature-based and other urban sustainability solutions: New York City study. *Land Degradation & Development*, 29(10), 3653–3662.

Ercolani, G., Chiaradia, E. A., Gandolfi, C., Castelli, F., & Masseroni, D. (2018). Evaluating performances of green roofs for stormwater runoff mitigation in a high flood risk urban catchment. *Journal of Hydrology*, 566, 830–845.

European Commission. (2015). *Report on the progress in implementation of the Water Framework Directive Programmes of Measures*. The Water Framework Directive and the Floods Directive: Actions towards the 'good status' of EU water and to reduce flood risks. ISBN 0060200901.

Faivre, N., Fritz, M., Freitas, T., De Boissezon, B., & Vandewoestijne, S. (2017). Nature-Based Solutions in the EU: innovating with nature to address social, economic and environmental challenges. *Environmental Research*, 159, 509–518.

Frantzeskaki, N. (2019). Seven lessons for planning nature-based solutions in cities. *Environmental Science & Policy*, 93, 101–111.

Fry, T. J., & Maxwell, R. M. (2017). Evaluation of distributed bmp s in an urban watershed-high-resolution modeling for stormwater management. *Hydrological Processes*, 31(15), 2700–2712.

Funk, A., Martínez-López, J., Borgwardt, F., Trauner, D., Bagstad, K. J., Balbi, S., & Hein, T. (2019). Identification of conservation and restoration priority areas in the Danube River based on the multi-functionality of river-floodplain systems. *Science of the Total Environment*, 654, 763–777.

Gasperi, J., Gromaire, M. C., Kafi, M., Moilleron, R., & Chebbo, G. (2010). Contributions of wastewater, run-off and sewer deposit erosion to wet weather pollutant loads in combined sewer systems. *Water Research*, 44(20), 5875–5886.

GCA. (2019). Adapt now: a global call for leadership on climate resilience. Rotterdam, The Netherlands: Global Commission on Adaptation; Washington, DC: World Resources Institute.

Gilvear, D. J., Spray, C. J., & Casas-Mulet, R. (2013). River rehabilitation for the delivery of multiple ecosystem services at the river network scale. *Journal of Environmental Management*, 126, 30–43.

Graves, R. A., Haugo, R. D., Holz, A., Nielsen-Pincus, M., Jones, A., Kellogg, B., & Schindel, M. (2020). Potential greenhouse gas reductions from natural climate solutions in Oregon, USA. PLoS *One*, 15(4), 0230424.

Grebel, J. E., Mohanty, S. K., Torkelson, A. A., Boehm, A. B., Higgins, C. P., Maxwell, R. M., & Sedlak, D. L. (2013). Engineered infiltration systems for urban stormwater reclamation. *Environmental Engineering Science*, 30(8), 437–454.

Guo, X., Du, P., Zhao, D., & Li, M. (2019). Modelling low-impact development in watersheds using the stormwater management model. *Urban Water Journal*, 16(2), 146–155.

Haase, A. (2017). Nature-based solutions to climate change adaptation in urban areas. In N. Kabisch, H. Korn, J. Stadler & A. Bonn (Eds.), *Nature-based solutions to climate change adaptation in urban areas linkages between science, policy and practice* (pp. 221–236). Cham: Springer.

Haines-Young, R., & Potschin, M. (2012). Common international classification of ecosystem services (CICES, Version 4.1). *European Environment Agency*, 33, 107.

Hansen, R., Olafsson, A. S., Van Der Jagt, A. P., Rall, E., & Pauleit, S. (2019). Planning multifunctional green infrastructure for compact cities: what is the state of practice? *Ecological Indicators*, 96, 99–110.

Harenda, K. M., Lamentowicz, M., Samson, M., & Chojnicki, B. H. (2018). The role of peatlands and their carbon storage function in the context of climate change. In: Zielinski, T., Sagan, I., Surosz, W. (eds.) *Interdisciplinary approaches for sustainable development goals*, Economic growth, social inclusion and environmental protection, 169–187. Springer, Cham.

Harrington, R., O'Donovan, G., & McGrath, G. (2013). Integrated constructed wetlands (ICW) working at the landscape scale: the Anne Valley project, Ireland. *Ecological Informatics*, 14, 104–107.

Her, Y., Jeong, J., Arnold, J., Gosselink, L., Glick, R., & Jaber, F. (2017). A new framework for modeling decentralized low impact developments using Soil and Water Assessment Tool. *Environmental Modelling & Software*, 96, 305–322.

Jamion, N. A., Lee, K. E., Mokhtar, M., Goh, T. L., Simon, N., Goh, C. T., & Bhat, I. U. H. (2022). The integration of nature values and services in the nature-based solution assessment framework of constructed wetlands for carbon-water nexus in carbon sequestration and water security. *Environmental Geochemistry and Health*, 45(5), 1201–1230.

Jones, P., & Macdonald, N. (2007). Making space for unruly water: sustainable drainage systems and the disciplining of surface runoff. *Geoforum*, 38(3), 534–544.

Kapos, V., Wicander, S., Salvaterra, T., Dawkins, K., & Hicks, C. (2019). *The role of the natural environment in adaptation, background paper for the global commission on adaptation*. Rotterdam and Washington, DC: Global Commission on Adaptation.

Karjalainen, S. M., & Heikkinen, K. (2005). The river life project and implementation of the water framework directive. *Environmental Science & Policy*, 8(3), 263–265.

Kolokotsa, D., Lilli, A. A., Lilli, M. A., & Nikolaidis, N. P. (2020). On the impact of nature-based solutions on citizens' health & well-being. *Energy and Buildings*, 229, 110527.

Kondo, M. C., Fluehr, J. M., McKeon, T., & Branas, C. C. (2018). Urban green space and its impact on human health. *International Journal of Environmental Research and Public Health*, 15(3), 445.

Kumar, S., & Dutta, V. (2019). Constructed wetland microcosms as sustainable technology for domestic wastewater treatment: an overview. *Environmental Science and Pollution Research*, 26(12), 11662–11673.

Kumar, S., Nand, S., Dubey, D., Pratap, B., & Dutta, V. (2020). Variation in extracellular enzyme activities and their influence on the performance of surface-flow constructed wetland microcosms (CWMs). *Chemosphere*, 251, 126377.

Kumar, S., Nand, S., Pratap, B., Dubey, D., & Dutta, V. (2021). Removal kinetics and treatment efficiency of heavy metals and other wastewater contaminants in a constructed wetland microcosm: does mixed macrophytic combinations perform better? *Journal of Cleaner Production*, 327, 129468.

Kumar, S., Pratap, B., Dubey, D., & Dutta, V. (2022a). Interspecific competition and their impacts on the growth of macrophytes and pollutants removal within constructed wetland microcosms treating domestic wastewater. *International Journal of Phytoremediation*, 24(1), 76–87.

Kumar, S., Pratap, B., Dubey, D., & Dutta, V. (2022c). Integration of Constructed Wetland Microcosms with Available Wastewater Treatment Technologies for the Polishing of Domestic Wastewater and Their Potential Reuses. *International Journal of Environmental Research*, 16(6), 99.

Kumar, S., Pratap, B., Dubey, D., Kumar, A., Shukla, S., & Dutta, V. (2022b). Constructed wetlands for the removal of pharmaceuticals and personal care products (PPCPs) from wastewater: origin, impacts, treatment methods, and SWOT analysis. *Environmental Monitoring and Assessment*, 194(12), 885.

Lafortezza, R., & Chen, J. (2016). The provision of ecosystem services in response to global change: evidences and applications. *Environmental Research*, 147, 576–579.

Lafortezza, R., Chen, J., Van Den Bosch, C. K., & Randrup, T. B. (2018). Nature-based solutions for resilient landscapes and cities. *Environmental Research*, 165, 431–441.

Lawrence, D., & Vandecar, K. (2015). Effects of tropical deforestation on climate and agriculture. *Nature Climate Change*, 5(1), 27–36.

Lennon, M. (2015). Green infrastructure and planning policy: a critical assessment. *Local Environment*, 20(8), 957–980.

Lieuw-Kie-Song, M., & Perez-Cirera, V. (2020). *Nature hires: How nature-based solutions can power a green jobs recovery*. Geneva: World Wide Fund for Nature.

Lorenz, K., & Lal, R. (2018). *Carbon sequestration in agricultural ecosystems* (pp. XII, 392. Springer Cham, Switzerland).

Luan, B., Yin, R., Xu, P., Wang, X., Yang, X., Zhang, L., & Tang, X. (2019). Evaluating Green Stormwater Infrastructure strategies efficiencies in a rapidly urbanizing catchment using SWMM-based TOPSIS. *Journal of Cleaner Production*, 223, 680–691.

Lundin, M., Bengtsson, M., & Molander, S. (2000). Life cycle assessment of wastewater systems: influence of system boundaries and scale on calculated environmental loads. *Environmental Science & Technology*, 34(1), 180–186.

Maes, J., Liquete, C., Teller, A., Erhard, M., Paracchini, M. L., Barredo, J. I., & Lavalle, C. (2016). An indicator framework for assessing ecosystem services in support of the EU Biodiversity Strategy to 2020. *Ecosystem Services*, 17, 14–23.

Masi, F., Rizzo, A., & Regelsberger, M. (2018). The role of constructed wetlands in a new circular economy, resource-oriented, and ecosystem services paradigm. *Journal of Environmental Management*, 216, 275–284.

McDonald, R., & Shemie, D. (2014). Urban water blueprint: Mapping conservation solutions to the global water challenge. Washington, DC: The Nature Conservancy.

McGonigle, D. F., Harris, R. C., McCamphill, C., Kirk, S., Dils, R., Macdonald, J., & Bailey, S. (2012). Towards a more strategic approach to research to support catchment-based policy approaches to mitigate agricultural water pollution: A UK case-study. *Environmental Science & Policy*, 24, 4–14.

Means, M. M., Ahn, C., Korol, A. R., & Williams, L. D. (2016). Carbon storage potential by four macrophytes as affected by planting diversity in a created wetland. *Journal of Environmental Management*, 165, 133–139.

Mell, I. C. (2008). Green infrastructure: concepts and planning. *FORUM ejournal*, 8(1), 69–80. Newcastle University.

Mitsch, W. J. (2017). Solving Lake Erie's harmful algal blooms by restoring the Great Black Swamp in Ohio. *Ecological Engineering*, 108, 406–413.

Nan, X., Lavrnić, S., & Toscano, A. (2020). Potential of constructed wetland treatment systems for agricultural wastewater reuse under the EU framework. *Journal of Environmental Management*, 275, 111219.

Nisbet, T., Silgram, M., Shah, N., Morrow, K., & Broadmeadow, S. (2011). Woodland for water: woodland measures for meeting water framework directive objectives. *Forest Research Monograph*, 4, 156.

Perni, A., & Martínez-Paz, J. M. (2013). A participatory approach for selecting cost-effective measures in the WFD context: The Mar Menor (SE Spain). *Science of the total Environment*, 458, 303–311.

Pineda Pinto, M. (2020). Environmental ethics in the perception of urban planners: A case study of four city councils. *Urban Studies*, 57(14), 2850–2867.

Potschin, M. B., & Haines-Young, R. H. (2011). Ecosystem services: Exploring a geographical perspective. *Progress in Physical Geography*, 35(5), 575–594.

Qiu, Y., Schertzer, D., & Tchiguirinskaia, I. (2021). Assessing cost-effectiveness of nature-based solutions scenarios: Integrating hydrological impacts and life cycle costs. *Journal of Cleaner Production*, 329, 129740.

Rajpar, M. N., & Zakaria, M. (2014). Effects of habitat characteristics on waterbird distribution and richness in wetland ecosystem of Malaysia. *Journal of Wildlife and Parks*, 28, 105–120.

Raymond, C. M., Breil, M., Nita, M. R., Kabisch, N., de Bel, M., Enzi, V., & Berry, P. (2017). An impact evaluation framework to support planning and evaluation of nature-based solutions projects. *Report prepared by the EKLIPSE expert working group on nature-based solutions to promote climate resilience in urban areas.* Centre for Ecology and Hydrology.

Rizzo, A., Tondera, K., Pálfy, T. G., Dittmer, U., Meyer, D., Schreiber, C., & Masi, F. (2020). Constructed wetlands for combined sewer overflow treatment: A state-of-the-art review. *Science of the Total Environment*, 727, 138618.

Rossa, L., Urbaniak, M., & Majewska, Z. (2019). Application of constructed wetland for treating runoff from the dairy cattle farm yard. *Journal of Ecological Engineering*, 20(10), 225–232.

Schindler, S., O'Neill, F. H., Biró, M., Damm, C., Gasso, V., Kanka, R., & Wrbka, T. (2016). Multifunctional floodplain management and biodiversity effects: a knowledge synthesis for six European countries. *Biodiversity and Conservation*, 25(7), 1349–1382.

Seddon, N., Chausson, A., Berry, P., Girardin, C. A., Smith, A., & Turner, B. (2020). Understanding the value and limits of nature-based solutions to climate change and other global challenges. *Philosophical Transactions of the Royal Society B*, 375(1794), 20190120.

Sherren, K., & Verstraten, C. (2013). What can photo-elicitation tell us about how maritime farmers perceive wetlands as climate changes? *Wetlands*, 33(1), 65–81.

Soana, E., Fano, E. A., & Castaldelli, G. (2021). The achievement of Water Framework Directive goals through the restoration of vegetation in agricultural canals. *Journal of Environmental Management*, 294, 113016.

Souliotis, I., & Voulvoulis, N. (2021). Natural Capital Accounting Informing Water Management Policies in Europe. *Sustainability*, *13*(20), 11205.

Souliotis, I., & Voulvoulis, N. (2022). Operationalising nature-based solutions for the design of water management interventions. *Nature-Based Solutions*, 2, 100015.

Spano, G., Dadvand, P., & Sanesi, G. (2021). The benefits of nature-based solutions to psychological health. *Frontiers in Psychology*, 12, 646627.

Stefanakis, A. I. (2019). The role of constructed wetlands as green infrastructure for sustainable urban water management. *Sustainability*, 11(24), 6981.

Stefanakis, A. I., Calheiros, C. S., & Nikolaou, I. (2021). Nature-based solutions as a tool in the new circular economic model for climate change adaptation. *Circular Economy and Sustainability*, 1(1), 303–318.

Stefanakis, A., Akratos, C. S., & Tsihrintzis, V. A. (2014). *Vertical flow constructed wetlands: eco-engineering systems for wastewater and sludge treatment*, Elsevier, Amsterdam, Netherlands, 392 pp. ISBN: 978-0-124-04612-2.

Stephens, A., Taket, A., & Gagliano, M. (2019). Ecological justice for nature in critical systems thinking. *Systems Research and Behavioral Science*, 36(1), 3–19.

Stutter, M. I., Chardon, W. J., & Kronvang, B. (2012). Riparian buffer strips as a multifunctional management tool in agricultural landscapes: introduction. *Journal of Environmental Quality*, 41(2), 297–303.

Tabari, H. (2020). Climate change impact on flood and extreme precipitation increases with water availability. *Scientific Reports*, 10(1), 1–10.

Teixeira, Z., Teixeira, H., & Marques, J. C. (2014). Systematic processes of land use/land cover change to identify relevant driving forces: Implications on water quality. *Science of the Total Environment*, 470, 1320–1335.

Thapa, K., Singh, C., Deshkar, S., & Shaw, R. (2021). Path towards sustainable water management: a case study of Shimla, India. In M. Mukherjee & R. Shaw (Eds.), *Ecosystem-based disaster and climate resilience. Disaster and risk research.* GADRI Book Series. Singapore: Springer.

UN-Water. (2013). What is water security? https://www.unwater.org/sites/default/files/app/uploads/2017/05/unwater_poster_Oct2013.pdf.

Versini, P. A., Jouve, P., Ramier, D., Berthier, E., & De Gouvello, B. (2016). Use of green roofs to solve storm water issues at the basin scale-Study in the Hauts-de-Seine County (France). *Urban Water Journal*, 13(4), 372–381.

Volk, M., Liersch, S., & Schmidt, G. (2009). Towards the implementation of the European Water Framework Directive? Lessons learned from water quality simulations in an agricultural watershed. *Land Use Policy*, 26(3), 580–588.

Wang, J., & Banzhaf, E. (2018). Towards a better understanding of Green Infrastructure: A critical review. *Ecological Indicators*, 85, 758–772.

Ward, N. D., Bianchi, T. S., Medeiros, P. M., Seidel, M., Richey, J. E., Keil, R. G., & Sawakuchi, H. O. (2017). Where carbon goes when water flows: carbon cycling across the aquatic continuum. *Frontiers in Marine Science*, 4, 7.

Washington, H., Chapron, G., Kopnina, H., Curry, P., Gray, J., & Piccolo, J. J. (2018). Foregrounding ecojustice in conservation. *Biological Conservation*, 228, 367–374.

Were, D., Kansiime, F., Fetahi, T., Cooper, A., & Jjuuko, C. (2019). Carbon sequestration by wetlands: a critical review of enhancement measures for climate change mitigation. *Earth Systems and Environment*, 3(2), 327–340.

Wise, S., Braden, J., Ghalayini, D., Grant, J., Kloss, C., MacMullan, E., & Yu, C. (2010). Integrating valuation methods to recognize green infrastructure's multiple benefits. In *Low impact development 2010: Redefining water in the city,* American Society of Civil Engineers, 1801 Alexander Bell Drive, Reston, VA (pp. 1123–1143).

Wong, C. P., Jiang, B., Kinzig, A. P., & Ouyang, Z. (2018). Quantifying multiple ecosystem services for adaptive management of green infrastructure. *Ecosphere*, 9(11), 02495.

Yaka, Ö. (2019). Rethinking justice: Struggles for environmental commons and the notion of socio-ecological justice. *Antipode*, 51(1), 353–372.

Yang, J., Pyrgou, A., Chong, A., Santamouris, M., Kolokotsa, D., & Lee, S. E. (2018). Green and cool roofs' urban heat island mitigation potential in tropical climate. *Solar Energy*, 173, 597–609.

3 Constructed Wetlands as an Emerging and Effective Wastewater Treatment Technology for Decentralized Locations

Jayaraman Sethuraman Sudarsan and Subramanian Nithiyanantham

3.1 INTRODUCTION

Water plays a significant part in human life. The best of all things is water (Birdie and Birdie, 1998). In European municipalities alone, more than 21 million tons of contaminated waste have to receive treatment annually (Vyas, 2004). Especially country like India with a huge population it is very difficult to manage the resources and effective treatment of wastewater also not happening. The use of conservative treatment plants includes more unit processes and unit operation and it makes the public's financial burden heavier, especially in a developing nation like India. This expense reflects a bigger percentage of the budget that is out of reach for small communities in particular (Susmita et al., 2015). But, obviously, a lot of this untreated water finds its route to the normal aquatic resources contaminating the various water resources and making them non-utilizable. Therefore, cost-effective, environment-friendly treatment for domestic wastewater is required. CWL promises a lot in this regard. To date, very inadequate studies on the functioning and efficiency of constructed wetlands, especially under tropical conditions like in India, have been reported (Juwarkar et al., 1995; Billore et al., 1999).

In the present situation managing the resources by nature-based treatment in the present case, the effective process is the root zone technique of domestic wastewater treatment which is the most suited to our climate and is economically viable. The effluents after employing Root Zone Treatment are reported to achieve the prescribed discharge limits for the safe disposal of the sewage. The added advantage of this technique is that it requires no operating cost with negligible maintenance cost. Moreover, it does not require the use of electricity, chemicals, or flocculation. The crops grown on domestic wastewater in root zone treatment are useful by-products and considerably enhance the economic viability of the process.

Wetlands: The wetlands are biodiversity-rich swampy marshlands and it is of two main types – NaturalWetlands (NWL) and Constructed Wetlands (CWL).

NWL: NWL such as swamplands, marshlands, and bayous have been conventionally used all over the world in some countries it is also used as the effective unit for conventional wastewater treatment (Hammer and Knigh, 1994).

CWL: CWL has been used to treat a variety of wastewater, and it has several benefits (Hammer and Knigh, 1994).

A planned system that is intended and built to use normal processes, such as wetland plants, soil, and their associated microbial assemblage to help treat various forms of wastewater is known as a

DOI: 10.1201/9781003441144-3

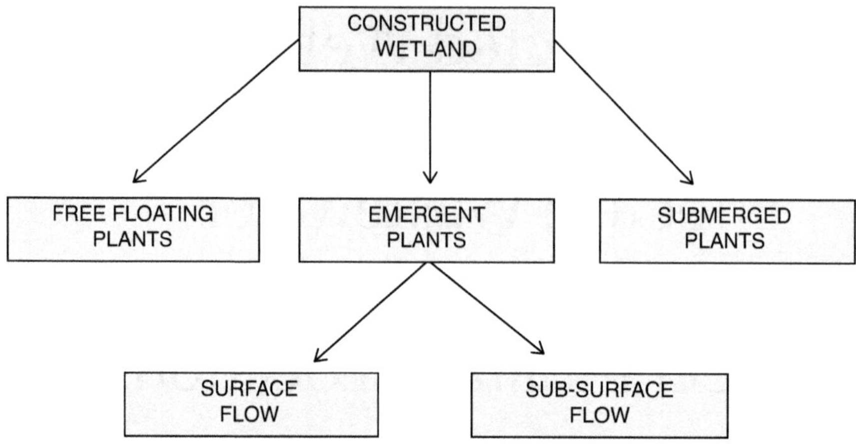

FIGURE 3.1 Classification of CWL.

CWL (Vymazal, 2006). These systems were built with the intention of utilizing microbial assemblages, wetland vegetation, and soil's natural processes to help treat various wastewater (Vymazal, 1996, 2010a).

Classification is done basically depending upon the flow of water and types of plants that are grown in the system, further classification is done as shown in Figure 3.1.

3.1.1 Need for the Research

The growth and development of developing countries like India with a huge explosion of human growth increase water stress and serious water pollution problems which in turn directives an eco-friendly system of Nature-based treatment process. CWL is a tangible and viable option.

3.1.2 Objective

- To check the functioning of the CWL system using emergent Macrophytes: Phragmites *Australis* and Typha *Latifolia* in treating domestic wastewater.
- To evaluate the efficiency and performance of the CWL in treating wastewater using native wetland species.

To accomplish these objectives, six constructed wetland systems were planned conceptualized, designed, operated, and tested in-house at Sri Ramasamy Memorial (SRM) University. All six wetland units were systems combining the features of Horizontal Flow (HF) and Vertical Flow (VF) with the help of baffles and holes.

3.2 LITERATURE REVIEW (A BRIEF SURVEY OF EARLIER WORKS)

In India, though constructed wetland research has not gained momentum, globally it has to its credit more literature. The review of earlier works and cited literature is limited to topics pertaining to the objectives of the present study.

Primitive Civilization employed a drainage system which is a mark of advanced civilization. The drainage wastewater was let out on vegetation and meadows unknowingly of the metabolism of the pollutant removal system that worked within latently. There is no documentation exists except an old note of hand in 1904 as the only evidence (Brix, 1994a). The use of CWL for wastewater treatment at Max Plank Institute in West Germany. Aquatic Macrophytes were used for water quality improvement (Seidel, 1953; Hammer and Bastian, 1989). This experiment was patented later (Seidel, 1973).

Reinhold Kickuth (1960) developed the horizontal surface flow wetland with the root zone method (Kickuth and Konmenn, 1987). Bastian et. Al., (1993) tested the use of wetlands for wastewater treatment. This led to a series of experiments in the USA and Canada in the use of CWL for wastewater treatment. Brix (1994b) presented the results through a worldwide database. Vymazal (1996) presented the treatment results of horizontal flow CWL treating domestic or municipal wastewater in the Czech Republic. The decade from 1970—1980 saw the blossoming of wetland research. The last decade of the 20th century and the beginning of the 21st century showed remarkable growth in constructed wetlands. The constructed wetland systems with low expenses over conventional treatment plants are cheaper by 70% (Narella et al., 2000; Sundaravadivel and Vigneshwaran, 2001). Jayakumar and Dandigi (2002) found that, for summer and rainy seasons, the constructed wetlands treatment for a well-settled community treating municipal wastewater was an efficient treatment technique, low cost, and easy to operate.

The constructed wetlands are generally reliable systems with no energy sources and chemical requirements, minimal operation, and large land requirements (Matcalf and Eddy, 2012). Reed et al. (1988) and Hammer and Bastian (1989) described the nature of wetlands and the conditions that prevail in wetlands. Brix (1994a) classified the Vegetative classification of constructed wetlands as free-floating, sub-emergent, and emergent. But constructed wetlands are yet to be commercialized in developing countries (Padma et al., 2011). The U.S. Environmental Protection Agency (US EPA, 1988) described the removal mechanism such as physical, chemical, and biological metabolism in constructed wetlands. The Typha *Latifolia* and Phragmites *Karka* plants in cement pipes and tested the treatment of wastewater. Following this, the country's first wetland of 90 m × 30 m was built in Sainik School Bhubaneshwar Orissa (Juwarkar et al., 1995; Billore et al., 1999). The removal efficiency was found to be above 50% for BOD, NH_4-N, TKN and a four-celled horizontal flow constructed wetland to evaluate the removal efficiency of wastewater and constituents of an Industrial effluent (Billore et al., 2001). The municipal wastewater in subsurface flows CWL in Spain using two hydraulic application rates and two species (Solono et al., 2003). A free surface wetland consisting of three cells marsh – pond – marsh. Cattails and Bulrushes were used as treatment species and found low removal efficiency for phosphorous (Cameron et al., 2003). The duckweed for the treatment of domestic primary effluent in a continuous flow-free surface-constructed wetland and obtained good results for TSS, BOD, and COD, but poor results for P and N (Ran et al., 2004).

The role of cattail accessed in the removal of BOD, COD, TSS, N, and P from municipal wastewater. The removal efficiencies were good for BOD, and NH_4-N but low for other parameters (Ciria et al., 2005). The N and bacterial removal in a combined system using Phragmites in the vertical flow and Typha in the horizontal flow system and compared with the unplanted combined flow system. There was no difference in bacterial removal (90% in both) but it was seen that nitrification predominates in the vertical flow system and denitrification in the horizontal flow system (Keffla and Gharbi, 2005). The effect of providing deep zones in constructed wetlands to prevent short-circuiting concluded that providing deep zones can improve the wetland performance (Light Body et al., 2007). According to Lee and Scholz, the removal of pollutants can be enhanced by increasing the macrophytes diversity in constructed wetlands (Lee and Scholz, 2007). Casseles-Osorio and Joan (2007) tested the municipal wastewater and obtained good results for BOD, COD, and TSS removal but poor results for NH_3-N removal. The need for primary treatment is recommended before the wastewater is led into a constructed wetland. The treatment efficiency of constructed wetland on wastewater from a milk processing plant. *Phragmites Karka Scirpus Cittorolis* and *Typha Latifolia* were used as wetland plants. *Typha* showed more efficiency in pollutant removal (Gosh and Gopal, 2007).

The potential usage of *Typha Latifolia* for phosphorus removal in a batch-constructed wetland and concluded that the technique is a viable option for the treatment of phosphorus-rich water (Sonavane et al., 2007). The use of constructed wetlands for the treatment of municipal wastewater using *Phragmites Australis* obtained good results (Ahmed et al., 2008). Indicating raw wastewater and treated wastewater in a pilot scale unit achieved the reducing concentration of TSS, TDS, TN, TP, BOD, and COD by 90%, 77%, 85%, 95%, 95%, and 69% respectively (Baskar, 2011). The nitrogen and phosphorus cycles in wetlands greatly influence their structure and function. The biological and chemical process transforms the majority of nutrients in wetlands (Zhou, 2010). Constructed wetlands have a number of removal mechanisms (Vymazal, 2010a, b). Based on the analysis of nutrient accumulation in *Typha Latifolia* and sediment in an integrated setup of constructed wetlands. It showed that though there was a reduction in the levels of BOD and COD, the removal efficiency depends largely upon the resident time of wastewater in the wetlands. The longer the resident time, the greater the removal efficiency of the plant (Csillaet al., 2005; Atiff and Niclas, 2011; Wu et al., 2014). CWL study with different analyses showed an outstanding purification effect of the system for BOD 91%, P 89%, N 63%, and NH_4-N 77% (Mander et al., 2005). From 1991 to 2020, the annual number of journal articles published and the number of articles cited in wetland research increased many folds. Most articles produced by a single country were USA followed by Canada and UK. 'Wetland' was the most active journal. The main issues in wetland research in the future might be Wetland Biodiversity and Constructed Wetland (Liang et al., 2010).

3.3 PROJECT SITE

The project site, Sri Ramaswamy Memorial Institute of Science and Technology (SRMIST) University, is located at Kattankulathur Block, Kanchipuram District, South Chennai, Tamil Nadu, India (Figure 3.2).

The six units of CWL were constructed at the unused free space near the existing conventional sewage treatment plant (STP) at the educational institute premises so that the wastewater from the pretreatment chamber could be utilized for the study.

3.4 METHODOLOGY

The block diagram (Figure 3.3) gives a pictorial representation of the methodology of the work done.

FIGURE 3.2 Project Site – SRM(SRMIST) University, Tamilnadu, India.

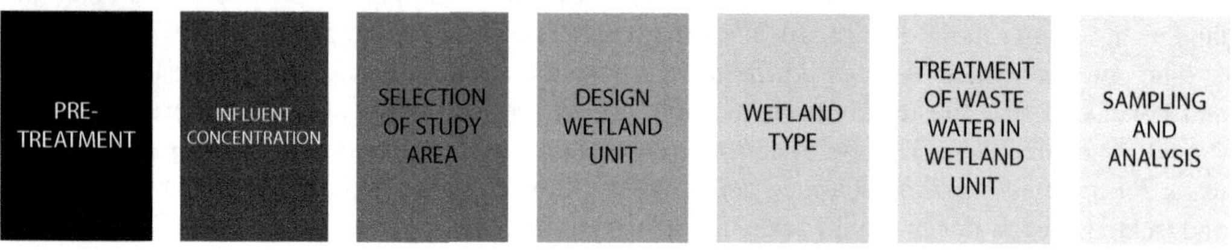

FIGURE 3.3 Step-by-Step methodologies.

3.4.1 Design of Constructed Wetland

Tennessee Valley Authority (TVA) technique is suitable for the plan of CWL units at community levels in the decentralized process of wastewater treatment (U.S. EPA, 2000). Domestic Wastewater: Based on this design criterion, the Final dimensions of the experimental wetland unit were designed as 2×1× 0.9 m. The depth is divided into three regions of 0.3 m gravel at the bottom, 0.3 m sand in the middle and 0.3 m native soil at the top on which the wetland plants are transplanted. The Design flow was estimated at 2.5 m^3 per day and the organic loading was estimated to be 0.03 kg BOD_5/person/day. The cross-sectional area based on organic loading and design inflow was estimated at 0.12 and 0.965 m^2 respectively. The treatment performance of constructed wetlands mainly depends on retention time. Based on the design criteria the retention time of 8 days was adopted in the study (Steiner and Watson, 1993; U.S. EPA, 1993; U.S. EPA, 2000).

3.4.2 Sampling and Analysis

Samples were collected at 24, 48, 72, 96, 144, and 192 hours. Twelve samples were collected each at the inlet and outlet varied for 12 months for one year at different seasons. The domestic wastewater samples were analyzed for the following parameters Chemical Oxygen Demand (COD), Biochemical Oxygen Demand (BOD), Total Nitrogen (TN), and Total Phosphorus (TP), and the dairy wastewater samples for BOD, COD, Total Suspended Solids (TSS), Total Volatile Solids (TVS), Phosphorus (P), pH, Total Nitrogen (TN), Total Solids (TS), Total Dissolved Solids (TDS), and Phosphate according to standard methods for water and wastewater examination derived by American Public Health Association (APHA) Manual 2005 (APHA, 1995).

3.4.3 Influent Characteristics

Domestic Wastewater: The influent to the wetland units was taken from the main sedimentation tank, where the wastewater from nine different areas of the University was collected for treatment. The characteristics of the influents are shown in Figures 3.4–3.7. The parameters in discussion fall within normal domestic wastewater limits (Garg, 2009). There is no striking trend seen in the variation either due to seasonality or due to population. Earlier work on this subject also indicates that there is no appreciable variation found in treatment efficiency owing to seasonal differences (Baskar, 2011). The treatment plants receive influent from different areas; hence a marginal fluctuation is observed in the influent characteristics BOD ranging from 391 maximum to 248 minimum mg/L,

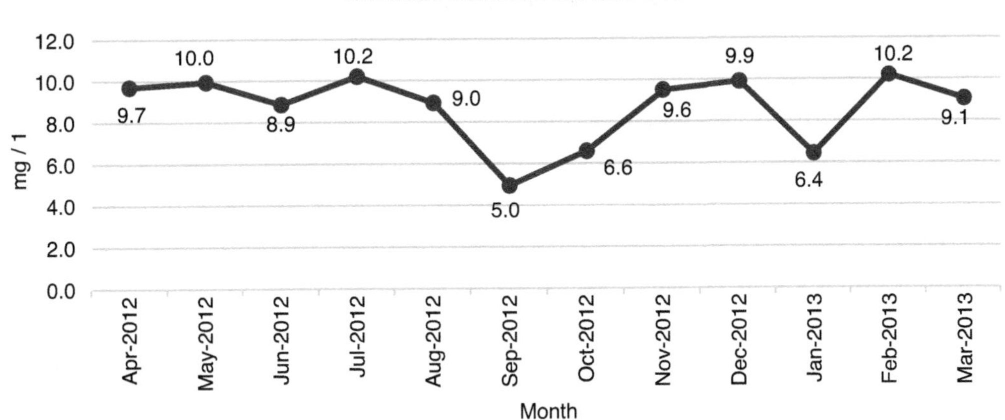

FIGURE 3.4 TP Influent characteristics over 12 months.

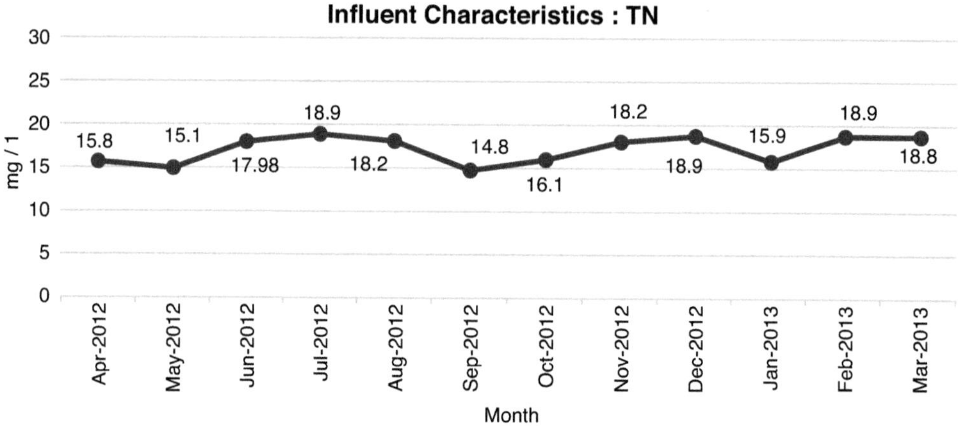

FIGURE 3.5 TN: Influent characteristics over 12 Months.

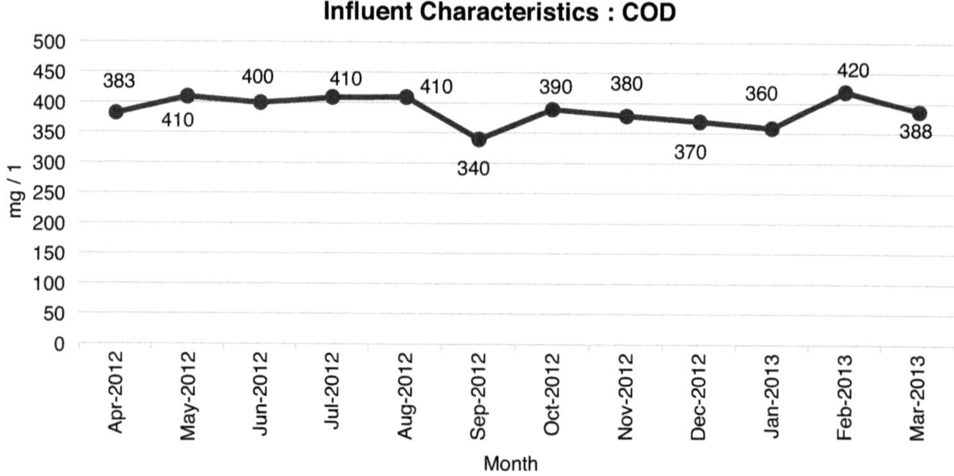

FIGURE 3.6 COD: Influent characteristics over 12 months.

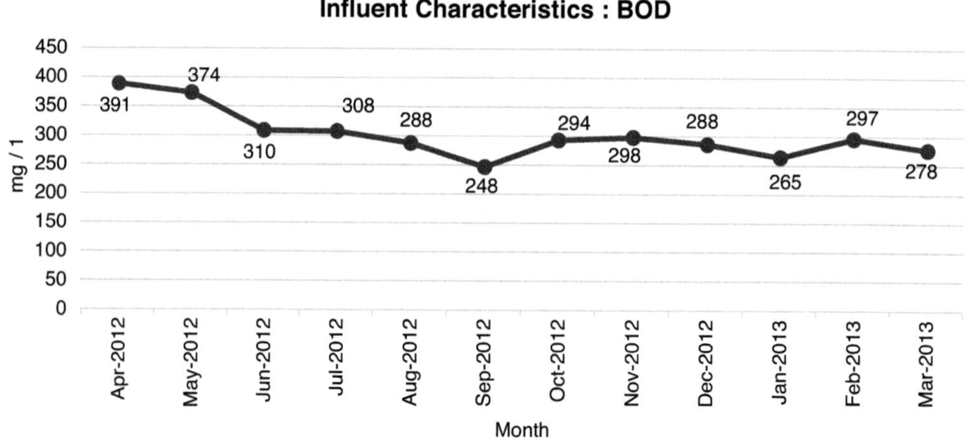

FIGURE 3.7 BOD: Influent characteristics over 12 months.

COD ranging from 420 maximum to 340 minimum mg/L, TP ranging from 10.2 maximum to 5 minimum mg/L and TN ranging from 18.5 maximum and 14.8 minimum over a period of 12 months which includes busy and lean activities of the institutional population(Kadlec, and Wallace, 2008). Target constituent removal proportion was used as a reference to access the CWL treatment performance (Netter, 1993; Patterson and Jones, 2001; Dias and Vymazal, 2011).

3.5 RESULTS AND DISCUSSION

3.5.1 Domestic Wastewater Treatment

Tests were conducted on six integrated flow units, of which two were Control units, two were planted units and the remaining two were planted units with hollow tubes. All six units were designed and constructed with uniform design criteria.

The Removal percentage was computed and analyzed for all six units. The results are discussed with respect to BOD, COD, TN, and TP removal. It was observed for individual parameters TP, TN, COD, and BOD over retention time of 24, 48, 72, 96, 144, and 192 hours. From Figures 3.8–3.11,

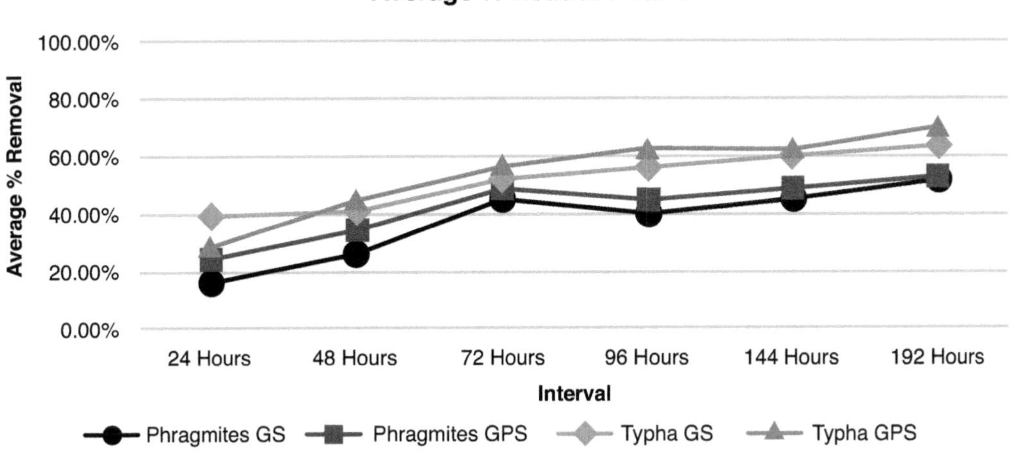

FIGURE 3.8 TP Reduction at different intervals.

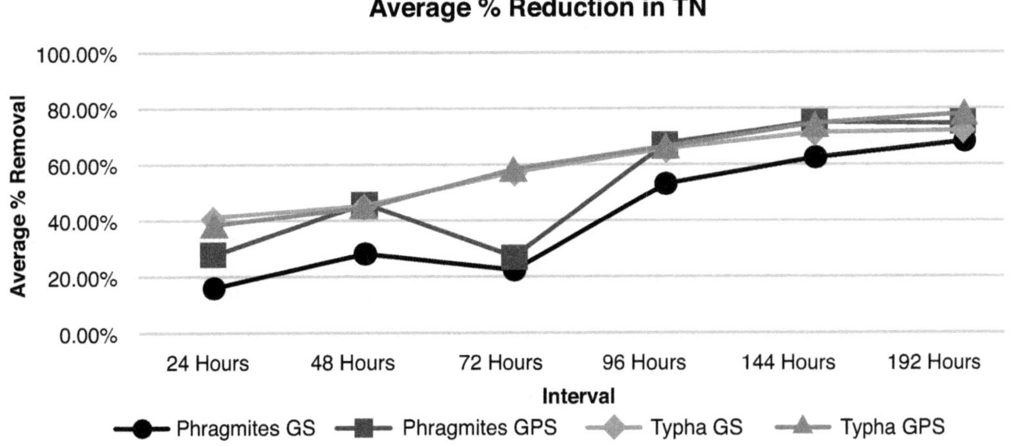

FIGURE 3.9 TN Reduction at different intervals.

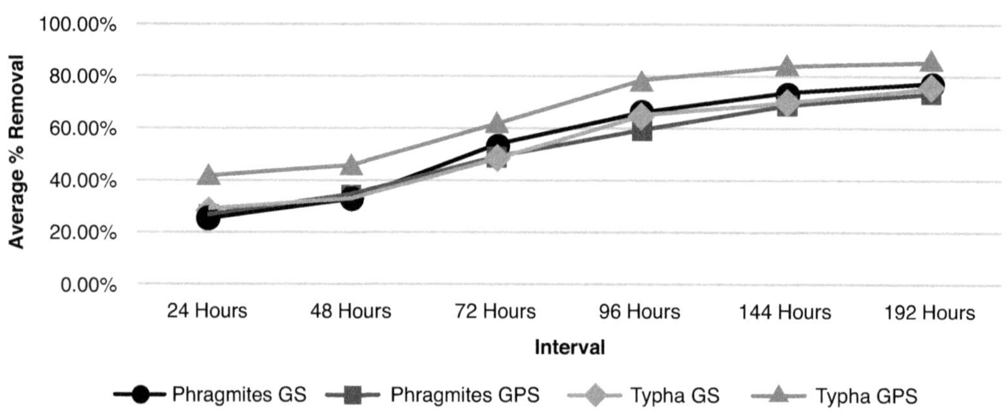

FIGURE 3.10 COD Reduction at different intervals.

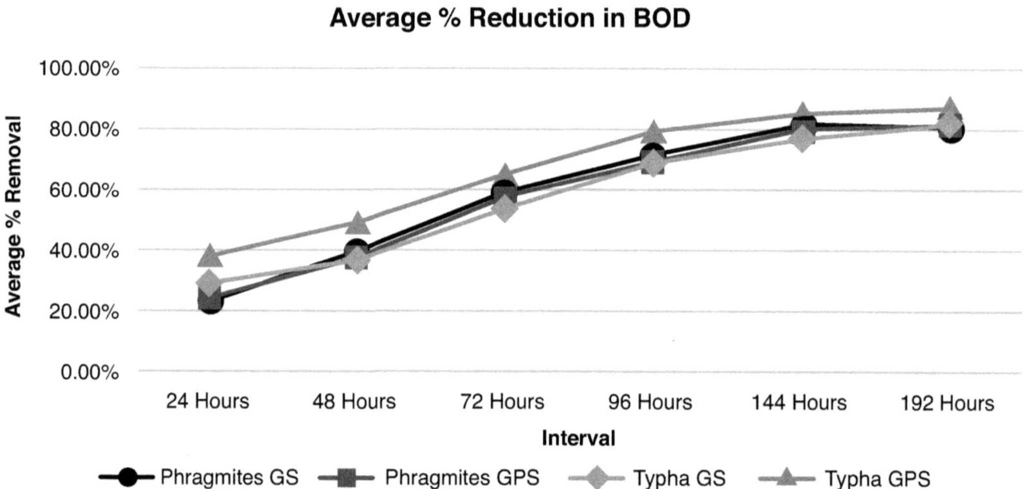

FIGURE 3.11 BOD Reduction at different intervals.

the average reduction percentage for TP, TN, BOD, and COD have progressively increased at different retention times and reached the optimum reduction efficiency at 192 hours. There is a good removal for all components by Day 8 i.e., around 60 to 80% for the inorganic components and close to 90% for the organic components. Among the wetland cells, *Typha* GPS showed better reduction efficiency of TP – 70%, TN – 78%, COD – 86%, and BOD – 87% than the other treatment units.

The higher percentage of TP removal is attributed to the following reasons. The gravel media, the high retention time, and the higher aeration capacity of the system which uses hollow plastic tubes (Netter, 1993).

Nitrogen removal is largely dependent on bacterial activity in the origin zone that is the roots. An extensive variety of organic processes such as Nitrification (Vertical flow) and Denitrification (Horizontal flow) contribute to the removal of TN and was also reported that 60 to 94% removal of TN was achieved from dairy wastewater (Tanner et al., 2002).

The reduction percentage of BOD is higher due to the biodegradable nature of the influent. The higher the biodegradability, the higher will be the reduction. BOD gets reduced by aerobic microorganisms attached to the media and to the plant roots. Higher BOD removal efficiency of 60% to 80% in summer was also observed (Thiesen and Martin, 1987).

Reduction in COD indicates the removal of organic components. The more organic and biodegradable compounds present in the influent account for better COD reduction percentage (Kadlec et al., 2000).

The statistical examination was carried out using SPSS and the results of the outcome represented in Table 3.1 depict the relationship between the actual values and the predicted values of the effluent characteristics. To assess the efficiency of organic removal the BOD reduction efficiency was correlated with reference to estimated value and predicted value the parameter BOD has shown a higher R-Value representing a higher % of degradation. The Constant of Determination R-square measures the goodness-of-fit of the estimated sample regression plane as follows, Y (BOD) = 157.34–7.269X_1. Here the coefficient of X_1 is 7.269 representing the effect of the retention period on influent characteristics. The estimated negative sign implies that such a result highlights that it would decrease by 7.27 for every unit along with the retention time starting from day one. Based on the standardized coefficient, T-GS and T-GPS sight a higher proportion of computation in all the parameters followed by P-GS and P-GPS. From this, it is obvious that plants play an important role in organic removal.

The uptake of nutrients of each component of the plant species was also studied during the period to ascertain the uptake of nutrients by root, stem, and leaf by *Typha* and *Phragmites* are discussed (Table 3.2).

Based on the findings it was detected that root had more acceptance followed by soil, stem, and leaf (Khayat Maesoon, 1986; Cooke, 1992; Arthur et al., 2003; Jason et al., 2004; Healy et al., 2007; Deepak et al., 2012; Rita et al., 2013; Vymazal, 2013; Patel Pratikand and Dharaiya Nishith, 2014). To ascertain the role and the treatment efficiency of wetland plant species, the physical properties of the plant like no of plants, the height of the plant grown, the color of the leaf, width of the stem and root length were estimated before and after treatment and the same is discussed in Table 3.3.

TABLE 3.1
Regression Analysis to Substantiate Organic Removal Efficiency

Parameter	Treatment Unit	R Value	R2	F Value	P Value	Regression Equation	Std Co-eff (b)
BOD	C-GS	0.65	0.42	80.525	<0.001	$Y = 278.48 - 9.008x$	–0.65
	C-GPS	0.77	0.6	167.83	<0.001	$Y = 301.40 - 8.340x$	–0.77
	P-GS	0.82	0.67	231.39	<0.001	$Y = 207.24 - 7.633x$	–0.82
	P-GPS	0.83	0.68	240.58	<0.001	$Y = 210.31 - 7.522x$	–0.83
	T-GS	0.84	0.71	270.16	<0.001	$Y = 195.97 - 6.390x$	–0.84
	T-GPS	0.78	0.62	169.58	<0.001	$Y = 157.34 - 7.269x$	–0.77

TABLE 3.2
Nutrient Budget of the Vegetated Wetland Units

	Storage (%)		Plant Absorption (%)		Soil Accumulation (%)	
Nutrient	T	P	T	P	T	P
N	78	75	47	45	9	8
P	63	53	41	35	44	37
K	54.5	51	4	3.7	46.9	43.9

Note: T represents *Typha Latifolia* and P represents *Phragmites Australis*.

TABLE 3.3
Growth Status of the Plant

Growth Status → Treatment Status ↓	No. of Plants	Height of the Plan (cm)	Width of the Stem (cm)	Average no. of Leaves/Plant	Color of Leaf	Root Length (cm)
Before treatment	12	7	0.6	5	Green	3
After treatment	16	12	0.7	9	Light green	4.7
Improve in yield (%)	33	71	16.6	80	—	57

TABLE 3.4
Projected Cost for a Full-Scale Unit (A Conventional Domestic Wastewater Treatment System on Campus Treating 0.5 MLD Necessitates the Following Costs)

Construction cost (~1500 m^2 in area)	INR. 18,47,755 (23,096$)/-
Equipment cost	INR. 26,15,158 (32,689$)/-
O&M (per year)	INR. 15,00,000 (18,750$)/-
Total	INR. 59,62,913 (74,536$)/-

3.5.2 Cost Estimation – Comparative Study

The estimated cost of constructing a pilot scale unit for treating 2.5 m^3 of water per day per wetland unit was INR 32,179 (402$). For constructing six wetland treatment units, it comes to around INR 180,000 (2,250$) (Table 3.4). However, the expenditure of land was not considered in the present study as the land cost would vary based on the location. For this study, the experimentation was steered on the land maintained by the University Management (Teng et al., 2012; Rita et al., 2013).

To ascertain the cost comparison analysis between a conventional domestic wastewater treatment system and a modified domestic wastewater treatment system, the following analysis was carried out (Sudarsan et al., 2012; Karodpati and Kote, 2013; Mirunaliniet al., 2014; Sudarsan et al., 2014).

If a modified wastewater treatment system is to be used for the same amount, the construction cost combined with the cost of media filling will be almost the same as that of a conventional Domestic wastewater treatment system. However, the equipment cost and O&M costs will drastically come down. This kind of nature-based CWL process will help to reduce 70%–80% in O&M by using a modified wastewater treatment unit (Sudarsan et al., 2012; Karodpati and Kote, 2013; Mirunaliniet al., 2014; Sudarsan et al., 2014; Deeptha et al., 2015; Sudarsan et al., 2015). Applying the same projection and assuming equal construction cost and 10% of equipment cost, the wetland for an equivalent influent discharge will turn out to be INR. 24,09,271 (30,115$)/- which is a 59% saving over the conventional design.

3.6 CONCLUSIONS

i. Though *Typha Latifolia* showed more BOD reduction in the effluent water than that of *Phragmites Australis*.
ii. Nutrient uptake was seen more by *Typha* than *Phragmites* which can be correlated to the reduction in the nitrate in the effluent water from the setup having Typha.
iii. The accumulation was in the order of root>soil> stem> leaf.
iv. The reduction of organics is found to be more than 70% in 12 days, without any chemical treatment with plant growth of more than 70%.

v. If the cost comparison is done with conventional STP and constructed wetland attached STP, the construction cost of constructed wetland combined STP is 59% less than Conventional STP.

3.7 RECOMMENDATIONS

i. Integrated (Hybrid) flow constructed wetland performs better in treating wastewater. Instead of going for fully vertical or fully horizontal. An Integrated (Hybrid) unit will be a better option as proved in the study.
ii. *Typha* and *Phragmites* perform well in treating wastewater. It is better to go for *Phragmites* in controlled field conditions and *Typha* in normal field conditions.
iii. Integrated (Hybrid) wetland is also suitable for treating industrial wastewater as proved by the study of dairy wastewater treatment.
iv. Water budget and Nutrient budget were also done to derive the Mass Balance relationship in the Integrated (Hybrid) wetland unit.
v. The detailed cost estimation was performed to ascertain the cost-effectiveness of the Integrated (Hybrid) constructed wetland unit. Based on the study, it is advisable for smaller communities to go in for a constructed wetland and not be dependent on power availability for wastewater treatment, mainly in developing and undeveloped countries.

ACKNOWLEDGMENT

The authors are thankful to The Management of SRMIST Chennai for providing support and facility in executing this study and the authors are thankful to the management of NICMAR University, Pune for their encouragement in the successful completion of this chapter.

REFERENCES

Ahmed, S., Popov, V., & Trevedi, R.C. (2008). Constructed Wetland as Territory Treatment of Municipal Wastewater. *Waste and Resource Management*, 16(2), 77–84.

APHA. (1995). *Standard Methods for the examination of Water and Wastewater*. A.E. Greenberg (Ed.) (19th edition). Washington, DC: American Public Health Association (APHA), American Water Works Association (AWWA) and the Water Environmental Federation (WEF).

Aracally-Caseless, O., & Joan, G. (2007). Effect of Physiochemical Treatment on the Removal Efficiency of Horizontal Subsurface Flow Constructed Wetland. *Journal of Environmental Pollution*, 146(1), 55–63.

Arthur, F.M., Meuleman-Richard van, L., Gerard, B.J., Rijs, J., & Verhoeven, T.A. (2003). Water and Mass Budgets of a Vertical-flow Constructed Wetland Used for Wastewater Treatment. *Ecological Engineering*, 20, 31–44.

Atiff, M., & Niclas, S. (2011). Nutrient Accumulation in *Typha Latifolia* and Sediment of a Representative – Integrated Constructed Wetland. *Water, Air and Soil Pollution*, 219, 329–341.

Baskar, G. (2011). Studies on Application of Subsurface Flow Constructed Wetland for Wastewater Treatment. Ph.D. Thesis, SRM University, Kattankulathur, and Tamilnadu, India.

Bastian, R.K., & Hammer, D.A. (1993). The Use of Constructed Wetland for Wastewater Treatment and Recycling. In G.A. Moshiri (Ed.), *Constructed Wetland for Water Quality Improvement* (pp. 59–68). Boca Ratan, FL: Lewis Publishers.

Billore, S.K., Singh, N., Ram, H.K., Sharma, J.K., Singh, V.P., Nelson, R.M., & Das, P. (2001). Treatment of Molasses based Distillery Effluent in Constructed Wetland in Central India. *Water Science and Technology*, 44 (11-12), 441–448.

Billore, S.K., Singh, N., Sharma, J.K., Dass, P., & Nelson, R.M. (1999). Horizontal Subsurface flow gravel bed Constructed Wetland with *Phragmites Karka* in Central India. *Water Science and Technology*, 40(3), 165–171.

Birdie, G.S., & Birdie, J.S. (1998). *Water Supply and Sanitary Engineering*. New Delhi: DhanpatRai Publishing Company (p) Limited.

Brix, H. (1994a). Functions of Macrophytes in Constructed Wetland. *Water Science and Technology*, 29(4), 71–78.

Brix, H. (1994b). Use of Constructed Wetland in Water Pollution Control Historical Development Present Status and Future Perspectives. *Water Science and Technology*, 30(8), 209–303.

Cameron, K., Madramootoo, C., Crolla, A., & Kinsley, C. (2003). Pollutant Removal from Municipal Sewage Lagoon Effluents with a Free Surface Wetland. *Journal of Water Science Research*, 3, 2803–2812.

Ciria, M.P., Solano, M.L., & Soriano, P. (2005). Role of Macrophytes – *Typha Latifolia* in Constructed Wetland for Wastewater Treatment and Assessment of Its Potential as a Bio Mass Fuel. *Journal of Bio-system Engineering*, 92, 535–544.

Cooke, James G. (1992). Phosphorus Removal Processes in a Wetland after a Decade of Receiving a Sewage Effluent. *Journal of Environmental Quality*, 21, 733–739.

Csilla, T., Tamás, M.G., & Zoltán, B. (2005). The use of Reed (*Phragmites Australis*) in Wastewater Treatment on Constructed Wetland. *Acta Biologica Szegediensis*, 49(1–2), 81–83.

Deepak, M., Sudarsan, J.S., Deeptha, V.T., & Baskar, G. (2012). Low-Cost Dairy Wastewater Treatment using Constructed Wetlands. *Journal of Institution of Public Health Engineers*, 3, 55–60.

Deeptha, V.T., Sudarsan, J.S., & Baskar, G. (2015). Performance and Cost Evaluation of Constructed Wetlands for Domestic Wastewater Treatment. *International Journal of Environmental Biology*, 36, 1071–1074.

Dias, V., & Vymazal, J. (2011). *Proceedings of the 10th International Conference on Wetland Systems for Water Pollution Control* (pp. 909–917). Lisbon: Ministerio de Ambiente, do Ordemamento do Territori e do Desenvolvimento Regional (MAOTDR) and IWA.

Garg, S.K. (2009). *Sewage Disposal and Air Pollution Engineering* Made easy Publications, New Delhi – 110016. (Vol. 2, pp. 194–203).

Gosh, D., & Gopal, P. (2007). Effect of Wetland Plant Species on Territory Treatment of Wastewater from a Milk Processing Unit. *International Journal of Ecology and Environmental Sciences*, 33(4), 273–292.

Hammer, D.A., & Bastian, R.K. (1989). Wetlands Ecosystems: Natural Water Purifiers? In D.A. Hammer (Ed.) *Construction Wetlands for Wastewater Treatment Municipal Industrial and Agricultural*, (pp. 5–20). Chelsea, MI: Lewis Publishers.

Hammer, D.A., & Knigh, R.L. (1994). Designing Constructed Wetland for Nitrogen Removal. *Water Science and Technology*, 29(4), 15–27.

Healy, M.G., Rodgers, M., &Mulqueen, J. (2007). Treatment of Dairy Wastewater Using Constructed Wetlands and Intermittent Sand Filters. *Bioresource Technology*, 98(12), 2268–2281.

Jason, A.K., William, R.W., & Joseph, J.D. (2004). Water Budget and Cost-Effectiveness Analysis of Wetland Restoration Alternatives: A Case Study of Levy Prairie, Alachua County, FL. *Ecological Engineering*, 22, 43–60.

Jayakumar, K.V., & Dandigi, M.N. (2002). *A Study on Use of Constructed Wetland for the Treatment of Municipal Wastewater during Summer and Rainy Seasons in Semi-arid City in India* (pp. 1–13). Global Solutions for Urban Drainage https://doi.org/10.1061/40644(2002)

Juwarkar, A.S., Oke, B., Juwarkar, A., & Patnaik, A.M. (1995). Domestic Wastewater Treatment through Constructed Wetland in India. *Water Science and Technology*, 32(3), 291–294.

Kadlec, R., Knight, R., Vymazal, J., Brix, H., Cooper, P., & Haberl, R. (2000). *Constructed Wetlands for Pollution Control: Processes, Performance, Design and Operation*. IWA Publishing.

Kadlec, R.H. & Wallace, S.D. (2008). *Treatment Wetlands*. CRC Press. (2nd Edition), Boca Raton, Florida.

Karodpati, S.M., & Kote, A.S. (2013). Energy-efficient and Cost-effective Sewage Treatment Using Phytroid Technology. *International Journal of Advanced Technology in Civil Engineering*, 2(1), 21–24.

Keffla, C., & Gharbi, A. (2005). Nitrogen and Bacteria Removal in Constructed Wetland: Treating Domestic Wastewater. *Journal of Desalination*, 185, 383–389.

Khayat, M.N. (1986). Some Problems Associated with Brewery and Dairy Wastewater Discharges. *Environmental International*, 12(5), 563–569.

Kickuth, R., & Konmenn, N. (1987). Patent: Method for Purification of Sewage Water. *United States and Germany*, 4, 793929.

Lee, B.H., & Scholz, M. (2007). What is the Role of Phragmites Australis in Experimental Constructed Wetland Filters Treating Urban Run-off. *Ecological Engineering*, 29(1), 87–95.

Liang, Z., Ming-Huang, W., Jie, H., & Yuh-Shan, H. (2010). A Review of Published Wetland Research, 1991-2008: Ecological Engineering and Ecosystems Restoration. *Ecological Engineering*, 36, 973–980.

Light Body, A.F., Nepf, H.M., & James, S.B. (2007). Mixing in Deep Zones within Constructed Wetland. *Ecological Engineering*, 29(2), 209–220.

Mander, Ü., Lohmus, K., Teiter, S., Nurk, K., Mauring, T., & Augustin, J. (2005). Gaseous Fluxes from Subsurface Flow Constructed Wetlands for Wastewater Treatment. *Journal of Environmental Science and Health*, 40(6-7), 1215–1226.

Matcalf & Eddy. (2012). Chapter 14 – Treatment, Reuse, and Disposal of Solids and Biosolids. In *Wastewater Engineering: Treatment and Reuse* (4th Edition). Tata McGraw Hill Publishers.

Mirunalini, V., Sudarsan, J.S., Deeptha, V.T., & Paramaguru, T. (2014). Role of Integrated Constructed Wetlands for Wastewater Treatment. *Asian Journal of Applied Sciences*, 10(3923), 1–5.

Narella, S., weaver, R.W., lesikar, B.J., & Persin, R.A. (2000). Improvement of Domestic Wastewater Quality by Subsurface Constructed Wetland. *Journal of Bio-Resource Technology*, 75, 19–25.

Netter, R. (1993). The Purification efficiency of Planted Soil Filter. In G.A. Moshini (Ed.), *Constructed Wetland for Water Quality Improvement* (pp. 249–254). Boca Raton, FL: Lewis Publishers.

Padma, V., Paul, G., Allan, W., Alka, T., & Mamta, T. (2011). Localized Domestic Wastewater Treatment: Part I - Constructed Wetlands - An Overview. *Journal of Scientific and Industrial Research*, 70, 583–594.

Patel Pratik, A., & Dharaiya Nishith, A. (2014). Constructed Wetland with Vertical Flow: A Sustainable Approach to Treat Dairy Effluent by Phyto Remediation. *International Journal of Engineering Science Innovative Technology (IJESIT)*, 3(1), 509–511.

Patterson, R.A., & Jones, M.J. (2001). *Peat Bed Filters for On-site Treatment of Septic tank Effluent* (pp. 315–322). Armidale: Lanfax Laboratories Armidale.

Ran, N., Agami, M., & Oron, G. (2004). A Pilot Study of Constructed Wetland Using Duckweed (Lemna-Jibba-L) for the Treatment of Domestic Primary Effluent in Israel. *Journal of Water Research*, 38, 2241–2248.

Reed, S.C., Middle, E., Brrok, S., & Crites, R.W. (1988). *Natural Systems for Waste Management and Treatment*, IInd edition. New Delhi: Tata McGraw Hill Publishers.

Rita, S.W.R, Chia, C.H., Tsang, J.C., &Wen, L.C. (2013). A Preliminary Investigation of Wastewater Treatment Efficiency and Economic Cost of Subsurface Flow Oyster-Shell-Bedded Constructed Wetland Systems. *Water*, 5, 893–916.

Seidel, K. (1953). *Pflanzungen zwischen Gewässern und land.* Mitteilungen Max-Planck Gesselschaft, pp. 17–20.

Seidal, K. (1973). Patent: Systems for Purification of Polluted Waters. United States US 3, 770623.

Solono, M.L., Soriano, P., & Ciria, M.P. (2003). Constructed Wetland as a sustainable solution for Wastewater Treatment in Small Villages. *Journal of Bio-Science Engineering*, 87, 109–118.

Sonavane, P.G., Ranade, S.V., & Munavalli, G.R. (2007). Potential of Typha Latifolia for Phosphorus removal in batch Constructed Wetland. Ecology Environment and Conservation, 1414, 553–557.

Steiner, G.R., & Watson, J.T. (1993). General Design, Construction and Operation Guidelines: Constructed Wetlands Wastewater Treatment Systems for Small Users Including Individual Residences / United States. Environmental Protection Agency. Tennessee Valley Authority. Morgantown, WV: National Small Flows Clearinghouse, West Virginia University, p. 42.

Sudarsan, J.S., Deeptha, V.T., & Das, A. (2012). Phyto-Remediation of Dairy Wastewater Using Constructed Wetland. *International Journal of Pharma and Bio Sciences*, 3(3), 745–755.

Sudarsan, J.S., Deeptha, V.T., Reenu, Lizbeth Roy., Prathap, M. Giri., & Kumar, S. (2014). Constructed Wetlands for Water Quality Improvement, Recycling and Reuse. *Journal of Aquatic Biology and Fisheries*, 2, 759–776.

Sudarsan, J.S, Reenu, Lizbeth Roy, Baskar, G., Deeptha, V.T., & Nithyanantham, S. (2015). Domestic Wastewater Treatment Performance Using Constructed Wetland. *Sustainable Water Resource Management*, 1(2), 89–96.

Sundaravadivel, M., &Vigneshwaran, S. (2001). Wastewater Collection and Treatment Technologies for Semi-urban Areas in India: A Case Study. *Water Science and Technology,* 93 (11) 329–336.

Susmita, M., Subhro, C., Surajit, S., Md Iqbal, A., & Arithra, G. (2015). Urban Wastewater: Treatment & Re-use: A Theoretical Perspective. *Journal of Basic and Applied Engineering Research*, 2(15), 1255–1259.

Tanner, C.C., Kadlec, R.H., Gibbs, M.M., Sukias, J.P.S., & Nguyen, M.L. (2002). Nitrogen Processing Gradients in Sub Surface Flow Treatment Wetlands – Influence of Wastewater Characteristics. *Ecological Engineering*, 18, 499–520.

Teng, C.J., Shao-Yuan, Leu, Chun-Han, Ko, Chihhao, Fan, Yiong-Shing, Sheu, Hui Yu, Hu. (2012). Economic and Environmental Analysis of using Constructed Riparian Wetlands to Support Urbanized Municipal Wastewater Treatment. *Ecological Engineering*, 44, 249–258.

Thiesen, A., & Martin, C.D. (1987). Aquatic Plant for Wastewater Treatment and Resource Recovery. Magnolia, Orlando FL, pp. 295–298.

U.S. EPA. (1988). Design Manual: Constructed Wetland and Aquatic Plant Systems for municipal Wastewater Treatments, EPA 625/1-88/022, USEPA Office of Water, Washington, DC.

U.S. EPA. (1993). Subsurface Flow Constructed Wetlands for Wastewater Treatment: A Technology Assessment. EPA 832-R-93-008, U.S. EPA Office of Water, Washington, DC.

U.S. EPA. (2000). Constructed Wetlands Treatment of Municipal Wastewaters. EPA 625/R-99/010, U.S. EPA Office of Research Development, Washington, DC.

Vyas, S.K. (2004). Non Energy Intensive Wastewater Treatment Systems. *Proceedings of National Seminar on Environmental Pollution in Punjab Agricultural University,* Ludhiana, pp. 46–54.

Vymazal, J. (1996). The Use of Subsurface Flow Constructed Wetland for Wastewater Treatment in Czech Republic. *Ecological Engineering*, 7, 1–14.

Vymazal, J. (2006). Removal of Nutrients in various types of Wetlands. *Science of Total Environment*, 380, 48–65.

Vymazal, J. (2010a). Constructed Wetland for Wastewater Treatment. *Water*, 2, 530–549.

Vymazal, J. (2010b). The Use of Constructed Wetland with Horizontal Subsurface Flow for Various Types of Wastewater. *Ecological Engineering*, 35, 1–17.

Vymazal, J. (2013). Emergent Plants Used in Free Water Surface Constructed Wetlands. *Ecological Engineering*, 61, 582–592.

Wu, S., Kuschk, P., Brix, H., Vymazal, J., & Dong, R. (2014). Development of Constructed Wetland in Performance Intensification for Wastewater Treatment: A Nitrogen and Organic Matter Targeted Review. *Water Research*, 57, 40–55.

Zhou, N.Q. (2010). Nitrogen Cycle Characterization in Wetland and its Influence in Natural Environment. www/eeexplore/org.

4 Use of Nature-based Technologies for the Treatment of Wastewater from Municipal Slaughterhouse

Jesus Castellanos Rivera, Florentina Zurita, and Luis Carlos Sandoval Herazo

4.1 INTRODUCTION

In developing countries due to the large population increase and lack of resources, freshwater pollution has been considerably suspected due to the lack of industrial wastewater treatment and its discharge into receiving water bodies. Municipalities and entities involved in the generation of effluents related to animal slaughter must propose new technologies to mitigate environmental impacts, derived not only from the effluent itself but also from the implemented technologies themselves; that is, more sustainable technologies and prone to care for the environment (Bustillo-Lecompte and Mehrvar, 2015).

It is important to mention that, in general, industrial effluents contaminate water bodies, therefore, they contribute to the exacerbation of water scarcity, for this reason, government entities and NGOs must work to issue environmental regulations and find solutions that empower the management and final disposal of wastewater. Properly treated wastewater can be used for different purposes (Lorenzo et al., 2012), which would also greatly benefit the well-being of the biosphere and individuals by keeping the water clean without the discharge of untreated wastewater (Saravanan et al., 2021). Survival without adequate services of the precious liquid is impossible since it greatly influences health, education, life expectancy, well-being, and social development (Bastian et al., 2020).

With regard to the meat industry, it is characterized by the generation of a considerable volume of waste and discharges derived from animal reception operations and slaughter stages. In developing countries such as Mexico and Colombia, although some slaughter centers are endorsed and regulated by the Ministry of the Environment, many of them currently operate without complying with quality and hygiene standards. This situation translates into a direct impact on natural resources, which are affected by the type of discharges that these profit centers generate into the environment, which have high loads of degradable organic matter and chemical residues (Perdigón, 2010).

The treatment of organic waste from slaughterhouses is a complex process, due to the components that make it up: proteins, fats, blood, and solids in suspension. The complexity for the processing, handling, and elimination or reusing of solid and liquid organic waste generated in slaughterhouses increases due to the huge volumes of waste that are generated daily (Singh et al., 2014; Qamar et al., 2022). Regularly, most of the biological treatments used in slaughterhouses include the activated sludge system, stabilization lagoons, and bioreactors (Vilvert et al., 2020). However, the problem is that many of these centers do not submit their effluents to any treatment, so they are discharged directly into the nearest body of water, violating current regulations.

DOI: 10.1201/9781003441144-4

Therefore, it is necessary to look for alternatives that facilitate the solution to manage this problem, considering different strategies such as green engineering and cleaner production (CP); looking for the economic and environmental cost benefit in guides for the construction and management of new technologies. So that the effects on the environment generated by this agro-industrial sector can be prevented and/or mitigated, as well as obtaining environmental and economic benefits with the reuse of water resources within the process or in other activities of the sector. Therefore, the objective of this work is to analyze the different nature-based technologies for the treatment of wastewater generated in municipal slaughterhouses.

4.2 MATERIALS AND METHODS

4.2.1 Information Search

This research is of a qualitative type, made up of articles integrated with a database of a large number of publications from the last decade, adding some research from the penultimate and antepenultimate decade to further strengthen the search, especially when the theoretical foundations.

The review was structured by articles, book chapters, books, and theses, which have been published in both English (90%) and Spanish (10%). Google Scholar was used for the search. (50%), Science Direct (10%), 1findr (10%), Springer (10%), Proquest (10%) and EBSCO (10%). The selected articles were reviewed by academic peers, who guaranteed the quality of the data collected. When starting the search, the following keywords were used: nature-based solutions, slaughterhouse wastewater, anaerobic/aerobic treatment, poultry slaughterhouse wastewater, activated sludge, slaughterhouse wastewater characterization, constructed wetlands, biological treatment, and operating conditions, among others.

4.2.2 Environmental Legislation on Municipal Slaughterhouses

To establish a correct discharge of wastewater from slaughterhouses into the environment, the current regulations have been consulted, as well as the maximum limits allowed for the discharge of this type of effluent in different jurisdictions around the world, including data provided by the Australian and New Zealand Environmental and Conservation Council, the Colombian ministry of Environment and Sustainable Development Colombia, the Council of the European Communities, the environment Canada, the Indian Central Pollution Control Board, the People's Republic of Chinese Ministry of environmental Protection, the US EPA and World Bank Group.

4.2.3 Theoretical Analysis

This document was organized with all the data collected, highlighting the main characteristics of wastewater from municipal slaughterhouses, as well as the main standards and environmental regulations for the discharge of effluents from municipal slaughterhouses. In addition, relevant information is presented on anaerobic and aerobic biological treatment used during the last 20–30 years in the handling and management of wastewater from slaughterhouses and of food and in the meat industry. For each of the nature-based solutions that is proposed (anaerobic, aerobic, or anaerobic/aerobic) it is explained what it consists of, how its internal bioremediation process is carried out, function, operation, design, and efficiency in the elimination of pollutants characteristic of this type of effluents.

4.2.4 Statistic Analysis

Descriptive statistics were applied, building frequency histograms, and tables describing the main parameters studied and the type of technology used.

4.3 RESULTS AND DISCUSSION

4.3.1 Characteristics of Wastewater from Municipal Slaughterhouses

Wastewater from municipal slaughterhouses (WWMS) contains a high level of proteins, fats, and carbohydrates that are generated from meat particles, visors, skin, and blood residues (Ziara et al., 2018). Residues of blood and meat particles, in addition to residues of disinfectants and cleaning products, are mainly sources of phosphorus that can be organic and inorganic phosphates (Baker et al., 2021). The characteristics of WWMS are shown in Table 4.1. Pollutants in this type of effluents are determined in terms of biochemical oxygen demand (BOD), chemical oxygen demand (COD), suspended solids (TSS), ammonium (NH_4-H), total phosphorus (PT), and pH. According to the studies in Table 4.1, the concentrations of BOD, COD, total suspended solids (TSS), NH_4-H, total nitrogen (TN), total organic carbon (TOC), TP, total sulfide (TS) and the pH exceeds the standard limit allowed by the World Bank (see section 3.2) for wastewater effluents (Mozhiarasi and Natarajan, 2022).

4.3.2 Environmental Regulation of Municipal Slaughterhouses in Developing Countries

WWMS has been considered by the United States Environmental Protection Agency (US EPA) as the most dangerous wastewater for nature and living beings (Yetilmezsoy et al., 2022).

TABLE 4.1
Typical Parameters of Wastewater from Slaughterhouses

Parameter[a]	Minimum value	Maximum value	Mean±SD
Alkalinity	12	433	222.30±210.20
VFA	833	1060	1306.67±473.33
pH	6.5	7.28	6.88±0.15
K	0.04		0.04±0.00
TSS	1462	7267	4530.25±1357.22
COD	5577	15385	9718.50±1716.41
FOG	600	666	633.00±33.00
NH_4-N	30	417.96	223.98±193.98
PT	26	43	36.10±5.36
TOC	862	5300	3081.00±2219.00
Total Solids	4558	27390	16055.33±6591.53
Pb	34.3		34.30±0.00
BOD	2300	10173	5955.38±2033.29
TVS	1310	18800	9685.25±3942.86
TS	29	1496	762.43±733.57
Turbidity	130	275	202.50±72.50
TN	156	1520	838.00±682.00
PO_4^{-3}	8	120	64.00±56.00

Source: Own Elaboration with Information Collected from Baker et al. (2021); Loganath and Senophiyah-Mary (2020).

[a] All parameters are expressed in mg/L with the exception of pH, SD: standard error.

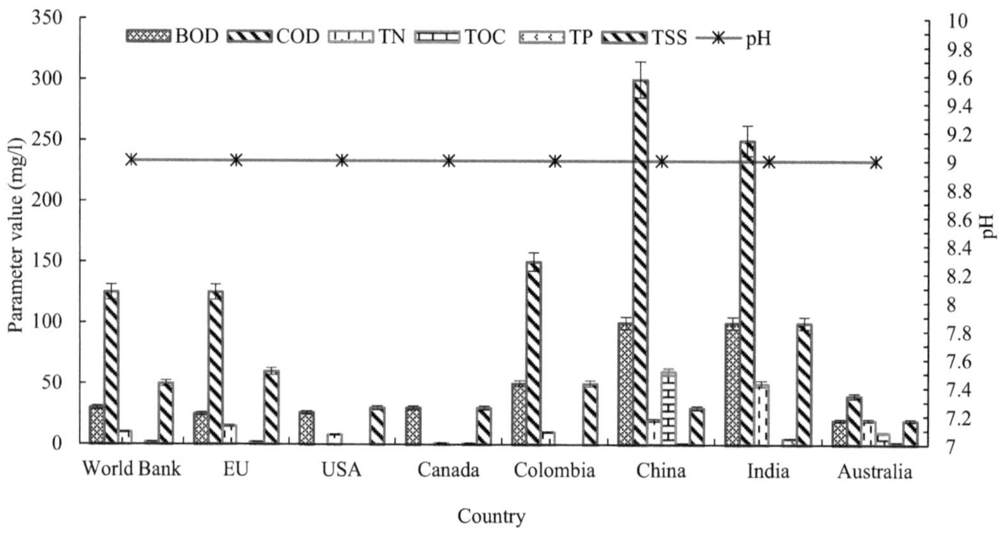

FIGURE 4.1 Comparison among the Standard limits of various countries for wastewater discharge from slaughterhouses.

Source: Own elaboration based on Farooq and Ahmad (2017).

The discharge of this type of effluent can cause the deoxygenation of rivers, lakes, and streams and the contamination of other natural receiving bodies (US EPA, 2024; Wizor and Nwankwoala, 2019). Anaerobic treatment is mostly used due to the high amount of organic matter contained in this type of effluent (Khawer et al., 2022). However, usually, the anaerobic treatment is not enough for a complete elimination of the organic load contained in the WWMS so a post-treatment based on aerobic systems is necessary (Aziz et al., 2022). Because of the physicochemical properties of this type of effluent, it is recommended not to use anaerobic or aerobic processes individually, since the discharge of effluents must comply with limits and standards established by the different world organizations on wastewater discharges (Musa and Idrus, 2021).

Standards and guidelines are an essential part of addressing the ecological effect of wastewater from municipal slaughterhouses in the industrial and business framework (Preisner et al., 2020). Figure 4.1 presents the standard limits allowed for the discharge of wastewater effluents from municipal slaughterhouses established by the World Bank, the Environmental Organization of Canada (2001–2012), the Australian Environmental Council, Malaysian Environmental Quality and the European Communities, in addition to other instances (Baker et al., 2021).

The limits of allowed wastewater pollutants and the corresponding legislation change according to the type of wastewater (food, agricultural, industrial, etc.) (Parida et al., 2021). According to the table above, the strictest level of the standard limits of pollutants allowed among the standards addressed are the guidelines of the Canadian regulations and the established by the United States, while the lowest requirements are established by the People's Republic of China Ministry of Environmental Protection and the Indian Central Pollution Control Board. The limits that have a greater tolerance interval are those established by Canada. The margin allowed by environmental standards provides an area of opportunity for the use of nature-based technologies (aerobic and/or anaerobic) with the purpose of minimizing the pollutants characteristic of this type of effluent and at the same time obtaining secondary products such as biogas (Baker et al., 2021).

4.3.3 Use and Application of Nature-Based Technologies for Municipal Wastewater Treatment

This section addresses information on different nature-based aerobic and anaerobic technologies used in the last 30 years for the sanitation and management of effluents from municipal slaughterhouses. The design, operation, and functioning for the elimination of contaminants present in these effluents are described in accordance with what was published by Aziz et al. (2019). This section addresses information on different aerobic and anaerobic technologies based on nature used in the last 30 years in the treatment and management of wastewater from municipal slaughterhouses, involving the design, operation and functioning in the elimination of contaminants present in these effluents from according to what was published by Aziz et al. (2019).

4.3.4 Natural Technologies Using Anaerobic Processes

A widespread technology for the biological treatment of organic waste is anaerobic digestion (AD) (Zamri et al., 2021; Meegoda et al., 2018). This technology could be considered an experienced technology and, at the same time, innovative and promising, since it can provide society with a sustainable energy source, and at the same time allow the efficient management of wastewater with a high organic matter content such as effluents from of municipal slaughterhouses (Basitere et al., 2020). The complex organic matter contained in this type of water is metabolized due to the action of different groups of bacteria and archaea under anaerobic conditions (no oxygen available) (Zhang et al., 2019). Four stages are involved in the AD process, namely (i) hydrolysis, (ii) acidogenesis, (iii) acetogenesis, and (iv) methanogenesis. In Figure 4.2, these processes are illustrated graphically, and the steps involved in them are explained below (Li et al., 2019; Aziz et al., 2019):

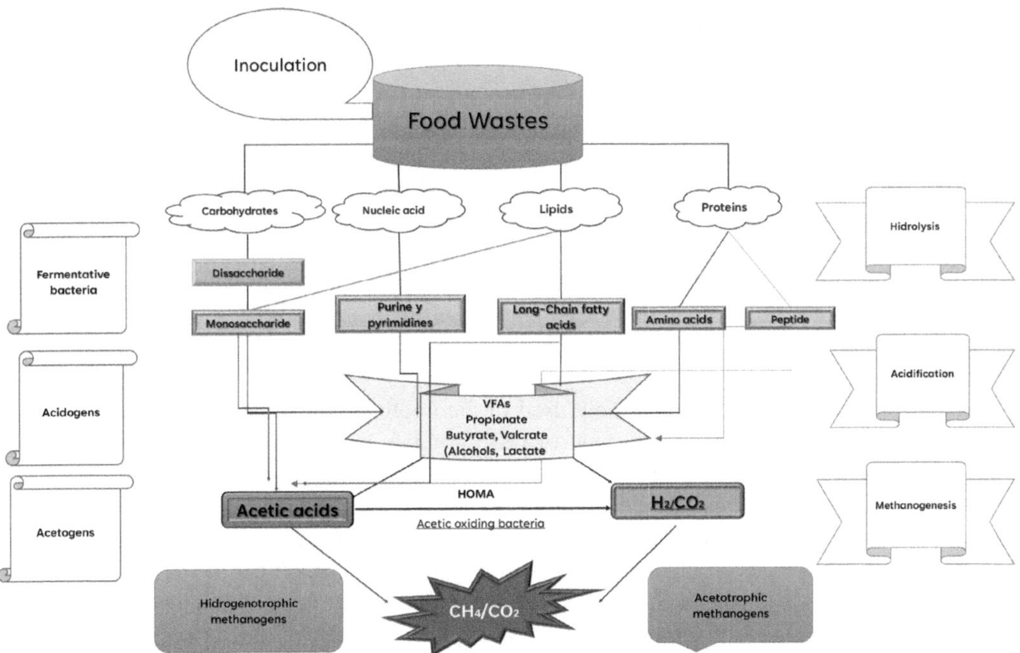

FIGURE 4.2 Diagram of biological processes involved in anaerobic digestion.

Source: Own elaboration with information taken from Li et al. (2019); Aziz et al. (2019); Wang et al. (2018).

1. In hydrolysis, which is the first step, long-chain organic macromolecules (polymers) that hydrolytic bacteria cannot metabolize are broken down into simpler molecules by the intervention of hydrolytic coenzymes so that they can pass through the cell walls of acidogenic bacteria without difficulty. Purines, pyrimidines, simple sugars, fatty acids, and alcohols are some of the products that are obtained in this first stage.
2. In the second stage called acidification (also called fermentation), the metabolism of compounds such as long-chain fatty acids (LCFAs), alcohols, amino acids, and monosaccharides is carried out. This step serves as an intermediate in the metabolism of the substrate as an oxidizing agent. Acidogenic bacteria, during this transformation ultimately metabolize the aforementioned compounds and generate larger (propionate, butyric acid) and smaller (acetic acid or acetate) organic molecules, normally called volatile fatty acids (VFAs).
3. In the third stage, the acetogens degrade the VFA produced, mainly to acetate and hydrogen. A limited number of bacteria called homoacetogenic bacteria use carbon dioxide (CO_2)/Hydrogen (H_2) as a base (substrate) to synthesize acetate (acetic acid).
4. In the fourth stage called methanogenesis, the acidification compounds (for example, methanoic acid, acetate, CO_2/H_2, among others) are transformed to methane by means of strictly anaerobic methanogens (as shown in Equations 4.1 to 4.5).

$$CH_3COO^- + H_2O \rightarrow CH_4 + HCO_3^- \tag{4.1}$$

$$HCO_3^- + H^+ \rightarrow CH_4 + 3H_2O \tag{4.2}$$

$$4CH_3OH \rightarrow CO_2 + 2H_2O \tag{4.3}$$

$$4HCOO^- + 2H^+ \rightarrow CH_4 + CO_2 + 2HCO_3^- \tag{4.4}$$

$$4H_2 + CO_2 \rightarrow CH_4 + 2H_2O \tag{4.5}$$

Since methanogenesis is a laborious phenomenon, the joint intervention of countless mesophilic bacteria is required to carry out this process. There are two groups of bacteria that transform hydrogen and acetate synthesized in the acetogenesis stage into methane. Although acetate-using bacteria are the main ones in charge of the greatest methane formation (around 70%), their growth in the medium is limited compared to hydrogen-using bacteria and this requires an increase in sludge concentration and longer retention times to promote their increase (Rabii et al., 2019).

4.3.4.1 Anaerobic ponds

Anaerobic ponds (Figure 4.3a) are widely used in the treatment of waste from the meat industry as a first stage of the secondary treatment of wastewater from municipal slaughterhouses with a high concentration of pollutants. This type of technology has been shown to reduce 90% COD and BOD (Schmidt et al., 2019). Anaerobic ponds are a viable option because the installation and operation costs are relatively low, in addition to the fact that their installation and handling are simple; however, there are two drawbacks associated with this type of technology, such as odor emissions and the generation of methane, a powerful greenhouse gas (Vilvert et al., 2020). Due to these drawbacks, it has been decided to use covered anaerobic ponds, which, although their installation and maintenance are higher compared to uncovered ponds, offer multiple significant benefits such as odor control, intensification of the organic matter degradation process, and the removal of BOD, the potential to capture methane-rich gas and the reduction of greenhouse gas emission, among others (Aziz et al., 2019; McCabe et al., 2013).

Anaerobic (covered) ponds have been applied with excellent results in slaughterhouse effluent waste sanitization currently and years ago (Farzadkia et al., 2016; McCabe et al., 2014; Mittal, 2006;

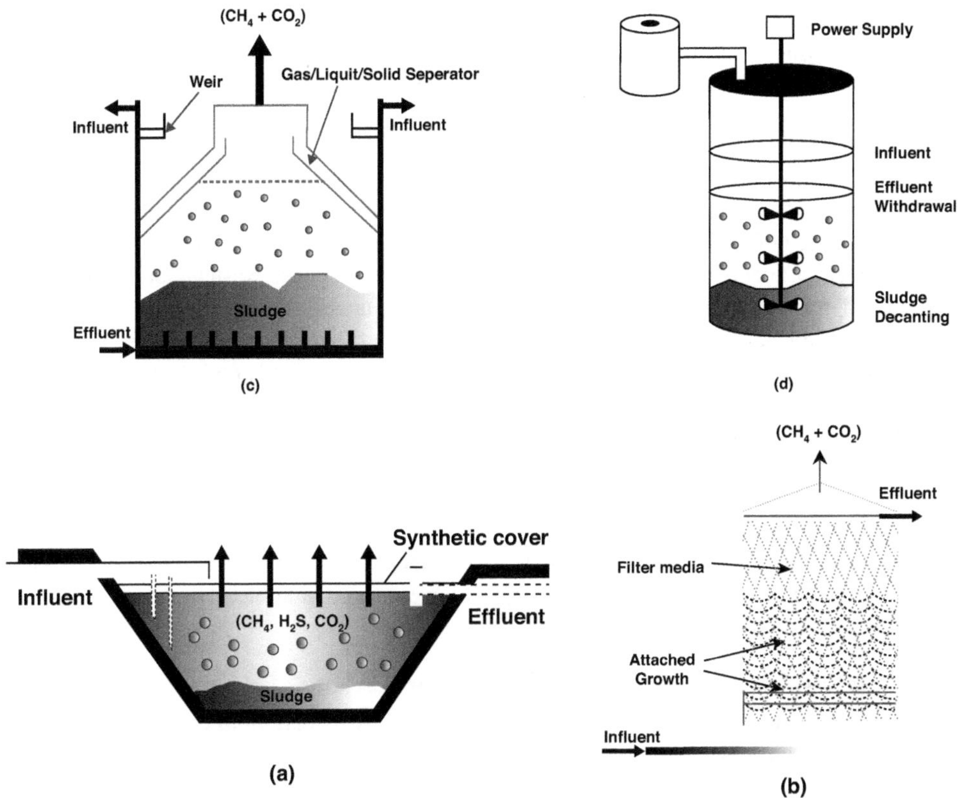

FIGURE 4.3 Main schemes of nature-based anaerobic technologies (a). Anaerobic pond, (b). Anaerobic filter (c). Upflow anaerobic sludge blanket (d). Anaerobics sequencing batch reactor.

Source: Own elaboration with information taken from Aziz et al. (2019)

Johns, 1995). Due to the nature of slaughterhouse wastewater and the high concentration of contaminants it contains, two important factors, namely land availability and a favorable climate are required for the successful use of anaerobic ponds for the removal of contaminants of this type of effluents (Aziz et al., 2019).

4.3.4.2 Anaerobic Contact Reactor

This type of reactor is a mechanically stirred vessel in which the content is totally mixed, and the generated sludge is reused (Morgan-Sagastume et al., 2019). The residue from the tank is diverted to a fluid separator (for example, a gravity settling bowl, utensil where the sludge is on the surface, or plate flocculant), where the waste is recovered and fed back into the anaerobic digester (Goel et al., 2022). Due to the large amount of contaminants contained in wastewater from municipal slaughterhouses, ACRs are very efficient in treating this type of effluent (Talaiekhozani et al., 2019). Changes in organic load are easily accomplished and hydraulic retention times (HRT) are relatively short (Patel et al., 2021). ACRs are less sensitive to acidity and to different inhibitors. Stirred bioreactors with cell retention based on membranes are very efficient in the generation of biogas (Basitere, 2017; Brindhadevi et al., 2021).

Waste from municipal slaughterhouses has been effectively treated with this type of reactor on several occasions (MASSE and Masse, 2000; Saddoud and Sayadi, 2007; Loganath and Senophiyah-Mary, 2020). These types of technologies have the competitive advantage that, if they are reinforced with the help of other techniques that allow the separation and reuse of sludge, they can retain it in an exceptional way. Additionally, because these reactors separate the HRT and sludge retention time (SRT) from the treatment system, it is very likely that the volume of the vessel will be considerably reduced (Loganath and Senophiyah-Mary, 2020).

4.3.4.3 Anaerobic Filter/Fixed Film Anaerobic Reactor

Slaughterhouse effluents, due to the high concentration of contaminants they have, have been an excellent incentive to be treated by means of an anaerobic fixed film reactor (AFFR) or also called an anaerobic filter (AF), which has been widely used in wastewater with a high concentration of pollutants. If the concentration of suspended solids in slaughterhouse wastewater is limited, higher extraction and contaminant removal efficiencies can be achieved. The configuration of the reactor is of vital importance, since, as a consequence of this, the effluents can flow toward the upper part or toward the lower part of the reactor. The microorganisms that are present in the filter medium are responsible for eliminating the concentration of organic contaminants that have adhered to the surface of the reactor film, which is normally made of stone and/or plastic (Figure 4.3b) (Ülgüdür et al., 2019). When the filter media become clogged, the performance of the reactor is affected, therefore, if an effective result is required, the effluents must be soluble for the treatment to be adequate (Khan et al., 2022; Akhbari et al., 2021; Armal and Shanta, 2015). For a correct design of an AF or a fixed film reactor, parameters such as HRT, depth, temperature, and the size/shape of the filter medium must be considered, since these will have a significant effect on the efficiency of the system (Rajakumar et al., 2011).

4.3.4.4 Upflow Anaerobic Sludge Blanket Reactor (UASB)

When a high-speed digester is modified it gives rise to a UASB (Chollom et al., 2020). Significant efficiencies have been achieved in the handling and management of wastewater from slaughterhouses in recent years (Saghir and Hajjar, 2022; Loganath and Senophiyah-Mary, 2020; Musa et al., 2020; Musa et al., 2019; Batubara et al., 2018). The operating mechanism is based on the entry of the effluent through the lower part of the reactor and its route through a sludge surface continuously until it reaches the upper part of the reactor, where the treated residual water is collected (Figure 4.3c). A UASB reactor consists of three basic elements, namely, the sludge, the effluent to be treated, and the gases generated such as methane and carbon dioxide (Arthur et al., 2022). The flow rate toward the top of the reactor makes the mixing of sludge and generation of biogas possible. Another parameter of crucial importance in the performance and operation of a UASB reactor is the size and shape of the sludge particle, which can be flocculent and/or granular (Rajagopal et al., 2019). It is important that the particles that participate in this process are retained inside the reactor (bottom) to inhibit their escape. In the same way, it is highly recommended that the ascending flow velocity be increased as the reactor performance requires it, which causes this vital design parameter in the construction of the UASB (Vashi et al., 2019; Bressani-Ribeiro et al., 2019). At the end of the 1980s, this type of technology was implemented with excellent results in New Zealand and Belgium (Mansourian et al., 2024). The amount of organic load that enters the UASB reactor determines the operation of these treatment systems, which also makes it a main drawback, since, if the amount of organic matter is high, it makes the UASB incapable of treating wastewater from slaughterhouses due to the amount of impurities present and substances such as lipids, proteins, and polysaccharides (Hollas et al., 2021).

4.3.4.5 Anaerobic SBR

One of the best nature-based technologies for the sanitation of effluents from slaughterhouses is the treatment by means of sequential batch biological reactors (Basheer et al., 2021; Aziz, 2022; Ripoll et al., 2022). This type of reactor was originally built for the first time in the year 1993 at Iowa State University in the United States Dague et al. (2018) as a result of the combination of the mechanisms of the anaerobic contact reactor (ACR) and the treatment by muds activated (Aziz et al., 2019). Feeding, reaction, sedimentation, and decantation are performed sequentially in a single batch reactor (Figure 4.3c). Due to these procedures, it is not necessary to use a settling tank,

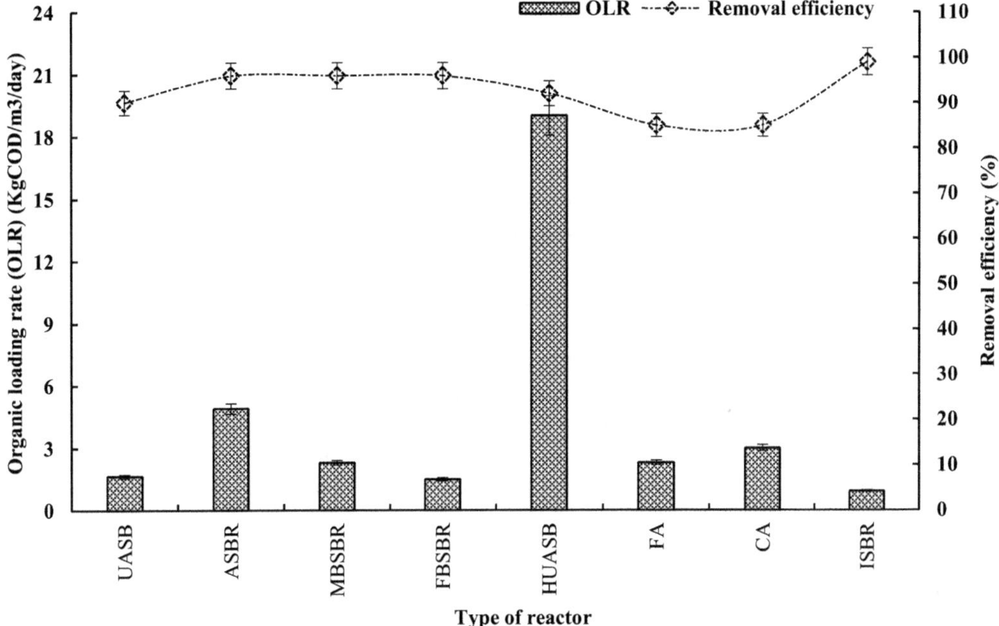

FIGURE 4.4 Main treatment systems used for the removal of organic load using anaerobic technologies.

Source: Own elaboration based on the data reported by Alayu and Yirgu (2018).

which makes it a simpler technology, in addition to the fact that its maintenance and operating costs are lower compared to other anaerobic reactors (Piotrowski et al., 2019). One of the most important factors in an anaerobic SBR is the retained biomass. A batch fixed volume reactor is used through which slaughterhouse effluents enter during the feeding stage. Subsequently, in the reaction stage and through the action of a mixer, the biomass and organic matter are combined for a determined period of time, continuously or in batches. This is where organic matter is transformed into biogas (Harada and Hidayat, 2022). The amount of biogas generated is directly proportional to the Feed/Mass ratio, therefore, in the initial stage of this phase there is a greater amount of biogas generated and as the organic matter decreases with respect to time, the amount of biogas that is produced is less and less (Guo et al., 2020). En la fase de sedimentación puesto que la carga orgánica (Feed/Mass) es mucho menor, la biomasa se separa del efluente (que es el agua residual tratada). If the amount of Feed/Mass is lower, a greater sedimentation of the biomass is favored (Shoukat et al., 2019). In the decantation, which is the last stage of the process, the treated effluent is removed from the reactor and the process cycle ends, giving rise to a new cycle similar to the previous one, where the number of times the process is carried out depends on HRT (Abd Nasir et al., 2019).

Figure 4.4 presents the removal of COD by different anaerobic technologies depending on the organic load of pollutants provided.

4.3.5 Natural Technologies Using Aerobic Processes

A large number of microorganisms through anabolic (autotrophs or food producers) and anabolic (heterotrophs or food consumers) functions carry out the AD process in the presence of oxygen (aerobic conditions). Organic compounds are used as an energy source by heterotrophs and autotrophs use carbonates such as carbon dioxide (CO_2) as an energy source. Mixed microorganisms break down biomass and ammonium gas into simpler molecules such as carbon dioxide, and nitrogenous compounds (nitrites (NO_3^-) and nitrates (NO_2^-), generating a significant number of cells. In aerobic metabolism two crucial steps are carried out, namely intracellular respiration and oxidation with synthesis (Ahmed et al., 2022; Yuan et al., 2022).

Technologies based on nature that are carried out under strict aerobic conditions are generally adequate if they are used after an anaerobic treatment, due to the large amount of oxygen and the treatment times necessary for the development of the process (Han et al., 2020; Bustillo-Lecompte and Mehrvar, 2017). For the processes to be efficient and feasible, effluents with relatively low amounts of COD must be used (concentrations less than 1,000 mg/L) (Maisuria et al., 2022; Tao et al., 2021). The efficiency of these systems in the elimination of pathogenic microorganisms and unpleasant pests, characteristic of slaughterhouse wastewater, has been demonstrated (Sib et al., 2020; Abdel-Mohsein et al., 2020). Some of the main disadvantages of these systems are the increased requirement of necessary oxygen, the constant maintenance of the system and a considerable amount of biomass generation. The generated sludge must have an additional treatment before being discharged. Despite these drawbacks, high removals of organic pollutants have been obtained by applying these aerobic technologies (Sandoval et al., 2022). Nitrogen is removed through nitrification/denitrification processes and phosphorus is removed through anaerobic and aerobic processes in a process called biological phosphorus removal (Tong et al., 2019; Hosseinlou, 2021; Buayoungyuen et al., 2022).

4.3.5.1 Aerobic Ponds/Lagoons

Lagoons and aerobic ponds are the most used procedure in the handling and management of slaughterhouse wastewater, although they are similar technologies, the type of aeration is what differentiates both treatment techniques. In the ponds, the supply of oxygen is carried out through the process of photosynthesis and in the lagoons artificial aeration is installed. With this type of ecotechnologies, excellent pollutant removals have been achieved, normally around 95% (Aziz et al., 2019). The characterization of the wastewater to be treated is of vital importance for oxygen demand and treatment times, in addition, during this type of treatment considerable suspended solids are generated if an efficient sedimentation system is not used (Vadiveloo et al., 2022). Although this type of treatment technique for slaughterhouse effluents has been used after the use of natural anaerobic technology, there is a limited number of studies on the application of these bioremediation techniques (Hernández et al., 2016; Vadiveloo et al., 2022).

4.3.5.2 Trickling Filters (TF)

It is a type of fixed bed reactor, where the development of microorganisms is carried out in highly porous media that are ideal for cell growth. The filter media used are usually rocks and plastics that have different areas. If it is desired to prevent agglomerations and achieve high organic load flows, it is important to use filter media of relatively large sizes. Subsequently, a light layer containing microorganisms forms on the surface of the medium. With this variation in particle size, the effluents that enter the reactor are filtered on the different layers used (Arthur et al., 2022). The bacterial consortium contained in the medium metabolizes organic matter under aerobic conditions. In the thinnest layer, the oxygen present is limited because there is not enough oxygen transfer, which results in the separation of bacterial consortia from the surface area. Generally, this type of biological technology is used as a secondary treatment in wastewater from slaughterhouses, however, although this type of bioreactor has very promising advantages (such as low operating cost, small spaces required, low energy cost, among others) have applied very little in the management of this type of effluents in developed and developing countries (Stefanakis et al., 2019).

4.3.5.3 Rotary Biological Contactor (RBC)

It is a type of fixed film bioreactor that contains several discs or plates, usually plastic, that are supported on a horizontal base (Figure 4.5b). The discs submerged up to a certain level rotate on the base in a container, the force that makes it possible for the discs to rotate is caused by a motor, that is, additional energy is required (Hamedi et al., 2021). The material with which the bioreactor discs are built is corrugated plastic, which allows the development and growth of microorganisms that

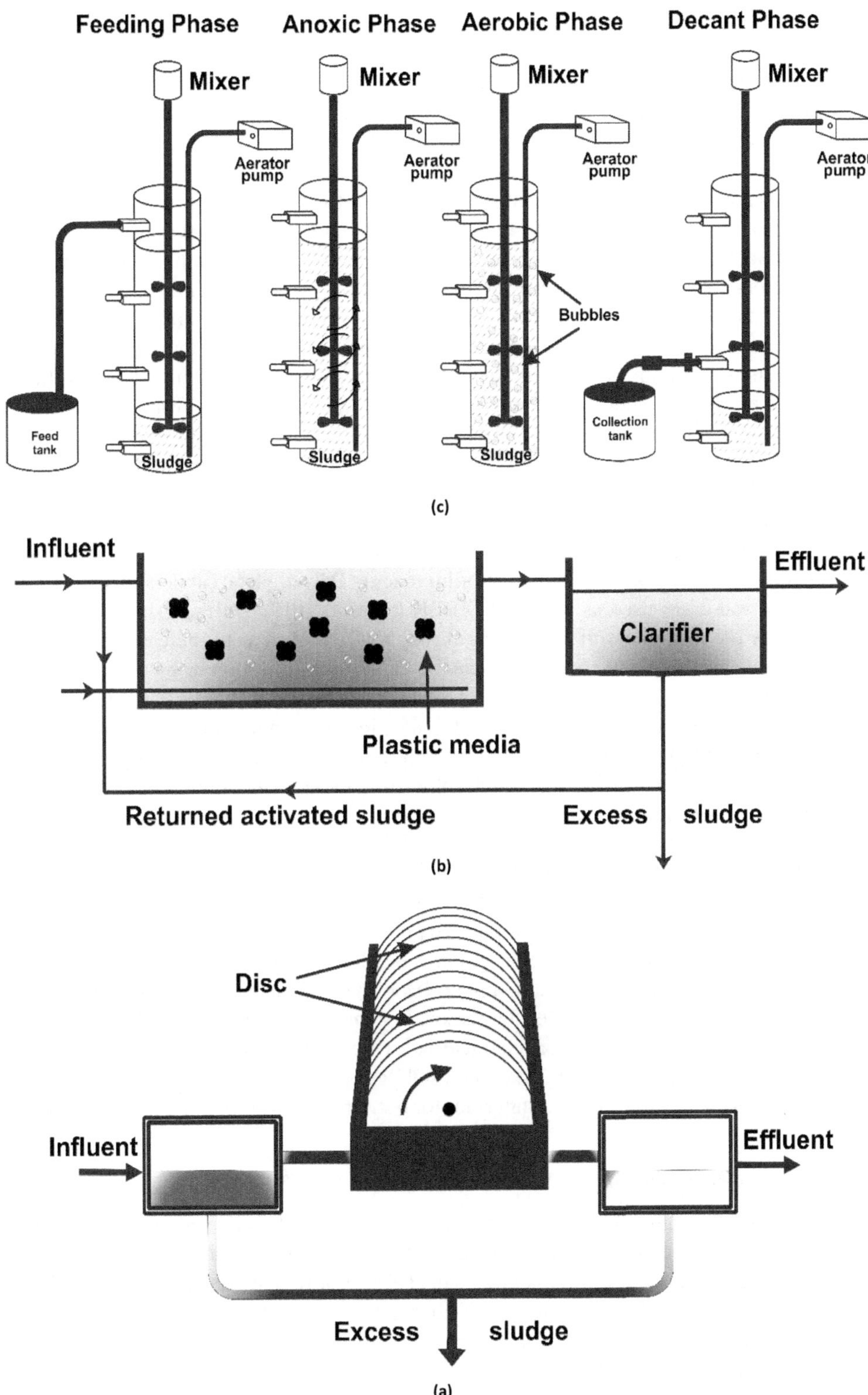

FIGURE 4.5 Main schemes of aerobic technologies based on nature (a) RBC (b) MBBR (c) ISBR

Source: Own elaboration with information taken from Aziz et al. (2019).

degrade organic matter, the average size of the discs is 1–2 m in radius. (Waqas et al., 2021; Hamedi et al., 2021). The space between the bioreactor plates is 0.3–0.4 cm, which allows the substrate, bacteria and oxygen to carry out the contact reactions. The microorganisms form a biofilm that is 0.1–0.4 cm thick on the plates, all of which can be identified in a relatively short time (approximately 7 days) (Waqas et al., 2021). The thickness of the film formed by microorganisms is a function of the speed of rotation of the discs and the strength of the wastewater used. Organic matter is metabolized by the bacterial consortium present in the effluents to be treated by absorbing atmospheric oxygen. The biofilm formed detaches from the superficial medium by the action of the shear stress caused by the wastewater, also inhibiting the growth of bacteria in the medium. Some advantages of rotary bioreactors are ease of construction of the reactor, low energy cost, simple maintenance, and freedom to handle unstable pollutant loads, among others (Hamedi et al., 2021). It also has energy advantages over other types of technologies that use pumps and reuse systems, not to mention the investment costs of mechanized systems, such as the activated sludge system. However, the application of these ecotechnologies in the bioremediation of residual effluents from slaughterhouses is scarce (Bustillo-Lecompte and Mehrvar, 2015; Guillén, 2017; Aziz et al., 2019).

4.3.5.4 Moving Bed Biofilm Reactor

The Moving Bed Biofilm Reactor (MBBR) is a modification of the activated sludge system. The process was invented by a Norwegian scientist in the early 1990s (Aziz et al., 2019). It combines the benefits of activated sludge system and the biofilm reactor. Polyethylene carriers that have a density close to that of water are used as media in MBBR (Safwat, 2019). Unlike fixed film bioreactors, MBBR involves the movement of biocarriers throughout the system to increase the contact area between substrates and biomass. Movement can be achieved by aeration or by mechanical mixing depending on the type of bioreactor (aerobic or anaerobic) (Figure 4.5b) (di Biase et al., 2019). The main benefits of MBBR are resistance to shock loads, lower volume requirement, no recycling or backwashing required, no mechanical intervention required in case of load fluctuations, and sufficient SRT for both heterotrophic and autotrophic microorganisms (Zinatizadeh and Ghaytooli, 2015; Aziz et al., 2019; Gzar et al., 2021).

4.3.5.5 Intermittent Sequencing Batch Reactor

An intermittent sequencing batch reactor (ISBR) is implemented not only for the removal of organic compounds but also for the removal of nitrogen and phosphorous (Wang et al., 2022). It has numerous advantages over other conventional aerobic systems. Feeding, reaction, sedimentation, and decantation take place one after the other in a complete cycle. Aerobic, anoxic, and anaerobic conditions can prevail in a single batch reactor for the removal of impurities, particularly N&P (Figure 4.5c) (Leonard et al., 2018). However, effluent characteristics are highly dependent on aeration rate, OLR, HRT, SRT, MLSS, specific oxygen uptake rate (SOUR), temperature, and F/M ratios. In addition, COD/TN and COD/TP ratios are also important for nutrient removal (Shafaei et al., 2022).

4.3.6 Natural Technologies Using Aerobic/Anaerobic Processes

Organic and inorganic compounds can be greatly reduced by combining anaerobic and aerobic processes. Numerous case studies in which sequential processes have been carried out for the treatment of both domestic and industrial wastewater (Saghafi et al., 2019; Komatsu et al., 2020; Komolafe et al., 2021; Terreros-Mecalco et al., 2022). Figure 4.7 shows some of the schematic diagrams of the sequential anaerobic and aerobic processes. The anaerobic-aerobic treatment provides an effluent that can meet international design standards and can be easily discharged to land or water bodies. The benefits of sequential treatment are the complete degradation of organic matter, the substantial production of biogas, the removal of nitrogen and phosphorus, the removal of heavy metals, phenols, and pharmaceutical compounds (Aziz et al., 2019). Furthermore, several studies have been carried out to evaluate the treatment of slaughterhouse wastewater by combining anaerobic and aerobic reactors (Del Pozo and Diez, 2005; Rajab et al., 2017; Tong et al., 2020; Gutu et al., 2021; Lopes et al., 2022).

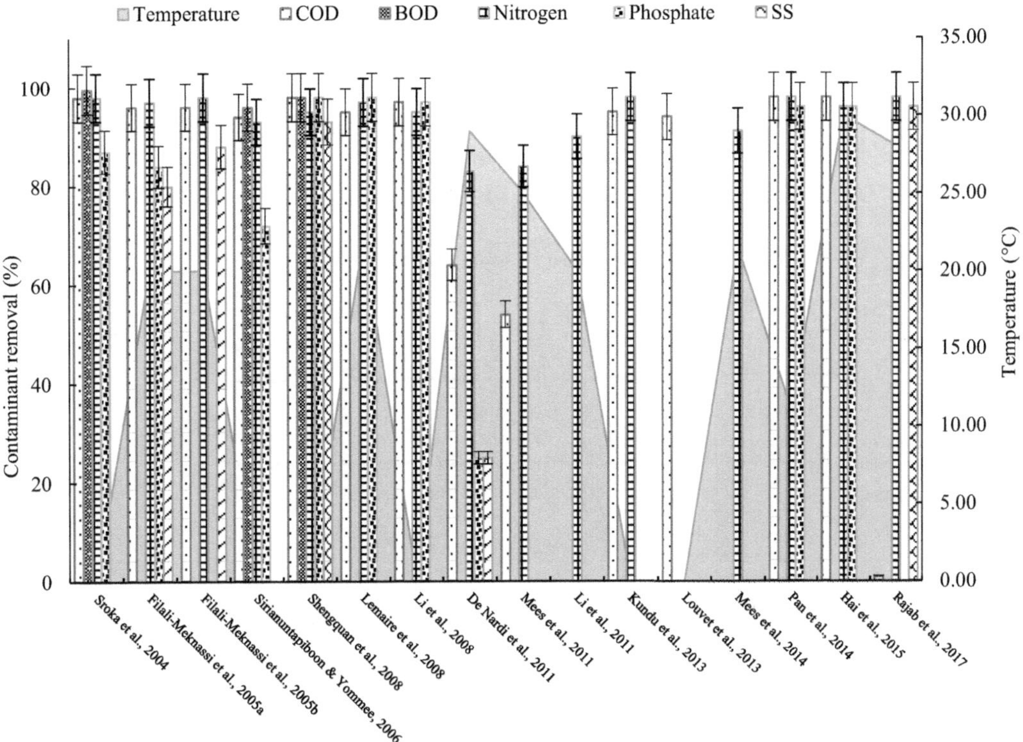

FIGURE 4.6 Removal of polluting from SBR under different operating conditions.

Source: Own elaboration based on the data reported by Aziz et al. (2019).

Figure 4.6 shows numerous parameters and operating conditions evaluated in the treatment of slaughterhouse waste by installing sequencing batch reactors (SBR) under different operating conditions, under aerobic and anaerobic conditions.

4.3.7 Constructed Wetlands

Constructed wetlands have been widely used for the management and management of wastewater, in general, this type of bioremediation technology is used after primary and secondary treatment, that is, as tertiary treatment. In this type of technology based on nature, vegetation, bacterial consortia, soil, porous material and other natural processes are used that together remove contaminants present in the effluents to be treated (Vymazal et al., 2021). The competitive advantages of these compared to other treatment technologies are: low operating and maintenance cost, easy to implement, low energy, and are harmless to nature and receiving bodies (Moreira and Dias, 2020; Rizzo et al., 2020). There are two types of constructed wetlands, and they are classified as surface flow and subsurface flow constructed wetlands. Taking into account the flow direction, they are divided into two groups: horizontal flow and vertical flow wetlands. Filter media used in wetlands, such as tezontle, river gravel, and stone, among others; contaminant removal in subsurface flow wetlands is much better compared to removals achieved with surface flow wetlands (Parde et al., 2021). To avoid effluent filtering through the bottom, the wetlands are covered with polyethylene plastic. In the late 1960s, the first large-scale implementation of constructed wetlands took place. In the first instance, domestic and municipal wastewater was treated, however, over the years it has been possible to treat industrial, agricultural, hospital, food, leachate, and slaughterhouse wastewater (Kataki et al., 2021).

Table 4.2 summarizes some of the main applications of slaughterhouse wastewater treatment over time.

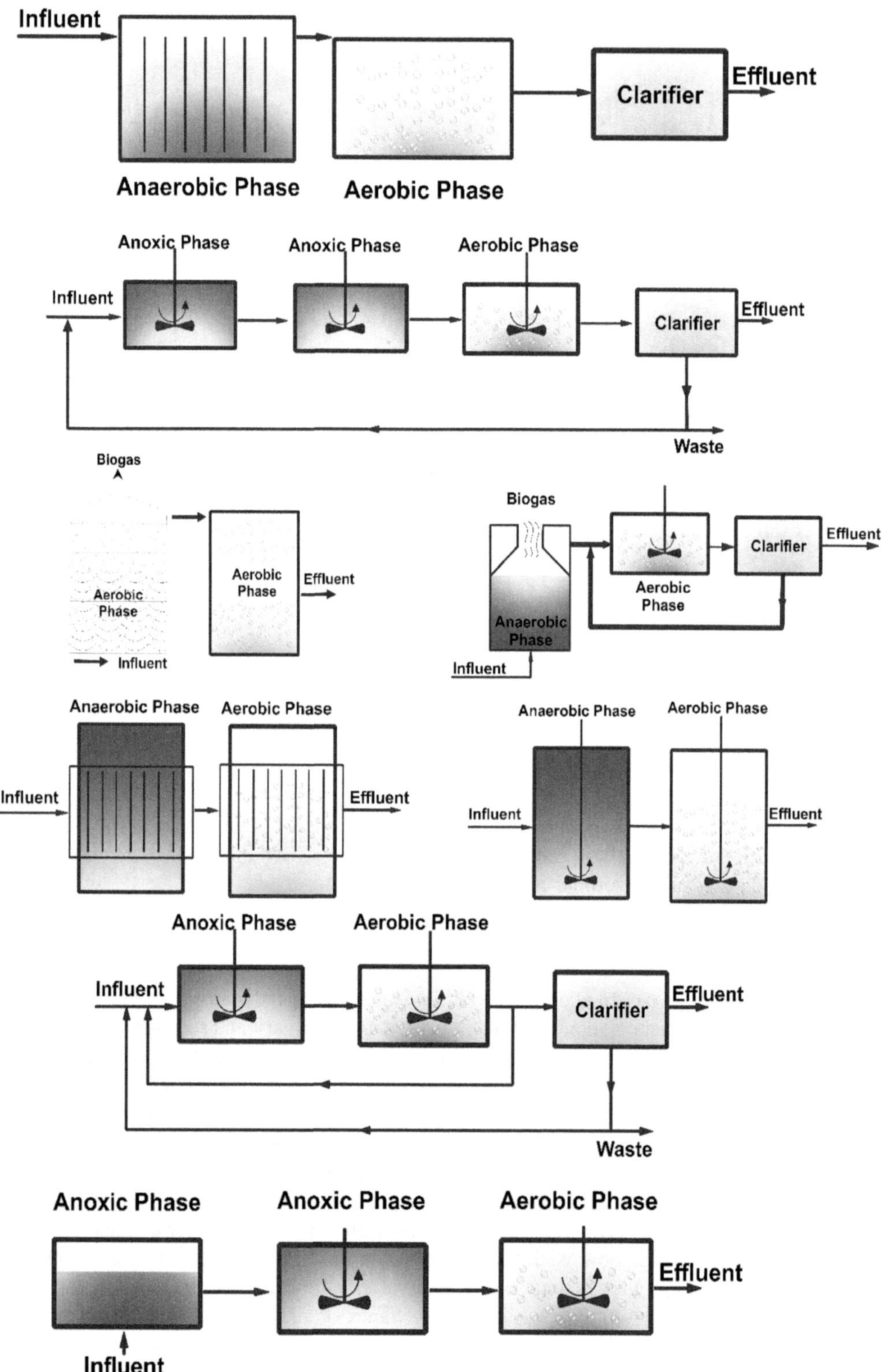

FIGURE 4.7 Representations of the main technologies based on nature anaerobic and sequential aerobic.

Source: Own elaboration based on the data reported by Aziz et al. (2019).

TABLE 4.2
Different Applications of Constructed Wetlands in Slaughterhouse Wastewater Treatment

Wetland Type	Treatment	Type of Water	Size	HRT	Contaminant Removal	Vegetation	Reference
Surface flow	Primary	Meat processing	Big scale	8.7 days	Nitrogen	*Maximum glyceria.*	Russell et al. (1994)
Subterranean flow	Primary	Dairy farm	Big scale	3 days	Nitrogen	*Schoenoplectus validus*	
Surface flow	Four wetlands in series	Meat processing	Big scale	7 days	N: 87%	*Maximum glyceria*	Van Oostrom (1995)
Primary sedimentation tank, an anaerobic lagoon, and a groundflow constructed wetland, in series	Tertiary	Slaughterhouse wastewater	Big scale	10.6 days	BOD_5: 91%; COD: 89%; SST: 85%; Coliforms: 2–3.5 logs	*Phragmites australis* (junco) and *Typha latifolia* (cattail)	Gutiérrez-Sarabia et al., 2004
Horizontal flow	Secondary	Sewage in slaughterhouse	NR	NR	Wastewater meets environmental protection requirements according to organic (BOD_5) and biogenic (Ntotal and Ptotal) pollutants	NR	Struseviciene and Strusevicius (2006)
Vertical flow	Secondary	Animal processing	Microcosm	NR	COD: 167 mg/l; Ammonia: 63 mg/L; TSS: 15 mg/L, in the effluent	*Typha latifolia*	Scholz (2006)
Vertical and horizontal flow	Serial and combined system	Wastewater from slaughterhouses	Microcosm	NR	BOD_5: 99.9%; COD: 97.4%; SST: 94.9; NH_4-N: 99.3%; NT: 78.2%	*Phragmites australis*	Soroko (2007)
Horizontal underground flow	Secondary	Wastewater of different chemical composition	NR	NR	N: 37%–44%	*Phragmites australis*	Gasiunas et al. (2005)
Built overland flow	Secondary	Wastewater from the slaughterhouse	Big scale	111 days	BOD_5: 95%; SST: 72%; TDS: 81%; PT: 88%; SRP: 97%; TKN: 87%; TAN: 87%	*Typha latifolia*	Carreau et al. (2012)
Vertical flow	Secondary	Sewage from the pig slaughterhouse	Big scale	5 days	COD: 36%: BOD: 66%; Coliform bacteria: 97%; Fecal bacteria: 99%	*Typha*	Pitaktunsakul et al., 2015
Horizontal underground flow	Tertiary	Slaughterhouse wastewater	Big scale	1.16 days	NH_4-N: 76%; NT: 48%; o-PO_4-P: 46%; TP: 74%; COD: 63%; fecal coliforms: 100%	*Cyperus papyrus*	Odong et al. (2015)
Vertical subsurface flow	Primary	Slaughterhouse wastewater	Mesocosm	5 days	BOD_5: 50%; COD: 55%; SST: 82%; NH_4-N: 26.5%	NR	Mburu et al. (2019)

4.4 MAIN CHALLENGES OF THE DEVELOPMENT AND IMPLEMENTATION OF NATURE-BASED TECHNOLOGIES FOR THE TREATMENT OF WASTEWATER FROM MUNICIPAL SLAUGHTERHOUSES

Although good results have been obtained in the handling, management and sanitation of wastewater from slaughterhouses, there are still some challenges related to these technologies used, therefore, an evaluation of these treatments is presented below, followed by recommendations that in turn become in excellent areas of opportunity for future research according to what was reported by Musa and Idrus (2021):

- Numerous studies on slaughterhouse wastewater treatment applying nature-based solutions have been reported. Although excellent removal of pollutants has been achieved with AD and at the same time biogas can be generated as a product with high added value, a disadvantage of this process is that high amounts of organic load are generated, which results in the inhibition biogas production and lower COD removals. It is recommended to investigate anaerobic co-digestion where slaughterhouse wastewater is combined with another organic substrate (Fat Oil & Grease) to achieve higher biogas production, compared to AD alone.
- The implementation of artificial wetlands has been used to treat wastewater from slaughterhouses, however, the effluents obtained do not comply with environmental regulations to be discharged.. Therefore, it is suggested to combine these methods with some other nature-based techniques to comply with the applicable regulations before being discharged.
- One of the most promising nature-based solutions for bioremediation of slaughterhouse water is the combination of aerobic/anaerobic techniques. These joint techniques are more effective in eliminating high concentrations of pollutants present in effluents such as slaughterhouse wastewater. If these combined technologies are complemented with some disinfection technique, the effluents generated can comply with the rules and regulations for wastewater discharge.
- An important area of opportunity regarding nature-based solutions in the bioremediation of slaughterhouse wastewater is the optimization of the techniques that have been reported so far, since until now there is limited literature on this subject. In addition, it is important to explore the removal of other types of organic components, such as pharmaceutical compounds and pathogenic microorganisms present. It is necessary to carry out more studies *in situ*, and the application of existing technologies on a large scale.
- It is very necessary to know the environmental factors that influence the removal of contaminants, more detailed studies are required to identify the mechanism of action of each environmental, physical-chemical, or biological factor in this process and its contribution to the efficiency of the treatment, for its subsequent escalation.
- The drawbacks of anaerobic technologies (temperature control, long HRT, sludge washing, among others) and aerobic technologies (energy cost, construction areas, inefficient scaling, sludge production, among others) make more research on waste necessary. Organic, inorganic, or cellulosic materials to make the nature-based technologies studied more sustainable.

4.5 CONCLUSIONS

The wastewater generated in municipal slaughterhouses is characterized by being of a recalcitrant nature due to the different residues that include blood, intestinal content, fat derived from the meat-trimming process, viscera, feces, and pathogens. The treatment of this effluent is challenging for slaughterhouses due to the complexity of its compounds since it is necessary to combine different technologies for the removal of contaminants, including physical pretreatments for the separation of solids and thermal pretreatments for the elimination of pathogens. It should be noted that it is not recommended to use anaerobic or aerobic processes individually since the effluent must comply with limits and standards established for its final disposal. Currently, with population growth, industrial development, and technological advances, there has been an increase in meat processing plants to meet the demands and, consequently, more and more areas of opportunity are opening up to apply science and research with the purpose of implementing nature-based technologies for the treatment of these effluents. These technologies allow not only for the removal of contaminants, but also for the treatment to be easy to operate, low investment and maintenance costs, and friendly to the environment. Specifically, AD is a recommended option for the treatment of WWMS due to its advantage of biogas production as an added value. This technology could be used in combination with ARCs to produce an effluent that meets standards for safe discharge into the environment or with reuse characteristics.

REFERENCES

Abd Nasir, M. A., Jahim, J. M., Abdul, P. M., Silvamany, H., Maaroff, R. M., & Yunus, M. F. M. (2019). The use of acidified palm oil mill effluent for thermophilic biomethane production by changing the hydraulic retention time in anaerobic sequencing batch reactor. *International Journal of Hydrogen Energy*, 44(6), 3373–3381. https://doi.org/10.1016/j.ijhydene.2018.06.149

Abdel-Mohsein, H. S., Feng, M., Fukuda, Y., & Tada, C. (2020). Remarkable removal of antibiotic-resistant bacteria during dairy wastewater treatment using hybrid full-scale constructed wetland. *Water, Air, & Soil Pollution*, 231(8), 1–12. https://doi.org/10.1007/s11270-020-04775-9

Ahmed, S. F., Mofijur, M., Parisa, T. A., Islam, N., Kusumo, F., Inayat, A., ... Ong, H. C. (2022). Progress and challenges of contaminate removal from wastewater using microalgae biomass. *Chemosphere*, 286, 131656.

Akhbari, A., Chuen, O. C., & Ibrahim, S. (2021). Start-up study of biohydrogen production from palm oil mill effluent in a lab-scale up-flow anaerobic sludge blanket fixed-film reactor. *International Journal of Hydrogen Energy*, 46(17), 10191–10204. https://doi.org/10.1016/j.ijhydene.2020.12.125

Alayu, E., & Yirgu, Z. (2018). Advanced technologies for the treatment of wastewaters from agro-processing industries and cogeneration of by-products: A case of slaughterhouse, dairy and beverage industries. *International Journal of Environmental Science and Technology*, 15(7), 1581–1596. https://link.springer.com/article/10.1007/s13762-017-1522-9

Armal, D. G. P., & Shanta, S. (2015). Slaughterhouse wastewater treatment by anaerobic fixed film fixed bed reactor packed with special media. *International Journal of Plant, Animal and Environmental Sciences*, 5(3), 151–156.

Arthur, P. M., Konaté, Y., Sawadogo, B., Sagoe, G., Dwumfour-Asare, B., Ahmed, I., & Williams, M. N. (2022). Performance evaluation of a full-scale upflow anaerobic sludge blanket reactor coupled with trickling filters for municipal wastewater treatment in a developing country. *Heliyon*, 8(8), e10129. https://doi.org/10.1016/j.heliyon.2022.e10129.

Aziz, A., Basheer, F., Sengar, A., Khan, S. U., & Farooqi, I. H. (2019). Biological wastewater treatment (anaerobic-aerobic) technologies for safe discharge of treated slaughterhouse and meat processing wastewater. *Science of the Total Environment*, 686, 681–708. https://www.sciencedirect.com/science/article/pii/S0048969719323368

Aziz, A., Rameez, H., Sengar, A., Sharma, D., Farooqi, I. H., & Basheer, F. (2022). Biogas production and nutrients removal from slaughterhouse wastewater using integrated anaerobic and aerobic granular intermittent SBRs-Bioreactors stability and microbial dynamics. *Science of the Total Environment*, 848, 157575. https://www.sciencedirect.com/science/article/pii/S0048969722046733

Baker, B. R., Mohamed, R., Al-Gheethi, A., & Aziz, H. A. (2021). Advanced technologies for poultry slaughterhouse wastewater treatment: A systematic review. *Journal of Dispersion Science and Technology*, 42(6), 880–899. https://www.tandfonline.com/doi/abs/10.1080/01932691.2020.1721007

Basheer, F., Aziz, A., Sharma, D., Sengar, A., & Farooqi, I. H. (2021). Bioenergy production and slaughterhouse wastewater treatment in a column-type anaerobic sequencing batch reactor without any external mixer or gas or liquid recirculation. *Journal of Environmental Engineering*, 147(3), 04021004. https://ascelibrary.org/doi/abs/10.1061/%28ASCE%29EE.1943-7870.0001859

Basitere, M. (2017). Performance evaluation of an up-and down-flow anaerobic reactor for the treatment of poultry slaughterhouse wastewater in South Africa (Doctoral dissertation, Cape Peninsula University of Technology). https://etd.cput.ac.za/handle/20.500.11838/2632.

Basitere, M., Njoya, M., Ntwampe, S. K. O., & Sheldon, M. S. (2020). Up-flow vs downflow anaerobic digester reactor configurations for treatment of fats-oil-grease laden poultry slaughterhouse wastewater: A review. *Water Practice and Technology*, 15(2), 248–260. https://iwaponline.com/wpt/article-abstract/15/2/248/73145.

Bastian, R. K., Shanaghan, P. E., & Thompson, B. P. (2020). Use of wetlands for municipal wastewater treatment and disposal-Regulatory issues and EPA policies. In *Constructed Wetlands for Wastewater Treatment* (pp. 265–278). CRC Press.

Batubara, F., Ritonga, N. A., & Turmuzi, M. (2018, December). Start-up of Upflow Anaerobic Sludge Blanket (UASB) reactor treating slaughterhouse wastewater. *Journal of Physics: Conference Series*, 1116(4), 042008. IOP Publishing. https://iopscience.iop.org/article/10.1088/1742-6596/1116/4/042008/meta

Bressani-Ribeiro, T., Chernicharo, C. A., Lobato, L. C., & Neves, P. N. (2019). Design of UASB reactors for sewage treatment. In C. A. L. Chernicharo & T. Bressani-Ribeiro (Eds.) *Anaerobic Reactors for Sewage Treatment: Design, Construction, and Operation*. https://doi.org/10.2166/9781780409238_0056

Brindhadevi, K., Shanmuganathan, R., Pugazhendhi, A., Gunasekar, P., & Manigandan, S. (2021). Biohydrogen production using horizontal and vertical continuous stirred tank reactor-a numerical optimization. *International Journal of Hydrogen Energy*, 46(20), 11305–11312. https://doi.org/10.1016/j.ijhydene.2020.06.155

Buayoungyuen, S., Panyapinyopol, B., Fongsatitkul, P., Patthanaissaranukool, W., & Warodomrungsimun, C. (2022). Simultaneous removal of nitrogen, phosphorus, and organic matter from slaughterhouse wastewater using AnA2/O2 SBR and its economic benefits. *EnvironmentAsia*, 15(1), 33–46.

Bustillo-Lecompte, C. F., & Mehrvar, M. (2015). Slaughterhouse wastewater characteristics, treatment, and management in the meat processing industry: A review on trends and advances. *Journal of Environmental Management*, 161, 287–302.

Bustillo-Lecompte, C. F., & Mehrvar, M. (2017). Treatment of actual slaughterhouse wastewater by combined anaerobic–aerobic processes for biogas generation and removal of organics and nutrients: An optimization study towards a cleaner production in the meat processing industry. *Journal of Cleaner Production*, 141, 278–289.

Carreau, R., VanAcker, S., VanderZaag, A. C., Madani, A., Drizo, A., Jamieson, R., & Gordon, R. J. (2012). Evaluation of a surface flow constructed wetland treating abattoir wastewater. *Applied Engineering in Agriculture*, 28(5), 757–766. https://elibrary.asabe.org/abstract.asp?aid=42416

Chollom, M. N., Rathilal, S., Swalaha, F. M., Bakare, B. F., & Tetteh, E. K. (2020). Comparison of response surface methods for the optimization of an upflow anaerobic sludge blanket for the treatment of slaughterhouse wastewater. *Environmental Engineering Research*, 25(1), 114–122. https://doi.org/10.4491/eer.2018.366

Dague, R. R., Urell, R. F., & Krieger, E. R. (2018, February). Treatment of pork processing wastewater in a covered anaerobic lagoon with gas recovery. In *Proceedings of the 44th Industrial Waste Conference*, May 9–11, 1989 (pp. 815–824). CRC Press. https://www.taylorfrancis.com/chapters/edit/10.1201/9781351076029-90/treatment-pork-processing-wastewater-covered-anaerobic-lagoon-gas-recovery-richard-dague-robert-urell-eugene-krieger

Del Pozo, R., & Diez, V. (2005). Integrated anaerobic-aerobic fixed-film reactor for slaughterhouse wastewater treatment. *Water Research*, 39(6), 1114–1122. https://www.sciencedirect.com/science/article/pii/S0043135405000060

di Biase, A., Kowalski, M. S., Devlin, T. R., & Oleszkiewicz, J. A. (2019). Moving bed biofilm reactor technology in municipal wastewater treatment: A review. *Journal of Environmental Management*, 247, 849–866. https://doi.org/10.1016/j.jenvman.2019.06.053

Farooq, R., & Ahmad, Z. (Eds.). (2017). *Physico-Chemical Wastewater Treatment and Resource Recovery*. IntechOpen. https://doi.org/10.5772/67803

Farzadkia, M., Vanani, A. F., Golbaz, S., Sajadi, H. S., & Bazrafshan, E. (2016). Characterization and evaluation of treatability of wastewater generated in Khuzestan livestock slaughterhouses and assessing of their wastewater treatment systems. *Global NEST Journal*, 18, 108–118. https://journal.gnest.org/sites/default/files/Submissions/gnest_01716/gnest_01716_published.pdf.

Gasiunas, V., Strusevicius, Z., & Struseviciene, M.-S., 2005. Pollutant removal by horizontal subsurface flow constructed wetlands in Lithuania. *Journal of Environmental Science and Health, Part A*, 40, 1467–1478. https://doi.org/10.1081/ESE-200055889.

Goel, S., Kumar, P., Sain, M., & Singh, A. (2022). Anaerobic treatment of food processing wastes and agricultural effluents. In *Food Processing Waste and Utilization* (pp. 313–328). CRC Press. https://www.taylorfrancis.com/chapters/edit/10.1201/9781003207689-16/anaerobic-treatment-food-processing-wastes-agricultural-effluents-simmi-goel-pankaj-kumar-mukul-sain-ajay-singh.

Guillén, J. A. S. (2017). *Autotrophic Nitrogen Removal from Low Concentrated Effluents: Study of System Configurations and Operational Features for Post-treatment of Anaerobic Effluents*. CRC Press. https://doi.org/10.1201/9781315115955.

Guo, M., Li, H., Baldwin, B., & Morrison, J. (2020). Thermochemical processing of animal manure for bioenergy and biochar. In *Animal Manure: Production, Characteristics, Environmental Concerns, and management* (Vol. 67, pp. 255–274). https://doi.org/10.2134/asaspecpub67.c21

Gutiérrez Sarabia, A., Fernández Villagómez, G., Martínez Pereda, P., Rinderknecht Seijas, N., & Poggi Varaldo, H. M. (2004). Slaughterhouse wastewater treatment in a full scale system with constructed wetlands. *Water Environment Research*, 76(4), 334–343. https://onlinelibrary.wiley.com/doi/abs/10.2175/106143004X141924

Gutu, L., Basitere, M., Harding, T., Ikumi, D., Njoya, M., & Gaszynski, C. (2021). Multi-Integrated Systems for Treatment of Abattoir Wastewater: A Review. *Water*, 13(18), 2462. https://doi.org/10.3390/w13182462

Gzar, H. A., Al-Rekabi, W. S., & Shuhaieb, Z. K. (2021, August). Applicaion of Moving Bed Biofilm Reactor (MBBR) for treatment of industrial wastewater: A mini review. *Journal of Physics: Conference Series*, 1973(1), 012024. IOP Publishing.

Hamedi, S., Babaeipour, V., & Rouhi, M. (2021). Design, construction and optimization a flexible bench-scale rotating biological contactor (RBC) for enhanced production of bacterial cellulose by Acetobacter Xylinium. *Bioprocess and Biosystems Engineering*, 44(6), 1071–1080.

Han, Y., Yang, L., Chen, X., Cai, Y., Zhang, X., Qian, M., ... Shen, G. (2020). Removal of veterinary antibiotics from swine wastewater using anaerobic and aerobic biodegradation. *Science of the Total Environment*, 709, 136094. https://doi.org/10.1016/j.scitotenv.2019.136094.

Harada, H., & Hidayat, E. (2022, January). The effects of F/M ratio on in treatment of wastewater from Brewery Slurry by an anaerobic sequencing batch reactor. In 6th International Conference of Food, Agriculture, and Natural Resource (IC-FANRES 2021) (pp. 418–422). Atlantis Press. https://doi.org/10.2991/absr.k.220101.057

Hernández, D., Riaño, B., Coca, M., Solana, M., Bertucco, A., & García-González, M. C. (2016). Microalgae cultivation in high rate algal ponds using slaughterhouse wastewater for biofuel applications. *Chemical Engineering Journal*, 285, 449–458. https://doi.org/10.1016/j.cej.2015.09.072

Hollas, C. E., Bolsan, A. C., Venturin, B., Bonassa, G., Tápparo, D. C., Cândido, D., ... Kunz, A. (2021). Second-generation phosphorus: Recovery from wastes towards the sustainability of production chains. *Sustainability*, 13(11), 5919. https://doi.org/10.3390/su13115919

Hosseinlou, D. (2021). Determination of design loading rates for simultaneous anaerobic oxidation/partial nitrification-denitrification process and application in treating dairy industry effluent. *Journal of Environmental Chemical Engineering*, 9(3), 105176. https://doi.org/10.1016/j.jece.2021.105176

Johns, M. R. (1995). Developments in wastewater treatment in the meat processing industry: A review. *Bioresource Technology*, 54(3), 203–216. https://www.sciencedirect.com/science/article/pii/0960852495001409

Kataki, S., Chatterjee, S., Vairale, M. G., Dwivedi, S. K., & Gupta, D. K. (2021). Constructed wetland, an eco-technology for wastewater treatment: A review on types of wastewater treated and components of the technology (macrophyte, biolfilm and substrate). *Journal of Environmental Management*, 283, 111986. https://doi.org/10.1016/j.jenvman.2021.111986

Khan, N. A., Bokhari, A., Mubashir, M., Klemeš, J. J., El Morabet, R., Khan, R. A., ... Show, P. L. (2022). Treatment of Hospital wastewater with submerged aerobic fixed film reactor coupled with tube-settler. *Chemosphere*, 286, 131838. https://doi.org/10.1016/j.chemosphere.2021.131838

Khawer, M. U. B., Naqvi, S. R., Ali, I., Arshad, M., Juchelková, D., Anjum, M. W., & Naqvi, M. (2022). Anaerobic digestion of sewage sludge for biogas & biohydrogen production: State-of-the-art trends and prospects. *Fuel*, 329, 125416. https://www.sciencedirect.com/science/article/pii/S0016236122022499

Komatsu, K., Onodera, T., Kohzu, A., Syutsubo, K., & Imai, A. (2020). Characterization of dissolved organic matter in wastewater during aerobic, anaerobic, and anoxic treatment processes by molecular size and fluorescence analyses. *Water Research*, 171, 115459. https://doi.org/10.1016/j.watres.2019.115459

Komolafe, O., Mrozik, W., Dolfing, J., Acharya, K., Vassalle, L., Mota, C. R., & Davenport, R. (2021). Occurrence and removal of micropollutants in full-scale aerobic, anaerobic and facultative wastewater treatment plants in Brazil. *Journal of Environmental Management*, 287, 112286. https://doi.org/10.1016/j.jenvman.2021.112286

Leonard, P., Finnegan, W., Barrett, M., & Zhan, X. (2018). Efficient treatment of dairy processing wastewater in a pilot scale Intermittently Aerated Sequencing Batch Reactor (IASBR). *Journal of Dairy Research*, 85(3), 384–387. https://doi.org/10.1017/S0022029918000596

Li, Y., Chen, Y., & Wu, J. (2019). Enhancement of methane production in anaerobic digestion process: A review. *Applied Energy*, 240, 120–137. https://www.sciencedirect.com/science/article/pii/S0306261919302715

Loganath, R., & Senophiyah-Mary, J. (2020). Critical review on the necessity of bioelectricity generation from slaughterhouse industry waste and wastewater using different anaerobic digestion reactors. *Renewable and Sustainable Energy Reviews*, 134, 110360. https://doi.org/10.1016/j.rser.2020.110360

Lopes, C. L., de Assis, T. M., Passig, F. H., de Lima Model, A. N., Mees, J. B. R., Cervantes, F. J., ... Gomes, S. D. (2022). Nitrogen removal from poultry slaughterhouse wastewater in anaerobic-anoxic-aerobic combined reactor: Integrated effect of recirculation rate and hydraulic retention time. *Journal of Environmental Management*, 303, 114162. https://www.sciencedirect.com/science/article/pii/S0301479721022246

Lorenzo, E. V., Bataller Venta, M., Fernández García, L. A., Tórrez, I. F., & Hung, Y.-T. (2012). Municipal wastewater treatment for reuse in agricultural irrigation. In *Handbook of Environment and Waste Management* (bll 347–381). World Scientific. https://doi.org/10.1142/9789814327701_0009

Maisuria, K. J., Shah, K. A., & Rana, J. K. (2022, September). Removal of Tannic acid and COD from synthetic Tannery wastewater. *IOP Conference Series: Earth and Environmental Science*, 1086(1), 012035. IOP Publishing.

Mansourian, R., Mousavi, S. M., & Rahimpour, M. R. (2024). Membrane technologies for the treatment of food. In *Current Trends and Future Developments on (Bio-) Membranes* (bll 3–23). Elsevier. https://doi.org/10.1016/B978-0-323-90258-8.00001-8 MASSE, O., & Masse, L. (2000). Treatment of slaughterhouse wastewater in anaerobic sequencing batch reactors. *Canadian Agricultural Engineering*, 42(3), 131. https://library.csbe-scgab.ca/docs/journal/42/42_3_131_ocr.pdf

Mburu, C., Kipkemboi, J., & Kimwaga, R. (2019). Impact of substrate type, depth and retention time on organic matter removal in vertical subsurface flow constructed wetland mesocosms for treating slaughterhouse wastewater. *Physics and Chemistry of the Earth, Parts A/B/C*, 114, 102792. https://doi.org/10.1016/j.pce.2019.07.005.

McCabe, B. K., Hamawand, I., Harris, P., Baillie, C., & Yusaf, T. (2014). A case study for biogas generation from covered anaerobic ponds treating abattoir wastewater: Investigation of pond performance and potential biogas production. *Applied Energy*, 114, 798–808. https://www.sciencedirect.com/science/article/pii/S0306261913008404.

McCabe, B. K., Harris, P., Baillie, C., Pittaway, P., & Yusaf, T. (2013). Assessing a new approach to covered anaerobic pond design in the treatment of abattoir wastewater. *Australian Journal of Multi-Disciplinary Engineering*, 10(1), 81–93. https://www.tandfonline.com/doi/abs/10.7158/14488388.2013.11464867

Meegoda, J. N., Li, B., Patel, K., & Wang, L. B. (2018). A review of the processes, parameters, and optimization of anaerobic digestion. *International Journal of Environmental Research and Public Health*, 15(10), 2224. https://www.mdpi.com/349758

Mittal, G. S. (2006). Treatment of wastewater from abattoirs before land application-a review. *Bioresource Technology*, 97(9), 1119–1135. https://www.sciencedirect.com/science/article/pii/S0960852405000076

Moreira, F. D., & Dias, E. H. O. (2020). Constructed wetlands applied in rural sanitation: A review. *Environmental Research*, 190, 110016. doi.org/10.1016/j.envres.2020.110016

Morgan-Sagastume, F., Jacobsson, S., Olsson, L. E., Carlsson, M., Gyllenhammar, M., & Horváth, I. S. (2019). Anaerobic treatment of oil-contaminated wastewater with methane production using anaerobic moving bed biofilm reactors. *Water Research*, 163, 114851. https://www.sciencedirect.com/science/article/pii/S0043135419306177.

Mozhiarasi, V., & Natarajan, T. S. (2022). Slaughterhouse and poultry wastes: Management practices, feedstocks for renewable energy production, and recovery of value added products. In *Biomass Conversion and Biorefinery* (pp. 1–24). https://link.springer.com/article/10.1007/s13399-022-02352-0

Musa, M. A., & Idrus, S. (2021). Physical and biological treatment technologies of slaughterhouse wastewater: A review. *Sustainability*, 13(9), 4656. https://www.mdpi.com/article/10.3390/su13094656

Musa, M. A., Idrus, S., Che Man, H., & Nik Daud, N. N. (2019). Performance comparison of conventional and modified upflow anaerobic sludge blanket (UASB) reactors treating high-strength cattle slaughterhouse wastewater. *Water*, 11(4), 806. https://doi.org/10.3390/w11040806

Musa, M. A., Idrus, S., Harun, M. R., Tuan Mohd Marzuki, T. F., & Abdul Wahab, A. M. (2020). A comparative study of biogas production from cattle slaughterhouse wastewater using conventional and modified Upflow Anaerobic Sludge Blanket (UASB) reactors. *International Journal of Environmental Research and Public Health*, 17(1), 283. https://doi.org/10.3390/ijerph17010283

Odong, R., Kansiime, F., Omara, J., & Kyambadde, J. (2015). Tertiary treatment of abattoir wastewater in a horizontal subsurface flow-constructed wetland under tropical conditions. *International Journal of Environment and Waste Management*, 15(3), 257–270. https://doi.org/10.1504/IJEWM.2015.069160

Parde, D., Patwa, A., Shukla, A., Vijay, R., Killedar, D. J., & Kumar, R. (2021). A review of constructed wetland on type, treatment and technology of wastewater. *Environmental Technology & Innovation*, 21, 101261. https://doi.org/10.1016/j.eti.2020.101261

Parida, V. K., Saidulu, D., Majumder, A., Srivastava, A., Gupta, B., & Gupta, A. K. (2021). Emerging contaminants in wastewater: A critical review on occurrence, existing legislations, risk assessment, and sustainable treatment alternatives. *Journal of Environmental Chemical Engineering*, 9(5), 105966. https://www.sciencedirect.com/science/article/pii/S221334372100943X

Patel, B. B., Gundaliya, P. J., & Amin, D. D. (2021). Optimized hybrid upflow anaerobic sludge blanket with post treatment processes for wastewater. *Current World Environment*, 16, 282–303. https://doi.org/10.12944/CWE.16.1.29

Perdigón, R. A. P. (2010). Una breve descripción del manejo de los residuos generados en los mataderos de Colombia y su inclusión en los procesos de las tecnologías limpias o apropiadas. *Boletín Semillas Ambientales*, 4(2), 4–6.

Piotrowski, R., Paul, A., & Lewandowski, M. (2019). Improving SBR performance alongside with cost reduction through optimizing biological processes and dissolved oxygen concentration trajectory. *Applied Sciences*, 9(11), 2268. https://doi.org/10.3390/app9112268

Pitaktunsakul, P., Chunkao, K., Dampin, N., & Poommai, S. (2015). Vertical-flow constructed wetlands in cooperating with oxidation ponds for high concentrated COD and BOD pig-slaughterhouse wastewater treatment system at Suphanburi-provincial municipality. *Modern Applied Science*, 9(8), 371. https://doi.org/10.5539/mas.v9n8p371

Preisner, M., Neverova-Dziopak, E., & Kowalewski, Z. (2020). An analytical review of different approaches to wastewater discharge standards with particular emphasis on nutrients. *Environmental Management*, 66(4), 694–708. https://link.springer.com/article/10.1007/s00267-020-01344-y

Qamar, M. O., Farooqi, I. H., Munshi, F. M., Alsabhan, A. H., Kamal, M. A., Khan, M. A., & Alwadai, A. S. (2022). Performance of full-scale slaughterhouse effluent treatment plant (SETP). *Journal of King Saud University-Science*, 34(3), 101891.

Rabii, A., Aldin, S., Dahman, Y., & Elbeshbishy, E. (2019). A review on anaerobic co-digestion with a focus on the microbial populations and the effect of multi-stage digester configuration. *Energies*, 12(6), 1106. https://doi.org/10.3390/en12061106

Rajab, A. R., Salim, M. R., Sohaili, J., Anuar, A. N., & Lakkaboyana, S. K. (2017). Performance of integrated anaerobic/aerobic sequencing batch reactor treating poultry slaughterhouse wastewater. *Chemical Engineering Journal*, 313, 967–974. https://www.sciencedirect.com/science/article/pii/S1385894716315613

Rajagopal, R., Choudhury, M. R., Anwar, N., Goyette, B., & Rahaman, M. S. (2019). Influence of pre-hydrolysis on sewage treatment in an up-flow anaerobic sludge BLANKET (UASB) reactor: A review. *Water*, 11(2), 372. https://doi.org/10.3390/w11020372

Rajakumar, R., Meenambal, T., Banu, J. R., & Yeom, I. T. (2011). Treatment of poultry slaughterhouse wastewater in upflow anaerobic filter under low upflow velocity. *International Journal of Environmental Science & Technology*, 8(1), 149–158. https://doi.org/10.1007/BF03326204

Ripoll, V., Agabo-García, C., Solera, R., & Perez, M. (2022). Anaerobic digestion of slaughterhouse waste in batch and anaerobic sequential batch reactors. *Biomass Conversion and Biorefinery*, 1–12. https://doi.org/10.1007/s13399-021-02179-1

Rizzo, A., Tondera, K., Pálfy, T. G., Dittmer, U., Meyer, D., Schreiber, C., ... Masi, F. (2020). Constructed wetlands for combined sewer overflow treatment: A state-of-the-art review. *Science of the Total Environment*, 727, 138618. https://doi.org/10.1016/j.scitotenv.2020.138618

Russell, J. M., Van Oostrom, A. J., & Lindsey, S. B. (1994). Denitrifying sites in constructed wetlands treating agricultural industry wastes: A note. https://www.tandfonline.com/doi/abs/10.1080/09593339409385408

Saddoud, A., & Sayadi, S. (2007). Application of acidogenic fixed-bed reactor prior to anaerobic membrane bioreactor for sustainable slaughterhouse wastewater treatment. *Journal of hazardous materials*, 149(3), 700–706. https://www.sciencedirect.com/science/article/pii/S030438940700516X

Safwat, S. M. (2019). Moving bed biofilm reactors for wastewater treatment: A review of basic concepts. *International Journal of Research*, 6(10), 85–90.

Saghafi, S., Ebrahimi, A., Mehrdadi, N., & Bidhendy, G. N. (2019). Evaluation of aerobic/anaerobic industrial wastewater treatment processes: The application of multi criteria decision analysis. *Environmental Progress & Sustainable Energy*, 38(5), 13166. https://doi.org/10.1002/ep.13166

Saghir, A., & Hajjar, S. (2022). Biological Treatment of Slaughterhouse Wastewater using Up flow Anaerobic Sludge Blanket (UASB)-anoxic-aerobic system. *Scientific African*, e01236. https://doi.org/10.1016/j.sciaf.2022.e01236

Sandoval, M. A., Espinoza, L. C., Coreño, O., García, V., Fuentes, R., Thiam, A., & Salazar, R. (2022). A comparative study of anodic oxidation and electrocoagulation for treating cattle slaughterhouse wastewater. *Journal of Environmental Chemical Engineering*, 10(5), 108306. https://doi.org/10.1016/j.jece.2022.108306

Saravanan, A., Senthil Kumar, P., Jeevanantham, S., Karishma, S., Tajsabreen, B., Yaashikaa, P. R., & Reshma, B. (2021). Effective water/wastewater treatment methodologies for toxic pollutants removal: Processes and applications towards sustainable development. *Chemosphere*, 280, 130595. https://doi.org/10.1016/j.chemosphere.2021.130595

Schmidt, T., Harris, P., Lee, S., & McCabe, B. K. (2019). Investigating the impact of seasonal temperature variation on biogas production from covered anaerobic lagoons treating slaughterhouse wastewater using lab scale studies. *Journal of Environmental Chemical Engineering*, 7(3), 103077. https://www.sciencedirect.com/science/article/pii/S2213343719302003

Scholz, M. (2006). Comparison of novel membrane bioreactors and constructed wetlands for treatment of pre-processed animal rendering plant wastewater in Scotland. *E-Water* (on-line publication of EWA), 14pp. https://www.ewa-online.eu/tl_files/_media/content/documents_pdf/Publications/E-WAter/documents/44_2006_03h.pdf

Shafaei, N., Bonakdarpour, B., & Faridizad, G. (2022). The application of membrane sequencing batch reactors in high rate activated sludge processes: The effect of the organic loading rate on the bioflocculation and fouling phenomenon. *Environmental Science: Water Research & Technology*, 8(7), 1561–1578. https://doi.org/10.1039/D1EW00882J

Shoukat, R., Khan, S. J., & Jamal, Y. (2019). Hybrid anaerobic-aerobic biological treatment for real textile wastewater. *Journal of Water Process Engineering*, 29, 100804. https://doi.org/10.1016/j.jwpe.2019.100804

Sib, E., Lenz-Plet, F., Barabasch, V., Klanke, U., Savin, M., Hembach, N., ... Bierbaum, G. (2020). Bacteria isolated from hospital, municipal and slaughterhouse wastewaters show characteristic, different resistance profiles. *Science of the Total Environment*, 746, 140894. https://doi.org/10.1016/j.scitotenv.2020.140894

Singh, A. L., Jamal, S., Baba, S. A., & Islam, M. M. (2014). Environmental and health impacts from slaughter houses located on the city outskirts: A case study. *Journal of Environmental Protection*, 05(06), 566–575..

Soroko, M. (2007). Treatment of wastewater from small slaughterhouse in hybrid constructed wetlands systems. *Ecohydrology and Hydrobiology*, 7(3–4), 339–343. https://doi.org/10.1016/S1642-3593(07)70117-9

Stefanakis, A. I., Bardiau, M., Trajano, D., Couceiro, F., Williams, J. B., & Taylor, H. (2019). Presence of bacteria and bacteriophages in full-scale trickling filters and an aerated constructed wetland. *Science of the Total Environment*, 659, 1135–1145. https://doi.org/10.1016/j.scitotenv.2018.12.415

Struseviciene, S. M., & Strusevicius, Z. (2006). Efficiency of wastewater treatment in slaughterhouse in two-stage constructed wetlands. In *International Scientific Conference: Research for Rural Development*, Jelgava (Latvia), 19–22 May 2006. Latvia University of Agriculture. https://agris.fao.org/search/en/providers/122652/records/6472467c08fd68d54600832d.

Talaiekhozani, A., Banisharif, F., Bazrafshan, M., Eskandari, Z., Heydari Chaleshtari, A., Moghadam, G., & Mohammad Amani, A. (2019). Comparing the ZnO/Fe(VI), UV/ZnO and UV/Fe(VI) processes for removal of Reactive Blue 203 from aqueous solution. *Environmental Health Engineering and Management*, 6(1), 27–39. https://doi.org/10.15171/EHEM.2019.04

Tao, C., Parker, W., & Bérubé, P. (2021). Characterization and modelling of soluble microbial products in activated sludge systems treating municipal wastewater with special emphasis on temperature effect. *Science of the Total Environment*, 779, 146471. https://doi.org/10.1016/j.scitotenv.2021.146471

Terreros-Mecalco, J., Guzmán-López, O., & García-Solorio, L. (2022). Simultaneous aerobic-anaerobic biodegradation of an industrial effluent of polymeric resins with high phenol concentration at different organic loading rates in a non-conventional UASB type reactor. *Chemical Engineering Journal*, 430, 133180. https://doi.org/10.1016/j.cej.2021.133180

Tong, S., Wang, S., Zhao, Y., Feng, C., Xu, B., & Zhu, M. (2019). Enhanced alure-type biological system (E-ATBS) for carbon, nitrogen and phosphorus removal from slaughterhouse wastewater: A case study. *Bioresource Technology*, 274, 244–251. https://doi.org/10.1016/j.biortech.2018.11.094

Tong, S., Zhao, Y., Zhu, M., Wei, J., Zhang, S., Li, S., & Sun, S. (2020). Effect of the supernatant reflux position and ratio on the nitrogen removal performance of anaerobic-aerobic slaughterhouse wastewater treatment process. *Environmental Engineering Research*, 25(3), 309–315. https://www.koreascience.or.kr/article/JAKO202019062607564.page

Ülgüdür, N., Ergüder, T. H., Uludağ-Demirer, S., & Demirer, G. N. (2019). High-rate anaerobic treatment of digestate using fixed film reactors. *Environmental Pollution*, 252, 1622–1632. https://doi.org/10.1016/j.envpol.2019.06.115

U.S. Environmental Protection Agency | US EPA. (n.d.). Retrieved July 4, 2024, from https://www.epa.gov/

Vadiveloo, A., Shayesteh, H., Bahri, P. A., & Moheimani, N. R. (2022). Comparison between continuous and daytime mixing for the treatment of raw anaerobically digested abattoir effluent (ADAE) and microalgae production in open raceway ponds. *Bioresource Technology Reports*, 17, 100981. https://doi.org/10.1016/j.biteb.2022.100981

Van Oostrom, A. J. (1995). Nitrogen removal in constructed wetlands treating nitrified meat processing effluent. *Water Science and Technology*, 32(3), 137–147. https://www.sciencedirect.com/science/article/pii/0273122395006141

Vashi, N. V., Shah, N. C., & Desai, K. R. (2019). Performance of UASB post treatment technologies for sewage treatment in Surat city. *Oriental Journal of Chemistry*, 35(4), 1352. https://doi.org/10.13005/ojc/350415

Vilvert, A. J., Junior, J. C. S., Bautitz, I. R., Zenatti, D. C., Andrade, M. G., & Hermes, E. (2020). Minimization of energy demand in slaughterhouses: Estimated production of biogas generated from the effluent. *Renewable and Sustainable Energy Reviews*, 120, 109613. https://www.sciencedirect.com/science/article/pii/S1364032119308214.

Vymazal, J., Zhao, Y., & Mander, Ü. (2021). Recent research challenges in constructed wetlands for wastewater treatment: A review. *Ecological Engineering*, 169, 106318. https://doi.org/10.1016/j.ecoleng.2021.106318

Wang, H., Miao, J., & Sun, Y. (2022). Aerobic nitrous oxide emission in anoxic/aerobic and intermittent aeration sequencing batch reactors. *Environmental Technology*, 1–10. https://doi.org/10.1080/09593330.2022.2144467

Wang, P., Wang, H., Qiu, Y., Ren, L., & Jiang, B. (2018). Microbial characteristics in anaerobic digestion process of food waste for methane production–A review. *Bioresource Technology*, 248, 29–36. https://doi.org/10.1016/j.biortech.2017.06.152

Waqas, S., Bilad, M. R., Aqsha, A., Harun, N. Y., Ayoub, M., Wirzal, M. D. H., ... Elma, M. (2021). Effect of membrane properties in a membrane rotating biological contactor for wastewater treatment. *Journal of Environmental Chemical Engineering*, 9(1), 104869. https://doi.org/10.1016/j.jece.2020.104869

Waqas, S., Bilad, M. R., & Man, Z. B. (2021). Performance and energy consumption evaluation of rotating biological contactor for domestic wastewater treatment. *Indonesian Journal of Science and Technology*, 6(1), 101–112. https://doi.org/10.17509/ijost.v6i1.31524

Wizor, C. H., & Nwankwoala, H. O. (2019). Effects of municipal Abattoir waste on water quality of Woji River in Trans-Amadi industrial area of Port Harcourt, Nigeria: Implication for sustainable urban environmental management. *International Journal of Geography and Geology*, 8(2), 44–57. https://archive.conscientiabeam.com/index.php/10/article/view/1996

Yetilmezsoy, K., Dinç-Şengönül, B., Ilhan, F., Kıyan, E., & Yüzer, N. (2022). Use of sheep slaughterhouse-derived struvite in the production of environmentally sustainable cement and fire-resistant wooden structures. *Journal of Cleaner Production*, 366, 132948. https://www.sciencedirect.com/science/article/pii/S0959652622025409

Yuan, H., Dang, Z., Li, C., Zhou, Y., Yang, B., & Huang, S. (2022). Simultaneous oxygen and nitrate respiration for nitrogen removal driven by aeration: Carbon/nitrogen metabolism and metagenome-based microbial ecology. *Journal of Water Process Engineering*, 50, 103196. https://doi.org/10.1016/j.jwpe.2022.103196

Zamri, M. F. M. A., Hasmady, S., Akhiar, A., Ideris, F., Shamsuddin, A. H., Mofijur, M., ... Mahlia, T. M. I. (2021). A comprehensive review on anaerobic digestion of organic fraction of municipal solid waste. *Renewable and Sustainable Energy Reviews*, 137, 110637. https://www.mdpi.com/508626.

Zhang, Q., Wang, M., Ma, X., Gao, Q., Wang, T., Shi, X., ... Yang, Y. (2019). High variations of methanogenic microorganisms drive full-scale anaerobic digestion process. *Environment International*, 126, 543–551. https://www.sciencedirect.com/science/article/pii/S0160412018323948

Ziara, R. M., Li, S., Subbiah, J., & Dvorak, B. I. (2018). Characterization of wastewater in two US cattle slaughterhouses. *Water Environment Research*, 90(9), 851–863. https://onlinelibrary.wiley.com/doi/abs/10.2175/106143017X15131012187971

Zinatizadeh, A. A. L., & Ghaytooli, E. (2015). Simultaneous nitrogen and carbon removal from wastewater at different operating conditions in a moving bed biofilm reactor (MBBR): Process modeling and optimization. *Journal of the Taiwan Institute of Chemical Engineers*, 53, 98–111. https://doi.org/10.1016/j.jtice.2015.02.034

5 Use of Plant-based Coagulants for Wastewater Treatment in Underdeveloped Communities

Sources, Processes, Effectiveness and Associated Mechanism

Priya Agarwal and Gaurav Saini

5.1 INTRODUCTION

One essential element for human life is water. It is the primary element of the hydrological cycle. Globally, water supplies are constantly being depleted as a result of population growth, industry, urbanization, and change in climate. More than 2 billion people worldwide inhabit underdeveloped nations where access to clean water is either very limited or nonexistent. On the other hand, the developing and wealthy nations are not doing so well either. It has been estimated that in such regions, as many as 4 billion people face some level of water scarcity for a span of at least one month every year (UNESCO, 2019). Poor sanitation and a lack of reliable water delivery systems are issues in developing nations (Mumbi et al., 2018). As per a World Health Organization's (WHO) research study, 844 million people suffer from a lack of access to potable drinking water. This includes around 159 million people who use one or the other type of surface water sources (such as rivers, streams, lakes, ponds, reservoirs, etc.) for their daily water consumption (WHO, 2018). Municipal wastewater (Natarajan et al., 2018; Maurya and Daverey, 2018), domestic wastewater (Vunain et al., 2019), and industrial wastewater (Kukić et al., 2018) are three primary sources of possible pollutants that could significantly worsen the quality of freshwater. The amount as well as the quality of domestic wastewater generated depends on the lifestyle and the population density in the given geography (Natarajan et al., 2018). Waterborne diseases like cholera, typhoid fever, hepatitis A, and amoebic dysentery are more likely to spread as a result of the indiscriminate discharge of industrial and/or municipal (domestic) wastewater, with no or partial treatment, into rivers and streams, especially in developing and underdeveloped nations (Vunain et al., 2019). This also has the potential to contaminate the groundwater and spread the infection to the populace that are dependent on groundwater sources for their water consumption.

Effective wastewater treatment is required in compliance with effluent requirements. The type of treatment to be used depends on the level of risk that water poses. Detergent-containing untreated wastewater poses serious environmental issues because it is hazardous to aquatic life (Mohamed et al., 2013). Additionally, it contains phosphates (Mohamed et al., 2014c) and elements: sodium (Na), chlorine (Cl), and Boron (B) (Mohamed et al., 2014e), that promote the growth of excessive algae in nearby waters like drains and others (Mohamed et al., 2014d). The wastewater from car washes, on the other hand, contains significant quantity of inorganics such as grits and sand which originally emanated from traffic, chemicals used in car washing/cleaning, and car usage. They also have oil and grease in them. Heavy metals from industrial waste are present in wastewater, which could be hazardous to the environment and people's health.

DOI: 10.1201/9781003441144-5

Because of the limited resources, infrastructure, and knowledge; development of wastewater treatment/management technologies to reduce water contamination has been a difficult and challenging undertaking. Research is being done on different processes and technologies to improve the quality of polluted water. They can be physical, chemical, or biological technologies (Sen, 2015; Jayalekshmi et al., 2021).

Physical methods employed for wastewater treatment include adsorption (Ali and Gupta, 2006), settling, medium & membrane filtration (Ezugbe and Rathilal, 2020), and Ultra-Violet (UV) radiation-based processes (O'Malley et al., 2020). Some examples of the chemical processes used for the treatment/management of different types of wastewaters include coagulation (Alibeigi-Beni et al., 2021), ion exchange (Ergunova et al., 2017), oxidation (Gogate and Pandit, 2004), disinfection (Collivignarelli et al., 2017), softening processes (Brastad and He, 2013), and catalytic reduction (Guo et al., 2020). A few biological methods include phytoremediation (Hu et al., 2020), anaerobic methods, bioreactor-based processes (Neoh et al., 2016), and microbial biodegradation (Huang, 2000). In underdeveloped countries, coagulation is considered a better choice because it is a simple and cost-effective process (Yin, 2010).

Coagulation is widely used to remove suspended solids, organic matter, turbidity, color (Vatvani, 2016), microorganisms, inorganic ions, metals, and contaminants to improve water quality (Teh et al., 2014). Mostly, aluminum salt (alum (Al_2 $(SO_4)_3.14H_2O$)) and ferric salt (ferrous sulfate ($FeSO_4.7H_2O$)) (Sćibanet al., 2009) are used in developing countries. The potential of these chemical coagulants is well known, but still, there are disadvantages associated such as ineffectiveness of chemical coagulants in low-temperature water, pH of treated water, and detrimental effects for the health of living beings (Vijayaraghavan et al., 2011). Moreover, voluminous and non-biodegradable sludge is generated after the treatment increasing the problem of disposal, and hence, the treatment costs (Desta and Bote, 2021). According to Flaten (2001), the use of chemical coagulants, based on aluminum, has neurotoxic and carcinogenic effects on humans and causes the development of neurodegenerative diseases like Alzheimer's disease (Bongiovani et al., 2015). As a result of these drawbacks of synthetic chemical coagulants, eco-friendly and sustainable alternatives, such as natural coagulants, are in great demand. These natural coagulants have several benefits, including being biodegradable, renewable, non-toxic, and affordable (Kukić et al., 2018; Maurya and Daverey, 2018). Natural coagulants have been successfully used in wastewater treatment, according to studies. Turbidity, Chemical Oxygen Demand (COD) (Mohamed et al., 2014a), Biochemical Oxygen Demand (BOD), hardness, Total Suspended Solids (TSS), Total Dissolved Solids (TDS), and color are the main parameters that are treated with natural coagulants (Desta and Bote, 2021). Some of the common ones include *Moringa Oleifera* (Chitra, and Muruganandam, 2020; Vunain et al., 2019; Mohamed et al., 2014a; Mohamed et al., 2014b), *Strychnos potatorum* (Mohamed et al., 2014a), Banana (Maurya and Daverey, 2018; Chitra and Muruganandam, 2020), *Phaseolus vulgaris* (common beans) (Kukić et al., 2018), Okra (Kaushal and Goyal, 2019), Neem (Maurya and Daverey, 2018), etc. Plant-based coagulants have been used to treat wastewater from several sources, including domestic wastewater (Vunain et al., 2019), industrial wastewater (Kukić et al., 2018; Gautam and Saini, 2020), and municipal wastewater (Natarajan et al., 2018; Maurya and Daverey, 2018).

Other than wastewater, treatment of water (Pritchard et al., 2009; Ang and Mohammad, 2020) and drilling slurries (Agarwal and Saini, 2020) have been carried out using natural coagulants.

5.2 COAGULATION MECHANISM BY NATURAL COAGULANTS

An essential process widely used in the treatment of both water and wastewater is coagulation (Jiang, 2015). The process of coagulation requires adding a coagulant, which enables small particles to aggregate to form larger flocs that can settle down during flocculation. Thus, coagulation is a chemical process that involves neutralization of particles in water and wastewater, whereas flocculation is a physical process that involves the formation of flakes from neutralized particles during the coagulation process. As a result, such impurities separate from the water suspension.

TABLE 5.1
Factors Affecting the Process of Coagulation

Coagulant Characteristics	Wastewater Characteristics	Physical Characteristics
Coagulant type	Water quality	Settling time
Coagulant dosage	Suspended solids	Mixing intensity
Coagulant quality	Temperature	Coagulant dosage end point
Coagulant lifespan	pH	
	Alkalinity	

The factors that determine the effectiveness of coagulation include, characteristics of coagulant used, characteristics of water to be treated, and characteristics of mixing process (Ang et al., 2020; Kumar et al., 2017). Table 5.1 presents the ideal operating conditions that significantly determine the most efficient coagulant needed for the treatment (Tetteh and Rathilal, 2005; Nimesha et al., 2022).

The types of equipment and reagents used, physical and chemical properties of pollutants (zeta potential), color, molecular weight of coagulant, the concentration of colloidal particles, and the presence or absence of impurities (trace elements and dissolved salts) are some additional factors that affect the coagulation process (Ang et al., 2020; Kumar et al., 2017).

Coagulation broadly depends on the operating conditions including the coagulant type, its dosage, rate of mixing, and time provided for settling of flocs. These factors determine the water quality. Moreover, the factors affecting the coagulation speed are pH, turbidity, amount/dose of the coagulant, temperature, mixing time, and speed. The optimal operating conditions are different for each type of coagulant and must be determined for the complete removal of contaminants.

The key constituents of natural coagulants include carbohydrates, protein, and lipids. Polysaccharides and amino acids are the main components of such coagulants. In a given suspension containing the coagulants, particle aggregation can take place through one or a combination of four different mechanisms: (i) double-layer compression; (ii) sweep flocculation; (iii) adsorption and charge neutralization; and (iv) adsorption and interparticle bridging (Miller et al., 2008; Bolto and Gregory, 2007; Crittenden et al., 2005). The coagulants compress the double layer, which can cause the destabilization of the suspended particles (Packham, 1965). Then, the coagulant encases suspended particles in a soft colloidal floc, which is the result of the sweep flocculation. The sorption of particles containing oppositely charged ions is termed as adsorption and results in surface charge neutralization; meanwhile in the mechanism of interparticle bridging, coagulants provide a polymeric chain for the binding of particles (Miller et al., 2008). It is generally believed that polymeric coagulants are associated with adsorption by neutralizing charges or bridging between particles due to their long-chained structures that increase adsorption sites. In the past, the dominant mechanisms observed for natural coagulants include charge neutralization and interparticle (or polymer) bridging.

Natural coagulants have diverse mechanisms of action. Table 5.2 presents the proposed mechanisms for commonly used natural coagulants and their functional groups. Water-soluble cationic coagulant proteins are present in *M. oleifera* extracts. These proteins have been reported to cause coagulation through a combination of charge neutralization and adsorption (Ndabigengesere et al., 1995; Theodoro et al., 2013) in which short protein chains with cationic surface charges adsorb and cause neutralization of the negatively charged colloidal particles present in water (Bolto and Gregory, 2007).

Anionic polyelectrolytes extracted from the seeds of a natural coagulant, *S. potatorum*, are used to destabilize suspended solid particles in water via the mechanism of interparticle bridging. Chemical bridging occurs in polymer chains where hydroxyl groups serve as adsorption sites, and the coagulation potency of -COOH and free -OH groups is increased (Yin, 2010). The functional

TABLE 5.2
Summary of Mechanisms of Natural Coagulants with their Functional Group

Natural Coagulants	Mechanism	Functional Group	References
Moringa oleifera	Adsorption and charge neutralization	Starch, cationic protein, glucose, fatty acids, alcoholic compounds, phenolic compounds, amines, and carboxylate groups	Vunain et al. (2019); Kurniawan et al. (2020)
Opuntia ficusindica	adsorption and bridging	D-xylose, galacturonic acid, l-rhamnose, l-arabinose, and d-galactose	Yin (2010); Vijayaraghavan et al. (2011)
Abelmoschus esculentus/Hibiscus esculentusas	Interparticle bridging	alkyl, ketone, aldehyde, carboxylic acids, esters, and amide	Zaharuddin et al. (2014); Durazzo et al. (2019); Kim et al. (2020)
Strychnos potatorum	Interparticle bridging	galactan and galactomannan	Vijayaraghavan et al. (2011)

groups present in Nirmali seeds are polysaccharides mixtures of galactanii and galactomannani that can result in about 80% turbidity reduction (Vijayaraghavan et al., 2011).

Cactus Polygalacturonicid found in cactus mucilage enables chemical bonding via hydrogen bonds (Miller et al., 2008). The particulates in solution do not approach directly but are generally connected to a polymer-like substance originating from such cactus species (Yin, 2010).

5.3 PLANT-BASED NATURAL COAGULANTS

In some regions within the developing world, traditional water treatment/management techniques have been using natural coagulants for centuries (Sutherland et al., 1994). Natural coagulants are divided into three groups based on their origins: plants, animals, and microbes (Nimesha et al., 2022).

Extracts from plants have long been used by humanity and date back to around 2000 BC. Evidence of plant products used as water clarifying agents has been inscribed by the Egyptians (Karnena and Saritha, 2022a; Saleem and Bachmann, 2019). Natural coagulants that are formed or produced from plants are known as plant-based natural coagulants. These are obtained from plant components such as seeds, leaves, fruits, stems, bark, and roots. Typically, plant-based coagulants are water-soluble, organic, and contain both ionic as well as non-ionic polymers (Bodlund et al., 2014). In recent years, natural coagulants have found significant attention in the treatment industry dealing with water and wastewater.

The commonly used natural coagulants with their form of usage are mentioned in Table 5.3. Several plant-based natural coagulants are widely accessible and have been proven to be safe (Desta and Bote, 2021; Nimesha et al., 2022). A number of research studies have investigated the use of coagulant aids or natural coagulants derived from *M. oleifera* seeds (Bina et al., 2010; Vigneshwaran et al., 2020). The application of natural coagulants is considered an environment-friendly and sustainable technology, as such substances are safe for consumption, and biodegradable in the environment. Moreover, natural coagulants are cost-effective, generate lesser sludge and do not alter the pH value of treated water.

Despite the fact that many plant-based coagulants have been reported, a few natural coagulants are known for their potential in treating wastewater within the scientific community and studies have generally focused on a niche of such plant-based materials, including *M. oleifera*, *Abelmoschus esculentus*, *S. potatorum*, and *Opuntia ficus-indica*.

TABLE 5.3
List of Natural Coagulants used for Wastewater Treatment

Scientific Name	Common Name	Sources of Extraction	Forms	References
Pisum sativum	Pea	Seed	Powder	Natarajan et al. (2018); Mbogo (2008)
Moringa oleifera	Drumstick	Seed	Powder	Vunain et al. (2019); Chitra and Muruganandam (2020); Desta and Bote (2021)
Tamarindus indica	Indian tamarind	Seed	Powder	Chitra and Muruganandam (2020); Kaur et al. (2012)
Phaseolus vulgaris	Common beans	Seed	Powder	Kukić et al. (2018)
Musa Cavendish	Banana	Peel/Stem	Powder/ Juice	Maurya and Daverey (2018); Chitra and Muruganandam (2020)
Azadirachta indica	Neem	Leaves/seeds	Powder	Maurya and Daverey (2018)
Carica papaya	Papaya	Seed	Powder	Yimer and Dame (2021); Maurya and Daverey (2018)
Arachis hypogea	Peanut	Seed	Powder	Subramani et al. (2018); Birima et al. (2013)
Aloe vera	Aloe vera	Leaves	Gel	Subramani et al. (2018)
Vigna radiata	Mung bean	Seeds	Powder	Subramani et al. (2018)
Opuntia ficusindica	Cactus	Pulp	Mucilage	Karnena et al. (2022b)
Cicer areitinum	Bengal gram	Seed	Powder	Kumar et al. (2017); Deosarkar et al. (2019)

5.3.1 Commonly Studied Natural Coagulants

The following section details a number of plant-based natural coagulants that have received si.

5.3.1.1 Moringa oleifera

M. oleifera belongs to the *Moringaceae* family (Okuda et al., 2001a). Its common names include horseradish tree and drumstick tree. It is a tropical plant that can be found across India as well as in many parts of Africa, Asia, and South America (Sutherland et al., 1994). Within the environmental scientific community, it is considered as the most extensively researched natural coagulant. Almost every part of *M. oleifera*, including the seeds, leaves, and roots, provides benefits. Moringa seeds are a source of both food and medicine and are commonly consumed by a large fraction of population in countries such as India. A water-soluble chemical and edible oil are present in the seeds (Anwar and Rashid, 2007). The chemical structures of several components found in *M. oleifera* (Sodvadiya et al., 2020) are illustrated in Figure 5.1.

It has been suggested to use *M. oleifera* seeds in rural parts of Africa and Asia for domestic/municipal wastewater treatment due to the growing interest in natural coagulants (Broin et al., 2002; Fatombi et al., 2013). According to studies by Katayon et al. (2006) and Beltrán-heredia et al. (2012), *M. oleifera* seeds can serve as coagulant, flocculant and absorbent for water treatment and can result in potable water. Reduction of heavy metals (copper, lead, zinc, and cadmium) using *M. oleifera* (Matouq et al., 2015; Basra et al., 2014). According to studies, both Gram-negative and Gram-positive bacteria are resistant to the antibacterial properties of *M. oleifera* seeds and leaves (Katata-Seru et al., 2018; Wan et al., 2016).

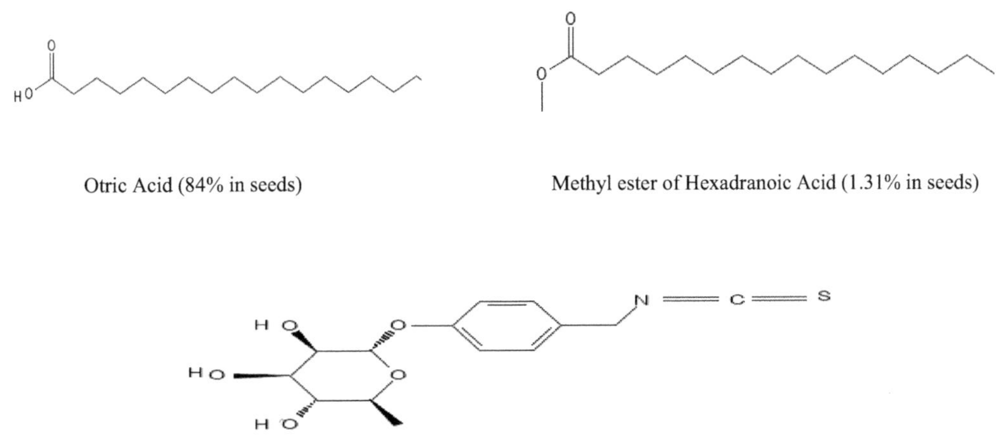

Otric Acid (84% in seeds)

Methyl ester of Hexadranoic Acid (1.31% in seeds)

FIGURE 5.1 Chemical structure of *Moringa* constituents and *Moringa* isothiocyanate.

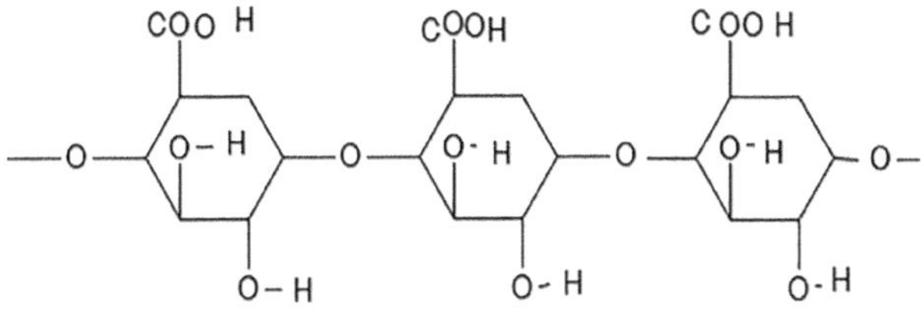

FIGURE 5.2 Polygalacturonic acid molecular chemical structure of *Opuntia* species.

M. oleifera seed is not known to have adverse effects on humans and has no significant disadvantages (Eman et al., 2014; Shan et al., 2017; Desta and Bote, 2021). Therefore, it can be used to treat wastewater both at household and industrial scales.

5.3.1.2 Opuntia ficus-indica

Opuntia ficus-indica plant is a member of the *Cactaceae* family (Prakash and Manikandan, 2012). Cacti species, known as opuntia, are made of mucilage that is kept in the plant's pulp (Karnena et al., 2022b). Cactus species are generally a resident of arid and semi-arid places. These grow in huge areas of Great Rift Valley and some parts of Northern Kenya. Various forms of cactus extract are raw juice, lyophilized powder or sun-dried powder (Mounir et al., 2014). In some regions, it is valued as a medicinal plant and a food source. Additionally, it is used to eliminate heavy metals (Mounir et al., 2014). Moreover, cactus is readily available without any adverse health effects, highly effective with biodegradable residues, produces toxic-free clean water, and generates a lower volume of sludge. Therefore, it can be used as an effective substitute for chemical coagulants. The chemical structure of Cactus (Othmani et al., 2020; Theodoro et al., 2013) is illustrated in Figure 5.2.

Cactus is made up of polygalacturonic acid, which has a long anionic chain with three primary groups: carboxyl (-COOH), carbonyl (-C=O), and hydroxyl (-OH). Hydroxyl group's deprotonation promotes the adsorption of different types of anionic pollutants and, consequently, their aggregation (Othmani et al., 2020; Theodoro et al., 2013).

5.3.1.3 Hibiscus esculentusas

H. esculentus belongs to the *Malvaceae* family (Kumar et al., 2010). This fruit is commonly known as ladyfinger, okra or gumbo okra, in different parts of the world. It is a common vegetable crop grown in swaths of Asia and Africa (Kumar et al., 2010). Okra is a versatile vegetable with several

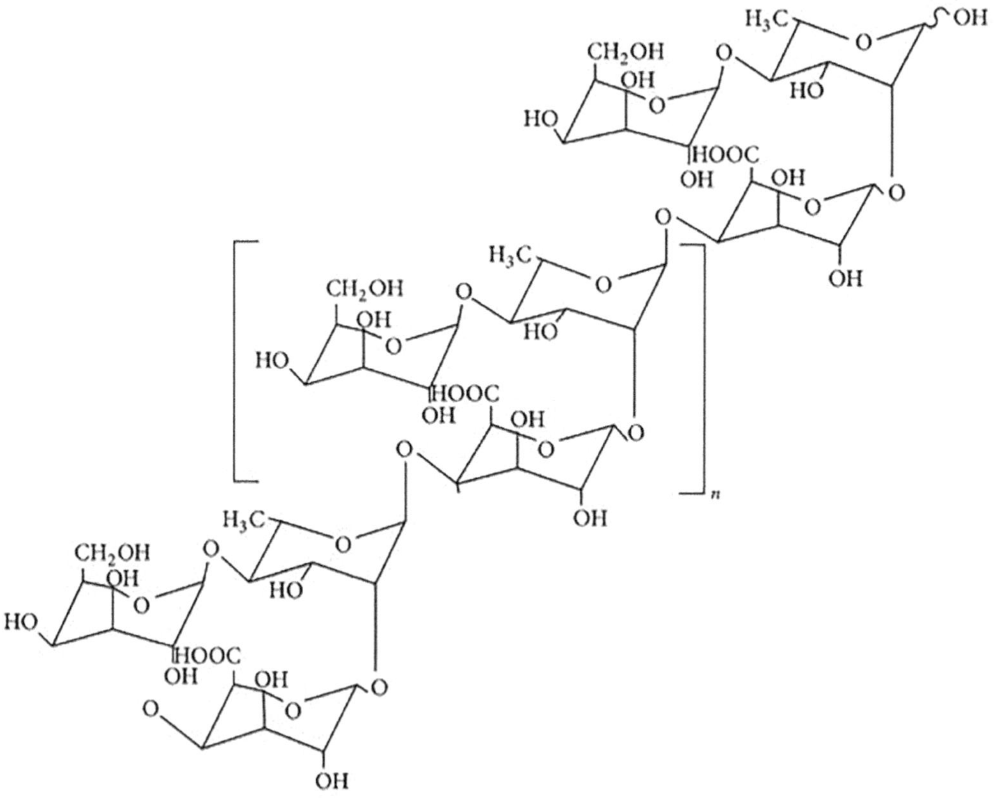

FIGURE 5.3 Okra gum polysaccharide molecular structure.

uses; a consumable in regions of Western Africa and Southeast Asia, while dried okra stems are often used as fuel for the production of paper in Malaysia (Lamont, 1999).

The polysaccharide structure of okra is depicted in Figure 5.3 (Othmani et al., 2020; Zaharuddin et al., 2014) and is made up of repeating units consisting of alternating rhamnose and galacturonic acid residues. In addition, disaccharide side chains comprised galactose connected to rhamnose residues are also present (Sengkhamparn et al., 2009).

5.3.1.4 Strychnos potatorum

S. potatorum, often called "Nirmali" belongs to the family *Loganiaceae* (Kumudhaveni et al., 2020). It can be found in vast regions of Asia, including Sri Lanka, Burma, and India (southern and central regions). It is used to make medications (Jayaram et al., 2009). Sanskrit scriptures from more than 4,000 years ago indicate that seeds were employed to clean turbid water obtained from surface sources (Shultz and Okun, 1984). It reveals that the Nirmali seeds are indeed the first ever plant-based coagulant that was used in water treatment (Swati and Govindan, 2005; Sarawgi et al., 2009).

The active coagulating agent for Nirmali is found to be polysaccharides (Othmani et al., 2020). These active compounds are generally known as galactomannans and are composed of two substances: galactose and mannose chains (Othmani et al., 2020).

5.4 PROCESSING OF NATURAL COAGULANTS

The plant-based natural coagulants are generally processed through one or a combination of three steps: primary, secondary, and tertiary, as shown in Figure 5.4 (Yin, 2010).

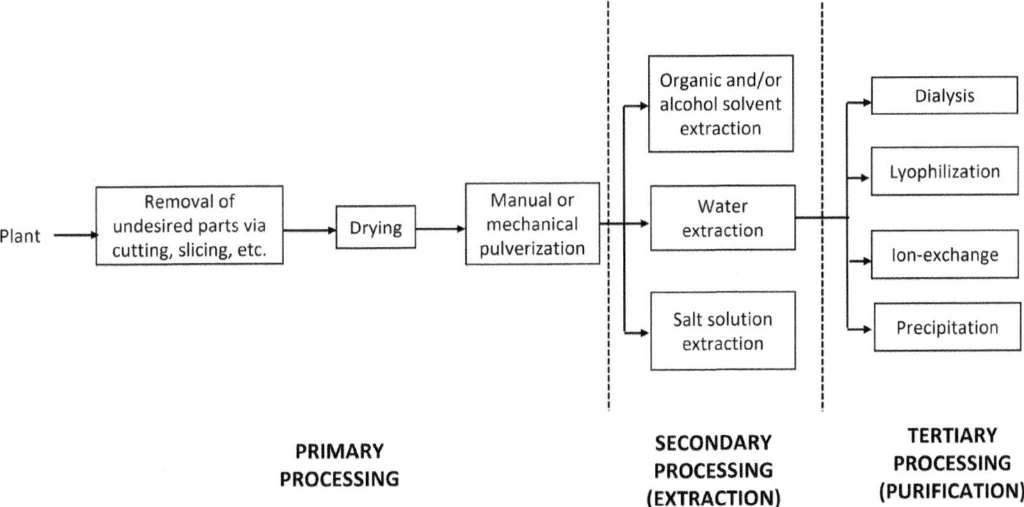

FIGURE 5.4 Processing steps of plant-based natural coagulants.

In the first or primary processing step, plant parts are cut or sliced, then dried in the oven or sun. After that, these parts are ground into a fine powder which is ready to use. This method is used in place of sophisticated processing equipment for research studies and domestic applications. This enables a low-cost and low-tech method of generating coagulants and is especially useful for resource-strapped underdeveloped and developing countries.

In the secondary processing stage, active agents are extracted by a variety of solvents, such as chloride-based salt solutions (sodium chloride (NaCl), potassium chloride (KCl), and magnesium chloride ($MgCl_2$), etc.), water, or organic solvents (acetone, alcohol) (Okuda et al., 2001a). In most of the cases, solvents are used together to extract active coagulants, useful and edible oils (Ghebremichael et al., 2005). Extraction with water is commonly used owing to the twin advantages it offers: easy availability and economics. In the case of *M. oleifera*, water-soluble protein is the active coagulating agent. However, salt solution (using NaCl) extraction proved more effective than water, resulting in greater coagulation activity at lower coagulant dosage (Okuda et al., 2001a; Okuda et al., 2001b). It was reported that the dosage of extracted components is 7.4 times less as compared to those extracted with distilled water and resulted in similar levels of turbidity reduction.

Tertiary processing is uncommon for coagulants derived from plants and is generally restricted only to academic/university research to purify extracts of *M. oleifera* because it raises processing costs (Ndabigengesere et al., 1995, Okuda et al., 2001b, Ghebremichael et al., 2005). Initial studies have suggested that lyophilization (Ndabigengesere et al., 1995), ion exchange (Ghebremichael et al., 2005), precipitation, and dialysis (Okuda et al., 2001a) are suitable purification techniques for *M. oleifera* extracts that can be incorporated with a upscaled system dealing with the treatment of high turbidity water.

While preparing the *M. oleifera* extracts, the seeds were first readily taken from the seed pods, dried, shelled, grinded and then sized appropriately using sieving. After forming the wet powder, either of the following methods is used for further processing. The obtained powder can be dried and used directly (Sethupathy, 2015) or extraction of active protein can be done by dissolving the seed powder in either of the solvents: salt solutions or organic solvents (Santos et al., 2012). On mixing the seed powder soluble proteins are yielded with net positive charge (Sutherland et al., 1994). The active ingredients for *M. oleifera* are proteins that are extracted using salt solution since it is regarded as the most effective option due to the salting-in process generally termed as the common-ion effect, which promotes the protein-protein dissociation. The solubility of proteins can be increased by enhancing the ionic strength of given solution through salt addition (Okuda et al., 2001a). Ultimately, increased protein solubility results in enhanced coagulating potential of *M. oleifera*. As an example, proteins extracted with a 1M sodium chloride solution removed 95% of the turbidity at 4 mL/L as compared to water. It removed 78% of turbidity at 32 mL/L for kaolin water (Okuda et al., 1999).

The cactus extracts are derived by oven-drying of the diced cladodes, then grinding them into powder form (Sellami et al., 2014). Moreover, the powder can be mixed with either of the solutions (plain water, saline water or organic solvents) to extract cactus coagulants. These are the methods for preparing commonly used natural coagulants for wastewater treatment.

5.5 TREATMENT OF WASTEWATER

In 1994, a study was conducted in southern Malawi (Thyolo treatment works). It comprises three main processes: flocculator-based clarifiers, rapid gravity sand filters, and disinfection through chlorine application. In this study, alum and soda ash were replaced by *M. oleifera* seed powder and the results obtained were comparable. It was for the first-time natural coagulant was applied as the primary alternative of synthetic chemical on an industrial scale (Sutherland et al., 1994).

Table 5.4 summarizes the studies that have used synthetic coagulants such as alum, ferric chloride ($FeCl_3$), poly-aluminum-silicate-chloride, and ferrous sulfate (Vunain et al., 2019; Chitra and Muruganandam, 2020; Mohamed et al., 2014a; Rodrigues et al., 2008) for wastewater treatment. The optimum dosages for chemical coagulants are traditionally 30–80 mg/L, depending on the source and strength of the water.

Table 5.5 summarizes the studies that investigated natural coagulants for the treatment of wastewater. As can be seen form the compiled results, the natural coagulants work at dosages ranging between 40 and 600 mg/L and are better suited to basic waters, with pH values ranging between 7 and 10.

TABLE 5.4
Treatment of Wastewater Using Chemical Coagulants

Chemical Coagulant	Sources of Wastewater	Dosage	Efficiency	References
Alum	Domestic wastewater	15 g/L	99.80% turbidity 96.70% *E. Coli*	Vunain et al. (2019)
$FeCl_3$	Domestic wastewater	15 g/L	99.70% Turbidity 96.40% *E. Coli*	Vunain et al. (2019)
Alum	Synthetic greywater Real greywater	1,000 mg/L	95.79% turbidity 93.31% turbidity	Chitra and Muruganandam (2020)
Alum Ferrous sulfate ($FeSO_4$),	Car wash wastewater	150 mg/L	87% turbidity, 74% COD, and 81% phosphorus 77% turbidity 71%COD 65% phosphorus	Mohamed et al. (2014)
Poly-aluminum-silicate-chloride	Wastewater from paper and pulp	40 mg/L	93.13% COD 91.12% turbidity	Rodrigues et al. (2008)
Alum	Domestic wastewater	30 mg/L	96% turbidity	Ugwu et al. (2017)
Alum	Municipal Wastewater	–	59% turbidity, 59% COD, 51% BOD, 30% hardness, 67%TSS and 65% TDS	Kaushal and Goyal (2019)

TABLE 5.5
Treatment of Wastewater Using Natural Coagulants

Coagulant	Sources of Wastewater	Optimum Conditions	Efficiency	Location	References
Pisum sativum	Municipal wastewater	4.0 g/L		Oman	Natarajan et al. (2018)
Moringa oleifera	Domestic wastewater	15 g/L	94.40% Turbidity 97.30% E.Coli	Zomba, Malawi	Vunain et al. (2019)
Moringa oleifera	Basic wastewater Acidic wastewater	0.4 g/500 ml	99.99% turbidity 95.34% color 59.99% COD 95.99% turbidity 90% color 55.99% COD		Desta and Bote (2021)
Tamarindus indica (tamarind) *Moringa oleifera* banana peels	Synthetic greywater	400 mg/L 600 1,000	61.33% turbidity 85.28% turbidity 90.42% turbidity		Chitra and Muruganandam (2020)
Tamarindus indica (tamarind), *Moringa oleifera*, banana peels	Actual greywater	400 600 1,000	63.75% turbidity 83.7% turbidity 88.88% turbidity		Chitra and Muruganandam (2020)
Phaseolus vulgaris (common beans)	Sugar production wastewater synthetic turbid water	0.4 mL/L	49.8% turbidity 90.1% turbidity		Kukić et al. (2018)
Musa Cavendish (banana peel), neem leaf powder, papaya seed powder, and banana stem juice	Municipal wastewater	0.4 g/L g/L 0.8 g/L 10mL/L	59.6%, 43.96%, 41.89%, and 18.78% turbidity 45.45%, 35.75%, 66.67%, 29.69% TSS 58%, 64%, 66.67%, 64% COD	Dehradun, India	Maurya and Daverey (2018)
Moringa oleifera *Strychnos potatorum*	Car wash wastewater	30 mg/L 50 mg/L	90% turbidity, 60% COD 75% phosphorus 96% turbidity 55% COD 65% phosphorus	Johor, Malaysia	Mohamed et al. (2014a)
Moringa oleifera	Car wash wastewater	60 mg/L	60% COD	Johor, Malaysia	Mohamed et al. (2014b)
Moringa oleifera	Domestic wastewater	200 mg/L	91% turbidity	Nigeria	Ugwu et al. (2017)
Moringa oleifera	Municipal wastewater	–	61% Turbidity, 65% COD, 55% BOD, 69% TSS, 68% TDS, 25% Hardness	India	Kaushal and Goyal (2019)

(Continued)

TABLE 5.5 (*Continued*)
Treatment of Wastewater Using Natural Coagulants

Coagulant	Sources of Wastewater	Optimum Conditions	Efficiency	Location	References
Okra	Municipal wastewater	–	65% Turbidity, 67% COD, 56% BOD, 30% hardness, 70% TSS, 69% TDS	India	Kaushal and Goyal (2019)
Arachis hypogea (Peanut) Mung bean Cactus		–	59.5% Cr, 70.5% Cd, 68.9% Zn, 12% hardness, 18.28% turbidity 57.3% Cr, 68.2% Cd, 64.8% Zn, 34% hardness, 62.28% turbidity 56.6% Cr, 69.4% Cd, 69.7% Zn, 20% hardness, 15.71% turbidity	India	Subramani et al. (2018)
Moringa oleifera	mixed domestic/ industrial wastewater	150 mg/L	40% BOD 80% Suspended Solids		Folkard et al. (1993)
Polygonum *Commelina,* *Typha* *Ipomoea,* *Limnophyton*	Turbid water sample		92% 91% 97% 93% 95% Turbidity	India	Patale et al. (2010)
Cicer areitinum Tamarind	Sewage wastewater	0.15 g/mL 0.2 g/mL	84.3% turbidity 80.0%	India	Deosarkar et al. (2019)

In another study, turbidity removal obtained for *M. oleifera*, alum, and ferric chloride were 94.4%, 99.8%, and 99.7%, respectively. The removal for *Escherichia coli* was 97.30% using *M. Oleifera*, which was more than conventional coagulants; alum and ferric chloride, 96.70% and 96.40%, respectively. Mohamed et al. (2014a), treated car wash wastewater using natural coagulants as it costs less and offers eco-friendly treatment. It leads to the formation of stronger flocs using natural coagulants through bridging effect as compared to alum (Vijayaraghavan et al., 2011). However, using *Moringa*, the coagulation is found to be more effective as compared to alum, because there is no need for post-treatment for pH (Al-Jadabi et al., 2021). Additionally, *M. oleifera* is locally available in various parts of the world and is an environment-friendly, non-toxic, and biodegradable coagulant that does not alter the pH of resulting water.

As far as observed in the studies, the use of natural coagulants for treating industrial wastewater is generally limited to lab research, as seen by scarcity of pilot or field scale reports. However, these experimental research studies have reported significant potential for using natural coagulants for treating wastewaters generated by different industries. It can be seen; natural coagulants perform better when used for wastewaters with less variety of pollutants.

5.6 USE OF TREATED WATER

It has been reported that final products post application of natural coagulants, including the treated water as well as the clarified sludge have the potential to be used further. There are many options for using treated water including irrigation purpose and watering. The treated water can be used in irrigation of cultivated plants used for the production of food as well as watering of areas for landscaping, gardens, parks, and sports facilities. The treated water can potentially be utilized for non-drinking applications, such as washing vehicles and flushing toilets.

In a study, the water, post-treatment, was used in four combinations to grow crops like okra, radish, and eggplant; 1:1 (groundwater: treated wastewater), 100% groundwater, 3:1 (treated wastewater: groundwater), and 100% treated wastewater. Not much difference was observed in physicochemical properties of soil using treated wastewater and groundwater (Ahmed et al., 2016). The physical analysis of crop showed a significant increase in plant productivity using treated wastewater. Further, the chemical characteristics confirmed to the applicable regulatory standards (Al-Busaidi and Ahmed, 2017). In addition, the treated wastewater was also used for growing plants with the potential to generate biofuel. In this study, the *Jatropha* plants gave the best growth (measured as plant height and yield) when irrigated with treated wastewater (Al-Busaidi, 2014).

5.7 MERITS OF PLANT-BASED NATURAL COAGULANTS

The use of *M. oleifera* permits avoiding various drawbacks of conventional coagulants like alum because it is a locally accessible naturally occurring substance that is environment-friendly, non-toxic, biodegradable in natural setting and does not alter the pH of water (Vunain et al., 2019). It is highlighted that the use of plant-based flocculants, such as *M. oleifera* seed, is strongly advised for domestic wastewater treatment in developing/underdeveloped communities, where residents, particularly in rural/non-urban communities, are accustomed to consuming contaminated and/or untreated turbid waters from surface sources. If appropriately applied (within the optimal dosage ranges and settling times), *M. oleifera* seed powder acts as a coagulant during the purification process and dramatically decreases suspended particles (flocculates) as well as the microbial count in water (Vunain et al., 2019).

Therefore, the usage of *M. oleifera* may lessen the spread of waterborne infections in developing countries, particularly in sub-Saharan Africa where there is a severe lack of conventional water treatment methods. The primary benefits of using naturally occurring plant-based coagulants as point-of-use (POU) water treatment materials are obvious; they are economical, unlikely to generate treated water with an extreme pH, and extremely biodegradable. These benefits are greatly enhanced if the plant used to make the coagulant is native to a rural area. The use of such coagulants is a necessary effort in accordance with the global sustainable development objectives in the age of climate change, resource depletion, and broad environmental deterioration. Moreover, plant-based coagulants were used to treat turbid water (Sanghi et al., 2002) and for this reason, scientists have identified a number of different plant varieties. *M. oleifera* should be used in developing countries unaltered in order to avoid additional costs that are not often allowed for these nations.

However, there are some issues that must be resolved before using natural coagulants further, esp. at an industrial scale. The inability of natural coagulants to regulate pH change is one of their key disadvantages. Although the use of *M. oleifera* seed powder does not ensure that raw water is entirely (100%) free of pathogenic microorganisms, it does clean the water and make it safer for domestic use than the present untreated/partially treated situation. *M. oleifera* seed powder is not the best choice for removing salt, potassium, and calcium from wastewater. The need for proper storage procedures for these coagulants adds to the overall expense of the treatment. The need for higher dosages of natural coagulants when compared to synthetic coagulants is another disadvantage of employing them for wastewater treatment (Aliyu et al., 2015).

TABLE 5.6
Advantages of Using Natural Coagulants

Natural Coagulant	Advantages	References
Moringa oleifera	Reduction in the quantity of microorganisms in raw water	Vunain et al. (2019)
Plant-based coagulants	Not dependent on chemicals Small amounts of sludge generated Biodegradable Less toxic Non-corrosive	Rocha et al. (2019)
Plant-based coagulants	Low price Abundant source Multipurpose Biodegradability	Othmani et al. (2020)
Plant-based coagulants	Not noxious to an aquatic environment	Nimesha et al. (2022)
Plant-based coagulants	Non-toxic Biodegradable Environment-friendly	Verma et al. (2012); Muralimohan et al. (2014)
Plant-based coagulants	Sludge obtained after the treatment with natural coagulants is non-toxic and biodegradable and can be used for soil improvement.	Prodanović et al. (2013)

5.8 CONCLUSION

Wastewater of all types (municipal and/or industrial) has been treated for a very long time using chemical coagulants. Chemical coagulants like aluminum salts and iron salts are commonly utilized. There are certain drawbacks to using chemical coagulants notwithstanding their high efficiency. These include expensive costs, the creation of sludge that isn't biodegradable, and poor health conditions brought on by the water's aluminum residue. A promising substitute for chemical coagulants is the use of plant-based natural coagulants. Natural coagulants made from plant-based sources are used, which is sustainable environmental technology because it prioritizes raising the standard of living in underprivileged places. Several plant-based species are known to serve as excellent coagulants for wastewater treatment. There are advantages and disadvantages associated with their application and need to be further studied and improved before their large-scale implementation. The application of natural coagulants is especially relevant for underdeveloped and developing regions due to their low cost, low-tech preparation, and wide availability. Further research in their storage, extraction of coagulant material, and pilot scale application is expected to yield natural alternatives to synthetic compounds in the field of municipal and industrial wastewater treatment.

REFERENCES

Agarwal, P., and Saini, G. (2020). Use of natural coagulants (Moringa oleifera and Benincasa hispida) for volume reduction of waste drilling slurries. Materials Today: Proceedings, 49. https://doi.org/10.1016/j.matpr.2020.12.924.

Ahmed, S., Saifullah, Ahmad, M., Swami, B.L., and Ikram, S. (2016). Green synthesis of silver nanoparticles using Azadirachta indica aqueous leaf extract. *Journal of Radiation Research and Applied Sciences*, 9(1), 1–7. https://doi.org/10.1016/j.jrras.2015.06.006

Al-Busaidi, A. (2014). Usage of treated wastewater for bio-fuel production. In *Conference: The 5th Joint GCC-Japan environment symposium, sustainable GCC environment: Challenge for our future.*

Al-Busaidi, A., and Ahmed, M. (2017). Maximum use of treated wastewater in agriculture. In *Water resources in arid areas: The way forward*, pp. 371–382. https://doi.org/10.1007/978-3-319-51856-5_21.

Ali, I., and Gupta, V. (2006). Advances in water treatment by adsorption technology. *Nature Protocols*, 1(6), 2661–2667.

Alibeigi-Beni, S., Habibi Zare, M., Chenar, M.P., Sadeghi, M., and Shirazian, S. (2021). Design and optimization of a hybrid process based on hollow-fiber membrane/coagulation for wastewater treatment. *Environmental Science and Pollution Research*, 28, 8235–8245. https://doi.org/10.1007/s11356-020-11037-y

Aliyu, L., Mukhtar, L.W., and Abba, Sani. (2015). Evaluation of coagulation efficiency of natural coagulants (Moringa oliefera, Okra) and alum, for Yamuna water treatment. *International Journal of Advance Research in Science and Technology*, 4(1), 5–833.

Al-Jadabi, N., Laaouan, M., Mabrouk, J., Fattah, G., and El-Hajjaji, S. (2021). Comparative study of the coagulation efficacy of Moringa oleifera seeds extracts to alum for domestic wastewater treatment of Ain Aouda City, Morocco. E3S Web of Conferences, 314, 08003. https://doi.org/10.1051/e3sconf/202131408003

Ang, T.H., Kiatkittipong, K., Kiatkittipong, W., Chua, S.C., Lim, J.W., Show, P.L., and Ho, Y.C. (2020). Insight on Extraction and characterisation of biopolymers as the green coagulants for microalgae harvesting. *Water*, 12(5), 1388. https://doi.org/10.3390/w12051388

Ang, W.L., and Mohammad, A.W. (2020). State of the art and sustainability of natural coagulants in water and wastewater treatment. *Journal of Cleaner Production*, 121267. https://doi.org/10.1016/j.jclepro.2020.121267

Anwar, F., and Rashid, U. (2007). Physiochemical characteristics of Moringa oleifera seeds and seed oil from a wild provenance of Pakistan. *Pakistan Journal of Botany*, 39, 1443–1453.

Basra, S.M.A., Iqbal, Z., ur-Rehman, Khalil, Ur-Rehman, Hafeez, and Ejaz, M.F. (2014). Time course changes in pH, electrical conductivity and heavy metals (Pb, Cr) of wastewater using *Moringa oleifera* Lam. seed and alum, a comparative evaluation. *Journal of Applied Research and Technology*, 12, 560–567. https://doi.org/10.1016/S1665-6423(14)71635-9

Beltrán-heredia, J., Sánchez-martín, J., Muñoz-Serrano, A., and Peres, J.A. (2012). Towards overcoming TOC increase in wastewater treated with *Moringa oleifera* seed extract. Chemical Engineering Journal, 188, 40–46. https://doi.org/10.1016/j.cej.2012.02.003.

Bina, B., Mehdinejad, M.H., Dalhammer, G., Rajarao, G., Nikaeen, M., and Attar, H.M. (2010). Effectiveness of Moringa oleifera Coagulant Protein as natural coagulant aid in removal of turbidity and bacteria from turbid waters. *World Academy of Science, Engineering and Technology*, 43(7), 618–620.

Birima, A.H., Hammad, H.A., Desa, M.N.M., and Muda, Z.C. (2013). Extraction of natural coagulant from peanut seeds for treatment of turbid water. IOP Conference Series: Earth and Environmental Science, 16, 012065. https://doi.org/10.1088/1755-1315/16/1/012065

Bodlund, I., Pavankumar, A.R., Chelliah, R., Kasi, S., Sankaran, K., and Rajarao, G.K. (2014). Coagulant proteins identified in Mustard: A potential water treatment agent. *International Journal of Environmental Science and Technology*, 11(4), 873–880.

Bolto, B., and Gregory, J. (2007). Organic polyelectrolytes in water treatment. *Water Research*, 41, 2301–2324.

Bongiovani, M.C., Camacho, F.P., Coldebella, P.F., Valverde, K.C., Nishi, L., and Bergamasco, R. (2015). Removal of natural organic matter and trihalomethane minimization by coagulation/flocculation/filtration using a natural tannin. *Desalination and Water Treatment*, 57(12), 5406–5415.

Brastad, K.S., and He, Z. (2013). Water softening using microbial desalination cell technology. *Desalination*, 309, 32–37. https://doi.org/10.1016/j.desal.2012.09.015

Broin, M., Santaella, C., Cuine, S., Kokou, K., Peltier, G., and Joe, T. (2002). Flocculent activity of a recombinant protein from Moringa oleifera Lam. Seeds. *Applied Microbiology and Biotechnology*, 60, 114–119. https://doi.org/10.1007/s00253-002-1106-5.

Chitra, D., and Muruganandam, L. (2020). Performance of natural coagulants on greywater treatment. *Recent Innovations in Chemical Engineering*, 13(1), 81–92.

Collivignarelli, M., Abbà, A., Alloisio, G., Gozio, E., and Benigna, I. (2017). Disinfection in wastewater treatment plants: Evaluation of effectiveness and acute toxicity effects. *Sustainability*, 9(10), 1704. https://doi.org/10.3390/su9101704

Crittenden, J.C., Trussell, R.R., Hand, D.W., Howe, K.J., and Tchobanoglous, G. (2005). *Water treatment principles and design* (2nd ed.). Hoboken, NJ: John Wiley & Sons.

Desta, W.M., and Bote, M.E. (2021). Wastewater treatment using a natural coagulant (Moringa oleifera seeds): Optimization through response surface methodology. *Heliyon*, 7(11), e08451.

Durazzo, A., Lucarini, M., Novellino, E., Souto, E.B., Daliu, P., and Santini, A. (2019). Abelmoschus esculentus (L.): Bioactive components; beneficial properties-focused on antidiabetic role-for sustainable health applications. *Molecules*, 24(1), 38. https://doi.org/10.3390/molecules24010038

Eman, N.A., Tan, C.S., and Makky, E.A. (2014). Impact of Moringa oleifera cake residue application on wastewater treatment: A case study. *Journal of Water Resource and Protection*, 6(7), 677–687.

Ergunova, O., Ferreira, R., Ignatenko, A., Lizunkov, V., and Malushko, E. (2017). Use of ion exchange filters in wastewater treatment. In K. Anna Yurevna, A. Igor Borisovich, W. Martin de Jong, and M. Nikita Vladimirovich (Eds.), *Responsible research and innovation,* vol. 26. *European proceedings of social and behavioural sciences,* pp. 541–549. https://doi.org/10.15405/epsbs.2017.07.02.69.

Ezugbe, E.O., and Rathilal, S. (2020). Membrane technologies in wastewater treatment: A review. *Membranes,* 10(5), 89. https://doi.org/10.3390/membranes10050089

Fatombi, J.K., Lartiges, B., Aminou, T., Barres, O., and Caillet, C. (2013). A natural coagulant protein from copra (*Cocos nucifera*): Isolation, characterization, and potential for water purification. *Separation and Purification Technology*, 116, 35–40. https://doi.org/10.1016/j.seppur.2013.05.015.

Flaten, T.P. (2001). Aluminium as a risk factor in Alzheimer's disease, with emphasis on drinking water. *Brain Research Bulletin*, 55, 187–96.

Folkard, G.K., Sutherland, J.P., and Grant, W.D. (1993). Natural coagulants at pilot scale. In J. Pickford (Ed.) Water, environment and management: Proc. of the 18th WEDC conference, Kathmandu, Nepal, 30 August–3 September 1992. Loughborough University Press, pp. 51–54.

Gautam, S., and Saini, G. (2020). Use of natural coagulants for industrial wastewater treatment. *Global Journal of Environmental Science and Management*, 6(4), 553–578. https://doi.org/10.22034/gjesm.2020.04.10

Ghebremichael, K.A., Gunaratna, K.R., Henriksson, H., Brumer, H., and Dalhammar, G.A. (2005). Simple purification and activity assay of the coagulant protein from Moringa oleifera seed. *Water Research*, 39, 2338–2344.

Gogate, P.R., and Pandit, A.B. (2004). A review of imperative technologies for wastewater treatment I: Oxidation technologies at ambient conditions. *Advances in Environmental Research*, 8(3-4), 501–551. https://doi.org/10.1016/s1093-0191(03)00032-7

Guo, Y., Ahmed Zelekew, O., Sun, H., Kuo, D.-H., Lin, J., and Chen, X. (2020). Catalytic reduction of organic and hexavalent chromium pollutants with highly active bimetal CUBiOS oxysulfide Catalyst under dark. Separation and Purification Technology, 116769. https://doi.org/10.1016/j.seppur.2020.116769

Hu, H., Li, X., Wu, S., and Yang, C. (2020). Sustainable livestock wastewater treatment via phytoremediation: Current status and future perspectives. Bioresource Technology, 123809. https://doi.org/10.1016/j.biortech.2020.123809

Huang, C. (2000). Optimal condition for modification of chitosan: A biopolymer for coagulation of colloidal particles. *Water Research*, 34(3), 1057–1062. https://doi.org/10.1016/s0043-1354(99)00211-0

Jayalekshmi, S.J., Biju, M., Somarajan, J., Ajas, P.E.M., and Sunny, D. (2021). Wastewater treatment technologies: A review. *International Journal of Engineering Research & Technology* (*IJERT*) *ICART*, 9(9), 1–5.

Jayaram, K., Murthy, I.Y.L.N., Lalhruaitluanga, H., and Prasad, M.N.V. (2009). Biosorption of lead from aqueous solution by seed powder of Strychnos potatorum L. *Colloids and Surfaces B*, 71, 248–254.

Jiang, J.Q. (2015). The role of coagulation in water treatment. *Current Opinion in Chemical Engineering*, 8, 36–44. https://doi.org/10.1016/j.coche.2015.01.008.

Karnena, M.K., and Saritha, V. (2022a). Contemplations and investigations on green coagulants in treatment of surface water: A critical review. *Applied Water Science*, 12, 150. https://doi.org/10.1007/s13201-022-01670-y

Karnena, M.K., Konni, M., Dwarapureddi, B.K., and Saritha, V. (2022b). Blend of natural coagulants as a sustainable solution for challenges of pollution from aquaculture wastewater. *Applied Water Science*, 12(3), 1–14.

Katata-Seru, L., Moremedi, T., Aremu, O.S., and Bahadur, I. (2018). Green synthesis of iron nanoparticles using Moringa oleifera extracts and their applications: Removal of nitrate from water and antibacterial activity against Escherichia coli. *Journal of Molecular Liquids*, 256 296–304. https://doi.org/10.1016/j.molliq.2017.11.093.

Katayon, S., Ng, S.C., Johari, M.M.N.M., and Ghani, L.A.A. (2006). Preservation of coagulation efficiency of Moringa oleifera, a natural coagulant. *Biotechnology and Bioprocess Engineering*, 11, 489–495, https://doi.org/10.1007/BF02932072.

Kaur, H., Yadav, S., Ahuja, M., and Dilbaghi, N. (2012). Synthesis, characterization and evaluation of thiolated tamarind seed polysaccharide as a mucoadhesive polymer. *Carbohydrate Polymers*, 90(4), 1543–1549.

Kaushal, R.K. and Goyal, H. (2019). Treatment of waste water using natural coagulants. *Proceedings of recent advances in interdisciplinary trends in engineering & applications* 2019. https://doi.org/10.2139/ssrn.3368088.

Kim, I.T.S., Sethu, V., Arumugasamy, S.K., and Selvarajoo, A. (2020). Fenugreek seeds and okra for the treatment of palm oil mill effluent (POME) – Characterization studies and modeling with backpropagation feedforward neural network (BFNN). *Journal of Water Process Engineering*, 37, 101500. https://doi.org/10.1016/j.jwpe.2020.101500

Kukić, D., Šćiban, M., Prodanović, J., Vasić, V., Antov, M., and Nastić, N. (2018). Application of natural coagulants extracted from common beans for wastewater treatment. *E-GFOS*, 9(16), 77–84.

Kumar, S., Dagnoko, S., Haougui, A., Ratnadass, A., Pasternak, D., and Kouame, C. (2010). Okra (Abelmoschus spp.) in West and Central Africa: Potential and progress on its improvement. *African Journal of Agricultural Research*, 5(25), 3590–3598.

Kumar, V., Othman, N., and Asharuddin, S. (2017). Applications of natural coagulants to treat wastewater - A review. MATEC Web of Conferences 103, ISCEE 2016. https://doi.org/10.1051/matecconf/201710306016

Kumudhaveni, B., Radha, R., and Suresh, A.J. (2020). Pharmacognostical and phytochemical standardization of seeds of *Strychnos potatorum* Linn. (Loganiaceae). *Journal of Pharmacognosy and Phytochemistry*, 9(2), 2054–2059.

Kurniawan, S.B., Abdullah, S.R.S., Imron, M.F., Said, N.S.M., Ismail, N., 'Izzati, Hasan H.A., Othman, A.R., and Purwanti, I.F. (2020). Challenges and opportunities of biocoagulant/bioflocculant application for drinking water and wastewater treatment and its potential for sludge recovery. *International Journal of Environmental Research and Public Health*, 17(24), 1–33.

Lamont, W.J. (1999). Okra – A versatile vegetable crop. *HortTechnology*, 9(21), 179184. https://doi.org/10.21273/HORTTECH.9.2.179

Matouq, M., Jildeh, N., Qtaishat, M., Hindiyeh, M., and Al Syouf, M.Q. (2015). The adsorption kinetics and modeling for heavy metals removal from wastewater by Moringa pods. *Journal of Environmental Chemical Engineering*, 3, 775–784. https://doi.org/10.1016/j.jece.2015.03.027.

Maurya, S., and Daverey, A. (2018). Evaluation of plant-based natural coagulants for municipal wastewater treatment. *3 Biotech*, 8(1), 77.

Mbogo, S.A. (2008). A novel technology to improve drinking water quality using natural treatment methods in rural Tanzania. *Journal of Environmental Health*, 70(7), 46–50.

Miller, S.M., Fugate, E.J., Craver, V.O., Smith, J.A., and Zimmerman, J.B. (2008). Toward understanding the efficacy and mechanism of Opuntia spp. as a natural coagulant for potential application in water treatment. *Environmental Science & Technology*, 42, 4274–4279.

Mohamed, R.M.S.R., Kassim, A.H.M., Anda, M., and Dallas, S. (2013). A monitoring of environmental effects from household greywater reuse for garden irrigation. *Environmental Monitoring and Assessment*, 185(10), 8473–8488.

Mohamed, R.M.S.R., Kutty, N.M.A.I., and Kassim, A.H.M. (2014a). Efficiency of using commercial and natural coagulants in treating car wash wastewater. *Australian Journal of Basic and Applied Sciences*, 8(16), 227–234.

Mohamed, R.M.S.R., Rahman, N.A., and Kassim, A.H.M. (2014b). *Moringa oleifera* and strychnos potatorum seeds as natural coagulant compared with synthetic common coagulants in treating car wash wastewater: Case study 1. *Asian Journal of Applied Sciences*, 2(5), 693–700.

Mohamed, R.M.S.R, Kassim, A.H.M., Anda, M., and Dallas, S. (2014c) The effects of elements mass balance from turf grass irrigated with laundry and bathtub greywater. *International Journal of Applied Environmental Sciences*, 9(4), 2033–2049.

Mohamed, R.M.S.R., Wurochekke, A.A., Chan, C.M., and Kassim, A.H.M. (2014d). The use of natural filter media added with peat soil for household greywater treatment. *GSTF International Journal of Engineering Technology (JET)*, 2(4), 33–38.

Mohamed, R.M.S.R., Kassim, A.H.M., Anda, M., and Dallas, S. (2014e). Elemental mass balance of Na, Cl and B for the turf grass irrigated with laundry and bathtub greywater. *Australian Journal of Basic and Applied Sciences*, 8(15), 158–165.

Mounir, B., Abdeljalil, Z., and Abdellah, A. (2014). Comparison of the efficacy of two bioflocculants in water treatment. *International Journal of Scientific Engineering and Technology*, 3(60), 734–737.

Mumbi, A.W., Fengting, L., and Karanja, A. (2018). Sustainable treatment of drinking water using natural coagulants in developing countries: A case of informal settlements in Kenya. *Water Utility Journal*, 18, 1–11.

Natarajan, R., Al Fazari, F., and Al Saadi, A. (2018). Municipal waste water treatment by natural coagulant assisted electrochemical technique-Parametric effects. *Environmental Technology & Innovation*, 10, 71–77. https://doi.org/10.1016/j.eti.2018.01.011.

Ndabigengesere, A., Narasiah, K.S., and Talbot, B.G. (1995). Active agents and mechanism of coagulation of turbid waters using Moringa oleifera. *Water Research*, 29, 703–710.

Neoh, C.H., Noor, Z.Z., Mutamim, N.S.A., & Lim, C.K. (2016). Green technology in wastewater treatment technologies: Integration of membrane bioreactor with various wastewater treatment systems. *Chemical Engineering Journal*, 283, 582–594. https://doi.org/10.1016/j.cej.2015.07.060

Nimesha, S., Hewawasam, C., Jayasanka, D.J., Murakami, Y., Araki, N., and Maharjan, N. (2022). Effectiveness of natural coagulants in water and wastewater treatment. *Global Journal of Environmental Science and Management,* 8(1), 101–116.

Othmani, B., Rasteiro, M.G., and Khadhraoui, M. (2020). Toward green technology: A review on some efficient model plant-based coagulants/flocculants for freshwater and wastewater remediation. *Clean Technologies and Environmental Policy*, 22(5), 1025–1040. https://doi.org/10.1007/s10098-020-01858-3

Okuda, T., Baes, A.U., Nishijima, W., and Okada, M. (1999). Improvement of extraction method of coagulation active components from Moringa Oleifera seed. *Water Research*, 33(15), 3373–3378. https://doi.org/10.1016/S0043-1354(99)00046-9

Okuda, T., Baes, A.U., Nishijima, W., and Okada, M. (2001a). Isolation and characterization of coagulant extracted from Moringa oleifera seed by salt solution. *Water Research*, 35, 405–410.

Okuda, T., Baes, A.U., Nishijima, W., and Okada, M (2001b). Coagulation mechanism of salt solution extracted active components in Moringa oleifera seeds. *Water Research*, 35, 830–834.

O'Malley, E., O'Brien, J.W., Verhagen, R., and Mueller, J.F. (2020). Annual release of selected UV filters via effluent from wastewater treatment plants in Australia. Chemosphere, 125887. https://doi.org/10.1016/j.chemosphere.2020.125887.

Packham, R.F. (1965). Some studies of the coagulation of dispersed clays with hydrolyzing salts. *Journal of Colloid and Interface Science*, 20, 81–92.

Patale, V., Chaudhari, P., Punita, P., Unadkat, K., and Payal, V. (2010). Application of the aquatic indigenous plant material coagulants for wastewater treatment. *Journal of Industrial Pollution Control*, 26, 15–18.

Prakash, M.J., and Manikandan, S. (2012). Response surface modeling and optimization of process parameters for aqueous extraction of pigments from prickly pear (Opuntia ficus-indica) fruit. *Dyes Pigments*, 95, 465–472. https://doi.org/10.1016/j.dyepig.2012.06.007

Pritchard, M., Mkandawire, T., Edmondson, A., O'Neill, J.G., and Kululanga, G. (2009). Potential of using plant extracts for purification of shallow well water in Malawi. *Physics and Chemistry of the Earth*, 34, 799–805.

Rodrigues, A.C., Boroski, M., Shimada, N.S., Garcia, J.C., Nozaki, J., and Hioka, N. (2008). Treatment of paper pulp and paper mill wastewater by coagulation-flocculation followed by heterogeneous photocatalysis. *Journal of Photochemistry and Photobiology A: Chemistry*. 194(1), 1–10.

Saleem, M., and Bachmann, R.T. (2019). A contemporary review on plant-based coagulants for applications in water treatment. *Journal of Industrial & Engineering Chemistry*, 72, 281–297.

Sanghi, R., Bhatttacharyaa, B., and Singh, V. (2002). Cassia angustifolia seed gum as an effective natural coagulant for decolourisation of dye solutions. *Green Chemistry*, 4, 252–254.

Santos, A.F., Paiva, P.M., Teixeira, J.A., Brito, A.G., Coelho, L.C., and Nogueira, R. (2012). Coagulant properties of Moringa oleifera protein preparations: Application to humic acid removal. *Environmental Technology*, 33(1), 69–75.

Sarawgi, G., Kamra, K., Suri, N., Kaur, A., and Sarethy, I.P. (2009). Effect of Strychnos potatorum Linn. seed extracts on water samples from different sources and with diverse properties. *Asian Journal of Water, Environment and Pollution*, 6(3), 13–17.

Sellami, M., Zarai, Z., Khadhraoui, M., Jdidi, N., Leduc, R., and Rebah, F.B. (2014). Cactus juice as biofoccu-lant in the coagulation-focculation process for industrial wastewater treatment: A comparative study with polyacrylamide. *Water Science and Technology*, 70(7), 1175–1181. https://doi.org/10.2166/wst.2014.328

Sen, T.K. (2015). Physical, chemical and biological treatment processes for water and wastewater. *Chemical Engineering Methods and Technology*, Nova Science Publishers, 1–454.

Sengkhamparn, N., Verhoef, V., Schols, H.A., Sajjaanantakul, T., and Voragen, A.G. (2009). Characterisation of cell wall polysaccharides from Okra (*Abelmoschus esculentus* L. *Moench*). *Carbohydrate Research*, 344(14), 1824–1832. https://doi.org/10.1016/j.carres.2008.10.012

Sethupathy, A. (2015). An experimental investigation of alum and *Moringa oleifera* Seed in water treatment. *International Journal of Advanced Research*, 3, 515–518

Shan, T.C., Matar, M. Al, Makky, E.A., and Ali, E.N. (2017). The use of *Moringa oleifera* seed as a natural coagulant for wastewater treatment and heavy metals removal. *Applied Water Science*, 7(3), 1369–1376.

Shultz, C.R., and Okun, D.A. (1984). Surface water treatment for communities in developing countries. *Prepared for the Water and Sanitation for Health (Wash) Proj Issue 29 of WASH technical report, Wiley-Interscience publication.*

Sodvadiya, M., Patel, H., Mishra, A., and Nair, S. (2020). Emerging insights into anticancer chemopreventive activities of nutraceutical *moringa oleifera*: Molecular mechanisms, signal transduction and in vivo efficacy. Current Pharmacology Reports. https://doi.org/10.1007/s40495-020-00210-z

Subramani, T., Kathirvel, C., Mohamed, H.H., Nowfis, M.M., and Niyasdeen, A. (2018). Plant based coagulant for waste water treatment. *International Journal of Emerging Trends & Technology in Computer Science*, 7(2), 177–185.

Sutherland, J.P., Folkard, G.K., Mtawali, M.A., and Grant, W.D. (1994). *Moringa oleifera* as a Natural Coagulant Affordable water supply and sanitation. In *Proceedings of the 20th WEDC International Conference,* Colombo, Sri Lanka, 22–26 August, pp. 297–299.

Swati, M., and Govindan, V.S. (2005). Coagulation studies on natural seed extracts. *Indian Water Works Association*, 37, 145–149.

Teh, C.Y., Wu, T.Y., and Juan, J.C. (2014). Optimization of agro-industrial wastewater treatment using unmodified rice starch as a natural coagulant. *Industrial Crops and Products*, 56, 17–26. https://doi.org/10.1016/j.indcrop.2014.02.018

Tetteh, E.K., and Rathilal, S. (2005). Application of organic coagulants in water and wastewater treatment. In A. Sand and E. Zaki (Eds.) *Organic Polymers* [Internet]. London: IntechOpen; 2019. https://www.intechopen.com/chapters/65706; https://dx.doi.org/10.5772/intechopen/84556.

Theodoro, J.D.P., Lenz, G.F., Zara, R.F., and Bergamasco, R. (2013). Coagulants and Natural Polymers: Perspectives for the Treatment of Water. *Plastic and Polymer Technology*, 2(3), 55–62.

Ugwu, S.N., Umuokoro, A.F., Echiegu, E.A., Ugwuishiwu, B.O., and Enweremadu, C.C. (2017). Comparative study of the use of natural and artificial coagulants for the treatment of sullage (domestic wastewater), *Cogent Engineering*, 4(1). https://doi.org/10.1080/23311916.2017.1365676.

UNESCO. (2019). The United Nations world water development report 2019: Leaving no one behind. *UNESCO Digital Library*.

Vatvani, C. (2016). The toxic waste that enters Indonesia's Citarum River, one of the world's most polluted. *Channel News Asia*. https://cleanwaterforpeople.com/2019/12/04/the-toxic-waste-that-enters-indonesias-citarum-river-one-of-the-worlds-most-polluted.

Vigneshwaran, S., Karthikeyan, P., Sirajudheen, P., and Meenakshi, S. (2020). Optimization of sustainable chitosan/*Moringa oleifera* as coagulant aid for the treatment of synthetic turbid water - a systemic study. *Environmental Chemistry and Ecotoxicology*, 2, 132–140.

Vijayaraghavan, G., Sivakumar, T., and Adichakkravarthy, V. (2011). Application of plant based coagulants for waste water treatment. *International Journal of Advanced Engineering Research and Studies*, 1(1), 88–92.

Vunain, E., Masoamphambe, E.F., Mpeketula, P.M.G., Monjerezi, M., and Etale, A. (2019). Evaluation of coagulating efficiency and water borne pathogens reduction capacity of *Moringa oleifera* seed powder for treatment of domestic wastewater from Zomba, Malawi. *Journal of Environmental Chemical Engineering*. https://doi.org/10.1016/j.jece.2019.103118

Wan, L., Chen, X., and Wu, A. (2016). Mini review on antimicrobial activity and bioactive compounds of *Moringa Oleifera, Medicinal Chemistry*, 6, 578–582. https://doi.org/10.4172/2161-0444.1000402.

World Health Organization (WHO), United Nations Children's Fund (UNICEF). (2018). Progress on drinking water, sanitation and hygiene: 2017 update and SDG baselines, 116. https://iris.who.int/bitstream/handle/10665/258617/9789241512893-eng.pdf sequence=1.

Yimer, A., and Dame, B. (2021). Papaya seed extract as coagulant for potable water treatment in the case of Tulte River for the community of Yekuset district, Ethiopia. *Environmental Challenges*, 4, 100198.

Yin, C.Y. (2010). Emerging usage of plant-based coagulants for water and wastewater treatment. *Process Biochemistry*, 45(9), 1437–1444.

Zaharuddin, N.D., Noordin, M.I., and Kadivar, A. (2014). The use of *Hibiscus esculentus* (Okra) gum in sustaining the release of propranolol hydrochloride in a solid oral dosage form. BioMed Research International. https://doi.org/10.1155/2014/735891.

6 Phytoremediation of Organic and Inorganic Contaminants from Industrial Wastewater

S.S. Rakesh, A. Nivetha, J. Ezra John, M. Prasanthrajan, and C. Sakthivel

6.1 INTRODUCTION

Plants are one of the most valuable resources on our planet which regulate the water cycle and provide food and habitat for living organisms. Many organic and inorganic pollutants released into water bodies from industrial wastewater constitute a threat to living creatures. Contaminants that are resistant in nature accumulate in soils, absorbed by crops, and enter into the food chain of humans and other living species (Tak et al., 2013). Heavy metals such as cadmium, mercury, selenium, manganese, nickel, copper, and lead pose a major impact on environment and living organisms (Allen, 2017). Plants species possess the capability to translocate the heavy metals in their internal organs and cells as part of their defence system which leads to alterations in metabolism and physiological changes caused by various contaminants. As a result, plants are used to the greatest extent possible to remediate the damaged environment. Plants possess antioxidants, hormonal signaling and plant growth promoting rhizobacteria (PGPR), phytochelatins, and function in signaling are suitable for phytoremediation. The phytoremedition techniques includes phytodegradation, phytotransformation, phytohydraulics, phytostabilization, phytoextraction/phytoaccumulation, rhizoremediation/phytostimulation and phytovolatilization. This chapter deals with the organic and inorganic pollutants in industrial wastewater, existing wastewater treatment methods, phytoremediation techniques for degradation of textile dyes, pesticides, pharmaceutical degradation, polycyclic aromatic compounds, and removal of heavy metals, microplastics and xenobiotics and integration in existing wastewater treatment with their limitations and challenges.

6.2 ORGANIC POLLUTANTS IN INDUSTRIAL WASTEWATER

The water is affected by numerous organic pollutants and the sources of organic pollutants are textile, paper, petrochemical, rubber industries, etc. In the last few decades, industries were increased in developing countries like India, China, etc (Dsikowitzky & Schwarzbauer, 2014). The halogenated contaminants are identified in dye manufacturing plants and textile mills and they are known as endocrine-disrupting chemicals. These compounds are utilized as a starting material for azo and anthraquinone dye preparation. Azo dyes are considered as the most important dye stuff used in the textile industries (Loos et al., 2007). Nonylphenol ethoxylates, diethyl phthalate, bis (ethyl hexyl) phthalate are found in the water discharged from tanneries industry as well as benzothiazoles are used as fungicides instead of chlorophenols in tannery field (Rodríguez et al., 2004).

During petroleum refining, crude oil like alkylated benzenes, indane, naphthalene and its derivatives, quinoline derivatives are discharged into the fresh water. In addition, ingredients of crude oil were identified including indoline and polycyclic aromatic hydrocarbons (Botalova et al., 2009). The paper and pulp industries discharges resin acids, lignin, catechol, terpenes, hydroxybenzaldehyde, etc. During bleaching of pulp chlorinated derivatives such as catechol, dehydroabietic acid, guaiacols,

 DOI: 10.1201/9781003441144-6

and syringol were discharged. The paper mill pollutants have the endocrine disruptors' bisphenol A (BPA), nonylphenol ethoxycarboxylates. Chlorine is used as a bleaching agent in the pulp and paper industries instead of chlorine dioxide, molecular oxygen, peroxide, etc. (Latorre et al., 2005).

The benzothiazoles, aniline derivatives are used as the catalyst for the process vulcanization which is identified in the tire and rubber manufacture plants. In rubber manufacture, phthalic acid esters are used as plasticizers and toluene is used as a solvent. Both di-tert-butylmethylphenol and diphenylamines are utilized as antioxidants. The nitroaromatic compounds were identified in the wastewater during the synthesis of trinitrotoluene (Ahmad et al., 2022).

6.3 INORGANIC POLLUTANTS IN INDUSTRIAL WASTEWATER

Wastewater contamination is a significant issue that has negative effects on people and ecosystems due to the discharge of inorganic pollutants from industrialization and globalization. The bulk of pollutants are produced by human activity as well as by natural processes including volcanic eruption, the release of industrial waste, mining, and agricultural chemical wastes, among others. Importantly, inorganic pollutants such as heavy metals (HMs) or metalloids, inorganic acids, metal complexes, and mineral acids are non-degradable and difficult to eradicate the contaminants from ecosystem. Even at low concentrations, heavy metals have a relatively high specific density of 5 g/ cm^3. The aggregation of heavy metals such as cadmium (Cd), arsenic (As), mercury (Hg), chromium (Cr), lead (Pb), copper (Cu), and selenium (Se), etc., as shown in Figure 6.1. These metals have a detrimental effect on aquatic life, and they can harm the liver, kidneys, blood vessels, nervous system, skin, bone marrow, and gastrointestinal system in humans (Borah et al., 2020). The wastewater originates from several areas and is divided into different categories, including domestic wastewater, industrial wastewater, sewer wastewater, and agricultural wastewater, among others.

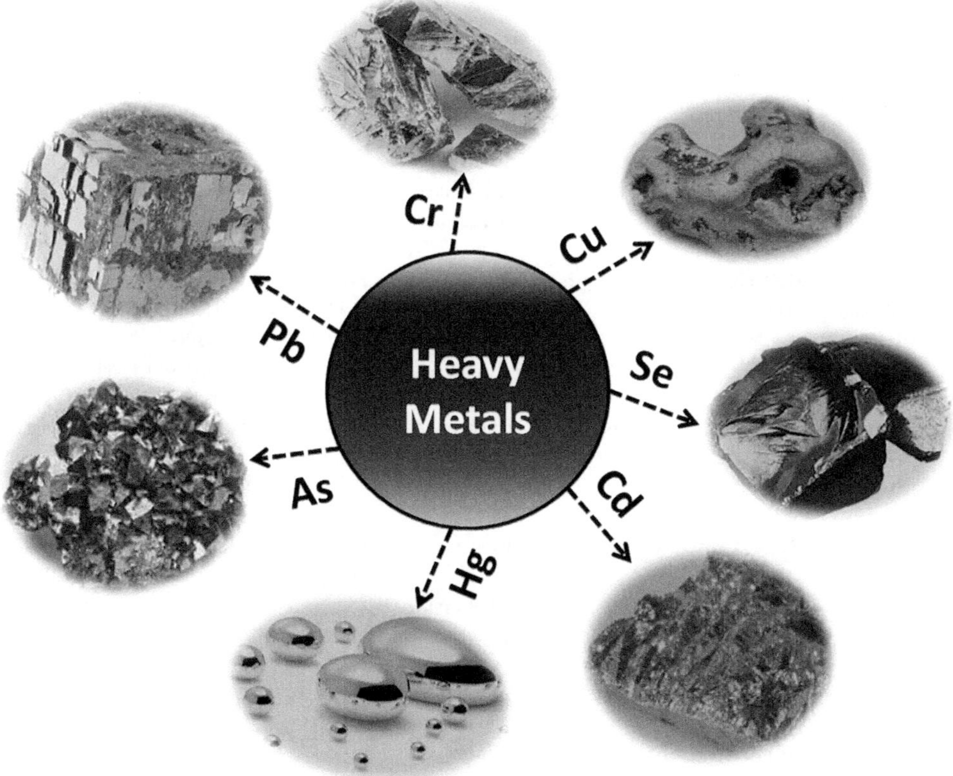

FIGURE 6.1 Pictorial representation of heavy metals.

Additionally, using industrial effluent to irrigate agriculture can have very harmful effects on the crops, leading to lasting ecosystem changes. The total amount of fresh water available from all water sources can be utilized by industry annually in excess of 20% (Ahmed et al., 2021). Additionally, it contributes to a number of environmental pollutions, including air contamination, ground water pollution, soil pollution, and pollution of the land. There are many water bodies that regulate and start the safety requirements for managing industrial wastewater, but they can only recommend the minimum acceptable standards (MINAS) for industrial and municipal contaminants. The pollutions control board is the Central Pollution Control Board (CPCB) in India, Integrated Pollution Prevention and Control (IPPC) in Europe, Total Maximum Daily Load (TMDL) in US (Wang & Yang, 2016). Nowadays, there are numerous industrial wastewater treatment techniques for controlling or recycling wastes. These techniques include electrochemical, chemical, physical, and biological processes to get rid of the pollutants. The heavy metal properties such as oxidation states, toxicity, and sources are given in Table 6.1.

TABLE 6.1
Overall View of Heavy Metal Properties

S.No.	Heavy Metals	Oxidation States	Maximum Permissible Limits (MPL)	Sources	pH Range	Health Issues	Ref.
1.	Copper (Atomic No. 29)	Cu^{+2} and Cu^{+1} (Cu^{+1} is toxic)	70–140 μg/dL	Industry and Agriculture	1.3 mg/L pH=7.4	Inflammatory disorders, biological systems, also Alzheimer disease	(Wang et al., 2015)
2.	Selenium (Atomic No. 34)	Se-2, Se^{0}, Se+4 and Se+6 (all forms are toxic)	10 μg/L	Drainage water, fossil fuals, metal ores and thermoelectric plants	pH>7.5	Human and aquatic diseases etc	(Balakrishnan et al., 2020)
3.	Cadmium (Atomic No. 48)	Cd^{+2} (toxic)	5 μg/m^{3}	Paints, battery and jewelry	pH=6.0	Kidney diseases, cancer and heart diseases etc	(Borah et al., 2020)
4.	Mercury (Atomic No. 80)	Hg^{-2}, Hg^{0} and Hg^{+2} (all forms are toxic)	0.001 mg/L	Natural rocks and coal environment	pH (5.0–11.0)	Reproduce system, metal disorder, memory loss and genetic diseases	(Carocci et al., 2014)
5.	Arsenic (Atomic No. 33)	As^{-3}, As^{0}, As^{+3}, and As^{+5} (As^{+3} is more toxic)	0.05 mg/L	Industry and ground water	pH (4.6–5.5)	DNA damage, chronic diseases, solar keratosis and palmar etc	(Wuana & Okieimen, 2011)
6.	Lead (Atomic No. 82)	Pb^{2+} and Pb^{4+} (Pb^{4+} is toxic)	0.01 mg/L	Batteries, weapon, paints, pipes, pigments amd paper etc	pH=5.6 (100 ppm)	Brain, kidney, liver and bone diseases	(Wani et al., 2015)
7.	Chromium (Atomic No. 24)	Cr^{-2}, Cr^{0}, Cr^{+3} and Cr^{+6} (Cr+6 is toxic)	0.05 mg/L	Ground water and industrial effluent etc	pH=8.3	Chronic diseases, pulmonary fibrosis, and bronchial asthma etc	(Priti Sharma et al., 2012)

6.4 EXISTING METHODS FOR INDUSTRIAL WASTEWATER TREATMENT

The primary function of wastewater treatment is to accelerate the natural processes that improve the water quality thereby it can be utilized by people. Water can be treated to any desired level of purity; it is determined by its intended purpose but, as required level of purity increases, the cost also increases. A Chemical, biological, or physical process or a combination of these, may be used in treatment. Pre-treatment, primary treatment, secondary treatment, and tertiary or advanced wastewater treatment are the several elements of wastewater treatment processes Figure 6.2. Pre and primary treatment are followed by secondary treatment stages in conventional treatment. Pre-treatment makes wastewater suitable for subsequent treatment operations (Von Sperling, 2007). Consideration is given to elements that pass through as sludge or otherwise incompatible with treatment processes. Common pre-treatment methods include equalization, neutralization for pH correction, toxics removal, oil and grease removal, and solids removal. Primary treatment processes consist of physical separation by screening, grit removal, and sedimentation. The majority of treatment plants were developed to purify wastewater to levels enough for discharged into streams or other receiving waterways. Primary treatment, depending on the amount of organics, remove a large fraction of the oxygen-demanding compounds (BOD). A well-designed and managed primary treatment plant can remove up to 35%–40% of BOD and 60%–65% of settleable solids.

The secondary stage employs biological methods, where bacteria and other microorganisms in the water digest organic waste, converting it into new bacterial cells, carbon dioxide, and other by-products. In secondary treatment a biological process, which is often activated sludge or trickling filtration or slightly modified suspended growth treatment systems. These biochemical activities are often aerobic, which is favorable for the oxidation of organics into carbon dioxide and water.

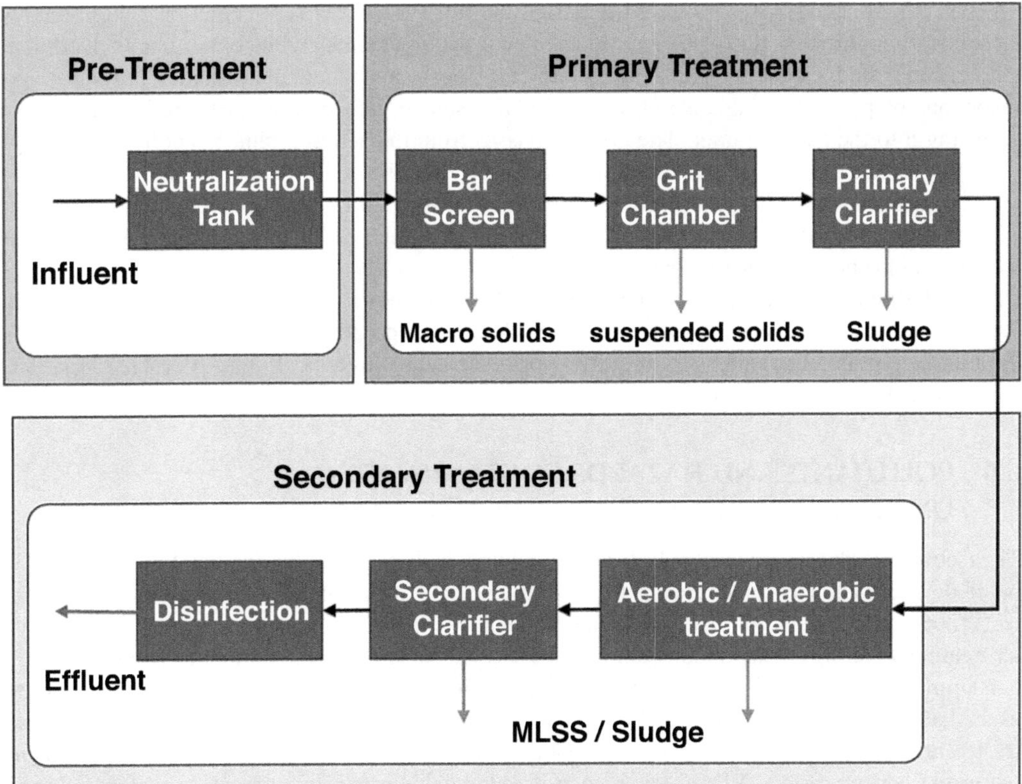

FIGURE 6.2 Schematics of conventional wastewater treatment process.

In secondary treatment a biological process, which is often activated sludge or trickling filtration or slightly modified suspended growth treatment systems. These biochemical activities are often aerobic, which is favorable for oxidation of organics into carbon dioxide and water. A well-designed secondary treatment process can remove 85%–95% of BOD and suspended particles (Von Sperling, 2007). All effluent must pass through secondary treatment before discharge under current laws in the majority of nations. Tertiary treatment methods are employed in the treatment sequence to decrease particular constituent or residue to very low levels.

6.4.1 Primary Treatment

Large floating debris that might block pipes or harm equipment are removed from the industrial effluent by passing it through a screen. Next to screening, it enters a grit chamber, where solids of small size drop to the bottom. Even after screening and grit removal, organic and inorganic contaminants remain in the effluent. When the flow rate through one of these tanks is lowered, the suspended particles gradually fall to the bottom, forming a mass of solids known as raw primary biosolids. Typically, biosolids are removed from tanks by pumping, and then they may undergo additional treatment for use in the soil application or co-combustion.

6.4.2 Secondary Treatment

Using microorganisms, the secondary treatment eliminates around 85% of the organic materials. The wastewater flows into secondary treatment after the sedimentation in the initial stage. The trickling filter and the activated sludge process are the two most used secondary treatment processes. A trickling filter is basically a 3–6-ft-deep bed of stones and polymers over which wastewater drips. Bacteria proliferate on these stones, eventually consuming the majority of the organic materials. The partially treated sewage travels from a trickling filter to another sedimentation tank to remove biosolids. By putting bacteria-laden air and sludge into close contact with contaminants in effluent, the activated sludge process speeds up treatment. The bacteria degrade organic substances into innocuous by-products (Englande et al., 2015). The sludge with billions of bacteria is reused by returning it to the aeration tank. The effluent passes from the aeration tank to another sedimentation tank for the removal of surplus sludge and microorganisms. Effluent from the sedimentation tank is normally treated with chlorine before being discharge to complete secondary treatment. Chlorination, when done correctly, will eliminate more than 99% of the dangerous bacteria after which de-chlorination is carried out.

This traditional treatment has been unable to fulfill demands for improved water quality over the years due to the introduction of wastewaters with varying properties and resistant contaminants. To eliminate pollutants, industries require hybrid secondary treatment and, in certain situations, sophisticated tertiary treatment.

6.5 POLLUTANTS AND HAZARDS CONTAMINANTS UNADDRESSED IN THE CONVENTIONAL APPROACH

The technology chosen for one application may not be the best choice for another. Since it should be based on site-specific considerations such as available resources, climate, land availability, economics, effluent type, and so on. New pollution challenges have placed extra constraints on wastewater treatment facilities (Englande et al., 2015). Heavy metals, chemical compounds, and hazardous contaminants are more difficult to remove from water nowadays. Increasing demand for water exacerbates the situation. The growing demand to reuse water necessitates improved wastewater treatment. Unfortunately, traditional treatment does not efficiently eliminate many of these elements of current concern. Many VOCs, toxics, non-biodegradable organics, persistent organic pollutants (POPs), nutrients, and emerging contaminants including EDCs and PPCPs are among them. As a result, more treatment with advanced technology is necessary. Precipitation, filtration, coagulation and flocculation, air stripping, ion exchange, adsorption, membrane processes, nitrification and/or denitrification. These procedures could be added to the secondary treatment process or integrated within it. Tertiary or advanced wastewater treatment can increase specific targeted pollutant removal efficiency (Gupta et al., 2012). The expense of achieving it, however, hinders implementation in many economic weaker countries. Because there is a demand in industrial facilities for the treatment of a high volume of flow with low contamination concentrations, this process requires pollutant-specific treatments. As a result, the removal of these newly developing contaminants from wastewaters fails. With the new state enforcement of zero liquid discharge (ZLD), companies are being pushed to use a cost-effective solution that addresses a broad range of pollutants without disrupting existing operations (Tong & Elimelech, 2016). In general, addressing the problem at the source, when volumes are low and certain contaminants are present in large quantities, is the most cost-effective technique.

This method employs plant or phyto-compound interactions (physical, biochemical, biological, chemical, and microbiological) to remove or degrade contaminants of interest. Some important factors to consider when choosing a plant as a phytoremediant include: root system, which may be fibrous or tap, above-ground biomass, toxicity of pollutant to plant, plant survival and its adaptability to prevailing environmental conditions, plant growth rate, site monitoring and above all, time availability (Macci et al., 2016). Most plants growing in contaminated areas are likely to be effective phytoremediators. As a result, the success of any phytoremediation strategy is dependent largely on improving the remediation potentials of native plants living in contaminated areas. Depending on the pollutant type phytoremediation involves a number of methods (extraction, degradation, filtering, stabilization, and volatilization). The removal of elemental contaminants (toxic heavy metals and radionuclides) is mostly accomplished by extraction, transformation, and sequestration.

6.6 PHYTOREMEDIATION

Phytoremediation is cost-effective *in situ* technology that utilizes the various species of plants to reduce the volume, mobility or toxicity of contaminants such as heavy metals, pesticides, fertilizers, and petroleum hydrocarbons in soil, groundwater, atmosphere, and industrial wastewater (EPA, 2020). The phytoremediation is classified as follows (Figure 6.3).

6.6.1 PHYTODEGRADATION/PHYTOTRANSFORMATION

The phytotransformation encompasses the uptake of harmful pollutants from soil, air, and water by various plants followed by accumulation and translocate in their shoot region. Plants absorb the pollutants based on their structure and hydrophilicity and store in plant tissues and in cellular structures or decomposes into CO_2 (deep oxidation) and water. The phytotransformation is assisted by enzymes which are used for the remediation of petroleum hydrocarbons, explosives, chlorinated solvents, and pesticides. The enzymes which assist phytodegradation of various compounds are listed in the table below (Table 6.2).

6.6.2 PHYTOHYDRAULICS

Phytohydraulics (Hydraulic control) utilizes plants with high transpiration rates that absorb significant volumes of water and pollutants such as organic and inorganic water-soluble contaminants, preventing them from further transport by its well-developed root system. The suitable plants for phytohydraulics are eucalyptus, willow, hybrid poplars, and birch. The plant absorption is driven mainly by means of temperature the transpiration rate is reduced in the winter season. The water requirement of a poplar tree was 100–200L per day (Shrirangasami et al., 2020). *Salix viminalis* requires 375L of water per day with the evapotranspiration rate of 22.7mm per day which suits for hydraulic control (Frédette et al., 2019).

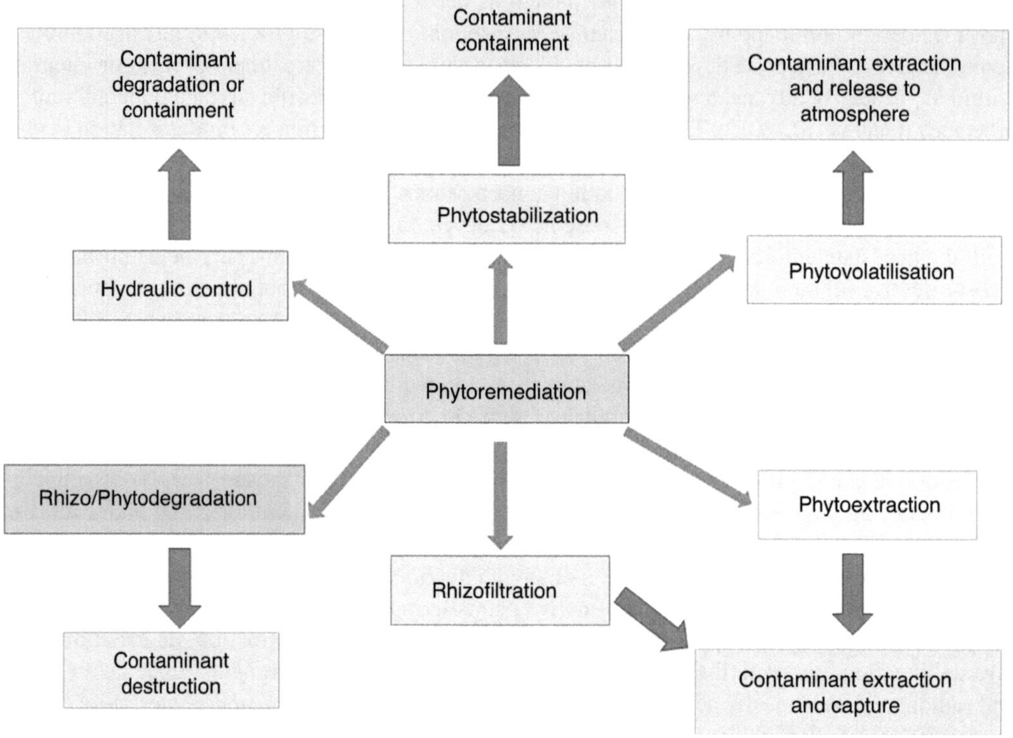

FIGURE 6.3 Existing phytoremediation techniques.

TABLE 6.2
Phytotransformation of Contaminants

Contaminants	Plant Species	Enzyme
Chlorinated solvents	Hybrid poplars (*Populus spp.*)	Dehalogenase
TNT, triaminotoluene	Parrot feather (*Myriophyllum aquaticum*), stonewort (*Nitella spp.*)	Laccase
Cyanide groups from aromatic rings	Willow (*Salix spp.*)	Nitrilase
Explosives and other nitroaromatic compounds	Hybrid poplars (*Populus spp.*), parrot feather (*Myriophyllum aquaticum*), stonewort (*Nitella spp.*)	Nitroreductase
Phenols	Horse radish (*Armoracia rusticana*)	Peroxidase
Phosphates, organophosphate pesticides	Giant duckweed (*Spirodela polyrhiza*)	Phosphatase

Source: Ncibi et al. (2017).

6.6.3 Phytostabilization

Phytostabilization involves the accumulation of pollutants principally heavy metals by root system of plants in to complex forms in the soil by means of organics present in the root exudates leading to immobilization and heavy metals leaching through plant transpiration which helps to reduce surface runoff followed by soil erosion reduction (McGrath et al., 2002). The factors affecting the phytoremediation are rhizosphere microfauna, rhizoexudates, and chelation of metal ions followed by vacuole compartmentation which gets controlled by soil factors such as pH, redox potential,

TABLE 6.3
Phytostabilization of Metals

S.No.	Plant Species	Element	Site of Stabilization	References
1	*Agrostis capillaris*	As	Wetlands ecosystem	Symeonidis et al. (1985)
2	*Microchloa altera*	Cu	Soils	Shutcha et al. (2010)
3	*Rhizophora mucronata*	Fe, Cu and Pb	Wetland	Pahalawattaarachchi et al. (2009)
4	*Pistia stratiotes* and *Eichhornia crassipes*	Cd	Wastewater	Sricoth et al. (2018)
5	*Typha latifolia*	Co, As, Cd, Cr and Zn	Industrial wastewater	Varun et al. (2011)
6	*Salix babylonica* and *Arachis pintoi*	Cu	Mine tailings	Andreazza et al. (2011)

organic matter, microorganisms, texture, and temperature (Chaignon et al., 2002). Sugars, polysaccharides, organic and amino acids, peptides, and proteins in root exudates play an important role in phytostabilization by promoting the accumulation, stability, or volatilization of pollutants from soil. The polysaccharides, sugars, organic acids, amino acids, and proteins in the root exudates assists phytostabilization by means of accumulation, stabilization, or volatilization of contaminants (Kushwaha et al., 2015). Graminaceous plants secrete amino acid compounds such as siderophore which bind to form stable complex with metals such as iron, cadmium, zinc, and copper (Xu et al., 2020). Plant cell walls are rich in pectin compounds that bind divalent and trivalent metal cations by homogalacturonans (HGA) which inactivates within the apoplast by reducing the toxicity of contaminants (Caffall & Mohnen, 2009). Heavy metal ions entered in the cytosol get sequestered into the vacuole by chelation with metalloproteins, organic acids, amino acids or resists toxicity of heavy metals and also by means of phytochelatins (PCs) and metallothioneins (MTs). The plant species such as *Halimione portulacoides, Sarcocornia perennis, Juncus maritimus,* and *Triglochin maritima* provide higher Hg stabilization (Table 6.3).

6.6.4 Phytoextraction

Phytoextraction is the process of extraction of metals from the soils by accumulating in shoot and root portion of the plants which was classified as induced continuous and phytoextraction (Salt et al., 1998). Induced phytoextraction utilizes the huge biomass producing plants with the addition of chemicals such as EDDS, EDTA, and NTA to enhance metal accumulation (Gómez-Garrido et al., 2018). Continuous phytoextraction uses hyperaccumulating plants to accumulate high levels of metals (McGrath et al., 2002). Various steps of phytoextraction process include the uptake of contaminants by the plants and translocation in various parts of the plants followed by vacuolar compartmentation (Kidd et al., 2009). The organics in the root exudates controls pH in the rhizosphere, for the stabilized absorption of metals by root hairs assisting metal penetration into root cells (Figure 6.4). The heavy metals are transferred from soil or aqueous solution to roots and to shoots through xylem tissues by symplast pathway resulting in the blockage of movement in apoplast pathway (Mahmood, 2010).

The phytoextraction uses 450–500 hyperaccumulator plants from various families which include *Lamiaceae, Caryophyllaceae, Leguminosea, Cunoniaceae Graminae, Asteraceae,* and *Cyperaceae.* The plants such as *Ipomea alpine, Euphorbia maeleocloela, Arabidopsis bisulcatus, Centella asiatica, Sesbaniia drumnondia, Sedum alfredei, Euphorbia marcoleida, Phragmites*

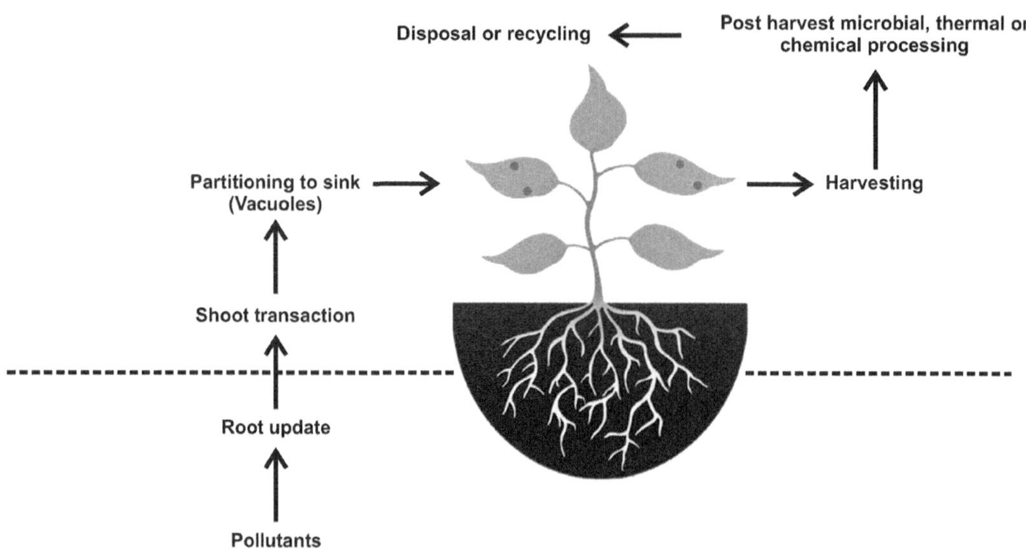

FIGURE 6.4 Phytoextraction of contaminants.

TABLE 6.4
Metal Hyperaccumulating Plants

Element	Threshold mg/kg	Plant Species	Percent (%)
Zinc	>3,000	*Noccaea caerulescens*	5.4
Thallium	>100	*Biscutella laevigata*	1.9
Selenium	>100	*Astragalus bisulcatus*	1.5
Lanthanum, Caesium	>1,000	*Dicranopteris linearis*	0.7
Lead	>1,000	*Noccaea rotondifolia subsp. cepaeifolia*	0.8
Nickel	>1,000	*Berkheya coddii*	7.6
Manganese	>10,000	*Virotia neurophylla*	5.5
Cobalt	>300	*Haumaniastrum robertii*	1.0
Copper	>300	*Aeolanthus biformifolius*	1.4
Cadmium	>100	*Arabidopsis halleri*	0.36
Arsenic	>1,000	*Pteris vittata*	2.3

Source: Reeves et al. (2018).

australis, Phytolacca Americana, Cardanine hupingshas and *Iberis intermedia* reported as hyper-accumulators for phytoextraction of copper, nickel, cadmium, zinc, chromium, arsenic, selenium, and titanium (Chaudhary et al., 2019) (Table 6.4).

6.6.5 Rhizofiltration/Rhizoremediation

Rhizofiltration employs plant roots to absorb or adsorb rhizosphere contaminants, metals, and organics from wastewater and soils, concentrate and precipitate them in the root zone (Olguín & Sánchez-Galván, 2011). Plant roots solubilize the heavy metals by acidifying the rhizosphere soil with protons released from the rhizo-exudates. The reduction in pH solubilizes metals in the soil

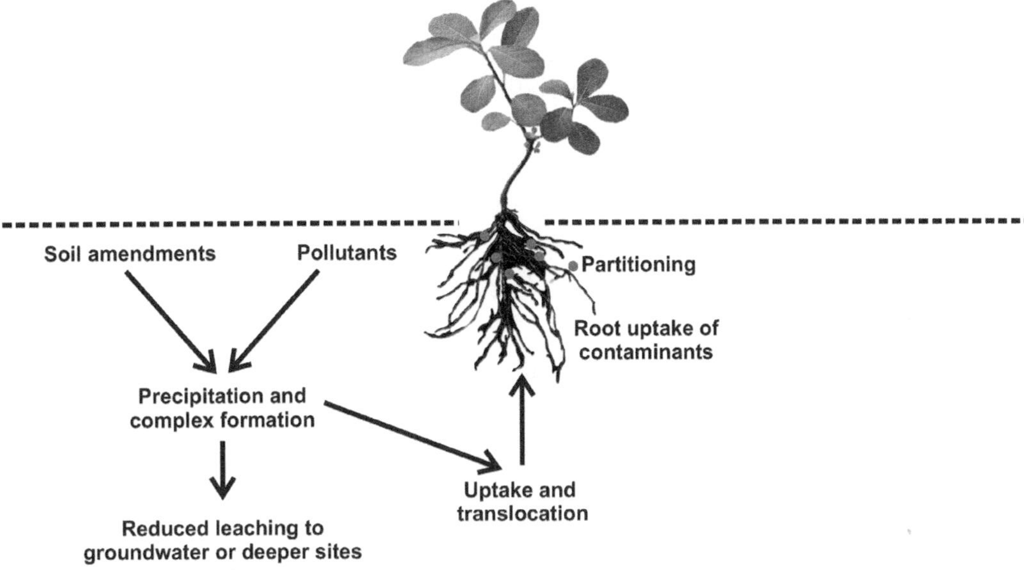

FIGURE 6.5 Rhizostabilization of soil pollutants.

enters into the root zone by means of extracellular or intracellular pathways leading to the accumulation of metal contaminants in the roots. Rhizofiltration is achieved by means of aquatic plants such as *Lemna minor, Phragmites australis, Hydrocotyle umbellate, Eichhornia crassipes,* and *Typha angustifolia.* Synthetic dyes are known to have complex structures that are difficult to degrade (Nilratnisakorn et al., 2007). The adsorption of textile dyes from textile effluent by plants such as *Typhonium flagelliforme, Typha angustifolia* absorbs Reactive Red 141 (diazo reactive dye), *Rheum rabarbarum* absorbs and accumulates synthetic anthraquinones (Figure 6.5).

6.6.6 Phytovolatilization

Phytovolatilization refers to the uptake and release of volatile compounds such as chlorinated solvents and metals into the atmosphere by plant species (Figure 6.6). Phytovolatilization has mainly been applied to groundwater, soil, sediments, and sludges. The contaminants treated by the technology are.

The steps involved in the phytovolatilization process are

1. Pollutants (organic or inorganic) are modified and translocated in the root zone
2. The modified compounds were translocated to foliage.
3. Pollutants are released into the atmosphere either by transpiration (volatile gaseous form) or evaporation

The GM plants with mercuric reductase (MerA) and bacterial organomercurial lyase (MerB) absorb divalent and methylmercury from the polluted habitats and release as elemental mercury (Hg^0) into the atmosphere (Rahman et al., 2008). De Souza et al. (2002) reported the selenium phytovolatilization is achieved by plant species such as *Chara canescens* and *Brassica juncea* from the contaminated sites and are used as supplementary cattle feed. The plants utilized for heavy metals accumulation include *Liriodendron tulipifera, Canna indica, Typha angustifolia, Azolla caroliniana, Cyperus papyrus, Arundo donax, Pteris vittata,* and *Colocasia esculenta* (Pilon-Smits et al., 1999).

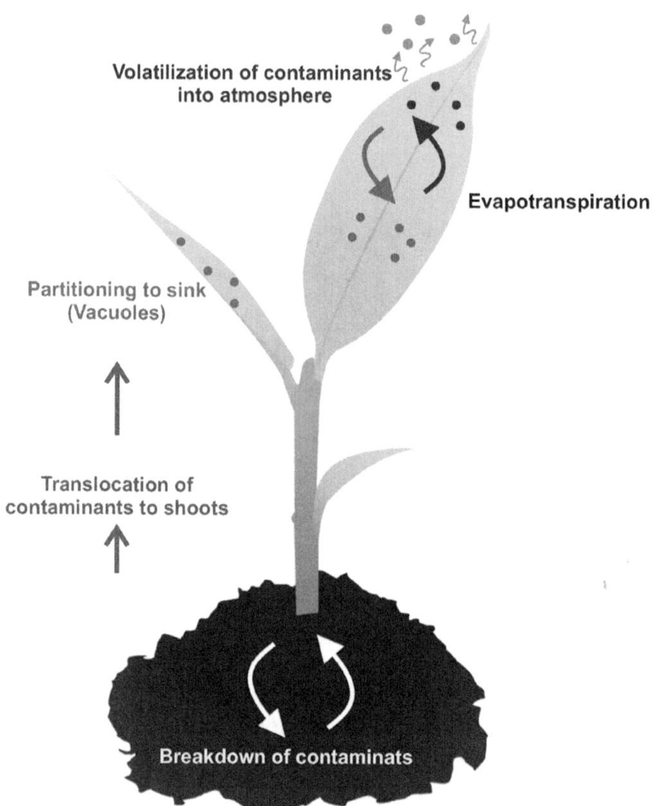

FIGURE 6.6 Phytovolatilization of heavy metals.

6.7 ORGANIC POLLUTANT ABATEMENT THROUGH PHYTOREMEDIATION TECHNIQUES

6.7.1 Dye Degradation

Each year, about one million tons of synthetic dyes are produced, of which 10% are dumped into the environment and natural resources as waste. The number of synthetic dyes currently made commercially and easily accessible on the market is close to 1,00,000. To fulfill the current demands of a growing population, manufacturing must be increased (M. M. Ahmad et al., 2022). This situation also results in a greater release of dye effluent, which is very detrimental to the environment. The removal of colored dyes from waste is critical because the presence of dyes with very tiny particle sizes of below 1 ppm is plainly apparent and has a substantial impact on the aquatic environment.

Adsorption on activated carbon, ion exchange, chemical precipitation, coagulation-flocculation, oxidation, electrochemical treatment, and membrane filtration are among the physical and chemical approaches used for the treatment of dye-contaminated wastewater (Ahmad et al., 2021). However, all of these approaches are expensive and inefficient, and they fail in the presence of different colors in the wastewater stream. Furthermore, the generation of harmful by-products is a major drawback of traditional methods. Due to its efficacy, effectiveness, and eco-friendliness, phytoremediation is one of the biological treatments receiving increased attention for the treatment of dye-contaminated environments (Figure 6.7). Utilizing live plants, phytoremediation removes dye pollutants from water or soil (Ahmad et al., 2015).

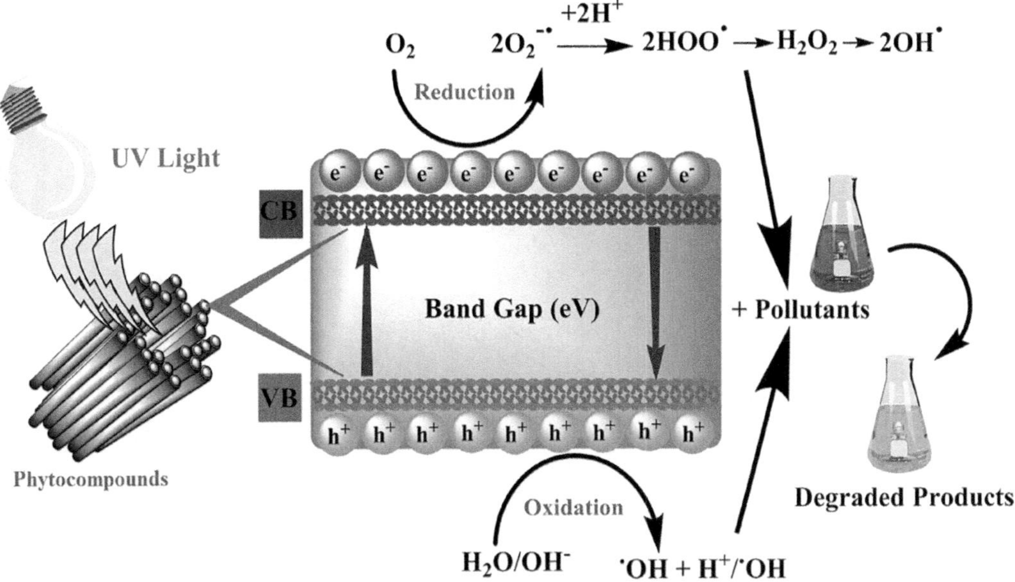

FIGURE 6.7 Mechanism of degradation of dye by phytocompounds.

For the elimination of polluted contaminants, several weeds, ferns, grasses, or agricultural wastes have been suggested. For the cleanup of textile effluents, and particularly with regard to the removal of the dye, Acid Orange 7, native populations of *Phragmites australis* have been extensively examined. *Lemna minor L.* plant with pond used to degrade the basic red 46 dye at low concentration (Yaseen & Scholz, 2017). Degradation, accumulation, dissipation, and immobilization of pollutants are common processes used in wastewater treatment facilities (Wang et al., 2017). Anamaria et al reported that *Lemna minor* plant has the potential to degrade crystal violet and malachite green up to 80% and 90% respectively which was owing to phytoextraction and phytodegradation (Török et al., 2015). *Tecoma stans var. angustata* and *Scirpus grossus* plant displayed better removal efficiency against synthetic dyes such as brilliant green, methylene blue (Almaamary et al., 2017; Rani & Abraham, 2016).

6.7.2 Pesticide Degradation

Agrochemicals like fertilizers, insecticides, herbicides, and fungicides are generated in large quantities every year. Agricultural chemicals contribute to environmental pollution once they have served their useful role in the production of food. Innovative technologies must be created in order to purify the polluted water and reduce this pollution. Due to its ability to effectively and inexpensively remediate various forms of pollution in situ, phytoremediation has grown in popularity (Figure 6.8). Due to their high photosynthetic activity, high pollutant absorption, ease of harvest, and high growth rates, certain aquatic plants, including *E. crassipes*, *L. minor*, and *Elodea canadensis*, are used to treat water pollution. Tront et al investigated the uptake and accumulation of 2,4-dichlorophenol by *Lemna minor*. Analysis of the plant revealed the presence of the herbicide and its metabolites in some degree of inhibited amounts in the plant tissues. Less than 10% of the original component was discovered in the plant, showing that *Lemna minor* sequestered almost 90% of the original compound. *L. minor* accumulates and destructs the agricultural chemicals into useful products for plant growth (Polińska et al., 2021) According to Olette et al., the dimethomorph was eliminated effectively by *L. minor* (115 μg/L). Both *L. minor* and macrophytes were effective in the elimination of fungicides (2010).

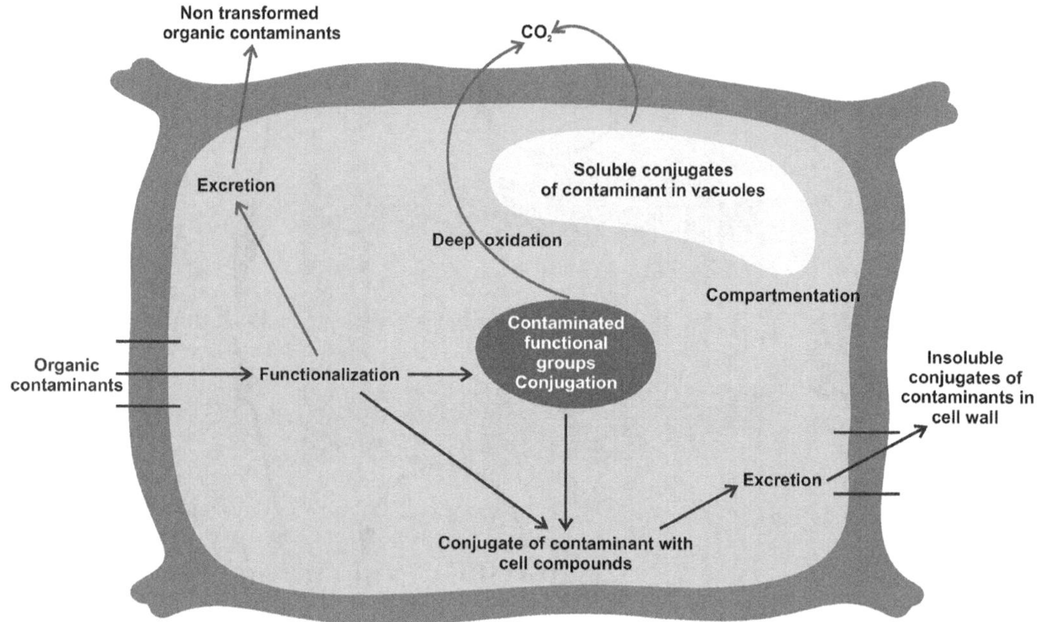

FIGURE 6.8 Transport and fate of organic contaminants in plant cell.

S. polyrhiza have displayed the better removal efficiency for metazachlor herbicide rather than *L. minor* (Müller et al., 2010). *L. minor* was active against atrazine compared with *M. aquaticum* (Teodorović et al., 2012). *E. canadensis* is found in the North America and have the potential to treat the pesticides. Olette et al. reported *E. canadensis* was used to remove the pesticides like dimethomorph (fungicides), copper sulfate (fungicides) and flazasulfuron (herbicides) (2008). *E. crassipes* have shown degradation efficiency up to 56% of 10 ppm of pesticide malathion.

6.7.3 Pharmaceutical Degradation

The most practical locations for disposing of industrial wastes are on land and in surface waters. A novel variety of chemically stable and persistent pharmaceutical pollutants that are disposed of with pharmaceutical effluents could potentially accumulate in ground and surface water. To safely dispose of the effluent into the environment, it becomes urgently necessary to make it free of harmful substances. *Lemna minor, Typha angustifolia L., Azolla filiculoides, Chrysopogon zizanioides (L.) Roberty, Eichhornia crassipes, Cyperus isoclaudus* have showed the degradation efficiency up to 100% for benzotriazole, hydrazine, 1,2-dichloroethane, diquat, cefadroxil, tetracycline, paracetomol, 17-α-ethinylestradiol, estrone, β-estrelestradiol, and levonorgestradiol.

The species *Cyperus alternifolius* was used to remove the oxybenzone and *Spirodela polyrhiza* effectively removed the DEET which is an insect repellent. The tetracycline was eliminated by *Phragmites australis* and *Phalaris arundinacea*. Dimethomorph was eliminated effectively by the plant species Lemna minor. Dordio et al. investigated the removal of ibuprofen, carbamazepine, and clofibric acid with the aid of *Typha spp.*, which is developed by using matrix of light expanded clay aggregates. The degradation efficiency was found to be 96%, 97%, and 75% with respect to ibuprofen, carbamazepine, and clofibric acid at 7 days (Dordio et al., 2010). *Typha, Phragmites, Iris*, and *Juncus* were used to remove the ibuprofen, iohexol. The ibuprofen completely removed within 24 days and iohexol removed upto 80% by all plant species (Zhang et al., 2016).

6.8 INORGANIC POLLUTANT ABATEMENT THROUGH PHYTOREMEDIATION

6.8.1 Heavy Metals Removal

In recent decades, the problem of HMs contamination has grown, making it even more important to remove the chemical from wastewater in order to transform it into usable water resources. Since the United Nations Organization for Education, Scientific, and Cultural Affairs has reported that 80% of wastewater produced from human and industrial activities is contaminated (UNESCO) (Arcipowski et al., 2017). There is a need for smart and systematic approaches to remove HMs from industrial effluent so that water sources can increase. Adsorption, photocatalytic degradation, chemical precipitation, chemical coagulation, electrochemical oxidation, ion exchange, flotation, reverse osmosis, and membrane filtration are a few of the approaches for water remediation highlighted in the earlier report (Agarwal & Singh, 2017; Du et al., 2020; Liu et al., 2019; Renu et al., 2017). Due to their vast surface area, which creates more active sites for accommodating the metal ions on their surface, adsorption and photocatalysis are the most effective ways for eliminating HMs at lower concentrations. The adsorption method has the advantages of being simple to use, economical, and environmentally beneficial. As in the previous report, environmentally friendly and chemically produced nanomaterials were considered to be effective adsorbents for the removal of HMs in industrial effluent. Adsorbents such as ZnO, MnO_2, MgO, TiO_2, Al_2O_3, CeO_2, and Fe_2O_3 have been used for the eradication of HMs (Kyzas & Matis, 2015). According to the literature, ZnO has superior adsorbent properties compared to other materials because of its strong ionic character, light sensitivity, abundance, affordability, and biocompatibility, all of which are encouraging for HMs' decontamination abilities (Kumar et al., 2013).

6.8.1.1 Mechanism for HMs Removal

The mechanism is the most key aspect in the removal of HMs because it explains the way in which HMs escape from industrial effluent. The mechanism is primarily divided into two types and used to remove HMs via adsorption or catalytic degradation; (i) photodegradation by redox reaction and (ii) physical adsorption. When examining the energy of UV light on the catalyst surface, the first redox reaction mechanism has evolved into the formation of excited electrons from the valence band to the conduction band, which results in the development of intermediate species like electrons in conduction band and holes in valence band is shown in Figure 6.9. The metal ions such as Cr^{+6}, Ag^{+}, and Pb^{2+} are more light-sensitive response that occurred under UV light comparatively Cu^{2+}, Cd^{+2}, Ni^{2+}, and Mn^{2+} ions where not response under light sources with the catalyst of ZnO NPs (Le et al., 2019). Second, the physical adsorption mechanism as follows; for example, ZnO NPs (M-O-H) has positive zeta potential which means the presence of positive charges on its surface and negative zeta potential value obtained during dispersed the catalyst in solution which clearly demonstrates that the presence of negative charged particles like OH^- groups on the catalyst surface. As a result, those negatively charged particles are allowing the positively charged HMs ions to congregate in clusters on the catalyst surface, which results in the adsorption of HMs from industrial effluent as shown in Figure 6.9.

6.8.2 Microplastics Removal

Microplastics (MPs) must range in size from less than 5 mm with different shapes. 350 billion tons of different types of plastics are accumulated annually from industrial trash, home garbage, agricultural waste, building construction, electrical goods, packaging, and automobile sources worldwide, etc. MPs take into account polyethylene (PE), polypropylene (PP), polyethylene terephthalate (PET), polystyrene (PS), and polyvinyl chloride (PVC) (Bratovcic, 2019). Recently, MPs found in human tissues and organs, food products, water resources like ocean, fresh water, lakes, and rivers.

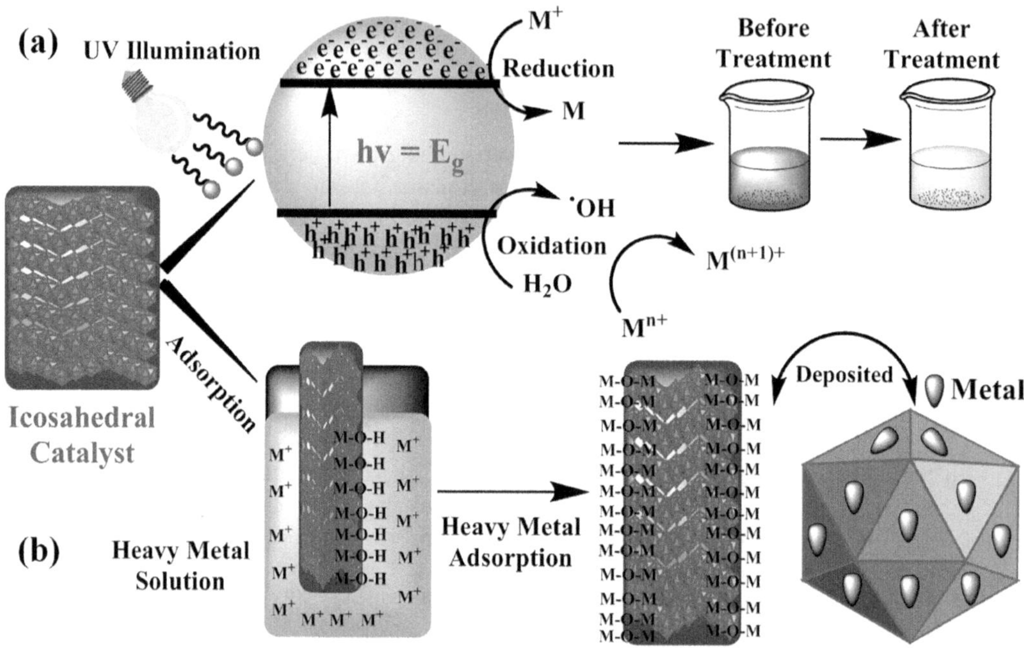

FIGURE 6.9 Schematic representation of mechanism for the removal of heavy metals (a) Photodegradation by redox reaction and (b) Physical adsorption.

(Abdullah et al., 2018). More than 6 billion metric tons of MPs were generated between 1950 and 2015, according to previous research. Among these, 4.9 billion metric tons of MPs were found in the surrounding areas and landfills. According to the scientific community, this will have increased by almost three times by 2050. The MPs are mainly deposited down in two ways; (i) Cosmetics are one of the primary sources of the accumulation of smaller dimension particles and (ii) The second source requires raising the aquatic medium's microbial activity as well as hydrolysis, mechanical forces, and microbial action to fragment bulk polymers under the influence of UV light energy. From different MPs, PP, PET, and PE are primarily dominant in the formation of residue with uncertain forms and fibers during the fragmentation process (Peiponen et al., 2019; X. Wang et al., 2020; Xia et al., 2020). The impacts of these MPs on large mammals, aquatic life, humans, and the environment, among other things, are numerous. The MPs in the contaminated solution can be removed or degraded using a variety of techniques. The techniques include the use of membranes, microbes, magnetic nanoparticles or metal organic composites for sequestration, dissolved air flotation, sedimentation, ultrafiltration, and coagulants. (Chen et al., 2020; Ma et al., 2019; Misra et al., 2020; Poerio et al., 2019). Water treatment facilities already in operation reportedly remove MPs up to 83% on average when MP concentrations were 3605 ± 497 particles L^{-1} which reduced after treatment was found to be 628 ± 28 particles L^{-1} (Krystynik et al., 2021). Due to the smaller size of the MPs, which were more difficult to connect with metal ions due to their inability to interact electrostatically, only 40% of the MPs could be removed using the coagulation approach using metal salts like Fe and Al. Additionally, the metal's insertion of phenolic compounds improves the effectiveness of MPs removal during coagulation. Recently, tannic acid from plant sources was identified to significantly increase the removal efficiency of MPs by more than 90% in just 5 minutes. Today's activated carbon materials, which are produced by green synthesis and then thermal activation, have good adsorption properties. Due to its high surface area and porosity, which allow MP particles to be accommodated on its surface and other forms of functionalized carbon materials have been obtained superior MPs removal effectiveness, it is utilized for the adsorption of MPs in industrial and domestic wastewater treatment. Basically, the mechanisms that led to the development of the adsorption process include electrostatic interaction, hydrogen bond interaction, π–π electron interaction and complexation, etc. (Chellasamy et al., 2022). The breakdown of MPs through photocatalysis has also been accomplished using bio-nanomaterials such carbon-enriched metallic particles and metal oxides doped with Ag. Furthermore, under certain circumstances, the biosynthesized compounds can improve the removal or degradation efficiency against MPs.

6.8.3 Xenobiotics

Greek words xenos and bios were used to create the term "xenobiotics," which refers to composite materials made up of a variety of organic and inorganic components such as phenolics, azodyes, synthetic polymers, pharmaceutical drugs, polycyclic aromatic hydrocarbons (PAHs), nirocompounds, halogen compounds, polychlorinated biphenyls (PCBs), pesticides, chlorinated derivatives, polybromonated biphenyls (PBBs), personal care products (PCPs), and triazines produces more effects to eco-systems by accumulation and not undergoes biodegradation process for a long-time and these compounds discharged from industrial revolution, and agriculture products (Figure 6.10) (Derby et al., 2021; Zhou et al., 2022). Therefore, it is mixed with the atmosphere either directly or indirectly before it reaches humans in the form of the food chain and it has a number of harmful impacts on both human and animal health, including mutagenic, carcinogenic, and teratogenic effects. The dichloro-diphenyl-trichloroethane (DDT) and methyl mercury chemical precipitate was found in fish in the aquatic system which creates attention among researchers to work on xenobiotic complexes in 1960 (Miglani et al., 2022). The primary

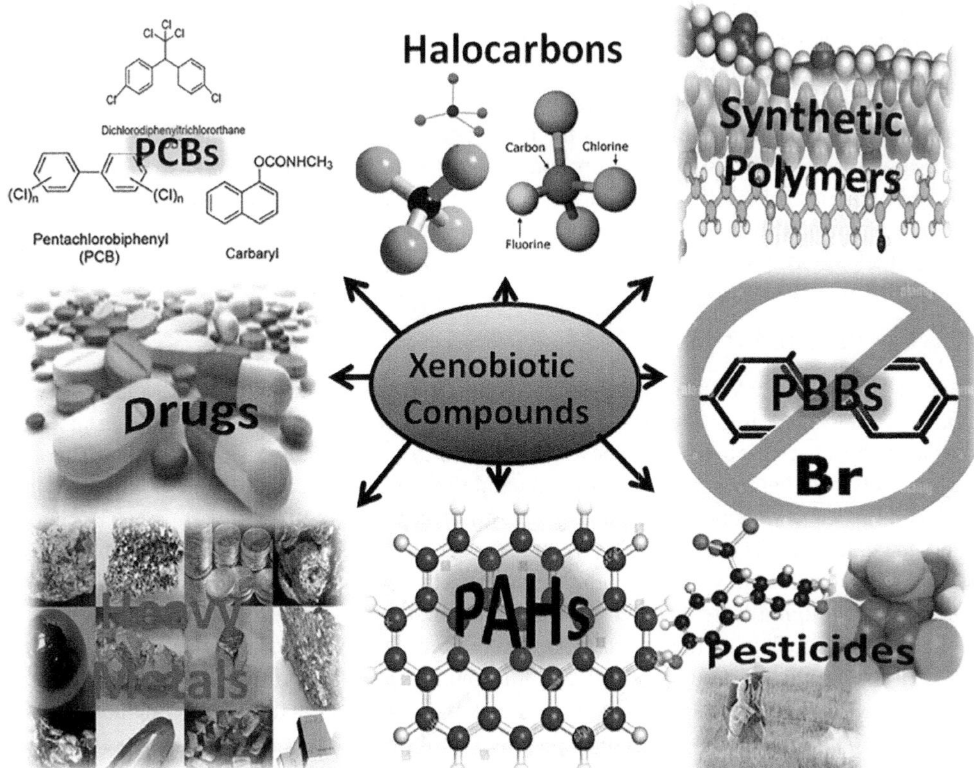

FIGURE 6.10 Schematic images of various xenobiotic compounds.

problem is the removal of xenobiotic substances from eco-systems, which must be accomplished through a variety of physical-chemical techniques as chemical precipitation, adsorption and filtration, electro-coagulation, and ozonation. However, these techniques are very expensive, involve a lengthy procedure, and result in secondary pollutants that may be more harmful than the original substances (Bhandari et al., 2021). The bioremediation technique, an alternative to the procedures mentioned above, assisted in eliminating xenobiotic elements that were present in water bodies in the form of sediments, soils, etc. The xenobiotic chemicals were successfully eliminated by the bioremediation method, which included living creatures such as bacteria, enzymes, fungi, plants, and others. Similar to this, xenobiotic contaminants have been degraded using microbes including Microbacterium, Alcaligenes, Aeromonas, Micrococcus, Penicillium, and Sphingobium, among others (Miglani et al., 2022). Moreover, the bioremediation process is more successful and ideal because it is affordable, environmentally benign, doesn't require any particular conditions, is simple to carry out, and is most prominent.

6.9 INTEGRATION IN EXISTING WASTEWATER TREATMENT

The technology seems promising and its easy incorporation in the conventional treatment process makes it an attractive solution. The plants root portion is mainly involved with the treatment process in general. The main objective of the process is to reduce the nutrient load primarily followed by the removal of contaminants that are at low concentrations. There are several methods to upscale the phytoremediation process to treat wastewater in industrial scale.

6.9.1 Land Farming

Land farming is a predominantly practiced bioremediation technique owing to its low cost and less equipment requirement for operation. This requires a large area in the treatment site. The effluent devoid of hazardous pollutant and with low nutrient are used as irrigation source for fast growing crops and trees. The nutrients are utilized by the plants preventing them from reaching water bodies. In this method the major drawback is the requirement of large area (Pooja Sharma et al., 2021). Regular monitoring of the soil and crop quality is essential in ensuring safety. Many industries with large premises follow this method for reducing nutrient load. Utilization of non-food crops is of utmost important if heavy metal is present in the effluent. The plants used are terrestrial and consume more water for their growth.

6.9.2 Constructed Wetland

Constructed wetlands (CWs) are manmade systems that have been engineered and built to use natural processes including wetland plants, soils, and the accompanying microbial communities to help treat wastewater (Figure 6.11). They are intended to mimic many of the processes that occur in natural wetlands, but in a more supervised condition. A constructed wetland is a shallow manmade basin loaded with substrate, generally topsoil or gravel, and planted with species tolerant to submerged conditions (Vymazal, 2011). Water is then routed into the system from one end and flows over the surface (surface flow) or through the substrate (subsurface flow) before being released from the other end through a weir or other device that regulates the depth of water. Surface flow wetlands and subsurface flow wetlands are two types of manmade wetlands that have been introduced. Because of its low cost, low energy need, and minimum operating optimization and expertise, wetland technology is preferred. Due to its numerous advantages and better environmental prospects, increasing amount of research are being conducted on its practical application in order to broaden our understanding of its operation. Furthermore, to provide more insight into its appropriate design, performance, operation, and maintenance for maximum environmental benefits. Despite the fact that CWs are sturdy and productive systems, their effectiveness depends on the regular adjustments

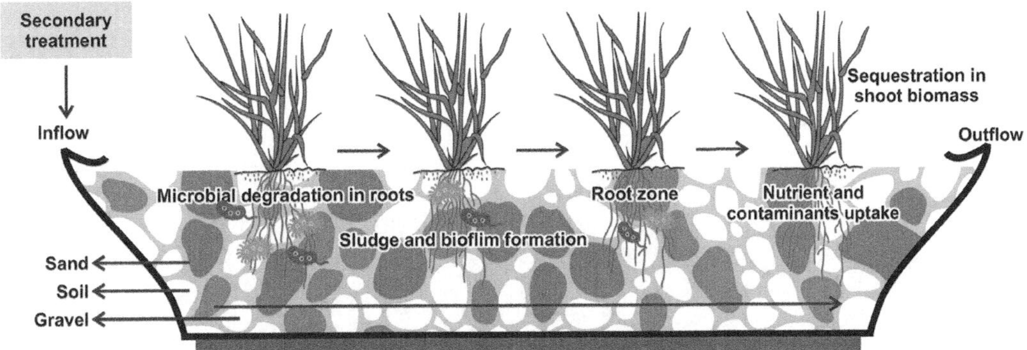

FIGURE 6.11 Integration of constructed wetland as a tertiary wastewater treatment.

to effectively manage emerging contaminants such as antibiotic and xenobiotics and for them to work effectively, they must be carefully designed and maintained (Reichenauer & Germida, 2008). A constructed wetland typically has three basic components.

An impervious layer is present at the bottom, generally polypropylene or clay. Above that a gravel layer that serves as a substrate for the root zone, providing nutrients and support is packed. Finally, an above-surface vegetation zone that has reed plants is the main component. The impermeable barrier in constructed wetland system avoids percolation of wastes down into subsurface aquifers. The gravel and root zone are important layer where water flows favoring bioremediation, and denitrification. The above-ground vegetative layer has well-adopted plant material. Within the layers, both the aerobic and anaerobic processes occur, and they may be partitioned into discrete zones. The removal processes might function singularly, sequentially, or concurrently on each contaminant group depending on the plants. For instance, in polluted groundwater, volatile organic compounds (VOCs) are predominantly removed by the integrative physical mechanism of diffusion-volatilization (Hu et al., 2020). Mechanisms such as adsorption of suspended materials, photochemical oxidation, and biological degradation may also be involved. Physical removal processes of pollutants are made possible by low flow rates thereby favoring settling, sedimentation, and volatilization.

6.9.3 Floating Wetlands

Floating treatment wetlands (FTWs) are engineered structures that allow aquatic emergent plants to thrive in water that would otherwise be too deep for them. Their roots stretch across the floating platforms forming thick columns of roots with a large surface area. FTWs have a structure that is similar to other conventional wetlands, besides the fact that plants in FTWs are anchored by man-made floating platforms. These buoyant mats keep the shoots of plants well above water surface, allowing them to grow roots into the treatment zone (Karthikeyan et al., 2022). Buoyancy is attained in FTWs by employing either low-density floating material for platforms or plants with aerenchymatous properties. In particular, materials, such as polyester sheets, PVC pipes, and bamboo having meshes, can be employed to construct the floating platforms (Figure 6.12). Helophytic plants are incorporated in the design of FTWs because of their natural capacity to encapsulate gases inside the rhizomes, causing them to float on the water. However, durability, utility, weight, environmental sensitivity, anchoring, adaptability, and affordability are the major issues in constructing floating wetlands. Plants in FTWs perform a variety of wastewater treatment activities like stabilization of the pond bed, reduction of water flow and turbulence, increased sedimentation, trapping/filtering suspended particles, and provision of habitat for the formation of microbial communities. Because of the free suspension of roots in the water column, which provides direct contact between pollutants and the rhizospheric microbial population, biological processes are more successful in FTWs

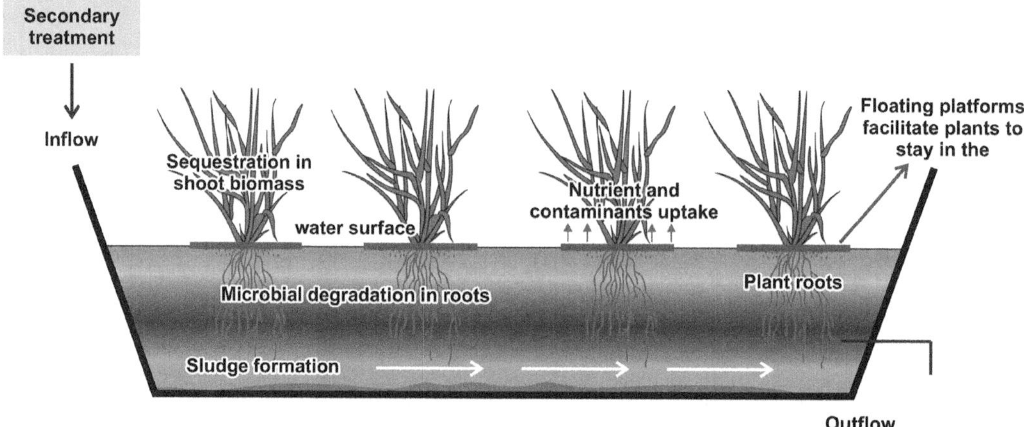

FIGURE 6.12 Integration of floating macrophytes as a tertiary wastewater treatment.

than in built wetlands. Microorganisms break down the organic stuff into essential nutrients, which are absorbed directly by the roots. Plants also produce a variety of organic chemicals (root exudates) that influence biochemical mechanisms including denitrification. This is especially successful for nutrient-rich wastewater in which nitrate reducers transform nitrates into N_2 gas and release it into the sky.

It has been found that plant in floating wetlands is able to accumulate more nitrogen than conventional wetland system (200 to 2,500 kg/ha/year). According to studies, treatment using native plants such as vetiver can restore water quality metrics to reusable norms in as little as two weeks (Karthikeyan et al., 2022). Consequently, plants increase the removal of suspended particles by attaching them to the rhizosphere and reduce the turbidity of water. Plants also create bioactive chemicals, which induces physicochemical changes of wastewater, resulting in improved sorption and sedimentation processes (Ansari et al., 2015). Interestingly, helophytes absorb atmospheric oxygen and discharge it into the rhizosphere. This release of oxygen occurs largely in daylight owing to photosynthesis. The oxygen generated by roots controls redox potential, which influences nitrogen transformation, destruction of certain phytotoxins, and decomposition of organic materials by microbes.

6.9.4 Hybrid Treatment Process

The floating wetland systems can also be combined with conventional treatment technologies to provide higher treatment efficiency. All the open tank treatment units can be mounted with a setup to facilitate the planting of macrophytes. The roots of the plants could uptake the pollutants without the need for separate space for treatment. This hydroponic system could also benefit by the aeration in the root zone which improves its removal of pollutant. The root zone provides an anchorage to the beneficial microbes that could degrade the pollutants in the wastewaters. The studies with the vetiver hydroponic hybrid systems have proven to be effective in reducing the time taken for treatment in addition to the removal of heavy metals from the effluent (Davamani et al., 2021).

Merits

- Phytoremediation is economically feasible since plants are autotrophic systems powered by solar energy, making them easy to maintain and with minimal installation and maintenance costs.
- Improves soil fertility by replenishment of organic matters to the soil

- It reduces exposure of the contaminants to the environment and ecosystem
- It is applicable over a large-scale and easily disposed
- Phytoremediation prevents metal leaching by stabilizing and immobilizing heavy metals
- Planting trees (Phytohydraulics) on the remediation sites possess aesthetic value.

Demerits

- Pollutant accumulation in edible parts of crop and vegetables.
- Low biomass production leads to repeated planting and disposal for decontamination
- The main limitation was selection of unique hyperaccumulator plant for each metallic element
- Chelate-enhanced phytoremediation pollutes the environment
- Handling and disposing contaminated plants are one of the major limitation
- Contaminants dissolved in groundwater are not suitable for aquatic phytoremediation

6.10 CHALLENGES AND FUTURE PERSPECTIVE

Since every part of a human being and all living things depend on water sources to survive, water cleanup is a serious concern. Because of this, the challenge today is to remove or degrade numerous types of organic and inorganic contaminants from industrial effluent. The production of biomaterials, including nanomaterials, bio-adsorbents, functionalized materials with naturally derived compounds, etc., depends on phyto-resources such as extracts and naturally derived compounds (Tripathi et al., 2019). Currently, a number of techniques are employed to recover the contaminants from wastewater, but they have several drawbacks, such as being unprofitable, creating secondary pollutants, complicating operations, and necessitating an advanced laboratory, and others. On the other hand, Bioremediation is an alternative conventional technique that can be utilized to eliminate industrial effluent utilizing biomaterials (Mustafa & Hayder, 2021). Due to the inclusion of numerous elements, natural chemicals, etc., these biomaterials can be made by using waste biomass and phytoconstituents, which also increases wastewater treatment efficiency. Particularly, 80% of biomass is present in agricultural waste discharged periodically, and it can be used as a precursor to producing biochar and bio composite materials. Biomass fuels are relatively affordable, biodegradable, regenerable, and reusable, among other things. Exploring the advances of biomaterials for the treatment of industrial and other wastewater is necessary;

i. Before looking into applications, it is important to study the physicochemical properties of the synthesized biomaterials, such as surface shape, compound purity, material stability, optical and electrical characteristics, etc.
ii. The biocatalysts must have their biocompatibility and cytotoxicity thoroughly examined in both in-vivo and in-vivo models.
iii. More essential, compared to existing used technologies, biomaterials should be inexpensive and simple to synthesize.
iv. The effectiveness of the biocatalysts against both industrial and commercial pollution needs to be further researched. The numerous characteristics, including the effects of pH, temperature, catalyst weight, quenchers, pollutants concentration, and catalyst reusability, must be researched in order to gain a basic understanding of the behavior of the catalyst.
v. Due to the likelihood that these materials could be used in real-time applications, researchers must focus on producing biomaterials on both a laboratory and industrial scale.
vi. It must be taken into consideration in the future to synthesis more biomaterials such as metallic nanoparticles and metal oxides from marine plants and algae.

6.11 CONCLUSION

Phytoremediation is an emerging green technology for decontaminating polluted environments. Its efficiency is highly dependent on the operation of the treatment process in conjunction with other post-treatment processes for faster pollutants removal by means of GM plants with pollutant removal ability as well as the appropriate application of hyperaccumulator plant species. The cost efficiency of phytoremediation, optimizing the treatment process, and the ideal combination connection with other treatment methods in phytoremediation have not been thoroughly addressed and may be considered as future study priorities. To summaries, phytoremediation being a specific technique is a tool for scientific research, and like all technologies, it has advantages and disadvantages; thus, the efficiency of this technique is highly dependent on proper selection and appropriate application with comfortable augmentation and amendments.

REFERENCES

Abdullah, M. M. S., Atta, A. M., Allohedan, H. A., Alkhathlan, H. Z., Khan, M., & Ezzat, A. O. (2018). Green synthesis of hydrophobic magnetite nanoparticles coated with plant extract and their application as petroleum oil spill collectors. *Nanomaterials*, *8*(10), 855.

Agarwal, M., & Singh, K. (2017). Heavy metal removal from wastewater using various adsorbents: a review. *Journal of Water Reuse and Desalination*, *7*(4), 387–419.

Ahmad, A., Mohd-Setapar, S. H., Chuong, C. S., Khatoon, A., Wani, W. A., Kumar, R., & Rafatullah, M. (2015). Recent advances in new generation dye removal technologies: novel search for approaches to reprocess wastewater. *RSC Advances*, *5*(39), 30801–30818.

Ahmad, M. M., Kotb, H. M., Mushtaq, S., Waheed-Ur-Rehman, M., Maghanga, C. M., & Alam, M. W. (2022). Green synthesis of Mn+ Cu bimetallic nanoparticles using vinca rosea extract and their antioxidant, antibacterial, and catalytic activities. *Crystals*, *12*(1), 72.

Ahmad, M. M., Mushtaq, S., Al Qahtani, H. S., Sedky, A., & Alam, M. W. (2021). Investigation of TiO2 nanoparticles synthesized by Sol-Gel method for effectual photodegradation, oxidation and reduction reaction. *Crystals*, *11*(12), 1456.

Ahmed, J., Thakur, A., & Goyal, A. (2021). Industrial wastewater and its toxic effects In: Shah, M. P. (eds) *Biological Treatment of Industrial Wastewater*, The Royal Society of Chemistry, pp. 1–14.

Allen, D., Haynes, H., & Arthur, S. (2017). Contamination of detained sediment in sustainable urban drainage systems. *Water*, *9*(5), 355.

Almaamary, E. A. S., Abdullah, S. R. S., Hasan, H. A., Rahim, R. A. A., & Idris, M. (2017). Treatment of methylene blue in wastewater using Scirpus grossus. *Malaysian Journal of Analytical Sciences*, *21*(1), 182–187.

Andreazza, R., Bortolon, L., Pieniz, S., Giacometti, M., Roehrs, D. D., Lambais, M. R., & Camargo, F. A. O. (2011). Potential phytoextraction and phytostabilization of perennial peanut on copper-contaminated vineyard soils and copper mining waste. *Biological Trace Element Research*, *143*(3), 1729–1739.

Ansari, A. A., Gill, S. S., Gill, R., Lanza, G. R., & Newman, L. (2015). *Phytoremediation: management of environmental contaminants*, Volume 2. Phytoremediation: Management of Environmental Contaminants (pp. 1–366). https://doi.org/10.1007/978-3-319-10969-5

Arcipowski, E., Schwartz, J., Davenport, L., Hayes, M., & Nolan, T. (2017). Clean water, clean life: promoting healthier, accessible water in rural Appalachia. *Journal of Contemporary Water Research & Education*, *161*(1), 1–18.

Balakrishnan, R. M., Uddandarao, P., Manirethan, V., & Raval, K. (2020). Insights on the advanced processes for treatment of inorganic water pollutants. In: Devi, P., Singh, P., Kansal, S. K. (eds) *Inorganic pollutants in water* (pp. 315–336). Elsevier.

Bhandari, S., Poudel, D. K., Marahatha, R., Dawadi, S., Khadayat, K., Phuyal, S., Shrestha, S., Gaire, S., Basnet, K., & Khadka, U. (2021). Microbial enzymes used in bioremediation. *Journal of Chemistry*, *2021*(1), 8849512.

Borah, P., Kumar, M., & Devi, P. (2020). Types of inorganic pollutants: metals/metalloids, acids, and organic forms. In *Inorganic pollutants in water* (pp. 17–31). https://doi.org/10.1016/B978-0-12-818965-8.00002-0

Botalova, O., Schwarzbauer, J., Frauenrath, T., & Dsikowitzky, L. (2009). Identification and chemical characterization of specific organic constituents of petrochemical effluents. *Water Research*, *43*(15), 3797–3812.

Bratovcic, A. (2019). Degradation of micro-and nano-plastics by photocatalytic methods. *Journal of Nanoscience and Nanotechnology Applications*, *3*, 206.

Caffall, K. H., & Mohnen, D. (2009). The structure, function, and biosynthesis of plant cell wall pectic polysaccharides. *Carbohydrate Research*, *344*(14), 1879–1900.

Carocci, A., Rovito, N., Sinicropi, M. S., & Genchi, G. (2014). Mercury toxicity and neurodegenerative effects. In: Whitacre, D. (eds) *Reviews of Environmental Contamination and Toxicology*, *229,* Springer, 1–18.

Chaignon, V., Bedin, F., & Hinsinger, P. (2002). Copper bioavailability and rhizosphere pH changes as affected by nitrogen supply for tomato and oilseed rape cropped on an acidic and a calcareous soil. *Plant and Soil*, *243*(2), 219–228.

Chaudhary, K., Khan, S., & Saraswat, P. K. (2019). Nano-phytoremediation technologies for groundwater contaminates: emerging research and opportunities: emerging research and opportunities. In: Chaudhary, K., Khan, S., Saraswat, P. K. (eds) *Advances in Environmental Engineering and Green Technologies*, IG Global, pp. 1–198.

Chellasamy, G., Kiriyanthan, R. M., Maharajan, T., Radha, A., & Yun, K. (2022). Remediation of microplastics using bionanomaterials: a review. *Environmental Research*, *208*, 112724.

Chen, Y.-J., Chen, Y., Miao, C., Wang, Y.-R., Gao, G.-K., Yang, R.-X., Zhu, H.-J., Wang, J.-H., Li, S.-L., & Lan, Y.-Q. (2020). Metal-organic framework-based foams for efficient microplastics removal. *Journal of Materials Chemistry A*, *8*(29), 14644–14652.

Davamani, V., Parameshwari, C. I., Arulmani, S., John, J. E., & Poornima, R. (2021). Hydroponic phytoremediation of paperboard mill wastewater by using vetiver (Chrysopogon zizanioides). *Journal of Environmental Chemical Engineering*, *9*(4), 105528.

De Souza, M. P., Pickering, I. J., Walla, M., & Terry, N. (2002). Selenium assimilation and volatilization from selenocyanate-treated Indian mustard and muskgrass. *Plant Physiology*, *128*(2), 625–633.

Derby, A. P., Fuller, N. W., Hartz, K. E. H., Segarra, A., Connon, R. E., Brander, S. M., & Lydy, M. J. (2021). Trophic transfer, bioaccumulation and transcriptomic effects of permethrin in inland silversides, Menidia beryllina, under future climate scenarios. *Environmental Pollution*, *275*, 116545.

Dordio, A., Carvalho, A. J. P., Teixeira, D. M., Dias, C. B., & Pinto, A. P. (2010). Removal of pharmaceuticals in microcosm constructed wetlands using Typha spp. and LECA. *Bioresource Technology*, *101*(3), 886–892.

Dsikowitzky, L., & Schwarzbauer, J. (2014). Industrial organic contaminants: identification, toxicity and fate in the environment. *Environmental Chemistry Letters*, *12*(3), 371–386.

Du, J., Zhang, B., Li, J., & Lai, B. (2020). Decontamination of heavy metal complexes by advanced oxidation processes: a review. *Chinese Chemical Letters*, *31*(10), 2575–2582.

Englande, A. J., Krenkel, P., & Shamas, J. (2015). Wastewater treatment &water reclamation. *Reference Module in Earth Systems and Environmental Sciences*. https://doi.org/10.1016/B978-0-12-409548-9.09508-7

EPA. (2020). *Introduction to Phytoremediation* (pp. 1–7). National Risk Management Research Laboratory, Office of Research and and Development, US Environmental Protection Agency. https://doi.org/10.4018/978-1-5225-9016-3.ch001

Frédette, C., Labrecque, M., Comeau, Y., & Brisson, J. (2019). Willows for environmental projects: a literature review of results on evapotranspiration rate and its driving factors across the genus Salix. *Journal of Environmental Management*, *246*, 526–537.

Gómez-Garrido, M., Mora Navarro, J., Murcia Navarro, F. J., & Faz Cano, Á. (2018). The chelating effect of citric acid, oxalic acid, amino acids and Pseudomonas fluorescens bacteria on phytoremediation of Cu, Zn, and Cr from soil using Suaeda vera. *International Journal of Phytoremediation*, *20*(10), 1033–1042.

Gupta, V. K., Ali, I., Saleh, T. A., Nayak, A., & Agarwal, S. (2012). Chemical treatment technologies for waste-water recycling-an overview. *RSC Advances*, *2*(16), 6380–6388. https://doi.org/10.1039/C2RA20340E

Hu, H., Li, X., Wu, S., & Yang, C. (2020). Sustainable livestock wastewater treatment via phytoremediation: current status and future perspectives. *Bioresource Technology*, *315*, 123809. https://doi.org/10.1016/j.biortech.2020.123809

Karthikeyan, G., Sara, K., Banu, P., Maheswari, M., Karthikeyan, S., John, J. E., Prabhu, R., & Poornima, R. (2022). Suitability of vetiver (Vetiveria zizanioides) for removal of Cr (III) from tannery effluent using floating bed and rhizofiltration systems. *Journal of Applied and Natural Science*, *14*(SI), 105–110. https://doi.org/10.31018/JANS.V14ISI.3575

Kidd, P., Barceló, J., Bernal, M. P., Navari-Izzo, F., Poschenrieder, C., Shilev, S., Clemente, R., & Monterroso, C. (2009). Trace element behaviour at the root-soil interface: implications in phytoremediation. *Environmental and Experimental Botany*, *67*(1), 243–259.

Krystynik, P., Strunakova, K., Syc, M., & Kluson, P. (2021). Notes on common misconceptions in microplastics removal from water. *Applied Sciences, 11*(13), 5833.

Kumar, K. Y., Muralidhara, H. B., Nayaka, Y. A., Balasubramanyam, J., & Hanumanthappa, H. (2013). Low-cost synthesis of metal oxide nanoparticles and their application in adsorption of commercial dye and heavy metal ion in aqueous solution. *Powder Technology, 246*, 125–136.

Kushwaha, A., Rani, R., Kumar, S., & Gautam, A. (2015). Heavy metal detoxification and tolerance mechanisms in plants: implications for phytoremediation. *Environmental Reviews, 24*(1), 39–51.

Kyzas, G. Z., & Matis, K. A. (2015). Nanoadsorbents for pollutants removal: a review. *Journal of Molecular Liquids, 203*, 159–168.

Latorre, A., Rigol, A., Lacorte, S., & Barceló, D. (2005). Organic compounds in paper mill wastewaters. *Water Pollution, 2*, 25–51.

Le, A. T., Pung, S.-Y., Sreekantan, S., & Matsuda, A. (2019). Mechanisms of removal of heavy metal ions by ZnO particles. *Heliyon, 5*(4), e01440.

Liu, C., Wu, T., Hsu, P.-C., Xie, J., Zhao, J., Liu, K., Sun, J., Xu, J., Tang, J., & Ye, Z. (2019). Direct/alternating current electrochemical method for removing and recovering heavy metal from water using graphene oxide electrode. *ACS Nano, 13*(6), 6431–6437.

Loos, R., Hanke, G., Umlauf, G., & Eisenreich, S. J. (2007). LC-MS-MS analysis and occurrence of octyl-and nonylphenol, their ethoxylates and their carboxylates in Belgian and Italian textile industry, waste water treatment plant effluents and surface waters. *Chemosphere, 66*(4), 690–699.

Ma, B., Xue, W., Hu, C., Liu, H., Qu, J., & Li, L. (2019). Characteristics of microplastic removal via coagulation and ultrafiltration during drinking water treatment. *Chemical Engineering Journal, 359*, 159–167.

Macci, C., Peruzzi, E., Doni, S., Poggio, G., & Masciandaro, G. (2016). The phytoremediation of an organic and inorganic polluted soil: a real scale experience. *International Journal of Phytoremediation, 18*(4), 378–386. https://doi.org/10.1080/15226514.2015.1109595

Mahmood, N. M. Q. (2010). Phytoremediation of arsenic (As) and mercury (Hg) contaminated soil. *World Applied Sciences Journal, 8*(1), 113–118.

McGrath, S. P., Zhao, J., & Lombi, E. (2002). Phytoremediation of metals, metalloids, and radionuclides. *Advances in Agronomy*, 75, 1–56.

Miglani, R., Parveen, N., Kumar, A., Ansari, M. A., Khanna, S., Rawat, G., Panda, A. K., Bisht, S. S., Upadhyay, J., & Ansari, M. N. (2022). Degradation of xenobiotic pollutants: an environmentally sustainable approach. *Metabolites, 12*(9), 818.

Misra, A., Zambrzycki, C., Kloker, G., Kotyrba, A., Anjass, M. H., Franco Castillo, I., Mitchell, S. G., Güttel, R., & Streb, C. (2020). Water purification and microplastics removal using magnetic polyoxometalate-supported ionic liquid phases (magPOM-SILPs). *Angewandte Chemie International Edition, 59*(4), 1601–1605.

Müller, R., Berghahn, R., & Hilt, S. (2010). Herbicide effects of metazachlor on duckweed (Lemna minor and Spirodela polyrhiza) in test systems with different trophic status and complexity. *Journal of Environmental Science and Health Part B, 45*(2), 95–101.

Mustafa, H. M., & Hayder, G. (2021). Recent studies on applications of aquatic weed plants in phytoremediation of wastewater: a review article. *Ain Shams Engineering Journal, 12*(1), 355–365. https://doi.org/10.1016/j.asej.2020.05.009

Ncibi, M. C., Mahjoub, B., Mahjoub, O., & Sillanpää, M. (2017). Remediation of emerging pollutants in contaminated wastewater and aquatic environments: biomass-based technologies. *CLEAN-Soil, Air, Water, 45*(5), 1700101.

Nilratnisakorn, S., Thiravetyan, P., & Nakbanpote, W. (2007). Synthetic reactive dye wastewater treatment by narrow-leaved cattails (Typha angustifolia Linn.): effects of dye, salinity and metals. *Science of the Total Environment, 384*(1–3), 67–76.

Olette, R., Couderchet, M., Biagianti, S., & Eullaffroy, P. (2008). Toxicity and removal of pesticides by selected aquatic plants. *Chemosphere, 70*(8), 1414–1421.

Olguín, E. J., & Sánchez-Galván, G. (2011). Phytofiltration of heavy metals. *Comprehensive Biotechnology, Second Edition, 6*, 207–213. https://doi.org/10.1016/B978-0-08-088504-9.00379-2

Pahalawattaarachchi, V., Purushothaman, C. S., & Vennila, A. (2009). Metal phytoremediation potential of *Rhizophora mucronata* (Lam.), *Indian Journal of Marine Sciences, 38*(2), 178–183.

Peiponen, K.-E., Räty, J., Ishaq, U., Pélisset, S., & Ali, R. (2019). Outlook on optical identification of micro-and nanoplastics in aquatic environments. *Chemosphere, 214*, 424–429.

Pilon-Smits, E. A. H., De Souza, M. P., Hong, G., Amini, A., Bravo, R. C., Payabyab, S. T., & Terry, N. (1999). *Selenium volatilization and accumulation by twenty aquatic plant species*. Wiley Online Library.

Poerio, T., Piacentini, E., & Mazzei, R. (2019). Membrane processes for microplastic removal. *Molecules, 24*(22), 4148.

Polińska, W., Kotowska, U., Kiejza, D., & Karpińska, J. (2021). Insights into the use of phytoremediation processes for the removal of organic micropollutants from water and wastewater: a review. *Water, 13*(15), 2065.

Rahman, R. A. A., Abou-Shanab, R. A., & Moawad, H. (2008). Mercury detoxification using genetic engineered Nicotiana tabacum. *Global Nest Journal, 10*, 432–438.

Rani, D. N., & Abraham, E. T. (2016). A potential tissue culture approach for the phytoremediation of dyes in aquaculture industry. *Biochemical Engineering Journal*, 115, 23–29.

Reeves, R. D., Baker, A. J. M., Jaffré, T., Erskine, P. D., Echevarria, G., & van der Ent, A. (2018). A global database for plants that hyperaccumulate metal and metalloid trace elements. *New Phytologist, 218*(2), 407–411.

Reichenauer, T. G., & Germida, J. J. (2008). Phytoremediation of organic contaminants in soil and groundwater. *ChemSusChem, 1*(8-9), 708–717. https://doi.org/10.1002/cssc.200800125

Renu, Agarwal, M., & Singh, K. (2017). Methodologies for removal of heavy metal ions from wastewater: an overview. *Interdisciplinary Environmental Review, 18*(2), 124–142.

Rodríguez, D. M., Wrobel, K., & Jimenez, M. G. G. (2004). Determination of 2-mercaptobenzothiazole (MBT) in tannery wastewater by high performance liquid chromatography with amperometric detection. *Bulletin of Environmental Contamination and Toxicology, 73*(5), 818–824.

Salt, D. E., Smith, R. D., & Raskin, I. (1998). Phytoremediation. *Annual Reviews of Plant Physiology and Plant Molecular Biology, 49*, 643–668.

Sharma, P., Bihari, V., Agarwal, S. K., Verma, V., Kesavachandran, C. N., Pangtey, B. S., Mathur, N., Singh, K. P., Srivastava, M., & Goel, S. K. (2012). Groundwater contaminated with hexavalent chromium [Cr (VI)]: a health survey and clinical examination of community inhabitants (Kanpur, India). *PloS One, 7*(10), e47877.

Sharma, P., Tripathi, S., Purchase, D., & Chandra, R. (2021). Integrating phytoremediation into treatment of pulp and paper industry wastewater: field observations of native plants for the detoxification of metals and their potential as part of a multidisciplinary strategy. *Journal of Environmental Chemical Engineering, 9*(4), 105547.

Shrirangasami, S. R., Rakesh, S. S., Murugaragavan, R. M., Ramesh, P. T., Varadharaj, S., Elangovan, R., & Saravanakumar, S. (2020). Phytoremediation of contaminated soils – A review. *International Journal of Current Microbiology and Applied Sciences, 9*(11), 3269–3283.

Shutcha, M. N., Mubemba, M. M., Faucon, M.-P., Luhembwe, M. N., Visser, M., Colinet, G., & Meerts, P. (2010). Phytostabilisation of copper-contaminated soil in Katanga: an experiment with three native grasses and two amendments. *International Journal of Phytoremediation, 12*(6), 616–632.

Sricoth, T., Meeinkuirt, W., Saengwilai, P., Pichtel, J., & Taeprayoon, P. (2018). Aquatic plants for phytostabilization of cadmium and zinc in hydroponic experiments. *Environmental Science and Pollution Research, 25*(15), 14964–14976.

Symeonidis, L., McNeilly, T., & Bradshaw, A. D. (1985). Interpopulation variation in tolerance to cadmium, copper, lead, nickel and zinc in nine populations of Agrostis capillaris L. *New Phytologist, 101*(2), 317–324.

Tak, H. I., Ahmad, F., & Babalola, O. O. (2013). Advances in the application of plant growth-promoting rhizobacteria in phytoremediation of heavy metals. In: Whitacre, D. (eds) *Reviews of Environmental Contamination and Toxicology, 223*, Springer, New York, pp. 33–52.

Teodorović, I., Knežević, V., Tunić, T., Čučak, M., Lečić, J. N., Leovac, A., & Tumbas, I. I. (2012). Myriophyllum aquaticum versus Lemna minor: sensitivity and recovery potential after exposure to atrazine. *Environmental Toxicology and Chemistry, 31*(2), 417–426.

Tong, T., & Elimelech, M. (2016). The global rise of zero liquid discharge for wastewater management: drivers, technologies, and future directions. *Environmental Science and Technology, 50*(13), 6846–6855.

Török, A., Buta, E., Indolean, C., Tonk, S., Silaghi-Dumitrescu, L., & Majdik, C. (2015). Biological removal of triphenylmethane dyes from aqueous solution by Lemna minor. *Acta Chimica Slovenica, 62*(2), 452–461.

Tripathi, S., Singh, V. K., Srivastava, P., Singh, R., Devi, R. S., Kumar, A., & Bhadouria, R. (2019). Phytoremediation of organic pollutants: current status and future directions. In *Abatement of environmental pollutants: trends and strategies.* Elsevier Inc. https://doi.org/10.1016/B978-0-12-818095-2.00004-7

Varun, M., D'Souza, R., Kumar, D., & Paul, M. S. (2011). Bioassay as monitoring system for lead phytoremediation through Crinum asiaticum L. *Environmental Monitoring and Assessment, 178*(1), 373–381.

Von Sperling, M. (2007). *Wastewater characteristics, treatment and disposal.* IWA Publishing (Vol. 6). https://doi.org/10.2166/9781780402086

Vymazal, J. (2011). Constructed wetlands for wastewater treatment: five decades of experience. *Environmental Science & Technology*, *45*(1), 61–69.

Wang, J., Yokokawa, M., Satake, T., & Suzuki, H. (2015). A micro IrOx potentiometric sensor for direct determination of organophosphate pesticides. *Sensors and Actuators B: Chemical*, *220*, 859–863.

Wang, L., Ji, B., Hu, Y., Liu, R., & Sun, W. (2017). A review on in situ phytoremediation of mine tailings. *Chemosphere*, *184*, 594–600.

Wang, Q., & Yang, Z. (2016). Industrial water pollution, water environment treatment, and health risks in China. *Environmental Pollution*, *218*, 358–365.

Wang, X., Zheng, H., Zhao, J., Luo, X., Wang, Z., & Xing, B. (2020). Photodegradation elevated the toxicity of polystyrene microplastics to grouper (Epinephelus moara) through disrupting hepatic lipid homeostasis. *Environmental Science & Technology*, *54*(10), 6202–6212.

Wani, A. L., Ara, A., & Usmani, J. A. (2015). Lead toxicity: a review. *Interdisciplinary Toxicology*, *8*(2), 55. https://doi.org/10.1515/INTOX-2015-0009

Wuana, R. A., & Okieimen, F. E. (2011). Heavy metals in contaminated soils: a review of sources, chemistry, risks and best available strategies for remediation. *International Scholarly Research Notices*, *2011*(1), 402647.

Xia, X., Sun, M., Zhou, M., Chang, Z., & Li, L. (2020). Polyvinyl chloride microplastics induce growth inhibition and oxidative stress in Cyprinus carpio var. larvae. *Science of the Total Environment*, *716*, 136479.

Xu, C., Yang, W., Wei, L., Huang, Z., Wei, W., & Lin, A. (2020). Enhanced phytoremediation of PAHs-contaminated soil from an industrial relocation site by Ochrobactrum sp. *Environmental Science and Pollution Research*, *27*(9), 8991–8999.

Yaseen, D. A., & Scholz, M. (2017). Comparison of experimental ponds for the treatment of dye wastewater under controlled and semi-natural conditions. *Environmental Science and Pollution Research*, *24*(19), 16031–16040.

Zhang, Y., Lv, T., Carvalho, P. N., Arias, C. A., Chen, Z., & Brix, H. (2016). Removal of the pharmaceuticals ibuprofen and iohexol by four wetland plant species in hydroponic culture: plant uptake and microbial degradation. *Environmental Science and Pollution Research*, *23*(3), 2890–2898.

Zhou, Y., Yao, L., Pan, L., & Wang, H. (2022). Bioaccumulation and function analysis of glutathione S-transferase isoforms in Manila clam Ruditapes philippinarum exposed to different kinds of PAHs. *Journal of Environmental Sciences*, *112*, 129–139.

7 Activity and Role of Different Extracellular Enzymes for Deterioration of Pollutant within the Root Zone of Macrophytes

A. Bharani, R. Jayashree, P.A. Shahidha, and R. Sunitha

7.1 INTRODUCTION

Bioremediation takes advantage of the inherent ability of living life forms to pristine climate that includes utilisation of microbes and other natural substances, for example, green growth, parasites and microbes to remove or eliminate contaminations from the climate attributable to their mixed digestion abilities (Volesky and Holan, 1995; Matheickal et al., 1999; Ho et al., 2002; Malik, 2004). Detoxification process using plants and microbes helps with change and debasement of pollutants into innoxious or harmless substances. Remediation using microbes is an innovation that takes advantage of relief cycles like biostimulation and bioaugmentation. Biostimulation uses native microbial populaces to treat polluted substrate through the addition of nutrients and different agents for fastening the normal constriction mechanisms. Bioaugmentation includes the presentation of microbes that are originating externally (obtained from outside the soil climate) fit for removing a specific toxin, some of the time utilising genetically modified microorganisms (Biobasics, 2006).

Degradation of organic pollutants utilising biodegradation component of microorganisms brings about ultimate biodegradation of natural impurities into CO_2, H_2O and non-organic mixtures, while conversion process turns into combination of compounds of natural foreign substances over completely to other easier natural mixtures. Biodegradation of poisonous natural contaminations like insecticides, organic compounds, compounds like halide natural mixtures and phenolic or anilinic compounds has been worked with by stimulants present in microorganisms. Stimulants separate the bonds between the atoms and help the exchange of electrons from a decreased natural contributor to some other product. Chemicals like oxidoreductase that catalyses the incorporation of molecular oxygen and multicopper oxidases found in plants, fungi, and bacteria intercede oxidative coupling through monomers, combination of two or more monomers with different substrates or restricting to fulvic like compound (Karigar and Rao, 2011). Bacteria that thrive in the anoxic environment are significant for degradation of halogens and nitrosamine, diminishing epoxides and nitro groups. Microbes, for example, *Pseudomonas, Bacillus, Neisseria, Moraxella, Trichoderma, Aerobacter, Micrococcus* and *Burkholderia* and *Acinetobacter* can degrade DDT, dieldrin and endrin. Anabaena (a cyanobacterium), *Pseudomonas spinosa, Pseudomonas aeruginosa* and *Burkholderia* can degrade endosulfan.

In healthy soils, microbes are encircled by carbon-rich material and other nutrients that are expected for cell upkeep and development. In any case, microorganisms can't straightforwardly move these biomolecules to protoplasm. Rather, they depend on the countless number of compounds that they integrate and deliver into their nearby ecosystem. These extracellular

DOI: 10.1201/9781003441144-7

chemicals depolarise natural mixtures and create solvent, less of oligosaccharides and amides that are then perceived by membrane receptors and taken across the external film and into the cell. Protein amalgamation, catalyst creation with emission is vigorously costly, requires nutrient source (Schimel and Weintraub, 2003), and is at last incapacitating except if there are bountiful incentives. In this manner, the designation of cell assets to chemical union and discharge should include a unique harmony between the speculation of the valuable assets dispensed for the formation of proteins with the energy and nutrients acquired because of their movement. In any case, the Earth is an innately contentious ecosystem for exoenzymes in light of the fact that exudation from the cytoplasm becomes dependent upon amalgamation, disintegration, and appease through both natural and artificial systems. From the outset, the syntrophic degradation of organic polysaccharides in soil appears as an unimaginable errand. Somewhat recently, progresses in molecular biology, microscopy, and scientific methods in addition to some innovative reevaluating have started to give new bits of knowledge into the environment of exoenzymes. The obtrusive, damaging and economically awful exercises of phytopathogens (Kikot et al., 2009) and the intricate enzymology of ruminative processing likewise require an itemised information on biopolymers dissolution and calcification. Numerous efficient natural contaminants are synthetically hard and / or ineffectively dissolvable dependent on exoenzymes (frequently by 'ligninases') prior to intake, ingestion followed by removal. Efficient remediation process using microbes of polluted soil will rely upon careful comprehension of these catalystic reactions. Microbial compounds to degrade adhesives, disturb microbial matrix and repulse colonisers seem important to the antifouling industry and pave way for innumerable patenting. Future of exoenzymes tends to be perpetual. Xenobiotic compounds form a portion of the earth acts as a drawn-out storehouse for carbon capturing as well as being the basic in soil structure, rigidness, crop nutrients and moisture maintenance, diversified microbial activity and numerous activities to enhance the fertility of soil for efficient crop growth. Expanding the terrestrial and aquatic heat, increased gaseous composition especially CO_2 (Finzi et al., 2006), and increasing regular moist and dry conditions will alter the microbes and speed up development and enzyme activities either straightforwardly or followed by the impacts on plants. The results of these progressions might remember a decay for the organic part: the C sequestration turns into a motion. Those endeavouring to survey the results of an unnatural weather change by creating prescient carbon cycle models should consider any anticipated expansions in the catalytic activities of the terrestrial systems and the related decrease in the then unmanageable organic matter.

What we aim at here is to highlight the complication and varied soil catalysts and the super molecules that they decompose. Also it focuses on the regulation, secretion and various consequences of these exoenzymes being transported from the protoplasm to the outer layers. The enzyme diffusion, survival and medium turnover are affected due to soil physico-chemical and biological properties, likewise the ratio of the material produced and accumulated by manufacturing cells. Pathways by which the bacteria and their exoenzymes try to oversee the generally degrading or influential characteristics of the land areas and the various methods they adopt for efficient medium detection and usage will be explained.

7.2 EXTRA CELLULAR ENZYMES – AS AN ELIXIR

Few bacteria namely heterotrophic microbes are potential to release extracellular enzymes required to bioremediate any degradable wastes. Decomposition of degradable substances comprises of different types of microorganisms *viz.*, proteolytic, cellulolytic, amylolytic, nitrification and denitrification. For heterotrophic bacteria cell regeneration, these organisms use ammonia as an energy source. Acetate, pyruvate and oxaloacetate compounds from lake waters use carbon as energy source. In aquaculture, organic wastes are generated from the wastes like residues from feed and the excreta released by the fishes. Some of the exoenzymes *viz.*, proteases, amylases and cellulases produced by heterotrophic microbes are employed in remediating organic wastes.

Microorganisms and their extracellular proteins should be equipped for identifying, moving towards, and changing natural trash to dissolvable smallest units (or short oligomers) that are in this manner shipped into the cytoplasm. As expressed already, the monomer parts of live and lysed plant and animal parts and microbes are frequently physically and chemically connected with one another and absorb or entangle with organic compounds having low atomic weight. The process leads to an obstruction that confines bacterial and exoenzyme access even to the generally weak dissolvable compounds of organic matter. The original medium may likewise capture inside soil parts and this will decrease restrict movement into the cell parts. In this way, debasement of macromolecules and all the more effectively refined non-chemical in soils need stimulants manufacture, yet in addition real concern of stimulants with their objective media and the target of impetus. Moreover, anoxic situations should be at reach where catalysts will thrive for actions to happen.

Exoenzymes might be related with the plasma layer of the microbe's cells, held inside and appended to the outer surface of the region between the inner and outer membrane, cell wall and glycocalyx, or delivered into the liquid phase of soil. The periplasm may give electronegative microbes a supply of movement held until an outer signal that emits is gotten – maybe a proficient and fast technique for answering the existence of a possible substrate. The enzymes of the periplasmic region may help in the transport of essential nutrients to the external ecosystem to face resisting outer world. For instance, glycosylation might happen in the periplasm. As to catalysis inside the periplasm, the target should be restricted to structures having low atomic mass that can be moved through the outer layer. Hence, stimulants that disintegrate disaccharides, for example, cellobiose and mannose, and a few phosphatases and enzyme that cleave nucleic acids are frequently (however not generally) confined to the periplasmic space and further change their substrates promptly preceding movement to the cytoplasmic region.

A few exoenzymes are held on the external walls of the cell membrane (i.e. frieze or fringe chemicals) that probably are designed so that their dynamic sites are uncovered and the regions that are generally liable to go after by proteases are safeguarded. Two different sorts of non-diffusing exoenzymes thrive: these inside the carbohydrate layer covering the cell or are important for a multicelled uncultured layer (Flemming and Wingender, 2002; Romani et al., 2008) and these coordinated inside minutely noticeable forms connected to, yet projecting from, the cell membrane. The latter, called polysaccharides with long chain were first depicted for anaerobic thermophile *Clostridium thermocellum*, an anaerobic bacteria coming under thermophiles and have been greatly explored (Bayer et al., 2004; Gold and Martin, 2007, Bayer et al., 2008). *Clostridium thermocellum*, produces many exoenzymes, including endoglucanases, exoglucanses, β-glucosidases, xylanases, lichenases, laminarinases, xylosidases, galactosidases, mannosidases, gelatin lyases, gelatin methylesterases, polygalacturonate hydrolases, cellobiose phosphorylases, and cellodextrin phosphorylases (Lamed et al., 1983; Demain et al., 2005) and a large number enzymes might be held inside cellulosomes.

In fact, since they are glycosylated or have disulfide linkages, many exoenzymes are fundamentally more steadier than their internal equivalence. Heat resistant, a broad pH range for action, and few proteolysis refusal are all provided by these changes. Other enzymes are stabilised by interacting with soil organic matter, clay minerals, or tannins (Burns, 1982; Joanisse et al., 2007). In contrast to being free in solution, some enzymes' activity is really largely linked to organic and inorganic colloids (Kandeler, 1990). The demise and death of their mother cell have caused many of these attached adjuvants to become externalised rather than strictly extracellular. In actuality, various places may contain the same enzyme. No matter where they came from, stabilised enzymes frequently have less internal process because complications can limit substrate availability, obstruct reactive regions, and alter composites (Allison and Jastrow, 2006; Nannipieri, 2006; Quiquampoix and Burns, 2007). Enzymes that have been preserved as humic matter and linked to organomineral complexes may still be active (Tate, 2002). But this will differ across various enzymes. For instance, (Gianfreda et al., 1995) discovered that the enzyme maintains its conformation and function when urease complexes with tannic acid are generated in the presence of ferric ions and aluminium hydroxide species. Clay-bound cellulose enzymes continued to function, according to Pflug (1982), but the activities of the enzymes that break down starch, -amylase and amyloglucosidase, were entirely suppressed. Depending on how it is distributed across the various adsorptive soil fractions, the same enzyme may even display distinct activity. Enzymes may also be protected by stabilisation from proteases and other enzymatic denaturants (Nannipieri et al., 1978; Nannipieri et al., 1988). If enzymes are removed from organomineral complexes, enzymatic function might be recovered. However, because of modifications in protein structure, irreversible deactivation of enzymes adsorbed to surfaces is frequent. The amount of energy needed to reverse the unfolding of the adsorbed enzyme may be more than the heat energy available if the enzyme is unravelled and the points that are in contact with cell surfaces increase (Quiquampoix et al., 2002). Therefore, although their ability to diffuse to distant substrates will be much diminished, extracellular enzymes forming a complex with organic molecules or adsorbed by clay minerals (or a mixture of both) are probably to have a lengthier turnover period than those free in the liquid phase (Hope and Burns, 1985). Unmeasured enzyme yield rates in soils, however, represent a critical knowledge gap in soil enzyme studies.

The ability of extracellular enzymes produced by microbial activity in the soil to break down composite organic substrates into less molecular weight compounds easily ingested by soil flora and fauna is well known. Although it is still up for debate, it is generally acknowledged that these extracellular enzymes' catalytic capability take part in an important role in the dynamics of humus-like substance (repository and yield) in soils. Therefore, any feature or set of features that change the synthesis and reactivity of soil catalysts may restrain the above acts.

7.3 TYPES OF EXTRACELLULAR ENZYMES AND THEIR ROLE IN DETOXIFICATION

No matter the trophic strategy, extracellular enzymes constitute a significant bacterial feature. Extracellular hydrolytic chemicals are located outside the cell membrane and classified into two groups: attached enzymes, which are kept in touch with the cell's outer layer by being anchored to or integrated into the layer, and free enzymes, which the cell releases into solution. Normal seawater contains both of the extracellular enzyme groups, but is hard to evaluate the ecological importance of the two protein architectures. It is widely acknowledged that the method for deploying enzymes has something to do with how microorganisms live. Due to the high nutrient concentrations on particles and the projected critical enzyme activity on particles, the free-enzyme system is advantageous for organisms that are associated with molecules. The surrounding substrate field is radically distinctive for free-living microorganisms utilising dissolved organic materials. Given that dissolved organic matter is so diluted, it has been projected that surface-attached enzymes will be the most economically productive technology in this situation. As far as anyone is concerned, the technologies currently in use for analyzing typical microbial networks in seawater are unable to determine the volume of extracellular proteins equivalent to the cell or to analyse the little amount of catalysts connected to the cell's outer layer. Thus, in size-fractionated water, just the ratio of dissolved enzymatic action to total protein action is often reported. It has been demonstrated that free enzymes make a remarkable contribution, contradicting the notion that surface-applied chemicals are the most expensive and effective method. However, a number of processes, including phytoplankton and zooplankton activities, enzymes that escaped from particles, dynamic compounds on cell fragments, or the release of enzymes as a result of cell lysis during sample filtration, may mask a potential sign of successfully released extracellular enzymes from microorganisms.

7.3.1 Intracellular Enzymes

Endoenzymes are another name for the substances found inside cells. They are protein-based cellular components. They also provide the main justification for the majority of biological processes.

This kind of enzyme supports prokaryotic and eukaryotic cells' internal metabolic processes. For instance, they can help the cell's respiration as well as photosynthesis by handling both processes. These proteins also allow for further intracellular processing.

7.3.2 Extracellular Enzymes

Extracellular chemicals aid in the vast majority of assimilation in the alimentary canal. An extracellular substance is a catalyst that is released and adheres to the cells outside. Exoenzymes are another term for these proteins. They play a crucial function when anything is degrading or decaying. They also break down complicated particles like cellulose and hemicellulose into smaller ones.

7.3.3 Extracellular and Intracellular Enzymes are Compared

Because, one functions inside the cell while the other acts outside, intracellular and extracellular differ significantly from one another (Table 7.1). The table compares and contrasts these enzymes' additional variations.

The cytoplasmic fluid or cellular organelles may be home to the intracellular enzymes. They perform functions for DNA replication and protein synthesis there. Additionally, they support the mitochondria's Krebs cycle and glycolysis processes. Additionally, they support the photosynthetic pathway inside the chloroplast where they are located. Lysosomes, which contain endoenzymes, also contribute significantly to intracellular digesting processes. Enzymes from extracellular sources are important in the breakdown process. Additionally, the majority of digestive enzymes, including trypsin, peptidase, and pancreatic amylase, are extracellular. As a result, digestion is facilitated. Exoenzymes are used by some of the bacteria to propagate dangerous diseases. Exoenzymes from microorganisms help in bioremediation as well. Some of them can be utilised in the paper industry and as biofuels.

7.3.4 Intracellular and Extracellular Enzymes have Similar Properties

- Both operate as catalysts for chemical processes and are enzymes.
- Both enzymes are active in living things.
- They're both proteins.
- Both varieties of enzymes are present in living things.
- The exo and endo enzymes constitute proteins made of amino acids seen in unicellular and multicellular organisms.

TABLE 7.1
Difference between Extracellular and Intracellular Enzymes

Intracellular Enzymes	Extracellular Enzymes
These enzymes operate inside the cell	It works outside of the cell.
In nature these are primarily inside the cell	Exoenzymes are quite rare.
Functions include intracellular digestion, protein synthesis, cellular respiration, DNA replication, and photosynthesis	Extracellular digestion, decomposition, and other functions.
Examples include ATP synthetase, DNA and RNA polymerase, and others. In addition, intracellular enzymes are present in the lysosome and peroxisome.	For instance, enzymes like peptidase, amylase, trypsin, collagenase, pepsin, sucrase, maltase, and kinases.

7.3.4.1 Protease

Protease production from Actinomycetes like members of the genera Streptomyces, Nocardia, and Nocardiopsis has been documented in a number of investigations. Most proteases exhibit tolerance to a variety of abiotic stressors like high pH, temperature, and salinity. Proteases from *Streptomyces* spp. can be utilised to digest a variety of agro-industrial wastes, including feathers, nails, hair, and plant wastes. Proteases are well known as significant industrial enzymes and have the potential to be widely exploited in the leather, baking, textile, detergent, brewery, cheese, and dehairing industries. The generations of proteases by more than 48 strains of soil-based Actinomycetes as well as their cytotoxic effects on cancer cells have been shown (Wariishi et al., 1992).

7.3.4.2 Cellulase

Cellulases are crucial industrial enzymes for the environmentally friendly generation of biofuel because they transform cellulose into fermentable sugars. Highly thermostable Streptomyces spp. cellulases include those from *S. ruber*, *S. lividans*, and *S. rutgersensis*. These enzymes are mostly employed as supplements in the paper and pulp business, textile industry, animal additives, and detergent industry. Cellulases that have the potential to be exploited economically are also produced by some members of the Thermobifida and Micromonospora genera. Cellulases from extremophiles, such as Ermobida, can degrade cotton and avicel because they are stable at high temperatures and pH levels. They can employ wheat, rice, and other crops as substrates (Fetzner, 1998; Jones et al., 2001).

7.3.4.3 Keratinase

Many Actinomycetes strains, including Streptomycetes spp. and Actinomadura, produce keratinases, an enzyme with significant industrial importance. Most often, these enzymes are employed in the hydrolysis of keratin. There is a high demand for creating biotechnological options for recycling keratin wastes, which involve using Actinomycetes keratinases to transform waste chicken feather, hair, nail, and wool into valuable items.

7.3.4.4 Amylases

Amylases, which can be divided into endoamylases and exoamylases, are thought to be a significant category of enzymes that hydrolyse starch into high fructose, glucose, and maltose syrups. Actinomycetes strains, such as *Streptomyces erumpens* and ermobidafusca, have the capacity to discharge amylases to the outside of the cells in order to perform extracellular digestion. These thermostable enzymes are useful in the paper and pulp, pharmaceutical, and baking industries. To increase the detergency of substances, several alkaliphilic *Actinomycetes* strains' amylases can be employed in detergent formulation. *Streptomyces* spp. amylases have a significant impact on biotechnological applications in a variety of industries and account for around 25% of the global enzyme market demand.

7.3.4.5 Xylanase

The majority of xylanases produced by Actinomycetes belong to the genus Streptomyces. Hemicelluloses' most prevalent component, xylan, is frequently employed in the upgrading of pulp and the biobleaching industries. Actinobacterial genera, *Actinomadura,* and *ermoactinomyces*, produce cellulases-free and thermostable xylanases with an optimal temperature of 70°C. The hydrolysis of several agricultural wastes, such as oil cake and straw waste, by some species of *Streptomyces* led to an increase in the production of biogas.

7.3.4.6 Lipases

Oils and fats can be hydrolysed by a number of Actinomycetes strains. Different classes of hydrolytic enzymes, such as lipases and esterases, catalyse the hydrolysis of lipids like triglycerides. Actinomycete members such as *Streptomyces exfoliates* and *Nocardiopsis alba* generate lipases that hydrolyse the ester bonds in triglycerides to create glycerol and fatty acids. Lipases have the potential to be used in the processing of oils and fats, cosmetics, diagnostics, and detergents.

7.3.4.7 Chitinases

The group of industrially significant enzymes known as chitinases also has the capacity to hydrolyse chitin. Some Actinomycetes produce chitinases, and since these enzymes are thermostable and functional over a wide pH range, they can be used in industrial settings. Actinomycetes strains that generate chitinases include *Streptomyces thermoviolaceus* and Microbispora species. The potential antioxidant chitibiose, which typically has uses in the biomedical and food industries, was recovered using chitinase from these Actinomycetes strains. Other than Streptomyces, Actinomycetes such as *Nocardiopsis prasina* have chitinases that are helpful in the hydrolysis of chitin oligosaccharides, which have the potential to be used as antioxidants, anti-microbials, anticancer, anti-coagulants, and antitumor agents. Chitinases are utilised to dispose of the waste products generated by the leather industry. Chitinases from Streptomyces species, including *S. aureofaciens*, *S. griseus*, and *S. griseoloalbus*, are helpful against phytopathogenic fungi.

7.3.4.8 Pectinases

Several species of *Streptomyces*, including *S. lydicus*, generate pectinases. These enzymes are employed in the food business for the clarification and extraction of wines, juices, oils, and flavouring compounds, as well as in the textile sector for the production of hemp and linen fibres. One of the most significant pectinases, polygalacturonase is employed extensively across numerous sectors.

7.4 RHIZOSPHERE OF MACROPHYTES – A DECONTAMINATION ZONE

A rhizosphere is a region that has an enormous number of rhizodeposits, making it the ideal place for microbial growth. Significant amounts of carbonaceous compounds that plants discharge into the rhizosphere boost and diversify the microbial community surrounding the root zone. The rhizosphere is a small area of soil that is home to a large number of bacteria and is directly influenced by plant root exudates. It is regarded as one of the planet's most dynamic interfaces (Philippot et al., 2013).

7.4.1 Role of Root Exudates

- Root exudates shield the plant from desiccation and feed nutrients to the microorganisms.
- It adds moisture to the soil, which helps microbes develop.
- It develops systemic resistance against abiotic stress like temperature, salt concentration, and pH.
- It defends the root against biotic stress, or pathogens, by releasing certain defence proteins and antimicrobial compounds.
- It aids in soil aggregation by secreting mucilage.

The following elements can alter the physicochemical properties of soil:

- osmotic pressure
- ionic equilibrium and
- oxidation reduction reactions

In the presence of adequate growth substrates, single-celled and multicellular microbes use synthetic pollutants as carbon and energy sources or co-metabolise them. Plant-derived carbon, which can be found in root exudates or other plant remnants, is an essential catalyst for these (co-metabolic) activities because of the oligotrophic nature of soil. Thus, contaminants can act as electron

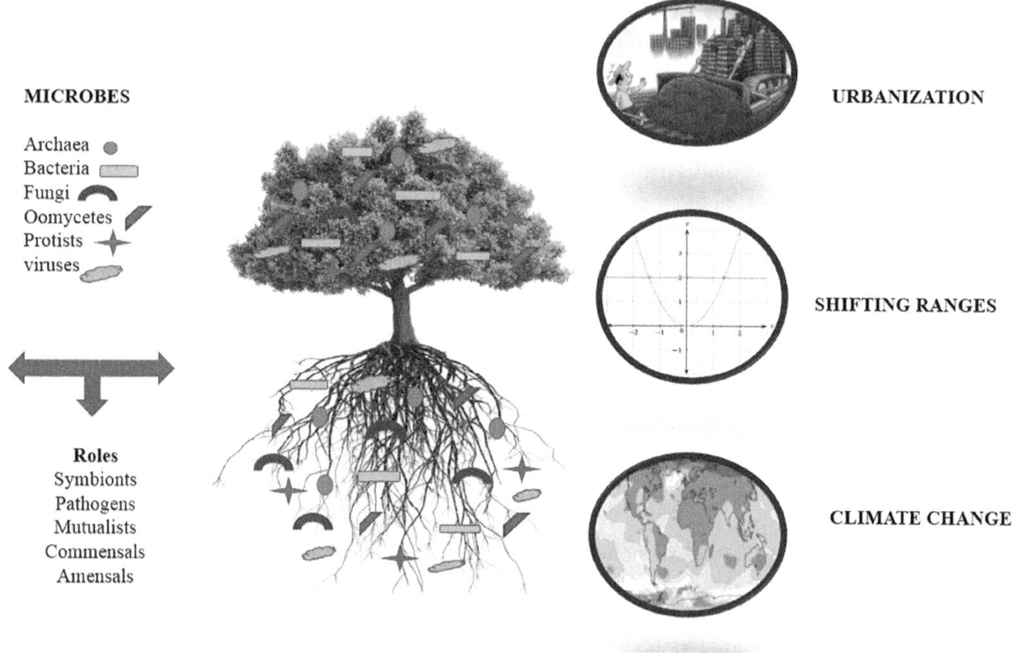

FIGURE 7.1 Rhizosphere of macrophytes and the influencing factors.

donors and undergo oxidation in both aerobic and anaerobic environments. For the transportation and transformation of organic pollutants as well as the stabilisation of impacted ecosystems interaction of plants with microorganisms are needed. Green liver idea (Fernandez-Luqueno et al., 2013) states that plants have the enzymatic capacity to biodegrade organic pollutants on their own. It is also evident that plants are obviously driving a number of significant procedures (water drift, carbon flux, oxidation reduction situations).

Numerous organisms feeding on dead and decayed matter or fungi or bacteria that feed on low nutrient medium, as well as mycorrhizal fungi, bacteria living in mutualism (Rhizobium, Frankia), bacterial and fungal organisms living within the plant, colonise the rhizosphere (Figure 7.1). Every time, benefits such as increased plant nutrition for minerals, increased plant forbearance for biotic and abiotic stress, and improved organic substance that regulate plant growth are traded for plant-derived carbon.

7.5 BACTERIAL AND FUNGAL EXTRACELLULAR ENZYMES

Phytoremediation is the process of cleaning up a polluted site using plants. It has been suggested that efficient phytoremediation depends on interactions between plants and microorganisms. Organisms that live within a plant can enhance plant-mediated remediation in polluted terrestrial and aquatic ecosystems as well as the suitability and revamping of related plants in these environments. Furthermore, endophytes containing catabolic genes can become more common in bacterial population in the presence of pollutants probably in contaminant-dependent manner. By introducing a transposon expressing a toluene-discrediting enzyme demonstrations showed how to intentionally enhance this process. When Sandhu et al. (2007), revealed concrete proof of vaporous non-chemical degradation by bacteria present inside the leaf region termed as "phylloremediation" Since then, numerous studies have described how bacteria support this process by boosting plant development or by degrading pollutants through specific metabolic pathways, as well as presented evidence of pollution uptake by leaf surfaces. By using toluene-degrading bacteria, De Kempeneer et al. (2004) reported that the phyllosphere microbiota performs toluene cleanup. Although there is growing evidence that leaf microbes may have an impact on urban pollution, much work needs to be done to determine (i) which plant species are best at degrading pollutants and lowering pathogenic genes? (ii) What microbial strains or plasmids might the urban phyllosphere use to best degrade air pollution? These are some of the questions that need to be answered. The biologically active soil region known as the rhizosphere, which is located around plant roots, serves as a nutrient sink. There is a strong connection going on between the microorganisms and the plant roots here. Rhizobacteria, the word for the microbes that live here and control the bionutrient cycles, break down humus material, and maintain the chemical nature of the soil. Effective root colonisation is confirmed to need the production of the O-antigen of lipopolysaccharide and cellulose, as well as the synthesis of thiamine and biotin, amino acid synthesis, an isoflavonoid inducible efflux pump, and a nine-polar flagellar shape. With the aid of in vivo expression technology (IVET), it was shown that roughly 20 genes are triggered in root-colonising pseudomonads. They create microcolonies, also known as biofilms that surround the roots and are protected by mucoid coating. Rhizoremediation, a method that uses rhizobacteria and mycorrhizae to promote plant growth, is one way to reduce the harmful impacts of trace element that some plants absorb from the environment. The biological removal of organic or inorganic pollutants from soils through microbial activity in the root region is known as rhizoremediation (Bharani and Senthilraja, 2022). Rhizoremediation is the process by which high- and low-molecular-weight exudates released by growing plants (such as proteins, phenolics, and organic acids) stimulate the viability and functionality of the plant growth-promoting rhizobacteria (PGPR), resulting in a more effective transformation/degradation of environmental pollutants. In addition, the microbial activity reduces heavy metal buildup in many plant organs. However, in most cases, heavy metals are poisonous to bacteria that have an impact on soil fertility. As a method of survival, some microbes have developed resistance or avoidance mechanisms that result in changes in metal speciation. The methods include the removal of heavy metals via efflux pumps, biosorption of metals by pigments, polysaccharides, and cell walls, and the development of proteins and peptides that bind to metals, such as metal MTs. As a result, the hazardous heavy metals are absorbed by the microbial biomass. Additionally, certain bacteria use enzymatic action to detoxify heavy metals as Cd, Hg, and Pb. Numerous studies have shown that heavy metals in soil can be effectively removed by rhizoremediation using rhizobacteria that are resistant to heavy metals and a variety of plant types. In metal-contaminated soil, metal-defiant root associated bacteria increase plant mycorrhizal and nodulation efficiency. For improved rhizoremediation efficiency, transgenic methods at the microbial and plant levels are documented in addition to natural bacteria. For instance, the synthesis of the metal-binding peptide (EC20) in the rhizobacterium *Pseudomonas putida* 06909 increased cadmium binding and cell proliferation in sunflower seedlings. The introduction of genetically modified microorganisms meant for rhizoremediation has been shown to change naturally occurring bacteria in the rhizosphere but not the adjacent soil.

The rhizoremediation method is visually appealing, inexpensive, uses solar energy, requires little upkeep, eliminates the need for additional recycling, and maintains soil fertility and ecology. As a result, this tactic is becoming more popular. In addition to repairing and making money, it guarantees human food security and shields people from numerous illnesses. To ensure that rhizoremediation is feasible, large-scale field tests and its evaluation are necessary. Scientists are currently faced with the task of figuring out how to apply this technique to the rehabilitation of nutrient-deficient or metal-contaminated non-agricultural soils. When numerous aspects of remediation techniques are considered, it becomes clear that all of these techniques, in general, only offer a short-term fix for the removal of contaminated lands and do not completely destroy the metals from the contaminated locations. Therefore, it is important to encourage both remedial technology and organic farming practices to stop metal poisoning of agricultural land.

7.6 DEGRADATION OF ORGANIC AND INORGANIC POLLUTANTS BY EXTRACELLULAR ENZYMES

In human bodies, enzymes operate as biological catalysts for metabolic reactions. All proteins, including enzymes, are comprised of sequences of amino acids. By lowering the activation energy of the chemical events, enzymes can either promote or block them. Enzymes have an active site for binding substrates. Enzyme and substrate have a unique interaction that functions like a lock and key. Enzymes come in two varieties: intracellular and extracellular, depending on where they function. Cells synthesise and store intracellular enzymes for use in intracellular biochemical processes. Enzymes that work outside of the cell are secreted. Endoenzymes function within the cell, whereas exoenzymes function from the outer layer of the cell. This is the primary distinction between intracellular and extracellular enzymes.

Recent research is constantly focused on developing novel enzymes or improving existing enzymes by engineering at the gene and protein levels in order to address the growing demand for robust, rapid turnover, conveniently accessible biocatalysts. Microorganism-produced enzymes are thought to be potential biocatalysts for a wide range of processes. Enzymes generated from microbial sources are typically recognised as safe and functional at a wide range of temperature, pH, salinity, or other harsh circumstances. One of the most diverse groups of microorganisms, actinomycetes are known for their adaptability in terms of metabolism.

It has been observed that different Actinomycetes genera produce a large range of prospective industrial enzymes that can be employed in biotechnological applications, and biomedical fields in particular (Hammel, 1997). Species of the phylum Actinomycetota, including gram-positive, filamentous, spore-forming bacteria, are actinomycetes of the genus Streptomyces. Streptomyces members are familiar in preparing a broad array of bioactive natural compounds, including antibiotics, antifungal, and anticancer agents using proteomics and metaproteomics, it is now possible to analyse the synthesis of microbial enzymes due to ongoing developments in sequencing technology and bioinformatics tools (Lehninger et al., 2004). The synthesis of amylases, cellulases, proteases, chitinases, xylanases, and pectinases by actinomycetes has been continuously studied and exploited. Bioremediation-related enzymes such oxido-reductases and hydrolases have had their mechanisms thoroughly researched.

The gram-positive filamentous bacteria known as "actinomycetes" are widely recognised for producing numerous extracellular enzymes. These enzymes are made within the cell but released to the outer layer of the cell where they work. Extracellular enzymes are crucial because they are used extensively in a variety of sectors, including agriculture, paper, textile, detergent, cosmetics, biosensors, medicines, and biorefineries. Wastes can be quickly converted into useful raw materials for direct industrial recycling.

Because these techniques are expensive, take a long time, and use substances with dangerous qualities, traditional chemical and physical treatments on biodegradable waste materials are ineffective for degradation. In place of conventional approaches, biodegradation using microbial extracellular enzymes is effective and affordable. By using enzymatic mechanisms, microorganisms can transform hazardous waste into non-toxic form. The process of biodegradation involves microbes like bacteria, fungus, and yeast that consume degradable waste as an energy source, completely degrading the waste.

7.6.1 Biodegradation of Organic Pollutants: Process Mechanism

Large, three-dimensional organic assemblies and mineral particles that include enzymes that restrict and affect their movement are found in soil.

Popular environmental pollutants that are toxic, genotoxic, mutagenic, and carcinogenic include aromatic hydrocarbons. In these chemicals, enzyme molecules may be adsorbed, immobilised, or stuck, resulting in the production of so-called spontaneously immobilised enzymes (Gianfreda and

Rao, 2004). It has been demonstrated that using microorganisms as a carbon source for microbe growth and production in the bioremedy of petroleum hydrocarbons is quite successful. It has been shown that Pseudomonas species isolated from soil contaminated with petroleum hydrocarbons can break down hydrocarbons. More than just pressure is needed for multiple petroleum hydrocarbon bacterial breakdown to take place. Microbial communities comprising strains from several genres were chosen in petroleum-contaminated soil or water based on metrics showing significant growth in crude oil, specific hydrocarbon compounds, or both. The ability of the strain to break down organic hydrocarbon contamination in soil samples that had been polluted with gasoline, fuel oil, or motor oil was studied. Biological methods were preferred to chemical and physical methods for textile performance upkeep. The main components of biological discoloration methods include plants, fungus, and microorganisms. To use the quantity of microorganisms, textile colours have been decolored.

7.7 REGULATION OF EXTRACELLULAR ENZYME ACTIVITY BY THE ENVIRONMENT

Enzymes may be lost from the producer cell and exposed to inhibitors, be denatured by physical and chemical factors, or be hydrolysed by protons. The substrate concentration is notoriously low and highly variable. Substrate complexation with humic substances, colloidal organic matter, and detritus can make it difficult for an enzyme to associate with its substrate. These conditions make it difficult for an enzyme to link with its substrate. The regulation of bacterial extracellular enzymatic activity is influenced by a complex web of interacting variables, which regulate enzyme expression and affect the kinetics of polymer degradation.

7.7.1 Biochemical Manipulation

Only when simple sources of organic C are insufficient can enzymes produced by microbes be used. Microbes can manufacture enzymes to release certain nutrients from organic matter when they are present in trace levels. When microbes grow on sources of easily utilisable dissolved organic materials, most aquatic microorganisms suppress the development of the majority of ectoenzymes. Ectoenzyme production only begins to resume until the concentration of readily accessible substrates in the water falls below a threshold level. Microorganisms can avoid the unnecessary manufacturing of inducible enzymes by employing the repression method for ectoenzyme synthesis. In aquatic conditions, the accumulation of the hydrolysis end product in the cell or the surrounding environment can also prevent the synthesis of numerous ectoenzymes. Despite the abundance of substrate in the environment, microbes can still create extracellular enzymes. Constitutive enzymes produce low amounts of reaction products that trigger the creation of more enzymes in the presence of a substrate. Enzyme synthesis is repressed and restored to constitutive levels once the product concentration is high enough to meet demand. Sometimes it is impossible to develop a definite response to nutrient availability. For instance, it has been found that adding dissolved inorganic nitrogen to natural and planktonic communities and semi-natural culture systems either results in a reduction in extracellular protease activity or has no effect. In certain situations, environmental regulation may be the primary factor dictating enzyme activity. The importance of organic N sources for bacterial growth has been hypothesised by the apparent positive association between amino peptidase activity and N limits.

7.7.2 Regulation of the Environment

According to a model of how the environment affects extracellular enzyme activity, these interactions between enzyme and substrate at the microenvironmental level—such as inhibition, adsorption, stabilisation, and humification—are what primarily control extracellular enzyme activity, even though environmental factors are still present. Enzyme synthesis is thus closely controlled by environmental variables like temperature at the ecosystem level.

Temperature: Since the affinity of enzyme systems reduces at low temperatures, one significant indirect consequence of temperature is its interference with that affinity. As the temperature increases, extracellular enzymes from arctic isolates and marine sediments become more active, with optimal values occurring well above the ambient environmental temperatures. The substrate affinities of mesophilic and psychrotolerant marine bacteria's membrane-bound transporters decrease in response to low temperatures within their specific temperature range. Studies have shown that extracellular enzymes' temperature sensitivity varies seasonally, which has been explained by the production of various isoenzymes across time, either by various microorganisms or by a single species that can create a variety of isoenzymes. Additionally, there is some proof of biogeographical trends in the temperature sensitivity of enzymes. For instance, numerous studies have found that the optimal temperature for some enzymes produced by bacteria that live in cold environments is exceptionally low.

Salinity: It appears that there is only a weak correlation between salinity and the metabolic activity of bacterioplankton. It has been determined that there are environments where the relationship between Leuamp activity and salinity is positive and environments where there is a negative relationship. Bacterioplankton often exhibits higher levels of glucosidase activity in low salinities, but in higher salinities, bacterioplankton appears to be better adapted to protein degradation. Less is known about how salinity regulates the microbial populations in sediment. Higher salinity sediments have lower extracellular enzymatic activity because osmolites demand more energy to produce than extracellular enzymes do to be released. The inverse tendency, however, has also been noted.

pH: Exoenzymes are directly impacted by the extracellular environment's pH, in contrast to endoenzymes that function in the cell's cytoplasmic buffer, because variations in the environment's hydrogen ion concentration change the ionisation amino acid condition and the active site's three-dimensional structure. The rate of enzyme activity decreases when the pH is off-balance. The pH optimum of extracellular enzymes is not always the same as the ambient pH, just like the enzymatic temperature optimum is not always the same. The pH changes brought on by photosynthesis can also impact the way enzymes function.

Other factors: UVB radiation and trace metals are two biologically significant elements that regulate *in-situ* activity.

7.8 CASE STUDIES

7.8.1 Enzymatic Activity in *Spartina maritima's* Rhizosphere:

Extracellular enzymatic activity (EEA) of different enzymes, including peroxidase, phenol oxidase, b-glucosidase, b-N-acetylglucosaminidase, and acid phosphatase, was examined in sediments colonised by *Spartina maritima* in two salt marshes of the Tagus estuary (Portugal). For a better understanding, the effects of the halophyte on microbial activity in the rhizosphere under varied site conditions, as well as its potential impacts on metal cycling and phytoremediation in salt marshes, were examined. The EEA of acid phosphatase and b-N-acetyl glucosaminidase was substantially higher than that of peroxidase. The fundamental explanation for this was that the two locations' organic matter compositions were different. The EEA of hydrolases was observed to positively correlate with root biomass, suggesting that the halophyte may have an impact on the function of the sediment microbial population. Through microbial interactions, this might have an impact on metal cycling in the rhizosphere (Reboreda & Caçador, 2008).

7.8.2 The Action of Enzymatic Antioxidants in Altering the Chlorpyrifos Stress - Exposed Plants

Plant protection products contribute to the massive environmental problem of water contamination. Chlorpyrifos, as one sort of an organophosphorus pesticide, was used for many years and its high proportion could harm environments. Macrophytes, for example, were among the plants that were subjected to a variety of stresses. In order to live, the macrophytes developed an efficient antioxidant system composed of enzymatic and non-enzymatic antioxidants. The restoration of aquatic habitats that have been harmed by plant protection products in our climate zone can be sped up if autochthonic macrophytes are present. The effects of various chlorpyrifos dosages on the enzyme systems of freshwater mint, needle spikerush, and Canadian waterweed were studied. In order to compare the activities of glutathione S-transferase and guaiacol peroxidase in leaves and roots, both enzyme activities appeared to be rising in plants exposed to harmful substances, according to research. The highest enzyme activity was impacted by the highest chlorpyrifos concentration. The water mint roots responded by exhibiting the highest amount of glutathione S-transferase activity during cultivation in a contaminated environment. Therefore, it was determined that an aquatic plant exposed to the harmful insecticide developed an enzymatic antioxidant system as a protective mechanism that was connected to the contaminant dosage and plant species (Sobiecka et al., 2022).

7.8.3 Various Exoenzyme Activities and their Influence on the Constructed Wetland Systems

Triple macrophytes have been used in the construction of a variety of manmade wetland systems. Within several constructed wetland microcosm units, an evaluation of enzyme activities was done in relation to time, soil depth, and their relationship to the efficiency of removing contaminants. The results of this investigation demonstrated that soil depth, sampling duration, and the kind of pollutants all exhibited a notable effect regarding the activity of enzymes and the effectiveness of pollutants removal. The majority of the created wetland microcosm units showed a strong correlation between phosphatase activity and the removal of soluble reactive phosphorus and total phosphorus. There was a strong positive correlation between the activity of urease and the plants *Typha latifolia*, *Phragmites karka*, and *Pistia stratiotes* in artificial wetland units planted with these plants. The removal of NO_3-N and NO_2-N was found to be favourably and negatively correlated with urease activity in a number of constructed wetland microcosm units. Similar to this, fluoresce during diacetate hydrolysis and BOD reduction have a moderately positive and negative connection. In the majority of the wetland units, there was a negative correlation between BOD removal and microbial biomass carbon. The surface layer of these wetland systems showed notable exoenzyme activity and differed from the lower layer with respect to vertical variation. Significant temporal fluctuations in enzyme activity were seen in constructed wetland systems (Kumar et al., 2020).

7.9 THE WAY FORWARD

Marine researchers are putting a lot of effort into understanding how the marine biota and biogeochemical cycles may be able to provide feedback to climate change since biological activities have a big impact on the ocean's capacity to absorb atmospheric CO_2. Heterotrophic bacteria's extracellular hydrolysis, further disintegration, and mineralisation of organic carbon all affect how air CO_2 is transported to the ocean. Despite the need for rigorous temporal and spatial sampling techniques and the challenge of extrapolating results from microcosm experiments to the natural world, there hasn't been much research on the combined effects of several causes of global change. Because of this, it is now impossible for scientists to foresee how EEA, the marine environment, and the subsequent carbon cycle activities might be impacted by the effects of global warming. Examining the current environments where such changes typically occur would be a potentially reliable technique to understand how microorganisms respond to environmental change on a global scale. Areas with values that are close to the upper and lower limits of the ranges of natural variation for these parameters, such as heterotrophic systems where respiration is noticeably higher than primary production or polar waters with lower calcium carbonate saturation rates, for example, could be studied to determine the effects of increased CO_2 and pH (Orr et al., 2005). Less buffered than the oceans freshwater lakes and estuaries show daily to seasonal pH changes. Additionally, coastal and estuary ecosystems demonstrate rapid regional pH changes.

Since they are UV-exposed environments, extreme elevation Andean lakes, for instance, have been proposed as examples of a larger spectrum of bacterial adaption mechanisms. For example, it has been suggested that Actinobacteria's importance in the microbial community of Tyrolean Alps lakes with varied UV transparency is related to their higher UV resilience (Warnecke et al., 2005). Similar to how bacteria living in surface microlayers are more exposed to sunlight and have increased solar radiation resistance. To accurately forecast how ecosystems will react to changing climatic conditions, it is vital to understand how microbial populations will change in response to various climate change sources. In multifactorial experiments under simulated climate settings, the impacts of atmospheric CO_2, temperatures, and rainfall on the composition of the soil microbes have been examined (Castro et al., 2010). To characterise how different climatic circumstances affect bacterial extracellular enzyme reactions in the aquatic environment, and finally, how those trends of organic matter recycling fluctuate in the context of broader environmental changes, similar experimental approaches should be very helpful. To assess how enzyme group sizes, stability, and activity are expected to change with differences in temperature and precipitation frequency, we urgently need to learn more about how microorganisms regulate enzyme production, secretion, and survival. There is growing evidence that higher atmospheric CO_2 will increase the rate of root exudation, affecting the relative availability of labile C and nutrients, especially N, in soils (Phillips et al., 2011; Kaiser et al., 2010). This could alter the heterotrophic soil respiration, which could lead to an increase in atmospheric CO_2. To estimate changes in the net ecosystem exchange of CO_2, we need further information on how variations in root exudation and plant debris deposition affect microbial enzyme production and cycling. Additionally, it is anticipated that fluctuations in microbial enzymes and root exudate activity may affect primal matter, necessitating immediate research (Schmidt et al., 2011).

7.10 CONCLUSIONS

Since biological processes have a significant impact on the ocean's ability to absorb atmospheric CO_2, marine researchers are working hard to understand how the marine biota and biogeochemical cycles may be able to offer feedback to climate change. Extracellular hydrolysis, further breakdown, and mineralisation of organic carbon by heterotrophic bacteria all have an impact on the transport of atmospheric CO_2 to the ocean. However, little research has been done on the combined effects of several drivers of global change, primarily due to the necessity for rigorous temporal and spatial sampling techniques and the difficulty of extrapolating microcosm experiment findings to the natural world. As a result, the scientific community still lacks the knowledge necessary to forecast how the effects of global warming will affect extracellular enzymatic processes, the aquatic ecosystems, and the subsequent carbon cycle processes. Analyzing current habitats where such changes naturally take place would be a reasonably reliable method to examine how microorganisms react to global environmental change. By examining areas where these parameters have values that are close to the upper and lower boundaries of the ranges of natural variation, such as heterotrophic systems where respiration is significantly higher than primary production or polar waters with lower calcium carbonate saturation rates, for instance, the effects of increased CO_2 and pH could be studied (Orr et al., 2005). Less stabilised than the oceans, freshwater lakes and estuaries show daily to seasonal pH fluctuations. Additionally, over short time intervals, coastal and estuarine habitats exhibit considerable pH regional fluctuation. Extreme altitude since Andean lakes are UV-exposed environments, it has been argued that they serve as examples of a larger spectrum of bacterial adaptive response techniques (Fernández Zenoff et al., 2006). For illustration, it is expected that Actinobacteria's significance in the microbial population of lakes in the Tyrolean Alps with varying UV transparency is connected to their better UV protection (Agogué et al., 2005). Similar to how bacteria living in surface microlayers are more exposed to sunlight and have increased solar radiation resistance (Castro et al., 2010). Making precise forecasts of ecosystem response to shifting climatic scenarios requires an understanding of how microbial populations will evolve in response to various climate change drivers. The influence of atmospheric CO_2, temperature, and precipitation on the makeup of the soil microbial community have been studied in multifactorial studies in simulated climate conditions. To characterise how different climatic circumstances affect the microbial exoenzyme activity in the aquatic ecosystem and, consequently, how the method of organic carbon reprocessing alters in the context of broader ecosystem modification, similar experimental approaches should be very helpful.

REFERENCES

Agogué, H., Joux, F., Obernosterer, I., & Lebaron, P. (2005). Resistance of marine bacterioneuston to solar radiation. *Applied and Environmental Microbiology*, *71*(9), 5282–5289.

Allison, S.D., & Jastrow, J.D. (2006). Activities of extracellular enzymes in physically isolated fractions of restored grassland soils. *Soil Biology and Biochemistry*, *38*, 3245–3256.

Bayer, E.A., Belaich, J.P., Shoham, Y., & Lamed, R. (2004). The cellulosomes: Multienzyme machines for degradation of plant cell wall polysaccharides. *Annual Review of Microbiology 58*, 521–554.

Bayer, E.A., Lamed, R., White, B.A., & Flint, H.J. (2008). From cellulosomes to cellulosomics. *Chemical Record*, *8*, 364–377.

Bharani, A., & Senthilraja, K. (2022). *A Textbook on Environmental Biotechnology* (p. 268). Jodhpur: Scientific Publishers Limited.

Burns, R.G. (1982). Enzyme-activity in soil – Location and a possible role in microbial ecology. *Soil Biology and Biochemistry*, *14*, 423–427.

Castro, H.F., Classen, A.T., Austin, E.E., Norby, R.J., & Schadt, C.W. (2010). Soil microbial community responses to multiple experimental climate change drivers. *Applied and Environmental Microbiology*, *76*(4), 999–1007.

De Kempeneer, L., Sercu, B., Vanbrabant, W., Van Langenhove, H., & Verstraete, W. (2004). Bioaugmentation of the phyllosphere for the removal of toluene from indoor air. *Applied Microbiology and Biotechnology*, *64*(2), 284–288.

Demain, A.L., Newcomb, M., & Wu, J.H.D. (2005). Cellulase, clostridia, and ethanol. *Microbiology and Molecular Biology Reviews*, *69*, 124–154.

Feldman Biobasics. (2006). The science and the issues. 9 February–24 November 2006.

Fernandez-Luqueno, F. et al. (2013). Microbial communities to mitigate contamination of PAHs in soil-possibilities and challenges: A review. *Environmental Science and Pollution Research*, *18*(1), 12–30.

Fetzner, S. (1998). Bacterial dehalogenation. *Applied Microbiology and Biotechnology*, *50*(6), 633–657.

Finzi, A.C, Sinsabaugh, R.L., Long, T.M., & Osgood, M.P. (2006). Microbial community responses to atmospheric carbon dioxide enrichment in a warm-temperate forest. *Ecosystems*, *9*, 215–226.

Flemming, H.C., & Wingender, J. (2002). What biofilms contain – Proteins, polysaccharides, etc? *Chemie in UnsererZeit*, *36*, 30–42.

Gianfreda, L., & Rao, M.A. (2004). Potential of extra cellular enzymes in remediation of polluted soils: A review. *Enzyme and Microbial Technology*, *35*, 339–354.

Gianfreda, L., Rao, M.A., & Violante, A. (1995). Formation and activity of urease-tannate complexes affected by aluminum, iron, and manganese. *Soil Science Society of America Journal*, *59*, 805–810.

Gold, N.D., & Martin, V.J.J. (2007). Global view of the Clostridium thermocellumcellulosome revealed by quantitative proteomic analysis. *Journal of Bacteriology*, *189*, 6787–6795.

Hammel, K.E. (1997). Fungal degradation of lignin. In G. Cadisch & K.E. Giller (eds) *Driven by Nature: Plant Litter Quality and Decomposition* (pp. 33–45). Wallingford: CAB International.

Ho, Y., Porter, J., & McKay, G. (2002). Equilibrium isotherm studies for the sorption of divalent metal ions onto peat: Copper, nickel and lead single component systems. *Water, Air, & Soil Pollution*, *141*, 1–33.

Hope, C.F.A., & Burns, R.G. (1985). The barrier-ring plate technique for studying extracellular enzyme diffusion and microbial-growth in model soil environments. *Journal of General Microbiology*, *131*, 1237–1243.

Joanisse, G.D., Bradley, R.L., Preston, C.M., & Munson, A.D. (2007). Soil enzyme inhibition by condensed litter tannins may drive ecosystem structure and processes: The case of *Kalmia angustifolia*. *New Phytologist*, *175*, 535–546.

Jones, J.P., O'Hare, E.J., & Wong, L.L. (2001). Oxidation of polychlorinated benzenes by genetically engineered CYP101 (cytochrome P450cam). *European Journal of Biochemistry*, *268*(5), 1460–1467.

Kaiser, C., Koranda, M., Kitzler, B., Fuchslueger, L., Schnecker, J., Schweiger, P., & Richter, A. (2010). Belowground carbon allocation by trees drives seasonal patterns of extracellular enzyme activities by altering microbial community composition in a beech forest soil. *New Phytologist*, *187*(3), 843–858.

Kandeler, E. (1990). Characterization of free and adsorbed phosphatases in soils. *Biology and Fertility of Soils*, *9*, 199–202.

Karigar, C.S., & Rao, S. (2011). Role of microbial enzymes in the bioremediation of pollutants: A review. *Enzyme Research*. Article ID 805187, 11 p. doi:10.4061/2011/805187

Kikot, G.E., Hours, R.A., & Alconada, T.M. (2009). Contribution of cell wall degrading enzymes to pathogenesis of Fusariumgraminearum: A review. *Journal of Basic Microbiology*, *49*, 231–241.

Kumar, S., Nand, S., Dubey, D., Pratap, B., & Dutta, V. (2020). Variation in extracellular enzyme activities and their influence on the performance of surface-flow constructed wetland microcosms (CWMs). *Chemosphere*, *251*, 126377.

Lamed, R., Setter, E., & Bayer, E.A. (1983). Characterization of a cellulose-binding, cellulase-containing complex in Clostridium thermocellum. *Journal of Bacteriology*, *156*, 828–836.

Lehninger, L., Nelson, D.L., & Cox, M.M. (2004). *Lehninger's Principles of Biochemistry* (4th edition). New York: W.H. Freeman.

Malik, A. (2004). Metal bioremediation through growing cells. *Environment International*, *30*(2), 261–278.

Matheickal, J., Yu, Q., & Woodburn, G. (1999). Biosorption of cadmium (II) from aqueous solutions by pre-treated biomass of marine alga *Durvillaeapotatorum*. *Water Research*, *33*, 335–342.

Nannipieri, P. (2006). Role of stabilised enzymes in microbial ecology and enzyme extraction from soil with potential applications in soil proteomics. *Nucleic Acids and Proteins in Soil*, *8*, 75–94.

Nannipieri, P., Ceccanti, B., & Bianchi, D. (1988). Characterization of humus phosphatase complexes extracted from soil. *Soil Biology and Biochemistry*, *20*, 683–691.

Nannipieri, P., Ceccanti, B., Cervelli, S., & Sequi, P. (1978). Stability and kinetic properties of humus urease complexes. *Soil Biology and Biochemistry*, *10*, 143–148.

Orr, J.C., Fabry, V.J., Aumont, O., Bopp, L., Doney, S.C., Feely, R.A., & Yool, A. (2005). Anthropogenic ocean acidification over the twenty-first century and its impact on calcifying organisms. *Nature*, *437*(7059), 681–686.

Pflug, W. (1982). Effect of clay-minerals on the activity of polysaccharide cleaving soil enzymes. *Soil Enzymes and Clay-minerals*, *145*(2), 493–502.

Philippot, L., Raaijmakers, J.M., Lemanceau, P., & van Der Putten, W.H. (2013). Going back to the roots: The microbial ecology of the rhizosphere. *Nature Reviews Microbiology*, *11*, 789–799.

Phillips, R.P., Finzi, A.C., & Bernhardt, E.S. (2011). Enhanced root exudation induces microbial feedbacks to N cycling in a pine forest under long-term CO2 fumigation. *Ecology Letters*, *14*(2), 187–194.

Quiquampoix, H., & Burns, R.G. (2007). Interactions between proteins and soil mineral surfaces: Environmental and health consequences. *Elements*, *3*, 401–406.

Quiquampoix, H., Servagent-Noinville, S., & Baron, M.H. (2002). Enzyme adsorption on soil mineral surfaces and consequences for the catalytic activity. In R. Dick & R. Burns (eds) *Enzymes in the Environment: Activity, Ecology, and Applications* (pp. 285–306). New York: Marcel Dekker.

Reboreda, R., & Caçador, I. (2008). Enzymatic activity in the rhizosphere of *Spartina maritima*: Potential contribution for phytoremediation of metals. *Marine Environmental Research*, *65*(1), 77–84.

Romani, A.M., Fund, K., Artigas, J., Schwartz, T., Sabater, S., & Obst, U. (2008). Relevance of polymeric matrix enzymes during biofilm formation. *Microbial Ecology*, *56*, 427–436.

Sandhu, A., Halverson, L.J., & Beattie, G.A. (2007). Bacterial degradation of airborne phenol in the phyllosphere. *Environmental Microbiology*, *9*(2), 383–392.

Schimel, J.P., and Weintraub, M.N. (2003). The implications of exoenzyme activity on microbial carbon and nitrogen limitation in soil: A theoretical model. *Soil Biology and Biochemistry*, *35*, 549–563.

Schmidt, M.W., Torn, M.S., Abiven, S., Dittmar, T., Guggenberger, G., Janssens, I.A., & Trumbore, S.E. (2011). Persistence of soil organic matter as an ecosystem property. *Nature*, *478*(7367), 49–56.

Sobiecka, E., Mroczkowska, M., & Olejnik, T.P. (2022). The enzymatic antioxidants activities changes in water plants tissues exposed to chlorpyrifos stress. *Antioxidants*, *11*(11), 2104.

Tate, R.L. (2002). Microbiology and enzymology of carbon and nitrogen cycling. Enzymes in the environment: Activity, ecology, and applications. In R. Dick and R. Burns (eds) *Enzymes in the Environment: Activity, Ecology, and Applications* (pp. 227–248). New York: Marcel Dekker.

Volesky, B., & Holan, Z. (1995). Biosorption of heavy metals. *Biotechnol Program, 11*, 235–250.
Wariishi, H., Valli, K., & Gold, M.H. (1992). Manganese (II) oxidation by manganese peroxidase from the basidiomycete *Phanerochaetechrysosporium*. Kinetic mechanism and role of chelators. *The Journal of Biological Chemistry, 267*(33), 23688–23695.
Warnecke, F., Sommaruga, R., Sekar, R., Hofer, J.S., & Pernthaler, J. (2005). Abundances, identity, and growth state of actinobacteria in mountain lakes of different UV transparency. *Applied and Environmental Microbiology, 71*(9), 5551–5559.

8 Integration of Nature-based Wastewater Treatment Technologies with Other Available Technologies as a Tertiary Treatment Unit

Mohamed Ali Wahab, Yasmin Cherni, and Gareth Griffiths

8.1 INTRODUCTION

Water is vital for life, but it is also one of the principal channels through which chemical and biological pollutants are disseminated and transmitted. It is a scarce resource that we must use wisely. Water supply is striving to keep up with the world's growing demands, which are impacted by several factors, including urban growth, decreasing water quality and climatic change. Thus, the 21st century's primary issue is the need for pure, clean water to support basic human needs (Pandey et al., 2021). The necessity for wastewater treatment has increased due to the depletion of freshwater supplies, which has shifted attention to water recycling and reuse (Qasem et al., 2021). Wastewater is a complicated effluent composed of biodegradable suspended and dissolved organic particles (Zhang et al., 2021). It contains trace metals, hazardous compounds and agricultural run-off, including fertilisers and pesticides are known to have environmental implications (Zhang et al., 2019). Thus, to be released into the receiving environment according to the regulations, wastewater must be appropriately treated (Vinardell et al., 2020; Xiang et al., 2020). In the context of sustainable development, which fundamentally includes protecting the environment and utilising renewable resources, wastewater reuse is beginning to attract significant interest from industries. Thus, wastewater treatment has become a priority and is defined as transforming into inorganic or stable organic molecules of highly complicated and biodegradable organic substances (Xiang et al., 2020). With clean maintenance and operation of the treatment plant, it attempts to reduce the concentration of pollutants in the wastewater (Wang et al., 2021). Typically, the treatment has three stages: primary, secondary and tertiary (Iervolino et al., 2020). These stages eliminate and reduce most of the total suspended solids. However, secondary treatment is typically insufficient to treat polluted effluent, characterised by expensive operation and higher concentrations of chemical reagents (Kehrein et al., 2020). Hence, integrating nature-based wastewater treatment technologies into wastewater treatment plants (WWTP) has been proposed to significantly remove organic, nutrients and toxic compounds (Boano et al., 2020). So, it can be concluded that countries and industries are trying to attempt alternatives to make one of the most crucial resources for life appropriate and available; nature-based wastewater treatment technologies have become an important focal point.

Integrating ecotechnologies for wastewater treatment provides comprehensive and vital applications in treating various contaminants and those discharged in wastewater from municipal, industrial and domestic activities. In recent decades, several investigations have developed major trends in wastewater treatment involving resource recovery, organic and inorganic matter degradation and nutrient removal and energy saving for environmental sustainability. To recover

DOI: 10.1201/9781003441144-8

resources, some innovative alternatives are being used and introduced for wastewater treatment (Nas et al., 2021). These technologies, such as oxidation ponds, microalgae, constructed wetlands (CW), green walls and solar treatment, with promising prospects and different applications, have been presented and discussed in this chapter. These nature-based wastewater technologies as tertiary units have been proven to be performant and suitable for removing various emerging contaminants in wastewater.

8.2 DIFFERENT TYPES OF WASTEWATER

As reported by the literature, four categories of wastewater are distinguished: (i) rainwater, (ii) agricultural wastewater, (iii) domestic wastewater and (iv) industrial wastewater (Pandey et al., 2021). Wastewater poses a severe issue for humans and the environment (Nas et al., 2021). Therefore, their composition, in turn, relies on their origin and source, increasing their toxicity. For instance, domestic and agricultural wastewater contains nutrient compounds that can lead to eutrophication which induces algae proliferation at the expense of other aquatic species (Boano et al., 2020). In the same vein, wastewater containing a high concentration of trace metals is also a severe source of toxicity for the environment, again for the population.

On the other hand, the industrial fields such as agro-food, textiles and pulp industry are mainly considered the principal consumers of water and highly chemically polluting sector (Singh et al., 2023). For instance, coloured effluents, like those from the textile industry or paper mills, have a significant visual impact due to their colour and are noted by researchers as an indication of the presence of dangerous pollution and toxic compounds (Pandey et al., 2021). Because the effluent contains various contaminants depending on its source, wastewater treatment performance is quite complex. Hence, there are many different types of wastewater to treat, and each has unique characteristics requiring specialised and sustainable treatment alternatives.

8.3 WASTEWATER TREATMENT

8.3.1 Primary Treatment

Wastewater may mechanically damage wastewater treatment equipment; it contains various solid materials, including clothing, shoes, plastics, sticks, rags and grits. Thus, primary wastewater treatment is the most crucial step in determining the overall performance of any WWTP (Arashiro et al., 2019). The preliminary process eliminates material that will either float or instantly settle due to the force of gravity. Furthermore, physical techniques such as screening, comminution and sedimentation are used. Concerning the screening system, long, small metal bars closely spaced together constitute the screens. Generally, screens are classified as fine or coarse, with openings between 0.1 and 6 mm. They minimise floating waste and other large objects from blocking pumps (Rashed et al., 2013). The screens are periodically cleaned manually or mechanically, and the material is immediately disposed of on the plant grounds. Concerning the coarse particles that pass through the screens, a comminutor and a grinder may be applied to grind and shred debris into small pieces. Then, the shredded materials are removed by flotation or sedimentation processes (Zhao et al., 2000). From the screening phase, wastewater moves into the grit chamber. It is defined as a long narrow tank designed to slow down the flow. Thereby grits, like sand, will settle out of the effluent. Grit represents the heavier inert materials like gravels, pebbles, sand, etc., in wastewater which will not decompose in the treatment. Grit induces excessive damage to plant equipment, essentially on pumps. The detention period is long enough to allow the higher specific gravity grit to settle relative to other materials. Also, the wastewater is purified of the grease that floats over. After the grit chamber, Sedimentation tanks remove suspended solids from the sewage that passes via screens and grit chambers (Sylwan and Thorin, 2021). The flow rate decreased to permit gravity's settling of suspended solids known as primary sludge. Using mechanical scrapers, the primary sludge is moved along the tank bottom. It is accumulated in a hopper and discharged using a pump (Crini and Lichtfouse, 2019). To reduce and remove the chemical pollution, the pre-treated effluent must move on to the secondary treatment using the best available alternative before being reused or discharged into the receiving environment.

8.3.2 Secondary Treatment

90% of BOD and 90% of suspended particles that escape primary treatment are reduced by secondary treatment. The removal is often completed by biological processes in which microorganisms use the organic compounds for their growth and enzyme activity. They transformed them into energy, water and carbon dioxide. Eliminating soluble organic matter at the treatment plant contributes to maintaining the receiving environment's dissolved oxygen balance. Previous works reported three basic processes applied as a biological treatment: the trickling filter, the activated sludge process and the rotating biological contactor (Gupta et al. 2012).

8.3.2.1 Trickling Filter

In a tank with a circular bed of stones, including gravel, pebbles and other materials, known as a trickling filter, wastewater is circulated by a rotating arm. The biofilm covering the bed is inhabited by many bacteria that degrade the dissolved organics when oxygen is available. Air circulating upward through the stone spaces provides sufficient oxygen for the metabolic processes. After that, the wastewater and detached biofilm are collected at the bottom of the trickling filter. It then enters the secondary settling chamber, where it is removed. Trickling filters' effectiveness is influenced by several variables, including features of wastewater depth of filtering and hydraulic loading. In addition, numerous investigations showed that different materials, such as plastic, lighter than gravel, for example, and have more surface area for microbial growth, might be used in place of the materials used in trickling filters (Tang et al., 2020).

8.3.2.2 Activated sludge Process

Comparing activated sludge to trickling filters is considered more effective and economical. An aeration tank follows the secondary clarifier in the active sludge treatment system. The aeration tank is regarded as the fundamental unit in this process. Microbial mass from the secondary settling tank is added to the aeration-activated sludge tank as inoculum when the wastewater passes. The mixture is then thoroughly stirred for 6–8 hours while receiving compressed air through porous diffusers at the bottom of the tank to maintain an aerobic condition. Later, wastewater is moved into the clarifier, where bacterial flocs are removed by settling from the effluent. A few quantities of this sludge are taken for inoculum, and the remaining solid is further treated and discharged (Delanka-Pedige et al., 2020). Activated sludge processes ensure maximum contact between the microbes and the effluent compared to trickling filters. These processes are performant and decrease BOD. Typically, their high operating costs are the only disadvantage.

8.3.2.3 Rotating Biological Contactor

The rotating biological contactor (RBC) is another biological secondary treatment method.RBC was constructed of parallel, closely spaced, circular plastic discs (PVC, polyethylene and expanded polystyrene), typically 3–4 m in diameter and 3 mm thick, that were attached to a rotating shaft. The medium rotates slowly, with the flowing wastewater covering approximately 40% of its surface area. The discs move into and out of the wastewater with the biomass attached to their surface. The microorganisms consume the oxygen, and they degrade the waste contained in the effluent. The process efficiency can be evolved by introducing several discs in the series to achieve the treatment goals (Waqas et al., 2021).

8.3.3 Tertiary Treatment

A final or tertiary treatment could also be required to remove the residual pollutants generated during the secondary treatment. The tertiary treatment stages are necessary for wastewater treatment to enhance effluent quality by removing residual solids, nutrients and harmful pollutants. Thus, the tertiary treatment is designed to produce a higher effluent rate than is usually attained after secondary treatment methods. It comprises several alternatives, such as advanced oxidation, activated carbons (AC) and membrane filtration (ultrafiltration, reverse osmosis) (Heidarinejad et al., 2020). This section has presented some of the tertiary processes that are most commonly used.

8.4 NATURE-BASED WASTEWATER TREATMENT TECHNOLOGIES

Natural wastewater treatment techniques have considerably increased potential than conventional treatment methods for reducing hazardous contaminants such nutrients, organic pollutants and trace metals below permitted limits. The top feature of nature-based wastewater treatment technologies is their ease combined with other alternatives. Nature-based wastewater treatment technologies are considered a revolutionary green wastewater treatment approach. Thus, several approaches for wastewater treatment using nature-based wastewater treatment technology are mentioned below.

8.4.1 Oxidation Pond

Waste stabilisation ponds or lagoons, also known as oxidation ponds, are designed to remediate wastewater by the interaction of microorganisms, mainly bacteria and algae, with sunlight. They are applied to treat different types of sewage, even complex industrial wastewater, and they function under other weather conditions. Moreover, they require easy maintenance and straightforward operations (Chen et al., 2022). Due to these factors, the pond offers an attractive method for environmentally friendly wastewater treatment with the following cautions: Operators should monitor the chemical and biological substances in the system to make sure the design requirements are satisfied, and the system is performing to regulatory standards, mainly if the effluent is to be reused (Hasan et al., 2019). According to the literature, aerobic (high-rate), anaerobic, facultative and maturation ponds are the four main categories of oxidation ponds (Butler et al., 2017).

8.4.1.1 Aerobic Ponds

High-rate ponds, also known as aerobic stabilisation ponds, have dissolved oxygen throughout their depth and include bacteria and algae in suspension. Thus, aerobic ponds are excellent for areas where land cost is low due to their high biochemical oxygen demand (BOD) removal capacity. Shallow and aerated ponds are the two different varieties of aerobic ponds (Dinh et al., 2020).

8.4.1.2 Anaerobic Ponds

In general, the anaerobic ponds are found to be between 2 and 5 m deep, with an ideal pH of less than 6.2, temperatures over 15°C, and retention times of between 1 and 1.5 days. These ponds have a 60% BOD removal rate. However, the effectiveness depends on the climate. The sedimentation is the driving force behind the treatment. The worm settles to the bottom of the pond, where microorganisms attach to settling particles to be removed or die due to a lack of food or the presence of a predator. Typically, these ponds are integrated along facultative ponds (Ho et al., 2021).

8.4.1.3 Facultative Ponds (Aerobic-anaerobic Ponds)

Typically, there are three zones in a pond as follows:

- Zone (1): there is a symbiotic interaction between bacteria and algae in an aerobic surface region;
- Zone (2): an area of the bottom that is anaerobic and where anaerobic bacteria actively deteriorate accumulated sediments;
- Zone (3): a zone where bacteria may survive both aerobic and anaerobic environments and thrive.

A facultative pond typically has a depth of 1–2 m and removes BOD at a rate of about 95%. Because these ponds use algae as decomposers, which is related to the photosynthetic processes inside the unit, the treatment duration might range between 2 and 3 weeks (Ho et al., 2021).

8.4.1.4 Maturation Ponds

These ponds ensure the removal of faecal coliforms, pathogens and nutrients. Compared to the other pond types, the depth extent between 1 and 1.15 m makes it more superficial than any pond other than the aerobic. Usually, the methanogenic conditions are maintained in maturation ponds (Ho et al., 2021).

In literature, due to its low maintenance costs and straightforward construction, numerous authors showed that oxidation ponds are a practical treatment technique, particularly in small communities and developing countries. Consequently, reusing treated wastewater, mainly from oxidation ponds, will expand as the global water problem intensifies (Daee et al., 2019).

8.4.2 Constructed Wetlands

The use of CWs to treat various types of wastewater, including sewage from urban, industrial effluents, and agricultural effluents, is a nature-based technology developing rapidly. As the term implies, CWs are made to represent natural wetlands, which are elaborate aquatic ecosystems that are either permanently or periodically saturated in water and play a significant role in the habitat creation for aquatic plants as well as the mitigation of extreme weather events like flood and droughts. Natural wetlands provide various ecosystem services, including the storage, management and purification of water, biodiversity, carbon sequestration, aesthetic value and recreational opportunities (Delle Grazie and Gill, 2022; Xu et al., 2020). The creation of artificial wetlands and their use in wastewater treatment are primarily motivated by the ability of natural wetlands to purify large quantities of water.

In Germany and North America, engineered and maintained wetland systems were created for the first time in the early 1960s (Vymazal, 2022). Since 1970, CWs have grown in popularity as one of the most effective nature-based solutions for treating different types of wastewater, and new CW configurations are constantly being developed to meet various treatment goals, such as the removal of contaminants of emerging concern and the efficient management of water pollution (Ji et al., 2022).

Additionally, CWs have very low energetic inputs and operating and maintenance costs. Most CW configurations have energy-free water flow since it depends mostly on gravity and is passive. The primary treatment processes are natural ones provided by plants, filter medium or substrate, and microorganisms, providing low-cost CWs. Therefore, the development of numerous CW configurations designed to treat various pollutants (organic, heavy metal, nitrogen, etc.) that can be applied to treat wastewater in both urban and rural regions have been inspired by the economic and ecological benefits of CWs and their implications for improving wastewater treatment sustainability (Kraiem et al., 2019).

Artificial wetlands can use the natural processes involving wetland vegetation, soils and their associated microbial assemblages to effectively treat wastewater because they are designed to be less complex than natural wetlands. CWs are more controlled than natural wetlands and offer greater

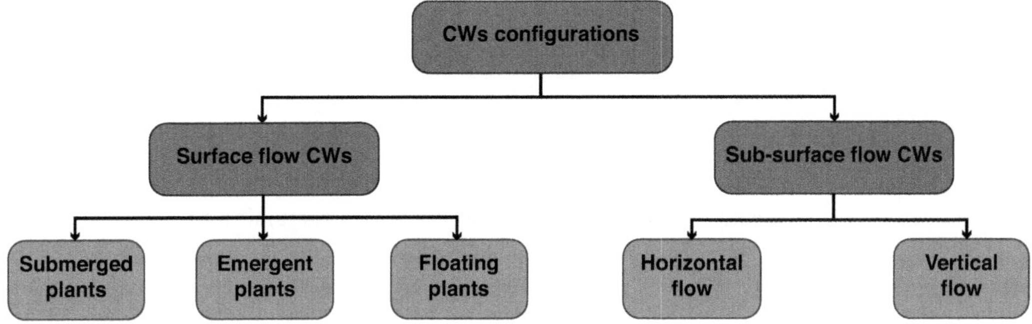

FIGURE 8.1 Classification of the main CW configurations.

flexibility in size, hydraulic conditions, substrate material composition and species of macrophytes planted there. CWs are divided into sub-surface flow CWs, and surface flow CWs based on the water flow pattern through the CW system (Figure 8.1).

8.4.2.1 Surface Flow CWs Systems

Surface flow (SF) CWs are the most basic CW arrangement and resemble natural wetlands in their features. The water surface is exposed to the atmosphere and shallow ponded depths (approximately 0.3 m), where treatment is primarily accomplished within the water column, are what identify SF CWs (Parde et al., 2021). SF CWs can be separated from wastewater pond systems by including macrophyte plants. Free-floating macrophyte systems, submerged macrophyte systems and emergent macrophyte systems are the categories of SF CWs based on the type of macrophytes. Large plants or floating leaves with well-developed submerged roots, like water hyacinth, to tiny surface-floating plants with few or no roots, like duckweed, are examples of free-floating macrophytes. The density and root system of free-floating macrophytes significantly impacts the biological removal of organic matter and nutrients because roots could provide a large surface area for attached microorganisms, increasing the potential for the decomposition of organic matter. Plant density regulates O_2 diffusion into the water, promoting nitrate denitrification and restricting algae growth by reducing sunlight passage. In the case of submerged (i.e. *Potamogeton pectinatus*, *Elodea canadansis*)and emergent (*Phragmites australis*, *Typha* sp.,) macrophyte, the CWs systems are more aerated, allowing for higher total suspended solid (TSS) removal as well as higher nitrification and removal biodegradation rates. However, the effectiveness of these removal rates is greatly influenced by the temperature and is higher in warm climates.

8.4.2.2 Sub-surface Flow CWs Systems

The substrate media is the most crucial element of sub-surface flow (SSF) CWs because it allows water flow and most biological and chemical processes. Indeed, SSF CWs are divided into Horizontal SSF (HSSF) and Vertical SSF categories based on the direction of the water flow (VSSF). The filter saturation rate and the type of biological activities (aerobic, anoxic, or anaerobic) that occur to remove pollutants are determined by the flow direction and the amount of water in the substrate. The biological processes occur in anoxic and anaerobic conditions, and HSSF is saturated. Therefore, macrophyte species play a significant part in ensuring the system's oxygenation through their root systems.

VSSF are unsaturated, in contrast to HSSF, and the filter is more aerated, allowing the growth of aerobic microorganisms involved in the aerobic removal of organic matter and the nitrification process to convert ammonia into nitrite. The adsorption of various contaminants, such as ammonium, phosphate and some heavy metals, as well as the filtration of suspended solids, depends heavily on the substrate media and the biological process. As a result, the CWs employed a wide variety of substrate materials. The macrophytes in VSSF prevent the medium from being blocked.

8.4.3 Green Walls and Roofs

Green infrastructures, such as green walls and roofs, can be used as an environmental system to reduce wastewater problems to achieve wastewater management and the main objective. Green roofs on building rooftops are an ancient practice (Pradhan et al., 2019). People built green roofs in the past to provide isolation and to mitigate the adverse effects of urbanisation. Any system that enables the growth of distinctive forms of flora on top of buildings is referred to as a green roof, often an eco-roof or a vegetative roof. It includes a set of layers containing the vegetation, substrate, filter and drainage layer that protect the support and enhance system efficiency. Typically, these solutions are applied over waterproofed roofs with an interlayer and a root barrier (Andric et al., 2020).

8.4.3.1 Green Roofs Classification

Based on substrate depth and applied plants, there are three green roofs: intensive, semi-intensive and extensive. A more significant substrate layer (around 15–40 cm) on intensive green roofs allows for integrating a wide range of plants. However, they involve regular maintenance and irrigation.

Semi-intensive solutions have a thicker base (12–25 cm) than extensive ones but require less irrigation and upkeep than intensive ones. All systems that permit greening a vertical surface of a structure (such as façades, walls, blind walls, partition walls, etc.) with a variety of plant species are referred to as "green walls," sometimes known as "vertical greening systems".

The aesthetics of inaccessible roofs can be improved with extensive green roofs and lightweight systems that can be put in existing flat or sloped roofs up to 30°. Vast solutions can limit the variety of plants due to their thickness (6–20 cm). According to the literature, Numerous studies have shown the value of using succulent plants, such as Sedum, because of their short roots, compatibility with scarce water resources and resistance to sun radiation. Nevertheless, other plant kinds, including mosses, grasses and wildflowers, can be used in vast systems since they are adapted to the local climate conditions and match the properties of the system (Andric et al., 2020).

8.4.3.2 Green Walls Categories

Green façades and living walls are the two primary divisions of green walls. Climbing plants that grow directly up the wall (much like using aerial roots) or have an indirect support system are frequently used to create green façades (e.g., wire). They have a small variety of plants and slow surface spread.

The more consistent growth of vegetation throughout the surface and the application of various plant species are made possible by living walls (whether continuous or modular). It may need nutrient supply and frequent irrigation (Semeraro et al., 2019).

To sum up, the effective installation of green roofs and green walls and the choice of the most suitable system depend essentially on building characteristics and climate conditions. Several researchers have studied the benefits of green roofs and green walls and the estimation of life cycle costs. They proved that green roofs are the most effective methods for enhancing water quality. Green roofs' vegetation layers and substrate are crucial for reducing runoff and adsorbing various pollutants. The substrate's absorption of heavy metals and contaminants from the effluent improves the water quality. However, green roofs' main drawbacks are their high cost and additional care requirements based on the vegetation type and irrigation requirements (Semeraro et al., 2019).

8.4.4 Algae cultivation Systems

Algae require a light source, an adequate temperature range, nutrients, carbon dioxide and water access to develop effectively. It depends on the species and how algae are grown. Most algae produce autotrophically, utilising light for photosynthesis and fixing dissolved CO_2 into organic molecules after 3-phosphoglycerate is formed by the activity of ribulose-5 phosphate carboxylase (Gross et al., 2015). In contrast, many algae can develop in low or no light conditions using an external carbon source, such as acetate or glucose, to support their growth, similar to animal cells. This process is known as heterotrophic growth (Kumar et al., 2021). Mixotrophic development involves simultaneously using autotrophic and heterotrophic methods to provide light and an external carbon supply. In combination with exposure to sunlight, Tris-acetate-phosphate (TAP) medium is frequently used to grow Chlamydomonas reinhardtii, for instance (Slade and Bauen, 2013). Traditional classifications of algae include green algae (Chlorophyta), brown algae (Phaeophyta), red algae (Rhodophyta) and others based on their pigment levels. However, although having minimal taxonomic significance, the terms "microalgae" and "macroalgae" are helpful concepts for classifying this varied group of organisms into functional groups. Microalgae (usually made up of a single or cluster of cells) have drawn significant attention because they may produce high-value products, including proteins, pigments and polyunsaturated fatty acids (Ubando et al., 2016). Open raceway ponds with stirrer paddles to promote circulation or enclosed photobioreactors constructed of polyethylene or tubular constructions are often used for more controlled microalgae development (Zhang et al., 2021). In general, the operating costs of photobioreactors are higher than those of open raceway ponds. The net energy ratio (NER) of biomass products is a valuable assessment technique defined as the sum of the energy consumed for cultivation, harvesting and drying divided by the energy content of the dry biomass (Low et al., 2021). A process with NER values less than 1 produces more energy than it uses. According to this method, raceway ponds often have NER1, while photobioreactors typically have >1, favouring the former as the preferred method. Open ponds, however, might not be practical, depending on the region. For instance, photobioreactors have been utilised in Iceland to produce polyunsaturated fatty acids. The cost of the product vs the cost of production will ultimately determine the decision. In the northern hemispheres, photobioreactors may be a more practical way to grow high-value products with values >£1,000 per Kg because prohibitive winter temperatures and light reduction would not negatively impact the production. Natural sunlight is favoured because it is cheaper and shouldn't be used more than 3,000 hours per year. This favours locations where plants can be grown between 30° north and 30° south of the equator, where cultivation temperatures are typically between 18° and 25°C (Elisabeth et al., 2021). Also, nutrient availability is essential. According to estimates, the EU would need 25 million tonnes of nitrogen and 4 million tonnes of phosphorus annually to replace fossil fuel with algal biomass. This is more than twice as much fertiliser production capacity as the EU. The base cost of production in an idealised raceway pond is predicted to be between 1.6 and 1.8 euros per kilogram (with CO_2 accounting for around 50% of the costs), with expected costs falling to between 0.3 and 0.4 euros per kilogram as a result of using cheaper sources of input (Qu et al., 2021). Nutrients can be found in lower-value sources like wastewater, such as poultry litter or milk waste, which can significantly lower algal culture's water and fertiliser requirements. Recently, their integration for biofuel production has been vigorously promoted to reduce costs further (Kim et al., 2021). Numerous algae have special equipment media modifications. For instance, silica must be added to the growth medium for diatoms since they have an exterior silica "shell" (Qu et al., 2021). Dunaliella and other algae that thrive in brackish or incredibly salty environments will need more NaCl. In addition to the macronutrients (N, P and K), various micronutrients may also need to be added for optimum growth. One such solution is Hutner's trace elements solution, which comprises H_3BO_3, $ZnSO_4$ ($7H_2O$), $MnCl_2$, $CoCl_2$, $FeSO_4$, $CuSO_4$ and $(NH_4)_6$ $(Mo_7O_2)_4$.

8.4.5 Solar Photochemical Processes

Photochemical processes for wastewater treatment have received significant interest as sustainable alternatives using a free, renewable and clean light source, like, as natural solar radiation (Tang et al., 2020). The most common and conventional wastewater treatment and disinfection technologies are chlorination, UV-C radiation, ozonation and filtration. Although AOPs have frequently been researched as prospective and promising wastewater alternatives in recent decades. Utilising AOPs is mainly influenced by their strong capacity to generate hydroxyl radicals (HO). However, their application faces its biggest obstacle due to the high running costs caused by the UV lamps' high energy consumption. Solar AOPs become a cost-effective and appealing alternative when replacing UV lights with natural solar radiation is emphasised. Thus, homogenous photocatalysis (photo-Fenton process) and heterogeneous photocatalysis using TiO_2 are the main explored alternatives, involving other study developments like doped semiconductors to enhance their catalytic activity for the treatment and purification of wastewater. However, these alternatives present a significant issue for their action, which require the addition of expensive products, the use of a pre-or post-treatment, or in some instances, both, which alter the physicochemical compositions of effluent (da Silva Brito et al., 2019; Berruti et al., 2022). In this regard, various oxidative substances have been investigated as prospective methods to enhance the sun photon action for wastewater treatment, including hydrogen peroxide, persulfate and peracetic acid.

8.4.5.1 Solar/Persulfate Process

Persulfate (PS) is a powerful oxidant (2.1 V) that can use two separate activation mechanisms to produce two SO4-• in water. like UV radiation, heat and transition metals. In the so-called sulphate radicals-based AOPs, PS has been investigated as a substitute for H_2O_2 due to its high solubility, stability, non-toxicity and low cost. Several experimental and theoretical studies have reported PS's effectiveness and efficiency in removing organic contaminants from wastewater. In addition, based on the strong oxidation capacity of the generated SO4 •- (2.5–3.1 V), its use has been extensive for microorganism disinfection and longer half-life compared to HO•. As reported by many researchers, the solar/PS process is a tremendously eco-friendly and cost-effective alternative for wastewater treatment applications. They successfully demonstrated the inactivation of *E. coli* and E. faecalis in effluent at various PS concentrations. They described the mechanism as the ability of the solar to activate PS, resulting in the formation of SO4•- radicals through the homolytic cleavage of the O-O link (Berruti et al., 2022). Thus, the effectiveness of the treatment is induced by the produced radicals. Furthermore, it's feasible that the effectiveness of the solar/PS process would depend simply on the interaction between the two variables and perhaps on the little thermolytic activation that results from the rise in water temperature during sun exposure. Therefore, additional research is necessary to comprehend the variables that control the fundamental process and evaluate treatment efficacy in more complex wastewater such as landfill leachate.

8.4.5.2 Solar/Peroxymonosulfate Process

Peroxymonosulfate (HSO_5^-, PMS) is the active component of triple potassium salt and a potent oxidising agent (1.8 V). Due to its effectiveness as a moderate oxidative agent that can oxidise various inorganic and organic materials, its uses in wastewater treatment have garnered considerable attention. In addition, the effectiveness of PMS in inhibiting prion proteins and degrading amino acids has been documented in numerous investigations. They established that its electrophilic nature and selective interaction with organic compounds having electron-rich moieties give it its oxidative capabilities. Today, Although the precise mechanisms governing the cell-death have not yet been determined, numerous studies focused on the combination of PMS and solar radiation for bacterial inactivation have been examined. The efficacy of PMS at very low doses for concurrent decontamination and disinfection of wastewater. Three microbiological pathogens (*E. coli*, *E. faecalis* and *P.aeruginosa*) were removed more effectively under natural solar radiation. Three mechanisms have been proposed to explain the synergistic effect primarily seen in microbial pathogens: (i) intracellular ROS imbalance caused by light inactivating key regulator enzymes (catalase, alkylhydroperoxide reductase, superoxide dismutases, hydroperoxidases and glutathione reductase); (ii) direct oxidation of cell wall components by PMS; and (iii) potential generation of SO_4-inside the cell after HSO_5^- or SO_4^{2-} diffusion and their reaction with transition metals naturally (Berruti et al., 2022).

Although additional research is required to thoroughly understand its mechanisms and capacity to inactivate various pathogens and wastewater degradation effectively, solar/PMS for water disinfection appears to be a promising technique.

8.5 WASTEWATER TREATMENT CONCEPTS INTEGRATING NATURE-BASED TECHNOLOGIES AS TERTIARY TREATMENT UNIT

8.5.1 Integrating Nature-based Technologies for Nutrient Removal

Nutrients in secondary wastewater are the main reason for many environmental issues, such as eutrophication and groundwater pollution with nitrate (Jaouadi et al., 2014; Wahab et al., 2010). For living things to grow, nitrogen and phosphorus must be integrated into various macromolecules, principally nucleic acids and proteins. However, excessive amounts in water systems can lead to a range of environmental problems. This has been particularly exacerbated in recent years, where arable crop soil fertilisers have run off into aquasystems leading to significant increases in their levels. To satisfy the growing demand for proteins, swine farms have multiplied exponentially worldwide, producing swine wastewater enriched in ammonia (Nagarajan et al., 2020). Because of the high ammonia content and chemical/biological oxygen requirement of swine wastewater, conventional wastewater treatment technologies are limited. Among the nature-based technologies, algae culture removes nutrients most effectively. The efficacy of algae-based remediation depends on their capacity to absorb and store nutrients in the biomass. The relationship between biomass productivity and removal efficiency (Nguyen et al., 2022). The average time intervals over which the nutrients are retained inside the algae is relatively long, >10 days, compared to a few hours in bacteria-based processes.

Nutrient uptake is linked to light levels (to support photosynthesis) and CO_2, the substrate for photosynthetic fixation. The properties of wastewater, light intensity, light-dark cycle, C/N and N/P ratio, CO_2 supply and cultivation method are only a few of the variables that could substantially impact how well microalgae remove nutrients and produce biomass (Li et al., 2019). The C/N and N/P ratios vary significantly with wastewater sources; however, their values are usually not compatible with optimal microalgal growth rates (Woertz et al., 2009). Biomass productivity generally decreases when the external N/P ratio increases. An N/P ratio higher than optimal could result in phosphorus restriction, whereas a below-optimal ratio could result in nitrogen limitation (Olguín, 2012). The possibility of utilising the interaction between microalgae and bacteria has been proposed to attenuate bioremediation (Fallahi et al., 2021). Oxygen produced by microalgae benefits nitrifying bacteria when the nitrogen source is ammonium. When nitrate is used as the nitrogen source, denitrifying bacteria may be inhibited by oxygen. While there is potential in combining bacterial and algal in consortia, significant challenges remain in practice since there is a complex interaction between the two sets of organisms. Algal-based treatment units are increasingly applied as a tertiary treatment to remove nutrients (ammonia and phosphate) and soluble organic matter from secondary-treated wastewater. Ansari et al. (2021) employed microalgae culture in outdoor pilot-scale pools run under natural environmental conditions to recover nutrients from secondary treated effluents. High removal efficiencies were obtained for nitrate, ammonia and phosphorus with 83.20%, 100% and 93.50%, respectively. In addition to high nutrient removal efficiencies, with lipid, protein and carbohydrate contents of 25.60%, 29.00% and 18.40% w/w dry cell weight basis, respectively, a high algal biomass concentration (0.79 g/L) was formed. The algal-based wastewater treatment project's economic calculation revealed a sufficient initial investment, reasonable operating costs and profits, with a payback period of approximately 14 years (Ansari et al., 2021). Boelee et al. (2011) investigated the capacity of microalgal biofilms as a tertiary treatment for the effluent of municipal wastewater treatment plants, and they found that the microalgal biofilm's maximum uptake capacity was reached at loading rates of 1.0 $g/m^2/day$ nitrogen and 0.13 $g/m^2/day$ phosphorus. Other researchers have shown that merging microalgae culture and membrane photobioreactor under actual environmental conditions can efficiently remove nutrients from the real wastewater treatment facility (Segredo-Morales et al., 2022; Solmaz and Işık, 2019). Recently, a novel sustainable technique known as microalgal-bacterial membrane photobioreactors (MPBRs) has been developed for the enhanced treatment of household secondary effluents, achieving significant nutrient removal (>95% and >70%, for nitrogen and phosphorus, respectively) (Sheng et al., 2017) and considerable biomass productivity (50–167 mg/L/d) (Praveen et al., 2018). Using natural microalgae consortium from a eutrophic lagoon for the tertiary treatment was also efficient in removing nutrients with 14 mg TN/L/d and 2 mg TP/L/d uptake rates in a photobioreactor (Segredo-Morales et al., 2022). In addition to their application for the treatment of secondary-treated wastewater, algal-based treatment units were widely applied to remove nutrients from anaerobic digestion effluents. Indeed, anaerobic digestion is known for its efficiency in converting organic matter into biogas (Wahab et al., 2016, 2014); however, the effluent generated is highly concentrated in nutrients. Wahab et al., (2022) revealed that 83% of ammonia and 85% of phosphorus had been removed from algal digestate. *Scenedesmus* sp. completely eliminated phosphate and ammonia from secondary effluent and microalgae digestate (Arias et al., 2018). Nevertheless, most studies have indicated the inhibition of microalgae growth when the ammonia concentration in the effluent exceeded 400 mg/L, especially in the lack of a high COD concentration (Wahab et al., 2022).

Green wall system has also demonstrated high nutrient removal from brewery wastewater with a 94% of reduction of total nitrogen. Before discharge to a municipal sewer system or an onsite treatment system, this green wall system has the potential to offer appropriate treatment of brewery effluent (Wolcott et al., 2022). It was discovered that the use of decorative species in green wall systems, such as *Canna lilies*, *Lonicera japonica* and ornamental grapevine, was crucial for the removal of high levels of nitrogen (>80%) and phosphorus (13%–99%) from greywater at the household scale. Greywater removal from green roofs planted with *Atriplex halimus* and stocked with perlite was 87% and 55% higher, respectively (Thomaidi et al., 2022). Green roof systems can accumulate 2.9 ± 1.1 g N/m^2/yr with no indication of levelling off even after 20 years (Mitchell et al., 2021).

Integrating CW as a tertiary unit for treating secondary wastewater has also demonstrated high nutrient removal efficiencies. To boost biodiversity and lower nitrogen (N) and phosphorus (P) inputs to Swedish waters, more than 13,000 ha of CWs have been put into place since 1990 (Djodjic et al., 2022). High nitrogen and phosphorus removal were achieved in mine waste-based wetlands (Wang et al., 2022). Nutrient removal efficiency in CWs depends mainly on the nature of the substrate media. Wu et al. (2022) investigated the tertiary treatment of municipal wastewater with CWs using different media substrates and revealed that the incorporation of biochar enhanced ammonia removal mainly by promoting the enrichment of nitrobacteria while the incorporation of zero-valent iron promoted nitrate removal by further enhancing denitrification and anammox. On the other hand, the saturation of the CWs seems to play an essential role in improving nitrogen removal from secondary treated wastewater through the development of anammox bacteria (Kraiem et al., 2019). Furthermore, the efficacy of the CWs to remove nutrients was greatly enhanced by adding artificial aeration, external carbon sources, a variety of plant species and substrates and bioaugmentation. (Kamilya et al., 2022).

8.5.2 Integrating Nature-based Technologies for Heavy Metal Removal

Heavy metal (HM) contamination of aquatic systems can have profound negative impacts on human health, flora, and fauna in affected habitats. This contamination arises from a combination of anthropogenic and natural sources. These include mining, industrial production (oil refining, smelting, pesticide production, chemical industry), traffic, sewage, industrial effluents and natural weathering of rocks.

Although many clean-up technologies, such as electrolytic technologies, ion exchange, precipitation, chemical extraction, hydrolysis, polymer micro-encapsulation and leaching, have been developed to sequester HMs, the majority are expensive and have low efficiency. Algae provide a different, environmentally responsible, and sustainable method for remediating HMs (Song et al., 2022). HM remediation mainly uses Chlorophyta from the genera Chlamydomonas, Chlorella and Scenedesmus. HM absorption can be influenced by several variables, including metal content, accessible biomass, pH, temperature, cations, anions and the organism's metabolic stage (Priya et al., 2022).

A practical and environmentally acceptable method for cleaning aquatic environments has been discovered: studying microalgae. In addition, microalgae are capable of efficiently removing N (90%–98.4%), P (66%–98%), Pb (75%–100%), Zn (15.6%–99.7%), Cr (52.54%–96%), Hg (77%–97%), Cu (45%–98%) and Cd (2%–93.06%) from contaminated aquatic environments (Singh et al., 2021). Microalgae can adsorb, accumulate, metabolise, or neutralise toxic substances into safer levels to remove heavy metals from wastewater treatment plants or polluted areas.

Tetradesmus obliquus, *Chlorella sorokiniana*, *Chlorella vulgaris*, *Arthrospira platensis* and *Arthrospira maxima* were tested for their tolerance to steel hot-rolling wastewater. The results revealed a reduction of hydrocarbons and iron of 75% and 97.9%, respectively, indicating that microalgae might be used as a novel and environmentally sustainable bioremediation tool for steel-industry wastewater (Blanco-Vieites et al., 2022). The microalgae demonstrated strong abilities to remove heavy metals from textile effluent, including Pb, Al, V, Cu and Se (Oyebamiji et al., 2019).

Desmodesmus sp. MAS1 and *Heterochlorella* sp. MAS3 were two acid-tolerant microalgae with the potential to remove heavy metals at pH 3.5. Results showed 40%–80% and 40%–60% removal of Fe and Mn, respectively, demonstrating the tolerance capacity of some microalgae species to treat metal-rich acid mine drainage (Abinandan et al., 2019).

Green microalgae biofilms were also used to remove heavy metals from mine effluent, and the results revealed that Fe, Zn and Cd had the highest removal rates of 85%, 95% and 99%, respectively. In comparison, copper and aluminium both had 100% removal rates (Makhanya et al., 2021). Microalgae may be a more effective wastewater treatment method than the current methods, and the technology may be utilised to remove heavy metals from contaminated wastewater.

The removal of heavy metals combined with biodiesel production by microalgae in a cost-effective way is a promising approach that has been proposed by (Hamed et al., 2022), who have revealed high removal efficiencies of Cu^{2+} (59.4, 98.1%) and Zn^{2+} (72.4%, 98.2%) in *C. sorokiniana* and *S. acuminatus*, respectively, indicating that microalgae are excellent remediators for industrial drainage and future sustainable algal-biofuels platform. The economic assessment of daily 1000-tonne biodiesel production was viable, with a return on high investment (16.4%) and a payback period of 5 years. A revolving algal biofilm (RAB) reactor's viability for the removal of HM was tested, and the system showed outstanding Ni removal performance with >90% Ni attributable to the high levels of EPS in RAB biofilm (Zhou et al., 2021).

Also, it was suggested that a fungi-microalgae symbiotic system would be effective in ensuring high Cd(II) elimination (98.89%) and microalgae harvest (Wang et al., 2021). This study showed that the essential EPS components, extracellular polysaccharides and extracellular proteins, were crucial in the symbiotic system's ability to handle heavy metal stress and resist its toxicity.

Using natural and synthetic algae-based microbiomes for heavy metal removal and recovery from wastewater has recently been proposed as an alternative to genetically modified organisms with inherent ecological consequences. By decreasing retention times and recovering HM from the microbial biomass, these microbiomes may boost effectiveness and further lower capital and operating costs of HM removal while promoting a closed-loop, circular, and sustainable economy (Greeshma et al., 2022).

The gradual use of omics-based techniques has enlightened our understanding of the heavy metal detoxification mechanism of microalgae by exposing some regulatory genes of microalgae engaged in cellular functions such as energy metabolism, photosynthesis, metal transport, antioxidant machinery, signalling, etc., under heavy metal stress (Tripathi and Poluri, 2021).

There has been extensive research into the use of CW to remove heavy metals from industrial effluent. The main processes for transforming heavy metals were adsorption and plant uptake, according to a dynamic simulation model that evaluated the transport and fate of heavy metals in vertical flow constructed wetland systems. The forcing functions considered were wastewater volume, temperature, heavy metals concentration, contact time, flow rate and adsorbent media (Mohammed and Babatunde, 2017).

The CW's performance was effective enough to remove heavy metals from the industrial effluent that was given to it, especially Cd, Fe and Cu (Khan et al., 2009). Additionally, *Typha Latifolia*-planted CW in Canada has shown a maximum selenium removal rate of 54.13% from mine effluent (Etteieb et al., 2021). HSSF-CW was also suggested as a successful passive bioremediation technique for highly acidic coal mine drainage from Assam, India. It achieved high average metal removal efficiency for Fe (73%), Al (79%), Zn (98%), Co (95%), Ni (99%) and Cr (100%) (Singh and Chakraborty, 2022). Using an economically rich substrate (cow manure and bamboo chips) planted with common cattail, the same researcher also assessed HSSF-CW for the remediation of acid mine drainage, indicating metal removal efficiency in the following order: Except manganese and cobalt, all heavy metals have poor bioaccumulation and translocation in common cattail: chromium (99.7%) > nickel (97.8%) > cobalt (93.7%) > iron (91.6%) >aluminium (59.7%) (Singh and Chakraborty, 2020).

The co-treatment of high-strength acid mine drainage and domestic wastewater using a multistage CW (mixing and sedimentation pond, vertical-flow constructed wetland and surface-flow constructed wetland) that was able to operate for 270 days revealed that this system was highly effective in removing dissolved metals (89.4% of Mn and 99% of Fe, Zn, Cd and Cu). The diverse microbe (such as sulphate-reducing bacteria, nitrifying bacteria and denitrifying bacteria) assisted in effectively remediating a situation where wetland plants absorbed only a small quantity of metal (Wang et al., 2021).

With a 5d-hydraulic retention period, CW combined with a micro-electric field (CW-MEF) has recently emerged as an innovative and effective water treatment technology to remove HM. Maximum removal efficiencies for Cu, Zn, Cd, Co, Ni and Pb were 95.6, 80.1, 74.0, 67.1, 69.8 and 99.6%, respectively (Si et al., 2019).

Constructed wetland-microbial fuel cells (CW-MFC) come in various designs, including up-flow, down-flow and hybrid up-flow/down-flow. CW-MFCs demonstrated outstanding removal performance for (Cu^{2+} and Zn^{2+}), resulting in concentrations below the first discharge standard for heavy metals (Cu and Zn) in mariculture effluent (Liu et al., 2022).

Various studies examining the removal of HM from wastewater show that CW medium retains >90% of the total removed metal mass (Chang et al., 2022; Singh and Chakraborty, 2021; Wdowczyk et al., 2022). According to Chang et al. (2022), the cooperation of organic-degrading bacteria primarily composed of Cellulomonas, Clostridium, Bacteroides and SRB consortia provides a mixture of organic solid waste and biochar, a preferable filler candidate for CW systems treating acid mine drainage. Wdowczyk et al. (2022) reported that the best reductions of HM were obtained on zeolite substrates in the case of municipal leachate landfills. Additionally, the nitrate-reduction-inhibiting effects of fluorine and heavy metals in wastewater were suggested to be reduced by using hydroxyapatite substrate. This research also sheds new light on the microbiota's structural and functional responses to various substrates in CWs (Wang et al., 2023). Recycling clamshells as a substrate in CW to remove heavy metals from acid mine drainage increased the length of the cattail root. They increased the population of sulphate-reducing bacteria, which led to the efficient removal of Zn, Cd, Cu, Pb and Mn (Nguyen et al., 2022). On the other hand, several researches have also shown the impact of vegetation on eliminating heavy metals in a built wetland. The excellent tolerance of *T. domingensis* to treat wastewater from an Argentine tool industry with high pH, conductivity and concentrations of heavy metals were established by Maine et al. (2009). Additionally, *E. crassipes* was chosen as the best free-floating macrophytes for removing heavy metals from wastewater (Rai, 2019). Another study assessing the amount of heavy metals sequestered in the aboveground biomass of *Phragmites australis* found that the amount of heavy metals accumulated in the plant biomass (aboveground standing stock) often only represent a small fraction of the annual inflow load. Still, in some studies, this fraction is relatively high, especially for zinc (up to 59%), more infrequently for cadmium (55%) and chromium (38%) (Vymazal and Březinová, 2016). Additionally, heavy metal deposition in stems decreases from the base to the top while increasing oppositely in leaves (Březinová and Vymazal, 2022). Without endangering the health of the plants, a catalytic combination of substrates, chelators and plants enabled the simultaneous removal of many metals from wastewater (Batool and Saleh, 2020).

Metals deposited in sediments were generally stable bound and not bioavailable, according to an examination into the removal, distribution and retention of metals in a created wetland over 20 years (Knox et al., 2021).

Additionally, green roofs are considered practical techniques for removing HM from wastewater. The substrate improved the effluent quality by absorbing the pollutants and trace metals in the wastewater (Liu et al., 2021). In this frame, to improve the wastewater quality, a large green roof was examined by Berndtsson et al. in 2009. By reducing trace metals like Cu, Zn, Pb and Cd, they were able to demonstrate the significance of green roofs on wastewater quality performance. The results showed that the extensive roof systems eliminated 97% Cu, 96% Zn, 99% Pb and 92% Cd over the summer (Berndtsson et al., 2009). As a result, green roofs offer a promising alternative to improve wastewater quality in urban areas. However, the trace metals in green roof runoff primarily rely on the age and kind of substrate. These elements are essential to improve wastewater quality.

8.5.3 Integrating Nature-based Technologies for Emerging Organic Pollutants Removal

A global threat to water quality and the environment today is emerging pollution. These pollutants are chemicals or substances that exist in low concentrations (ng/L to g/L) and have an impact on people, flora and fauna both directly and indirectly. Pharmaceuticals, insecticides and hormones are the broad categories into which they can be separated. Emerging contaminants can also be made of synthetic or natural materials. The list of developing contaminants is, therefore, quite significant because it is growing in relation to human needs to improve quality of life. Moreover, it is worth studying and investigating emerging pollutants' removal using appropriate and suitable treatments. However, the main goals of traditional sewage treatment facilities, which are still in use and were created years ago, are organic and bacteriological removal rather than the elimination of emerging organic compounds (de Oliveira et al., 2020).

In this frame, natural-based treatment systems were recently investigated as tertiary units to remove emerging pollutants in secondary wastewater. CWs are the most efficient in removing emerging organic pollutants among these natural systems. Indeed, CWs involve different processes simultaneously to remove emerging contaminants such as aerobic/anoxic/anaerobic biodegradation, sorption, plant uptake and photodegradation (AL Falahi et al., 2022; Bai et al., 2022; Kamilya et al., 2023). Many studies revealed that HSSF-CW is more efficient in treating pharmaceuticals (such as ibuprofen, naproxen, diclofenac, ketoprofen, clofibric acid and carbamazepine) than conventional WWTP (AL Falahi et al., 2022). A high removal rate of ibuprofen (>99%) was recorded in VSSF-CW treating real domestic wastewater (AL Falahi et al., 2021). Recently, the effectiveness of CWs in eliminating steroids and antibiotics was reported to be 90% (Kamilya et al., 2023). Vertical up-flow CWs, with birnessite-coated sand, achieved a removal efficiency of 98.5% of diclofenac from secondary effluent of the wastewater treatment plant (Cheng et al., 2022).

It was recommended to design a green wall system with more than 30 minutes of greywater residence time using a mixture of coco coir and zeolite for the effective removal of xenobiotic organic compounds from domestic greywater based on the results of a recent study examining the performance of five lightweight green wall media types to remove the emerging contaminants from greywater (Abd-ur-Rehman et al., 2022). Several studies have highlighted the capacity of microalgae to eliminate newly emerging pollutants from wastewater through acclimation, co-metabolism and algal-bacterial consortia. Biosorption, bioaccumulation, biodegradation, photolysis, hydrolysis and volatilisation are some methods used by microalgae-based systems to remove contaminants (Zhou et al., 2022). Three algal-bacterial consortia systems—algal-bacterial activated sludge, algal-bacterial biofilm reactor and algal-bacterial built wetland system achieved high removal performance of emerging pollutants from wastewater. A semi-closed, tubular horizontal photobioreactor (PBR) was used to test the ability of microalgae-based treatment systems to remove a variety of priority pesticides under real/environmental conditions. The results showed variable removals between 100% and negative values (García-Galán et al., 2020). *Scenedesmus* sp.-dominated microalgae growing in a semi-closed environment has also exhibited significant surfactant removal efficiency (90%–97%) from municipal wastewater (Serejo et al., 2020). Additionally, a high-rate algae pond has shown a high removal rate (>90) of developing organic pollutants from urban wastewater, including caffeine, acetaminophen, ibuprofen, methyl dihydrojasmonate and hydrocinnamic acid (Matamoros et al., 2015). Similar to this, *Chlorella sorokiniana* was used to remove 60%–100% of Diclofenac, ibuprofen, Paracetamol and Metoprolol Trimethoprim from anaerobically treated black water (de Wilt et al., 2016). On the other hand, the biodegradation, photodegradation and sorption processes, the majority of which occur in the initial phases of treatment (for example, in the anaerobic or facultative ponds), have been linked to the effectiveness of wastewater stabilisation ponds in the removal of organic micropollutants (Gruchlik et al., 2018). According to K'oreje et al. (2018), wastewater stabilisation ponds can potentially extract pharmaceutically active chemicals. Although the removal rate varied depending on the molecule and the bearing phase (from 28.4% for boscalid to 89.4% for aclonifen in the dissolved phase and from 22.1% for pendimethalin to 96.8% for metolachlor in the particulate fraction), the pond also demonstrated satisfactory efficiency in pesticide attenuation (Chaumet et al., 2022). On average, domestic wastewater treatment efficiency in wastewater stabilisation ponds was 71% for monobutyltin, 47% for dibutyltin and 55% for tributyltin (Sabah et al., 2016).

A promising technique for removing recalcitrant contaminants from water is photocatalytic oxidation. It is important to note that some emerging pollutants capable of photocatalytic degradation due to their molecular composition can be degraded using an external energy source, such as UV or solar light. The Fenton process is one of the AOPs frequently researched for tertiary wastewater treatment due to its effectiveness in oxidising refractory organic compounds, reasonable cost and straightforward implementation. This alternative combines ferrous ions (Fe^{2+}) and hydrogen peroxide (H_2O_2). Multiple studies have demonstrated that several variables, including contact time, pH, reagent concentrations and beginning pollutant concentrations, impact the Fenton process (Cai et al., 2021). The solar photo-Fenton process' operational viability has also been evaluated in continuous flow at neutral pH, resulting in a treatment capacity of 900 $L/m^2/d$ with micropollutant removal percentages of about 60% (Arzate et al., 2020). According to Prieto-Rodríguez et al. (2013), the solar photo-Fenton process was successfully used as a tertiary treatment for MWTP effluents, eliminating 80% of all micropollutants. Additionally, solar photo-Fenton was used to remove 30 pollutants of emerging concern from hospital wastewater, with a 78% overall removal rate (Lofrano et al., 2021). The design of the hydrodynamics in the large Raceway Pond Reactor must be carefully examined to reduce power consumption while improving mixing performance, according to a computational fluid dynamics model investigating the effect of mixing and hydrodynamics on the removal of contaminants of emerging concern found in a secondary WWTP effluent by the solar photo-Fenton process (Peralta Muniz Moreira et al., 2021).

8.5.4 Integrating Nature-based Technologies for Wastewater Disinfection

Wastewater contains a number of dangerous pathogens, including bacteria, protozoa and viruses, that can lead to cholera, gastroenteritis, fever and hepatitis, among other illnesses. The most prevalent and traditional methods for disinfecting water, in addition to chlorination, are UV-C radiation, ozonation and filtration; however, in recent years, nature-based technologies like oxidation ponds and AOPs have also received significant attention as potential and promising tertiary treatments (Achouri et al., 2021; Berruti et al., 2022).

One of the most effective natural approaches for treating infections is oxidation ponds. For treating diseases, many variables are necessary, including pH, light intensity, time and temperature, and dissolved concentration. *Faecal enterococci*, Ascaris eggs and other faecal coliforms can all be removed from various ponds. Other pathogens, like *Campylobacter jejuni* and *Salmonella enterica* in particular, can be reduced by 96.4% in the oxidation pond (Butler et al., 2017). Generally, a treatment system with oxidation ponds may efficiently remove pathogens and reduce germs when detention is between a few days and several weeks (Butler et al., 2017).

In waste stabilisation pond wastewater treatment systems, maturation ponds use sunshine and other naturally occurring phenomena to disinfect pathogens and reduce carbon and nutrients. Due to sunlight's high attenuation in turbid water, the effectiveness of UV disinfection in the water column depends significantly on a microorganism's vertical position and proximity to the water surface (Dahl et al., 2017). *E. coli* were not susceptible to exogenous sensitisers found in waste stabilisation pond water. However, dissolved oxygen concentrations and the presence of UVB wavelengths had a significant impact on *E. coli* inactivation (Kadir and Nelson, 2014). *E. coli* and viruses are rendered inactive by UVB due to direct sunlight inactivation. UVB disinfection may, however, be less essential than other removal methods when averaged over time and full-scale pond depth, as shown by the extremely high attenuation with a depth of UVB in wastewater treatment pond (99% elimination in the top 8 cm) (Park et al., 2021). Designing pond systems and predicting inactivation rates in ponds and other natural treatment systems that use sunlight for disinfection requires a thorough understanding of the mechanisms of sunlight-mediated inactivation, including the factors that control the mechanisms (such as the impact of wavelength, water quality, water depth, etc.). (Kadir and Nelson, 2014). According to a process-based model, sun disinfection and daphnia predation account for 65% and 25% of the elimination in the study pond, respectively (Hernández-Crespo et al., 2022). Over a wide variety of water temperatures, solar radiation intensities and hydraulic residence periods, it was found that an innovative wastewater tertiary treatment based on Daphnia filtration increased particle removal and *E. coli* inactivation (Serra et al., 2022).

Implementing CW, which achieves pathogen elimination efficiency up to as high as 4–5 log eduction, is another viable alternative for pathogen treatment in a sustainable manner (Shingare et al., 2019). Physicochemical processes like sedimentation, unfavourable water chemistry, oxidation, temperature, exposure to sunlight or UV radiation, biological interactions like the secretion of antibiotics or biocides, interaction with biofilms, protection from predators, attachment of microbes to macrophytes, etc., were used to treat pathogens in CW.

Hybrid wetland systems were shown to be the most effective because of a longer retention duration, and horizontal SSF CWs have superior reduction capacity than free water surface flow CWs (Wu et al., 2016).

By removing up to 99.99% of *Pythium ultimum* and *Fusarium oxysporum*, Gruyer et al. (2013) shown that CW can provide an effective and secure alternative to treat and then reuse greenhouse effluent. Pathogen removal is assisted by microbes and enzymes that degrade cell walls. Furthermore, the elimination of pathogens in CW changed with the season. In fact, CW in Leon, Spain, achieved maximum values for E. coli and total coliforms in the spring and the fall at 99.9% and 99.9%, respectively. For *faecal streptococci* in the winter at 97.0%, for *Clostridium* in the summer at 100%, and for *Giardia cysts*, *Cryptosporidium oocysts* and helminth eggs throughout the year at 100%, as well (Molleda et al., 2008).

After 15 days of treatment, high rates of faecal indicator bacteria removal were seen, with enterococci removal reaching >95% and enterobacteria removal reaching >98%. This reduced the amount of dangerous biological contaminants that were initially present in the cattle effluent (Bôto et al., 2023). Similarly, bacteriological parameters were reduced to 99.9%, and complete removal of *Salmonella* sp was achieved during three days by *Cyperus papyrus* macrophyte (Hamad, 2020).

The best candidate partners to *Typha latifolia* for the highest purification efficiency, of above 97% for removal of faecal indicator bacteria colonies and functional genes, and of more than 75% for removal of faecal bacterial strains, pathotypes, virulent and well as resistant genes, were also suggested to be *Elodea nuttallii* and *Myriophyllum spicatum* macrophytes (Omondi Donde et al., 2020).

Recently, CWs have also been used to reduce the number of bacteria that are resistant to antibiotics (ARB). To test the removal of antibiotic-resistant bacteria from hospital wastewater, eight horizontal SSF pilot scale artificial wetlands were created. In vegetated wetlands, there was a significantly (P 0.005) greater removal of antibiotic-resistant bacteria (80.8% to 93.2%), and the CW system removed total and faecal coliforms by 7.1 logs10 and 5.1 logs10, respectively (Dires et al., 2018).

On the other hand, treatment for municipal wastewater treatment plant effluent. Solar photo-Fenton has been demonstrated to be a successful method for disinfecting wastewater that is supported by radiation alone and oxidative radicals produced in the majority of the wastewater (Achouri et al., 2021). Utilising Raceway Pond Reactors (RPR), solar photo-Fenton application to tertiary treatment of urban wastewater has sped up application while lowering investment costs from 400 € m^{-2} in reactors based on compound parabolic collectors to 10 € m^{-2}. (Fiorentino et al., 2022).

In RPR, the use of the solar photo-Fenton process led to the inactivation of *E. coli* below the detection limit (DL) in less than 15 minutes at an acidic pH and in 60 minutes with 0.2 mM of Fe^{3+} and 4.41 mM of H_2O_2 at a neutral pH. Trimethoprim and antibiotic resistance genes were also removed (Fiorentino et al., 2022). With no regrowth after treatment, solar photo-Fenton disinfection in an RPR showed 100% elimination of wild enteric bacteria, total coliforms and *E. coli* within 90 min of treatment. (Freitas et al., 2017).

Pilot-scale solar photo-Fenton in an RPR with H2O2 as an oxidant eliminated 2.5 log units of *E. coli*, whereas solar photo-Fenton mediated by persulfate eliminated 3.0 log units. Additionally, the number of bacteria strains resistant to ampicillin, chloramphenicol, erythromycin, amoxicillin, sulfadiazine, sulfamethoxazole and trimethoprim decreased in both procedures (solar photo-Fenton and persulfate mediated solar photo-Fenton) (Starling et al., 2021).

The investigation of the inactivation of Total Coliforms, *Escherichia coli* and *Enterococcus sp.* in WWTP secondary effluents by the solar photo-Fenton process in continuous flow at neutral pH in open reactors revealed that 30 min of HRT is the most effective operation condition capable of producing 305 m^3/m^2/year of disinfected water while complying with the Spanish Law (RD 1620/2007) for water reuse (De la Obra Jiménez et al., 2019).

In comparison to photo-Fenton and TiO_2-photocatalysis, H_2O_2/solar process demonstrated the best inactivation results at near neutral pH. Solar AOPs at pilot scale were also effective for actual wastewater disinfection in the presence of *Curvularia* sp (Aguas et al., 2017). The best disinfection efficiency was found when solar disinfection was compared to solar-driven AOPs (such as H_2O_2/sunlight, TiO_2/sunlight, H_2O_2/TiO_2/sunlight and natural photo-Fenton) for the inactivation of a multidrug-resistant *E. coli* strain isolated from real biologically treated wastewater. All processes resulted in a complete inactivation (5-log decrease) of bacteria until DL (Ferro et al., 2015).

Another comparison of various AOP systems (UV/H_2O_2, UV/Cl_2, O_3, O_3/UV, H_2O_2/O_3/UV and Cl_2/O_3/UV) operated at pilot scale as tertiary treatment of municipal wastewater revealed that AOPs using UV irradiation were the most effective processes for wastewater disinfection, resulting in a complete inactivation of selected indicator organisms by low ozone dose (1.5 mg/L) and UV fluence (191–465 mJ/cm^2) (Sgroi et al., 2021). The use of advanced oxidation processes (AOPs), specifically O_3, O_3/H_2O_2 and UV/H_2O_2, in combination with granular activated carbon (GAC) to polish secondary effluents were also efficient to disinfect the samples, promoting nearby 5-log reduction of *E. coli* and total coliforms (Antonio da Silva et al., 2021). GAC and AOPs, specifically O_3, O_3/H_2O_2 and UV/H_2O_2, when used to polish secondary effluents were effective in disinfecting the samples and generating close to 5-log reductions in E. coli and total coliforms (Antonio da Silva et al., 2021).

In order to produce treated water suitable for reuse, the combination of coagulation/flocculation and Fenton processes was also investigated. The results showed that when using hydrogen peroxide and added iron concentrations of 100 and 7 mg/L, respectively, presented the maximum enterobacteria inactivation after 120 min at 25°C. Real recirculating aquaculture systems wastewater disinfection along with low concentration peroxymonosulfate for *E. coli* (CGMCC 1.3373) inactivation improved *E. coli* inactivation over UVA-LED alone (Qi et al., 2020).

8.6 CONCLUSION

Water contamination has evolved into a serious environmental problem on a worldwide scale, especially in light of the presence of various pollutants. Any activity, including domestic, agricultural and industrial, discharges effluents containing harmful contaminants, some of which are toxic and affect both humans and the environment. These problems highlight the need for a novel, reliable alternative to protect water resources and improve water quality. Physical, chemical and biological techniques are typically used in conventional wastewater treatment to remove soluble and insoluble pollutants from effluents. In fact, the efficient integration of powerful and reducing technology can help to improve the effluent quality. New progress in wastewater treatment technology has included the detoxification of resistant organic and inorganic compounds, energy conservation and the creation of bioenergy for environmental sustainability. Due to the high efficacy of pollutants removal achieved by nature-based wastewater treatment methods, it is important to highlight that these alternatives may maximum, innovative ways to demonstrate their considerable applicability in various fields.

REFERENCES

Abd-ur-Rehman, H.M., Deletic, A., Zhang, K., & Prodanovic, V. (2022). The comparative performance of lightweight green wall media for the removal of xenobiotic organic compounds from domestic greywater. *Water Research*, 221, 118774.

Abinandan, S., Subashchandrabose, S.R., Panneerselvan, L., Venkateswarlu, K., & Megharaj, M. (2019). Potential of acid-tolerant microalgae, Desmodesmus sp. MAS1 and Heterochlorella sp. MAS3, in heavy metal removal and biodiesel production at acidic pH. *Bioresource Technology*, 278, 9–16.

Achouri, F., Said, M.B., Wahab, M.A., Bousselmi, L., Corbel, S., Schneider, R., & Ghrabi, A. (2021). Effect of photocatalysis (TiO2/UVA) on the inactivation and inhibition of Pseudomonas aeruginosa virulence factors expression. *Environmental Technology*, 42, 4237–4246.

Aguas, Y., Hincapie, M., Fernández-Ibáñez, P., & Polo-López, M.I. (2017). Solar photocatalytic disinfection of agricultural pathogenic fungi (Curvularia sp.) in real urban wastewater. *Science of The Total Environment*, 607–608, 1213–1224.

AL Falahi, O.A., Abdullah, S.R.S., Hasan, H.A., Othman, A.R., Ewadh, H.M., Al-Baldawi, I.A., Kurniawan, S.B., Imron, M.F., & Ismail, N. (2021). Simultaneous removal of ibuprofen, organic material, and nutrients from domestic wastewater through a pilot-scale vertical sub-surface flow constructed wetland with aeration system. *Journal of Water Process Engineering*, 43, 102214.

AL Falahi, O.A., Abdullah, S.R.S., Hasan, H.A., Othman, A.R., Ewadh, H.M., Kurniawan, S.B., & Imron, M.F. (2022). Occurrence of pharmaceuticals and personal care products in domestic wastewater, available treatment technologies, and potential treatment using constructed wetland: A review. *Process Safety and Environmental Protection*, 168, 1067–1088.

Andric, I., Kamal, A., & Al-Ghamdi, S.G. (2020). Efficiency of green roofs and green walls as climate change mitigation measures in extremely hot and dry climate: Case study of Qatar. *Energy Reports*, 6, 2476–2489.

Ansari, F.A., Nasr, M., Rawat, I., & Bux, F. (2021). Artificial neural network and techno-economic estimation with algae-based tertiary wastewater treatment. *Journal of Water Process Engineering*, 40, 101761.

Antonio da Silva, D., Pereira Cavalcante, R., Batista Barbosa, E., Machulek Junior, A., César de Oliveira, S., & Falcao Dantas, R. (2021). Combined AOP/GAC/AOP systems for secondary effluent polishing: Optimization, toxicity and disinfection. *Separation and Purification Technology*, 263, 118415.

Arashiro, L.T., Ferrer, I., Rousseau, D.P., Van Hulle, S.W., & Garfi, M. (2019). The effect of primary treatment of wastewater in high rate algal pond systems: Biomass and bioenergy recovery. *Bioresource Technology*, 280, 27–36.

Arias, D.M., Solé-Bundó, M., Garfí, M., Ferrer, I., García, J., & Uggetti, E. (2018). Integrating microalgae tertiary treatment into activated sludge systems for energy and nutrients recovery from wastewater. *Bioresource Technology*, 247, 513–519.

Arzate, S., Campos-Mañas, M.C., Miralles-Cuevas, S., Agüera, A., García Sánchez, J.L., & Sánchez Pérez, J.A. (2020). Removal of contaminants of emerging concern by continuous flow solar photo-Fenton process at neutral pH in open reactors. *Journal of Environmental Management*, 261, 110265.

Bai, Y., Wang, Z., & van der Hoek, J.P. (2022). Remediation potential of agricultural organic micropollutants in in-situ techniques: A review. *Ecological Informatics*, 68, 101517.

Batool, A., & Saleh, T.A. (2020). Removal of toxic metals from wastewater in constructed wetlands as a green technology; catalyst role of substrates and chelators. *Ecotoxicology and Environmental Safety*, 189, 109924.

Berndtsson, J.C., Bengtsson, L., & Jinno, K. (2009). Runoff water quality from intensive and extensive vegetated roofs. *Ecological Engineering*, 35(3), 369–380.

Berruti, I., Nahim-Granados, S., Abeledo-Lameiro, M.J., Oller, I., & Polo-López, M.I. (2022). Recent advances in solar photochemical processes for water and wastewater disinfection. *Chemical Engineering Journal Advances*, 10, 100248.

Blanco-Vieites, M., Suárez-Montes, D., Delgado, F., Álvarez-Gil, M., Battez, A.H., & Rodríguez, E. (2022). Removal of heavy metals and hydrocarbons by microalgae from wastewater in the steel industry. *Algal Research*, 64, 102700.

Boano, F., Caruso, A., Costamagna, E., Ridolfi, L., Fiore, S., Demichelis, F., & Masi, F., (2020). A review of nature-based solutions for greywater treatment: Applications, hydraulic design, and environmental benefits. *Science of the Total Environment*, 711, 134731.

Boelee, N.C., Temmink, H., Janssen, M., Buisman, C.J.N., & Wijffels, R.H. (2011). Nitrogen and phosphorus removal from municipal wastewater effluent using microalgal biofilms. *Water Research*, 45, 5925–5933.

Bôto, M.L., Dias, S.M., Crespo, R.D., Mucha, A.P., & Almeida, C.M.R. (2023). Removing chemical and biological pollutants from swine wastewater through constructed wetlands aiming reclaimed water reuse. *Journal of Environmental Management*, 326, 116642.

Březinová, T.D., & Vymazal, J. (2022). Distribution of heavy metals in Phragmites australis growing in constructed treatment wetlands and comparison with natural unpolluted sites. *Ecological Engineering*, 175, 106505.

Butler, E., Hung, Y.T., Suleiman Al Ahmad, M., Yeh, R.Y.L., Liu, R.L.H., & Fu, Y.P. (2017). Oxidation pond for municipal wastewater treatment. *Applied Water Science*, 7(1), 31–51.

Cai, Q.Q., Jothinathan, L., Deng, S.H., Ong, S.L., Ng, H.Y., & Hu, J.Y. (2021). Fenton-and ozone-based AOP processes for industrial effluent treatment. In M.P. Shah (Ed.), *Advanced Oxidation Processes for Effluent Treatment Plants* (pp.199–254). Elsevier.

Chang, J., Deng, S., Li, X., Li, Y., Chen, J., & Duan, C. (2022). Effective treatment of acid mine drainage by constructed wetland column: Coupling walnut shell and its biochar product as the substrates. *Journal of Water Process Engineering*, 49, 103116.

Chaumet, B., Probst, J.-L., Payré-Suc, V., Granouillac, F., Riboul, D., & Probst, A. (2022). Pond mitigation in dissolved and particulate pesticide transfers: Influence of storm events and seasonality (Auradé agricultural catchment, SW-France). *Journal of Environmental Management*, 320, 115911.

Chen, J., Jiang, X., Zhang, Y., Zhang, Y., Sun, Y., & Zhang, L. (2022). Organic matter conversion and contributors to bioavailability in biogas slurry treated by biochar/persulfate during the oxidation pond process. *Journal of Cleaner Production*, 355, 131770.

Cheng, C., Zhang, J., Xu, J., Yang, Y., Bai, X., & He, Q. (2022). Enhanced removal of nutrients and diclofenac by birnessite sand vertical flow constructed wetlands. *Journal of Water Process Engineering*, 46, 102656.

Crini, G., & Lichtfouse, E. (2019). Advantages and disadvantages of techniques used for wastewater treatment. *Environmental Chemistry Letters*, 17(1), 145–155.

da Silva Brito, G.F., Oliveira, R., Grisolia, C.K., Guirra, L.S., Weber, I.T., & de Almeida, F.V. (2019). Evaluation of advanced oxidative processes in biodiesel wastewater treatment. *Journal of Photochemistry and Photobiology A: Chemistry*, 375, 85–90.

Daee, M., Gholipour, A., & Stefanakis, A.I. (2019). Performance of pilot Horizontal Roughing Filter as polishing stage of waste stabilization ponds in developing regions and modelling verification. *Ecological Engineering*, 138, 8–18.

Dahl, N.W., Woodfield, P.L., Lemckert, C.J., Stratton, H., & Roiko, A. (2017). A practical model for sunlight disinfection of a subtropical maturation pond. *Water Research*, 108, 151–159.

De la Obra Jiménez, I., Esteban García, B., Rivas Ibáñez, G., Casas López, J.L., & Sánchez Pérez, J.A. (2019). Continuous flow disinfection of WWTP secondary effluents by solar photo-Fenton at neutral pH in raceway pond reactors at pilot plant scale. *Applied Catalysis B: Environmental*, 247, 115–123.

de Oliveira, M., Frihling, B.E.F., Velasques, J., Filho, F.J.C.M., Cavalheri, P.S., & Migliolo, L. (2020). Pharmaceuticals residues and xenobiotics contaminants: Occurrence, analytical techniques and sustainable alternatives for wastewater treatment. *Science of the Total Environment*, 705, 135568.

de Wilt, A., Butkovskyi, A., Tuantet, K., Leal, L.H., Fernandes, T.V., & Langenhoff, A., Zeeman, G. (2016). Micropollutant removal in an algal treatment system fed with source separated wastewater streams. *Journal of Hazardous Materials*, 304, 84–92.

Delanka-Pedige, H.M., Munasinghe-Arachchige, S.P., Zhang, Y., & Nirmalakhandan, N., (2020). Bacteria and virus reduction in secondary treatment: Potential for minimizing post disinfectant demand. *Water Research*, 177, 115802.

Delle Grazie, F.M., & Gill, L.W. (2022). Review of the ecosystem services of temperate wetlands and their valuation tools. *Water*, 14, 1345.

Dinh, T.T.U., Soda, S., Nguyen, T.A.H., Nakajima, J., & Cao, T.H. (2020). Nutrient removal by duckweed from anaerobically treated swine wastewater in lab-scale stabilization ponds in Vietnam. *Science of the Total Environment*, 722, 137854.

Dires, S., Birhanu, T., Ambelu, A., & Sahilu, G. (2018). Antibiotic resistant bacteria removal of subsurface flow constructed wetlands from hospital wastewater. *Journal of Environmental Chemical Engineering*, 6, 4265–4272.

Djodjic, F., Geranmayeh, P., Collentine, D., Markensten, H., & Futter, M. (2022). Cost effectiveness of nutrient retention in constructed wetlands at a landscape level. *Journal of Environmental Management*, 324, 116325.

Elisabeth, B., Rayen, F., & Behnam, T. (2021). Microalgae culture quality indicators: A review. *Critical Reviews in Biotechnology*, 41(4), 457–473.

Etteieb, S., Zolfaghari, M., Magdouli, S., Brar, K.K., & Brar, S.K. (2021). Performance of constructed wetland for selenium, nutrient and heavy metals removal from mine effluents. *Chemosphere*, 281, 130921.

Fallahi, A., Rezvani, F., Asgharnejad, H., Nazloo, E.K., Hajinajaf, N., & Higgins, B. (2021). Interactions of microalgae-bacteria consortia for nutrient removal from wastewater: A review. *Chemosphere*, 272, 129878.

Ferro, G., Fiorentino, A., Alferez, M.C., Polo-López, M.I., Rizzo, L., & Fernández-Ibáñez, P. (2015). Urban wastewater disinfection for agricultural reuse: Effect of solar driven AOPs in the inactivation of a multidrug resistant E. coli strain. *Applied Catalysis B: Environmental, Photocatalysis: Science and Applications*, 178, 65–73.

Fiorentino, A., Soriano-Molina, P., Abeledo-Lameiro, M.J., de la Obra, I., Proto, A., Polo-López, M.I., Pérez, J.A.S., & Rizzo, L. (2022). Neutral (Fe3+-NTA) and acidic (Fe2+) pH solar photo-Fenton Vs chlorination: Effective urban wastewater disinfection does not mean control of antibiotic resistance. *Journal of Environmental Chemical Engineering*, 10, 108777.

Freitas, A.M., Rivas, G., Campos-Mañas, M.C., Casas López, J.L., Agüera, A., Sánchez & Pérez, J.A. (2017). Ecotoxicity evaluation of a WWTP effluent treated by solar photo-Fenton at neutral pH in a raceway pond reactor. *Environmental Science and Pollution Research*, 24, 1093–1104.

García-Galán, M.J., Monllor-Alcaraz, L.S., Postigo, C., Uggetti, E., López de Alda, M., Díez-Montero, R., & García, J. (2020). Microalgae-based bioremediation of water contaminated by pesticides in peri-urban agricultural areas. *Environmental Pollution*, 265, 114579.

Greeshma, K., Kim, H.-S., & Ramanan, R. (2022). The emerging potential of natural and synthetic algae-based microbiomes for heavy metal removal and recovery from wastewaters. *Environmental Research*, 215, 114238.

Gross, M., Jarboe, D., & Wen, Z. (2015). Biofilm-based algal cultivation systems. *Applied Microbiology and Biotechnology*, 99(14), 5781–5789.

Gruchlik, Y., Linge, K., & Joll, C. (2018). Removal of organic micropollutants in waste stabilisation ponds: A review. *Journal of Environmental Management*, 206, 202–214.

Gruyer, N., Dorais, M., Zagury, G.J., & Alsanius, B.W. (2013). Removal of plant pathogens from recycled greenhouse wastewater using constructed wetlands. *Agricultural Water Management*, 117, 153–158.

Gupta, V.K., Ali, I., Saleh, T.A., Nayak, A., & Agarwal, S. (2012). Chemical treatment technologies for wastewater recycling – An overview. *Rsc Advances*, 2(16), 6380–6388.

Hamad, M.T.M.H. (2020). Comparative study on the performance of Typha latifolia and Cyperus Papyrus on the removal of heavy metals and enteric bacteria from wastewater by surface constructed wetlands. *Chemosphere*, 260, 127551.

Hamed, S.M., El Shimi, H.I., van Dijk, J.R., Osman, A.I., Korany, S.M., & AbdElgawad, H. (2022). A novel integrated system for heavy metals removal and biodiesel production via green microalgae: A techno-economic feasibility assessment. *Journal of Environmental Chemical Engineering*, 10, 108804.

Hasan, M.M., Saeed, T., & Nakajima, J. (2019). Integrated simple ceramic filter and waste stabilization pond for domestic wastewater treatment. *Environmental Technology & Innovation*, 14, 100319.

Heidarinejad, Z., Dehghani, M.H., Heidari, M., Javedan, G., Ali, I., & Sillanpää, M. (2020). Methods for preparation and activation of activated carbon: A review. *Environmental Chemistry Letters*, 18(2), 393–415.

Hernández-Crespo, C., Fernández-Gonzalvo, M.I., Miglio, R.M., & Martín, M. (2022). Escherichia coli removal in a treatment wetland - pond system: A mathematical modelling experience. *Science of the Total Environment*, 839, 156237.

Ho, L., Jerves-Cobo, R., Morales, O., Larriva, J., Arevalo-Durazno, M., Barthel, M., & Goethals, P. (2021). Spatial and temporal variations of greenhouse gas emissions from a waste stabilization pond: Effects of sludge distribution and accumulation. *Water Research*, 193, 116858.

Iervolino, G., Zammit, I., Vaiano, V., & Rizzo, L. (2020). Limitations and prospects for wastewater treatment by UV and visible-light-active heterogeneous photocatalysis: A critical review.In M.J. Muñoz-Batista, A. Navarrete Muñoz & R. Luque (Eds.), Heterogeneous *Photocatalysis*, (pp. 225–264). Cham: Springer.

Jaouadi, S., Wahab, M.A., Anane, M., Bousselmi, L., & Jellali, S. (2014). Powdered marble wastes reuse as a low-cost material for phosphorus removal from aqueous solutions under dynamic conditions. *Desalination and Water Treatment*, 52, 1705–1715.

Ji, Z., Tang, W., & Pei, Y. (2022). Constructed wetland substrates: A review on development, function mechanisms, and application in contaminants removal. *Chemosphere*, 286, 131564.

Kadir, K., & Nelson, K.L. (2014). Sunlight mediated inactivation mechanisms of Enterococcus faecalis and Escherichia coli in clear water versus waste stabilization pond water. *Water Research*, 50, 307–317.

Kamilya, T., Majumder, A., Yadav, M.K., Ayoob, S., Tripathy, S., & Gupta, A.K. (2022). Nutrient pollution and its remediation using constructed wetlands: Insights into removal and recovery mechanisms, modifications and sustainable aspects. *Journal of Environmental Chemical Engineering*, 10, 107444.

Kamilya, T., Yadav, M.K., Ayoob, S., Tripathy, S., Bhatnagar, A., & Gupta, A.K. (2023). Emerging impacts of steroids and antibiotics on the environment and their remediation using constructed wetlands: A critical review. *Chemical Engineering Journal*, 451, 138759.

Kehrein, P., Van Loosdrecht, M., Osseweijer, P., Garfí, M., Dewulf, J., & Posada, J. (2020). A critical review of resource recovery from municipal wastewater treatment plants-market supply potentials, technologies and bottlenecks. *Environmental Science: Water Research & Technology*, 6(4), 877–910.

Khan, S., Ahmad, I., Shah, M.T., Rehman, S., & Khaliq, A. (2009). Use of constructed wetland for the removal of heavy metals from industrial wastewater. *Journal of Environmental Management*, 90, 3451–3457.

Kim, G.Y., Heo, J., Kim, K., Chung, J., & Han, J.I. (2021). Electrochemical pH control and carbon supply for microalgae cultivation. *Chemical Engineering Journal*, 426, 131796.

Knox, A.S., Paller, M.H., Seaman, J.C., Mayer, J., & Nicholson, C. (2021). Removal, distribution and retention of metals in a constructed wetland over 20 years. *Science of the Total Environment*, 796, 149062.

K'oreje, K.O., Kandie, F.J., Vergeynst, L., Abira, M.A., Van Langenhove, H., Okoth, M., & Demeestere, K. (2018). Occurrence, fate and removal of pharmaceuticals, personal care products and pesticides in wastewater stabilization ponds and receiving rivers in the Nzoia Basin, Kenya. *Science of the Total Environment*, 637–638, 336–348.

Kraiem, K., Kallali, H., Wahab, M.A., Fra-vazquez, A., Mosquera-Corral, A., & Jedidi, N. (2019). Comparative study on pilots between ANAMMOX favored conditions in a partially saturated vertical flow constructed wetland and a hybrid system for rural wastewater treatment. *Science of the Total Environment*, 670, 644–653.

Kumar, B.R., Mathimani, T., Sudhakar, M.P., Rajendran, K., Nizami, A.S., Brindhadevi, K., & Pugazhendhi, A. (2021). A state of the art review on the cultivation of algae for energy and other valuable products: Application, challenges, and opportunities. *Renewable and Sustainable Energy Reviews*, 138, 110649.

Li, K., Liu, Q., Fang, F., Luo, R., Lu, Q., Zhou, W., & Ruan, R. (2019). Microalgae-based wastewater treatment for nutrients recovery: A review. *Bioresource Technology*, 291, 121934.

Liu, L., Cao, J., Ali, M., Zhang, J., & Wang, Z. (2021). Impact of green roof plant species on domestic wastewater treatment. *Environmental Advances*, 4, 100059.

Liu, F., Lu, T., & Zhang, Y. (2022). Performance assessment of constructed wetland-microbial fuel cell for treatment of mariculture wastewater containing heavy metals. *Process Safety and Environmental Protection*, 168, 633–641.

Lofrano, G., Faiella, M., Carotenuto, M., Murgolo, S., Mascolo, G., Pucci, L., & Rizzo, L., (2021). Thirty contaminants of emerging concern identified in secondary treated hospital wastewater and their removal by solar Fenton (like) and sulphate radicals-based advanced oxidation processes. *Journal of Environmental Chemical Engineering*, 9, 106614.

Low, S.S., Bong, K.X., Mubashir, M., Cheng, C.K., Lam, M.K., Lim, J.W., & Show, P.L. (2021). Microalgae cultivation in palm oil mill effluent (POME) treatment and biofuel production. *Sustainability*, 13(6), 3247.

Maine, M.A., Suñe, N., Hadad, H., Sánchez, G., & Bonetto, C. (2009). Influence of vegetation on the removal of heavy metals and nutrients in a constructed wetland. *Journal of Environmental Management*, 90, 355–363.

Makhanya, B.N., Nyandeni, N., Ndulini, S.F., & Mthembu, M.S. (2021). Application of green microalgae biofilms for heavy metals removal from mine effluent. Physics and Chemistry of the Earth, Parts A/B/C. *Integrated Water Resources Management for Sustainable Development in Eastern and Southern Africa*, 124, 103079.

Matamoros, V., Gutiérrez, R., Ferrer, I., García, J., & Bayona, J.M. (2015). Capability of microalgae-based wastewater treatment systems to remove emerging organic contaminants: A pilot-scale study. *Journal of Hazardous Materials*, 288, 34–42.

Mitchell, M.E., Emilsson, T., & Buffam, I. (2021). Carbon, nitrogen, and phosphorus variation along a green roof chronosequence: Implications for green roof ecosystem development. *Ecological Engineering*, 164, 106211.

Mohammed, A., & Babatunde, A.O. (2017). Modelling heavy metals transformation in vertical flow constructed wetlands. *Ecological Modelling*, 354, 62–71.

Molleda, P., Blanco, I., Ansola, G., & de Luis, E. (2008). Removal of wastewater pathogen indicators in a constructed wetland in Leon, Spain. *Ecological Engineering*, 33, 252–257.

Nagarajan, D., Lee, D.J., Chen, C.Y., & Chang, J.S. (2020). Resource recovery from wastewaters using microalgae-based approaches: A circular bioeconomy perspective. *Bioresource Technology*, 302, 122817.

Nas, B., Dolu, T., Argun, M.E., Yel, E., Ateş, H., & Koyuncu, S. (2021). Comparison of advanced biological treatment and nature-based solutions for the treatment of pharmaceutically active compounds (PhACs): A comprehensive study for wastewater and sewage sludge. *Science of the Total Environment*, 779, 146344.

Nguyen, T.T., Huang, H., Nguyen, T.A.H., & Soda, S. (2022). Recycling clamshell as substrate in lab-scale constructed wetlands for heavy metal removal from simulated acid mine drainage. *Process Safety and Environmental Protection*, 165, 950–958.

Olguín, E.J. (2012). Dual purpose microalgae-bacteria-based systems that treat wastewater and produce biodiesel and chemical products within a Biorefinery. *Biotechnology advances*, 30(5), 1031–1046.

Omondi Donde, O., Makindi, S.M., Tian, C., Tian, Y., Hong, P., Cai, Q., Yang, T., Wang, C., Wu, X., & Xiao, B. (2020). A novel integrative performance evaluation of constructed wetland on removal of viable bacterial cells and related pathogenic, virulent and multi-drug resistant genes from wastewater systems. *Journal of Water Process Engineering*, 33, 101060.

Oyebamiji, O.O., Boeing, W.J., Holguin, F.O., Ilori, O., & Amund, O. (2019). Green microalgae cultured in textile wastewater for biomass generation and biodetoxification of heavy metals and chromogenic substances. *Bioresource Technology Reports*, 7, 100247.

Pandey, A.K., Kumar, R.R., Kalidasan, B., Laghari, I.A., Samykano, M., Kothari, R., & Tyagi, V.V. (2021). Utilization of solar energy for wastewater treatment: Challenges and progressive research trends. *Journal of Environmental Management*, 297, 113300.

Parde, D., Patwa, A., Shukla, A., Vijay, R., Killedar, D.J., & Kumar, R. (2021). A review of constructed wetland on type, treatment and technology of wastewater. *Environmental Technology & Innovation*, 21, 101261.

Park, J.B.K., Weaver, L., Davies-Colley, R., Stott, R., Williamson, W., Mackenzie, M., McGill, E., Lin, S., Webber, J., & Craggs, R.J. (2021). Comparison of faecal indicator and viral pathogen light and dark disinfection mechanisms in wastewater treatment pond mesocosms. *Journal of Environmental Management*, 286, 112197.

Peralta Muniz Moreira, R., Cabrera Reina, A., Soriano Molina, P., Sánchez Pérez, J.A., & Li Puma, G. (2021). Computational fluid dynamics (CFD) modeling of removal of contaminants of emerging concern in solar photo-Fenton raceway pond reactors. *Chemical Engineering Journal*, 413, 127392.

Pradhan, S., Al-Ghamdi, S.G., & Mackey, H.R. (2019). Greywater recycling in buildings using living walls and green roofs: A review of the applicability and challenges. *Science of The Total Environment*, 652, 330–344.

Praveen, P., Guo, Y., Kang, H., Lefebvre, C., Loh, K.-C., 2018. Enhancing microalgae cultivation in anaerobic digestate through nitrification. *Chemical Engineering Journal*, 354, 905–912.

Prieto-Rodríguez, L., Spasiano, D., Oller, I., Fernández-Calderero, I., Agüera, A., & Malato, S. (2013). Solar photo-Fenton optimization for the treatment of MWTP effluents containing emerging contaminants. *Catalysis Today*, 209, 188–194.

Priya, A.K., Jalil, A.A., Vadivel, S., Dutta, K., Rajendran, S., Fujii, M., & Soto-Moscoso, M. (2022). Heavy metal remediation from wastewater using microalgae: Recent advances and future trends. *Chemosphere*, 305, 135375.

Qasem, N.A., Mohammed, R.H., & Lawal, D.U. (2021). Removal of heavy metal ions from wastewater: A comprehensive and critical review. *Npj Clean Water*, 4(1), 1–15.

Qi, W., Zhu, S., Shitu, A., Ye, Z., & Liu, D. (2020). Low concentration peroxymonosulfate and UVA-LED combination for E. coli inactivation and wastewater disinfection from recirculating aquaculture systems. *Journal of Water Process Engineering*, 36, 101362.

Qu, W., Zhang, C., Chen, X., & Ho, S.H. (2021). New concept in swine wastewater treatment: Development of a self-sustaining synergetic microalgae-bacteria symbiosis (ABS) system to achieve environmental sustainability. *Journal of Hazardous Materials*, 418, 126264.

Rai, P.K. (2019). Heavy metals/metalloids remediation from wastewater using free floating macrophytes of a natural wetland. *Environmental Technology & Innovation*, 15, 100393.

Rashed, I.G.A.A., Afify, H.A., Ahmed, A.E.M., & Ayoub, M.A.E.S. (2013). Optimization of chemical precipitation to improve the primary treatment of wastewater. *Desalination and Water Treatment*, 51(37–39), 7048–7056.

Sabah, A., Bancon-Montigny, C., Rodier, C., Marchand, P., Delpoux, S., Ijjaali, M., & Tournoud, M.-G. (2016). Occurrence and removal of butyltin compounds in a waste stabilisation pond of a domestic waste water treatment plant of a rural French town. *Chemosphere*, 144, 2497–2506.

Segredo-Morales, E., González, E., González-Martín, C., & Vera, L. (2022). Secondary wastewater effluent treatment by microalgal-bacterial membrane photobioreactor at long solid retention times. *Journal of Water Process Engineering*, 49, 103200.

Semeraro, T., Aretano, R., & Pomes, A. (2019). Green roof technology as a sustainable strategy to improve water urban availability. In *IOP Conference Series: Materials Science and Engineering,* 471(9), 092065.

Serejo, M.L., Farias, S.L., Ruas, G., Paulo, P.L., & Boncz, M.A. (2020). Surfactant removal and biomass production in a microalgal-bacterial process: Effect of feeding regime. *Water Science and Technology*, 82, 1176–1183.

Serra, T., Barcelona, A., Pous, N., Salvadó, V., & Colomer, J. (2022). Disinfection and particle removal by a nature-based Daphnia filtration system for wastewater treatment. *Journal of Water Process Engineering*, 50, 103238.

Sgroi, M., Snyder, S.A., & Roccaro, P. (2021). Comparison of AOPs at pilot scale: Energy costs for micropollutants oxidation, disinfection by-products formation and pathogens inactivation. *Chemosphere*, 273, 128527.

Sheng, A.L.K., Bilad, M.R., Osman, N.B., & Arahman, N. (2017). Sequencing batch membrane photobioreactor for real secondary effluent polishing using native microalgae: Process performance and full-scale projection. *Journal of Cleaner Production*, 168, 708–715.

Shingare, R.P., Thawale, P.R., Raghunathan, K., Mishra, A., & Kumar, S. (2019). Constructed wetland for wastewater reuse: Role and efficiency in removing enteric pathogens. *Journal of Environmental Management*, 246, 444–461.

Si, Z., Wang, Y., Song, X., Cao, X., Zhang, X., & Sand, W. (2019). Mechanism and performance of trace metal removal by continuous-flow constructed wetlands coupled with a micro-electric field. *Water Research*, 164, 114937.

Singh, D.V., Bhat, R.A., Upadhyay, A.K., Singh, R., & Singh, D. (2021). Microalgae in aquatic environs: A sustainable approach for remediation of heavy metals and emerging contaminants. *Environmental Technology & Innovation*, 21, 101340.

Singh, S., & Chakraborty, S. (2020). Performance of organic substrate amended constructed wetland treating acid mine drainage (AMD) of North-Eastern India. *Journal of Hazardous Materials,* 397, 122719.

Singh, S., & Chakraborty, S. (2021). Bioremediation of acid mine drainage in constructed wetlands: Aspect of vegetation (Typha latifolia), loading rate and metal recovery. *Minerals Engineering*, 171, 107083.

Singh, S., & Chakraborty, S. (2022). Biochemical treatment of coal mine drainage in constructed wetlands: Influence of electron donor, biotic-abiotic pathways and microbial diversity. *Chemical Engineering Journal*, 440, 135986.

Singh, B.J., Chakraborty, A., & Sehgal, R. (2023). A systematic review of industrial wastewater management: Evaluating challenges and enablers. *Journal of Environmental Management,* 348, 119230.

Slade, R., & Bauen, A. (2013). Micro-algae cultivation for biofuels: Cost, energy balance, environmental impacts and future prospects. *Biomass and bioenergy*, 53, 29–38.

Solmaz, A., & Işık, M. (2019). Polishing the secondary effluent and biomass production by microalgae submerged membrane photo bioreactor. *Sustainable Energy Technologies and Assessments*, 34, 1–8.

Song, Y., Wang, L., Qiang, X., Gu, W., Ma, Z., & Wang, G. (2022). The promising way to treat wastewater by microalgae: Approaches, mechanisms, applications and challenges. *Journal of Water Process Engineering*, 49, 103012.

Starling, M.C., Costa, E.P., Souza, F.A., Machado, E.C., de Araujo, J.C., & Amorim, C.C. (2021). Persulfate mediated solar photo-Fenton aiming at wastewater treatment plant effluent improvement at neutral PH: Emerging contaminant removal, disinfection, and elimination of antibiotic-resistant bacteria. *Environmental Science and Pollution Research*, 28, 17355–17368.

Sylwan, I., & Thorin, E. (2021). Removal of heavy metals during primary treatment of municipal wastewater and possibilities of enhanced removal: A review. *Water*, 13(8), 1121.

Tang, W., Li, X., Liu, H., Wu, S., Zhou, Q., Du, C., & Yang, C. (2020). Sequential vertical flow trickling filter and horizontal flow multi-soil-layering reactor for treatment of decentralized domestic wastewater with sodium dodecyl benzene sulfonate. *Bioresource Technology*, 300, 122634.

Thomaidi, V., Petousi, I., Kotsia, D., Kalogerakis, N., & Fountoulakis, M.S. (2022). Use of green roofs for greywater treatment: Role of substrate, depth, plants, and recirculation. *Science of The Total Environment*, 807, 151004.

Tripathi, S., & Poluri, K.M. (2021). Heavy metal detoxification mechanisms by microalgae: Insights from transcriptomics analysis. *Environmental Pollution*, 285, 117443.

Ubando, A.T., Cuello, J.L., El-Halwagi, M.M., Culaba, A.B., Promentilla, M.A.B., & Tan, R.R. (2016). Application of stochastic analytic hierarchy process for evaluating algal cultivation systems for sustainable biofuel production. *Clean Technologies and Environmental Policy*, 18(5), 1281–1294.

Vinardell, S., Astals, S., Peces, M., Cardete, M.A., Fernández, I., Mata-Alvarez, J., & Dosta, J. (2020). Advances in anaerobic membrane bioreactor technology for municipal wastewater treatment: A 2020 updated review. *Renewable and Sustainable Energy Reviews,* 130, 109936.

Vymazal, J. (2022). The historical development of constructed wetlands for wastewater treatment. *Land*, 11, 174.

Vymazal, J., & Březinová, T. (2016). Accumulation of heavy metals in aboveground biomass of Phragmites australis in horizontal flow constructed wetlands for wastewater treatment: A review. *Chemical Engineering Journal*, 290, 232–242.

Wahab, M.A., Habouzit, F., Bernet, N., Jedidi, N., & Escudié, R. (2016). Evaluation of a hybrid anaerobic biofilm reactor treating winery effluents and using grape stalks as biofilm carrier. *Environmental Technology*, 37, 1676–1682.

Wahab, M.A., Habouzit, F., Bernet, N., Steyer, J.-P., Jedidi, N., Escudié, R. (2014). Sequential operation of a hybrid anaerobic reactor using a lignocellulosic biomass as biofilm support. *Bioresource Technology*, 172, 150–155.

Wahab, M.A., Jellali, S., & Jedidi, N. (2010). Ammonium biosorption onto sawdust: FTIR analysis, kinetics and adsorption isotherms modeling. *Bioresource Technology*, 101, 5070–5075.

Wahab, M.A., Kebelmann, K., Schartel, B., & Griffiths, G. (2022). Valorization of macroalgae digestate into aromatic rich bio-oil and lipid rich microalgal biomass for enhanced algal biorefinery performance. *Journal of Cleaner Production*, 341, 130925.

Wang, H., Zhang, M., Xue, J., Lv, Q., Yang, J., & Han, X. (2021). Performance and microbial response in a multi-stage constructed wetland microcosm co-treating acid mine drainage and domestic wastewater. *Journal of Environmental Chemical Engineering*, 9, 106786.

Wang, J., Chen, R., Fan, L., Cui, L., Zhang, Y., Cheng, J., Wu, X., Zeng, W., Tian, Q., Shen, L. (2021). Construction of fungi-microalgae symbiotic system and adsorption study of heavy metal ions. *Separation and Purification Technology*, 268, 118689.

Wang, J.-F., Zhou, H.-Z., Tang, G.-H., Huang, J.-W., Liu, H., Cai, Z.-X., He, Z.-W., Zhu, H., & Song, X.-S. (2023). Reducing the inhibitive effect of fluorine and heavy metals on nitrate reduction by hydroxyapatite substrate in constructed wetlands. *Journal of Hazardous Materials*, 446, 130692.

Wang, R., Zhao, X., Wang, T., Guo, Z., Hu, Z., Zhang, J., Wu, S., & Wu, H. (2022). Can we use mine waste as substrate in constructed wetlands to intensify nutrient removal? A critical assessment of key removal mechanisms and long-term environmental risks. *Water Research*, 210, 118009.

Waqas, S., Bilad, M.R., Aqsha, A., Harun, N.Y., Ayoub, M., Wirzal, M.D.H., & Elma, M. (2021). Effect of membrane properties in a membrane rotating biological contactor for wastewater treatment. *Journal of Environmental Chemical Engineering*, 9(1), 104869.

Wdowczyk, A., Szymańska-Pulikowska, A., & Gałka, B. (2022). Removal of selected pollutants from landfill leachate in constructed wetlands with different filling. *Bioresource Technology*, 353, 127136.

Woertz, I., Feffer, A., Lundquist, T., & Nelson, Y. (2009). Algae grown on dairy and municipal wastewater for simultaneous nutrient removal and lipid production for biofuel feedstock. *Journal of Environmental Engineering*, 135(11), 1115.

Wolcott, S., Endreny, T.A., & Newman, L.A. (2022). Green walls as a novel wastewater treatment option for craft breweries. *Ecological Engineering*, 184, 106783.

Wu, J., Zheng, J., Ma, K., Jiang, C., Zhu, L., & Xu, X. (2022). Tertiary treatment of municipal wastewater by a novel flow constructed wetland integrated with biochar and zero-valent iron. *Journal of Water Process Engineering*, 47, 102777.

Wu, S., Carvalho, P.N., Müller, J.A., Manoj, V.R., & Dong, R. (2016). Sanitation in constructed wetlands: A review on the removal of human pathogens and fecal indicators. *Science of the Total Environment*, 541, 8–22.

Xiang, W., Zhang, X., Chen, J., Zou, W., He, F., Hu, X., & Gao, B. (2020). Biochar technology in wastewater treatment: A critical review. *Chemosphere*, 252, 126539.

Xu, X., Chen, M., Yang, G., Jiang, B., & Zhang, J. (2020). Wetland ecosystem services research: A critical review. *Global Ecology and Conservation*, 22, e01027.

Zhang, L., Shen, Z., Fang, W., & Gao, G. (2019). Composition of bacterial communities in municipal wastewater treatment plant. *Science of the Total Environment*, 689, 1181–1191.

Zhang, Z., Guo, L., Liao, Q., Gao, M., Zhao, Y., Jin, C, & Wang, G. (2021). Bacterial-algal coupling system for high strength mariculture wastewater treatment: Effect of temperature on nutrient recovery and microalgae cultivation. *Bioresource Technology*, 338, 125574.

Zhao, W., Ting, Y.P., Chen, J.P., Xing, C.H., & Shi, S.Q. (2000). Advanced primary treatment of waste water using a bio-flocculation-adsorption sedimentation process. *Acta biotechnologica*, 20(1), 53–64.

Zhou, H., Zhao, X., Kumar, K., Kunetz, T., Zhang, Y., Gross, M., & Wen, Z. (2021). Removing high concentration of nickel (II) ions from synthetic wastewater by an indigenous microalgae consortium with a Revolving Algal Biofilm (RAB) system. *Algal Research*, 59, 102464.

Zhou, J.-L., Yang, L., Huang, K.-X., Chen, D.-Z., & Gao, F. (2022). Mechanisms and application of microalgae on removing emerging contaminants from wastewater: A review. *Bioresource Technology*, 364, 128049.

9 Reuse of Treated Water for Agricultural Activities

Situation, Opportunities and Environmental Risks and Associated Health Problems

Humberto Raymundo Gonzalez Moreno, Lidia Elena Requena Hernández, Alan Antonio Rico Barragan, and Georgina Martínez-Reséndiz

9.1 INTRODUCTION

The natural resource most used by organisms is water. This resource is essential for the growth and development of human and industrial settlements, which is why they generally develop on the river banks of bodies of water. Agriculture is one of the main sources of food generation, which needs water to maintain food security. The water used for this purpose, in general, is fresh water, a resource that in recent decades is scarce, since the greater the population, the greater the demand. Therefore, it is essential to use alternative sources of water. One of these alternatives is the use of treated wastewater mainly in agriculture. Treated wastewater is defined as water that has received at least secondary treatment and basic disinfection and is reused after leaving a domestic wastewater treatment facility.

The use of wastewater for irrigation is a valuable resource and an option, in dry regions, because it is considered an inexhaustible, reliable additional source of water and nutrients. Thus, in several water-scarce countries, reusing wastewater is considered very important. Some of the wastewater reuse applications are agricultural irrigation, groundwater recharge, industrial reuse, urban applications such as street and fire cleaning and ecological and recreational uses. Until now, the reuse of wastewater for agricultural irrigation has been more widely accepted than in other fields, mainly in arid regions, where water resources are scarce.

9.2 REUSE OF WASTEWATER

In arid and semi-arid regions, water is a scarce resource, so policymakers are forced to consider all possible sources of water that can be used economically and efficiently to increase productivity and fill the scarcity of a major resource in agriculture. Worldwide, the reuse of wastewater in agriculture is known to contribute to water quality conservation and the use of water for purposes other than irrigation. However, it is important to plan and optimize resource disposal and reuse, because it must have certain characteristics to be usable and have no impact on the environment, mainly in the soil.

9.2.1 Characteristics of Wastewater After Treatment for Agriculture

The characteristics of the treated wastewater varied, depending on the source of origin (e.g., industrial, domestic, or municipal). Likewise, there is a great variety of systems that can be applied for

DOI: 10.1201/9781003441144-9

the treatment of wastewater; the treatment technology to be installed to reuse wastewater for agricultural irrigation must take into account (Rosillo et al., 2017):

1. Type of crop, since each one has the capacity to assimilate different qualities of water.
2. Irrigation techniques and systems, since the particle content could block or cover the nozzles or outlet holes.
3. Required nutrient content, to eliminate or reduce the use of agrochemicals.
4. Labor management of wastewater and irrigation for the protection of farmers, and pathogen content.

Some important agronomic parameters to consider in irrigation water quality are Electrical conductivity (CE), cations: Sodium (Na), Calcium (Ca), Magnesium (Mg) and Potassium (K); anions: Carbonate (CO_3^{-2}), Bicarbonates (HCO_3^{-2}), Chloride (Cl^-), Sulphate (SO_4^{-2}) y Nitrate (NO^{3-}) (Fipps, 2007). When it is suspected that the quality of the water is a problem related to the presence of heavy metals, it is taken into account the provisions of current regulations or also taking as reference the most important, for their effects on human health, among them are: Arsenic, Cadmium, Cyanides, Copper, Chromium, Mercury, Nickel, Lead and Zinc (Instituto Mexicano de Tecnología del Agua y el Ministerio de Medio Ambiente y Agua de Bolivia, 2018).

Another of the main important parameters to consider in treated wastewater is the bacteriological load, important in the quality of irrigation water for agricultural crops. The microorganisms to be evaluated are: Fecal Coliforms and Helminth Eggs; as they are international indicators used to establish water contamination by enteric bacteria and parasites. Determining this parameter is necessary, because the sanitary quality of those irrigated products would be affected (Instituto Mexicano de Tecnología del Agua y el Ministerio de Medio Ambiente y Agua de Bolivia, 2018).

9.2.2 Reasons for Reusing Treated Wastewater

About 380,000 million m^3 (380 billion liters) of wastewater are generated each year in the world, it is believed that in 2030, it will be 470,000 million m^3 and in 2050 they will be 51% more than at present (Qadir et al., 2020). Evidence has been found for the effects of treated wastewater irrigation on soil fertility and chemical properties, where it has been found that treated wastewater can improve soil fertility (Kesari et al., 2021). In addition, the reuse of treated wastewater for agricultural irrigation can be beneficial for several reasons: reducing water scarcity due to abundant wastewater year-round, reducing the risk of pollution from directly to the environment, and bringing economic benefits, mainly due to the nutrient content of wastewater, which reduces fertilizer costs for farmers (Irénikatché Akponikpè et al., 2011). Emphasizing the nutrient content and its richness in elements necessary for plants, when used, wastewater generates a higher crop yield, since it accelerates plant growth and therefore a significant increase in yield, clearly associated with the contribution of nutrients by wastewater (Ungureanu et al., 2020).

9.3 CURRENT SITUATION OF TREATED WASTEWATER

Currently, the quantities of fresh water are decreasing, however, the demographic expansion does not wait in its growth rate, and the needs that it demands for its functioning as a society. Currently, water is used for agri-food, textile, or any other type of industry. In addition, it is also used for domestic use, where aquifers and springs gain strength. Despite this, freshwater reserves are not sufficient for current demands and especially those of the future (Badii et al., 2008). In the context of humanity, some strategies have been proposed to ensure the proper use of freshwater throughout its hydrological scheme (Figure 9.1), i.e., from the time it precipitates, drains, and infiltrates, however, final drainage can be a worldwide problem, especially in underdeveloped or third world countries.

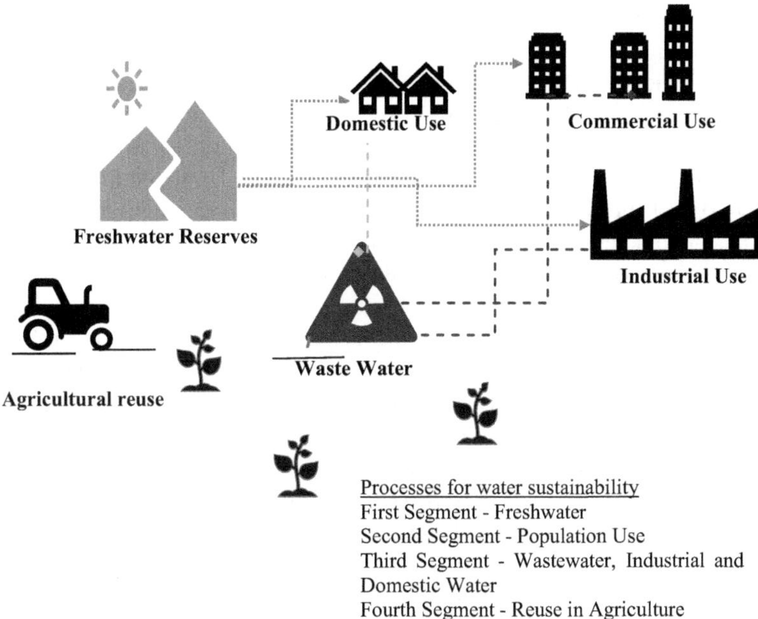

FIGURE 9.1 Common process of freshwater use, occupation, and crop reuse.

One of the strategies maintained by the UN in its Goal 6 is "clean water and sanitation", where the wastewater used in all operations of the population plays an important role. Wastewater treatment ranges from artificial wetlands to aerobic or anaerobic wastewater treatment plants.

It is therefore important to clarify that wastewater sanitation is not a luxury, but a necessity, and currently also a potential advantage to be able to different uses of the same treated water in different processes, an example of this is the agricultural plane Although agricultural irrigation accounts for only 10% of water use, it is the largest use of freshwater in the world, from here could be derived some circumstances of danger for the use of treated water, when using in crops (Fao y Fida, 2006), the positive aspects of using treated water in many crops generate absolute advantages, as long as they have at least a primary treatment, where organic matter and suspended solids can be eliminated (Mara and Carnicross, 1990). Commonly, irrigation is localized, and it has to do with the type of crops to be irrigated when the crops have large foliage and height, it is advisable to use localized irrigation; this is not the case if the crops are small and low, which could be irrigated by sprinkler irrigation. Some countries and regions that have irrigation percentages are: Israel, 67%; India, 25%, and South Africa, 24%. In Latin America, approximately 400 m^3/s of raw wastewater is discharged to surface sources and the soil is irrigated mainly with untreated wastewater; more than half of that is produced in Mexico (Post, 2006).

9.3.1 Main Countries Where Treated Wastewater is Used

Currently, the quantities of fresh water are decreasing, however, the demographic expansion does not wait in its growth rate, and the needs that it demands for its functioning as a society. Currently, water is used for agri-food, textile, or any other type of industry. In addition, it is also used for domestic use, where aquifers and springs gain strength. Despite this, freshwater reserves are not sufficient for current demands and especially those of the future (Badii et al., 2008). In the context of humanity, some strategies have been proposed to ensure the proper use of freshwater throughout its hydrological scheme (Figure 9.1), i.e., from the time it precipitates, drains, and infiltrates, however, final drainage can be a worldwide problem, especially in underdeveloped or third world countries.

One of the strategies maintained by the UN in its Goal 6 is "clean water and sanitation", where the wastewater used in all operations of the population plays an important role. Wastewater treatment ranges from artificial wetlands to aerobic or anaerobic wastewater treatment plants.

It is therefore important to clarify that wastewater sanitation is not a luxury, but a necessity, and currently also a potential advantage to be able to different uses of the same treated water in different processes, an example of this is the agricultural plane Although irrigation for agricultural purposes represents only 10% of the water used, this is the activity with the highest consumption of fresh water on the planet, from here could be derived some circumstances of danger for the use of treated water, when using in crops (Fao y Fida, 2006), the positive aspects of using treated water in many crops generate absolute advantages, as long as they have at least a primary treatment, where organic matter and suspended solids can be eliminated (Mara and Carnicross, 1990; Bakker et al., 2000). Commonly, irrigation is localized, and it has to do with the type of crops to be irrigated when the crops have large foliage and height, it is advisable to use localized irrigation; this is not the case if the crops are small and low, which could be irrigated by sprinkler irrigation. Some countries and regions that have irrigation percentages are: Israel, 67%; India, 25%, and South Africa, 24%. In Latin America, about 400 m^3/s of raw wastewater is delivered to surface sources and the areas are irrigated, most of the time, with untreated wastewater; more than half of this amount is generated in Mexico (Post, 2006).

In 2006, the WHO published new guidelines for the use of wastewater, excretes, and graywater (WHO, 2006), which are a tool for the preventive management of wastewater in agriculture to maximize public health safety. The quality processes of irrigated fruits or crops should be supported by safety standards that maintain the quality at the time of ingestion.

In fact, treated water depends to a large extent on the functionality of the sanitation systems, first on the supply, since it is largely from reliable sources and the supply is from potable water, although on several occasions the water of origin is contaminated. Therefore, to a large extent, river basins play an important role in the interaction between the supply at the source and the consequence of its drainage or final destination, since, depending on the amount of population, the quality of sanitation depends on the quantity of population (Figure 9.2). In this sense, the current population has caused more than 80% of wastewater to be discharged into rivers or directly into the sea without any treatment.

Between 1990 and 2015, the share of the world's population using an improved source of drinking water increased from 76% to 91%, but water scarcity affects more than 40% of the world's population and is expected will increase. Additionally, more than 1,700 million people currently live in river basins where water use exceeds replenishment capacity, often as human settlements destroy vegetation, destroy forests, and affect processes resurfacing and depth. Approximately 70% of all water extracted from rivers, lakes, and aquifers is used for agricultural irrigation. Regarding sanitation, 2.4 million people do not have access to basic sanitation facilities such as latrines or latrines,

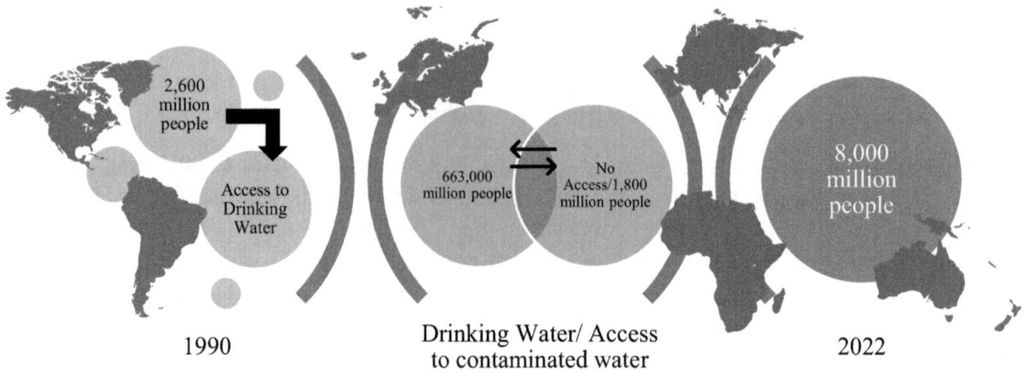

FIGURE 9.2 Predictions of treated water, drinking water, and amount of population in the world.

resulting in more than 80% of man-made wastewater being dumped directly into rivers or into the sea without any treatment. As a result, about 1,000 children die every day from diarrhea-related illnesses (Un, 2016).

9.3.2 Reuse of Treated Water in Agriculture Worldwide

The reuse of wastewater for crop irrigation should be understood as a planned process with different regulations depending on the requirements of each country, infrastructure, and management for the purposes required by the crop, as shown in Table 9.1, in Latin America the use of treated wastewater is mostly used for irrigation of vegetables, the authors in Mexico report an additional disinfectant to secondary treatments, which would lead to greater confidence in its consumption. In Europe, irrigation is reported mainly in trees where the fruit that is consumed is not directly irrigated, due to the policies applied in the country. In Greece, the use of gray wastewater is limited, unlike in Spain, where black water is used. According to the countries mentioned in the table that are part of the MENA region, the authors emphasize the use of domestic wastewater where, through secondary and tertiary treatments, different species of fruit trees, cereals, and crops in home gardens are irrigated; it is important to mention that being one of the areas with the lowest supply of clean water, there is a constant need to make use of resources that are more available, such as wastewater.

The information reported for Australia, unlike the countries mentioned above, no treatment system is used, so the water is discharged directly from the different existing reservoirs or pipes.

9.4 TREATED WATER FOR USE IN CROPS FROM A SUSTAINABLE ECONOMIC PERSPECTIVE

9.4.1 Circular Perspective

The scarcity of water resources due to the increase in the world population puts in perspective the need to improve its management, water is of vital importance for the survival of the human being in addition to sustaining a large part of the items of the global economy (Unidas, 2021). There are two main problems to be faced, one of them is the physical scarcity of water, that is to say, those regions where there are not enough natural resources to satisfy the demand and the other, economic water scarcity, that is to say, the lack of infrastructure for an extraction efficiently and comply with the necessary supply, for this reason, the adoption of a circular approach arises to achieve, in the first instance, sufficient supplies in a safe and sustainable process (Cansi and Cruz, 2020) and in the second instance, a new approach for analyzing the relationship between markets, consumers and the resources obtained, promoting in the same way sustainable and efficient policies and practices (Voulvoulis, 2018).

The concept of a circular economy is being implemented in different areas and places and has regenerative and restorative characteristics based on three fundamental principles, the first is the preservation of resources and the reduction in their use, the second is the optimization of resources including the re-elaboration, renovation and recycling and the third and last one the promoter of the effectiveness in the implementation of the system watching over a reduction of damages and negative impacts (Cerdá and Khalilova, 2016).

In this order of ideas, when applying this sustainable economic perspective to treated water for use in crops, reference should be made to the transition of the water extracted towards the actions of a reduction in the use of the resource, a recycling cycle, and a treatment from wastewater, protecting the environment. To achieve the transition in a more fluid way (Table 9.2), strategies must be established where it must be educated and recognize that water can no longer be considered an unlimited resource that can only be obtained through extraction from the environment, apply the knowledge and practices from in other sectors, such as the energy sector, considering the use of digital technologies as a support tool, must operate from transparency to develop appropriate and innovative infrastructure and must provide access to data and the necessary information in real time

TABLE 9.1
Use of Treated Crop Water Around the World

	País	Cultivo	Tipo de agua Residual	Tratamiento	% de uso	Referencia
Latin America	Mexico	Vegetables	Domestic waste with secondary treatment with disinfection	Extended aeration treatment plant	50,50	(Luciano and Yoval, 2003)
			Raw domestic waste	NA	100	
	Cuba	Radish, carrot, and Marigold flower	Raw domestic waste	NA	100	(Méndez et al., 2006)
			Raw and pure aqueduct water	NA	50,50	
			Aqueduct water	NA	100	
Europe	Cartagena, Spain	Lemon rootstock Macrophyla	Black waste with secondary treatment	Treatment plant	100	(Pedrero and Alarcón, 2009)
	Campotejar, Spain	Lemon rootstock Macrophyla	Sewage with tertiary treatment and well water	Treatment plant	50,50	
	Greece	Olives	Graywater with secondary treatment	Filter packed bed compact	Typical applied by farmers	(Petousi et al., 2015)
			Graywater with tertiary treatment	Treatment of packed bed filter effluent using a sand filter followed by chlorination. packed using a sand filter followed by chlorination.	Typical applied by farmers	
Mena	Tunisia	Vineyards, citrus, olive trees, peaches, pears, apples, pomegranates, alfalfa, sorghum olive trees, peaches, pears, apples, pomegranates, alfalfa, sorghum, and cereals.	Domestic waste with secondary treatment.	Mostly activated sludge; ponds; some tertiary treatment.	100	(Qadir et al., 2010)
	Jordan	Home gardens	Domestic wastewater, secondary and tertiary treatment.	Lagoons and activated sludge; frequent overloading	100	(Redwood, 2008)
Oceania	Australia	Silverbeet	Untreated gray wastewater and drinking water	NA	50,50	(Pinto et al., 2010)
			Graywater	NA	100	
			Drinking water	NA	100	

Source: Own elaboration.

TABLE 9.2
Transition Strategies

Education and Recognition	Water is Not an Unlimited Resource
Apply other contexts	Apply the knowledge and practices from in other sectors
Transparency	Using transparency to develop innovative and appropriate infrastructure
Easy access	Access to data and the necessary information in real time to contribute with efficient public policies

Source: World Economic Forum (Fortuno, 2017).

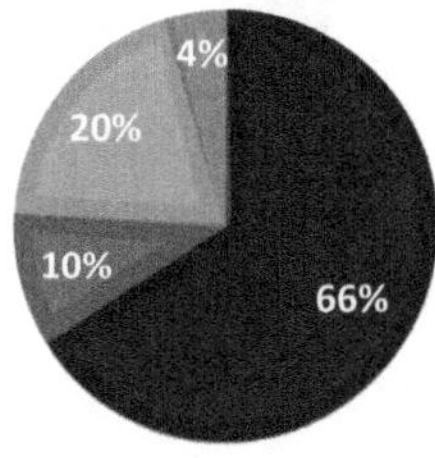

FIGURE 9.3 Water extracted percentages.

Source: World Economic Forum (Fortuno, 2017).

to contribute with efficient public policies and can connect the market with the clients and consumers of the resource (Sarni and Alexander, 2022).

According to data presented by the World Economic Forum, 36% of the total population resides in areas with water stress and of the water extracted on the planet, 66% of said extraction is used in the irrigation of crops, and the other 34% is consumed in domestic work (10%), industrial activities (20%) and the rest is eliminated in the evaporation of deposits (4%) (Figure 9.3) (Fortuño, 2017). Now, according to UNESCO reports, economically rich countries treat approximately 70% of wastewater, this figure is reduced to 38% in countries with medium-high incomes and 28% in countries with medium-low incomes (Figure 9.4). Leaving behind poor countries where only 8% of wastewater from either industrial or municipal sources is treated in any way. This supports the rough estimate that more than 80% of total wastewater is likely to be released into the environment without

The above percentages are important to identify the current situation worldwide and from there to move towards a circular economic policy in favor of water, which includes the emergence of innovative products and technologies and thus adopt effective models in which the collaboration of different actors is involved, achieving the reuse of water in compliance with norms, criteria, and quality standards to promote a new water market, transforming the management of water resources from a linear perspective to a circular perspective (AQUASTAT, 2022)

9.4.2 Comparative from a Global, Regional, and Domestic Approach

Proper management of water and related resources is an objective part of many programs around the world because, as mentioned earlier, such management supports the well-being of the world's people and ecosystem integrity in an integrated economy. By 2050, the demand for water resources is expected to increase by 55% due to population growth, leading to an increase in demand for

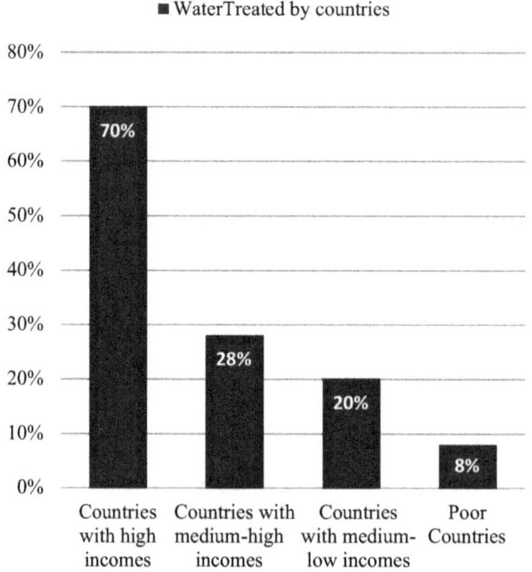

FIGURE 9.4 Water treated by countries.

Source: UNESCO adequate treatment (2017).

industry, domestic use, electricity generation, agriculture, cultivation, and consumption habits (United Nations World Water Assessment Programme, 2015).

The feasibility of water reuse will depend on local circumstances, which will affect the balance between costs and benefits (Winpenny et al., 2013). Although there are only approximate data on the current situation regarding the management of water resources, abstracted, residual, and treated in the world, the regions can be grouped (due to economic development and access to monetary and infrastructure resources) as follows:

Europe and North America: Many of the countries that make up these regions of the planet have high levels of economic development, both in terms of gross domestic product and per capita domestic product, which means that the demand for resources is greater for the population, including natural resources, with the understanding that the greater the wealth, the greater the demand for products and services, which has led governments to design policies that are adapted to care for the environment with respectful practices, implementing green technologies and a circular economy.

Asia and the Pacific: This region is densely populated and continuously faces challenges with respect to climate change, accelerating urbanization, and available water supplies. It also has a large industrial development which increases the demand for resources of all kinds as well as policies with a sustainable approach, and this sustainability is directly linked to water resources, whose main problems are: lack of access to drinking water and sanitation.

The Arab Region: The main challenge in this region of the world is water scarcity and this hinders sustainable development, the least developed countries often obtain the resource from unreliable and unsafe sanitation sources, and better public policies are required to improve water resources management. The agricultural sector remains the largest consumer of water, however, this can vary significantly from country to country, an example being Djibouti which uses 16% of its freshwater withdrawals on crops, while Somalia uses 99% of its total withdrawals (United Nations World Water Assessment Programme, 2015). This urgent need for water supplies has prompted Arab governments to embark on programs to improve irrigation efficiency, reuse wastewater, and harvest water as ways to obtain the precious liquid.

Latin America and the Caribbean: This is a very diverse region in terms of ecology and development, and despite economic growth over the past decade, it remains the most unequal region in the world in terms of resource distribution. Infrastructure plays an important and investment-constrained role, so the priority for countries in this geo-block is to strengthen public policies to improve water management and economic linkages with sustainable socio-economic development, as they are not currently in the foreground of public order.

Africa: Africa's primary goal is to ensure its participation in the global economy, and perhaps more than any other region of the world, the challenge of sustainable development is the development of natural resources and people without repeating the negative impact patterns that have been established throughout its history to grow further and learn more from these negative experiences and ensure sustainability, for example by supporting 'clean' industrial processes that reduce waste and water pollution.

For some researchers, Mexico belongs to the North American region, and for others, it belongs to Latin America, depending on the geographic or economic focus. According to information provided by the National Institute of Statistics and Geography (INEGI), the American continent is divided into three parts: North America (consisting of the United States of America, Canada, and Mexico), Central America, and South America, so for the purposes of this paper, it will be considered as such (Geografía, 2022). In this context, by 2021, Mexico had 2,050 water quality monitoring sites operated by the National Water Commission (Conagua) throughout the country, with fresh water withdrawn exclusively (Agua, 2022).

Mexico is a country with a great cultural, ideological, material, and population diversity, which can be divided into regions based on natural, geographical, ideological, ethnic, and economic relationships (Salazar, 2020). Regarding the management, use and distribution of water resources, Mexico, according to CONAGUA, allocates water to three main sectors: agriculture, public supply and industry. For the country, the sector that most consume the liquid resource is agriculture in 2017 received 76% of the concessioned volume, which is mainly used for irrigation, representing an approximate 66,799 cubic hectometers (also including livestock and aquaculture activities), so one of the main challenges facing resource management is the sustainable use trying to increase efficiency in the agricultural sector, Similarly, optimized management in irrigation districts is an indicator that allows us to evaluate the efficiency with which water is used in agricultural production, which will depend on various factors such as the conduction from the supply source to the plots, the type of crop in which it is used and the way in which the resource is used (Naturales, 2022).

From this issue arises the need to take advantage of the wastewater supply through reuse, paying special attention to the quality of the resource in accordance with the regulations in force in the country (NOM-001-SEMARNAT-1996) or those indicated by the World Health Organization (WHO). It is estimated that the total municipal wastewater generated in the country is 228.7 m^3/s, and the flow that receives some type of treatment is 113.3 m^3/s, which represents an average of approximately 48.66%, which evidently shows that there is still a long way to go with respect to sustainable management (Estrada and Rojas, 2016).

In view of the above, it is vitally important to lay the groundwork for a sustainable approach to the use of treated water for crops in Mexico, which can be categorized as follows (Figure 9.5):

1. Encouraging the reuse of wastewater for agricultural uses in areas with scarce water resources.
2. Verifying that wastewater comes from non-industrialized areas.
3. Comply with the minimum water quality standards established by applicable norms to receive treatment.
4. Analyze that the irrigation area complies with the characteristics of permeability, drainage, etc.
5. Validate that the crop pattern is tolerant to the use of treated water for irrigation.
6. Conclude that it is a profitable project for the producers considering the necessary financial analysis for the sanitation and utilization processes.

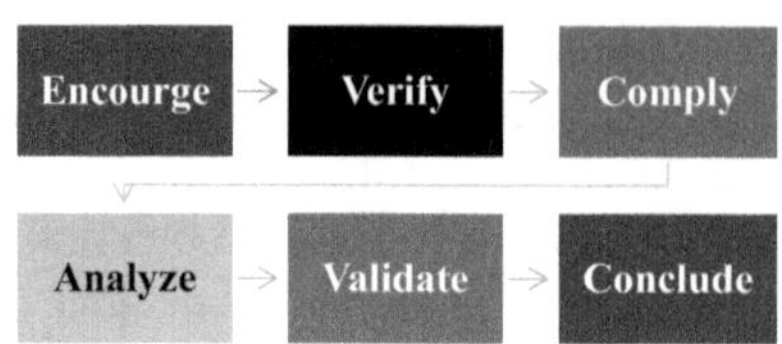

FIGURE 9.5 General guidelines for reuse projects in favor of a sustainable approach.

Source: Own source.

The aforementioned guidelines are necessary to provide a sustainable economic approach with certainty in the process, necessary to migrate to more efficient scenarios in the management of water resources, which, as has already been repeatedly pointed out, are the trend in a world that seeks to ensure continuous development without neglecting the environment and without leading to a gradual shortage of the precious liquid that is necessary for any human activity for economic, industrial, energy or even mere survival purposes.

9.5 IMPACTS OF TREATED WATER AND ITS USE IN AGRICULTURE

In 2008, Mexico had 2,101 municipal wastewater (MWW) treatment plants treating only 84 m^3/s of wastewater, despite collecting 208 m^3/s in sewer systems. Many plants operated below capacity due to poor maintenance and insufficiently trained personnel. The primary treatment systems used were activated sludge (44.3%) and waste stabilization ponds (18%) for the MWW collected (Zurita et al., 2012). This type of water is used for different purposes, of which the most important is for agricultural irrigation; 35,000 ha are irrigated with this type of water around the country (Olvera, 2017). Unsustainable use of water will lead to the use of untreated wastewater; This is mainly due to the lack of infrastructure to collect, treat and dispose of wastewater. On the other hand, when treated wastewater is mainly used for agricultural purposes. The level of treatment depends on the capacity of the treatment plants: in 56% of cities in low- and middle-income countries, treatment plants are operating or only partially functioning (Raschid-Sally, 2013).

Treated wastewater improves its quality when diluted or mixed with fresh water. Salinity has been shown to reduce soil respiration, enzyme activity, soil microbial biomass, and bacterial growth rates, all of which affect biogeochemical cycles (Zhang et al., 2019). On the other hand, the use of treated wastewater can also improve soil fertility because it contains organic matter; elements such as N, P, and K; and trace elements such as Fe, Mn, Zn, and Cu necessary for plant growth (Hassan et al., 2022). However, if wastewater is not properly treated, nutrients can accumulate in the soil and present a threat to the environment (Elgallal et al., 2016).

9.5.1 Environmental Risks

Treated wastewater is a safe and promising alternative to agricultural irrigation, and effective planning and management can make it more cost-effective (Khan et al., 2022). While irrigation with treated wastewater and treated wastewater can reduce the use of natural water resources, it can also lead to environmental problems.

9.5.2 Soil Impacts

Agriculture is one of the main agro-industrial activities generating high demand for water, therefore the reuse of treated wastewater has increased to protect already scarce water resources and maintain quality. environment by reducing the amount of wastewater discharged into freshwater bodies. However, despite the advantages and benefits of reuse, the use of treated wastewater has limitations, mainly in terms of physical and chemical parameters, as they can affect thoroughly on soil quality and plant growth. Undesirable substances in wastewater, such as high concentrations of salts, dissolved solids, toxic substances, microorganisms, and heavy metals, can be present in wastewater even after it has been treated (Iloms et al.2020; Yang et al., 2019; Abedinzadeh et al., 2019).

Abedi-Koupai et al. (2011) mention that irrigating the soil with wastewater with salt content and suspended solids affects the physical properties of the soil. When wastewater is treated, two large percentages of what is considered polluting are eliminated, however, there are some that cannot be removed in their entirety and, when these wastewaters are used, mainly in agriculture, they impact on the soil and the crops themselves.

9.5.2.1 Microorganisms

Wastewater from treated wastewater can contain bacteria beneficial to various soil functions as well as a number of potential pathogens (Ibekwe et al., 2018). Total coliforms, fecal, and *Escherichia coli* are the bacteria with the highest representation in treated water and, in other cases, Salmonella, Shigella and legionella (Bonetta et al., 2022). However, the presence of these pathogens does not always affect crops (Farhadkhani et al., 2018). Mkhinini et al. (2020) say that irrigation with long-term treated wastewater increases the metabolic activity of soil microorganisms; In the same way, it also increases the metabolic activity of soil microorganisms. Meanwhile, the accumulation of elements can seriously threaten biological processes in the soil and is a limiting factor for the reuse of wastewater after treatment in agriculture.

Because the microbial safety of products that are irrigated by treated wastewater is affected by the type and concentration of microorganisms in them, climatic conditions, irrigation method, and plant type; the impact of wastewater reuse needs to be properly assessed. However, in treated water applied to different crops, the absence of coliforms has been observed.

9.5.2.2 Salts

The quality of the water that is applied to irrigation not only depends on the total salt content but also on the type of salt and could damage the soil for a long time. The most common soil problems associated with water salinity are the rate of water infiltration through the soil and is one of the most brutal environmental factors limiting the productivity of crop plants (Shrivastava & Kumar, 2015). Salt water can cause salinity impacting soil cause impacts structure by limiting organic matter accumulation and plant growth. This affects hydraulic properties like water retention and hydraulic conductivity, which in turn influence soil structure during wetting and drying cycles through infiltration and evaporation processes (Tang et al., 2021).

The saturated hydraulic conductivity of the soil characterizes the porosity of the soil and can be divided into permeable, semi-permeable, and poorly permeable (Mantovani et al., 2009). The effect of the use of treated wastewater on the chemical properties of the soil can only be noticed after a long period of use and can vary depending on the physicochemical composition of the soil, the atmospheric conditions climate, and soil type (Heidarpour et al., 2007). The permeability loss can be lost when the electrical conductivity is approximately the same as that of domestic wastewater. Plants absorb the Pb and Cd of the soil, which makes plants potential sources of contamination for humans and animals.

9.5.2.3 Metals and Other Compounds

Through the use of treated wastewater for cropland irrigation, organic micropollutants are introduced into soils and potentially transferred to groundwater (Helmecke et al., 2020). Pharmaceutical compounds are applied to agricultural land by irrigation with treated wastewater. Carbamazepine, lamotrigine, caffeine, metoprolol, sulfamethoxazole, and sildenafil are stable in soil after discharge into treated wastewater. Several reports have shown that many pharmaceutical compounds are not completely degraded during wastewater treatment and therefore residues are found in sludge and/or treated wastewater (Grossberger et al., 2014; Williams and McLain, 2012). Treated water can also increase levels of antibiotic resistance (Gatica and Cytryn, 2013).

Chemical accumulation can pose a serious threat to biological processes in the soil and limit the reuse of treated wastewater in agriculture (Mkhinini et al., 2020). Rezapour et al. (2019), found that irrigation with treated wastewater results in a representative accumulation of metal concentrations in the soil and as expected were found in wheat roots, shoots, and grains. Plants have the ability to absorb elements such as Pb or Cd from the soil, elements that are highly toxic, making them potential sources of contamination for humans and animals (consumers). On the other hand, treated wastewater can also have a positive impact on soil and plant nutrition by improving soil fertility by reducing the need for mineral fertilizers due to its richness in organic matter and minerals such as N, P, and K (Perulli et al., 2019). The nutritional efficiency of treated wastewater is still underestimated because it can result in significant fertilizer savings due to its high nutrient content. Figure 9.6 shows the irrigation route with treated wastewater containing pathogens/contaminants.

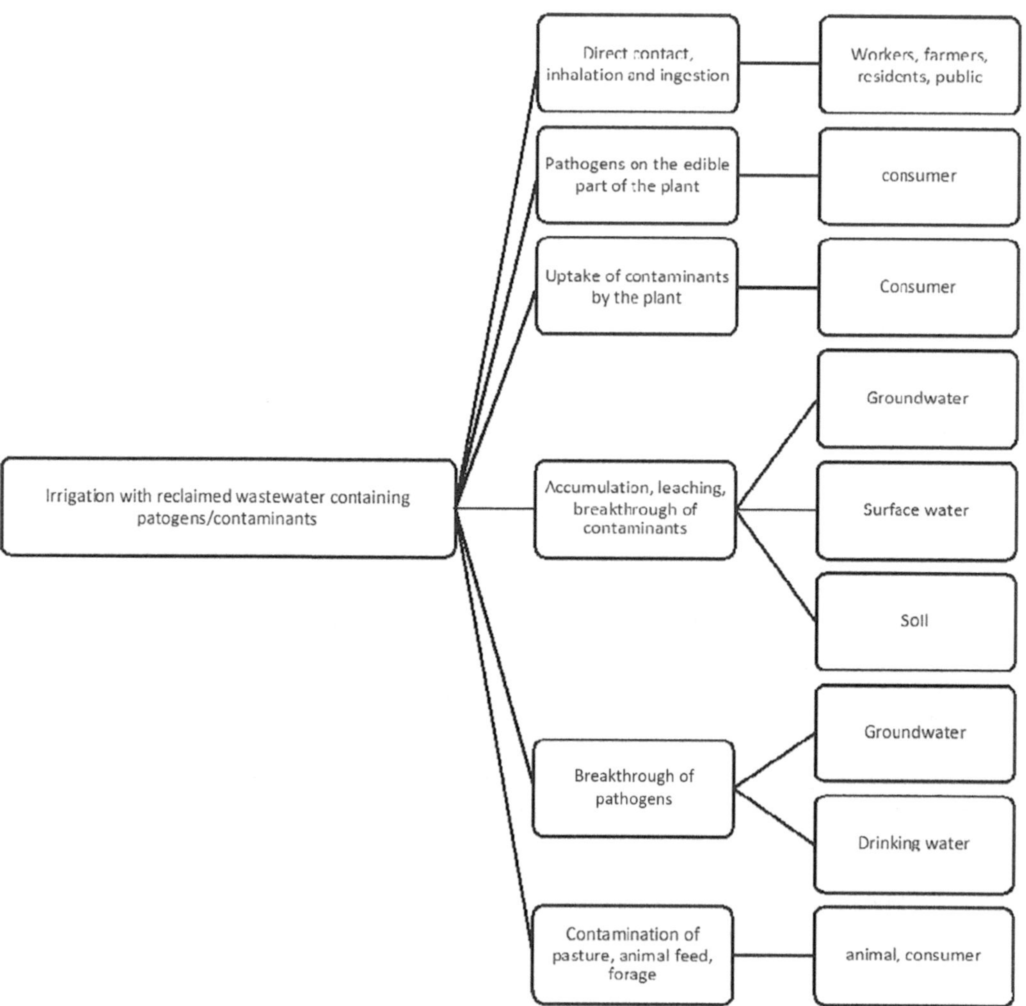

FIGURE 9.6 Irrigation with treated wastewater containing pathogens

Source: Taken from Helmeke et al. (2020).

9.6 HEALTH PROBLEMS ASSOCIATED WITH AGRICULTURAL WASTEWATER

Public health safety is one of the main issues with applying agricultural wastewater (Leonel and Tonetti, 2021). In this sense, reusing wastewater inadequately treated or not at all is connected with health risks that depend on the pollutant, concentration, and exposure time (Adegoke et al., 2018). The contaminants that can be fin in agricultural wastewater are distributed into three groups: inorganic, organic chemicals, and microbiological pollutants (Dickin et al., 2016).

9.6.1 Inorganic and Organic Chemicals Pollutants

The wastewater discharged from several sources as industrial, residential, or agricultural areas, is typically not treated in emerging nations or is subjected to basic treatments. As a result, the effluents under these conditions are frequently contaminated with several pollutants. Therefore, numerous negative health impacts have been reported on a farming community that applied wastewater for irrigation (Wu, 2020).

A significant part of inorganic pollutants easily found in wastewater are heavy metals, which can be arranged in the following order based on how toxically they are to organisms: mercury (Hg) > copper (Cu) > zinc (Zn) > nickel (Ni) > lead (Pb) > cadmium (Cd) > chromium (Cr) > iron (Fe) > manganese (Mn) > aluminum (Al) (Zwolak et al., 2019).

Heavy metals in wastewater can be harmful and pose a health concern if they are consumed and present in enough concentrations; also, the level of risk is related to the type of pollutant. In the case of Hg, the nervous system is susceptible to all types; meanwhile, EPA has declared two forms of Hg carcinogenic: mercuric chloride and methyl mercury. Other problems related to Hg exposure are irritability, changes in vision or hearing, memory problems, and altered brain functions and memory problems (Jaishankar et al., 2014). Exposure to Cu presents gastrointestinal side effects, headache, and neurological symptoms. Meanwhile, high concentrations can produce cardiac and renal failure, methemoglobinemia, rhabdomyolysis, intravascular hemolysis, hepatic necrosis, encephalopathy, and death (Royer and Sharman, 2022). On the other hand, levels of Pb can affect cognitive development in children; for example, Pb toxicity is related to adverse effects on the children's IQ in children (<6 years old) (Acosta-Saavedra et al., 2011). Another metal is Cr; a significant dose may adversely affect the gut, throat, heart, liver, kidneys, and mouth. In addition, lung cancer, respiratory tract irritation, nose, and skin ulcers are all caused by the exposure (ATSDR, 2016).

The prevailing vegetable and crop cultivation situation use wastewater heavily contaminated with inorganic compounds. Furthermore, heavy metals can accumulate in vegetables that are irrigated by polluted wastewater; for example, Ullah et al (2022) determined the concentration of heavy metals in some vegetables; as a result, the values of inorganic pollutants (Fe, Cr, Pb) were higher than the WHO limits, presenting a health risk, as commented before.

Other chemicals in agricultural wastewater are persistent organic pollutants (POPs); these substances can move over long distances, resist deterioration, and bioaccumulate due to their physical and chemical features (Fitzgerald and Wikoff, 2014). Table 9.1 resumed some of the POPs, their toxicity, and their health effects.

DDT: 1,1,1-trichloro-2,2-bis[p-chlorophenyl]-ethane, PCBs: polychlorinated biphenyls.

The primary sources of POPs are agricultural and industrial activities, which negatively affect environmental and public health (Aravind Kumar et al., 2022). The IARC (International Agency for Research on Cancer) has categorized PCBs as probably carcinogenic to humans. PCBs may also have noncarcinogenic health effects, such as dermal lesions, immune system issues, hepatotoxicity, immunosuppression, reproductive and developmental toxicity, endocrine disruption, neurotoxicity, and carcinogenicity endocrine disruptors (Wu, 2020). Furthermore, low levels of PCB exposure are associated with subtle neurodevelopmental effects in newborns and children (Ross, 2004).

9.6.2 Microbiological Pollutants

Several reports have demonstrated adverse health public risks from microbial pollutants in untreated wastewater, including agricultural use, diarrhea, giardiasis, campylobacteriosis, encephalitis, salmonellosis, typhoid, hepatitis A, salmonellosis, and gastroenteritis (Kesari et al., 2021). In addition, human parasites such as protozoa and helminth eggs are of particular importance; They are the most difficult to eliminate by treatment processes and cause a number of gastrointestinal infections in both developed and developing countries (Hussain et al., 2002).

A study developed by Muñoz et al. (2010), estimated the exposure risk of bacterial concentrations in effluents commonly used for irrigation. The annual probability of infection by bacteria, a value below 10^{-4}, is considered acceptable. The level observed in an effluent exceeds the limit commented before even considering the minimum concentration of pathogens. Compared to developed countries, the concentration of viruses, protozoan parasites, and helminths in wastewater in developing countries might range from 10 to 1,000 times higher (Blanca Jiménez et al., 2009).

Controlling the diversity of pollutants (inorganic, organic, or microbiological) is indispensable to preventing human health risks. However, the evaluation of the public health risks associated and the economic evaluation has to be applied to determine a realistic posture of wastewater use.

9.7 CHALLENGES POSED BY WASTEWATER AND THE SUCCESSFUL IMPLEMENTATION OF CONSTRUCTED WETLANDS

One of the alternative methods of treating wastewater and returning it uses for agricultural irrigation is to implement constructed wetlands; a sustainable and cost-effective technology that is applicable for the removal of pollutants and pathogens from wastewater. At present, wastewater treatment is still a subject that has challenges, it remains one of the problems of the last decades. That is why there is a need to develop cost-effective and high-performance treatment systems since inadequate treatment poses major health, social, and environmental problems. An alternative method for wastewater treatment and reuse for agricultural irrigation is the construction of wetlands; sustainable and cost-effective technology is applied to remove contaminants and pathogens from wastewater (Shingare et al., 2019).

The main challenge of wastewater treatment is the presence of new contaminants in water, which of course require new treatment systems. Micropollutants (Kosek et al., 2020), nano and microplastics (Enfrin et al., 2019; Iroegbu et al., 2020), radioactive substances (Zhang et al., 2019), persistent substances (Berg et al., 2022), fluorinated organic compounds (Kurwadkar et al., 2022), endocrine disrupting compounds belonging to the group of emerging pollutants (Vieira et al., 2021) are some examples of pollutants that are currently present in water, are some examples of contaminants that are currently present in water, their removal becomes complex, and if this water is used without prior treatment, there is a risk of contaminating the soil and groundwater when it is used as irrigation water in agriculture.

Constructed wetlands are identified as a cost-effective and environmentally friendly ecotechnology to remediate polluted water, especially in decentralized communities and rural areas, where due to economic limitations, establishing a conventional water treatment plant is not viable. Figure 9.7 shows the classification of artificial wetland systems for the treatment of liquid effluents. Adapted from Wallace and Knight (2006). In this sense, it is also promoted to design, apply, and evaluate new technologies such as biochar as a novel alternative substrate for wastewater treatment (Deng et al., 2021).

9.8 CONCLUSIONS

Reusing wastewater after treatment is one of the options to solve water shortages in agricultural activities, as well as reduce health risks and reduce the amount of wastewater entering freshwater sources. In this case, water resources can be used to irrigate crops, the edible part

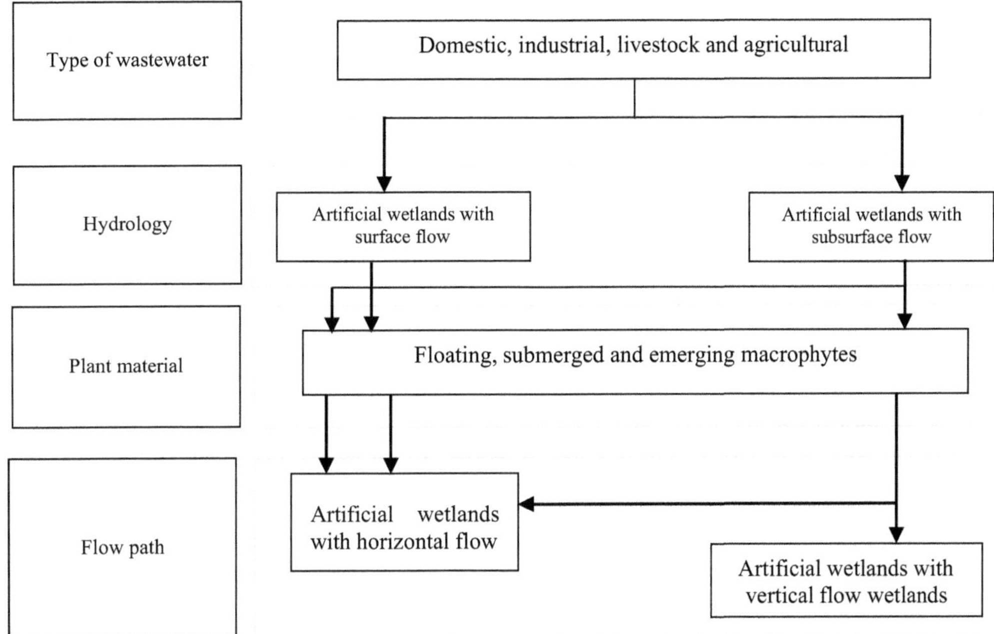

FIGURE 9.7 Classification of constructed wetlands.

of which is not in contact with irrigation water, to avoid microbial contamination of the final product. A major challenge for both conventional wastewater treatment plants and wetlands is to find the point where the water quality obtained after treatment does not cause any impact on any components. of the environment, as well as human health and agricultural crops itself, irrigated with purified water.

REFERENCES

Abedi-Koupai, J., Mostafazadeh-Fard, B., Afyuni, M., & Bagheri, M. R. (2011). Effect of treated wastewater on soil chemical and physical properties in an arid region. *Plant, Soil and Environment*, *52*(8), 335–344. https://doi.org/10.17221/3450-PSE

Abedinzadeh, M., Etesami, H., & Alikhani, H. A. (2019). Characterization of rhizosphere and endophytic bacteria from roots of maize (Zea mays L.) plant irrigated with wastewater with biotechnological potential in agriculture. *Biotechnology Reports*, 21, e00305. https://doi.org/10.1016/j.btre.2019.e00305

Acosta-Saavedra, L. C., Moreno, M. E., Rodríguez-Kessler, T., Luna, A., Gómez, R., Arias-Salvatierra, D., & Calderon-Aranda, E. S. (2011). Environmental exposure to lead and mercury in Mexican children: A real health problem. *Toxicology Mechanisms and Methods*, *21*(9), 656–666. Taylor & Francis. https://doi.org/10.3109/15376516.2011.620997

Adegoke, A. A., Amoah, I. D., Stenström, T. A., Verbyla, M. E., & Mihelcic, J. R. (2018). Epidemiological evidence and health risks associated with agricultural reuse of partially treated and untreated wastewater: A review. *Frontiers in Public Health*, *6*, 337. https://doi.org/10.3389/FPUBH.2018.00337/BIBTEX

Agua, C. N. (November 2022). CONAGUA. https://www.gob.mx/conagua/articulos/calidad-del-agua#:~:text=En%202021%20se%20contaba%20con,Conagua%20en%20todo%20el%20pa%C3%ADs.&text=Calidad%20del%20agua.,estuvo%20constituida%20por%20788%20sitios.

Akponikpè, P. B. I., Wima, K., Yacouba, H., & Mermoud, A. (2011). Reuse of domestic wastewater treated in macrophyte ponds to irrigate tomato and eggplant in semi-arid West-Africa: Benefits and risks. *Agricultural Water Management*, *98*(5), 834–840. https://doi.org/10.1016/j.agwat.2010.12.009

AQUASTAT. (Occtober 2022). Organizaci6n de las Naciones Unidas para la Alimentaci6n y la Agricultura. https://www.fao.org/aquastat/es/

Aravind kumar, J., Krithiga, T., Sathish, S., Renita, A. A., Prabu, D., Lokesh, S., Geetha, R., Namasivayam, S. K. R., & Sillanpaa, M. (2022). Persistent organic pollutants in water resources: Fate, occurrence, characterization and risk analysis. *Science of the Total Environment, 831*, 154808. https://doi.org/10.1016/j.scitotenv.2022.154808

ATSDR. (2016). ToxFAQsTM: Cromo (Chromium). https://www.atsdr.cdc.gov/es/toxfaqs/es_tfacts7.html

Badii, M., Landeros, J., & Cerna, E. (2008). El recurso de agua y sustentabilidad. *Revista Daena International Journal of Good Conscience, 3*(1), 661–671.

Berg, C., Crone, B., Gullett, B., Higuchi, M., Krause, M. J., Lemieux, P. M., Martin, T., Shields, E. P., Struble, E., Thoma, E., & Whitehill, A. (2022). Developing innovative treatment technologies for PFAS-containing wastes. *Journal of the Air & Waste Management Association, 72*(6), 540–555. https://doi.org/10.1080/10962247.2021.2000903

Bonetta, S., Pignata, C., Gasparro, E., Richiardi, L., Bonetta, S., & Carraro, E. (2022). Impact of wastewater treatment plants on microbiological contamination for evaluating the risks of wastewater reuse. *Environmental Sciences Europe, 34*(1), 20. https://doi.org/10.1186/s12302-022-00597-0

Cansi, F., & Cruz, P. M. (15 July 2020). Sostenibilidad: económica, social y ambiental. https://doi.org/10.14198/Sostenibilidad2020.2.04

Cerdá, E., & Khalilova, A. (2016). Ministerios de Industria, Comercio y Turismo. https://www.mincotur.gob.es/Publicaciones/Publicacionesperiodicas/EconomiaIndustrial/RevistaEconomiaIndustrial/401/CERD%C3%81%20y%20KHALILOVA.pdf

Deng, S., Chen, J., & Chang, J. (2021). Application of biochar as an innovative substrate in constructed wetlands/biofilters for wastewater treatment: Performance and ecological benefits. *Journal of Cleaner Production, 293*, 126156. https://doi.org/10.1016/j.jclepro.2021.126156

Dickin, S. K., Schuster-Wallace, C. J., Qadir, M., & Pizzacalla, K. (2016). A review of health risks and pathways for exposure to wastewater use in agriculture. *Environmental Health Perspectives, 124*(7), 900–909. Public Health Services, US Dept of Health and Human Services. https://doi.org/10.1289/ehp.1509995

Elgallal, M., Fletcher, L., & Evans, B. (2016). Assessment of potential risks associated with chemicals in wastewater used for irrigation in arid and semiarid zones: A review. *Agricultural Water Management, 177*, 419–431. https://doi.org/10.1016/j.agwat.2016.08.027

Enfrin, M., Dumée, L. F., & Lee, J. (2019). Nano/microplastics in water and wastewater treatment processes - Origin, impact and potential solutions. *Water Research, 161*, 621–638. https://doi.org/10.1016/j.watres.2019.06.049

Estrada, O. X., & Rojas, H. S. (2016). Reúso de aguas residuales en la agricultura. Jiutepec, Morelos: Instituto Mexicano de Tecnología del Agua.

FAO y FIDA. (2006). El agua para la alimentación, la agricultura y los medios de vida rurales. En: El agua, una responsabilidad compartida. 2º Informe de las Naciones Unidas sobre el desarrollo de los recursos hídricos en el mundo. Executive Summary. 47 p. www.unesco.org/water/wwap/index_es.shtml [Accessed September 2006].

Farhadkhani, M., Nikaeen, M., Yadegarfar, G., Hatamzadeh, M., Pourmohammadbagher, H., Sahbaei, Z., & Rahmani, H. R. (2018). Effects of irrigation with secondary treated wastewater on physicochemical and microbial properties of soil and produce safety in a semi-arid area. *Water Research, 144*, 356–364. https://doi.org/10.1016/j.watres.2018.07.047

Fipps, G. (2007). Irrigation Water Quality Standards and Salinity Management Strategies. *Texas Cooperaive Extension Services*. https://twon.tamu.edu/wp-content/uploads/sites/3/2021/06/irrigation-water-quality-standards-and-salinity-management-strategies-1.pdf

Fitzgerald, L., & Wikoff, D. S. (2014). Persistent organic pollutants. In *Encyclopedia of toxicology* (3rd ed., pp. 820–825). Elsevier. https://doi.org/10.1016/B978-0-12-386454-3.00211-6

Fortuño, M. (31 March 2017). World Economic Forum. https://es.weforum.org/agenda/2017/03/la-economia-del-agua-cada-vez-sera-mas-importante

Gatica, J., & Cytryn, E. (2013). Impact of treated wastewater irrigation on antibiotic resistance in the soil microbiome. *Environmental Science and Pollution Research, 20*(6), 3529–3538. https://doi.org/10.1007/s11356-013-1505-4

Geografía, I. N. (November 2022). Cuéntame de México. https://cuentame.inegi.org.mx/territorio/mexico.aspx?tema=T

Grossberger, A., Hadar, Y., Borch, T., & Chefetz, B. (2014). Biodegradability of pharmaceutical compounds in agricultural soils irrigated with treated wastewater. *Environmental Pollution, 185*, 168–177. https://doi.org/10.1016/j.envpol.2013.10.038

Hassan, J., Rajib, Md. M. R., Sarker, U., Akter, M., Khan, Md. N.-E.-A., Khandaker, S., Khalid, F., Rahman, G. K. M. M., Ercisli, S., Muresan, C. C., & Marc, R. A. (2022). Optimizing textile dyeing wastewater for tomato irrigation through physiochemical, plant nutrient uses and pollution load index of irrigated soil. *Scientific Reports*, *12*(1), 10088. https://doi.org/10.1038/s41598-022-11558-1

Heidarpour, M., Mostafazadeh-Fard, B., Abedi Koupai, J., & Malekian, R. (2007). The effects of treated wastewater on soil chemical properties using subsurface and surface irrigation methods. *Agricultural Water Management*, *90*(1–2), 87–94. https://doi.org/10.1016/j.agwat.2007.02.009

Helmecke, M., Fries, E., & Schulte, C. (2020). Regulating water reuse for agricultural irrigation: Risks related to organic micro-contaminants. *Environmental Sciences Europe*, *32*(1), 4. https://doi.org/10.1186/s12302-019-0283-0

Hussain, I., Raschid, L., Hanjra, M. A., & van der Hoek, W. (2002). *Wastewater use in Agriculture Review of Impacts and Methodological Issues in Valuing Impacts*. Colombo: International Water Management Institute.

Ibekwe, A. M., Gonzalez-Rubio, A., & Suarez, D. L. (2018). Impact of treated wastewater for irrigation on soil microbial communities. *Science of The Total Environment*, *622–623*, 1603–1610. https://doi.org/10.1016/j.scitotenv.2017.10.039

Iloms, E., Ololade, O. O., Ogola, H. J. O., & Selvarajan, R. (2020). Investigating industrial effluent impact on municipal wastewater treatment plant in Vaal, South Africa. *International Journal of Environmental Research and Public Health*, *17*(3), 1096. https://doi.org/10.3390/ijerph17031096

Instituto Mexicano de Tecnología del Agua y el Ministerio de Medio Ambiente y Agua de Bolivia. (2018). Guía Técnica para el Reúso de Aguas Residuales en la Agricultura. In *Proyecto de Cooperación Triangular México Bolivia y Alemania*. Edición y Fotografías: IMTA y GIZ en Bolivia. SEGUNDA EDICIÓN, p. 118.

Iroegbu, A. O. C., Sadiku, R. E., Ray, S. S., & Hamam, Y. (2020). Plastics in municipal drinking water and wastewater treatment plant effluents: Challenges and opportunities for South Africa-a review. *Environmental Science and Pollution Research*, *27*(12), 12953–12966. https://doi.org/10.1007/s11356-020-08194-5

Jaishankar, M., Tseten, T., Anbalagan, N., Mathew, B. B., & Beeregowda, K. N. (2014). Toxicity, mechanism and health effects of some heavy metals. *Interdisciplinary Toxicology*, *7*(2), 60–72. Slovak Toxicology Society. https://doi.org/10.2478/intox-2014-0009

Jiménez, B., Mara, D., Carr, R., & Brissaud, F. (2009). Wastewater treatment for pathogen removal and nutrient conservation: Suitable systems for use in developing countries. *Wastewater Irrigation and Health*, 175–196. https://doi.org/10.4324/9781849774666-19

Kesari, K. K., Soni, R., Jamal, Q. M. S., Tripathi, P., Lal, J. A., Jha, N. K., Siddiqui, M. H., Kumar, P., Tripathi, V., & Ruokolainen, J. (2021). Wastewater treatment and reuse: A review of its applications and health implications. *Water, Air, & Soil Pollution*, 232(5), Article no. 208. https://doi.org/10.1007/s11270-021-05154-8

Khan, M. M., Siddiqi, S. A., Farooque, A. A., Iqbal, Q., Shahid, S. A., Akram, M. T., Rahman, S., Al-Busaidi, W., & Khan, I. (2022). Towards sustainable application of wastewater in agriculture: A review on reusability and risk assessment. *Agronomy*, 12(6), 1397. https://doi.org/10.3390/agronomy12061397

Kosek, K., Luczkiewicz, A., Fudala-Książek, S., Jankowska, K., Szopińska, M., Svahn, O., Tränckner, J., Kaiser, A., Langas, V., & Björklund, E. (2020). Implementation of advanced micropollutants removal technologies in wastewater treatment plants (WWTPs) - Examples and challenges based on selected EU countries. *Environmental Science & Policy*, *112*, 213–226. https://doi.org/10.1016/j.envsci.2020.06.011

Kurwadkar, S., Dane, J., Kanel, S. R., Nadagouda, M. N., Cawdrey, R. W., Ambade, B., Struckhoff, G. C., & Wilkin, R. (2022). Per- and polyfluoroalkyl substances in water and wastewater: A critical review of their global occurrence and distribution. *Science of the Total Environment*, *809*, 151003. https://doi.org/10.1016/j.scitotenv.2021.151003

Leonel, L. P., & Tonetti, A. L. (2021). Wastewater reuse for crop irrigation: Crop yield, soil and human health implications based on giardiasis epidemiology. *Science of The Total Environment*, *775*, 145833. https://doi.org/10.1016/j.scitotenv.2021.145833

Luciano, M. I., & Yoval, S. (2003). Cultivo De Hortalizas Con Aguas Residuales Tratadas. *1996*, 211–248. http://documentacion.ideam.gov.co/openbiblio/bvirtual/018834/MEMORIAS2004/CapituloIV/4CultivodeHortalizasLuciano.pdf

Mara, D. y Carnicross, S. (1990). Directrices para el uso sin riesgos de aguas residuales y excretas en agricultura y acuicultura. Organización Mundial de la Salud (OMS), Ginebra.

Mantovani, E. C., Bernardo, S., & Palaretti, L. F. (2009). *Irrigação: princípios e métodos*. 3. ed. Viçosa: UFV. 355p.

Méndez, M., Pérez, J., Hernández, G., & Campos, O. (2006). Uso de las aguas residuales para el riego de cultivos agrícolas, en la agricultura urbana. *Revista Ciencias Técnicas Agropecuarias*, *15*(3), 17–21.

Mkhinini, M., Boughattas, I., Alphonse, V., Livet, A., Giusti-Miller, S., Banni, M., & Bousserrhine, N. (2020). Heavy metal accumulation and changes in soil enzymes activities and bacterial functional diversity under long-term treated wastewater irrigation in East Central region of Tunisia (Monastir governorate). *Agricultural Water Management*, *235*, 106150. https://doi.org/10.1016/j.agwat.2020.106150

Muñoz, I., Tomàs, N., Mas, J., García-Reyes, J. F., Molina-Díaz, A., & Fernández-Alba, A. R. (2010). Potential chemical and microbiological risks on human health from urban wastewater reuse in agriculture. Case study of wastewater effluents in Spain. *Journal of Environmental Science and Health, Part B*, *45*(4), 300–309. https://doi.org/10.1080/03601231003704648

Naturales, S. d. (30 November 2022). SEMARNAT.GOB.MX. https://apps1.semarnat.gob.mx:8443/dgeia/informe18/tema/cap6.html

Olvera, B. I. (2017). *Efectos del agua tratada, en suelo y cultivos agrícolas: caso de la planta tratadora de aguas residuales del municipio de atlixco, puebla*. Tesis. Colegio de Postgraduados.

Pedrero, F., & Alarcón, J. J. (2009). Effects of treated wastewater irrigation on lemon trees. *Desalination*, *246*(1–3), 631–639. https://doi.org/10.1016/j.desal.2008.07.017

Perulli, G. D., Bresilla, K., Manfrini, L., Boini, A., Sorrenti, G., Grappadelli, L. C., & Morandi, B. (2019). Beneficial effect of secondary treated wastewater irrigation on nectarine tree physiology. *Agricultural Water Management*, *221*, 120–130. https://doi.org/10.1016/j.agwat.2019.03.007

Petousi, I., Fountoulakis, M. S., Saru, M. L., Nikolaidis, N., Fletcher, L., Stentiford, E. I., & Manios, T. (2015). Effects of reclaimed wastewater irrigation on olive (Olea europaea L. cv. 'Koroneiki') trees. *Agricultural Water Management*, *160*, 33–40. https://doi.org/10.1016/j.agwat.2015.06.003

Pinto, U., Maheshwari, B. L., & Grewal, H. S. (2010). Effects of greywater irrigation on plant growth, water use and soil properties. *Resources, Conservation and Recycling*, *54*(7), 429–435.

Post, J. (2006). Wastewater treatment and reuse in the eastern medi-terranean region. *Water*, *21*, 36–41.

Qadir, M., Bahri, A., Sato, T., & Al-Karadsheh, E. (2010). Wastewater production, treatment, and irrigation in Middle East and North Africa. *Irrigation and Drainage Systems*, *24*(1–2), 37–51. https://doi.org/10.1007/s10795-009-9081-y

Qadir, M., Drechsel, P., Jiménez Cisneros, B., Kim, Y., Pramanik, A., Mehta, P., & Olaniyan, O. (2020). Global and regional potential of wastewater as a water, nutrient and energy source. *Natural Resources Forum*, *44*(1), 40–51. https://doi.org/10.1111/1477-8947.12187

Raschid-Sally, L. (2013). City Waste for Agriculture: Emerging Priorities which Influence Agenda Setting. *Aquatic Procedia*, *1*, 88–99. https://doi.org/10.1016/j.aqpro.2013.07.008

Redwood, M. (2008). Greywater irrigation: challenges and opportunities. *CABI Reviews*, *2008*, 1–7. https://doi.org/10.1079/PAVSNNR20083063

Rezapour, S., Atashpaz, B., Moghaddam, S. S., & Damalas, C. A. (2019). Heavy metal bioavailability and accumulation in winter wheat (Triticum aestivum L.) irrigated with treated wastewater in calcareous soils. *Science of the Total Environment*, *656*, 261–269. https://doi.org/10.1016/j.scitotenv.2018.11.288

Rosillo Martínez, J., García Martínez, M. B., & Moreno Mata, F. A. (2017). El agua residual tratada y el agricultor de Santa María del río (modular-temporary housing: tradition and recycling modular)(temporária de habitação: tradição e reciclagem). *H+D HÁBITAT MÁS DISEÑO* 18(9), 17–26. https://doi.org/10.58493/habitat.2017.18.01

Ross, G. (2004). The public health implications of polychlorinated biphenyls (PCBs) in the environment. *Ecotoxicology and Environmental Safety*, *59*(3), 275–291. https://doi.org/10.1016/J.ECOENV.2004.06.003

Royer, A., & Sharman, T. (2022). Copper toxicity. StatPearls Publishing. https://www.ncbi.nlm.nih.gov/pubmed/22332094

Salazar, E. M. (2020). El agua residual como generadora del espacio de la actividad agrícola en el Valle del Mezquital, Hidalgo, México. Hermosillo: CONACYT.

Sarni, W., & Alexander, A. (22 April 2022). World Economic Forum. https://www.weforum.org/agenda/2022/04/water-sector-renewable-circular/

Shingare, R. P., Thawale, P. R., Raghunathan, K., Mishra, A., & Kumar, S. (2019). Constructed wetland for wastewater reuse: Role and efficiency in removing enteric pathogens. *Journal of Environmental Management*, *246*, 444–461. https://doi.org/10.1016/j.jenvman.2019.05.157

Shrivastava, P., & Kumar, R. (2015). Soil salinity: A serious environmental issue and plant growth promoting bacteria as one of the tools for its alleviation. *Saudi Journal of Biological Sciences*, *22*(2), 123–131. https://doi.org/10.1016/j.sjbs.2014.12.001

Tang, S., She, D., & Wang, H. (2021). Effect of salinity on soil structure and soil hydraulic characteristics. *Canadian Journal of Soil Science*, *101*(1), 62–73. https://doi.org/10.1139/cjss-2020-0018

Ullah, N., Rehman, M. U., Ahmad, B., Ali, I., Younas, M., Aslam, M. S., Rahman, A. U., Taheri, E., Fatehizadeh, A., & Rezakazemi, M. (2022). Assessment of heavy metals accumulation in agricultural soil, vegetables and associated health risks. *PLoS ONE, 17*(6 June). https://doi.org/10.1371/JOURNAL.PONE.0267719

UN. (2016). *Sustainable Development Goal 6: Ensure Availability and Sustainable Management of Water and Sanitation for All.* https://sustainabledevelopment.un.org/sdg6

UNESCO. (2017). Informe Mundial sobre el desarrollo de los recursos hídricos de las Naciones Unidas 2017: Las aguas residuales: el recurso desaprovechado, cifras y datos. Colombella, Perugia: UNESCO.

Ungureanu, N., Vlăduţ, V., & Voicu, G. (2020). Water scarcity and wastewater reuse in crop irrigation. *Sustainability*, 12(21), 9055. https://doi.org/10.3390/su12219055

Unidas, N. (2021). Informe Mundial de las Naciones Unidas sobre el Desarrollo de los Recursos Hídricos 2021: El valor del agua. París.

United Nations World Water Assessment Programme. (2015). The United Nations world water development report 2015: Water for a sustainable world. Paris: UNESCO.

Vieira, W. T., de Farias, M. B., Spaolonzi, M. P., da Silva, M. G. C., & Adeodato Vieira, M. G. (2021). Endocrine-disrupting compounds: Occurrence, detection methods, effects and promising treatment pathways-A critical review. *Journal of Environmental Chemical Engineering*, *9*(1), 104558. https://doi.org/10.1016/j.jece.2020.104558

Voulvoulis, N. (2018). Water reuse from a circular economy perspective and potential. *Current Opinion in Environmental Science & Health*, *2*, 32–45. https://doi.org/10.1016/j.coesh.2018.01.005

WHO (World Health Organization). (2006). Guidelines for the safe use of wastewater excreta and greywater. *Wastewater Use in Agriculture*, *2*, 191. https://doi.org/10.1016/j.resconrec.2009.09.007

Williams, C. F., & McLain, J. E. T. (2012). Soil persistence and fate of carbamazepine, lincomycin, caffeine, and ibuprofen from wastewater reuse. *Journal of Environmental Quality*, *41*(5), 1473–1480. https://doi.org/10.2134/jeq2011.0353

Winpenny, J., Heinz, I., & Koo-Oshima, S. (2013). *Reutilización del agua en la agricultura: ¿Beneficios para todos?* Roma: Oficina de Intercambio de Conocimientos, Investigación y Extensión, FAO.

Wu, B. (2020). Human health hazards of wastewater. In *High-risk pollutants in wastewater*, pp. 125–139. https://doi.org/10.1016/B978-0-12-816448-8.00006-X

Yang, J., Hou, B., Wang, J., Tian, B., Bi, J., Wang, N., Li, X., & Huang, X. (2019). Nanomaterials for the removal of heavy metals from wastewater. *Nanomaterials*, *9*(3), 424. https://doi.org/10.3390/nano9030424

Zhang, X., Gu, P., & Liu, Y. (2019). Decontamination of radioactive wastewater: State of the art and challenges forward. *Chemosphere*, *215*, 543–553. https://doi.org/10.1016/j.chemosphere.2018.10.029

Zhang, W., Wang, C., Xue, R., & Wang, L. (2019). Effects of salinity on the soil microbial community and soil fertility. *Journal of Integrative Agriculture*, *18*(6), 1360–1368. https://doi.org/10.1016/S2095-3119(18)62077-5

Zurita, F., Roy, E. D., & White, J. R. (2012). Municipal wastewater treatment in Mexico: current status and opportunities for employing ecological treatment systems. *Environmental Technology*, *33*(10), 1151–1158. https://doi.org/10.1080/09593330.2011.610364

Zwolak, A., Sarzyńska, M., Szpyrka, E., & Stawarczyk, K. (2019). Sources of soil pollution by heavy metals and their accumulation in vegetables: A review. *Water, Air, and Soil Pollution*, *230*(7), 1–9. https://doi.org/10.1007/S11270-019-4221-Y/TABLES/3

10 PGPR Assisted Phytoremediation of Industrial Wastewater for Environmental Sustainability

Sheel Ratna, Monika Singh, Adarsh Kumar, and Rajesh Kumar

10.1 INTRODUCTION

Industrialization has led to the generation of ample waste, which is growing at an alarming rate and needs to be worked upon. Jones et al. (2021) study estimates global wastewater production at 359.4 × 109 m^3/yr, of which 63% (225.6 × 109 m^3/yr) is collected and 52% (188.1 × 109 m^3/yr) is treated. In India, a total of 764 units of Chemical, Distillery, Food, Dairy & Beverage, Pulp & Paper, Sugar, Textile, Bleaching & Dyeing, Tannery, and other industries generated 501 MLD (millions of liters per day) wastewater during different steps (ENVIS, 2020–2021). Industrial wastewaters have both degradable and non-degradable organic and inorganic pollutants. These pollutants are present in the form of persistent dyes, biochemical oxygen demand (BOD), chemical oxygen demand (COD), total dissolved solid (TDS), total nitrogen, heavy metals such as lead (Pb), zinc (Zn), mercury (Hg), nickel (Ni), cadmium (Cd), copper (Cu), chromium (Cr), arsenic (As), etc. In the textile industry, wastewater contains a variety of persistent coloring pollutants (dyes), formaldehyde, phthalates, phenols, surfactants, perfluorooctanoic acid (PFOA), pentachlorophenol chlorides, and sulfates with heavy metals. Bleaching processes in pulp and paper industries release dangerous chloroorganics and other highly colored compounds that have harmful impacts on the environment. Out of total water consumed by pulp and paper industries, 75% is discharged as contaminated water containing high amounts of sulfur compounds, sodium, calcium, and carbonates (Kumar and Chandra, 2021) In addition, tannery industries discharge about 400 mg/L sulfide-containing effluents into the nearby water, where sulfate transforms into hydrogen sulfide (H_2S) which is highly detrimental for living organisms (Mishra et al., 2019). Various industries discharge volatile organic compounds (VOCs) like gasoline, benzene, formaldehyde, toluene, and xylene into the environment. Distillery wastewater contains a mixture of organic and inorganic pollutants such as melanoidins, di-n-octyl phthalate, di-butyl phthalate, benzenepropanoic acid and 2-hydroxysocaproic acid, and toxic metals (Chowdhary et al. 2018). Due to the pollutants, IWW causes carcinogenic, mutagenic, genotoxic, cytotoxic, and allergenic health hazards to terrestrial and aquatic living organisms including human health (Kishor et al., 2021; Mishra et al., 2019). On the other hand, in agricultural land, it causes inhibition of seed germination and depletion of vegetation by reducing the soil alkalinity and manganese availability, if discharged without adequate treatment. Aquatic ecosystem pollutants cause eutrophication, reduction of photosynthesis activity, and accumulation of heavy metals in organisms (Ratna et al., 2021a; Kumar and Chandra, 2021; Kishor et al., 2021). Continually improving and affordable wastewater management provides opportunities for both pollution reduction and clean water supply augmentation, while simultaneously promoting sustainable development and supporting the transition to a circular economy (Jones et al., 2021). So, industrial wastewater treatment is obligatory before discharge or reuse. Various physiochemical processes are used for wastewater treatment, but they have limitations like high cost and are less environmental friendly. However

DOI: 10.1201/9781003441144-10

these strategies are expensive, demand intensive labor, alter soil properties, and disturb the soil native microbiome. Chemical remediation may also lead to very harmful side effects (Alves et al., 2022; Vocciante et al., 2022). Biological methods are ecofriendly and sustainable approaches for wastewater treatment. Various biological agents such as algae, fungi, bacteria, microbial fuel cells, and plants are reported for wastewater treatment.

Phytoremediation is being looked upon as a sustainable approach to wastewater treatment. Phytoremediation refers to the use of plants and associated microbes to reduce the concentrations or toxic effects of contaminants in the environment. But toxic pollutants of wastewater affect the growth and phytoremediation efficiency. So some microbes present in rhizosphere of plant enhance the plant growth through various metabolic activities called plant growth-promoting microbes. So this chapter focused on the PGPR-assisted phytoremediation of various types of industrial wastewaters.

10.2 TYPES OF INDUSTRIAL WASTEWATER AND ITS TOXICITY

Industrial wastewater contains a variety of organic and inorganic pollutants in the form of BOD, COD, TDS, salts and coloring compounds, and heavy metals (Ratna et al., 2021a; Kishor et al., 2021; Singh et al., 2021). The most contaminating metals found are arsenic (As), cadmium (Cd), chromium (Cr), copper (Cu), lead (Pb), mercury (Hg), nickel (Ni), and zinc (Zn) in industrial wastewater (Vocciante et al., 2022). These elements pile up and enter our food chain, consequently accumulating in body tissues of living organisms (bioaccumulation) and it also leads to magnification of their concentrations when they move from lower to higher level in a food chain via biomagnification (Ali and Khan, 2019; Vocciante et al., 2022). Metals have deleterious effects even at lower concentrations, hence their accumulation is very much risky to plants, animals, and humans (Kleckerová and Dočekalová, 2014; Srivastava et al., 2017). Metal from wastewater leads to lack of vegetation, consequently leading to soil erosion and off-site pollution, i.e., contamination of water reservoirs (Kumar et al., 2019). Metals are even toxic at very low concentrations, and hence they are harmful to plants, animals, and humans (Lal et al., 2018, Kumar et al., 2024). Human exposure to metals can also lead to health hazards such as skin lesions, lar disease, behavioral and attention deficit disorders, nervous and immune system impairment, gastrointestinal and renal dysfunction, and cancer, among many other complications (Althomali et al., 2024).

10.3 TREATMENT TECHNIQUES FOR INDUSTRIAL WASTEWATER

Several cleaning techniques are used to treat industrial wastewater in developed and developing countries such as physical, chemical, and biological techniques.

10.3.1 Physiochemical Treatments

Although a number of physicochemical methods such as adsorption, coagulation, oxidation, ozonation, electrolysis, membrane filtration, and thermolysis are being employed for industrial wastewater treatment but all of these are costly, use a large amount of chemicals, and generate a large amount of sludge after treatment, which also acts as a secondary pollutant in the environment. So these treatment methods are less environmentally, costly process for industrial wastewater treatment. So globally research studies focused on the development of alternatives to physiochemical treatment methods.

10.3.2 Biological Technique

Biological methods using microbes (bacteria, microalgae, fungi, etc.) are gaining attention for the treatment of industrial wastewaters. Pollutants are degraded by acclimated microorganisms during wastewater treatment, but some remain after treatment due to their persistent or recalcitrant nature.

10.3.2.1 Phytoremediation of Industrial Wastewater

Remediation of toxic substances or conversion of toxic substances into non-toxic substances is called remediation. Using potential microorganisms and their consortia of microbes to lower the toxicity or degrade or accumulate toxic pollutants is called bioremediation. Bioremediation in the rhizosphere of those plants which remediate toxic compounds of industrial polluted sites and grow normally in these waste effluents of industries is called Phytoremediation/botanical-remediation/agro-remediation/vegetative-remediation. Phytoremediation low cost, efficient, and ecofriendly remediation strategy that has good public acceptance sustainable, solar-driven a novel type of bioremediation, in which plants are used as a remediator or accumulating agents for the removal/lower the toxicity or degradation or accumulation of toxic pollutants. Phytoremediation technologies are available for various environments and types of contaminants. These involve different processes such as in situ stabilization or degradation and removal (i.e., volatilization or extraction) of contaminants. They are classified according to their type of contaminant and substrate, i.e., phytostabilization and phytoextraction (Liu et al., 2019; Khan et al., 2022).

10.3.2.1.1 Phytostabilization

Phytostabilization reduces the mobility and bioavailability of metals and therefore limits their leaching and entry into groundwaters and food chain, respectively (Shah and Daverey, 2020). It retains contaminants and can be stabilized in the roots or within the rhizosphere (Verma et al., 2017; Alves et al., 2022). In plants, metal tolerance is generally increased by symbiotic, root-colonizing, arbuscular mycorrhizal fungi (AMF), through metal sequestration in the AMF hyphae (Baghaie and Aghilizefreei, 2019). Plants can also convert contaminants into less toxic forms, or decrease their bioavailability. Roots of plants secrete certain natural chelates which can form complexes with metals in the rhizosphere. These include siderophores, organic acids, and phenolics. Metals such as the toxic can be converted to the much less toxic by enzymes found on the roots of wetland plants. In addition, plants, and their associated soil microbes, can release chemicals that act as biosurfactants in the soil that increase the uptake of contaminants (Lal et al., 2018; Rastogi et al., 2021a). Contaminants can be stabilized in natural and constructed wetlands through a process called phytofiltration. This involves rhizofiltration where metals are precipitated within the rhizosphere (Lal et al., 2018; Liu et al., 2022). Metal-plaque forms typically on the roots of wetland plants through the release of oxygen via the aerenchyma of roots (Prakash et al., 2021). Iron oxides precipitate along with other metals into the metal plaque. Metal plaque on roots acts as a reservoir for active iron (Fe^{2+}), which in turn increases the tolerance of plants to other toxic metals (Tangahu et al., 2022; Ustiatik et al., 2022).

10.3.2.1.2 Phytoextraction

Phytoextraction refers to the uptake of metals from wastewater and their translocation and accumulation in plant shoots, which is followed by harvesting and safe disposal of biomass (Mukherjee et al., 2015; Harmon, 2022). The shoot metal concentrations and biomass yield reflect the phytoextraction potential (Yang et al., 2022; Harmon, 2022). Phytoextraction uses the ability of plants to accumulate contaminants in the aboveground, harvestable biomass. This process involves repeated harvesting of the biomass in order to lower the concentration of contaminants in wastewater. Phytoextraction is either a continuous process (using metal hyperaccumulating plants, or fast growing plants), or an induced process (using chemicals to increase the bioavailability of metals in the contaminated environment). The effectiveness of phytoextraction is a function of a plant's biomass production and the content of contaminants in the harvested biomass. Therefore, fast-growing crops that accumulate metals have a great potential in phytoextraction. The use of crops in phytoextraction can be improved by manipulation of their associated soil microbes. Inoculation of plant growth-promoting rhizobacteria and arbuscular mycorrhizal fungi increases plant biomass (Baghaie and Aghilizefreei, 2019). The AMF-plant symbiosis usually results in reduced accumulation of metals in the aboveground biomass of plants. Therefore, suppressing AMF activity, by using specific soil fungicides, has resulted in increased metal accumulation in plants. The role of AMF in regulating metal uptake by plants appears to vary depending on numerous factors, such as AMF populations, plant species, nutrient availability, and metal content in the soil. Also, this regulation of AMF is usually metal-specific; where the uptake of essential metals is generally increased, but the uptake of non-essential metals is inhibited (Wang et al., 2024).

10.3.2.1.3 Phytovolatilization

Phytovolatilization used to transform pollutants into a volatile form, and it is released into the atmosphere through stomata (Shah and Daverey, 2020; Vocciante et al., 2022). However, here arises a problem of shifting of contaminants from soil to the atmosphere, and is likely to be redeposited into the original environment (Ratna et al., 2021b; Kumar et al., 2021). Also, the success of phytoremediation depends on the plant growth rate, root system, depth and distribution, soil characteristics, resistance to pests and pathogens and environmental and climatic adaptations (Khan et al., 2022; Tripathi et al., 2021). Phytovolatilization involves the uptake of contaminants by plant roots and their conversion to a gaseous state, and release into the atmosphere through evapotranspiration.

10.3.2.1.4 Phytodegradation

Phytodegradation involves the degradation of organic contaminants directly, through the release of enzymes from roots, or through metabolic activities within plant tissues (Singh et al., 2023). In phytodegradation, organic contaminants are taken up by roots and metabolized in plant tissues to less toxic substances. Phytodegradation of hydrophobic organic contaminants has been particularly successful. Poplar trees (*Populus* spp.) have been used successfully in phytodegradation of toxic and recalcitrant organic compounds. Rhizodegradation involves the attenuation of organic contaminants into less toxic substances within the rhizosphere through the biodegradation of soil microbes. This process is facilitated by root exudates (organic molecules) that sustain populations of soil microbes. To enhance this process a specific inocula of bacteria can be added to contaminated soils. Bacterial inocula contain strains that have desired metabolic activity to degrade targeted contaminants. Inoculating plants with genetically engineered strains of bacteria that degrade a specific contaminant has shown promising results (Thijs et al., 2016; Kumar et al., 2022). Phytoremediation of various types of industrial wastewater is listed in Table 10.1.

TABLE 10.1
Phytoremediation of Various Type Industrial Wastewaters

Phytoremediation	Pollutants Reductions	Wastewater	References
Vetiveria zizanioides (L.) Nash and *Cyperus alternifolius* L.	Reduction in TOC (89.3%–95.3%), TN (40.86%–82.7%), and PC (96.8%–98.8%) by *V. zizanioides* and TOC (84.9%–89.1%), TN (23.7%–92.7%), and PC (38.9%–92.1%) by *C. alternifolius*	Olive mill wastewater	Goren et al. (2021)
Pistia stratiotes	SCOD, NH_4-N, NO_3-N, and Sol. P reduction were up to 65%, 98%, 70%, and 65%, respectively	Parboiled Rice Mill Wastewater	Mukherjee et al. (2015)
Vetiver (*Chrysopogon zizanioides*)	EC, TSS, TDS, BOD, COD, TN, P, K, Pb, and Cd reduction were 39.39%, 81.19%, 56.19%, 58.01%, 59.24%, 47.78%, 54.64%, 55.68%, 65.63%, and 64.29% respectively	Paperboard mill wastewater	Davamani et al. (2021)

(Continued)

TABLE 10.1 (*Continued*)
Phytoremediation of Various Type Industrial Wastewaters

Phytoremediation	Pollutants Reductions	Wastewater	References
Stenotrophomonas species, *Sphingobium* species and *Phragmites australis*	4-n-nonylphenol, mono-ethoxylated nonylphenols and di-ethoxylated nonylphenols reduction were 87%, 70% and 87% by *P. australis* and 88%, 84%, 71% respectively	Tannery wastewater	Di Gregorio et al. (2015)
Eichhornia crassipes	Cyanide, BOD, and COD reduction were 95%, 89%, and 76% respectively	Steel industrial wastewater	Saha et al. (2018)
Typha angustifolia and *Eichhornia crassipes*	BOD reduction (91%)	Wastewater	Sricoth et al. (2018)
Spirogyra	Methylene blue and Cr(VI) reduction were 91.3% and 91.4% respectively	Simulated wastewater	Singh, et al. (2021)
Chrysopogon zizanioides	Cr(VI) reduction (82%–100%)	Wastewater	Masinire et al. (2021)
Typha angustifolia	Organics (92%), ammonia (52%), phosphorus (60%), and ibuprofen (76%)	Pharmaceutical-contaminated wastewater	Li et al. (2020)
Para grass (*Brachiara mutica*) and *papyrus* (*Cyprus papyrus*)	Cr (VI) reduction were 83.08% and 73.77% respectively by para grass, papyrus	Tannery wastewater	Kassaye et al. (2017)
Azolla pinnata	Reduction in pH (9.41%), EC (61.42%), TDS (71.56%), TKN (73.25%), and TP (65.37%)	Dairy wastewater	Goala et al. (2021)
Ipomoea Aquatica and *Centella asiaticas*	Reduction in Heavy metals by *I. aquatic* and *C. asiatica* up to 87% and 83% respectively	Industrial Wastewater	Hanafiah et al. (2020)
Typha latifolia and water Hyacinth	*T. latifolia* in removing turbidity (90.03%), EC (82.31%), Color (95.98%), Fe (92.01%), Cu (87.78%), and Zn (75.81%); water hyacinth reduced turbidity (64.15%), EC (62.19%), color (50.29%), Fe (54.15%) and Cu (70.17%) and Zn (85.97%)	Industrial wastewater	Abbas et al. (2021)
Pistia stratiotes L	Color (86%), TDS (66%), COD (77%), and chloride (61.33%) reduction	Dye-Containing Wastewater	Ahila et al. (2020)
Chrysopogon zizanoides L.	Cr (61.10%) and Ni (95.65%) reduction	Electroplating wastewater	Nugroho et al. (2021)
Lemna minor L.	Chloride (30%), sulfate (16%) and TDS (14%) reduction	Steel Wastewater	Saha et al. (2015)
Spirodela polyrhiza (L.) Schleid and bacteria consortium *Pseudomonas stutzeri, Janibacter anophelis, Bacillus safensis, Bacillus pumilus, Bacillus thuringiensis,* and *Bacillus cereus*	Reduction of COD (77.36%), color (91.70%), Ca (61.65%), Fe (69.41%), Ni (89.30%), Cd (88.37%), nitrate (70.83%), phosphate (73.11%), and sulfate (75.49%)	Textile wastewater	Parihar and Malaviya (2023)

(*Continued*)

TABLE 10.1 (*Continued*)
Phytoremediation of Various Type Industrial Wastewaters

Phytoremediation	Pollutants Reductions	Wastewater	References
Brachiara mutica and *Phragmites australis*, consortium of hydrocarbon-degrading bacteria (*Bacillus subtilis strain* LORI66, *Klebsiella sp. strain* LCRI87, *Acinetobacter Junii strain* TYRH47, *Acinetobacter sp. strain* LCRH81)	Oil content (97%), COD (93%), and BOD (97%) reduction	Oil field wastewater	Rehman et al. (2018)
Al–Fe electrode and *Canna indica*	Electrocoagulation and *Canna indica* with electrocoagulation reduced COD up to 86.4% and 97% respectively	Dairy industry wastewaters	Akansha et al. (2020)
Eichhornia crassipes	Cr (VI), BOD and COD reduction in 99.5% 50% and 34% respectively	Industrial mines wastewater	Saha et al. (2017)
Vetiveria zizanioides L.	Reduction in BOD_5 and NO_3^- were 78.47% and 90.53%	Wastewater	Parnian and Furze (2021).
Salvinia molesta	Phosphate (95%) and COD (39%) reduction	Palm Oil Mill Effluent	Ng and Chan (2017)

Total organic carbon (TOC), total nitrogen (TN), and phenolic compounds (PC), soluble COD (SCOD), ammoniacal nitrogen (NH_4-N), nitrate nitrogen (NO_3-N), electrical conductivity (EC), total soluble salts (TSS), total dissolved solids (TDS), Biological oxygen demand (BOD), chemical oxygen demand (COD), total nitrogen (TN), total Kjeldahl's nitrogen (TKN), and total phosphorus (TP).

10.4 PLANT GROWTH PROMOTING RHIZOSPHERIC MICROBES

Hiltner coined the term "Rhizosphere" that made up of two words: rhizo (root) and sphere (surrounding zone) (Hiltner, 1904). In the rhizosphere zone, microbes (bacteria, fungi, mycorrhiza, and cyanobacteria) present that enhance its growth directly and indirectly called plant growth-promoting rhizospheric (PGPR) microbes (Curl and Truelove, 2012). PGPR microbes are from broader genera, some of them are free-living, or in symbiotic association with the plants, and some are bacterial endophytes that can colonize plant interior tissues (Glick, 2012). The Rhizospheric zone contains an overcrowded number of microorganisms having specific activity (Glick, 2012; Vocciante, 2022). However, root exudates serve as constant source of energy and nutrients which promotes microbial growth, activity, and diversity. They serve as high source of carbon (C), nitrogen (N) which determines the presence of microbial community (Sharma et al., 2021). These exudates serve as repellant and attractant according to their composition. They are rich sources of amino acids, vitamins, hormones, and toxins and also the antagonistic compounds, that promote the growth of beneficial bacterial species (Alves et al., 2022). Complementing this promotes plant growth and well-being by providing phytoavailable nutrients, mitigating biotic and abiotic stresses and therefore protecting plants against pathogens (Verma et al., 2017; Ratna et al., 2021). They help by modifying root patterns and metabolizing root exudates that are reabsorbed by plants (Jach et al., 2022). Metal-tolerant strains, complement plants by enhancing their metal tolerance, detoxifying metalliferous soils thereby enhancing metal uptake in plants (Rastogi et al., 2021b). These microbes have developed such metabolism that they can sustain under toxic pollutants and they are capable of accumulating, transforming, and detoxifying metallic toxicity (Mishra et al., 2019 Lal et al., 2018). Deep understanding of these bacterial metabolisms and formulating different strategies can play a crucial role in improving and sustaining phytoremediation systems (Vocciante, 2022; Verma et al., 2017). These PGPRs can be broadly categorized as biofertilizers, phytostimulators, phizoremediators, and biopesticides (Tak et al., 2013; Verma et al., 2017). Plant growth-promoting bacteria affect plant growth through direct and indirect mechanisms. Direct mechanisms acquire the resources and lead to the synthesis of phytohormones that directly affect plant growth (Prakash, 2021). Indirect mechanisms secrete certain antagonistic compounds that promote systemic resistance and thereby promote plant growth (Saraf et al., 2014; Vocciante et al., 2022).

10.4.1 Direct Mechanisms

Microbe provides plants with mineral resources that are crucial for their survival those are phosphorus (P), nitrogen (N), carbon (C), and iron (Fe) (Glick, 2012). Nitrogen, which is a key element for plant growth and development, is present in the atmosphere as N_2 which is unavailable to plants. Nitrogenase enzyme which is a Nitrogen-fixing microorganism continuously converts atmospheric N_2 into phytoavailable forms such as ammonia and nitrate (Hoffman et al., 2014). Phosphorous is the plant macronutrient (P) which is often found in an insoluble form, which remains unavailable to plants (Glick, 2012). Phosphate solubilizing microbes secrete enzymes such as phosphatases, phytases, and lyases and low molecular weight organic acids (such as acetic, malic, succinic, and citric acids, to list but a few) that play a vital role in increasing the bioavailable levels of P (Goswami et al., 2016; Gupta et al., 2022). Iron (Fe) also plays a vital role in plant metabolism; however, it is mostly present in insoluble forms. Plant microorganisms can convert Fe^{3+} into Fe^{2+}, which becomes readily available for plants (Manoj et al., 2020). PGPR microbes synthesize and secrete the phytohormone indole-3-acetic acid (IAA), some of which is taken up by the plant. IAA stimulates plant cell proliferation and/or plant cell elongation or leads to the induction of the transcription of the plant enzyme ACC synthase that catalyzes the formation of ACC. IAA loosens plant cell walls, which facilitates cell elongation and consequently increases the extent of root exudation. ACC (a non-ribosomal amino acid) is exuded from seeds, roots, or leaves along with other small molecule components of root exudates, some of the plant and may be taken up by the bacteria associated with these tissues, and subsequently cleaved by ACC deaminase (Glick, 2014). Siderophores are basically low molecular weight, ferric ion-specific chelating agents secreted by bacteria and fungi growing under low iron stress. Their role is basically to scavenge iron from the environment and to make the mineral, available to the microbial cell which is primarily important for various metabolic activities in the cell. Interest in it has accrued about five decades ago, with the realization that most aerobic and facultative anaerobic microorganisms synthesize at least one siderophore. Siderophores are the result of virulent mechanisms in microorganisms pathogenic to both animals and plants. Iron in the Environment and in the aerobic atmosphere of the planet has become converted to oxyhydroxide polymers which are sparingly soluble. Microorganisms growing under aerobic conditions need iron for a variety of functions including the reduction of oxygen for synthesis of ATP, and for the reduction of ribotide which is precursors of DNA, it is also required in the formation of heme, and also for certain other essential purposes. At least one micromolar iron is needed for optimum growth. These microorganisms secrete some specific molecule which competes with hydroxyl ion for the ferric state of iron and make it available. However, not all microbes require iron, and siderophores can be dispensed with in these rare cases. Some lactic acid bacteria have no heme enzymes, and the crucial iron-containing ribotide reductase (4) has been replaced with an enzyme using adenosylcobalamin as the radical generator. Other microbes grow anaerobically on Fe (II). Siderophores are only associated to microbes and are not products of plant or animal metabolism, which have their own pathways for uptake of iron. Siderophore ligands are said to be "virtually specific" for Fe (III) among the naturally occurring metal ions of abundance. Validating the fact that the siderophore ligand shows strong affinity for only the higher oxidation state of iron sets this natural complexing agent apart from molecules such as heme, which serve effectively as electron shuttles. At the same time, the relatively weak complexing of Fe (II) affords an efficient means of release, via reduction, inside the cell. This large discrepancy in the binding constants for Fe (II) and Fe (III) drives down the oxidation-reduction potential, and there has been some discussion that the actual value may be beyond the range of natural reducing agents. This aspect of the problem requires clarification and elucidation at the enzyme level. Probably the significant feature is the oxidation-reduction potential of the enzymeferric siderophore complex rather than the potential of the free ferric chelate. With few exceptions, the "hard" acid ion, Fe (III), is linked to hard base atoms, such as oxygen, which accounts for the preference for ferric ions. Siderophores are the, are the proteins that are synthesized by bacteria, that can bind to a wide range of metals, thus enhancing their availability for plant uptake (Sheoran et al., 2016). Additionally, microbes also supply plants with phytohormones such as auxins, cytokinins and gibberellins, that are helpful for enhancing plant growth and also preventing metal phytotoxicity (Goswami et al., 2016). These phytohormones play different mechanisms in plants, they can be responsible for stimulating seed germination, root growth, shoot growth, and leaf expansion by cell elongation, division, and differentiation, among other beneficial effects (Wang et al., 2016). The most important auxin synthetized by bacteria is indole-3-acetic acid (IAA). Microbes synthesize the enzyme 1-aminocyclopropane-1-carboxylate (ACC) deaminase that regulates ethylene levels in plants (Ma et al., 2003). Ethylene is a multi-role plant hormone that controls plant growth and senescence. However, under higher concentrations it becomes detrimental. This problem is overshoot by ACC deaminase which plays a pivotal role in normalizing the levels of this phytohormone (Ma et al., 2003; Iqbal et al., 2017).

10.4.2 Indirect Mechanisms

Microbes can also indirectly enhance plant growth by producing certain antibiotic compounds known as allelochemicals, which provide defense against pathogens. Allelochemicals include antibiotics, hydrogen cyanide (HCN), lytic enzymes, and siderophores (Glick, 2012). PGPR-synthesized antibiotics such as pyrrolnitrin, phenazines, phloroglucinols, cyclic lipopeptides, and lipopeptides play an important role in preventing the growth of other bacteria and fungi (Saraf et al., 2014). Some PGPBs also exhibit cyanogenic activity (HCN release), which in addition to often operating as a biocontrol strategy, also enhances the action of bacterial antibiotics (Beneduzi et al., 2012). Lytic enzymes are known to deter pathogenic fungi propagation through the hydrolysis of cell wall components (Glick, 2012). Besides biocontrol, these lytic enzymes like cellulases, chitinases, glucanases, and proteases, also play a role in nutrient recycling through organic matter decomposition, increasing nutrient availability (Karthik et al., 2017). Lastly, siderophores contribute to biocontrol by preventing pathogens from acquiring Fe, limiting their proliferation (Rajkumar et al., 2010). Bacteria also prompt a process called induced systemic resistance that is responsible for a more efficient and faster reaction against a great variety of pathogens (Beneduzi et al., 2012; Novo et al., 2018). Furthermore, some PGPBs conduct biocontrol by simply outcompeting pathogens for resource acquisition and rhizosphere colonization (Whipps, 2001). Although plant growth-promoting bacteria promote the growth of plants (Glick, 2014) in a number of ways, the key bacterial trait that facilitates plant growth is the possession of the enzyme 1-aminocyclopropanel-carboxylate (ACC) deaminase. This enzyme cleaves the plant ethylene precursor, ACC, into ammonia and α-ketobutyrate. Thereby decreasing ethylene levels in plants (Glick, 2014), which if present in higher concentrations can inhibit plant growth or even death. Plants exude a large fraction of their photosynthetically fixed carbon (estimated to generally be in the range of 5%–30%) through their roots in the form of sugars, organic acids, and amino acids since these compounds act as a bacterial food source, this makes bacteria attracted around the roots of plants (i.e., the rhizosphere) are 10 to 1,000 times higher than in the bulk soil. The exuded ACC is cleaved by bacterial ACC deaminase which is seen that the bacterium is de facto acting as a sink for ACC. Consequently, lowering either the endogenous or the IAA-stimulated ACC level, the amount of ethylene that could potentially form in the plant is reduced. As a result of lowering plant ethylene levels, ACC deaminase-containing plant growth-promoting bacteria can reduce a portion of the ethylene inhibition, which is produced following a wide range of abiotic and biotic stresses. Consequently, plants growing in association with ACC deaminase-containing plant growth-promoting bacteria generally have longer roots and shoots and are also more resistant to growth inhibition by a variety of ethylene-inducing stresses (Glick, 2014). However, after the immediate effect, following abiotic or biotic stress, the pool of ACC in the plant is lower than that of the level of ACC deaminase in the associated bacterium. Since stress induces the induction of ACC oxidase in the plants leading to the increased flux through ACC oxidase resulting in the first (small) peak of ethylene that in turn induces the transcription of protective/defensive genes in the plant. Simultaneously, bacterial ACC deaminase is induced by the increasing amounts of ACC that occur from the induction of ACC synthase in the plant so that the magnitude of the second, deleterious, ethylene peak is decreased significantly (typically by 50%–90%). Since, ACC oxidase has a greater affinity for ACC to that of ACC deaminase, when ACC deaminase-producing bacteria are present, plant ethylene levels are dependent upon the ratio of ACC oxidase to ACC deaminase (Glick, 2014). That is, to effectively reduce plant ethylene levels, ACC deaminase must function before any significant amount of ACC oxidase is induced. A naive view of the interaction of IAA-producing bacteria with plants might posit that since IAA activates the transcription of ACC synthease, these bacteria should all ultimately result in the production of relatively high concentrations of ACC and subsequently inhibitory levels of ethylene. However, this is in fact not the case because as plant ethylene levels increase, the ethylene that is produced feedback inhibits IAA signal transduction thereby limiting the extent that IAA can activate ACC synthase transcription (Stearns et al., 2012). In the presence of ACC deaminase, there is much less ethylene and subsequent ethylene feedback inhibition of IAA signal transduction so that the bacterial IAA can continue to both promote plant growth and increase ACC synthase transcription. However, in this case, a large portion of the additional ACC that is synthesized is cleaved by the bacterial ACC deaminase. With plant growth-promoting bacteria that both secrete IAA and synthesize ACC deaminase, plant ethylene levels do not become elevated to the same extent as when plants interact with bacteria that secrete IAA but do not synthesize Adeaminase. In the presence of ACC deaminase, there is much less ethylene and subsequent ethylene feedback inhibition of IAA signal transduction so that the bacterial IAA can continue to both promote plant growth and increase ACC synthase transcription. However, in this case, a large portion of the additional ACC that is synthesized is cleaved by the bacterial ACC deaminase. The net result of this cross-talk between IAA and ACC deaminase is that by lowering plant ethylene levels, ACC deaminase facilitates the stimulation of plant growth by IAA.

10.5 PGPR ASSISTED PHYTOREMEDIATION

In recent years, bio-assisted phytoremediation or rhizoremediation has played an important role in decontamination of the wastewater (Manoj et al., 2020). Rhizoremediation is the most emerging, eco-friendly, and potentially effective process of biodegradation of contaminants from wastewater (Tangahu et al., 2022; Baghaie and Aghilizefreei, 2019). It involves the removal of specific contaminants from contaminated sites by mutual interaction of plant roots and suitable microbial species. Rhizosphere is a micro-environment where microorganisms (PGPR) form special types of communities with plant growth-promoting capabilities, and remove the toxic contaminants (Danyal et al., 2022). Phytoremediation of industrial effluents assisted by plant growth-promoting bacteria. Environmental Science and Pollution Research, 1–16. Interactions between plants and PGPR enhanced the phytoremediation efficiency of wastewater (Verma et al., 2017; Raklami et al., 2022). Recent findings on PGPR-assisted phytoremediation of industrial wastewater contaminated sites are shown in Figure 10.1 and Table 10.2.

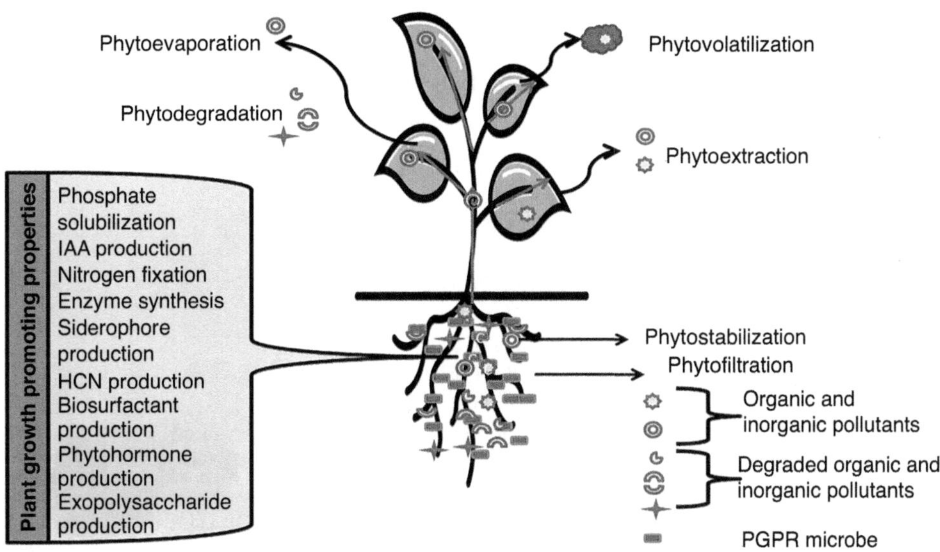

FIGURE 10.1 PGPR-assisted phytoremediation.

TABLE 10.2
PGPR-Assisted Phytoremediation of Various Types of Wastewaters

PGPR-Assisted Phytoremediation	Pollutants Reductions	Wastewater	References
Spirodela polyrhiza (L.) Schleid and bacteria consortium *Pseudomonas stutzeri, Janibacter anophelis, Bacillus safensis, Bacillus pumilus, Bacillus thuringiensis,* and *Bacillus cereus*	Reduction of COD (77.36%), color (91.70%), calcium (61.65%), iron (69.41%), nickel (89.30%), cadmium (88.37%), nitrate (70.83%), phosphate (73.11%), and sulfate (75.49%)	Textile wastewater	Parihar and Malaviya (2023)
Brachiara mutica and *Phragmites australis,* consortium of hydrocarbon-degrading bacteria (*Bacillus subtilis strain* LORI66, *Klebsiella sp. strain* LCRI87, *Acinetobacter Junii strain* TYRH47, *Acinetobacter sp. strain* LCRH81)	Reduction in oil content (97%), COD (93%), and BOD (97%)	Oil field wastewater	Rehman et al. (2018)
Dracaena sanderiana endophytic bacteria and *Bacillus cereus* NI	Inoculated plants and B. cereus NI shows highest BPA removal at 99.30±0.97% while only inoculated plants or *B. cereus NI* alone were 97.98±1.98% and 90.05±4.14%, respectively	Plastic industry wastewater	Suyamud et al. (2020)
Cynodon dactylon L. and *Bacillus cereus* (RCS-4 MZ520573.1)	All physico-chemical parameters of the sludge significantly improved up to 70%–75%	Distillery sludge	Tripathi et al. (2022)

(Continued)

TABLE 10.2 (*Continued*)
PGPR-Assisted Phytoremediation of Various Types of Wastewaters

PGPR-Assisted Phytoremediation	Pollutants Reductions	Wastewater	References
Cannabis sativa and *Bacillus thuringiensis* (MW887525), *Bacillus cereus* (MW887524), *Achromobacter denitrificans* (MW886333), *Bacillus subtilis* (MW886231)	Enhanced the metal accumulation pattern of Fe, Cu, Zn (162.15±0.085)> Mn (63.92±0.093)> Pb (28.619±0.192)> Ni (5.02±0.078)> Cd (2.53±0.085)> Cr (1.87±0.079) mg kg^{-1} in their shoot, root followed by leaf		Singh et al. (2023)
Oryza sativa and *Cd/Zn-tolerant PGPR strain Bacillus sp. ZC3-2-1*	Increase the phytoextraction of Cd^{2+} and Zn^{2+} by rice to up to 48.2% and 8.0%	Heavy metal contaminated soil	Liu et al. (2022)
Oryza sativa with PGPR consortium (*Sphingomonas, Massilia, Nitrospira*)	Reduction up to di (2-ethylhexyl) phthalate (DEHP) (86.1%) Cd^{2+} (76.0%) and Zn^{2+} (92.2%)	DEHP and heavy metals co-contaminated soil	Liu et al. (2022)
Jeotgalicoccus huakuii and *Bacillus amyloliquefaciens* isolated from *Cynodon dactylon isolated* and *Eleusine indica* respectively	Hg bioaccumulation up to 85.04%	Mercury-contaminated soil	Ustiatik et al. (2022)
constructed wetland planted with *Scirpus grossus* bioaugmented with rhizobacteria (*Bacillus cereus, B. pumilus, B. subtilis, Brevibacillus choshinensis,* and *Rhodococcus rhodochrous*)	Pb removal up to 99.99%	Contaminated water	Tangahu et al. (2022)
Sorghum inoculated with PGPR and Arbuscular Mycorrhizal Fungi (AMF)	PGPR and AMF significantly increased sorghum phytoremediation efficiency by 8% and 13.4%, respectively.	Soil treated with Pb-polluted cow manure	Baghaie and Aghilizefreei (2019)
Alternanthera philoxeroides in association with PGP (*Klebsiella sp.* VITAJ23)	Decolorization up to 79%	Reactive green dye (RGD)	Sinha et al. (2019)
Brevundimonas diminuta MYS6 associated *Helianthus annuus* L.	Cu removal (94.8%)	Cu contaminated soil	Rathi and Yogalakshmi (2021)

10.6 CONCLUSION AND FUTURE PROSPECTS

Industrial wastewaters have a variety of organic and inorganic pollutants in the form of BOD, COD, TDS, salts, minerals and heavy metals, and coloring compounds. If industrial wastewater is released into the environment without or improper treatment, it causes hazardous impact on terrestrial and aquatic ecosystems and human health. So wastewater treatment through Phytoremediation is one of the oldest techniques to remove pollutants from wastewater through interaction between plant and microorganisms. In rhizosphere of plant group of microbes live that enhances the plant growth directly and indirectly known as plant growth promoting rhizospheric (PGPR) microbes. PGPR microbes assisted/enhanced the phytoremediation efficiency of plants by various metabolic activities. PGPR directly enhances plant growth by releasing growth hormones (auxin, cytokinins, and gibberellins production) supplying nutrients (nitrogen, phosphorus iron through nitrogen fixation phosphate solubilization, siderophore production) and indirectly through lowering of stress (biocontrol through hydrogen cyanide and biosurfactants production). So, PGPR-assisted phytoremediation can be applied as ecofriendly and cost-effective approach for sustainable management of industrial wastewaters.

REFERENCES

Abbas, N., Butt, T., Ahmad, M., Deeba, F., & Hussain, N. (2021). Phytoremediation potential of *Typha latifolia* and water hyacinth for removal of heavy metals from industrial wastewater. *Chemistry International*, 7(2), 103–111.

Ahila, K. G., Ravindran, B., Muthunarayanan, V., Nguyen, D. D., Nguyen, X. C., Chang, S. W., ... Thamaraiselvi, C. (2020). Phytoremediation potential of freshwater macrophytes for treating dye-containing wastewater. *Sustainability*, 13(1), 329.

Akansha, J., Nidheesh, P. V., Gopinath, A., Anupama, K. V., & Kumar, M. S. (2020). Treatment of dairy industry wastewater by combined aerated electrocoagulation and phytoremediation process. *Chemosphere*, 253, 126652.

Ali, H., & Khan, E. (2018). Assessment of potentially toxic heavy metals and health risk in water, sediments, and different fish species of River Kabul, Pakistan. *Human and Ecological Risk Assessment: An International Journal*, 24(8), 2101–2118.

Althomali, R. H., Abbood, M. A., Saleh, E. A. M., Djuraeva, L., Abdullaeva, B. S., Habash, R. T., ... Najafi, M. L. (2024). Exposure to heavy metals and neurocognitive function in adults: a systematic review. *Environmental Sciences Europe*, 36(1), 18.

Alves, A. R. A., Yin, Q., Oliveira, R. S., Silva, E. F., & Novo, L. A. B. (2022). Plant growth-promoting bacteria in phytoremediation of metal-polluted soils: Current knowledge and future directions. *Science of the Total Environment*, 838, 156435.

Baghaie, A. H., & Aghilizefreei, A. (2019). Neighbor presence of plant growth-promoting rhizobacteria (PGPR) and arbuscular mycorrhizal fungi (AMF) can increase sorghum phytoremediation efficiency in a soil treated with Pb polluted cow manure. *Journal of Human, Environment, and Health Promotion*, 5(4), 153.

Beneduzi, A., Ambrosini, A., & Passaglia, L. M. (2012). Plant growth-promoting rhizobacteria (PGPR): their potential as antagonists and biocontrol agents. *Genetics and Molecular Biology*, 35, 1044–1051.

Chowdhary, P., Raj, A., & Bharagava, R. N. (2018). Environmental pollution and health hazards from distillery wastewater and treatment approaches to combat the environmental threats: A review. *Chemosphere*, 194, 229–246.

Curl, E. A., & Truelove, B. (2012). The rhizosphere, 15. Springer Science & Business Media.

Danyal, Y., Mahmood, K., Ullah, S., Rahim, A., Raheem, G., Khan, A. H., & Ullah, A. (2022). Phytoremediation of industrial effluents assisted by plant growth promoting bacteria. *Environmental Science and Pollution Research*, 30, 5296–5311.

Davamani, V., Parameshwari, C. I., Arulmani, S., John, J. E., & Poornima, R. (2021). Hydroponic phytoremediation of paperboard mill wastewater by using vetiver (*Chrysopogon zizanioides*). *Journal of Environmental Chemical Engineering*, 9(4), 105528.

Di Gregorio, S., Giorgetti, L., Ruffini Castiglione, M., Mariotti, L., & Lorenzi, R. (2015). Phytoremediation for improving the quality of effluents from a conventional tannery wastewater treatment plant. *International Journal of Environmental Science and Technology*, 12, 1387–1400.

ENVIS (Environmental Information System). (2020–2021). National status of waste water generation & treatment. https://sulabhenvis.nic.in/Database/Nationalstatusofwastewatergenerationandtreatment_2090.aspx?format=Print.

Glick, B. R. (2012). Plant growth-promoting bacteria: mechanisms and applications. *Scientifica*, 2012(1), 963401.

Glick, B. R. (2014). Bacteria with ACC deaminase can promote plant growth and help to feed the world. *Microbiological Research*, 169(1), 30–39.

Goala, M., Yadav, K. K., Alam, J., Adelodun, B., Choi, K. S., Cabral-Pinto, M. M., & Shukla, A. K. (2021). Phytoremediation of dairy wastewater using Azolla pinnata: Application of image processing technique for leaflet growth simulation. *Journal of Water Process Engineering*, 42, 102152.

Goren, A. Y., Yucel, A., Sofuoglu, S. C., & Sofuoglu, A. (2021). Phytoremediation of olive mill wastewater with Vetiveria zizanioides (L.) Nash and Cyperus alternifolius L. *Environmental Technology & Innovation*, 24, 102071.

Goswami, D., Thakker, J. N., & Dhandhukia, P. C. (2016). Portraying mechanics of plant growth promoting rhizobacteria (PGPR): a review. *Cogent Food & Agriculture*, 2(1), 1127500.

Gupta, R., Kumari, A., Sharma, S., Alzahrani, O. M., Noureldeen, A., & Darwish, H. (2022). Identification, characterization and optimization of phosphate solubilizing rhizobacteria (PSRB) from rice rhizosphere. *Saudi Journal of Biological Sciences*, *29*(1), 35–42.

Hanafiah, M. M., Zainuddin, M. F., Mohd Nizam, N. U., Halim, A. A., & Rasool, A. (2020). Phytoremediation of aluminum and iron from industrial wastewater using Ipomoea aquatica and *Centella asiatica. Applied Sciences*, 10(9), 3064.

Harmon, S. M. (2022). Biodegradable chelate-assisted phytoextraction of metals from soils and sediments. In *Current Opinion in Green and Sustainable Chemistry* (Vol. 37). Elsevier B.V. https://doi.org/10.1016/j.cogsc.2022.100677.

Hiltner, L. (1904). Ober neuter erfahrungen und probleme auf dem gebiete der bodenbakteriologie unter besonderer berucksichtigung der grundungung und brache. *Arbeiten der Deutshen Landwirschafthchen Gesellschaft*, 98, 59–78.

Hoffman, B. M., Lukoyanov, D., Yang, Z. Y., Dean, D. R., & Seefeldt, L. C. (2014). Mechanism of nitrogen fixation by nitrogenase: the next stage. *Chemical Reviews*, 114(8), 4041–4062.

Iqbal, N., Khan, N. A., Ferrante, A., Trivellini, A., Francini, A., & Khan, M. I. R. (2017). Ethylene role in plant growth, development and senescence: interaction with other phytohormones. *Frontiers in Plant Science*, 8, 475.

Jach, M. E., Sajnaga, E., & Ziaja, M. (2022). Utilization of legume-nodule bacterial symbiosis in phytoremediation of heavy metal-contaminated soils. *Biology*, 11(5), 676.

Jones, E. R., Van Vliet, M. T., Qadir, M., & Bierkens, M. F. (2021). Country-level and gridded estimates of wastewater production, collection, treatment and reuse. *Earth System Science Data*, 13(2), 237–254.

Karthik, C., Elangovan, N., Kumar, T. S., Govindharaju, S., Barathi, S., Oves, M., & Arulselvi, P. I. (2017). Characterization of multifarious plant growth promoting traits of rhizobacterial strain AR6 under Chromium (VI) stress. *Microbiological Research*, 204, 65–71.

Kassaye, G., Gabbiye, N., & Alemu, A. (2017). Phytoremediation of chromium from tannery wastewater using local plant species. *Water Practice & Technology*, 12(4), 894–901.

Khan, A. U., Khan, A. N., Waris, A., Ilyas, M., & Zamel, D. (2022). Phytoremediation of pollutants from wastewater: A concise review. *Open Life Sciences*, 17(1), 488–496.

Kishor, R., Purchase, D., Saratale, G. D., Saratale, R. G., Ferreira, L. F. R., Bilal, M., ... Bharagava, R. N. (2021). Ecotoxicological and health concerns of persistent coloring pollutants of textile industry wastewater and treatment approaches for environmental safety. *Journal of Environmental Chemical Engineering*, 9(2), 105012.

Kumar, A., & Chandra, R. (2021). Biodegradation and toxicity reduction of pulp paper mill wastewater by isolated laccase producing Bacillus cereus AKRC03. *Cleaner Engineering and Technology*, 4, 100193.

Kumar, A., Kumar, G., Ratna, S., & Kumar, S. (2024). Microbiological characterization of plant growth promoting bacteria isolated from vermiwash and their possible utilization in chromium contaminated fields. *Annals of Plant and Soil Research*, 26, 56–63.

Kumar, G., Bhatt, P., & Lal, S. (2021). Phytoremediation: A synergistic interaction between plants and microbes for removal of petroleum hydrocarbons. In *Soil Contamination-Threats and Sustainable Solutions*. IntechOpen.

Kumar, M., Goswami, L., Singh, A. K., & Sikandar, M. (2019). Valorization of coal fired-fly ash for potential heavy metal removal from the single and multi-contaminated system. *Heliyon*, 5(10), e02562.

Kumar, R., Gupta, K., & Bordoloi, N. (2022). The potential of engineered endophytic bacteria to improve phytoremediation of organic pollutants. In *Advances in Microbe-Assisted Phytoremediation of Polluted Sites* (pp. 477–496). Elsevier.

Lal, S., Ratna, S., Said, O. B., & Kumar, R. (2018). Biosurfactant and exopolysaccharide-assisted rhizobacterial technique for the remediation of heavy metal contaminated soil: An advancement in metal phytoremediation technology. *Environmental Technology & Innovation*, 10, 243–263.

Li, Y., Lian, J., Wu, B., Zou, H., & Tan, S. K. (2020). Phytoremediation of pharmaceutical-contaminated wastewater: Insights into rhizobacterial dynamics related to pollutant degradation mechanisms during plant life cycle. *Chemosphere*, 253, 126681.

Liu, A., Wang, W., Chen, X., Zheng, X., Fu, W., Wang, G., ... Guan, C. (2022). Phytoremediation of DEHP and heavy metals co-contaminated soil by rice assisted with a PGPR consortium: Insights into the regulation of ion homeostasis, improvement of photosynthesis and enrichment of beneficial bacteria in rhizosphere soil. *Environmental Pollution*, 314, 120303.

Liu, A., Wang, W., Zheng, X., Chen, X., Fu, W., Wang, G., ... Guan, C. (2022). Improvement of the Cd and Zn phytoremediation efficiency of rice (Oryza sativa) through the inoculation of a metal-resistant PGPR strain. *Chemosphere*, 302, 134900.

Liu, Z., Lin, H., Cai, T., Chen, K., Lin, Y., Xi, Y., & Chhuond, K. (2019). Effects of phytoremediation on industrial wastewater. In *IOP Conference Series: Earth and Environmental Science,* 371(3), 032011 IOP Publishing.

Ma, W., Sebestianova, S. B., Sebestian, J., Burd, G. I., Guinel, F. C., & Glick, B. R. (2003). Prevalence of 1-aminocyclopropane-1-carboxylate deaminase in Rhizobium spp. *Antonie Van Leeuwenhoek*, 83, 285–291.

Manoj, S. R., Karthik, C., Kadirvelu, K., Arulselvi, P. I., Shanmugasundaram, T., Bruno, B., & Rajkumar, M. (2020). Understanding the molecular mechanisms for the enhanced phytoremediation of heavy metals through plant growth promoting rhizobacteria: A review. *Journal of Environmental Management*, 254, 109779.

Masinire, F., Adenuga, D. O., Tichapondwa, S. M., & Chirwa, E. M. (2021). Phytoremediation of Cr (VI) in wastewater using the vetiver grass (Chrysopogon zizanioides). *Minerals Engineering*, 172, 107141.

Mishra, S., Chowdhary, P., Bharagava, R.N. (2019). Conventional methods for the removal of industrial pollutants, their merits and demerits. In *Emerging and Eco-friendly Approaches for Waste Management* (pp. 1–31). Singapore: Springer.

Mukherjee, B., Majumdar, M., Gangopadhyay, A., Chakraborty, S., & Chaterjee, D. (2015). Phytoremediation of parboiled rice mill wastewater using water lettuce (Pistia stratiotes). *International Journal of Phytoremediation*, 17(7), 651–656.

Novo, L. A., Castro, P. M., Alvarenga, P., & da Silva, E. F. (2018). Plant growth–promoting rhizobacteria-assisted phytoremediation of mine soils. In, Prasad, M. N. V., de Campos Favas, P. J., & Maiti, S. K. (Eds.), *Bio-geotechnologies for Mine Site Rehabilitation* (pp. 281–295). Elsevier.

Nugroho, A. P., Butar, E. S. B., Priantoro, E. A., Sriwuryandari, L., Pratiwi, Z. B., & Sembiring, T. (2021). Phytoremediation of electroplating wastewater by vetiver grass (Chrysopogon zizanoides L.). *Scientific Reports*, 11(1), 14482.

Parihar, A., & Malaviya, P. (2023). Textile wastewater phytoremediation using Spirodela polyrhiza (L.) Schleid. Assisted by novel bacterial consortium in a two-step remediation system. *Environmental Research*, 221, 115307.

Parnian, A., & Furze, J. N. (2021). Vertical phytoremediation of wastewater using *Vetiveria zizanioides* L. *Environmental Science and Pollution Research*, 28(45), 64150–64155.

Prakash, J. (2021). Plant growth promoting rhizobacteria in phytoremediation of environmental contaminants: Challenges and future prospects. In, *Bioremediation for Environmental Sustainability*, 191–218. Elsewere.

Rajkumar, M., Ae, N., Prasad, M. N. V., & Freitas, H. (2010). Potential of siderophore-producing bacteria for improving heavy metal phytoextraction. *Trends in Biotechnology*, 28(3), 142–149.

Raklami, A., Meddich, A., Oufdou, K., & Baslam, M. (2022). Plants-microorganisms-based bioremediation for heavy metal cleanup: Recent developments, phytoremediation techniques, regulation mechanisms, and molecular responses. *International Journal of Molecular Sciences*, 23(9), 5031.

Rastogi, S., Ratna, S., & Kumar, R. (2021a). Screening of biosurfactant producing bacteria isolated from hydrocarbon contaminated site and their potential in biosorption of Pb (II) and oil biodegradation. *Tenside Surfactants Detergents*, 58(6), 435–441.

Rastogi, S., Ratna, S., Said, O. B., & Kumar, R. (2021b). Physiological and molecular aspects of retrieving environmental stress in plants by microbial interactions. In, Sharma, A. (Eds.), *Microbes and Signaling Biomolecules Against Plant Stress: Strategies of Plant-Microbe Relationships for Better Survival* (pp. 107–125). Singapur, Springer.

Rathi, M., & Yogalakshmi, K. N. (2021). Brevundimonas diminuta MYS6 associated *Helianthus annuus* L. for enhanced copper phytoremediation. *Chemosphere*, 263, 128195.

Ratna, S., Rastogi, S., & Kumar, R. (2021a). Current trends for distillery wastewater management and its emerging applications for sustainable environment. *Journal of Environmental Management,* 290, 112544.

Ratna, S., Rastogi, S., & Kumar, R. (2021b). Phytoremediation: A synergistic interaction between plants and microbes for removal of unwanted chemicals/contaminants. In, Sharma, A (Ed.), *Microbes and Signaling Biomolecules against Plant Stress: Strategies of Plant-Microbe Relationships for Better Survival* (pp. 199–222). Singapore, Springer.

Rehman, K., Imran, A., Amin, I., & Afzal, M. (2018). Inoculation with bacteria in floating treatment wetlands positively modulates the phytoremediation of oil field wastewater. *Journal of Hazardous Materials*, 349, 242–251.

Saha, P., Banerjee, A., & Sarkar, S. (2015). Phytoremediation potential of Duckweed (*Lemna minor* L.) on steel wastewater. *International Journal of Phytoremediation*, 17(6), 589–596.

Saha, P., Mondal, A., & Sarkar, S. (2018). Phytoremediation of cyanide containing steel industrial wastewater by *Eichhornia crassipes. International Journal of Phytoremediation,* 20(12), 1205–1214.

Saha, P., Shinde, O., & Sarkar, S. (2017). Phytoremediation of industrial mines wastewater using water hyacinth. *International Journal of Phytoremediation*, 19(1), 87–96.

Saraf, M., Pandya, U., & Thakkar, A. (2014). Role of allelochemicals in plant growth promoting rhizobacteria for biocontrol of phytopathogens. *Microbiological Research*, 169(1), 18–29.

Shah, V., & Daverey, A. (2020). Phytoremediation: A multidisciplinary approach to clean up heavy metal contaminated soil. *Environmental Technology & Innovation*, 18, 100774.

Sharma, P., Pandey, A. K., Udayan, A., & Kumar, S. (2021). Role of microbial community and metal-binding proteins in phytoremediation of heavy metals from industrial wastewater. *Bioresource Technology*, 326, 124750.

Sheoran, V., Sheoran, A. S., & Poonia, P. (2016). Factors affecting phytoextraction: A review. *Pedosphere*, 26(2), 148–166.

Singh, A. K., & Chandra, R. (2019). Pollutants released from the pulp paper industry: Aquatic toxicity and their health hazards. *Aquatic Toxicology*, 211, 202–216.

Singh, A., Pal, D. B., Kumar, S., Srivastva, N., Syed, A., Elgorban, A. M., ... Gupta, V. K. (2021). Studies on Zero-cost algae based phytoremediation of dye and heavy metal from simulated wastewater. *Bioresource Technology*, 342, 125971.

Singh, A. K., & Chandra, R. (2019). Pollutants released from the pulp paper industry: Aquatic toxicity and their health hazards. *Aquatic Toxicology*, 211, 202–216.

Singh, K., Tripathi, S., & Chandra, R. (2023). Bacterial assisted phytoremediation of heavy metals and organic pollutants by *Cannabis sativa* as accumulator plants growing on distillery sludge for ecorestoration of polluted site. *Journal of Environmental Management*, 332, 117294.

Sinha, A., Lulu, S., Vino, S., & Osborne, W. J. (2019). Reactive green dye remediation by *Alternanthera philoxeroides* in association with plant growth promoting Klebsiella sp. VITAJ23: A pot culture study. *Microbiological Research*, 220, 42–52.

Sricoth, T., Meeinkuirt, W., Pichtel, J., Taeprayoon, P., & Saengwilai, P. (2018). Synergistic phytoremediation of wastewater by two aquatic plants (*Typha angustifolia* and *Eichhornia crassipes*) and potential as biomass fuel. *Environmental Science and Pollution Research*, 25, 5344–5358.

Suyamud, B., Thiravetyan, P., Gadd, G. M., Panyapinyopol, B., & Inthorn, D. (2020). Bisphenol A removal from a plastic industry wastewater by Dracaena sanderiana endophytic bacteria and *Bacillus cereus* NI. *International Journal of Phytoremediation*, 22(2), 167–175.

Stearns, J. C., Woody, O. Z., McConkey, B. J., & Glick, B. R. (2012). Effects of bacterial ACC deaminase on Brassica napus gene expression. *Molecular Plant-Microbe Interactions*, 25(5), 668–676.

Tak, H. I., Ahmad, F., & Babalola, O. O. (2013). Advances in the application of plant growth-promoting rhizobacteria in phytoremediation of heavy metals. *Reviews of Environmental Contamination and Toxicology*, 223, 33–52.

Tangahu, B. V., Abdullah, S. R. S., Basri, H., Idris, M., Anuar, N., & Mukhlisin, M. (2022). Lead (Pb) removal from contaminated water using constructed wetland planted with Scirpus grossus: Optimization using response surface methodology (RSM) and assessment of rhizobacterial addition. *Chemosphere*, 291, 132952.

Thijs, S., Sillen, W., Rineau, F., Weyens, N., & Vangronsveld, J. (2016). Towards an enhanced understanding of plant-microbiome interactions to improve phytoremediation: Engineering the metaorganism. *Frontiers in Microbiology*, 7, 341.

Tripathi, S., Sharma, P., Purchase, D., & Chandra, R. (2021). Distillery wastewater detoxification and management through phytoremediation employing Ricinus communis L. *Bioresource Technology*, 333, 125192.

Tripathi, S., Yadav, S., Sharma, P., Purchase, D., Syed, A., & Chandra, R. (2022). Plant growth promoting strain *Bacillus cereus* (RCS-4 MZ520573. 1) enhances phytoremediation potential of Cynodon dactylon L. in distillery sludge. *Environmental Research*, 208, 112709.

Ustiatik, R., Nuraini, Y., Suharjono, S., Jeyakumar, P., Anderson, C. W., & Handayanto, E. (2022). Mercury resistance and plant growth promoting traits of endophytic bacteria isolated from mercury-contaminated soil. *Bioremediation Journal*, 26(3), 208–227.

Verma, C., Das, A. J., & Kumar, R. (2017). PGPR-assisted phytoremediation of cadmium: An advancement towards clean environment. *Current Science*, 113(4), 715.

Vocciante, M., Grifoni, M., Fusini, D., Petruzzelli, G., & Franchi, E. (2022). The role of plant growth-promoting rhizobacteria (PGPR) in *mitigating pPlant's Environmental Stresses. Applied Sciences*, 12(3), 12031231.

Wang, Q., Dodd, I. C., Belimov, A. A., & Jiang, F. (2016). Rhizosphere bacteria containing 1-aminocyclopropane-1-carboxylate deaminase increase growth and photosynthesis of pea plants under salt stress by limiting Na+ accumulation. *Functional Plant Biology*, 43(2), 161–172.

Wang, Q., Liu, M., Wang, Z., Li, J., Liu, K., & Huang, D. (2024). The role of arbuscular mycorrhizal symbiosis in plant abiotic stress. *Frontiers in Microbiology*, 14, 1323881.

Whipps, J. M. (2001). Microbial interactions and biocontrol in the rhizosphere. *Journal of Experimental Botany*, 52(1), 487–511.

Yang, Z., Yang, F., Liu, J. L., Wu, H. T., Yang, H., Shi, Y., ... Chen, K. M. (2022). Heavy metal transporters: Functional mechanisms, regulation, and application in phytoremediation. *Science of the Total Environment*, 809, 151099.

11 Role of Biosurfactant in Wastewater Treatment
Current Research and Future

Talita Corrêa Nazareth Zanutto, Conrado Planas Zanutto, and Danielle Maass

11.1 INTRODUCTION

Organic and metallic contaminants are harmful to the environment, being the major causes of pollution of water bodies. The contamination of aquatic systems occurs mainly through improperly treated effluent discharges and accidental oil spills. The chemical and pharmaceutical industries, battery production industry, metallurgy, oil refineries, and agriculture are the main polluting sources. In this sense, efficient methods of removal and/or degradation of wastewater pollutants need to be implemented, being currently carried out through chemical, physical, and biological processes (Elumalai et al., 2021; Nafi and Taseidifar, 2022).

The biosurfactant-facilitated remediation of hydrocarbon and metal compounds from wastewater has been reported by several works in the literature. Due to their eco-friendly nature, biosurfactants are an attractive solution for remediation procedures. The biosurfactant enables the remediation process due to the amphipathic structure of the molecule, which acts as an adsorbent of metals in the liquid phase, being subsequently separated by foam flotation. Regarding the removal of organic pollutants, the biosurfactant alters the cell surface hydrophobicity and membrane permeability of microorganisms. At low concentrations, biosurfactant monomers are deposited on the surface of the organic contaminant, causing a reduction in interfacial tension and a decrease in repulsive forces between the two phases (Zeng et al., 2018).

In addition, the capacity of biosurfactants to aggregate in the form of micelles allows the micelles to capture target molecules in an aqueous phase. This occurs due to the structure of the micelle, which is hydrophilic on its surface and lipophilic in its core, being able to increase the solubility of the contaminant in a liquid medium, and also to enable its elimination by microorganisms due to better conductivity and mobility of dissolved pollutants (Liu et al., 2017; Manga et al., 2021; Rizzo et al., 2021).

From this perspective, some of the approaches used in the remediation of water systems by biosurfactants were presented in this chapter. In addition, the challenges related to the use of biosurfactants in wastewater treatment, which mainly involve their high production cost, were also addressed. Thus, this chapter intends to contribute to the dissemination of knowledge about biosurfactants and their application in the bioremediation of the main pollutants of wastewater through a review and discussion of works reported in recent literature.

11.2 BIOSURFACTANTS: CLASSIFICATION, PROPERTIES AND PRODUCTION

Biosurfactants are molecules produced by microorganisms that have interfacial activity due to their molecular structure composed of a polar and a non-polar part. The hydrophilic moiety comprises amino acid, peptide, phosphate, carbohydrate,alcohol, or carboxylic acidwhether the hydrophobic part is typically composed of hydrocarbons of varying lengths. Due to their amphipathic nature,

DOI: 10.1201/9781003441144-11

biosurfactants tend to preferentially distribute between interfaces of different polarities, being able to solubilize hydrophobic compounds, and increasing their bioavailability (Kashif et al., 2022).

Distinct microorganisms are ableto synthesize biosurfactants since these molecules play an important physiological role in the survival of the microorganism. Thus, some yeast, fungi, and bacteria can excrete biosurfactants into the culture medium or integrate them into the cell wall. According to Nitschke et al. (2002), microorganisms produce biosurfactants to regulate cell surface properties (cell adhesion-release to surfaces), facilitate the transport of nutrients, and use themas antibiotics. Furthermore, biosurfactants improve the emulsification and solubilization of hydrocarbon substrates, thus facilitating the survival of the microorganism (Nitschke et al., 2002).

The main physicochemical properties of biosurfactants are reduction of surface tension, emulsification, dispersion, and foaming. The performance of a biosurfactant is measured through its capacity to lower surface tension, specifically through the critical micelle concentration (CMC), which depends on the molecular structure of the biosurfactant. The diverse molecular structures of different congeners of biosurfactants give them varied functional activities due to the formation of micelles with different morphological structures, which guarantee greater specificity. Micelles normally have a spherical shape with various nanometers in diameter, where their aggregation comprises vesicles (Pβ), lamellar, crystalline, cubic, and hexagonal phases (Kashif et al., 2022).

Biosurfactants present variousbenefits over chemical surfactants obtained from petroleum derivatives. Among these advantages, high biodegradability, excellent surface activity, low toxicity, and stability against temperature and pH variations are the most prominent. Furthermore, biosurfactants can be obtained from agro-industrial waste, which contributes to a circular economy.

11.2.1 Classification and Chemical Structure

Biosurfactants are classified based on their chemical nature and producing microorganisms. Basically, they are divided into groups of low or high molecular weight, being distributed into five classes:(i) Glycolipids, (ii) Lipopeptides, (iii) Polymeric, (iv) Particulate, (v) Fatty acids, phospholipids, and neutral lipids (Desai and Banat, 1997). According to Kashif et al. (2022),high molecular weight biosurfactants present high emulsifying activity and strong adhesion to surfaces, although are not as efficient in reducing surface tension. On the other hand, low molecular weight biosurfactants present high surface activity, being useful for lowering surface and interfacial tension (Jahan et al., 2020). Once low molecular weight biosurfactants such as glycolipids and lipopeptides are the best well-studied in terms of production and application, they will be prioritized in the following sections.

11.2.1.1 Glycolipids

Glycolipids are biosurfactants composed of a fatty acid linked to a carbohydrate fraction, being divided according to the nature of the carbohydrate portion (i.g. sophorose lipids, rhamnose lipids, mannosylerythritol lipids, trehalose lipids, diglycosyl diglycerides, galactosyl-diglyceride, etc.) (Inès and Dhouha, 2015).The chemical structures of some glycolipid-type biosurfactants can be visualized in the article of Inès and Dhouha (2015).

The best known glycolipids are the sophorolipids produced by yeasts, the rhamnolipids produced by *Pseudomonas* sp, and the trehalolipids produced by *Rhodococcus* sp. (Inès and Dhouha, 2015). The micellar structure of glycolipids varies when the headgroup is affected by variations in pH, electrolytes, etc (Jahan et al., 2020). Rhamnolipids, sophorolipids, and mannosylerythritol lipids are the only biosurfactants produced on an industrial scale (Henkel et al., 2017).

11.2.1.1.1 Rhamnolipids

Rhamnolipids are biosurfactants widely reported in the literature and are composed of one or two rhamnoses (glycon part) linked to hydroxy fatty acids (aglycon part) through an α-1,2- glycosidic bond (Jahan et al., 2020). The aglycon portion is mostly composed of one to three β-hydroxy fatty acid chains linked by an ester bond formed between the β-hydroxyl group of the distal chain and the carboxyl group of the proximal chain (Chebbi et al., 2021). Rhamnolipids are efficient anionic surfactants, ableto reduce the surface tension of water to 27 mN/m. Furthermore, rhamnolipids promote the formation of emulsions, facilitate the solubilization of organic compounds in the aqueous phase, and modify the hydrophobicity of the cell surface (Parus et al., 2023).

11.2.1.1.2 Sophorolipids

Sophorolipids are molecules composed of a hydrophilic portion of sophorose (D-glucopyranose) linked by a O-β-glicosídica bond to a fatty acid tail (17-L-hydroxy-D9-octadecenoic acid). The head of sophorose can contain up to two acetyl groups and the tail usually has between 16 and 20 carbon atoms arranged in saturated to polyunsaturated chains. These molecules can be observed in lactonic or acidic form, where they can be lactonized due to a condensation reaction between the carboxyl group of the fatty acid and the 4" hydroxyl group of sophorose. An ester linkage attaches a second sophorose moiety to the fatty acid tail (Dierickx et al., 2022; Qazi et al., 2022).

11.2.1.1.3 Mannosylerythritol Lipids

Mannosylerythritol lipids (MEL)comprisea combination of a partially acylated derivative of 4-O-β-D-mannopyranosyl-D-erythritol and a hydrophobic tail, normally composed of two fatty acids with a variety of the degree of acetylation. The acetylation takes place at C4 and/or C6, whereas the two fatty acids are connected to the C2 and C3 positions of mannose via ester bonds. The degree of acetylation defines the classification of MELs, where diacetylation at positions C4 and C6 corresponds to MEL-A, monoacetylation at C4 or C6 corresponds to MEL-C and MEL-B, respectively, and MEL-D has no acetylation (Coelho et al., 2020).

11.2.1.1.4 Trehalolipids

Trehalolipids are glycolipids composed of a group of carbohydrates (non-reducing disaccharide) connected to a hydrophobic tail (fatty acid), which may be hydroxylated fatty acids with branched chains or aliphatic types. The numbers of hydrophobic chains in each molecule of trehalose lipids vary and can form mono-, di- and tetra esters. Bacteria such as *Rhodococcus, Nocardia, Micrococcus, Gordonia, Mycobacterium*, among others, are the most studied producers (Salek et al., 2022).

11.2.1.2 Lipopeptides

Lipopeptides are the second largest group of biosurfactants, consisting of surfactins, iturins, fengycins, lichenysins, among others. *Bacillus* and *Pseudomonas* genera are the most reported producing strains in the literature. Lipopeptides are composed of fatty acids (chains between 12 and 18 carbons) connected to a peptide chain (containing between 4 and 12 amino acids). The molecule has a hydrophilic head (linear) or a lactone ring (cyclic lipopeptides).(Vecino et al., 2021). The chemical structures of some lipopeptides can be visualized in the articles of Mongkolthanaruk (2012), Chen et al. (2015), and Yao et al. (2021).

Surfactin is one of the most prominent and well-studied biosurfactants of the lipopeptide class, being a cyclic heptapeptide linked to a β-OH fatty acid (lactone) chain containing from 12 to16 carbons.Severalvariations in the chiral sequence of amino acids in the peptide chain may appear, where amino acids from the aliphatic group, Val, Leu, and Ile have already been observed at positions 2nd, 4th, and 7th (Zanotto et al., 2019). In addition, the physicochemical properties can be altered depending on the length of the fatty acid chain, as well as the replacement of the amino acid component of the peptide ring (Jahan et al., 2020).

11.2.2 Biosurfactant Production

Biosurfactant production can be carried out under aerobic and anaerobic conditions and in submerged or solid-state fermentations. The biosurfactants-producing microorganismis capable of synthesizing tensoatives in a medium containing water-soluble and water-insoluble substrates, where the amount of production and physicochemical properties are strongly influenced by the carbon source, nitrogen source, ion concentration, temperature, pH, bacterial strains and culture condition. Oils/fats, hydrocarbons, and carbohydrates are the main carbon sources employed, although glucose is the most used by the glycolytic pathway for metaboliteproduction. Inorganic (ammonium nitrates and sulfates) and organic (urea, peptone, yeast extract) sources of nitrogen have been used in the production of biosurfactants, influencing the synthesis of metabolites and microbial growth. Furthermore, the presence of metallic compounds, such as manganese, phosphorus, iron, and others, has positively influenced the production of microbial surfactants (Sarubbo et al., 2022).

Even though microorganisms are able to produce biosurfactants under unfavorable conditions in terms of substrates and nutrients, process optimization plays a key role in achieving higher production yields, where several statistical approaches (Full/Fractional Factorial and Response surface designs) have been successfully employed to investigated and optimize process variables (e.g. agitation, aeration, etc) and components of the culture medium (e.g. concentration of carbon source, nitrogen, micronutrients, etc) for biosurfactant production (Nazareth et al., 2021a, b). In addition, artificial intelligence combined with genetic algorithm approaches has recently been used to maximize production in a short time (Manga et al., 2021).

The Carbon/Nitrogen (C/N) ratio is also relevant for biosurfactant production, where high C/N ratios strongly contribute to the production of biosurfactants since microbial growth is reduced and the metabolism is directed towards metabolites production. In addition, the literature confirms the need for the presence of nitrogen in the culture medium in smaller amounts to achieve higher productivity (Kashif et al., 2022).

Although biosurfactants have several advantages compared to chemical surfactants, their production on an industrial scale is not competitive yet, mainly concerning low yields, high costof carbon source (30%–50%), and separation/purification processes (Zanotto et al., 2019). In terms of reducing substrate-related costs, the use of low-cost carbon sources, such as agro-industrial waste, has been evaluated by several scientists around the world in the last decades, as canbe seen in Table 11.1. Renewable feedstock such as waste glycerol (Zhao et al., 2019; Janek et al., 2021), kitchen waste, sugarcane molasses (Takahashi et al., 2011; Minucelli et al., 2017) petroleum oil waste (Mostafa et al., 2019), and brewery waste (Khan et al., 2009; Nazareth et al., 2021a) have been reported in the literature as a potential carbon source in the production of biosurfactants.

11.3 BIOSURFACTANTS FOR REMOVAL OF HEAVY METALS FROM WASTEWATER

Heavy metals (HM) have a density greater than 5 g/m^3 and are defined according to their atomic weight, which can vary between 63.5 and 200.6. Although some HMs play an essential role in biochemical processes, excessive exposure to these metal ions can harm ecosystems and human health (Fu and Wang, 2011; Srivastava and Majumder, 2008), as described in Table 11.2. Its high solubility in aquatic systems, low biodegradability, and high migration rate between environmental complexes (e.g. water, soil), make heavy metals easily bioaccumulate in these systems, and may be dangerous for living beings (e.g. plants, animals) (Abidli et al., 2022; Morosanu et al., 2017). Therefore, the World Health Organization (WHO) has determined the maximum contaminant levels (MCLs) to ensure that no metal or an established minimum concentration is allowed in an aquatic system (Abdullah et al., 2019).

Although the contamination of ecosystems by HMs can be caused by natural processes, the majority are caused by industrial and urban processes (see Figure 11.1), making their removal a global environmental challenge. In this sense, several techniques have been establishedfor the removal or treat HMs from wastewater aiming to diminish their impact on environmental and human health. These techniques include adsorption, chemical precipitation, electrochemical, membrane filtration, ion exchange, treatment technologies, etc. (Fu and Wang, 2011; Abdullah et al., 2019). However,

TABLE 11.1
Biosurfactant Production from Renewable Resources

Biosurfactant (BS) Type	Low-cost Substrate	Microorganism	Fermentation Conditions	Bs Concentration	Productivity	Reference
Rhamnolipid	Distillery wastewater (20% v/v)	*Pseudomonas aeruginosa* SRRBL1	Batch, 37°C, 120 rpm, 120h	2.90 g/L	0.040 g/L/h	Ratna et al. (2022)
Rhamnolipid	waste frying oil (20 g/L)	*Pseudomonas aeruginosa*	Batch, 30°C, 150 rpm pH -, 96h	1.25 g/L, CMC: 150 mg/L	0.013 g/L/h	Pathania and Jana (2020)
Rhamnolipid	Glycerol (2% v/v)	*Pseudomonas* sp. CH1	Batch, 30°C, 180 rpm pH 7.0, 192h	5.07 g/L, CMC: 80 mg/L	0.026 g/L/h	Huang et al. (2020)
Rhamnolipid	Glycerol (40 g/L)	*Pseudomonas aeruginosa* TGC01	Batch, 30°C, 150 rpm pH 7.2, 96 h	11.0 g/L, CMC: 100 mg/L	0.114 g/L/h	Bezerra et al. (2019)
Sophorolipid	Rice husk (RH) and Wheat straw (WS)	*Starmerella bombicola* ATCC 22214	Solid-state fermentation, 30°C, 0.30 L air kg^{-1} total wet mass min^{-1}, 168h	RH (g SLs g^{-1}DM); WS (0.20 g SLs g^{-1}DM)	WS (0.16 g SLs g^{-1}DM), RH (0.09 g SLs g^{-1}DM)	Rodríguez et al. (2021)
Sophorolipid	Food waste hydrolysate	*Starmerella bombicola*	Fed batch, 30°C, pH 3.5, 3.48 vvm, 92h	115.2 g/L	1.25 g/L/h	Kaur et al. (2019)
Mannosylerythritol lipids	Cassava wastewater	*Pseudozyma tsukubaensis*	Batch, 30°C, 100–150 rpm, 0.4–0.8 vvm, pH -, 84h	1.26 g/L	0.015 g/L/h	Andrade et al. (2017)
Surfactin	Orange peel extract (10% v/v)	*Bacillus haynesii* E1	Batch, 35°C, 130 rpm, pH 6.0, 120 h	3.7 g/L, CMC: 50 mg/L	0.030 g/L/h	Rastogi et al. (2021)
Surfactin	Sugarcane molasses (5 % w/v)	*Bacillus subtilis* RSL-2	Batch, 41°C, 130 rpm, pH 6.6, 240 h	12.34 g/L, CMC: 80 mg/L	0.057 g/L/h	Verma et al. (2020)
Surfactin	Bagasse of sugarcane (4%)	*Bacillus safensis* J2	Batch, 30°C, 150 rpm, pH -, 72 h	0.92 g/L	–	Das and Kumar. (2019)

TABLE 11.2
Harmful Effects Caused by Heavy Metal(loid)s

Metal(loid)	Harmful Effects
Arsenic (As)	Long-term As exposure can cause different types of cancer (e.g. bladder, lungs, and skin) (Nafi and Taseidifar, 2022).
Cadmium (Cd)	Renal dysfunction and death (high level exposure) are possible situations caused by chronic exposure to Cadmium, which is also classified as a possible carcinogen by U.S. Environmental Protection Agency (Fu and Wang, 2011).
Copper (Cu)	Excessive ingestion may cause vomiting, convulsions, or death (Paulino et al., 2006).
Chromium (Cr)	Cr(VI) can cause skin irritation and lung carcinoma (Khezami and Capart, 2005).
Lead (Pb)	Can damage brain functions, the reproductive system, the kidney, and the liver (Naseem and Tahir, 2001).
Mercury (Hg)	High concentrations of Hg cause damage to the central nervous system, chest pain, dyspnoea, and impairment of kidney and pulmonary functions (Namasivayam and Kadirvelu, 1999).
Nickel (Ni)	Excess of Ni can cause skin dermatitis, lung (pulmonary fibrosis) and kidney damage, as well as gastrointestinal discomfort (Borba et al., 2006).
Zinc (Zn)	Excess Zn can cause anemia, skin irritations, stomach cramps, and vomiting (Oyaro et al., 2007).

Source: Adapted from Fu and Wang (2011).

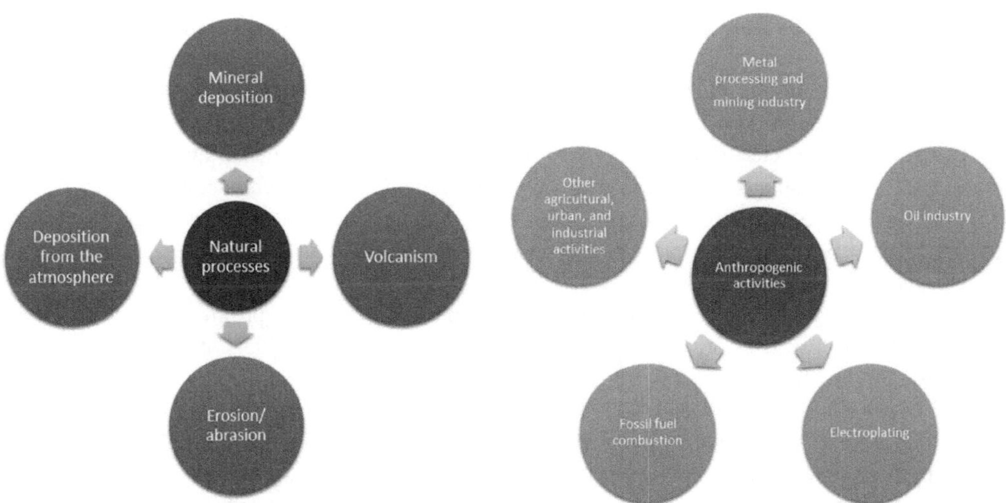

FIGURE 11.1 Sources of heavy metal contamination.

Source: Adapted from Fu and Wang (2011).

these traditional techniques have some limitations. For example, in an electrochemical process, the removal of HMs is limited when treating contaminants at low concentrations. The ion exchange process is considered an expensive method since there arecosts with large maintenance. Also, in the oxidation processes, secondary pollution is formed due to the addition of chemicals (Nafi and Taseidifar, 2022).

Biological treatment methods have been the most widely used in the treatment of different types of wastewater. Methods that use biosurfactants have been promising in biological treatments since they play an important role in mitigating ionic pollution by heavy metals since they are excellent surface-active molecules (Kashtiaray et al., 2021). Some of these methods will be discussed below as well their advantages and limitations.

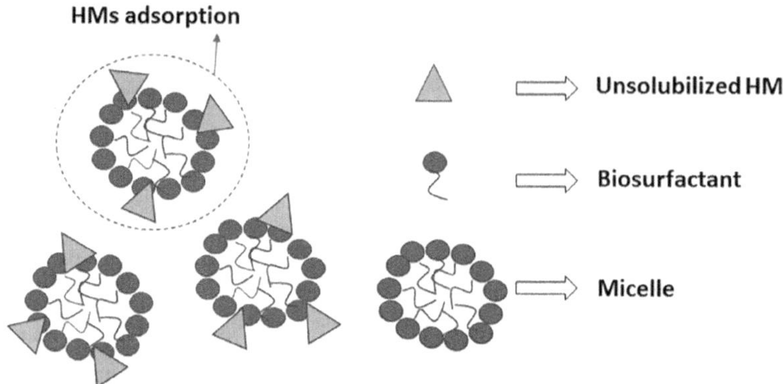

FIGURE 11.2 Scheme of heavy metal adsorption on biosurfactant micelle.

Source: Adapted from Rasheed et al. (2020).

The use of biosurfactants in the removal of HMs has been studied by using different methodologies such as phytoremediation, and soil-washing, among others. Moreover, biosurfactants can increase the efficiency of metal ion removal when the ultrafiltration method is used, being applied in membrane pretreatment. As can be seen in Figure 11.2, the micelle formed by the biosurfactant is capable of adsorbing the unsolubilized HMs. The use of biosurfactants, in this case, is interesting due to their excellent foaming capacity and stability against pH and temperature variations (Rasheed et al., 2020). For example, Tabagari et al. (2022) studied the remediation of water contaminated with Pb and Cu through *Spirulina platensis* (algae) in the presence of biosurfactant. (Rhamnolipids and Trehalose) as a chelating agent. They observed an enhancement in the absorption of copper between 42%–68% using Rhamnolipid and 73% when Trehalose was used in comparison to EDTA, a toxic chelating agent of heavy metals. These results suggest that the use of biosurfactants is an economically viable and eco-friendly approach for treating wastewater contaminated with lead and copper ions.

A new technology called Emulsion Liquid Membrane (ELM) is very promising to remove/separate HMs ions from wastewater, since being an efficient method with high selectivity, low operational cost, and low time demand.

The ELM technique is divided into three parts, which are(i) the membrane phase, (ii) the internal, and (iii) the external phase, as described in Figure 11.3. The ELM technique becomes especially interesting when biosurfactant is used as a stabilizer of the emulsion due to its biodegradability and low toxicity, very distinct characteristics when a chemical surfactant is used (Ferreira et al., 2019).

Another possibility to apply biosurfactants into wastewater treatment is through the mechanism of ion flotation. In this technique, the biosurfactant forms a complex with the HM ion, when the polar head of the biosurfactant bounds with the HM ion. Subsequently, air bubbles are added to the solution in order to make the complex float, forming a foam. Then, the foam is removed through a scum board. Thus, biosurfactants play both roles: collector and frother, being essential to this type of process since they facilitate the adsorption of the colligend species onto the surface of an air bubble (Peng et al., 2019).

11.4 BIOSURFACTANTS FOR REMOVAL OF ORGANIC POLLUTANTS FROM WASTEWATER

Organic compounds are commonly present in wastewater, such as dyes, oil residues, phenolic compounds, and pesticides, among others. These compounds represent a high risk to society and the environment, as they are toxic to the soil and water bodies, compromising human, fauna and flora

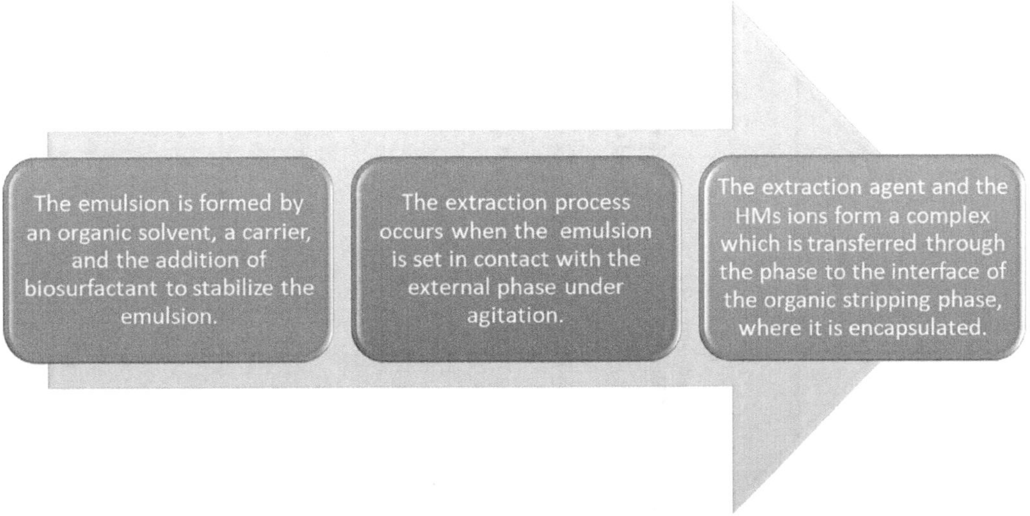

The emulsion is formed by an organic solvent, a carrier, and the addition of biosurfactant to stabilize the emulsion.

The extraction process occurs when the emulsion is set in contact with the external phase under agitation.

The extraction agent and the HMs ions form a complex which is transferred through the phase to the interface of the organic stripping phase, where it is encapsulated.

FIGURE 11.3 Emulsion liquid membrane phases.

Source: Adapted from Ferreira et al. (2019).

integrity. In this scenario, research has been carried out with the aim of developing efficient, low-cost, and eco-friendly technologies in order to solve this serious challenge for the global community.

Removal of pollutants from wastewater is commonly performed by coagulation, adsorption, and photocatalysis(Ajiboye et al., 2021). For instance, chemically aided dissolved air flotation has high efficiency, but has a high cost due to coagulating agents, in addition to generating sludge that is difficult to dispose of (Damasceno et al., 2014). Moreover, synthetic surfactants are commonly used for the remediation of contaminants from water and soil through the solubilization of hydrophobic compounds through micelles or metal removal when acting as chelating agent.Nonetheless, the intensive use of chemical surfactants can cause novel pollution due to their accumulation in the environment. Thus, biosurfactants are a promising alternative for applications in bioremediation since these compounds behave similarly to their synthetic counterparts, but with high biodegradability, low toxicity, and a wide variety of functional groups compared to chemical surfactants (Liu et al., 2017). The use of biosurfactants in the remediation of organic compounds was reported in the works presented in Table 11.3.

11.4.1 Oily Wastewater

Large amounts of petroleum hydrocarbons and other hazardous compounds (e.g. phenols, naphthalenes, and sulfides) are present in wastewater from petroleum refineries (Patel and Patel, 2020), where aromatic and aliphatic compounds represent 75% of wastewater content (Varjani et al., 2020). According to Yan et al. (2014), the amount of petroleum effluents from refineries represents between 0.4 and 1.6 timesthe total crude oil processed. Furthermore, it is estimated that by 2030 the global demand for oil will reach more than 107,000 barrels per day.

In this sense, many scientists investigated the potential use of biosurfactants in the remediation of petroleum wastewater, such as Patel and Patel (2020), who evaluated the production of biosurfactant by *Stenotrophomonas*sp. S1VKR-26 and its potential to remove contaminants (e.g. polycyclic aromatic hydrocarbons) present in petroleum wastewater. A 5L bioreactor was used for the 21-day biodegradation experiments in triplicate, containing 2,700 mL of refinery wastewater and 300 mL of bacterial culture. After 3 weeks, the maximum degradation achieved was 98.3% for pyrene, followed by 93%, 92%, and 86% for naphthalene, fluoranthene, and phenanthrene, respectively. After

TABLE 11.3
Biosurfactants for the Removal of Organic Compounds

Biosurfactant (Bs)	Pollutant	Removal Mechanism	Biosurfactant Concentration	Time	Removal Efficiency	Reference
Rhamnolipid	Bisphenol A (50 ppm)	Enzymatic degradation + Bs	1 ppm	120 min	65%	Alshabib and Onaizi (2020)
Lipopetide	Triclosan (0.05–5 mg/L)	Bs degradation	7.8 g/L	16 h	100%	Jayalatha and Devatha (2019)
Sophorolipid	Hexachlorocycloh-exane isomers (200 μg/mL)	Biodegradation in shake flask containing *Sphingomonas sp.*	40 μg/mL	2–4 days	70%	Manickam et al. (2012)
Rhamnolipid	Polyaromatic hydrocarbon PAHs (65.36 ng/mL)	Biodegradation in aerobic completely stirred tank reactor (CSTR)	15 mg/L	25 days	87%	Sponza and Gök (2010)
Rhamnolipid	Chlorpyrifos (0.1 g/L)	Biodegradation in shake flask containing *Pseudomonas sp.*	0.1 g/L	120 h	98%	Singh et al. (2009)
Mannosyleryth-ritol lipids (MEL)	Kerosene (8% v/v)	Biodegradation in bioreactor containing immoblized *C. antarctica*	1%	15 h	87%	Hua et al. (2004)

17 days, the Total petroleum hydrocarbons weresignificantly bioremediated from the wastewater after the treatment of strain S1VKR-26, achieving 72% of removal. According to the authors, the biosurfactant produced improved biodegradation due to the availability of substrate through emulsification, and also by modification of the cell surface of bacteria.

In another study, Huang et al. (2020) investigated the remediation of hydraulic fracturing flowback wastewater (HFFW) by biosurfactant (glycolipid). The HFFW consists of organic compounds, such as polycyclic aromatic hydrocarbons and n-alkanes, being generated from the fracking process used to extract oil and gas. The biosurfactant was produced by *Pseudomonas*sp. CH1 was later used to enhance the removal of the organic pollutants from HFFW. The results indicated that the addition of biosurfactant improved the removal of n-alkanes, achieving 80.21%, 79.09%, and 78.17% of removal for treatments with concentrations above-CMC (critical micelle concentration), sub-CMC, and CMC (80 mg/L), respectively. Concerning the removal of polycyclic aromatic hydrocarbons, the maximum removal efficiency achieved was 73.46% for biosurfactant concentration below CMC. Although higher concentrations of biosurfactants increase the solubilization of organic compounds and, consequently, their degradation, the authors believe that excess biosurfactant (concentrations above the CMC) was used as a substrate for microorganisms. Thus, the authors also concluded that low concentrations of biosurfactants (below the CMC) are effective in removing organic compounds from residue HFFW.

The bioremediation of polycyclic aromatic hydrocarbons by *Pseudomonas aeruginosa* S5, which is a biosurfactant-producing strain, was also evaluated by Sun et al. (2019). Coking wastewater holds high quantities of hydrophobic organics (e.g. pyridine and quinoline), where most PAHs are transferred to the sludge in the course of the wastewater treatment. The potential solubilization of fluoranthene, phenanthrene, and benzo[a]pyrene by the glycolipid was also investigated, where the results indicated that, for concentrations above the CMC (96.5 mg/L), there was a significant increase in the solubility of the components. Concerning the biodegradation experiments, the authors found that the removal of PAHs from the sludge phase, especially high-weight molecular, was enhanced by biosurfactant-producing microorganisms, achieving 44% removal after 15 days.

Several studies of oily wastewater bioremediation by biosurfactants have been reported in the literature, such as the work of Rizzo et al. (2021), where the authors investigated the production of biosurfactants by bacteria isolated from bilge water (BW) and their role in the bioremediation of hydrocarbon-enriched effluents. The bilge water (BW) is a complex mixture of several compounds accumulated on the bottom of ships resulting from leaks from machinery and washing water. According to the authors, these compounds, which include seawater, lubricating oil, fuels, oily sludge, and other chemical compounds, are more harmful to the environment than accidental oil spills. The isolated strains were able to produce biosurfactant with high emulsifying activity and a good capacity to reduce surface tension. In addition, the authors evaluated the efficiency of BW purification by biosurfactant-producing strains through microcosm-scale experiments, where the impact of the addition of extracted biosurfactant on the purification activity was evaluated. The authors found that the addition of biosurfactant caused an improvement in the biodegradation processes, in which the initial toxicity decreased in both microcosms evaluated and to a greater extent with the addition of biosurfactant.

Interesting studies of effluent pre-treatments combined with biosurfactants have been reported in the literature, for instance, the use of the system of Dissolved air flotation (DAF) – biosurfactant (BS) for oily wastewater treatment. Silva et al. (2018) investigated the efficiency of DAF-BS system for synthetic oily effluent treatment, where the biosurfactants were produced by *Bacillus cereus* UCP1615 and *Pseudomonas cepacia* CCT6659 using low-cost substrates (waste soy frying oil (2%), residual canola oil (2%) and corn steep liquor (3%)). A central composite rotatable design (CCRD) was utilized to enhance the oil separation efficiency, where the results indicated an increase in oil separation efficiency from 53.74% to 94.11% and 80.01% for biosurfactant from *P. cepacian* and *B. cereus*, respectively. Statistical analysis showed that the factor with the greatest impact on the separation efficiency was the oily water flow, followed by the concentration of biosurfactant. The authors also evaluated the treatment of a real effluent from a thermoelectric plant using DAF - BS produced by *P. cepacian*, achieving 95.7% removal. Furthermore, the authors concluded that when using the DAF system alone, it was not possible to reach the maximum limits determined by Brazilian legislation and that the DAF-BS system ensured adequate treatment of the effluent.

In another study of effluent pre-treatment combined with biosurfactant, Damasceno et al. (2014) evaluated the production of biosurfactant (rhamnolipid) and enzyme pool and their combined use for solubilization of fat present in thewastewater from a poultry processing plant, where the biodegradability was estimated by COD removal efficiency and biogas concentration. To investigate the interactive effect ofbiosurfactant and enzyme pools on the pre-treatment of effluent, several parameters (enzyme concentration, enzyme pool type, and biosurfactant concentration) were analyzed in a CCRD. Analysis of residual oils and greases indicated removals from 51% to 90%. Furthermore, the optimized condition corresponded to 24 mg/L of biosurfactant and 0.5% (w/v) *P. brevicompactum* pool.

11.4.2 Dye Wastewater

The textile industry has a significant role in the world economy. Nevertheless, it contributes to the increase of environmental pollution, being responsible for the pollution of 20% of global water and the consumption of around 93 billion cubic meters of water per year (Chen et al., 2021). The main source of pollution is due to the inefficient process of dyeing the fabric, or the residual dye (15%–50%) not fixed to the fabric that is discarded in the effluent. In addition, around 10,000 tons of the total dyes produced annually (7×10^7tons/yr) are used in the textile industry, with azo dyes being the most widely used (60%) (Al-Tohamy et al., 2022).

In this perspective, biosurfactants have been shown to be effective in the bioremediation of dyes from textile effluents, as can be seen in the work of Nor et al. (2022), who evaluated the role of extremophilic bacterium and its biosurfactant in the treatment of organic pollutant in wastewater. The mechanism behind the decolorization by bacteria in the presence of biosurfactant occurs due to the enzymes activity which improves the bacterial membrane permeability and facilitates dye diffusion. In the experiments, the isolated *Bacillus cereus* was capable to produce lipopeptide, with high emulsifying activity, by using molasses as the sole carbon source. In addition, the potential of dye discoloration of textile effluent by *B. cereus* was evaluated considering variations in salinity, temperature, pH,and concentration of molasse. The authors found that the decolorization efficiency of *B. cereus* in the presence of biosurfactant (0.075% w/v) was 87% while without it was 83%.

Other types of dyes present in wastewater have been removed through the use of biosurfactants, such as methyl violet dye, which was removed through photocatalytic discoloration by biosurfactant-modified iron oxide nanoparticles (Bhosale et al., 2019). According to the authors, the nanostructured material improves the degradation of organic pollutants through chemical reduction. In addition, the functionalization of magnetic nanoparticles with biosurfactant prevents oxidation and reduces the toxicity of nanoparticles due to the low toxicity nature of the biosurfactant. The nanoparticles were synthesized by co-precipitation and functionalized by glycolipid from *Pseudomonas aeruginosa* ATCC 9027, showing an effective removal (92.72%) of cationic dye in the presence of dodecyl sulfate (SDS) as a binding agent. However, without SDS, the efficiency was close to 28%.

Perez-Ameneiro et al. (2015) evaluated the removal efficiency of dye compounds in the winery wastewater by lignocellulosic biocomposite encapsulated in calcium alginate beads, containing biosurfactant. First, corn steep liquor was used as a carbon source in the production of biosurfactants, which werelater integrated into the biocomposite through the formation of an emulsion between sodium alginate and lignocellulosic composite. The results indicated that the presence of biosurfactant in the formula improved the adsorption of pollutants in the polymeric matrix, where removal of

76% of the colored compounds in the wastewater treated with the combination of biocomposite and biosurfactant was achieved, while without the presence of the biosurfactant, the removal achieved was approximately 70%.

11.4.3 Other Wastewater

Onaizi (2021) investigated the remediation of Bisphenol A (BPA) from wastewater through an enzymatic process containing biosurfactant. BPA is a hazardous phenolic compound thatpresents several applications in industry, mainly in the manufacturing of various plastics (e.g. paints, pipe linings, and papers). Different conditions were evaluated in order to identify factors with a statistically significant effect on BPA removal, such as reaction time, initial BPA concentration, salinity, pH, and temperature. The author's findings were that the addition of rhamnolipid and increasing the time of reaction improved BPA removal efficiency due to the favorable conformational changes of the laccase induced by the biosurfactant, contact time between biocatalyst and BPA, and the ability of rhamnolipid to prevent access to the active site of the enzyme by reaction products. Also, the pH was the factor that had the greatest significant effect on the response variable. The maximum BPA removal achieved was 84% when adding 1 ppm of biosurfactant, where a positive correlation between the addition of rhamnolipid and the removal of BPA was proved.

Another relevant organic residue with the potential to contaminate the surface/groundwater is pesticides, where organochlorine and organophosphate present low solubility in water.Hence, Gaur et al. (2019) investigated the potential of rhamnolipid in dissolving hydrophobic pesticides (HCH and endosulfan). *Lysinibacillus sphaericus* IITR51 was able to produce 1.6 g/L of biosurfactant from 1.5% glycerol in 72 h when incubated at 30°C and 160 rpm. The biosurfactant showed stability against temperature variations (4°C–100°C), salt concentration (2%–14%) and pH (4.0–10.0), and increased solubility by 1.18 mg/L (7.2 times) for α-endosulfan, 0.76 mg/L (2.9 times) for β-endosulfan and 13.37 mg/L (1.8 times) for γ-HCH. The authors also found that the solubilization forβ-endosulfan andα-endosulfan was greater than when using commercial surfactants (Triton X100).

In another similar work, García-Reyes et al. (2018) reported, for the first time, the potential solubilization of pesticides by biosurfactant produced by *Pseudomonas* sp. B0406. The biosurfactant was produced in batch, where the strain was incubated for 144 h, at 37°C and 120 rpm, with glycerol (20 g/L) as a carbon source. The authors found that endosulfanpesticidesand methyl parathion were effectively solubilized by rhamnolipid, which increased the solubility of both pesticides from 34.58 at 48.10 mg/L for methyl parathion and 0.41 at 0.92 mg/L for endosulfan.

Triclosan (TCS) is a compound belonging to the group of polychlorinated phenoxyphenols and is used in a variety of industrial, personal care, and household products. Due to its extensive consumption, TCS is an environmental pollutant frequently detected in aquatic environments (10–5160 ng/L) and can be convertedto toxic compounds (e.g. chlorophenol, chloroform, and chlorodioxins) through reaction with oxidants or photodegradation. The bioremediation of TCS by microorganisms is proven, however, it is a slow process due to the low biodegradability and poor degradation competence of microbes (Guo et al., 2016).

In this sense, Guo et al. (2016) investigated the potential of rhamnolipids (RL) to improve the degradation of TCS, after 56 days of treatment, by indigenous microbes in a water-sediment system. The authors investigated several parameters (one factor at a time) capable of impacting the efficiency of TCS biodegradation, such as initial TCS concentration, ion strength, temperature, pH, RL congeners and concentration, and agitation.The addition of di-RL congener enhanced the biodegradation efficiency of TCS, achieving 93.87%. For mono-RL, 47.3% of removal was achieved. According to the authors,the better results for di-RL can be attributed to di-RL micelles having core-shell interfaces and larger micellar hydrophobic core volumes,which makes it possible to solubilize larger amounts of TCS when compared to mono-RL micelles. Furthermore, the efficiency of TCS biodegradation by rhamnolipid was strongly influenced by variations in pH, decreasing from 91.47% to 63.92%, for pH values of 9 and 6, respectively.

11.5 RESEARCH NEEDS AND FUTURE DIRECTIONS TO BIOSURFACTANTS IN WASTEWATER TREATMENT

The potential use of biosurfactants in the remediation of wastewater containing metallic and organic residues has been widely reported in the literature in recent years. However, a better understanding of the mechanisms involved in the role of the biosurfactant in the removal of pollutants needs to be clarified. Thus, in-depth research to investigate the interactions at the micellar and sub-micellar levels between biosurfactants andpollutants in wastewater treatment is needed (Kashif et al., 2022). Factors such as pH, temperature, operational conditions, properties, and concentration of the biosurfactant affect its performance in the removal of contaminants in wastewater (Malkapuram et al., 2021), requiring further investigation to achieve a full understanding since the number of research in this aspect remains modest.

Despite the attractiveness of biosurfactant application for remediation purposes, most of the works reported in the literature were carried out on a laboratory scale, showing a gap in applied research on a larger scale for wastewater treatment. The key concern regarding the employment of biosurfactants in wastewater treatment methods is their production, which is expensive due to the high cost of carbon sources and extraction/purification processes, and poor yields. Moreover, the separation of biosurfactants from the culture medium is subject to contamination at a certain level that can result in a compromised performance of the biosurfactant since CMC is intimately correlated to the purity of the biosurfactant (Malkapuram et al., 2021).

Many studies have proved that biosurfactants can be produced from low-cost substrates, such as agro-industrial residues. Furthermore, optimized processes in terms of culture medium and operational parameters have been widely reported in the literature. However, there is still a gap in terms of processes and strategies aimed at reducing the costs concerning purification processes. The total costs of life cycle assessment (LCA), which addresses issues related to climate change, environmental degradation, and resource depletion must also be considered, which is still an underexplored and essential approach toward a sustainable and circular economy (Johnson et al., 2021).

Even though the use of biosurfactants in wastewater remediation showspromising results, continued research efforts to enable the production of biosurfactants and their large-scale application in wastewater treatment systems are essential.

11.6 CONCLUSION

In this chapter, the possibility of using biosurfactants in the treatment of wastewater containing metallic and organic pollutants was presented. Studies have indicated that biosurfactants are efficient in wastewater remediation processes, in which rhamnolipids have stood out, showing efficiency in the removal of organic compounds and heavy metals, although the inherent mechanisms of increased removal are different. The main obstacle to using biosurfactants to remove pollutants from wastewater is their high cost, unlike synthetic surfactants which are cheap to produce and well-established. Despite this, biosurfactants have great environmental appeal due to their many advantages over chemical surfactants, such as high biodegradability and low toxicity. Furthermore, it has already been proven that biosurfactants can be obtained from renewable sources, which would make their production and use feasible. However, deeper investigations of large-scale production from renewable substrates, improvement of production yield, development and/or improvement of cheaper purification techniques, LCA, and technical-economic evaluation studies are the biggest challenges.

REFERENCES

Abdullah, N., Yusof, N., Lau, W. J., Jaafar, J., & Ismail, A. F. (2019). Recent trends of heavy metal removal from water/wastewater by membrane technologies. *Journal of Industrial and Engineering Chemistry, 76*, 17–38. Korean Society of Industrial Engineering Chemistry. https://doi.org/10.1016/j.jiec.2019.03.029

Abidli, A., Huang, Y., ben Rejeb, Z., Zaoui, A., & Park, C. B. (2022). Sustainable and efficient technologies for removal and recovery of toxic and valuable metals from wastewater: Recent progress, challenges, and future perspectives. Chemosphere,*292*. Elsevier Ltd. https://doi.org/10.1016/j.chemosphere.2021.133102

Ajiboye, T. O., Oyewo, O. A., & Onwudiwe, D. C. (2021). Simultaneous removal of organics and heavy metals from industrial wastewater: A review. Chemosphere,*262*. 128379

Alshabib, M., & Onaizi, S. A. (2020). Enzymatic remediation of bisphenol a from wastewaters: Effects of biosurfactant, anionic, cationic, nonionic, and polymeric additives. Water, Air, and Soil Pollution, 231(8). https://doi.org/10.1007/s11270-020-04806-5

Al-Tohamy, R., Ali, S. S., Li, F., Okasha, K. M., Mahmoud, Y. A. G., Elsamahy, T., Jiao, H., Fu, Y., & Sun, J. (2022). A critical review on the treatment of dye-containing wastewater: Ecotoxicological and health concerns of textile dyes and possible remediation approaches for environmental safety. *Ecotoxicology and Environmental Safety* 231, 113160

Andrade, C. J. de, Andrade, L. M. de, Rocco, S. A., Sforça, M. L., Pastore, G. M., & Jauregi, P. (2017). A novel approach for the production and purification of mannosylerythritol lipids (MEL) by Pseudozyma tsukubaensis using cassava wastewater as substrate. Separation and Purification Technology, *180*, 157–167. https://doi.org/10.1016/j.seppur.2017.02.045

Bezerra, K. G. O., Gomes, U. V. R., Silva, R. O., Sarubbo, L. A., & Ribeiro, E. (2019). The potential application of biosurfactant produced by Pseudomonas aeruginosa TGC01 using crude glycerol on the enzymatic hydrolysis of lignocellulosic material. *Biodegradation*, *30*(4), 351–361. https://doi.org/10.1007/s10532-019-09883-w

Bhosale, S. S., Rohiwal, S. S., Chaudhary, L. S., Pawar, K. D., Patil, P. S., & Tiwari, A. P. (2019). Photocatalytic decolorization of methyl violet dye using Rhamnolipid biosurfactant modified iron oxide nanoparticles for wastewater treatment. *Journal of Materials Science: Materials in Electronics*, *30*(5), 4590–4598. https://doi.org/10.1007/s10854-019-00751-0

Borba, C. E., Guirardello, R., Silva, E. A., Veit, M. T., & Tavares, C. R. G. (2006). Removal of nickel(II) ions from aqueous solution by biosorption in a fixed bed column: Experimental and theoretical breakthrough curves. *Biochemical Engineering Journal*, *30*(2), 184–191. https://doi.org/10.1016/j.bej.2006.04.001

Chebbi, A., Franzetti, A., Duarte Castro, F., Gomez Tovar, F. H., Tazzari, M., Sbaffoni, S., & Vaccari, M. (2021). Potentials of winery and olive oil residues for the production of rhamnolipids and other biosurfactants: A step towards achieving a circular economy model. *Waste and Biomass Valorization*, *12*(8), 4733–4743. https://doi.org/10.1007/s12649-020-01315-8

Chen, W. C., Juang, R. S., & Wei, Y. H. (2015). Applications of a lipopeptide biosurfactant, surfactin, produced by microorganisms. *Biochemical Engineering Journal*, *103*, 158–169. Elsevier. https://doi.org/10.1016/j.bej.2015.07.009

Chen, X., Memon, H. A., Wang, Y., Marriam, I., & Tebyetekerwa, M. (2021). Circular economy and sustainability of the clothing and textile industry. Materials Circular Economy, 3(1). https://doi.org/10.1007/s42824-021-00026-2

Coelho, A. L. S., Feuser, P. E., Carciofi, B. A. M., de Andrade, C. J., & de Oliveira, D. (2020). Mannosylerythritol lipids: antimicrobial and biomedical properties. *Applied Microbiology and Biotechnology*,*104*(6), 2297–2318. Springer. https://doi.org/10.1007/s00253-020-10354-z

Damasceno, F. R. C., Freire, D. M. G., & Cammarota, M. C. (2014). Assessing a mixture of biosurfactant and enzyme pools in the anaerobic biological treatment of wastewater with a high-fat content. Environmental Technology, 35(16), 2035–2045. https://doi.org/10.1080/09593330.2014.890249

Das, A. J., & Kumar, R. (2019). Production of biosurfactant from agro-industrial waste by Bacillus safensis J2 and exploring its oil recovery efficiency and role in restoration of diesel contaminated soil. Environmental Technology and Innovation, 16. https://doi.org/10.1016/j.eti.2019.100450

Desai, J. D., & Banat, I. M. (1997). Microbial production of surfactants and their commercial potential. *Microbiology and Molecular Biology Reviews*, *61*(1). https://doi.org/10.1128/mmbr.61.1.47-64.1997

Dierickx, S., Castelein, M., Remmery, J., de Clercq, V., Lodens, S., Baccile, N., de Maeseneire, S. L., Roelants, S. L. K. W., & Soetaert, W. K. (2022). From bumblebee to bioeconomy: Recent developments and perspectives for sophorolipid biosynthesis. Biotechnology Advances,*54*. Elsevier Inc. https://doi.org/10.1016/j.biotechadv.2021.107788

Elumalai, P., Parthipan, P., Huang, M., Muthukumar, B., Cheng, L., Govarthanan, M., & Rajasekar, A. (2021). Enhanced biodegradation of hydrophobic organic pollutants by the bacterial consortium: Impact of enzymes and biosurfactants. Environmental Pollution, 289. https://doi.org/10.1016/j.envpol.2021.117956

Ferreira, L. C., Ferreira, L. C., Cardoso, V. L., & Filho, U. C. (2019). Mn(II) removal from water using emulsion liquid membrane composed of chelating agents and biosurfactant produced in loco. *Journal of Water Process Engineering*, *29*. https://doi.org/10.1016/j.jwpe.2019.100792

Fu, F., & Wang, Q. (2011). Removal of heavy metal ions from wastewaters: A review. *Journal of Environmental Management*, *92*(3), 407–418. https://doi.org/10.1016/j.jenvman.2010.11.011

García-Reyes, S., Yáñez-Ocampo, G., Wong-Villarreal, A., Rajaretinam, R. K., Thavasimuthu, C., Patiño, R., & Ortiz-Hernández, M. L. (2018). Partial characterization of a biosurfactant extracted from Pseudomonas sp. B0406 that enhances the solubility of pesticides. Environmental Technology, *39*(20), 2622–2631. https://doi.org/10.1080/21622515.2017.1363295

Gaur, V. K., Bajaj, A., Regar, R. K., Kamthan, M., Jha, R. R., Srivastava, J. K., & Manickam, N. (2019). Rhamnolipid from a Lysinibacillus sphaericus strain IITR51 and its potential application for dissolution of hydrophobic pesticides. *Bioresource Technology*, *272*, 19–25. https://doi.org/10.1016/j.biortech.2018.09.144

Guo, Q., Yan, J., Wen, J., Hu, Y., Chen, Y., & Wu, W. (2016). Rhamnolipid-enhanced aerobic biodegradation of triclosan (TCS) by indigenous microorganisms in water-sediment systems. *Science of the Total Environment*, *571*, 1304–1311. https://doi.org/10.1016/j.scitotenv.2016.07.171

Henkel, M., Geissler, M., Weggenmann, F., & Hausmann, R. (2017). Production of microbial biosurfactants: Status quo of rhamnolipid and surfactin towards large-scale production. *Biotechnology Journal*, *12*(7). Wiley-VCH Verlag. https://doi.org/10.1002/biot.201600561

Hua, Z., Chen, Y., Du, G., & Chen, J. (2004). Effects of biosurfactants produced by Candida Antarctica on the biodegradation of petroleum compounds. *World Journal of Microbiology and Biotechnology*, *20*, 25–29. https://doi.org/10.1023/B:WIBI.0000013287.11561.d4

Huang, X., Zhou, H., Ni, Q., Dai, C., Chen, C., Li, Y., & Zhang, C. (2020). Biosurfactant-facilitated biodegradation of hydrophobic organic compounds in hydraulic fracturing flowback wastewater: A dose-effect analysis. Environmental Technology and Innovation, 19. https://doi.org/10.1016/j.eti.2020.100889

Inès, M., & Dhouha, G. (2015). Glycolipid biosurfactants: Potential related biomedical and biotechnological applications. *Carbohydrate Research*, *416*, 59–69. Elsevier Ltd. https://doi.org/10.1016/j.carres.2015.07.016

Jahan, R., Bodratti, A. M., Tsianou, M., & Alexandridis, P. (2020). Biosurfactants, natural alternatives to synthetic surfactants: Physicochemical properties and applications. In *Advances in Colloid and Interface Science* (Vol. 275). Elsevier B.V. https://doi.org/10.1016/j.cis.2019.102061

Janek, T., Gudiña, E. J., Połomska, X., Biniarz, P., Jama, D., Rodrigues, L. R., Rymowicz, W., & Lazar, Z. (2021). Sustainable surfactin production by bacillus subtilis using crude glycerol from different wastes. Molecules, 26(12). https://doi.org/10.3390/molecules26123488

Jayalatha, N. A., & Devatha, C. P. (2019). Degradation of Triclosan from Domestic Wastewater by Biosurfactant Produced from Bacillus licheniformis. *Molecular Biotechnology*, *61*(9), 674–680. https://doi.org/10.1007/s12033-019-00193-3

Johnson, P., Trybala, A., Starov, V., & Pinfield, V. J. (2021). Effect of synthetic surfactants on the environment and the potential for substitution by biosurfactants. In *Advances in Colloid and Interface Science* (Vol. 288). Elsevier B.V. https://doi.org/10.1016/j.cis.2020.102340

Kashif, A., Rehman, R., Fuwad, A., Shahid, M. K., Dayarathne, H. N. P., Jamal, A., Aftab, M. N., Mainali, B., & Choi, Y. (2022). Current advances in the classification, production, properties and applications of microbial biosurfactants - A critical review. In *Advances in Colloid and Interface Science*, 306, 102718.

Kashtiaray, A., Khadir, A., Ardestani, A. N., Salehpour, N. (2021). Biosurfactants and sustainable multifunctional biocompounds for wastewater remediation. In: Inamuddin & Charles Oluwaseun Adetunji (eds.) *Green Sustainable Process for Chemical and Environmental Engineering and Science* (pp. 395–417). Elsevier. B.V. https://doi.org/10.1016/B978-0-12-822696-4.00015-2.

Kaur, G., Wang, H., To, M. H., Roelants, S. L. K. W., Soetaert, W., & Lin, C. S. K. (2019). Efficient sophorolipids production using food waste. *Journal of Cleaner Production*, *232*, 1–11. https://doi.org/10.1016/j.jclepro.2019.05.326

Khan, A. W., Rahman, M. S., & Ano, T. (2009). Application of malt residue in submerged fermentation of Bacillus subtilis. *Journal of Environmental Sciences*, *21*(Suppl. 1). https://doi.org/10.1016/S1001-0742(09)60030-9

Khezami, L., & Capart, R. (2005). Removal of chromium(VI) from aqueous solution by activated carbons: Kinetic and equilibrium studies. *Journal of Hazardous Materials*, *123*(1-3), 223–231. https://doi.org/10.1016/j.jhazmat.2005.04.012

Liu, Z., Li, Z., Zhong, H., Zeng, G., Liang, Y., Chen, M., Wu, Z., Zhou, Y., Yu, M., & Shao, B. (2017). Recent advances in the environmental applications of biosurfactant saponins: A review. *Journal of Environmental Chemical Engineering*, *5*(6), 6030–6038. Elsevier Ltd. https://doi.org/10.1016/j.jece.2017.11.021

Malkapuram, S. T., Sharma, V., Gumfekar, S. P., Sonawane, S., Sonawane, S., Boczkaj, G., & Seepana, M. M. (2021). A review on recent advances in the application of biosurfactants in wastewater treatment. *Sustainable Energy Technologies and Assessments*, *48*. https://doi.org/10.1016/j.seta.2021.101576

Manga, E. B., Celik, P. A., Cabuk, A., & Banat, I. M. (2021). Biosurfactants: Opportunities for the development of a sustainable future. In *Current Opinion in Colloid and Interface Science*, 56, 101514.

Manickam, N., Bajaj, A., Saini, H. S., & Shanker, R. (2012). Surfactant mediated enhanced biodegradation of hexachlorocyclohexane (HCH) isomers by Sphingomonas sp. NM05. *Biodegradation*, *23*(5), 673–682. https://doi.org/10.1007/s10532-012-9543-z

Minucelli, T., Ribeiro-Viana, R. M., Borsato, D., Andrade, G., Cely, M. V. T., de Oliveira, M. R., Baldo, C., & Celligoi, M. A. P. C. (2017). Sophorolipids production by Candida bombicola ATCC 22214 and its potential application in soil bioremediation. *Waste and Biomass Valorization*, *8*(3), 743–753. https://doi.org/10.1007/s12649-016-9592-3

Mongkolthanaruk, W. (2012). Classification of bacillus beneficial substances related to plants, humans and animals. *Journal of Microbiology and Biotechnology*, *22*(12), 1597–1604. https://doi.org/10.4014/jmb.1204.04013

Morosanu, I., Teodosiu, C., Paduraru, C., Ibanescu, D., & Tofan, L. (2017). Biosorption of lead ions from aqueous effluents by rapeseed biomass. *New Biotechnology*, *39*, 110–124. https://doi.org/10.1016/j.nbt.2016.08.002

Mostafa, N. A., Tayeb, A. M., Mohamed, O. A., & Farouq, R. (2019). Biodegradation of petroleum oil effluents and production of biosurfactants: Effect of initial oil concentration. *Journal of Surfactants and Detergents*, *22*(2), 385–394. https://doi.org/10.1002/jsde.12240

Nafi, A. W., & Taseidifar, M. (2022). Removal of hazardous ions from aqueous solutions: Current methods, with a focus on green ion flotation. *Journal of Environmental Management*, 319, 115666.

Namasivayam, C., & Kadirvelu, K. (1999). Uptake of mercury (II) from wastewater by activated carbon from an unwanted agricultural solid by-product: coirpith. Carbon, *37,* 79–84.

Naseem, R., & Tahir, S. S. (2001). Removal of Pb(II) from Aqueous/Acidic Solutions by Using Bentonite as an Adsorbent. *Water Research*, *35*(16).

Nazareth, T. C., Zanutto, C. P., Maass, D., de Souza, A. A. U., & Guelli Ulson de Souza, S. M. de A. (2021a). A low-cost brewery waste as a carbon source in bio-surfactant production. *Bioprocess and Biosystems Engineering*, 44, 2269–2276.

Nazareth, T. C., Zanutto, C. P., Maass, D., Ulson De Souza, A. A., & Ulson De Souza, S. M. D. A. G. (2021b). Impact of oxygen supply on surfactin biosynthesis using brewery waste as substrate. *Journal of Environmental Chemical Engineering*, *9*(4). https://doi.org/10.1016/j.jece.2021.105372

Nitschke, M., Gláucia, E., & Pastore, M. (2002). Biossurfactantes: propriedades e aplicações. *Química Nova*, *25*(5). https://doi.org/10.1590/S0100-40422002000500013

Nor, F. H. M., Abdullah, S., Ibrahim, Z., Nor, M. H. M., Osman, M. I., al Farraj, D. A., AbdelGawwad, M. R., & Kamyab, H. (2022). Role of extremophilic Bacillus cereus KH1 and its lipopeptide in treatment of organic pollutant in wastewater. Bioprocess and Biosystems Engineering. https://doi.org/10.1007/s00449-022-02749-1

Onaizi, S. A. (2021). Statistical analyses of the effect of rhamnolipid biosurfactant addition on the enzymatic removal of Bisphenol A from wastewater. Biocatalysis and Agricultural Biotechnology, 32. https://doi.org/10.1016/j.bcab.2021.101929

Oyaro, N., Juddy, O., Murago, E., & Gitonga, E. (2007). The contents of Pb, Cu, Zn and Cd in meat in Nairobi, Kenya. *Journal of Food, Agriculture & Environment, 5*, 119–121.

Parus, A., Ciesielski, T., Woźniak-Karczewska, M., Ślachciński, M., Owsianiak, M., Ławniczak, Ł., Loibner, A. P., Heipieper, H. J., & Chrzanowski, Ł. (2023). Basic principles for biosurfactant-assisted (bio)remediation of soils contaminated by heavy metals and petroleum hydrocarbons – A critical evaluation of the performance of rhamnolipids. *Journal of Hazardous Materials*, *443*. Elsevier B.V. https://doi.org/10.1016/j.jhazmat.2022.130171

Patel, K., & Patel, M. (2020). Improving bioremediation process of petroleum wastewater using biosurfactants producing Stenotrophomonas sp. S1VKR-26 and assessment of phytotoxicity. Bioresource Technology, 315. https://doi.org/10.1016/j.biortech.2020.123861

Pathania, A. S., & Jana, A. K. (2020). Utilization of waste frying oil for rhamnolipid production by indigenous Pseudomonas aeruginosa: Improvement through co-substrate optimization. *Journal of Environmental Chemical Engineering*, *8*(5). https://doi.org/10.1016/j.jece.2020.104304

Paulino, A. T., Minasse, F. A. S., Guilherme, M. R., Reis, A. v., Muniz, E. C., & Nozaki, J. (2006). Novel adsorbent based on silkworm chrysalides for removal of heavy metals from wastewaters. *Journal of Colloid and Interface Science*, *301*(2), 479–487. https://doi.org/10.1016/j.jcis.2006.05.032

Peng, W., Chang, L., Li, P., Han, G., Huang, Y., & Cao, Y. (2019). An overview on the surfactants used in ion flotation. *Journal of Molecular Liquids*, *286*. Elsevier B.V. https://doi.org/10.1016/j.molliq.2019.110955

Perez-Ameneiro, M., Vecino, X., Cruz, J. M., & Moldes, A. B. (2015). Wastewater treatment enhancement by applying a lipopeptide biosurfactant to a lignocellulosic biocomposite. *Carbohydrate Polymers*, *131*, 186–196. https://doi.org/10.1016/j.carbpol.2015.05.075

Qazi, M. A., Wang, Q., & Dai, Z. (2022). Sophorolipids bioproduction in the yeast Starmerella bombicola: current trends and perspectives. *Bioresource Technology*, 346, 126593.

Rasheed, T., Shafi, S., Bilal, M., Hussain, T., Sher, F., & Rizwan, K. (2020). Surfactants-based remediation as an effective approach for removal of environmental pollutants-A review. *Journal of Molecular Liquids*, *318*. Elsevier B.V. https://doi.org/10.1016/j.molliq.2020.113960

Rastogi, S., Tiwari, S., Ratna, S., & Kumar, R. (2021). Utilization of agro-industrial waste for biosurfactant production under submerged fermentation and its synergistic application in biosorption of Pb2+. Bioresource Technology Reports, 15. https://doi.org/10.1016/j.biteb.2021.100706

Ratna, S., & Kumar, R. (2022). Production of di-rhamnolipid with simultaneous distillery wastewater degradation and detoxification by newly isolated *Pseudomonas aeruginosa* SRRBL1. *Journal of Cleaner Production*, 336, 130429.

Rizzo, C., Caldarone, B., de Luca, M., de Domenico, E., & Giudice, A. lo. (2021). Native bilge water bacteria as biosurfactant producers and implications in hydrocarbon-enriched wastewater treatment. *Journal of Water Process Engineering*, *43*. https://doi.org/10.1016/j.jwpe.2021.102271

Rodríguez, A., Gea, T., Sánchez, A., & Font, X. (2021). Agro-wastes and inert materials as supports for the production of biosurfactants by solid-state fermentation. *Waste and Biomass Valorization*, *12*(4), 1963–1976. https://doi.org/10.1007/s12649-020-01148-5

Sałek, K., Euston, S. R., & Janek, T. (2022). Phase behaviour, functionality, and physicochemical characteristics of glycolipid surfactants of microbial origin. In *Frontiers in Bioengineering and Biotechnology*, 10, 816613. Frontiers Media S.A. https://doi.org/10.3389/fbioe.2022.816613

Sarubbo, L. A., Silva, M. da G. C., Durval, I. J. B., Bezerra, K. G. O., Ribeiro, B. G., Silva, I. A., Twigg, M. S., & Banat, I. M. (2022). Biosurfactants: Production, properties, applications, trends, and general perspectives. *Biochemical Engineering Journal*, *181*(108377). https://doi.org/10.1016/j.bej.2022.108377

Silva, E. J., Almeida, D. G., Luna, J. M., Rufino, R. D., Santos, V. A., & Sarubbo, L. A. (2018). Use of bacterial biosurfactants as natural collectors in the dissolved air flotation process for the treatment of oily industrial effluent. *Bioprocess and Biosystems Engineering*, *41*(11), 1599–1610. https://doi.org/10.1007/s00449-018-1986-0

Singh, P. B., Sharma, S., Saini, H. S., & Chadha, B. S. (2009). Biosurfactant production by Pseudomonas sp. and its role in aqueous phase partitioning and biodegradation of chlorpyrifos. *Letters in Applied Microbiology*, *49*(3), 378–383. https://doi.org/10.1111/j.1472-765X.2009.02672.x

Sponza, D. T., & Gök, O. (2010). Effect of rhamnolipid on the aerobic removal of polyaromatic hydrocarbons (PAHs) and COD components from petrochemical wastewater. *Bioresource Technology*, *101*(3), 914–924. https://doi.org/10.1016/j.biortech.2009.09.022

Srivastava, N. K., & Majumder, C. B. (2008). Novel biofiltration methods for the treatment of heavy metals from industrial wastewater. *Journal of Hazardous Materials*, *151*(1), 1–8. https://doi.org/10.1016/j.jhazmat.2007.09.101

Sun, S., Wang, Y., Zang, T., Wei, J., Wu, H., Wei, C., Qiu, G., & Li, F. (2019). A biosurfactant-producing Pseudomonas aeruginosa S5 isolated from coking wastewater and its application for bioremediation of polycyclic aromatic hydrocarbons. *Bioresource Technology*, *281*, 421–428. https://doi.org/10.1016/j.biortech.2019.02.087

Tabagari, I., Varazi, T., Chokheli, L., Kurashvili, M., Pruidze, M., Khatisashvili, G., Karpenko, O., Lubenets, V., & Fragsteinund Niemsdorff, P. von. (2022). Enhancement of Spirulina platensis remediation action using biosurfactants for wastewater treatment. *International Journal of Environmental Research*, *16*(1). https://doi.org/10.1007/s41742-022-00392-y

Takahashi, M., Morita, T., Wada, K., Hirose, N., Fukuoka, T., Imura, T., & Kitamoto, D. (2011). Production of sophorolipid glycolipid biosurfactants from sugarcane molasses using Starmerella bombicola NBRC 10243. *Journal of Oleo Science*, *60*(5):267–273. https://doi.org/10.5650/jos.60.267

Varjani, S., Joshi, R., Srivastava, V. K., Huu, & , Ngo, H., & Guo, W. (2020). Treatment of wastewater from petroleum industry: Current practices and perspectives. *Environmental Science and Pollution Research, 27*, 27172–27180. https://doi.org/10.1007/s11356-019-04725-x

Vecino, X., Rodríguez-López, L., Rincón-Fontán, M., Cruz, J. M., & Moldes, A. B. (2021). Nanomaterials synthesized by biosurfactants. Comprehensive Analytical Chemistry, *94*, 267–301. Elsevier B.V. https://doi.org/10.1016/bs.coac.2020.12.008

Verma, R., Sharma, S., Kundu, L. M., & Pandey, L. M. (2020). Experimental investigation of molasses as a sole nutrient for the production of an alternative metabolite biosurfactant. *Journal of Water Process Engineering, 38*. https://doi.org/10.1016/j.jwpe.2020.101632

Yan, L., Wang, Y., Li, J., Ma, H., Liu, H., Li, T., & Zhang, Y. (2014). Comparative study of different electrochemical methods for petroleum refinery wastewater treatment. *Desalination, 341*(1), 87–93. https://doi.org/10.1016/j.desal.2014.02.037

Yao, L., Selmi, A., & Esmaeili, H. (2021). A review study on new aspects of biodemulsifiers: Production, features and their application in wastewater treatment. Chemosphere, 284. https://doi.org/10.1016/j.chemosphere.2021.131364

Zanotto, A. W., Valério, A., de Andrade, C. J., & Pastore, G. M. (2019). New sustainable alternatives to reduce the production costs for surfactin 50 years after the discovery. *Applied Microbiology and Biotechnology, 103*(21-22), 8647–8656. Springer Verlag. https://doi.org/10.1007/s00253-019-10123-7

Zeng, Z., Liu, Y., Zhong, H., Xiao, R., Zeng, G., Liu, Z., Cheng, M., Lai, C., Zhang, C., Liu, G., & Qin, L. (2018). Mechanisms for rhamnolipids-mediated biodegradation of hydrophobic organic compounds. Science of the Total Environment, *634*, 1–11. Elsevier B.V. https://doi.org/10.1016/j.scitotenv.2018.03.349

Zhao, F., Jiang, H., Sun, H., Liu, C., Han, S., & Zhang, Y. (2019). Production of rhamnolipids with different proportions of mono-rhamnolipids using crude glycerol and a comparison of their application potential for oil recovery from oily sludge. RSC Advances, *9*(6), 2885–2891. https://doi.org/10.1039/c8ra09351b

12 Microbial Adsorbent for Heavy Metal Removal from Wastewater for Environmental Safety

Athar Hussain and Richa Madan

12.1 INTRODUCTION

As a consequence of rapid industrialization and technological advancement, augmented quantities of heavy metals, metalloids as well as organic contaminants have caused significant harm to the ecosystem. Due to the toxicity, widespread occurrence, and recalcitrant nature of heavy metals, pollution by them has become a main cause of concern in the past few years. Heavy metals are demarcated as metals with a range of atomic weight between 63.5 and 200.6 g/mol having a density of more than 5 g/mL (Lapedes, 1974; Briffa et al., 2020). Living organisms undergo various metabolic processes which are assessed by several heavy metals, like Cd, Zn, Cr, Co, Pb, Cu, Fe, Se, Ni, Hg, etc. There happens to be a range of safe, sufficient concentrations for ingestion by humans for each important element. Low concentrations are likely to result in deficiencies while high concentrations can be harmful to the organism.

Both manmade and natural sources contribute to the influx of hazardous heavy metals into the natural environment. Volcanic eruptions, mineral erosion and rocks, weathering, etc. are the chief natural sources of these heavy metals. Industrial activities such as electroplating, painting and polishing industries, mining, smelting, fertilizer factories, etc. are the main anthropogenic causes. Severely polluted wastewater discharged from countless manmade activities is hazardous to humans along with the environment, hence, prescribed treatment of wastewater is essential before discharge to any water body. These heavy metals pose grave concerns due to their toxic nature, nonbiodegradability, bioaccumulation, and bio-magnification (Kumar, 2006; Wang and Chen, 2009; Alengebawy et al., 2021). Because they cannot be broken down by physical or chemical processes, heavy metals are able to persist for an extremely long time in the environment. Depending on suitable environmental conditions, some of them might change to more poisonous variants. The biomagnification and bioaccumulation of these metals along the food chain prove to be detrimental to living organisms. Contact with these heavy metals in extreme concentrations becomes the major cause of adverse effects including organ damage, growth inhibition, nervous system damage, cancer, and in extreme cases, death. Table 12.1 shows the sources and toxic effects of heavy metals.

12.1.1 Conventional Wastewater Methods of Treatment

For the treatment of wastewater contaminated with hazardous heavy metals, many traditional treatment techniques are frequently utilized, including membrane separation, chemical precipitation electrochemical treatment, coagulation and flocculation, adsorption, etc. (Chai et al., 2021; Shrestha et al., 2021). The drawbacks of these methods include a short lifespan and high replacement cost of membrane in case of membrane process. The processes of coagulation and flocculation produce excess scum and sludge leading to disposal problems. Many studies agree that adsorption is a very

 DOI: 10.1201/9781003441144-12

TABLE 12.1
Sources of Heavy Metals and their Toxic Effects

S.No.	Heavy Metals	Major Sources	Toxic Effects
1.	Cadmium	Plastic, mining, fertilizer, refining, welding, pesticide	Kidney damage, gastrointestinal disorder, bronchitis, lung insufficiency, itai–itai disease cancer, hypertension
2.	Copper	Plating, paint, copper polishing, printing	neurotoxicity, dizziness, diarrhea
3.	Chromium	Textile dyeing, paints, alloy manufacturing	Carcinogenic, teratogenic, mutagenic, severe diarrhea
4.	Arsenic	Smelting, pesticides, mineral sedimentation, mining	Dermatitis, bronchitis, bone marrow depression
5.	Lead	Mining, manufacturing of batteries, paint, pigments, electroplating, coal burning	Anemia, loss of appetite, anorexia, brain, liver, kidney damage, intestinal damage
6.	Mercury	Batteries, mining, paper, and paint industries	Nervous system damage, dermatitis, kidney damage corrosive to skin, eyes
7.	Nickel	Enameling, electroplating, paint making	Bronchitis, lung cancer
8.	Zinc	Mining activities, refineries, brass manufacturing, plumping	Gastrointestinal distress

Source: Godt et al. (2007); Ozturk (2007); Gadd (2010).

effective method for treating wastewater that has been contaminated with harmful heavy metals and other contaminants. Diverse inorganic and organic wastes are used in the adsorption process as possible adsorbents to eliminate pollutants from contaminated water (Qasem et al., 2021; Singh et al., 2022). Adsorption is an easy and affordable wastewater treatment method with the majority of adsorbents being regenerated and reused. Heavy metal adsorption usually involves the mechanisms of ion exchange, physisorption, and chemisorption. Functional groups that are present on the adsorbent's surface are in charge of accumulating the heavy metal ions from the surrounding solution (Bradl, 2004; Chakraborty et al., 2022; Chen et al., 2022). Common functional groups found on solid surfaces include carboxyl, carbonyl, phenolics, and hydroxyl. A change in pH has a substantial impact on these functional groups. The electronegativity of the cation species is a crucial factor in chemisorption. Strong covalent bond formation is encouraged by high electronegativity (Beni and Esmaeili, 2020; Priya et al., 2022). In their study, which involved the examination of the behavior of surface functional groups and pores on adsorbents, Alfarra et al. (2004) came to the conclusion that functional groups on the surface are firm sites that fix rigid metal ions while surface pores happen to be soft spots that have the ability to trap soft ions.

Adsorbents with high metal adsorption capacities must be employed for efficient wastewater treatment polluted with toxic heavy metals. Researchers have utilized a variety of surface modification methods, comprising physical and chemical methods. While physical variations include heat treatment, microwave treatment, etc., chemical modifications involve metal/metal oxide impregnation and acid/base treatment.

12.1.2 Characterization of Adsorbent

It is vital to precisely characterize an adsorbent's features in order to comprehend its potential and practical usefulness. To appropriately analyze the properties of an adsorbent, it is crucial to specify its composition, surface area, surface morphology, particle size surface chemistry, and size distribution. Many constituents have been thoroughly explored to assess their ability as adsorbents for the removal of pollutants from wastewater and water. Utilizing low-cost adsorbents made from waste materials provides a number of benefits, primarily economic and environmental advantages. Microbes and waste products of agricultural, industrial, municipal, and industrial origins are frequently utilized as adsorbents in the adsorption process (Kanamarlapudi et al., 2018; Ramírez Calderón et al., 2020; Ubando et al., 2021).

12.2 HEAVY METAL ADSORPTION BY MICROBIAL BIOMASS

The exploration of agricultural and biological substances as prospective heavy metal adsorbents has resulted from the quest for inexpensive and conveniently accessible adsorbents. It has been suggested that using biological resources to remove metals—a process known as biosorption—is a straightforward, affordable, effective, and ecologically acceptable method (Javanbakht et al., 2011). One of the key benefits of biosorption is the cheap cost and comparatively better efficiency of metal ions removal from the diluted medium. In comparison to other procedures, including those used to clean drinking water, the procedure has an increased selectivity along with being inexpensive for low concentrations of heavy metals (Kordialik-Bogacka and Diowksz, 2014; Salehi et al., 2014).

It has been shown that a variety of microorganisms, comprising yeast, bacteria, fungus, and algae are capable of effectively aggregating heavy metals in their dead and live stages. The capacity of live biomass to bind heavy metal ions is influenced by the age of the cells as well as nutritional and environmental conditions. Additionally, the detrimental effects of hazardous heavy metals on living cells can accumulate to a certain degree and cause cell death. In addition to not requiring maintenance or sustenance, the biosorbents may be kept for an extended time without suffering many negative effects on their functionality, and also generation of metabolic toxins is nil (Rangabhashiyam et al., 2013; Yan and Viraraghavan, 2003). Nevertheless, adequate consideration should also be given to the eventual disposition of these materials, as some processes, like incineration or landfilling have technical drawbacks. Temperature, cationic and anionic metal type, oxidation state, concentration, pH, wastewater composition, removal mechanism, concentration of biosorbents, concurrent existence of organic and inorganic components, and dissolved and volatile constituents are key environmental aspects that may affect the capacity of microorganisms of metal-binding (Aryal, 2021; Li et al., 2022).

12.2.1 Structure of Microorganisms

One of the earliest living forms on Earth was bacteria. They are prokaryotic organisms that are spherical, rod-shaped, spiral-shaped, or filamentous in shape. A lipid membrane surrounds the cell wall of bacteria, which is situated beyond the cytoplasmic membrane and is made up of polysaccharide chains-based peptidoglycan. According to the makeup of the cell wall, bacteria are separated into Gram-negative and Gram-positive and groups. Gram-positive bacteria have a thick peptidoglycan coating linked by amino acid bridges, while gram-negative bacteria happen to have a thin peptidoglycan layer in their cell walls (Pankratova et al., 2019). The biosorption tendency of heavy metals relies on the bacterial cell walls. Metal binding is carried out by bacterial functional groups such as peptidoglycan, teichuronic and teichoic acids, lipopolysaccharides, phospholipids, and different proteins.

Aquatic plants called algae are devoid of real stems and roots. Being autotrophic in nature, they can produce a significant amount of biomass even with little food. Algae are regarded to be good biosorbent resources owing to their substantial size, strong biosorption capacity, and being devoid of harmful substance generation. They are often divided into three categories: macroalgae (marine or brown algae), microalgae (fresh water or green algae), and red algae. The category of brown algae happens to have the highest heavy metal absorption capability among all three types. The ionic charge of the metal, the type of algae, and the chemical makeup of the metal ion solution are what cause heavy metal ions to attach to algal surfaces (Abbas et al., 2014; Khani, 2011; Flouty and Estephane, 2012; Trinelli et al., 2013).

Fungi are living eukaryotic creatures which include mushrooms, yeasts, moulds and other species. Fungi are believed to have excellent biosorption tendencies due to their unique cell wall structure. Fungi can be employed as biosorbent materials in both states of living and dead. The mechanism of heavy metal biosorption by fungi is separated into two stages: (i) first is dynamic metal uptake, likewise identified as bioaccumulation, and (ii) second is biosorption, which comprises heavy metal ion adsorption to the surface of the fungal cell wall. The heavy metal absorption ability might be affected by metabolic inhibitors, temperature, pH, and some other variables (Bano et al., 2018).

12.2.2 Factors Affecting Biosorption

Ion exchange, chemisorption, adsorption-complex formation, condensation of heavy metal with hydroxide, and surface adsorption are some of the mechanisms that contribute to the biosorption capacity (Figure 12.1). The total uptake of metal is influenced by a variety of processes, each of which operates independently. The cellular components of biomass include imidazole, phosphate hydroxyl, carboxyl, amino, sulfate, amide, sulfhydryl, imine, phenol, thioether, carbonyl, phosphodiester, and phosphonate have properties of binding heavy metals in biomass. In fact, the majority of microbes have negatively charged surfaces due to the ionized functional groups, which aid in the binding of metals. The biosorption capacity and capability of bacteria for adsorbing heavy metal ions is affected by many factors such as pH, contact duration, temperature, the concentration of heavy metal ions, and biomass and makeup of aquatic environment (Figure 12.2).

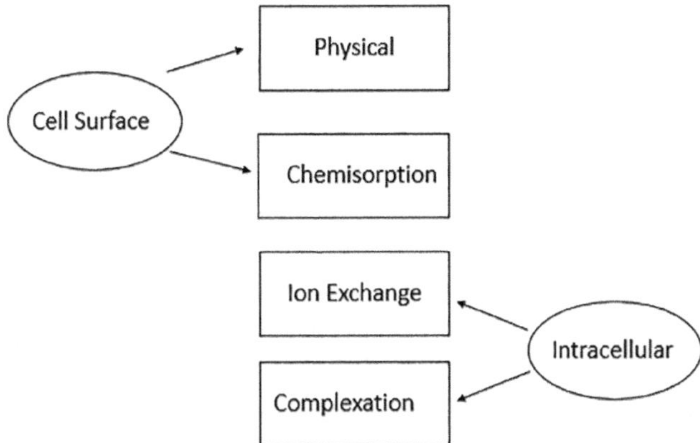

FIGURE 12.1 Mechanisms of biosorption.

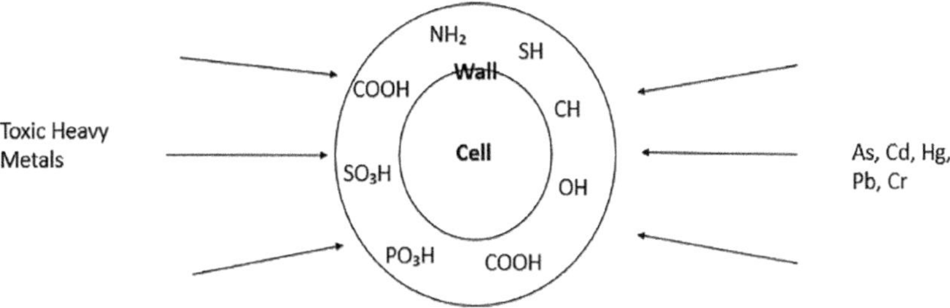

FIGURE 12.2 Structure of cell wall of microorganisms.

12.2.2.1 Effect of pH

The parameter of pH is a significant feature that influences the surface functional groups that are present on the bacterial cell wall and the solution chemistry of heavy metal ions. It also affects the metal ions' solubility in the solution because some positive ions that are on the biomass surface are replaced by H^+ ions. The additional negative binding sites that are exposed on metal cations when pH values rise may be the cause of this increase in biosorption ability (Long et al., 2014). Low pH levels limit the binding sites which are on the cell wall being associated with hydrogen ions, which prevent metal cations from accessing the surface functional groups due to repulsive forces. Conversely, when pH values rise, the repulsive interactions between negatively charged biomass surfaces and metal anions reduce the capacity to absorb the metals. But when pH values fall, the biosorption efficiency of metal anions tends to rise as a result of positively charged surface groups on biomass. It has been reported that with rising pH from 1.0 to 5.0, the biosorption efficacy of Cr(III) also rose but decreased with pH further reaching 7.0. While for Cr(VI), biosorption increased from pH 1.0 to 2.0 and declined up to 7.0 pH (Aryal and Liakopoulou-Kyriakides, 2014). Farhan and Khadom (2015) reported rapid uptake of Cr by *Saccharomyces Cerevisiae* from aqueous solutions at pH 5.0–6.0.

12.2.2.2 Effect of Temperature

The effectiveness of the sorption of metallic ions on biosorbents depends on the temperature of the medium. As of the solute's increased surface activity and kinetic energy, biosorption of heavy metals is often altered as temperature rises. However, at higher temperatures, certain binding sites for metal ions may be destroyed. Whether metal ions interact with binding sites exothermically or endothermically determines the capacity of bacteria to adsorb metal ions. According to studies, *A. niger*, *Paecilomyces* sp., and *Penicillium* sp. had the highest Co (II) adsorption capacities at 50°C during a 24-hour period: 94.1%, 100%, and 97.1% respectively (Cárdenas González et al., 2019). The optimal temperature for elimination of Pb(II) for A. niger was assessed to be 37°C with biosorption capacity ranging from 3.25–172.25 mg/g with 200–1400 mg/L initial concentration (Iram et al., 2015).

12.2.2.3 Effect of Concentration of Biomass

The amount of biomass employed as the sorption medium affects how well heavy metals are biosorbed. Biosorption efficiency often rises as biomass concentration increases, most likely as a result of an increase in binding sites. However, the adsorption capability of heavy metals might also diminish as biomass content rises, leaving certain binding sites for metal ions unfilled due to restrictions of mobility of ionic species in the adsorbing medium.

12.2.2.4 Effect of Equilibrium Time

One of the most essential characteristics of wastewater treatment using biosorbents is the equilibrium time. The parameter of equilibrium time of biosorption process denotes the sorption-desorption procedures that occur on the biomass surface after the metal ion fully saturates. When the optimum time is reached, the biosorption capacities are nearly constant, indicating that the sorption process is in equilibrium.

12.2.2.5 Effect of Initial Metal Concentration

The amount of heavy metal ions that are present at the commencement of the biosorption process is crucial. During the start of the biosorption process, when the amount of metal ions is high, the process is faster as the initial amount of metal ions delivers the required driving energy to overcome the metal ion mass transfer between aqueous and solid phases. When the amount of heavy metal ions is lower, the interface between binding sites on the bacteria and the heavy metal ions increases, resulting in greater biosorption efficiency. Absorption capacity is practically constant at higher dosages, which might be attributable to saturation of potential binding sites.

12.3 BIOSORPTION KINETICS

The adsorption of metal ions by the biosorbent may well be estimated by comparing the initial and final amounts of metal ions in the medium before and after biosorption. According to Abdi and Kazemi (2015), the biosorption capability of a biosorbent should be considered while observing the kinetics of biosorption of a heavy metal. A biosorption isotherm is formulated as a plot of adsorption of metal (q) vs solute concentration at equilibrium given by Equation 12.1.

$$q = \frac{V(C_i - C_e)}{M} \quad (12.1)$$

where q is the concentration of adsorbed metal ion (mg/g), C_i is the metal concentration at the beginning (mg/L), C_e is the metal concentration at equilibrium (mg/L), *V* is the total volume of the solution (L), *M* is the bioadsorbent mass.

Temperature, pH, and ionic strength are held constant while metal concentration is altered to examine the isotherm plots. Biosorption isotherms are commonly characterized using one of two models: Freundlich and Langmuir (Wang and Guo, 2020).

12.3.1 FREUNDLICH ISOTHERM

This is a non-linear kinetic model involving monolayer sorption of heavy metal ions on the dynamic sites of the adsorbent. It is expressed using Equation 12.2.

$$q_e = K_F C_e^{1/n} \quad (12.2)$$

where q_e is the concentration of adsorbed metal ions (mg/g), *K* is L/mg, C_e is the concentration of metal ions in solution at equilibrium (mg/L), *n* is Freundlich constant related to adsorption capacity.

12.3.2 LANGMUIR ISOTHERM

It is a linear model describing the monolayer adsorption of heavy metal ions on the dynamic spots of the adsorbent. It is depicted using Equation 12.3.

$$q_e = \frac{q_{max}\, bC_e}{1 + bC_e} \quad (12.3)$$

where q_e is the concentration of adsorbed metal ions (mg/g), q_{max} is the maximum metal uptake (mg/g), *b* is Langmuir constant, C_e is the concentration of metal ions in solution at equilibrium (mg/L).

12.4 BIOSORPTION BY MICROORGANISMS OF HEAVY METALS

The hazardous heavy metals may be acquired involving different processes by both living and dead biomasses. The advantages of dead biomass include inexpensive collection and supply as waste, no vulnerability to metal toxicity and can be easily used in any condition. Biosorption, on the other hand, is the totality of all interactions between cell wall with metal ions, comprising mechanisms of adsorption ion exchange and surface complexation reactions.

12.4.1 ADSORPTION BY BACTERIA

One of the most diverse, prolific, and useful microorganisms is bacteria which portray an important part in the ecosphere. They are widely utilized as biosorbents for wastewater treatment due to their minute size, ubiquity, and adaptability to reproduce in even unfavorable environments along with under controlled circumstances. Bacterial biomass is also produced as a byproduct in industries such as fermentation and food processing. These bacterial byproducts, as well as a purposefully cultivated population, are used in the bioremediation process. Some species of bacteria commonly utilized in heavy metal removal are *Bacillus, Micrococcus, Escherichia, Streptomyces, and Pseudomonas*. Metal binding is facilitated by abundant metal complexing agents and functional groups on the bacterial cell surface as well as sulfate on the polysaccharide thin layer.

Gawali Ashruta et al. (2014) studied heavy metals removal from aqueous solution using EPS produced using gram-negative bacteria in India. The extracellular polymeric substance produced by gram-negative bacteria consortium was demonstrated to be a promising adsorbent, capable of removing copper zinc, chromium lead, cobalt, nickel, and cadmium from an aqueous system with removal efficiencies of 71.0%, 77.15%, 74.48%, 78.18%, 76.12%, 66.63%, and 72.71% respectively. Saranya et al. (2018) assessed the removal of toxic heavy metals using *Cronobacter muytjensii* KSCAS2 at 100 and 200 mg/L initial metal concentrations. At 100 mg/L concentration, the removal efficacy of heavy metals (Cr, Cu, Cd, Zn,) achieved were 72.45%, 76.51%, 55.23%, and 61.51% respectively. However, at the initial concentration of 200 mg/L, the removal efficacy declined to 44.62%, 52.80% 63.1%, 67%, and for Cd, Zn, Cr, Cu respectively. Rizvi et al. (2020) studied three bacterial strains namely *P. aeruginosa, B. subtilis,* and *A. chroococcum* for biosorption of heavy metals recovered from polluted soil in Seoul, South Korea. At a concentration of 25 μg/mL, *B. subtilis* was observed to biosorb 96% chromium with optimum pH 8 and 60 minutes contact time. Kurniawan et al. (2022) studied *Aeromasss hydrophyla* and *Branhamella spp* for the elimination of toxic heavy metals from synthetic effluent with the initial amount of 100 mg/L with optimum pH at 6 and adsorbent dose of 3 g/L, 97% of Cd, Pb, Cu was removed.

12.4.2 Adsorption by Algae

Algae, which is abundant in all bodies of fresh water, have been employed as biosorbents for the removal of hazardous heavy metals. The categories of algae that have been used in biosorption studies are red, green, and brown algae. But brown algae have received the most spotlight due to their higher sorption capability. Microalgae have a high probability of eliminating toxic heavy metals from effluent via the biosorption process owing to the distribution of functional groups on the exterior cell wall (Spain et al., 2021).

Hadiyanto et al. (2014) assessed Cu^{2+} and Cr^{2+} removal from textile effluent using immobilized algae in Indonesia. By utilising *Chlorella* sp., the maximum Cu and Cr removed were 89% and 90%. Cu absorption capacity was maximized at 16.2 and 13.6 mg/g using *Spirulina* and *Chlorella*, respectively, whereas Cr absorption capacity was maximized at 1.95 and 1.45 mg/g using *Spirulina* and *Chlorella*, respectively. Goher et al. (2016) carried out a study on Cu^{2+}, $Cd^{2+,}$ and Pb^{2+} removal using dead cells of *Chlorella vulgaris* from aqueous solution in Egypt. The removal efficiency of Cu^{2+}, $Cd^{2+,}$ and Pb^{2+} achieved were 97.7%, 95.5%, and 99.4% at equilibrium pH 5–6 and contact time of 120 minutes. Ibrahim et al. (2018) compared different marine algae for the elimination of heavy metals from synthetic effluent in Egypt. Four distinct species of dried marine macroalgae were employed to eradicate heavy metals from synthetic effluent namely: *Ulva lactuca, Colpomenia sinosa Pterocladia capillacea,* and *Jania rubens*. The red alga *J. rubens* had the highest heavy metal removal efficiency, followed by *C. sinosa*, and *U. lactuca* with average removal efficiency of 91%, 89%, and 85%, respectively. The best adsorption occurred at pH of 5.0 with a contact period of 60 minutes using the adsorbent dosage of 20 g/L with the initial concentration of 40 mg/L. Bahaa et al. (2019) assessed macro algae for biosorption for the elimination of heavy metals from synthetic wastewater in Iraq. Synthetic wastewater with 10 mg/L concentrations of chromium, copper, and cadmium was employed 1 g of macroalgae powder. Copper, cadmium, and chromium removal efficiencies achieved were 70%, 85%, and 80%, respectively. Tavana et al. (2020) carried out a study on the elimination of Cu^{2+} using dead green algae *Schizomeris leibleinii* and recorded maximum biosorption capacity of 55.06 mg/g with optimum pH 6 at a contact time of 60 minutes using the adsorbent dosage of 0.4 g at 25°C. Rajivgandhi et al. (2021) studied marine algae for the elimination of toxic heavy metal ions from marine environment. The isolated strain was *Nocardiopsis* sp. GRG 3 (KT235642) which eliminated 74.7% Pb, 5.90% Cd, 51.90% Hg, and 85.90% Cr. Jayakumar et al. (2022) studied Cd^{2+} and Zn^{2+} elimination from aqueous solution using brown algae, *Sargassum polycystum*. The equilibrium pH for Cd^{2+} and Zn^{2+} was 4.65 and 5.7 at biosorbents dosage of 1.8 g/L and 1.2 g/L with 105.26 mg/g and 116.2 mg/g maximum adsorption capacity respectively.

12.4.3 Adsorption by Fungi

Metal ion biosorption is caused by ionic interactions and complexation between ions of heavy metals and functional groups on the surface of fungi. The primary functional groups involved in fungal biosorption include amine, amide, carboxyl, and phosphate.

Huang et al. (2012) studied the biosorption capacity of jelly fungus *Auricularia polytricha* for Cd^{2+}, Pb^{2+}, and Cu^{2+} removal from aqueous solution in China. The maximum removal efficiency recorded for Cd^{2+}, Pb^{2+}, and Cu^{2+}, and of 63.3 mg/g, 221.0 mg/g, and 73.7 mg/g, respectively with optimum pH 5 and biosorbents dosage of 4 g/L at 60 minutes contact time. Rao and Bhargavi (2013) assessed Pb^{2+} and Ni^{2+} removal from aqueous medium using pretreated *Aspergillus Niger* in India. The removal efficiency recorded for Pb^{2+} and Ni^{2+} was 80% and 60% at pH 6–7 and 5–8 respectively. Smily and Sumithra (2017) conducted chromium biosorption using Trichoderma sp. BSCR02 from aqueous medium in India. The equilibrium pH assessed was 5 with an initial metal ion amount of 200 mg/L using biosorbents dosage of 1.6 mg/mL for contact time of 120 minutes. Rodriguez et al. (2018) evaluated the biosorption capacity of *Aspergillus Niger* with initial heavy metal concentration of 100 mg/L using 1 g of fungus. The removal efficiency of Zn^{2+}, Co^{2+}, Hg^{2+}, and Cu^{2+} recorded was 100%, 71.4%, 83.2%, and 37% with pH ranging from 4.0–5.5. Noormohamadi et al. (2019) studied the bioremoval of Ni^{2+} and Cd^{2+} by living *Phanerochaete chrysosporium* from an aqueous solution in Iran. The removal efficiency recorded for Ni^{2+} and Cd^{2+} were 89.48% and 96.23% at pH 6 and 36°C. The data was in accordance with Langmuir isotherm with 46.50 mg/g and 71.43 mg/g maximum biosorption capacity for Ni^{2+} and Cd^{2+} respectively. Sundararaju et al. (2020) assessed the biosorption of Ni^{2+} using *Penicillium* sp. MRF1 from electroplating effluent in India. The removal efficiency of Ni^{2+} recorded was 74.6% with optimum pH of 5.5, biosorbent dosage of 7.5 g/L at 140 minutes contact time. Yadav and Mishra (2022) studied the biosorption capacity of Penicillium rubens for Cr^{6+} and Ni^{2+} removal from an aqueous medium in India. The removal efficiency recorded was 93.8% and 95.6% for Cr^{6+} and Ni^{2+}.

12.5 CONCLUSION

The practice of using microorganisms as adsorbents in the treatment of water and wastewater has numerous appealing qualities, including their contribution to waste disposal cost reduction, which leads to environmental protection. To adequately address the main issues involved, a sustainable method must be established to identify the most suitable biosorbent, operational environments, and effective process of elimination of heavy metals. Furthermore, additional study is needed in biosorbent characteristics, like surface area and shape, zeta potential, particle size, and surface functional groups, as they are significant in biosorption tests and are modified by biosorbent pretreatment. A deeper understanding and detailed examination of the biosorption mechanism of eradication of heavy metals, along with the efficient recovery and reuse of biosorbent materials in usable condition is essential.

REFERENCES

Abbas, S. H., Ismail, I. M., Mostafa, T. M., & Sulaymon, A. H. (2014). Biosorption of heavy metals: a review. *Journal of Chemical Science and Technology*, *3*(4), 74–102.

Abdi, O., & Kazemi, M. (2015). A review study of biosorption of heavy metals and comparison between different biosorbents. *Journal of Materials and Environmental Science*, *6*(5), 1386–1399.

Alengebawy, A., Abdelkhalek, S. T., Qureshi, S. R., & Wang, M. Q. (2021). Heavy metals and pesticides toxicity in agricultural soil and plants: Ecological risks and human health implications. *Toxics*, *9*(3), 42.
Alfarra, A., Frackowiak, E., & Béguin, F. (2004). The HSAB concept as a means to interpret the adsorption of metal ions onto activated carbons. *Applied Surface Science*, 228(1-4), 84–92.
Aryal, M. (2021). A comprehensive study on the bacterial biosorption of heavy metals: materials, performances, mechanisms, and mathematical modellings. *Reviews in Chemical Engineering*, *37*(6), 715–754.
Aryal, M., & Liakopoulou-Kyriakides, M. (2014). Characterization of Mycobacterium sp. strain Spyr1 biomass and its biosorption behavior towards Cr (III) and Cr (VI) in single, binary and multi-ion aqueous systems. *Journal of Chemical Technology & Biotechnology*, *89*(4), 559–568.
Bahaa, S., Al-Baldawi, I. A., Yaseen, S. R., & Abdullah, S. R. S. (2019). Biosorption of heavy metals from synthetic wastewater by using macro algae collected from Iraqi Marshlands. *Journal of Ecological Engineering*, *20*(11), 18–22.
Bano, A., Hussain, J., Akbar, A., Mehmood, K., Anwar, M., Hasni, M. S., ... Ali, I. (2018). Biosorption of heavy metals by obligate halophilic fungi. *Chemosphere*, *199*, 218–222.
Beni, A. A., & Esmaeili, A. (2020). Biosorption, an efficient method for removing heavy metals from industrial effluents: a review. *Environmental Technology & Innovation*, *17*, 100503.
Bradl, H. B. (2004) Adsorption of heavy metal ions on soils and soils constituents.
Briffa, J., Sinagra, E., & Blundell, R. (2020). Heavy metal pollution in the environment and their toxicological effects on humans. *Heliyon*, *6*(9), e04691.
Cárdenas González, J. F., Rodríguez Pérez, A. S., Vargas Morales, J. M., Martínez Juárez, V. M., Rodríguez, I. A., Cuello, C. M., ... Muñoz Morales, A. (2019). Bioremoval of Cobalt (II) from aqueous solution by three different and resistant fungal biomasses. *Bioinorganic Chemistry and Applications*, 2019, (1) 8757149.
Chai, W. S., Cheun, J. Y., Kumar, P. S., Mubashir, M., Majeed, Z., Banat, F., ... Show, P. L. (2021). A review on conventional and novel materials towards heavy metal adsorption in wastewater treatment application. *Journal of Cleaner Production*, *296*, 126589.
Chakraborty, R., Asthana, A., Singh, A. K., Jain, B., & Susan, A. B. H. (2022). Adsorption of heavy metal ions by various low-cost adsorbents: a review. *International Journal of Environmental Analytical Chemistry*, *102*(2), 342–379.
Chen, X., Hossain, M. F., Duan, C., Lu, J., Tsang, Y. F., Islam, M. S., & Zhou, Y. (2022). Isotherm models for adsorption of heavy metals from water-A review. *Chemosphere*, *135545*.
Farhan, S. N., & Khadom, A. A. (2015). Biosorption of heavy metals from aqueous solutions by Saccharomyces Cerevisiae. *International Journal of Industrial Chemistry*, *6*(2), 119–130.
Flouty, R., & Estephane, G. (2012). Bioaccumulation and biosorption of copper and lead by a unicellular algae Chlamydomonas reinhardtii in single and binary metal systems: a comparative study. *Journal of Environmental Management*, *111*, 106–114.
Gadd, G. M. (2010). Metals, minerals and microbes: geomicrobiology and bioremediation. *Microbiology*, *156*, 609–643.
Gawali Ashruta, A., Nanoty, V., & Bhalekar, U. (2014). Biosorption of heavy metals from aqueous solution using bacterial EPS. *International Journal of Life Sciences*, *2*, 373–377.
Godt, F. J., Scheidig, G. C., Esche, V., Brandenburg, P., Reich, A., Papandreou, A., Stournaras, C. J., & Panias, D. (2007). Copper and cadmium adsorption on pellets made from fired coal fly ash, *Journal of Hazardous Materials*, *148*, 538–547.
Goher, M. E., Abd El-Monem A., Abdel-Satar, A. M., Ali, M. H., Hussian, A. E., & Napiórkowska-Krzebietke, A. (2016). Biosorption of some toxic metals from aqueous solution using non-living algal cells of Chlorella vulgaris. *Journal of Elementology*, *21*(3): 703–714.
Hadiyanto, A. G., Pradana, L., Buchori, C., & Budiyati, S. (2014). Biosorption of heavy metal Cu 2+ and Cr 2+ in textile wastewater by using immobilized algae. *Research Journal of Applied Sciences, Engineering and Technology*, *7*(17), 3539–3543.
Huang, H., Cao, L., Wan, Y., Zhang, R., & Wang, W. (2012). Biosorption behavior and mechanism of heavy metals by the fruiting body of jelly fungus (Auricularia polytricha) from aqueous solutions. *Applied Microbiology and Biotechnology*, *96*(3), 829–840.
Ibrahim, W. M., Abdel Aziz, Y. S., Hamdy, S. M., & Gad, N. S. (2018). Comparative study for biosorption of heavy metals from synthetic wastewater by different types of Marine Algae. *J Bioremediat Biodegrad*, *9*(425), 2.
Iram, S., Shabbir, R., Zafar, H., & Javaid, M. (2015). Biosorption and bioaccumulation of copper and lead by heavy metal-resistant fungal isolates. *Arabian Journal for Science and Engineering*, *40*(7), 1867–1873.

Javanbakht, V., Zilouei, H., & Karimi, K. (2011). Lead biosorption by different morphologies of fungus Mucor indicus. *International Biodeterioration & Biodegradation*, *65*(2), 294–300.

Jayakumar, V., Govindaradjane, S., Rajamohan, N., & Rajasimman, M. (2022). Biosorption potential of brown algae, Sargassum polycystum, for the removal of toxic metals, cadmium and zinc. *Environmental Science and Pollution Research*, *29*(28), 41909–41922.

Kanamarlapudi, S. L. R. K., Chintalpudi, V. K., & Muddada, S. (2018). Application of biosorption for removal of heavy metals from wastewater. *Biosorption*, *18*(69), 70–116.

Khani, M. H. (2011). Uranium biosorption by Padina sp. algae biomass: kinetics and thermodynamics. *Environmental Science and Pollution Research*, *18*(9), 1593–1605.

Kordialik-Bogacka, E., & Diowksz, A. (2014). Metal uptake capacity of modified Saccharomyces pastorianus biomass from different types of solution. *Environmental Science and Pollution Research*, *21*(3), 2223–2229.

Kumar, U. (2006). Agricultural products and by-products as a low cost adsorbent for heavy metal removal from water and wastewater: A review. *Scientific Research and Essays*, *1*(2), 033–037.

Kurniawan, T. A., Lo, W., Othman, M. H. D., Goh, H. H., & Chong, K. K. (2022). Biosorption of heavy metals from aqueous solutions using activated sludge, Aeromasss hydrophyla, and Branhamella spp based on modeling with GEOCHEM. *Environmental Research*, *214*, 114070.

Lapedes, D. N. (1974). *Dictionary of scientific and technical terms*. McGraw Hill, New York, 674

Li, C., Yu, Y., Fang, A., Feng, D., Du, M., Tang, A., ... Li, A. (2022). Insight into biosorption of heavy metals by extracellular polymer substances and the improvement of the efficacy: A review. *Letters in Applied Microbiology*, *75*, 1064–1073.

Long, J., Luo, D., & Chen, Y. (2014). Identification and biosorption characterization of a thallium-resistant strain. *The Chinese Journal of Applied and Environmental Biology*, *20*, 426–430.

Noormohamadi, H. R., Fat'hi, M. R., Ghaedi, M., & Ghezelbash, G. R. (2019). Potentiality of white-rot fungi in biosorption of nickel and cadmium: Modeling optimization and kinetics study. *Chemosphere*, *216*, 124–130.

Ozturk, A. (2007). Removal of nickel from aqueous solution by the bacterium Bacillus thuringiensis, *Journal of Hazardous Materials*, *147*, 518–523.

Pankratova, G., Hederstedt, L., & Gorton, L. (2019). Extracellular electron transfer features of Gram-positive bacteria. Analytica chimica acta, *1076*, 32–47.

Priya, A. K., Gnanasekaran, L., Dutta, K., Rajendran, S., Balakrishnan, D., & Soto-Moscoso, M. (2022). Biosorption of heavy metals by microorganisms: Evaluation of different underlying mechanisms. *Chemosphere*, *307*, 135957.

Qasem, N. A., Mohammed, R. H., & Lawal, D. U. (2021). Removal of heavy metal ions from wastewater: A comprehensive and critical review. *Npj Clean Water*, *4*(1), 1–15.

Rajivgandhi, G., Vimala, R. T. V., Maruthupandy, M., Alharbi, N. S., Kadaikunnan, S., Khaled, J. M., ... Li, W. J. (2021). Enlightening the characteristics of bioflocculant of endophytic actinomycetes from marine algae and its biosorption of heavy metal removal. *Environmental Research*, *200*, 111708.

Ramírez Calderón, O. A., Abdeldayem, O. M., Pugazhendhi, A., & Rene, E. R. (2020). Current updates and perspectives of biosorption technology: an alternative for the removal of heavy metals from wastewater. Current Pollution Reports, *6*(1), 8–27.

Rao, P. R., & Bhargavi, C. (2013). Studies on biosorption of heavy metals using pretreated biomass of fungal species. *International Journal of Chemical Engineering*, *3*(3), 171–180.

Rangabhashiyam, S. Anu, N., & Selvaraju, N. (2013). Sequestration of dye from textile industry wastewater using agricultural waste products as adsorbents. *Journal of Environmental Chemical Engineering*, *1* (4), 629–641.

Rizvi, A., Ahmed, B., Zaidi, A., & Khan, M. (2020). Biosorption of heavy metals by dry biomass of metal tolerant bacterial biosorbents: an efficient metal clean-up strategy. *Environmental Monitoring and Assessment*, *192*(12), 1–21.

Salehi, P., Tajabadi, F. M., Younesi, H., & Dashti, Y. (2014). Optimization of lead and nickel biosorption by Cystoseira trinodis (brown algae) using response surface methodology. *Clean-Soil, Air*, Water, *42*(3), 243–250.

Saranya, K., Sundaramanickam, A., Shekhar, S., Meena, M., Sathishkumar, R. S., & Balasubramanian, T. (2018). Biosorption of multi-heavy metals by coral associated phosphate solubilising bacteria Cronobacter muytjensii KSCAS2. *Journal of Environmental Management*, *222*, 396–401.

Shrestha, R., Ban, S., Devkota, S., Sharma, S., Joshi, R., Tiwari, A. P., ... Joshi, M. K. (2021). Technological trends in heavy metals removal from industrial wastewater: A review. *Journal of Environmental Chemical Engineering*, *9*(4), 105688.

Singh, A., Pal, D. B., Mohammad, A., Alhazmi, A., Haque, S., Yoon, T., ... Gupta, V. K. (2022). Biological remediation technologies for dyes and heavy metals in wastewater treatment: New insight. *Bioresource Technology*, *343*, 126154.

Smily, J.R.M.B., & Sumithra, P.A. (2017). Optimization of Chromium Biosorption by Fungal Adsorbent, *Trichoderma* sp. BSCR02 and its Desorption Studies. HAYATI *Journal of Biosciences*, *24*(2), 65–71.

Spain, O., Plöhn, M., & Funk, C. (2021). The cell wall of green microalgae and its role in heavy metal removal. *Physiologia Plantarum*, *173*(2), 526–535.

Sundararaju, S., Manjula, A., Kumaravel, V., Muneeswaran, T., & Vennila, T. (2020). Biosorption of nickel ions using fungal biomass Penicillium sp. MRF1 for the treatment of nickel electroplating industrial effluent. Biomass Conversion and Biorefinery, 12, 1059–1068.

Tavana, M., Pahlavanzadeh, H., & Zarei, M. J. (2020). The novel usage of dead biomass of green algae of Schizomeris leibleinii for biosorption of copper (II) from aqueous solutions: Equilibrium, kinetics and thermodynamics. *Journal of Environmental Chemical Engineering*, *8*(5), 104272.

Trinelli, M. A., Areco, M. M., & dos Santos Afonso, M. (2013). Co-biosorption of copper and glyphosate by Ulva lactuca. *Colloids and Surfaces B: Biointerfaces*, *105*, 251–258.

Ubando, A. T., Africa, A. D. M., Maniquiz-Redillas, M. C., Culaba, A. B., Chen, W. H., & Chang, J. S. (2021). Microalgal biosorption of heavy metals: a comprehensive bibliometric review. *Journal of Hazardous Materials*, *402*, 123431.

Wang, J., & Chen, C. (2009). Biosorbents for heavy metals removal and their future. *Biotechnology Advances*, *27*(2), 195–226.

Wang, J., & Guo, X. (2020). Adsorption isotherm models: Classification, physical meaning, application and solving method. *Chemosphere*, *258*, 127279.

Yadav, S., & Mishra, A. (2022). Fungal biosorption of the heavy metals chromium (VI) and nickel from industrial effluent-contaminated soil. *Journal of Applied and Natural Science*, *14*(1), 233–239.

Yan, G., & Viraraghavan, T. (2003). Heavy-metal removal from aqueous solution by fungus Mucor rouxii. *Water Research*, *37*(18), 4486–4496.

13 Microbial-assisted Phytoremediation of Industrial Wastewater Recycling and Reuses for Overcoming Water Scarcity

Anupriya Verma and Gaurav Saini

13.1 INTRODUCTION

Industrial sectors are crucial to the development and prosperity of any economy. Despite this, the natural ecosystem is seriously threatened by industrial wastewater discharge which contaminates water with various pollutants such as plastics, heavy metals, pharmaceuticals, organic compounds,and xenobiotic compounds. Unfortunately, most of this waste is improperly handled and/or receives little to no treatment and is generally disposed ofin the receiving water bodies. Another concerning aspect is the wide variability in the types and composition of the waste that is generated by different industries, which results in a lack of appropriate treatment techniques that can be applied on a broader scale (such as a city scale). The problems are further compounded in the developing and underdeveloped sections of a country, where a lack of resources forces the industries to indiscriminately dispose of their wastewater leading to environmental issues. In Mexico,81% of industrial wastes are discharged directly into water bodies with no or inadequate treatment along with 58% of urban wastewater, resulting in the contamination of ~73% of the water bodies (Vargas-González,2014). Various pollutants, such as organic pollutants and toxic metals, are currently making their way into the rivers and soils from a wide variety of sources likeindustries, municipalities, agriculture, sewage sludge,and landfill leachate (Yang et al., 2020).

Plastics, heavy metals, pharmaceuticals, polycyclic aromatic hydrocarbons (PAHs), phenolic allelochemicals, and polychlorinated biphenyls (PCBs) are some of the harmful contaminants, generated by multiple industries (Ansari, 2016). For several aquatic and terrestrial ecosystems, environmental pollution brought on mostly by toxins released from industries has become a severe concern. The increase of total dissolved solids (TDS),total suspended solids (TSS), Chemical Oxygen demand, and Biochemical Oxygen demand (BOD)in industrial effluent makes it harmful to zooplankton andphytoplankton (Kumar et al., 2021c). Additionally, the Environmental Protection Agency (EPA), United States has classified some metals as contaminants, including cadmium (Cd), lead (Pb), Copper (Cu), Mercury (Hg),Arsenic (As) Chromium (Cr), and Nickel (Ni) (Kumar et al., 2020c; Lardieri, 2021; Nadar, 2018).

Wastewater discharged from industries contains a wide range of pollutants, including metals, mutagenic,and carcinogenic chemicals; most of which are harmful even after secondary treatment (Kumar et al., 2021c). In addition to secondary treatments, tertiary treatments can be used to safely dispose of industrial effluent into water bodies and /or recycled/reused. However,

DOI: 10.1201/9781003441144-13

proper tertiary treatment techniques are not available in developing and underdeveloped countries, and also these engineered techniques (primary, secondary, and tertiary) are not sustainable and environment-friendly.

The engineered treatment process itself uses certain harmful chemicals like potassium chloride, chlorine dioxide, hydrogen oxide, etc. Hydrogen peroxide is rapidly decomposed in the body and repeated exposure causes chronic respiratory tract irritation, burning, skin peeling, and partial or complete lung collapse (Agency for Toxic Substances and Disease Registry, 2005). Chlorine dioxide can cause throat irritation, breath shortness, watery eyes, and skin irritation (Lardieri, 2021). The risks posed to humans and other living beings by wastewater pollution havebeen substantial (Hoang, 2020). However, wastewater generated from industries is region, and industry-specific resulting in a wide range of pollutants. Due to the wide range of pollutants present in industrial effluent, it has become difficult for researchers and environmental engineers to build a sustainable, cost-competitive, effective, eco-friendly technology with steady remediation performance.

While considering the above problem environmental biotechnology has seen immense interestin the last few decades in the investigation of some alternatives to the current physio-chemical industrial effluent management systems. Due to technological and economic limitations, conventional wastewater clean-up techniques are not sustainable and environment-friendly. As an alternative to engineered treatments, phytoremediation is a quick, simple, environmentally friendly method that has been extensively researched to break down and mineralize various xenobiotic and resistant organic pollutants (Nadar, 2018). Figure 13.1 illustrates the schematic diagram for the treatment of industrial effluent via phytoremediation

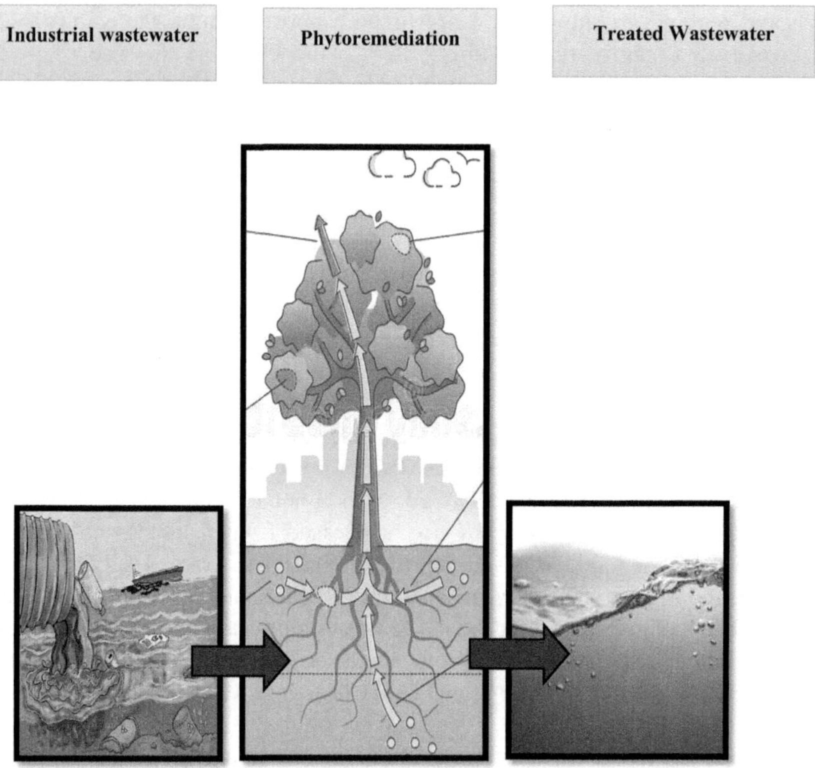

FIGURE 13.1 Industrial effluent treatment via phytoremediation.

13.2 PHYTOREMEDIATION: DEFINITION, TYPES, ADVANTAGES AND DISADVANTAGES

Phytoremediation is a low-cost, environment-friendly method for treating industrial and municipal wastewater. Using plants or associated microbes to clean up polluted soil, sediment, and water is known as phytoremediation (Dietz and Schnoor, 2001). Greek *phyto* (meaning plant) and Latin *remedium* (meaning to repair or eliminate an ill) are the two words that make up the term"phytoremediation" (Wang, 2020). Based on recent studies, this technique has garnered a significant amount of attentionespecially for the treatment of industrial effluents.

Numerous studies have demonstrated that phytoremediation is highly effective in removing a variety of pollutants including persistent organic pollutants (POPs), metals, and hydrocarbons (Ekta and Modi, 2018; Shah and Achlesh, 2020; Abhilash and Tripathi, 2016; Pandey, 2017; Kassaye, 2017; Mehmood, 2019). Additionally, the possibility of phytoremediation has also been evaluated in prior outstanding studies (Mehmood, 2019; Man, 2020).

Phytoremediation uses various processes such as phytoextraction, phytovolatilization, phytostabilization, and phytofiltration, to detoxify polluted water, metal-contaminated lands, and groundwater (Wang, 2020). Phytodegradation and rhizodegradation are two processes that are frequently utilized in phytoremediation procedures and are used to break down both organic and inorganic pollutants. The interaction between plants and microbes involves several biological, physiological, endogenous, genetic, and secondary metabolite production mechanisms that aid in the breakdown and detoxification of wastewater (Ozyigit and Ilhan, 2015). Natural and plant-based materials are commonly used for industrial wastewater and waste remediation (Agarwal and Saini, 2020; Gautam and Saini, 2020).

Biochemical processes are used by plants and microorganisms to interact with and absorb heavy metals. For instance, a nutrient transport system allows different types of elemental contaminants to enter plants. By using particular enzymes, organic pollutants are digested inside the plant cell during phytodegradation or phytotransformation. Dehalogenase and nitroreductase are the enzymes that help break down the nitroaromatic molecules as well as insecticides, anilines, and other chlorinated chemicals. Figure 13.2 illustrates the schematic model of various phytoremediation methods.

- Phytostabilization: Mobility and bioavailability of contaminants are eliminated or diminished.
- Phytoextraction: Biomass like shootthat can be harvested accumulates contaminants.
- Phytovolatization: includes pollutants being absorbed by plant roots, being transformed into a gas, and then being released into the atmosphere.
- Rhizofilteration:process by which pollutants in solution around the root zone are adsorbed onto plant roots or absorbed into plant roots (rhizosphere).

There are some more phytoremediation methods

- Phytodegradation: Inside plant tissues, plant enzymes break down organic pollutants.
- Phytodesalination: Halophytes extract more salt from saline soils.
- Phytofiltration: Plants remove pollutants from aquatic systems.
- Rhizodegradation: Microbes connected to plant roots breakdown contaminants.

Figure 13.3 illustrates a life cycle assessment and types of phytoremediation approaches used for the treatment of wastewater.

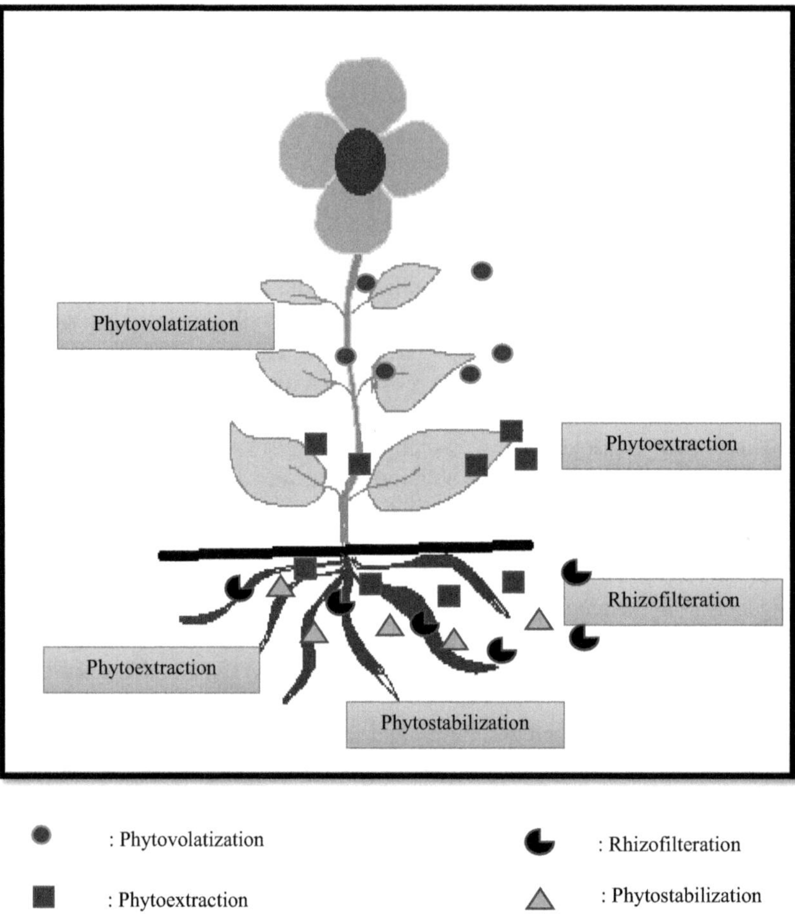

FIGURE 13.2 Schematic diagram of phytoremediation methods.

It is significant to remember that each one of these techniques works with various pollutants. Numerous studies demonstrate the excellent phytoremediation of persistent organic pollutant (POPs), polycyclic aromatic hydrocarbon (PAHs), and other toxins using various species of grasses likevetiver grass *(Vetiveria zizanioides)*(Ekperusi, 2019; Hu et al., 2019; Kumar et al., 2019; Lu, 2018). In one of the studies,*V. zizanioides* was grown in industrial wastewater in order to treat metal-contaminated soil and showedsignificant reduction in BOD (Biochemical Oxygen Demand) (Kafil, 2019). This suggests that the use of grasses is both widespread and efficient.

Phytoremediation has become very popular and well-known in recent years. Performance, viability, and an environmentally friendly approach are the main advantages. An effective method for cleaning up contaminated regions, phytoremediation also has long-term environmental advantages. Keeping an eye on the plants is simple. Given that it uses microorganisms and plants that are found naturally, it may be the least harmful option. It enhances soil healthwhile protecting soil fertility and maintaining the surface soil. Phytoremediation requireslow or no technical skills and hence can provide employment to rural communities. It is easy to maintain and requires less routine work. In addition to this,the phytoremediation technique can also be used in combination with engineered technologies (Ansari et al., 2020). Despite having advantages over other treatment methods, phytoremediation still has several drawbacks.

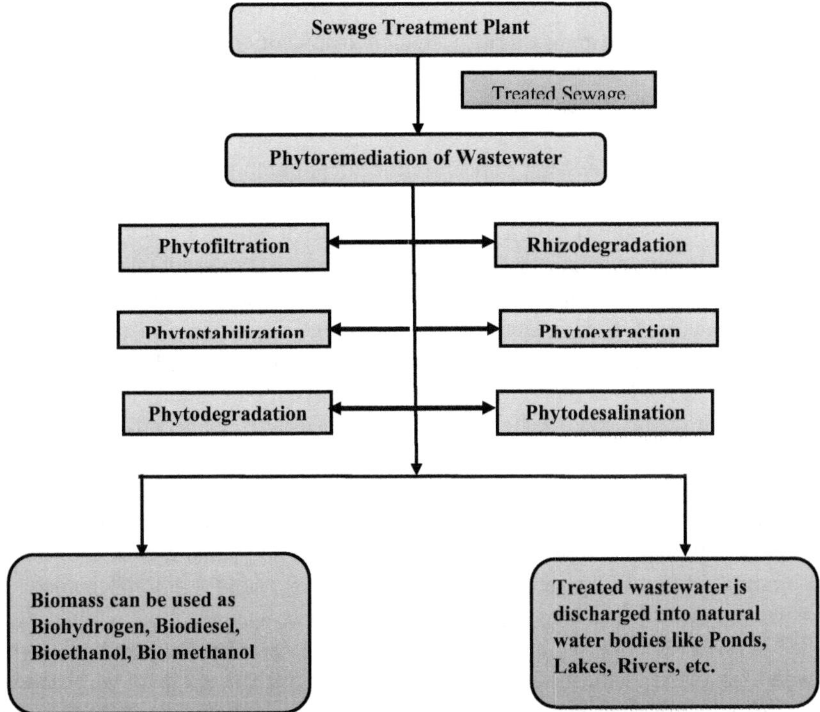

FIGURE 13.3 Life-cycle assessment and types phytoremediation approaches.

There are a few drawbacks, including slow plant growth, time consumption, reducedbiomass, and heavy metal sensitivity. Typically, toxic element phytoremediation uses plants that can be picked and removed from the contaminated site but usually arenot removedfrom the surrounding environment. Phytoremediation is limited to the surface and depth of the roots. Plant-based cleanup techniques cannot completely prevent the liquidation of contaminants in groundwater. It can take a few months to many years for phytoremediation to take place because it is a slow, biologically cyclically dependentprocess (Cameselle, 2019). Favorable weather and plant-friendly weather conditions are also necessary for sustainable plant restoration (Shah and Achlesh, 2020). Employing plants and microorganisms separately to remediate contaminated waterand soilis not particularly successful or sufficient. The harmful substances/pollutants can build up in the tissues of the plants, where they can eventually infiltrate the food chain (Abhilash and Tripathi, 2016). Phytoremediation is insufficient in areas where a variety of contaminants have been introduced. In comparison to other clean-up strategies, the remediation by plants processes slowly (Ma, 2010). The microorganisms, on the other hand, are restricted to the soil's surface and are unable to go deep into the soil. Apart from this they also cease the process when they run out of food. The release of air pollutants when plant-based materials are disposed of by burning, as if the plants involved in phytoremediation are burned at the end of life cycle will further lead to air pollution (Abdurrahmanet al., 2020). Consequently, some of the drawbacks of the phytoremediation process continue to be significant problems and may hinder its widespread application.

13.3 INDUSTRIAL WASTEWATER CHARACTERIZATION

Industrialization is essential for society's sustainability, but environmental pollution is a significant problem (Pandey, 2017). While considering the growth of the ecosystem, its impact on that development should be evident (Kassaye, 2017). Additionally, nearly 829,000 people die each year from illnesses caused by contaminated water (Mehmood, 2019). The emission of heavy metals from industrial sources is one of the main threats to the environment and living things. After secondary treatment, the massive amount of wastewater released by industries can directly impact marine and land life (Müller, 2019). Recent discovery of emerging contaminants in industrial effluents is another cause of concern.

Wastewater has an unusually high concentration of pollutants, which could be a sign of the complexity and cause for endangering aquatic life. The high BOD and/or COD concentration in industrial effluent indicates significant toxicity. The release of numerous wood extracts and chemical compounds into the wastewater utilizedin the pulping also raises COD values (Man, 2020). The effects of chromium (Cr) compounds on people, animals, plants, and microorganisms are well-documentedto be toxic, mutagenic,genotoxic, and carcinogenic. Because of ongoing human activities, there is a significant concentration of metals in industrial locations (Man, 2020). The amount of metals in wastewater depends on the industry's raw materials, their capacity to accumulate, and how a plant reacts biochemically under particular stress conditions. Bioconcentration and translocation factors were used to gauge the plant's capacity and reaction to metal stress (Man, 2020).

Due to several unique characteristics likereactive oxygen species (ROS), biomass, antioxidant activity, etc., plants can tolerate complicated contaminants from various contaminated areas. To better understand how metal accumulation in various parts of plants and microbes is changing, plant metal-binding protein-like genes were extracted from a variety of microorganisms and plants, including maize, soybean, tobacco, wheat, brassica, and rice (Duan, 2012; Hejna, 2020). Native plants utilized in phytoremediation are viewed as an appealing technique for the clean-up of industrial wastewater (Shahand Achlesh, 2020; Ekperusi,2019). Instead of chemical procedures, heavy metal environmental remediation by phytoremediation has recently been the focus of research (Kumar et al., 2020c).

13.4 INDUSTRIAL WASTEWATER TREATMENT VIA PHYTOREMEDIATION

There have been several studiesconducted on the possible applications of plants in wastewater treatment. Through static and dynamic studies, it was established that water hyacinth, water lettuce, and *Myriophyllum spicatum* were successful at removing water nutrients (Wang, 2017). Plants from families like *Gramineae*, *Ceratophyllaceae*, and*Pontederiaceae*have a relatively higher capacity to absorb heavy metals (Wuang, 2016). The most effective approach for removing advanced nutrients is microalgae-based processes (Wang, 2017). These processes exhibit remarkable qualities such as simultaneous nitrogen (N) and phosphorous (P) removal, and additional chemical-free treatment. According to one of the studies platensis can be used as agricultural fertilizer and to purify fish water (Wuang, 2016). Water hyacinth is effective at treating industrial effluent. It was also shown that Water hyacinth is a good option for nutrient uptake and enhancing water quality among aquatic plants (Rezania, 2015).

Chenopodium album demonstrated the highest accumulation of manganese (Mn) in leaves while growing in wastewater from the paper and pulp sectors (Kumar et al., 2020c). *Argemone mexicana* had the greatest concentration of iron (Fe) in the root from distillery waste based on sugarcane molasses. Additionally, the hairy roots of *Brassica napus* can eliminate 80% of chromium (Cr) generatedfrom tannery wastewater (Perotti, 2020). High levels of toxic contaminants in wastewater, including manganese (Mn), cadmium (Cd), copper (Cu), chromium (Cr),arsenic (As), zinc (Zn), andlead (Pb), make it difficult to use the herb in fields where it is grown (Luo, 2012). Additionally, some species, including *Mentha suaveolens*, *Cistus salvifolius*,*Phytolacca americana*, *Hypochaeris radicata*, and *Pinus pinaster* can detoxify metals and metalloids,making them suitable for the treatment of wastewater (Pratas et al., 2014). According to reports, some additional plants, such as *Solanum lycopersicum*, *Erato polymnioides*, *Festuca arundinacea*, *Populus canescens* are appropriate for phytoremediation (Liu, 2020). The application of metals detoxification phytoremediation technique for wastewater treatment is cost-effective since metals are regarded as a unique category of pollutants (Cameselle, 2019). The most often reported global contaminants are cadmium (Cd),lead (Pb), nickel (Ni), and mercury (Hg) (Udayan, 2017).

Chromium (Cr) exists in a wide range of oxidation states, from 2 to +6, but only the trivalent (III) and hexavalent (VI) kinds of Cr are the most common (Mishra and Bharagava, 2016). *Calotropis gigantea* extracts were used to study Chromium (Cr) biosorption (Udayan, 2017). Large amounts of agricultural land and freshwater ecosystems are contaminated by hazardous metals released from the pulp and paper industries (Kumar and Chandra, 2021a).

The growth and metabolism of many aquatic species, including *Selenastrum capricornutum, Eisenia fetida,Chlorella vulgaris, Daphnia magna*, and *Oryzias latipes*, are significantly impacted by industrial effluent (Miyazaki, 2002). *Trigonella foenumgraecum L Triticum aestivum, Allium cepa,Aestivum sativum*, and *Phaseolus mungo* wereused to test the phytotoxic and cytotoxic effects of wastewater from the pulp and paper sector (Haq, 2017).

Pollutant elimination is influenced by the length of exposure, the number of pollutants present, environmental variables like pH and temperature, and plant features like species and root systems (Anand, 2017). Along with these, there are a variety of aquatic plants, including those that are free-floating (*Pistia stratiotes, Lemna spp., Azolla pinnata,*and *Riccia fluitans*), emergent plants (*Distichlis spicata, Justicia Americana, Diodia Virginiana, Typha spp.*, and *Hydrochloa caroliniensis*), and submerged plants (*Hygrophilla corymbose, Ruppia Maritima, Vallisneria Americana, Hydrilla verticillata*, and *Myriophyllum aquaticum*) that were used for phytoremediation processes (Ekperusi, 2019). *Egeria densa*was used to treat industrial wastewater (Mustafa and Gasim, 2021). *Lemna minor*was used to treat textile and distillery wastewater (Saleh, 2017). Phytoremediation potentials of these plants for different industrial wastewater are listed below in Table 13.1.

13.4.1 Microbial-assisted Phytoremediation

The potential for remediation of heavy metals, organic chemicals, and xenobiotic substances was greatly boosted by the rhizospheric or endophytic interaction of bacteria and plants (Barea, 2015).

TABLE 13.1
Plants Used for Treating Industrial Wastewater

Industrial Wastewater Type	Plants	Treatment Duration	References
Heavy metal effluent	*E. densa*	7 days	Marques (2013)
Textile wastewater	*Typha angustifolia L.*	7 days/constructed wetlands	Seroja (2018)
Tofu wastewater	*Vertiveria zizaniodes*	15 days	Al-Baldawi (2018)
Wastewater (stable or radioactive Cobalt (Co) and Cesium (Cs)	*M. spicatum*	20 days	Saleh (2020)
Petrochemical industry	*Typha angustifolia L.*	42 days/subsurface constructed wetlands	Al-Baldawi (2018)
Pulp and paper mill final effluent	*Azolla pinnata; Scirpus grossus*	28 days	Seroja (2018)
Textile, distillery, and institutional wastewater;	*L. minor*	28 days	Amare (2018)
Wastewater (stable or radioactive Cobalt (Co) and Cesium (Cs)	*Ludwigia stolonifera*	20 days	Saleh (2017)
Palm oil mill effluent	*Ipomeo aquatica*	25 days/ bucket treatment system	Md Sa'at and Nastaein (2017)

Endophytic or symbiotic relationships between plants and microorganisms in the rhizosphere of plants is the best way to treat highly contaminated wastewater (Wu, 2006). Endophytic bacteria enable the breakdown of contaminants within the plants, while plant roots can assist microorganisms in reaching deeper within the soil and enhancing aeration and nutrition supply (Kuiper et al., 2004).

When combined with plants, numerous microbial cultures have been employed to reduce synthetic plastic pollutants,heavy metal contamination, and agrochemical pollutants like herbicides and pesticides from contaminated locations. These microorganisms can withstand high pollutant concentrations and have been effective in bioremediation, which demonstrates the synergistic benefits of plants and their root exudates. When barley was grown with *white rot* fungi, the number of dioxins in the soil was reduced by 35% (Oh, 2014). When *Pseudomonas sp.*Lk9 was inoculated into *Solanum nigrum*, bioaccumulation of Cd was increased by 28.9% (Chen, 2014). When infected with *Pseudomonas putida*, plants like *Eruca sativa* increase the uptake of nickel by 46% from the soil and lower the nickel level in the soil (Kamran et al., 2016). According to several studies, interactions between plants and microbes can improve the effectiveness of bioremediation (Nagata, 2015; Amare, 2018). A list of the remediation of heavy metal pollutants by using plant-microbe interaction through microbial-assisted phytoremediation is summarized in Table 13.2.

Explosives are extremely dangerous substances that are difficult to clean up. At the locations of munitions manufacturing and military action these substances severely pollute the environment (Singh, 2012). Nitrate esters (R–ONO2, where R is any organic residue), nitrosamines ($R_2N–N=O$, where R is an alkyl group), and nitroaromatics ($C_6H_5NO_2$) are the three main types of explosives (Panzand Korneliusz, 2012). These substances contaminate the soil and quickly drain into the groundwater, which causes cations in both terrestrial and marine organisms to bioaccumulate and biomagnify (Panzand Korneliusz, 2012). In some experiments, plants and bacteria worked in concert to successfully remediate soil and water that had been contaminated by explosives. When *Agrostis capillaries*wereused to inoculate the bacterial species of *Acer pseudoplatanus*, it was demonstrated that trinitrotoluene (TNT) both promotes plant growth and detoxifies ion ability from the soil (Thijs, 2014).

Plastic is a material that is widely employed in the creation of many things because it is strong, water-resistant,lightweight, and reasonably inexpensive. However, due to plastic's slow

TABLE 13.2
Plant-Microorganisms Interactions for Pollutant Remediation

Micro-Organisms	Plant	Pollutants	References
Pseudomonas putida	*Eruca sativa*	Ni	Kamran et al.(2016)
Kocuria fl ava	*Cicer arietinum* L.	Cr	Singh (2014)
Pseudomonas sp.	*Solanum nigrum* L.	Cd, Zn, and Cu	Kamran et al. (2016)
Bacillus safensis + Kocuria rosea	*Helianthus annuus* L.	Ni	Mohammadzadeh (2014)
Pseudomonas sp. Lk9	*Solanum nigrum* L.	Cd, Zn, Cu	Chen (2014)
Burkholderia cepacia SCAUK0330	Maize	P	Zhao et al. (2014)
Pseudomonas putida and *Rhodopseudomonas* sp.	*Cicuta virosa* L.	Zn	Nagata (2015)
Enterobacter sp. and *Klebsiella* sp.	*Brassica napus*	Cd, Pb, and Zn	Jing (2014)
Ralstonia eutropha, Chryseobacterium humi	*Helianthus annuus*	Zn, Cd	Chandanshive (2017)
Bradyrhizobium	Soybean	Cu	Chandanshive (2017)
Ralstonia eutropha and Chryseobacterium humi	*Zea mays* L.	Cd	Chandanshive (2017)

decomposition, the rapid buildup of plastic throughout the environment poses a major threat to the ecosystem. Plastics are available in various chemical forms including nylon, polyethylene, propylene, polystyrene, and polyurethane (PUR) (Kale et al., 2015). Researchers focus on the use of endophytes' capacity for plastic breakdown since the chemical linkages in certain plastics resemblethose of natural plant polymers (Carraher,2013). There have been reports of the breakdown of plastic (polyethylene and polypropylene) by the endophytic fungus *Lasiodiplodia theobromae* and *Paecilomyces lilacinus* from the plants *Humboldtia brunoise* and *Psychotria flavida*, respectively (Sheik, 2015). A 2-month period was required by the endophytic bacteria *Nocardiopsis sp. mrinalini* in combination with *Hibiscus rosa sinensis* to break down 22% of polythene and 10% of plastic (Singh and Padmavathy, 2015).

Today's increased use of a variety of prescription medications has contaminated drinking water, groundwater, surface water, and even seawater (Fent, 2006). For instance, benzodiazepine-class synthetic chemical carbamazepine is used as an antiepileptic and mood stabilizer. Itsoveruse poses a severe threat to the ecosystem because of its environmental persistence and potential for toxicity (Sauvêtre and Schröder, 2015). Utilizing interactions between plants and microbes, such pollutants can be remediated. In one study, the carbamazepine was successfully removed from the wastewater by the endophytic interaction of the *Chryseobacterium taeanense* and *Phragmites australis* plant (Sauvêtre and Schröder, 2015).

Pesticides are insect killers or repellents that control insects that are thought to be damaging to crops. These substances are poisonous and difficult for natural mechanisms to break down. These substances pollute both soil and water. They eventually become part of the food chain and accumulate at various tropical levels. Organochlorine insecticides have been prohibited in the past due to environmental concerns, yet contaminated areas still exist in abandoned plants and dump sites (Sun, 2015). Plant-microbial interactions have been proven to be the most successful in removing these organochlorine pesticides in numerous studies.

Withania somnifera and *Staphylococcus cohnii* are two examples of plant-microbe combinations that can accelerate the uptake of lindane by 76% from soil in 90 days (Abhilash and Srivastava, 2011). It was found that when the soil was bioaugmented with the plant *Orychophragmus violaceus*, bacteria *Gammaproteobacteria* and *Flavobacteria* were able to digest 85.4% dichlorodiphenyltrichloroethanes (DDTs) and 81.18% hexachlorocyclohexanes (HCHs) (Sun, 2015). The root and rhizosphere of *Lolium multiflorum (ryegrass)* are successfully used by the root endophytic bacteria *Mesorhizobium sp.* to digest chlorpyrifos and its metabolite 3,5,6 trichloro-2-pyridinol (Jabeen, 2016).

Ferulic acid (4-hydroxy-3-methoxycinnamic acid) is a very hazardous chemical made by agro-industrial processes that, even at low concentrations, inhibits the growth of organisms (Fiorentino, 2003). By using it as its only carbon source, the endophytic fungus *Phomopsis liquidambari* from a polycarp stem of *Bischofi* can break down 97% of ferulic acid in 48 hours (Xie and Chuan, 2015). Cinnamic acid is another phenolic allelochemical that is commonly present in soils used for continuous cropping and itsbuild-up can reduce plant productivity. Cinnamic acid is degraded by plant-microbe interactions like those between peanut and the endophytic fungus *P. liquidambari*, which also promotesseedling growth (Xie and Chuan, 2015). Another study found that the presence of the endophytic fungus *P. liquidambari B3* increased the production of peanut pods and the numberof root nodules (Faroon et al., 1998). This interaction demonstrates how the allelochemicals can be removed from the soil environment using microbial-assisted phytoremediation.

PCBs are a class of 209 structurally related synthetic organic compounds known as polychlorinated biphenyls (PCBs). PCBs are a variety of waxy solids and liquids that range in hue from colorless to light yellow (Faroon et al., 1998). These compounds have been collected in the environment for a long time since they are highly poisonous and difficult for natural processes to break down. PCBs can seriously harm human health and damage practically every element of the environment, including the air, water, and soil (Vorrink, 2014). To bioremediate PCBs, researchers make use of the potential of plant-microbe interaction (Faroon et al., 1998).

13.5 MICROALGAL SPECIES

Numerous species of microalgae are remarkably effective at removing pathogens, heavy metals, pesticides, heavy metals, nitrogen, and phosphorus. *Chlorella*, *Scenedesmus*, *Phormidium*, *Botryococcus*, *Arthrospira*, and *Chlamydomonas* are a few examples of these microalgae (Molazadeh, 2019). By eliminating phosphorus and nitrogen, microalgae control the eutrophication process. Microalgae can effectively replace biological therapy, which turns undesired organic and inorganic components into useful biomass. Numerous microalgal species were employed for wastewater treatment. The algae that are taken out of the treatment pond could be used to make valuable goods and food. Algae can store harmful substances like arsenic, and selenium in their cells and can remove them from aquatic environments (Ali et al., 2013).

Chlorella, Scenedesmus, *Euglena*, *Chlamydomonas*, and *Ankistrodesmus* are some examples of microalgae species that have been shown to thrive well in sewage. Only a few microbial algae are capable of removing heavy metals (Molazadeh, 2019). *Chlorella vulgaris* (for the removal of cadmium, copper, and zinc), *Chlamydomonas spp.* (for lead removal), and *Scenedesmus chlorelloides* are some examples. Additionally, different species' susceptibilities to organic contaminants in wastewater have been observed (Molazadeh, 2019).

A further benefit of using algae in phytoremediation is the high biomass production by these species, which results in significant heavy metal absorption and accumulation. A wide range of pollutants,such as pentachlorophenol (PCP), metals, radionuclides, fertilizers, petroleum hydrocarbons, and chlorinated solvents, are potentially amenable to phytoremediation (Baghour et al.,2001). Many heavy metals have been found to accumulate in plants. Heavy metal immobilization in the cell wall or compartmentalization in vacuoles are two ways that heavy metals might be tolerated by plants (Davis et al., 2003). As a result of tolerance mechanisms, some algae exhibit a high potential for heavy metal accumulation, and many algae produce phytochelatins and metallothioneins, which can bind to heavy metals and transport them into vacuoles (Suresh and Ravishankar, 2004).

Around the world, macroalgae have been widely utilizedto assess marine habitats and heavy metal pollution. Many different regions of the world have recently used different species of green algae to detect the levels of heavy metals like *Enteromorpha* (Al-Homaidan et al., 2011). Macroalgae are frequently used as biomonitors of metal availability in marine environments because of their capacity to store metals inside their tissues (Chekroun and Mourad, 2013).

13.6 PLANT-MICROBIAL INTERACTIONS

The biogeochemical cycling of nutrients and materials, the preservation of plant health, and soil quality are all exhibited as major environmental processes that are regulated by interactions between microbes and particular molecules (Barea et al., 2004). Large populations of metabolically active populations of advantageous soil-borne bacteria must form the basis for excellent bioremediation (Metting, 2013). High adaptability in a wide range of conditions, rapid growth, and biochemical plasticity to metabolize a variety of natural and xenobiotic compounds for a well-established ecosystem are the driving forces in these soil-borne bacteria (Lee, 2013).

In addition to supporting life, soil offers ecosystem services that are crucial for maintaining biodiversity, regulating biogenic gases and the earth's climate, and producing primary crops (Welbaum et al., 2004). Numerous human activities have contaminated a sizeable portion of the entire amount of land, and it is anticipated that this level will gradually rise in the years to come. Phytoremediation is a novel practical and economical plant-based technology that employs plants and the microbial flora that are linked to clean up the environment (Salt et al., 1998). Utilizing a plant-based remediation technology that functions as a transformation system to metabolize organic compounds while utilizing plants' nutrient utilization activities can result in the efficient breakdown of organic compounds (Kassel, 2002). Alternately, using this plant-based method, hazardous substances including heavy metals can be eliminated from the environment through absorption and bioaccumulation (Ali et al., 2013). Due to their involvement in key metabolic processes, soil-borne microbial communities with diverse genetic and functional capacities can vary greatly in soils and have an impact on soil function (Weyens et al., 2009).

Plants and the related rhizospheric microbial populations have a complex mutualistic connection. Plants' ability to absorb trace elements can be improved by soil-borne microbes such as rhizospheric bacteria (Weyens et al., 2009). Generally, a plant species needs a dense and highly branching root system to support numerous bacteria, primary and secondary metabolism, as well as establishment, survival, and ecological interactions with other organisms for rhizoremediation to be successful (Kuiper et al., 2004). Many organic contaminants can be directly absorbed by plants from the soil through their roots in addition to being affected by microbial interactions. It's crucial for soil microbes to break down organic leftovers so that plant nutrients like carbon (C), potassium (K), nitrogen (N), sulfur (S), and phosphate can be released (Macek et al., 2000). However, the employment of these biological remediation technologies to clean up heavy metal-contaminated areas destroys the native biotic components of soils, and their implementation is complex and somewhat expensive (Baker et al., 2000).

13.7 CONCLUSION

The industrial revolution has enhanced the convenience and comfort of human beings, unfortunately at the expense of resources and the environment. Industrial effluents, partially treated or untreated find their way to the water bodies, both surface and subsurface, causing contamination through a variety of pollutants, such as heavy metals, PAHs, organic and inorganic substances, etc. Such contamination can be treated with physicochemical methods, however, the same may be technologically and economically prohibitive esp. to developing and underdeveloped world. Phytoremediation, a subset of bioremediation, may offer hope for such places. The use of plants to decontaminate effluents and soil has been known for some time. Specific microbes can be used to enhance the potential of plants for the removal of pollutants. These microbes include bacteria, fungi, microalgae, etc. A number of studies have shown the potential of such microbial-assisted phytoremediation for the treatment of industrial wastewater. This process is not without its own drawbacks and needs further research to find the most appropriate plant-microbe set and optimum conditions for the removal of particular contaminant/s at a steady enough pace. It is clear that such a low-cost, low-tech sustainable solution is the way forward in a world full of carbon-intensive operations and holds the promise for appropriate management of industrial wastewater. It will also reduce the load on wastewater treatment processes and can provide partially or completely treated water for further uses including recycling and reuse.

LIST OF ABBREVIATIONS

BOD Biochemical oxygen demand
COD Chemical oxygen demand
DDT Dichlorodiphenyltrichloroethanes
HCH Hexachlorocyclohexanes
PAH Polycyclic aromatic hydrocarbon
PCB Polychlorinated biphenyls
PCP Pentachlorophenol
TNT Trinitrotoluene

REFERENCES

Abhilash, P.C., & Srivastava, S. (2011). Influence of Rhizospheric Microbial Inoculation and Tolerant Plant Species on the Rhizoremediation of Lindane. *Environmental and Experimental Botany*, 74, 1, 127–130. https://doi.org/10.1016/j.envexpbot.2011.05.009.

Abhilash, P.C., & Tripathi, V. (2016). Sustainability of Crop Production from Polluted Lands. *Energy, Ecology and Environment*, 1, 1, 54–65. https://doi.org/10.1007/s40974-016-0007-x.

Abdurrahman, M.I., Chaki, S., & Saini, G. (2020). Stubble Burning: Effects on Health &Environment, Regulations and Management Practices. *Environmental Advances*, 2. https://doi.org/10.1016/j.envadv.2020.100011

Agarwal, P., & Saini, G. (2020). Use of Natural Coagulants (Moringa oleifera and Benincasa hispida) for Volume Reduction of Waste Drilling Slurries. *Materials Today: Proceedings*, 49, 3274–3278. https://doi.org/10.1016/j.matpr.2020.12.924

Agency for Toxic Substances and Disease Registry. (2005). Medical Management Guidelines for Hydrogen Sulfide. Health San Francisco, p. 2, https://www.osha.gov/OshDoc/data_Hurricane_Facts/hydrogen_sulfide_fact.pdf.

Al-Baldawi, I.A.(2018). Removal of 1,2-Dichloroethane from Real Industrial Wastewater Using a Sub-Surface Batch System with Typha Angustifolia L. *Ecotoxicology and Environmental Safety*, 147, 260–265. https://doi.org/10.1016/j.ecoenv.2017.08.022.

Al-Homaidan, A.A., Al-Ghanayem, A.A., & Areej, A.H. (2011). Green Algae as Bioindicators of Heavy Metal Pollution in Wadi Hanifah Stream, Riyadh, Saudi Arabia. *International Journal of Water Resources and Arid Environments*, 1, 10.

Ali, H., Khan, E., & Sajad, M.A. (2013). Phytoremediation of Heavy Metals – Concepts and Applications. *Chemosphere*, 91, 869–881.

Amare, E. (2018). Wastewater Treatment by Lemna Minor and Azolla Filiculoides in Tropical Semi-Arid Regions of Ethiopia. *Ecological Engineering*, 120, 464–473. https://doi.org/10.1016/j.ecoleng.2018.07.005.

Anand, S. (2017). Macrophytes for the Reclamation of Degraded Waterbodies with Potential for Bioenergy Production. *Phytoremediation Potential of Bioenergy Plants*, 333–351. https://doi.org/10.1007/978-981-10-3084-0_13.

Ansari, A. A. (2016). Phytoremediation. In: S. S. Gill, R. Gill, G. Lanza, & L. Newman (eds.). Berlin, Germany: Springer International Publishing (pp. 12–18).

Ansari, A.A., Naeem, M. Gill, S.S., & AlZuaibr, F.M. (2020). Phytoremediation of contaminated waters: An eco-friendly technology based on aquatic macrophytes application, *Egyptian Journal of Aquatic Research*, 46, 4, 371–376.

Baghour, M., Moreno, D.A., Hernandez, J., Castilla, N., & Romero, L. (2001). Influence of Root Temperature on Phytoaccumulation of As, Ag, Cr, and Sb in Potato Plants (Solanum tuberosum L. var. Spunta). *Journal of Environmental Science and Health, Part A*, 36, 1389–1401.

Baker, Alan J.M., & Reeves, Roger D. (2000). Phytoremediation of Toxic Metals: Using Plants to Clean the Environment. *Journal of Plant Biotechnology*, 1, 1, 304.

Barea, J.M. (2015). Future Challenges and Perspectives for Applying Microbial Biotechnology in Sustainable Agriculture Based on a Better Understanding of Plant-Microbiome Interactions. *Journal of Soil Science and Plant Nutrition*, 15, 2, 261–282. https://doi.org/10.4067/s0718-95162015005000021.

Barea, J.M., Azcón, R., & Azcón-Aguilar, C. (2004). Mycorrhizal Fungi and Plant Growth Promoting Rhizobacteria. In A. Varma, L. Abbott, D. Werner, & R. Hampp (Eds.), *Plant Surface Microbiology* (pp. 351–371). Heidelberg: Springer.

Carraher, C.E, Jr. (2013). *Carraher's Polymer Chemistry* (6th Edn; J.J.Lagowski,Ed.). Basel: CRC Press.

Chandanshive, V.V. (2017). Co-Plantation of Aquatic Macrophytes Typha angustifolia and Paspalum scrobiculatum for Effective Treatment of Textile Industry Effluent. *Journal of Hazardous Materials*, 338, 47–56. https://doi.org/10.1016/j.jhazmat.2017.05.021.

Chekroun, K.B., & Mourad, B. (2013). The Role of Algae in Phytoremediation of Heavy Metals: A Review. *Journal of Materials and Environmental Science*, 4, 6, 873–880.

Chen, L. (2014). Interaction of Cd-Hyperaccumulator Solanum Nigrum L. and Functional Endophyte Pseudomonas Sp. Lk9 on Soil Heavy Metals Uptake. *Soil Biology and Biochemistry*, 68, 300–308. https://doi.org/10.1016/j.soilbio.2013.10.021.

Davis, T.A., Volesky, B., & Mucci, A. (2003). A Review of the Biochemistry of Heavy Metal Biosorption by Brown Algae. *Water Research*, 37, 4311.

Dietz, A., & Schnoor, J.L. (2001). Advances in Phytoremediation. *Environ Health Perspect. Environ Health Perspect*, 109, 163–168.

Duan, G. (2012). Expressing ScACR3 in Rice Enhanced Arsenite Efflux and Reduced Arsenic Accumulation in Rice Grains. *Plant and Cell Physiology*, 53, 154–163. https://doi.org/10.1093/pcp/pcr161.

Ekperusi, A.O. (2019). Application of Common Duckweed (Lemna Minor) in Phytoremediation of Chemicals in the Environment: State and Future Perspective. *Chemosphere*, 223, 285–309. https://doi.org/10.1016/j.chemosphere.2019.02.025.

Ekta, P., & Modi, N.R. (2018). A Review of Phytoremediation. *Journal of Pharmacognosy and Phytochemist ry*, 7, 4, 1485–1489.

Faroon, O., Llados, F., & George, J. (1998). Toxicological Profi Le for Polychlorinated Biphenyls. Department of Health and Human Services Agency for Toxic Substances and Disease Registry, Atlanta.

Fent, K. (2006). Ecotoxicology of Human Pharmaceuticals. *Aquatic Toxicology*, 76, 2, 122–159. https://doi.org/10.1016/j.aquatox.2005.09.009.

Fiorentino, A. (2003). Environmental Effects Caused by Olive Mill Wastewaters: Toxicity Comparison of Low-Molecular-Weight Phenol Components. *Journal of Agricultural and Food Chemistry*, 51, 4, 1005–1009. https://doi.org/10.1021/jf020887d.

Gautam, S., & Saini, G. (2020). Use of Natural Coagulants for Industrial Wastewater Treatment. *Global Journal of Environmental Science and Management*, 6, 4, 553–578. https://doi.org/10.22034/gjesm.2020.04.10

Haq, I. (2017). Genotoxicity Assessment of Pulp and Paper Mill Effluent before and after Bacterial Degradation Using Allium Cepa Test. *Chemosphere*, 169, 642–650. https://doi.org/10.1016/j.chemosphere.2016.11.101.

Hejna, M. (2020). Bioaccumulation of Heavy Metals from Wastewater through a Typha Latifolia and Thelypteris Palustris Phytoremediation System. *Chemosphere*, 241. https://doi.org/10.1016/j.chemosphere.2019.125018.

Hoang, H.G. (2020). Heavy Metal Contamination Trends in Surface Water and Sediments of a River in a Highly-Industrialized Region. *Environmental Technology and Innovation*, 20. https://doi.org/10.1016/j.eti.2020.101043.

Howard, S., Donald, R., & Tchobanoglous, G. (1985). *Environmental Engineering* (International Ed). New York: McGraw-Hill. Inc.

Jabeen, H. (2016). Enhanced Remediation of Chlorpyrifos by Ryegrass (Lolium multiflorum) and a Chlorpyrifos Degrading Bacterial Endophyte Mezorhizobium Sp. HN3. *International Journal of Phytoremediation*, 18, 2, 126–133. https://doi.org/10.1080/15226514.2015.1073666.

Jing, Yuan Xiao (2014). Characterization of Bacteria in the Rhizosphere Soils of Polygonum Pubescens and Their Potential in Promoting Growth and Cd, Pb, Zn Uptake by Brassica Napus. *International Journal of Phytoremediation*, 16, 4, 321–333.https://doi.org/10.1080/15226514.2013.773283.

Kafil, M. (2019). Phytoremediation Potential of Vetiver Grass Irrigated with Wastewater for Treatment of Metal Contaminated Soil. *International Journal of Phytoremediation*, 21, 2, 92–100. https://doi.org/10.1080/15226514.2018.1474443.

Kale, S.K., Deshmukh, A.G., Dudhare, M.S., & Patil, V.B. (2015). Microbial Degradation of Plastic: A Review. *Journal of Biochemical Technology*, 6, 2, 952–961.

Kamran, M.A. (2016). Bioaccumulation of Nickel by E. Sativa and Role of Plant Growth Promoting Rhizobacteria (PGPRs) under Nickel Stress. *Ecotoxicology and Environmental Safety*, 126, 256–263. https://doi.org/10.1016/j.ecoenv.2016.01.002.

Kassaye, G. (2017). Phytoremediation of Chromium from Tannery Wastewater Using Local Plant Species. *Water Practice and Technology*, 12, 4, 894–901, https://doi.org/10.2166/wpt.2017.094.

Kassel, A.G. (2002). Phytoremediation of Trichloroethylene Using Hybrid Poplar. *Physiology and Molecular Biology of Plants*, 8, 1, 3–10.

Kuiper, I., Lagendijk, E.L., Bloemberg, G.V., & Lugtenberg, B.J.J. (2004). Rhizoremediation: A Beneficial Plant-Microbe Interaction. *Molecular Plant-Microbe Interactions*, 17, 1, 6–15.

Kumar, A., & Chandra, R. (2021). Biodegradation and Toxicity Reduction of Pulp Paper Mill Wastewater by Isolated Laccase Producing Bacillus cereus AKRC03. *Cleaner Engineering and Technology*, 4, 100193. https://doi.org/10.1016/j.clet.2021.100193.

Kumar, A., Saxcena, G., Kumar, V., & Chandra, R. (2020). Environmental Contamination, Toxicity Profile and Bioremediation Approaches for Treatment and Detoxification of Pulp Paper Industry Wastewater. In G. Saxcena, V. Kumar, & M. Shah (Eds.), *Bioremediation for Environmental Sustainability Toxicity, Mechanisms of Contaminants Degradation, Detoxification, and Challenges*, (pp. 375–402). https://doi.org/10.1016/C2019-0-01053-9.

Kumar, A., Singh, A.K., & Chandra, R. (2021). Recent Advances in Physicochemical and Biological Approaches for Degradation and Detoxification of Industrial Wastewater. In I. Haq & A. Kalamdhad (Eds.), *Emerging Treatment Technologies for Waste Management*. 1, 28. DOI:10.1007/978-981-16-2015-7.

Lardieri, A. (2021). Harmful Effects of Chlorine Dioxide Exposure. *Clinical Toxicology*, 59, 5, 448–449. https://doi.org/10.1080/15563650.2020.1818767.

Lee, Y.J. (2013). Enhancement of Plant-Microbe Interactions Using Rhizosphere Metabolomics-Driven Approach and Its Application in the Removal of Polychlorinated Biphenyls. *Molecular Microbial Ecology of the Rhizosphere*, 2, 1191–98, https://doi.org/10.1002/9781118297674.ch114.

Liu, S. (2020). Prospect of Phytoremediation Combined with Other Approaches for Remediation of Heavy Metal-Polluted Soils. *Environmental Science and Pollution Research*, 27, 14, 16069–16085. https://doi.org/10.1007/s11356-020-08282-6.

Lu, B. (2018). Removal of Water Nutrients by Different Aquatic Plant Species: An Alternative Way to Remediate Polluted Rural Rivers. *Ecological Engineering*, 110, 18–26. https://doi.org/10.1016/j.ecoleng.2017.09.016.

Luo, X.S. (2012). Trace Metal Contamination in Urban Soils of China. *Science of the Total Environment*, 421–422, 17–30. https://doi.org/10.1016/j.scitotenv.2011.04.020.

Macek, T., Macková, M., & Kás, J. (2000). Exploitation of Plants for the Removal of Organics in Environmental Remediation. *Biotechnology Advances*, 18, 1, 23–34. https://doi.org/10.1016/S0734-9750(99)00034-8.

Man, Y.B., Lei, K.M., & Chow, K.L. et al. (2020). Ecological risks of heavy metals/metalloid discharged from two sewage treatment works to Mai Po Ramsar site, South China. *Environ Monit Assess* 192, 466. https://doi.org/10.1007/s10661-020-08397-w.

Marques, Ana P.G.C. (2013). Inoculating *Helianthus annuus* (Sunflower) Grown in Zinc and Cadmium Contaminated Soils with Plant Growth Promoting Bacteria - Effects on Phytoremediation Strategies. *Chemosphere*, 92, 1, 74–83. doi.org/10.1016/j.chemosphere.2013.02.055.

Md Sa'at, S.K., & Nastaein, Q.Z. (2017). Phytoremediation Potential of Palm Oil Mill Effluent by Constructed Wetland Treatment. *Engineering Heritage Journal*, 1, 49–50. https://doi.org/10.26480/gwk.01.2017.49.54.

Mehmood, K. (2019). Treatment of Pulp and Paper Industrial Effluent Using Physicochemical Process for Recycling. *Water (Switzerland)*, 11. https://doi.org/10.3390/w11112393.

Metting, F.B. Jr. (2013). Structure and Physiological Ecology of Soil Microbial Communities. In *Soil Microbial Ecology: Applications in Agricultural and Environmental Management* (pp. 3–26). https://api.semanticscholar.org/CorpusID:81440534.

Mishra, S., &Bharagava, R.N. (2016). Toxic and Genotoxic Effects of Hexavalent Chromium in Environment and Its Bioremediation Strategies. *Journal of Environmental Science and Health - Part C Environmental Carcinogenesis and Ecotoxicology Reviews*, 34, 1, 1–32, https://doi.org/10.1080/10590501.2015.1096883.

Miyazaki, A. (2002). Acute Toxicity of Chlorophenols to Earthworms Using a Simple Paper Contact Method and Comparison with Toxicities to Fresh Water Organisms. *Chemosphere*, 47, 1, 65–69. https://doi.org/10.1016/S0045-6535(01)00286-7.

Mohammadzadeh, A. (2014). Effects of Nickel and PGPBs on Growth Indices and Phytoremediation Capability of Sunflower (Helianthus annuus L.). *Archives of Agronomy and Soil Science*, 60, 12, 1765–17678. https://doi.org/10.1080/03650340.2014.898839.

Molazadeh, M. (2019). The Use of Microalgae for Coupling Wastewater Treatment with CO2 Biofixation. *Frontiers in Bioengineering and Biotechnology*, 7. https://doi.org/10.3389/fbioe.2019.00042.

Müller, J.B. (2019). Comparative Assessment of Acute and Chronic Ecotoxicity of Water Soluble Fractions of Diesel and Biodiesel on Daphnia Magna and Aliivibrio Fischeri. *Chemosphere*, 221, 640–646. https://doi.org/10.1016/j.chemosphere.2019.01.069.

Mustafa, H.M., & Gasim, H. (2021). Recent Studies on Applications of Aquatic Weed Plants in Phytoremediation of Wastewater: A Review Article. *Ain Shams Engineering Journal*, 12, 1, 355–365. https://doi.org/10.1016/j.asej.2020.05.009.

Nadar, S.S. (2018). Enzyme Assisted Extraction of Biomolecules as an Approach to Novel Extraction Technology: A Review. *Food Research International*, 108, 309–330. https://doi.org/10.1016/j.foodres.2018.03.006.

Nagata, S. (2015). Root Endophytes Enhance Stress-Tolerance of Cicuta Virosa L. Growing in a Mining Pond of Eastern Japan. *Plant Species Biology*, 30, 2, 116–125. https://doi.org/10.1111/1442-1984.12039.

Oh, K. (2014). Study on Application of Phytoremediation Technology in Management and Remediation of Contaminated Soils. *Journal of Clean Energy Technologies*, 216–220. https://doi.org/10.7763/jocet.2014.v2.126.

Ozyigit, İ. İ., & Dogan, İ. (2014). Plant-microbe interactions in phytoremediation. In: K. Hakeem, M. Sabir, M. Öztürk and A. Murmet (eds.), *Soil Remediation and Plants: Prospects and Challenges*, (pp. 255–285). doi:10.1016/B978-0-12-799937-1.00009-7.

Pandey, A.K., Singh, B., Upadhyay, S., Pandey, S., Mishra, V.K., & Jain, P.A. (2017). Solution for Sustainable Development for Developing Countries: Waste Water Treatment by Use of Membranes (Review). *International Journal of Current Microbiology and Applied Sciences*, 6(8): 1212–1228. doi: https://doi.org/10.20546/ijcmas.2017.608.149.

Panz, K., &Korneliusz, M. (2012). Phytoremediation of Explosives (TNT, RDX, HMX) by Wild-Type and Transgenic Plants. *Journal of Environmental Management*, 113, 85–92. https://doi.org/10.1016/j.jenvman.2012.08.016.

Perotti, R. (2020). CR(VI) Phytoremediation by Hairy Roots of Brassica napus: Assessing Efficiency, Mechanisms Involved, and Post-Removal Toxicity. *Environmental Science and Pollution Research*, 27, 9, 9465–9474. https://doi.org/10.1007/s11356-019-07258-5.

Pratas J.C. P., Varun J., & DSouza M, S. M. R. (2014). Phytoremediation of Soils Contaminated with Metals and Metalloids at Mining Areas: Potential of Native Flora. *Environmental Risk Assessment of Soil Contamination*. InTech; 2014. http://dx.doi.org/10.5772/57469.

Rezania, S. (2015). Perspectives of Phytoremediation Using Water Hyacinth for Removal of Heavy Metals, Organic and Inorganic Pollutants in Wastewater. *Journal of Environmental Management*, 163, 125–133. https://doi.org/10.1016/j.jenvman.2015.08.018.

Saleh, H.M. (2017). Uptake of Cesium and Cobalt Radionuclides from Simulated Radioactive Wastewater by Ludwigia Stolonifera Aquatic Plant. *Nuclear Engineering and Design*, 315, 194–199. https://doi.org/10.1016/j.nucengdes.2017.02.018.

Saleh, H.M. (2020). Potential of the Submerged Plant Myriophyllum spicatum for Treatment of Aquatic Environments Contaminated with Stable or Radioactive Cobalt and Cesium. *Progress in Nuclear Energy*, 118. https://doi.org/10.1016/j.pnucene.2019.103147.

Salt, D.E., Smith, R.D., & Raskin, I. (1998). Phytoremediation. *The Annual Review of Plant Biology*, 49, 643–668.

Sauvêtre, A., & Schröder, P. (2015). Uptake of Carbamazepine by Rhizomes and Endophytic Bacteria of Phragmites Australis. *Frontiers in Plant Science*, 6. https://doi.org/10.3389/fpls.2015.00083.

Seroja, R. (2018). Tofu Wastewater Treatment Using Vetiver Grass (Vetiveria zizanioides) and Zeliac. *Applied Water Science*, 8. https://doi.org/10.1007/s13201-018-0640-y.

Shah, V., & Achlesh, D. (2020). Phytoremediation: A Multidisciplinary Approach to Clean up Heavy Metal Contaminated Soil. *Environmental Technology and Innovation*, 18. https://doi.org/10.1016/j.eti.2020.100774.

Sheik, S. (2015). Biodegradation of Gamma Irradiated Low Density Polyethylene and Polypropylene by Endophytic Fungi. *International Biodeterioration and Biodegradation*, 105, 21–29, https://doi.org/10.1016/j.ibiod.2015.08.006.

Singh, B. (2012). Microbial Remediation of Explosive Waste. *Critical Reviews in Microbiology*, 38, 2, 152–167. https://doi.org/10.3109/1040841X.2011.640979.

Singh, M.J., & Padmavathy, S. (2015). Biosurfactant, Polythene, Plastic, and Diesel Biodegradation Activity of Endophytic Nocardiopsis Sp. Mrinalini9 Isolated from Hibiscus rosasinensis Leaves. *Bioresources and Bioprocessing*, 2, 1. https://doi.org/10.1186/s40643-014-0034-4.

Singh, R. (2014). Microorganism as a Tool of Bioremediation Technology for Cleaning Environment: A Review. *Proceedings of the International Academy of Ecology and Environmental Sciences*, 4, 1, 1.

Sun, G. (2015). Biodegradation of Dichlorodiphenyltrichloroethanes (DDTs) and Hexachlorocyclohexanes (HCHs) with Plant and Nutrients and Their Effects on the Microbial Ecological Kinetics. *Microbial Ecology*, 69, 2, 281–292. https://doi.org/10.1007/s00248-014-0489-z.

Suresh, B., & Ravishankar, G.A. (2004). Phtoremediation a Novel and Promising Approach for Environmental Cleanup. *Critical Reviews in Biotechnology*, 24, 97–124. https://www.tandfonline.com/action/journalInformation?journalCode=ibty20.

Thijs, S. (2014). Exploring the Rhizospheric and Endophytic Bacterial Communities of Acer Pseudoplatanus Growing on a TNT-Contaminated Soil: Towards the Development of a Rhizocompetent TNT-Detoxifying Plant Growth Promoting Consortium. *Plant and Soil*, 385, 1–2, 15–36. https://doi.org/10.1007/s11104-014-2260-0.

Udayan, A., Arumugam, M., & Pandey, A. (2017). Nutraceuticals From Algae and Cyanobacteria. In: Rajesh Prasad Rastogi, Datta Madamwar, Ashok Pandey (eds.) *Algal Green Chemistry, Elsevier*, pp, 65–89. https://doi.org/10.1016/B978-0-444-63784-0.00004-7.

Vargas-González, H.H. (2014). Effects of Sewage Discharge on Trophic State and Water Quality in a Coastal Ecosystem of the Gulf of California. *The Scientific World Journal*. https://doi.org/10.1155/2014/618054.

Vorrink, S.U. (2014). Hypoxia Perturbs Aryl Hydrocarbon Receptor Signaling and CYP1A1 Expression Induced by PCB 126 in Human Skin and Liver-Derived Cell Lines. *Toxicology and Applied Pharmacology*, 274, 3, 408–416. https://doi.org/10.1016/j.taap.2013.12.002.

Wang, J.H. (2017). Microalgae-Based Advanced Municipal Wastewater Treatment for Reuse in Water Bodies. *Applied Microbiology and Biotechnology*, 101, 7, 2659–2675. https://doi.org/10.1007/s00253-017-8184-x.

Wang, L. (2020). Field Trials of Phytomining and Phytoremediation: A Critical Review of Influencing Factors and Effects of Additives. *Critical Reviews in Environmental Science and Technology*, 50, 24, 2724–2774. https://doi.org/10.1080/10643389.2019.1705724.

Welbaum, G., Sturz, A.V., Dong, Z., & Nowak, J. (2004). Fertilising Soil Microorganisms to Improve Productivity of Agroecosystems. *Critical Reviews in Plant Sciences*. 23, 175–193.

Weyens, N., Van der Lelie, D., Taghavi, S., & Vangronsveld, J. (2009). Phytoremediation: Plant-- Endophyte Partnerships Take the Challenge. *Current Opinion in Biotechnology*, 20, 248–254.

Wu, C.H. (2006). Engineering Plant-Microbe Symbiosis for Rhizoremediation of Heavy Metals. Applied and Environmental Microbiology, 72, 2, 1129–1134. https://doi.org/10.1128/AEM.72.2.1129-1134.2006.

Wuang, S.C. (2016). Use of Spirulina Biomass Produced from Treatment of Aquaculture Wastewater as Agricultural Fertilizers. *Algal Research*, 15, 59–64. https://doi.org/10.1016/j.algal.2016.02.009.

Xie, X.G., & Chuan C.D. (2015). Biodegradation of a Model Allelochemical Cinnamic Acid by a Novel Endophytic Fungus Phomopsis Liquidambari. *International Biodeterioration and Biodegradation,* 104, 498–507. https://doi.org/10.1016/j.ibiod.2015.08.004.

Yang, Y., Liu, Y., Li, Z., Wang, Z., & Wei, H. (2020). Significance of Soil Microbe in Microbial-Assisted Phytoremediation: An Effective Way to Enhance Phytoremediation of Contaminated Soil. *International Journal of Environmental Science and Technology*, 17, 4, 2477–2484. https://doi.org/10.1007/s13762-020-02668-2.

Zhao, Ke, Penttinen, P., Zhang, X., Ao, X., Liu, M., Yu, X., & Chen, Q. (2014). Maize Rhizosphere in Sichuan, China, Hosts Plant Growth Promoting Burkholderia Cepacia with Phosphate Solubilizing and Antifungal Abilities. *Microbiological Research*, 169, 1, 76–82. https://doi.org/10.1016/j.micres.2013.07.003.

14 Wastewater Treatment through Exopolysaccharides-producing Microbes

Priyanka, S.K. Dwivedi, and Pavan Kumar

14.1 INTRODUCTION

Industrialization, urbanization, intensive agricultural practices, mismanagement of solid waste, and improper wastewater disposal, leading to an increase in pollution and water resources becoming more scarce every day (Kumar and Dutta, 2019; Siddharth et al., 2021). Conventional methods of wastewater remediation, such as filtration, sedimentation, and disinfection are often pricy and generate secondary contaminants and other harmful byproducts (Dey et al., 2016; Khan et al., 2019). Anaerobic sludge beds and aerobic thermophilic bioreactors are being used for biological wastewater treatment although having several disadvantages they produce sludge in huge quantities and insufficient aggregation, respectively (Zhao et al., 2019; Banerjee et al., 2021). Additionally, the use of inorganic and organic materials increases the concentrations of dissimilar pollutants in receiving water bodies (Karri et al., 2021). Activated sludge is the primary component among all biological processes used for the treatment of all kinds of wastewater (Wanner, 2021; de Sousa et al., 2019). The majority of microbes exist as microbial aggregations like biofilms, granules, and sludge flocs (Huang et al., 2022). The electron microscopy techniques have confirmed the occurrence of extracellular polymeric substances (EPS) in these microbial aggregations (Sheng et al., 2010; Ding et al., 2015; Trivedi, 2020). EPS are high molecular weight, complex natural polymers and metabolic products of microbes either present on the cell surface or present in extracellular space constituting 50%–90% organic matter (Mahapatra and Banerjee, 2013; Siddharth, et al., 2021). EPS constitutes monosaccharide units that are linked together by glycosidic linkages. They can be homopolysaccharides (single type of sugar monomers) or heteropolysaccharides (different types of sugar monomers). EPS plays a key role in the cohesion and adhesion of cells by binding them to other cell constituents such as proteins and polysaccharides (Siddharth et al., 2021). They are also reported to play a crucial role in biofilm development, protection against harmful substances, bacteriophages, and biotic abiotic stress (Dave et al., 2016; Costa et al., 2018). Polysaccharides and proteins are the significant ingredients of the EPS while other macromolecules including lipids, uronic acids, nucleic acids, and humic compounds are also present in certain proportions (Zhao et al., 2019). Intracellular, structural, and extracellular polysaccharides are the three primary categories of microbial polysaccharides. As a carbon reserve, the intracellular polysaccharides support the cell when it is starving. Structural polysaccharides play a vital role in the progress of cellular structures. Polysaccharides present on the cell surface or in extracellular space are called exopolysaccharides (Mehta et al., 2021). Exopolysaccharides have a wide range of biotechnological uses including thickening, stabilizing, and texturizing agents in food, textiles, paint, detergents, beverages, pharmaceutical industries, anti-aging compounds in skin care products, and flocculating agents in the wastewater treatment processes (Figure 14.1) (Han et al., 2016; Trivedi, 2020; Ateş, 2021; Debnath et al., 2021; Topal et al., 2022). Wastewater discharge from industries possesses high organics and nutrients, dark in color, and has an unpleasant smell.

DOI: 10.1201/9781003441144-14

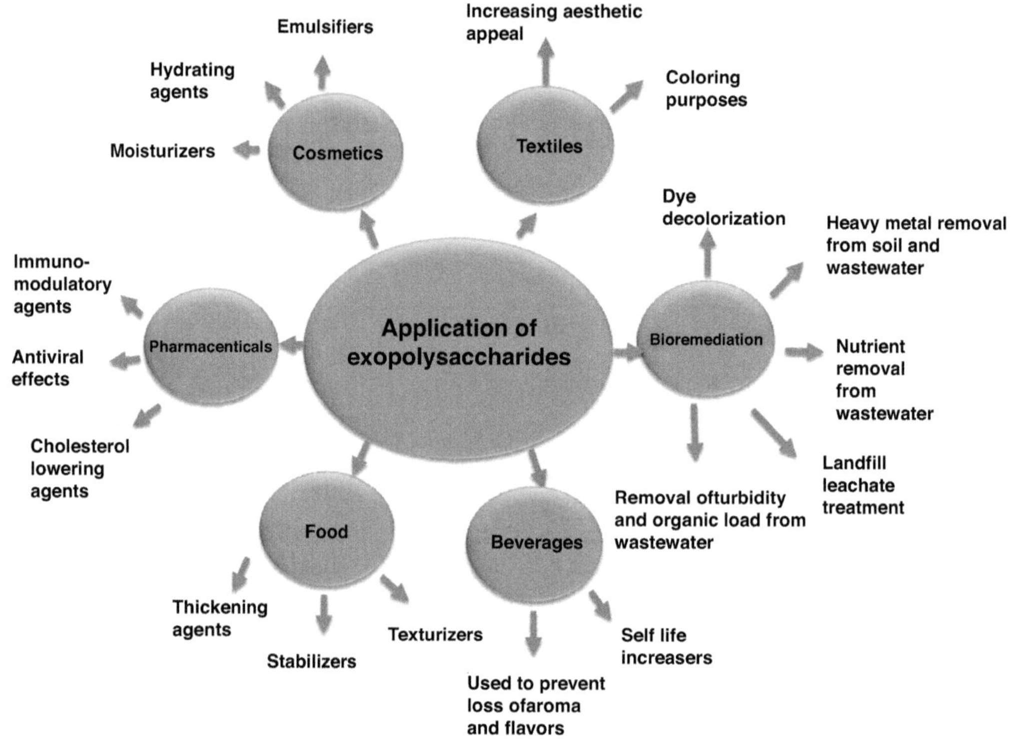

FIGURE 14.1 Applications of Exopolysaccharides in various sectors.

Source: Modified from Debnath et al. (2021).

Exopolysaccharides are natural polymers that have the ability to combat drawbacks related to conventional methods used for wastewater treatment as these are environmentally benign, cost-effective biological waste treatment alternatives. These are being used for the detoxification of metals, turbidity, nutrient removal, oil spills, hydrocarbon removal and dye removal from wastewater. This chapter discusses the exopolysaccharides of microbial origin, extraction methods, involved mechanisms, affecting factors and their applications in wastewater treatment

14.2 EXOPOLYSACCHARIDE-PRODUCING MICROBES

A wide variety of fungi, bacteria, and algae are described as exopolysaccharide producers. Numerous bacterial isolates, including *Solibacillus silvestris* (Wan et al., 2013) from activated sludge, *Bacillus subtilis* from wastewater sludge, and soil (Shahnavaz et al., 2015), *Bacillus cereus* from soil (Abdel-Wahab et al., 2022), *Bacillus agaradhaerens* from an alkaline lake, and *Alteromonas* (Bacosa et al., 2018) from surface waterbodies, have been identified to produce exopolysaccharide with effective flocculation and coagulation activity. Tyagi et al. (2020) isolated exopolysaccharide from *Parapedobacter sp.* ISTM3 isolated from heavy metals mixture solution. Amongst mesophilic bacteria, *Lactobacillus spp., Streptococcus spp., and Leuconostoc mesenteroides* are considered excellent candidates for EPS development. In addition to these, several other probable EPS producers are *Bacillus spp., Acetobacter spp., Pseudomonas spp., Escherichia spp., Sinorhizobium spp.,* and *Aureobasidium spp.,* etc. (Ibrahim et al., 2022; Habib et al., 2022).

Exopolysaccharides such as scleroglucan and pullulan produced by different fungal species have also been stated such as *Aureobasidium Pullulans,* (Cheng et al., 2012; Wang et al., 2015; Wu et al., 2016) *Ganoderma lucidum* (Tang et al., 2009; Fazenda et al., 2010; Wan et al., 2016) and *Sclerotium rolfsii* (Desai et al., 2008; Taskin et al., 2010).

The exopolysaccharides formed by microalgae serve as flocculants in wastewater treatment systems. Besides this, algal exopolysaccharide has antibacterial, anticancer, antioxidant, immunomodulatory, anti-inflammatory, and anticoagulant attributes (Guzmán et al., 2003; Najdenski et al., 2013; Moreira et al., 2022). Algal species of *Chlamydomonadaceae, Desmidiaceae, Porphyridiaceae, Chlorellaceae,* and *Glaucosphaeraceae* are the most significant families for exopolysaccharide production (Gaignard et al., 2019). Different exopolysaccharide-producing bacterial, fungal and algal species are listed in Table 14.1.

TABLE 14.1
Different Exopolysaccharide-Producing Microbial Species with their Composition

Microbial Species	Exopolysaccharide Composition (Sugar Monomers)	Produced Exopolysaccharide	References
	Bacteria		
Agrobacterium sp.	Glucose	Curdlan	Yang et al. (2016)
Bacillus subtilis	Fructose	Levan	Öner et al. (2016)
Streptococcus sp *Lactobacillus* sp. *Leuconostoc* sp.	Glucose	Dextran	Patel and Prajapat (2013)
Azotobacter vinelandii	Glucose, mannose	Alginate	Cimini et al. (2012)
Xanthomonas sp.	Glucose, mannose, galactose A	Xanthan	Palaniraj and Jayaraman (2011)
Klebsiella pneumoniae	Galactose A, fucose, galactose	Fucogel	Guetta et al., 2003
Enterobacter A47	Glucose A, fucose, glucose-galactose,	FucoPol	Freitas et al. (2011)
	Fungi		
Botryosphaeria rhodina	Glucose	Botryosphaeran	Dekker et al. (2019)
Diaporthe sp.	Glucose, mannose galactose	Glucan	Orlandelli et al. (2017)
Aureobasidium Pullulans	Glucose	pullulan	Wu et al. (2016)
Fusarium solani	Rhamnose, galactose	–	Mahapatra and Banerjee (2012)
Aspergillus sp. Y16	Mannose, galactose,	–	Chen et al. (2011)
Pleurotus tuberregium	Mannose, glucose,	–	Zhang and Cheung (2011)
	Algae		
Phaeodactylum tricornutum		–	–
Spirulina sp. LEB-18	Glucose, galactose, fructose	–	de Jesus et al. (2019)
Porphyridium sp.	Xylose, galactose, Glucose, glucose A	–	Soanen et al. (2016)
Scenedesmus obliquus AS-6-1	Glucose, galactose, mannose, rhamnose, fructose.	–	Guo et al. (2013)
Chlorella vulgaris	Glucose, mannose, galactose.	–	Alam et al. (2014)

14.3 EXTRACTION, PURIFICATION AND CHARACTERIZATION OF EXOPOLYSACCHARIDES

Exopolysaccharide-producing microorganisms are known to flourish in habitats with a high carbon/nitrogen ratio such as wastewater treatment plants of domestic from sugar, paper, and food manufacturing. Microbes are more effective and less expensive producers of exopolysaccharides than plants due to rapid growth, capacity to grow in relatively cheap media, lack of need for extensive space, and ease of manipulation. Hence, the isolation and identification of microbial exopolysaccharides to compete with conventional exopolysaccharides emerged as an important area of research (Raza et al., 2011; Ibrahim et al., 2022). Isolation of exopolysaccharides from plants requires rough methods like acid precipitation, alkali treatment, and bleaching, which may change its stability and structural composition as they have a cell wall. Bacteria produce polysaccharides extracellularly in a short time than any other microorganism (Kanmani and Yuvapriya, 2018). Assessing the physicochemical characteristics of cells from various sources requires a suitable approach to extract exopolysaccharides without interfering with cell lysis or the EPS biofilm matrix's structural integrity. There are two types of EPS that can be further divided into bound EPS (sheaths, capsular polymers, loosely bound polymers, condensed gels, and integrated organic components) and soluble EPS (soluble macromolecules, colloids, and slimes) (Laspidou and Rittmann, 2002; Huang et al., 2022). The soluble one is reported to have effects on microbial actions and sludge characteristics despite the weak connection with cells. A dynamic structure of bound EPS is likely to present around bacteria or in sludge flocs, with tightly bound EPS (TB-EPS) establishing the inner layer and lightly bound EPS (LB-EPS) dispersing in the external layer (Zanna et al., 2022). Extraction of exopolysaccharides from dissimilar sources like biofilm, sludge, cell suspensions, and solid surfaces is accomplished by physical and chemical methods. Heating, centrifugation, and ultrasonic are the most often physical extraction methods used to separate and subsequent dissolution of exopolysaccharides. While in chemical extraction, chemicals are utilized to weaken the bonds between the cell and EPS for rapid dissolution of exopolysaccharide. Loosely bound exopolysaccharides can be separated by adopting mild extraction methods like heating at low temperatures and centrifugation while tightly bound ones require more harsh combinations of physical and chemical methods. The supernatant is typically precipitated with chilled organic solvents like methanol and ethanol (95%) or methanol or acetone, for 12 – 24 hours at 4°C in order to separate crude exopolysaccharide. Ethanol precipitation is used to extract fungal exopolysaccharides using various proportions of the culture/water solution and alcohol. Exopolysaccharides are then retained as a vacuum-dried or lyophilized powder after being dialyzed against water to remove extra salt. The exopolysaccharides are then deproteinized via gel filtration, and ion exchange filtration methods using prescribed chemical reagents. Several fundamental parameters such as protein, carbohydrate, sulfated sugar, and uronic acid should be examined for the characterization of extracted exopolysaccharide (Harimawan and Ting, 2016; Mohamed et al., 2018; Wang et al., 2018). Fourier transform infrared spectroscopy (FTIR), High-performance liquid chromatography (HPLC), and nuclear magnetic resonance (NMR), X-ray diffraction (XRD), time-of-flight mass spectrometer (TOF) are a few techniques that have been used for the qualitative analysis of extracted exopolysaccharide (Mishra et al., 2011; Saravanan et al., 2017; Lynch et al., 2018). Numerous physicochemical extraction methods have been utilized for the extraction of exopolysaccharides are exhibited in Figure 14.2 and Table 14.2.

14.4 FACTORS AFFECTING THE PRODUCTION OF MICROBIAL EXOPOLYSACCHARIDES

Environmental factors have an impact on microorganisms' exopolysaccharide production ability as they may impede or stimulate enzyme activity and cellular structure and protein synthesis. Therefore, culture parameters such as temperature, pH, dissolved oxygen (DO), and agitation are

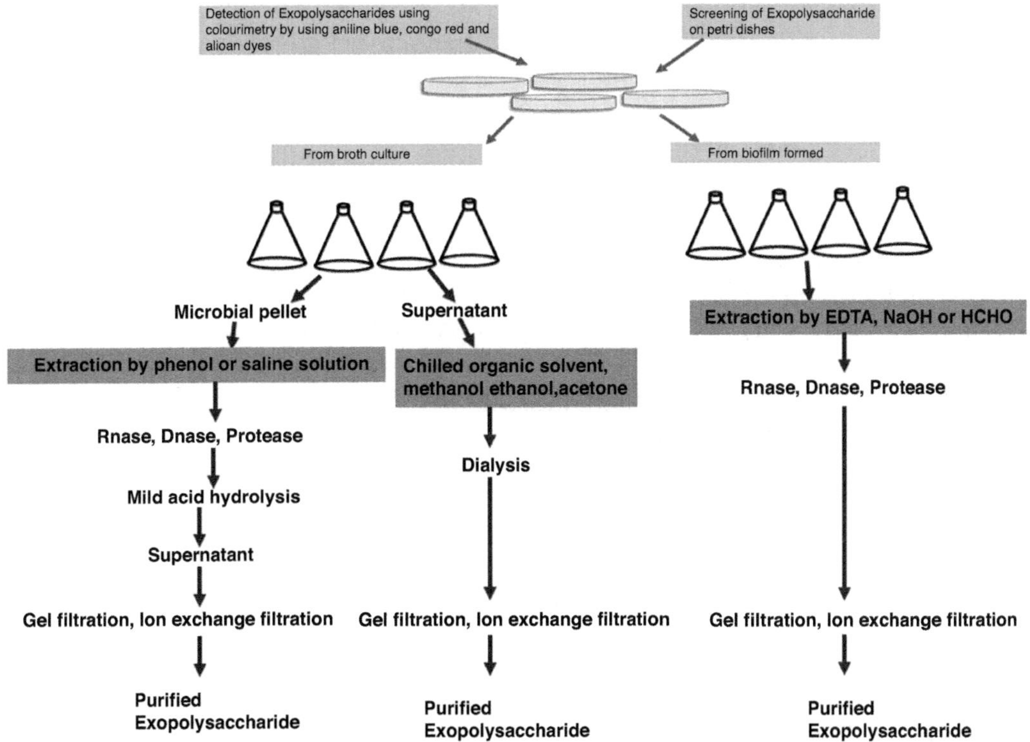

FIGURE 14.2 Steps involved in extraction and characterization of exopolysaccharide.

Source: Adapted from Mehta et al. (2021).

TABLE 14.2
Extraction Methods Employed for EPS

Method	Process	Mechanism
Physical	Heating	The acceleration of EPS dissolution will be achieved by heating to promote molecular mobility.
	Centrifugation	The impulsive pressure from the centrifugal force induces EPS to dissolve into the solution.
	Sonication	Extraction occurred under the force of sound energy
Chemical	Acidic treatment	A stronger repulsive effect causes EPS to separate from the cell surface.
	Alkali treatment	The ionization of the carboxylic groups is caused by mixing with an alkaline NaOH solution which induces repulsion between the cells and EPS.
	EDTA treatment	EDTA causes EPS to disintegrate by removing the charged divalent cation from the matrix
	Cation exchange Resin (CER)	EPS is recovered from the biofilm by eliminating the divalent positive ion
	Ethanol extraction	Ethanol weakens the binding affinity among EPS and cells triggering denaturing of EPS
	Glutaraldehyde	Glutaraldehyde causes fixation of biofilm matrix and EPS
	NaCl	The exchange of cations helps to break up the EPS when used in high concentrations of NaCl
Enzymatic extraction	Using enzymes	The sludge's structure is disturbed and the EPS subsequently dissolved using proteins and carbohydrates hydrolyze enzymes.

Source: Modified from Sheng et al. (2010) and Siddharth et al. (2021).

required to assess and regulate within prescribed ranges to obtain stable and reproducible bioprocess performance. In Enterobacter A47No sufficient growth and exopolysaccharide production was observed while decreasing pH from 7 to 5.6 but at pH 8.4 reduction in growth and exopolysaccharide was observed (Torres et al., 2012). The availability of nutrients and oxygen is determined by mixing and aeration. Several microbes have been seen to produce a high exopolysaccharide in the presence of vigorous agitation and/or aeration such as gellen (Prajapati et al., 2013). However high agitation may alter the properties of produced exopolysaccharide like a reduction in molecular weight of alginate as reported in *Azotobactor vinelandii* (Galindo et al., 2007) and (Radchenkova et al., 2014) optimize agitation conditions in a bioreactor for maximum (yield of 170 µg/ml) exopolysaccharide production from *Aeribacillus pallidus*. Low DO concentration favors the high yield of alginate and FucoPol in *Enterobacter* strain A47 DSM 23139 (Freitas et al., 2011).

14.5 ROLE OF MICROBIAL EXOPOLYSACCHARIDES IN WASTEWATER REMEDIATION

14.5.1 Removal of Nutrients

It is necessary to immobilize microbial strains with sufficient carbon, nitrogen, and phosphorus sources in agar beads or as biofilms for better nutrient removal from wastewater (Guarin and Pagilla, 2021). Immobilization facilitates microbial cells to be preserved, proliferated, and keep maintained during the entire filtration process. It has been reported that the versatility and characteristics of the EPS generated inside the system are essential for the development of biofilm and directly affect the efficacy of biofiltration (Guarin and Pagilla, 2021). There are numerous biofilm-dependent reactor designs (such as packed bed, trickle bed, fluidized bed, up-flow anaerobic sludge blanket, rotating biofilm, airlift suspensions, etc.) have been established. In these designs, EPS material encompasses the colony of native microbial communities to carry out a filtration process for nutrient removal (Li et al., 2015).

14.5.2 Dye Decolorization

Wastewater from textile, food, pharmaceutical, and leather industries is majorly responsible for the addition of different kinds of dyes, organic load, and turbidity causing toxicity to aquatic life and higher animals as dyes are carcinogenic in nature (Siddharth et al., 2021). Due to the charge density of various functional groups such as phosphate, amino, hydroxyl, and carboxyl, microbial exopolysaccharides have been proven potential dye-removing agents (Banerjee et al., 2021). Charge density assists in determining treatment efficiency. As it increases the viscosity of the polymeric solution increases as well resulting in high dye removal efficiency (Bratby, 2016). Zhou et al. (2010) reported a psychrotolerant bacterial exopolysaccharide (55 mg/L) from *Pseudoalteromonas sp.* Bsi20310 (antarctic sea-ice bacterium) exhibited high coagulation capacity towards brilliant red X-3B dye and improved the decolorization process by up to 90% in the occurrence of 150 mg/L ferric chlorides. Moreover, the use of exopolysaccharide-stabilized nanoparticles has appeared as a new keystone in dye removal (Li et al., 2020). Saravanan et al. (2017) reported the characterization of exopolysaccharide-stabilized AgNPs for the removal of Congo red (CR) and methyl orange (MO). Nanoparticles are more reactive towards different chemical compounds due to their high surface area hence considered as effective wastewater treatment tools. Combined treatment of chitosan-stabilized AgNPs with an advanced oxidation process (AOP), successfully degraded dyes (Santhanalakshmi and Dhanalakshmi, 2012). Ladnorg et al. (2019) extracted Alginate-like exopolysaccharide using aerobic granular sludge, and they found 69% removal of methylene blue dye in 60 min at 35°C.

14.5.3 Removal of Heavy Metals

Different bacteria have been found to release a variety of extracellular polysaccharides (EPS) having the capacity to bind with cations by biosorption from wastewater (Choi and Yun, 2006; Dave

TABLE 14.3
Exopolysaccharides Producing Microbes and their Heavy Metal Removal Potential

Strain	Type of Wastewater/ Medium	Metals and their Concentration	Removal (%)/Uptake (mg/g)	References
Pseudomonas veronii 2E	Synthetic wastewater	1 mM, Cu	82.8%	Busnelli et al. (2021)
Bacillus licheniformis	Synthetic solution	25 mg/L, Cu, Zn	Cu-86%, Zn-81%	Biswas et al. (2020)
Enterobacter A47	Synthetic solution	10 mg/L, Pb	Pb^{2+} 93.9±5.3%	Concórdio-Reis et al. (2020)
Paenibacillus peoriae	Synthetic solution	Pb (100 mg/L)	277.54 mg/g	Fella-Temzi et al. (2018)
Bacillus and *sphaericus Rhizobium radiobacter*	Molasses wastewater	Cu, Zn, Cr	Cu, Zn- >90% Cr- 30%	Zhao et al. (2017)
Paenibacillus polymyxa.	Synthetic wastewater	0.5 mg/L Arsenate, 0.5 mg/L Arsenite	98.9% 84.6%	Zhao et al. (2016)
Bacillus licheniformis	Synthetic solution	100 mg/L Cr^{+6}	88%	Natarajan (2015)
Athelia rolfsii	Synthetic solution	Cd, Cu, Zn (100 mg/L)	Cd-116.28, Cu-103.09, Zn- 153.85 mg/g	Li et al. (2016)

et al., 2020). Chemical modifications such as methylation, acetylation, sulfonylation, and phosphorylation can enhance the biosorption ability of exopolysaccharides (Desbrieres et al., 2018). It has been reported that 1–10 mg/L of exopolysaccharide concentration can attain more than 50% of the removal of heavy metals (Siddharth et al., 2021). In a study, *Bacillus firmus* efficiently removed 98.3%, 74.9%, and 61.8% of Pb, Cu, and Zn respectively at pH 7 (Salehizadeh and Shojaosadati, 2003). Exopolysaccharides produced from *Paenibacillus peoriae* uptake 277.54 mg/g of 100 ppm Pb after 180 min at pH 6.8 (Fella-Temzi et al., 2018). However, after lead treatment, the densities of biofilm-linked polysaccharide constituents in *Pseudomonas aeruginosa* N6P6 were found to decrease, resulting in a lower biofilm density (Kumari and Das, 2019). High-volumetric-density biofilms typically have greater sorption capacities than low-density biofilms, distinguished by having lower levels of exopolysaccharide content (Kumari and Das, 2019). Abinaya -Sindu and Gautam (2017) found similar results with *P. aeruginosa* MTCC 2297 under Cd exposure. Exopolysaccharides macromolecules have negatively charged functional groups in their structures allowing them to cause electrostatic connections to bind the positively charged metal ions (Concórdio-Reis et al., 2020; Gupta and Diwan, 2017). The heavy metal removing exopolysaccharides producing microbes from different mediums are listed in Table 14.3.

14.5.4 Removal of Turbidity and Organic Content by Flocculation Activity of Exopolysaccharide

Wastewater from the paper, tannery, and textile industries solely accounts for approximately 20% of the world's water contamination (Hasanbeigi and Price, 2015. These industries' wastewater discharges contain a high quantity of suspended solids, which enhances turbidity. Flocculation is the most often used technique to enhance the effluent characteristics through ion bridging by multivalent cations such as Mg^{2+} and Ca^{2+} (Chen et al., 2022). Numerous Microbial strains including *Klebsiella terrigena, Pseudomonas plecoglossicida Bacillus subtilis, Pseudomonas pseudoalcaligenes, Staphylococcus aureus,* and *Exiguobacterium acetylium* have been used as natural bio-flocculant (Buthelezi et al., 2009; Nie et al., 2011; Nontembiso et al., 2011; Kanmani and Yuvapriya, 2018). *Paenibacillus sp.* M21629 produced exopolysaccharides, floccuronic acid has removed 99.8%, 98.8%, and 89.2%

turbidity from kaolin suspension, coal wastewater, high-turbidity, and drinking water, respectively. *Paenibacillus mucilaginosus* produced exopolysaccharide and removed 81.5%–88% of suspended solids from peppermill wastewater (Tang et al., 2014). A fungus, *Talaromyces trachyspermus* OU5 produced an exopolysaccharide that removed 75% of turbidity from swine wastewater (Fang and Shi, 2016). The exopolysaccharide-producing bacterial strains *Pseudomonas aeruginosa* and *putida* have shown the capability to eliminate organics from aqueous environments due to their excellent flocculation and biosorption properties (Wang et al., 2012). *Scenedesmus sp.* strain SD07 was grown in 75% diluted effluent, it produced biomass in high amount (1.93±0.10 g/L), alongwith 378 mg/L of exopolysaccharides. Moreover, it removes 96.32%, 93.26%, 91.36%, 88.41% of total phosphate, total nitrogen, chemical oxygen demand (COD), and ammonium, respectively from 75% diluted effluent (Silambarasan et al., 2023). These microbial species have been studied to examine the possible applications of their exopolysaccharides for the removal of turbidity (biological and non-biological) and organic content from wastewater. During the flocculation process, the reduction in protein existing on the surface of exopolysaccharide tends to disintegrate the formed aggregates.

14.6 FUTURE PERSPECTIVES AND CHALLENGES

It is critical to improving the identification of promising bacterial strains for developing exopolysaccharides for adsorption, dewatering, and flocculation during wastewater treatment. Furthermore, new extraction techniques should be established in combination with both physicochemical techniques to obtain improved efficacy without disrupting the structural organization of exopolysaccharides. Additionally, inexpensive nutrient sources should be selected for media to save costs. The use of exopolysaccharides stated in several studies is still limited, hence more studies are necessary to address the drawbacks of variations in EPS features in diverse field situations. To successfully commercialize exopolysaccharides, there is need to need to optimize several factors such as physio-chemical conditions of the microbial growth, contact time, dose, and other related factors. To prevent pathogenic pollution, studies on microbial growth are also needed. However, reusability and regeneration studies must be carried out. Furthermore, it is frequently challenging to enforce precise control of crucial elements on a commercial scale since numerous environmental regulating factors, such as pH, ionic strength and temperature can alter the efficiency of exopolysaccharides. Exopolysaccharides immobilized on appropriate carriers might be used as an alternative to successfully scale up wastewater treatment.

14.7 CONCLUSION

EPS are effective, environmentally benign, natural polymers developed by microbial populations such as bacteria, fungi, algae (alginate, carrageenan, and agar), and plants (cellulose, starch, and pectin). These are traditionally utilized in several areas such as food, paint, medicines, and cosmetics as a thickening, stabilizing, and texturizing agents. Various natural polymers used in wastewater treatment have the ability to combat drawbacks related to conventional methods. Exopolysaccharides are significant constituents of EPS. Exopolysaccharide generally released in adverse conditions to promote their survival in harsh environments. The use of exopolysaccharide for detoxification of pollutants and bioremediation of wastewater is free from the pathogenicity related to the usage of entire organisms as it is a non-living entity. From the above discussions, it has been proved that these can remediate all kinds of wastewater by eliminating various pollutants present in wastewater such as dye and heavy metals. The distinctive structural configurations of exopolysaccharides allow them to such pollutants in a cost-effective, convenient, and environment-friendly manner.

ACKNOWLEDGMENT

The authors are grateful to the Head of the Department of Environmental Science, Babasaheb Bhimrao Ambedkar University, Lucknow and UGC, New Delhi, India for furnishing the necessary facilities and granting a non-NET fellowship respectively to the first author.

REFERENCES

Abdel-Wahab, B. A., Abd El-Kareem, H. F., Alzamami, A., Fahmy, C. A., Elesawy, B. H., Mostafa Mahmoud, M., ... Saied, E. M. (2022). Novel exopolysaccharide from marine bacillus subtilis with broad potential biological activities: insights into antioxidant, anti-inflammatory, cytotoxicity, and anti-Alzheimer activity. *Metabolites*, *12*(8), 715.

Abinaya Sindu, P., & Gautam, P. (2017). Studies on the biofilm produced by *Pseudomonas aeruginosa* grown in different metal fatty acid salt media and its application in biodegradation of fatty acids and bioremediation of heavy metal ions. *Canadian Journal of Microbiology*, *63*(1), 61–73.

Alam, M. A., Wan, C., Guo, S. L., Zhao, X. Q., Huang, Z. Y., Yang, Y. L., Chang, J. S., & Bai, F. W. (2014). Characterization of the flocculating agent from the spontaneously flocculating microalga *Chlorella vulgaris* JSC-7. *Journal of Bioscience and Bioengineering*, *118*, 29–33.

Ateş Duru, O. (2021). General overview of bacterial exopolysaccharides focused on medical applications. *Academia Letters*, Article 405. https://doi.org/10.20935/AL405."

Bacosa, H. P., Kamalanathan, M., Chiu, M. H., Tsai, S. M., Sun, L., Labonté, J. M., & Quigg, A. (2018). Extracellular polymeric substances (EPS) producing and oil degrading bacteria isolated from the northern Gulf of Mexico. *PLoS One, 13*(12), e0208406.

Banerjee, A., Sarkar, S., Govil, T., González-Faune, P., Cabrera-Barjas, G., Bandopadhyay, R., ... Sani, R. K. (2021). Extremophilic exopolysaccharides: biotechnologies and wastewater remediation. *Frontiers in Microbiology*, *12*, 721365.

Biswas, J. K., Banerjee, A., Sarkar, B., Sarkar, D., Sarkar, S. K., Rai, M., & Vithanage, M. (2020). Exploration of an extracellular polymeric substance from earthworm gut bacterium (Bacillus licheniformis) for bioflocculation and heavy metal removal potential. *Applied Sciences*, *10*(1), 349.

Bratby, J. (2016). *Coagulation and Flocculation* in *Water* and *Wastewater Treatment*. IWA Publishing, London.

Busnelli, M. P., Lazzarini Behrmann, I. C., Ferreira, M. L., Candal, R. J., Ramirez, S. A., & Vullo, D. L. (2021). Metal-pseudomonas veronii 2E interactions as strategies for innovative process developments in environmental biotechnology. *Frontiers in Microbiology*, *12*, 622600.

Buthelezi, S. P., Olaniran, A. O., & Pillay, B. (2009). Turbidity and microbial load removal from river water using bioflocculants from indigenous bacteria isolated from wastewater in South Africa. *African Journal of Biotechnology*, *8*(14), 3261–3266.

Chen, S., Cheng, R., Xu, X., Kong, C., Wang, L., Fu, R., ... Zhang, J. (2022). The structure and flocculation characteristics of a novel exopolysaccharide from a Paenibacillus isolate. *Carbohydrate Polymers*, *291*, 119561.

Chen, Y., Mao, W., Tao, H., Zhu, W., Qi, X., Chen, Y., ... Li, N. (2011). Structural characterization and antioxidant properties of an exopolysaccharide produced by the mangrove endophytic fungus Aspergillus sp. Y16. *Bioresource Technology*, *102*(17), 8179–8184.

Cheng, J. J., Chang, C. C., Chao, C. H., & Lu, M. K. (2012). Characterization of fungal sulfated polysaccharides and their synergistic anticancer effects with doxorubicin. Carbohydrate *Polymers*, *90*(1), 134–139.

Choi, S. B., & Yun, Y. S. (2006). Biosorption of cadmium by various types of dried sludge: an equilibrium study and investigation of mechanisms. *Journal of Hazardous Materials*, *138*(2), 378–383.

Cimini, D., De Rosa, M., & Schiraldi, C. (2012). Production of glucuronic acid-based polysaccharides by microbial fermentation for biomedical applications. *Biotechnology Journal*, *7*(2), 237–250.

Concórdio-Reis, P., Reis, M. A., & Freitas, F. (2020). Biosorption of heavy metals by the bacterial exopolysaccharide FucoPol. *Applied Sciences*, *10*(19), 6708.

Costa, O. Y., Raaijmakers, J. M., & Kuramae, E. E. (2018). Microbial extracellular polymeric substances: ecological function and impact on soil aggregation. *Frontiers in Microbiology*, 9, 1636.

Dave, S. R., Upadhyay, K. H., Vaishnav, A. M., & Tipre, D. R. (2020). Exopolysaccharides from marine bacteria: production, recovery and applications. *Environmental Sustainability*, *3*, 139–154.

Dave, S. R., Vaishnav, A. M., Upadhyay, K. H., & Tipre, D. R. (2016). Microbial exopolysaccharide-an inevitable product for living beings and environment. *Journal of Bacteriology & Mycology: Open Access*, *2*(4), 109–111.

de Jesus, C. S., de Jesus Assis, D., Rodriguez, M. B., Menezes Filho, J. A., Costa, J. A. V., de Souza Ferreira, E., & Druzian, J. I. (2019). Pilot-scale isolation and characterization of extracellular polymeric substances (EPS) from cell-free medium of Spirulina sp. LEB-18 cultures under outdoor conditions. *International Journal of Biological Macromolecules*, *124*, 1106–1114.

de Sousa Rollemberg, S. L., de Barros, A. N., Lira, V. N. S. A., Firmino, P. I. M., & Dos Santos, A. B. (2019). Comparison of the dynamics, biokinetics and microbial diversity between activated sludge flocs and aerobic granular sludge. *Bioresource Technology*, 294, 122106.

Debnath, A., Das, B., Devi, M.S., & Ram, R.M. (2021). Fungal exopolysaccharides: types, production and application. In: Vaishnav, A., & Choudhary, D.K. (eds) *Microbial Polymers* (pp. 45–68). Springer, Singapore.

Dekker, R. F., Queiroz, E. A., Cunha, M. A., & Barbosa-Dekker, A. M. (2019). Botryosphaeran-A fungal exopolysaccharide of the (1→ 3)(1→ 6)-β-D-glucan kind: structure and biological functions. In: Cohen, E., & Merzendorfer, H. (eds) *Extracellular Sugar-Based Biopolymers Matrices. Biologically-Inspired Systems* (vol. 12, pp. 433–484). Springer, Cham.

Desai, K. M., Survase, S. A., Saudagar, P. S., Lele, S. S., & Singhal, R. S. (2008). Comparison of artificial neural network (ANN) and response surface methodology (RSM) in fermentation media optimization: case study of fermentative production of scleroglucan. *Biochemical Engineering Journal*, *41*(3), 266–273.

Desbrieres, J., Peptu, C. A., Savin, C. L., & Popa, M. (2018). Chemically modified polysaccharides with applications in nanomedicine. In: *Biomass as Renewable Raw Material to Obtain Bioproducts of High-Tech Value* (pp. 351–399). Elsevier, Amsterdam.

Dey, P., Gola, D., Mishra, A., Malik, A., Kumar, P., Singh, D. K., ... Jehmlich, N. (2016). Comparative performance evaluation of multi-metal resistant fungal strains for simultaneous removal of multiple hazardous metals. *Journal of Hazardous Materials*, *318*, 679–685.

Ding, Z., Bourven, I., Guibaud, G., van Hullebusch, E. D., Panico, A., Pirozzi, F., & Esposito, G. (2015). Role of extracellular polymeric substances (EPS) production in bioaggregation: application to wastewater treatment. *Applied Microbiology and Biotechnology*, *99*(23), 9883–9905.

Fang, D., & Shi, C. (2016). Characterization and flocculability of a novel proteoglycan produced by Talaromyces trachyspermus OU5. *Journal of Bioscience and Bioengineering*, *121*(1), 52–56.

Fazenda, M. L., Harvey, L. M., & McNeil, B. (2010). Effects of dissolved oxygen on fungal morphology and process rheology during fed-batch processing of Ganoderma lucidum. *Journal of Microbiology and Biotechnology*, *20*(4), 844–851.

Fella-Temzi, S., Yalaoui-Guellal, D., Rodriguez-Carvajal, M. A., Belhadi, D., Madani, K., & Kaci, Y. (2018). Removal of lead by exopolysaccharides from Paenibacillus peoriae strainTS7 isolated from rhizosphere of durum wheat. *Biocatalysis and Agricultural Biotechnology*, *16*, 425–432.

Freitas, F., Alves, V. D., Torres, C. A., Cruz, M., Sousa, I., Melo, M. J., ... Reis, M. A. (2011a). Fucose-containing exopolysaccharide produced by the newly isolated Enterobacter strain A47 DSM 23139. *Carbohydrate Polymers*, *83*(1), 159–165.

Gaignard, C., Laroche, C., Pierre, G., Dubessay, P., Delattre, C., Gardarin, C., ... Michaud, P. (2019). Screening of marine microalgae: investigation of new exopolysaccharide producers. *Algal Research*, *44*, 101711.

Galindo, E., Peña, C., Núñez, C., Segura, D., & Espín, G. (2007). Molecular and bioengineering strategies to improve alginate and polydydroxyalkanoate production by Azotobacter vinelandii. *Microbial Cell Factories*, *6*(1), 1–16.

Guetta, O., Mazeau, K., Auzely, R., Milas, M., & Rinaudo, M. (2003). Structure and properties of a bacterial polysaccharide named Fucogel. *Biomacromolecules, 4*(5), 1362–1371.

Guarin, T. C., & Pagilla, K. R. (2021). Microbial community in biofilters for water reuse applications: a critical review. *Science of the Total Environment*, *773*, 145655.

Guo, S. L., Zhao, X. Q., Wan, C., Huang, Z. Y., Yang, Y. L., Alam, M. A., ... Chang, J. S. (2013). Characterization of flocculating agent from the self-flocculating microalga Scenedesmus obliquus AS-6-1 for efficient biomass harvest. *Bioresource Technology*, *145*, 285–289.

Gupta, P., & Diwan, B. (2017). Bacterial exopolysaccharide mediated heavy metal removal: a review on biosynthesis, mechanism and remediation strategies. *Biotechnology Reports*, *13*, 58–71.

Guzmán, S., Gato, A., Lamela, M., Freire-Garabal, M., & Calleja, J. M. (2003). Anti-inflammatory and immunomodulatory activities of polysaccharide from Chlorella stigmatophora and Phaeodactylum tricornutum. *Phytotherapy Research*, *17*(6), 665–670.

Habib, B., Vaid, S., Bangotra, R., Sharma, S., & Bajaj, B. K. (2022). Bioprospecting of probiotic lactic acid bacteria for cholesterol lowering and exopolysaccharide producing potential. *Biologia*, *77*(7), 1931–1951.

Han, X., Yang, Z., Jing, X., Yu, P., Zhang, Y., Yi, H., & Zhang, L. (2016). Improvement of the texture of yogurt by use of exopolysaccharide producing lactic acid bacteria. *BioMed Research International*, 2016.

Harimawan, A., & Ting, Y.-P. (2016). Investigation of extracellular polymeric substances (EPS) properties of P. aeruginosa and B. subtilis and their role in bacterial adhesion. *Colloids and Surfaces B: Biointerfaces*, *146*, 459–467.

Hasanbeigi, A., & Price, L. (2015). A technical review of emerging technologies for energy and water efficiency and pollution reduction in the textile industry. *Journal of Cleaner Production*, *95*, 30–44.

Huang, L., Jin, Y., Zhou, D., Liu, L., Huang, S., Zhao, Y., & Chen, Y. (2022). A review of the role of extracellular polymeric substances (EPS) in wastewater treatment systems. *International Journal of Environmental Research and Public Health*, *19*(19), 12191.

Ibrahim, H. A., Abou Elhassayeb, H. E., & El-Sayed, W. M. (2022). P+otential functions and applications of diverse microbial exopolysaccharides in marine environments. *Journal of Genetic Engineering and Biotechnology*, *20*(1), 151.

Kanmani, P., & Yuvapriya, S. (2018). Exopolysaccharide from Bacillus sp. YP03: its properties and application as a flocculating agent in wastewater treatment. *International Journal of Environmental Science and Technology*, *15*, 2551–2560.

Karri, R. R., Ravindran, G., & Dehghani, M. H. (2021). Wastewater-sources, toxicity, and their consequences to human health. In: *Soft Computing Techniques in Solid Waste and Wastewater Management* (pp. 3–33). Elsevier.

Khan, I., Aftab, M., Shakir, S., Ali, M., Qayyum, S., Rehman, M. U., ... Touseef, I. (2019). Mycoremediation of heavy metal (Cd and Cr)-polluted soil through indigenous metallotolerant fungal isolates. *Environmental Monitoring and Assessment*, *191*(9), 1–11.

Kumari, S., & Das, S. (2019). Expression of metallothionein encoding gene bmtA in biofilm-forming marine bacterium Pseudomonas aeruginosa N6P6 and understanding its involvement in Pb (II) resistance and bioremediation. *Environmental Science and Pollution Research*, *26*(28), 28763–28774.

Kumar, S., & Dutta, V. (2019). Constructed wetland microcosms as sustainable technology for domestic wastewater treatment: an overview. *Environmental Science and Pollution Research*, *26*(12), 11662–11673.

Ladnorg, S., Junior, N. L., Dall, P., Domingos, D. G., Magnus, B. S., Wichern, M., ... da Costa, R. H. R. (2019). Alginate-like exopolysaccharide extracted from aerobic granular sludge as biosorbent for methylene blue: thermodynamic, kinetic and isotherm studies. *Journal of Environmental Chemical Engineering*, *7*(3), 103081.

Laspidou, C. S., & Rittmann, B. E. (2002). A unified theory for extracellular polymeric substances, soluble microbial products, and active and inert biomass. *Water Research*, *36*(11), 2711–2720.

Li, C., Chen, D., Ding, J., & Shi, Z. (2020). A novel hetero-exopolysaccharide for the adsorption of methylene blue from aqueous solutions: isotherm, kinetic, and mechanism studies. *Journal of Cleaner Production*, *265*, 121800.

Li, H., Wei, M., Min, W., Gao, Y., Liu, X., & Liu, J. (2016). Removal of heavy metal Ions in aqueous solution by Exopolysaccharides from Athelia rolfsii. *Biocatalysis and Agricultural Biotechnology*, *6*, 28–32.

Li, W. W., Zhang, H. L., Sheng, G. P., & Yu, H. Q. (2015). Roles of extracellular polymeric substances in enhanced biological phosphorus removal process. *Water Research*, *86*, 85–95.

Lynch, K. M., Zannini, E., Coffey, A., & Arendt, E. K. (2018). Lactic acid bacteria exopolysaccharides in foods and beverages: isolation, properties, characterization, and health benefits. *Annual Review of Food Science and Technology*, *9*, 155–176.

Mahapatra, S., & Banerjee, D. (2012). Structural elucidation and bioactivity of a novel exopolysaccharide from endophytic *Fusarium solani* SD5. *Carbohydrate Polymers*, *90*, 683–689.

Mahapatra, S., & Banerjee, D. (2013). Fungal exopolysaccharide: production, composition and applications. *Microbiology Insights*, *6*, MBI-S10957.

Mehta, K., Shukla, A., & Saraf, M. (2021). Articulating the exuberant intricacies of bacterial exopolysaccharides to purge environmental pollutants. *Heliyon*, *7*(11), e08446.

Mishra A, Kavita K, Jha B (2011) Characterization of extracellular polymeric substances produced by micro-algae Dunaliella salina. *Carbohydrate Polymers*, *83*, 852–857.

Mohamed, S. S., Amer, S. K., Selim, M. S., & Rifaat, H. M. (2018). Characterization and applications of exopolysaccharide produced by marine Bacillus altitudinis MSH2014 from Ras Mohamed, Sinai, Egypt. *Egyptian Journal of Basic and Applied Sciences*, *5*(3), 204–209.

Moreira, J. B., Kuntzler, S. G., Bezerra, P. Q. M., Cassuriaga, A. P. A., Zaparoli, M., da Silva, J. L. V., ... de Morais, M. G. (2022). Recent advances of microalgae exopolysaccharides for application as bioflocculants. *Polysaccharides*, *3*(1), 264–276.

Najdenski, H. M., Gigova, L. G., Iliev, I. I., Pilarski, P. S., Lukavský, J., Tsvetkova, I. V., ... Kussovski, V. K. (2013). Antibacterial and antifungal activities of selected microalgae and cyanobacteria. *International Journal of Food Science & Technology*, *48*(7), 1533–1540.

Natarajan, K. A. (2015). Production and characterization of bioflocculants for mineral processing applications. *International Journal of Mineral Processing*, *137*, 15–25.

Nie, M., Yin, X., Jia, J., Wang, Y., Liu, S., Shen, Q., ... Wang, Z. (2011). Production of a novel bioflocculant MNXY1 by Klebsiella pneumoniae strain NY1 and application in precipitation of cyanobacteria and municipal wastewater treatment. *Journal of Applied Microbiology*, *111*(3), 547–558.

Nontembiso, P., Sekelwa, C., Leonard, M. V., & Anthony, O. I. (2011). Assessment of bioflocculant production by Bacillus sp. Gilbert, a marine bacterium isolated from the bottom sediment of Algoa Bay. *Marine Drugs*, *9*(7), 1232–1242.

Öner, E. T., Hernández, L., & Combie, J. (2016). Review of levan polysaccharide: from a century of past experiences to future prospects. *Biotechnology Advances*, *34*(5), 827–844.

Orlandelli, R. C., da Silva, M. D. L. C., Vasconcelos, A. F. D., Almeida, I. V., Vicentini, V. E. P., Prieto, A., ... Pamphile, J. A. (2017). β-(1→3, 1→6)-D-glucans produced by Diaporthe sp. endophytes: purification, chemical characterization and antiproliferative activity against MCF-7 and HepG2-C3A cells. *International Journal of Biological Macromolecules*, *94*, 431–437.

Palaniraj, A., & Jayaraman, V. (2011). Production, recovery and applications of xanthan gum by *Xanthomonas campestris*. *Journal of Food Engineering*, *106*, 1–12.

Patel, A., & Prajapat, J. B. (2013). Food and health applications of exopolysaccharides produced by lactic acid bacteria. *Advances in Dairy Research, 1*(2), 1–7.

Prajapati, V. D., Jani, G. K., Zala, B. S., & Khutliwala, T. A. (2013). An insight into the emerging exopolysaccharide gellan gum as a novel polymer. *Carbohydrate Polymers*, *93*(2), 670–678.

Radchenkova, N., Vassilev, S., Martinov, M., Kuncheva, M., Panchev, I., Vlaev, S., & Kambourova, M. (2014). Optimization of the aeration and agitation speed of Aeribacillus palidus 418 exopolysaccharide production and the emulsifying properties of the product. *Process Biochemistry*, *49*(4), 576–582.

Raza, W., Makeen, K., Wang, Y., Xu, Y., & Qirong, S. (2011). Optimization, purification, characterization and antioxidant activity of an extracellular polysaccharide produced by Paenibacillus polymyxa SQR-21. *Bioresource Technology*, *102*(10), 6095–6103.

Salehizadeh, H., & Shojaosadati, S. A. (2003). Removal of metal ions from aqueous solution by polysaccharide produced from Bacillus firmus. *Water Research*, *37*(17), 4231–4235.

Santhanalakshmi, J., & Dhanalakshmi, V. (2012). Chitosan silver nanoparticles assisted oxidation of textile dyes with H2O2 aqueous solution: kinetic studies with pH and mass effect. *Indian Journal of Science and Technology*, *5*, 3834–3838.

Saravanan, C., Rajesh, R., Kaviarasan, T., Muthukumar, K., Kavitake, D., & Shetty, P. H. (2017). Synthesis of silver nanoparticles using bacterial exopolysaccharide and its application for degradation of azo-dyes. *Biotechnology Reports*, *15*, 33–40.

Shahnavaz, B., Karrabi, M., Maroof, S., & Mashreghi, M. (2015). Characterization and molecular identification of extracellular polymeric substance (EPS) producing bacteria from activated sludge. *Journal of Cell and Molecular Research*, *7*(2), 86–93.

Sheng, G. P., Yu, H. Q., & Li, X. Y. (2010). Extracellular polymeric substances (EPS) of microbial aggregates in biological wastewater treatment systems: a review. *Biotechnology Advances*, *28*(6), 882–894.

Siddharth, T., Sridhar, P., Vinila, V., & Tyagi, R. D. (2021). Environmental applications of microbial extracellular polymeric substance (EPS): a review. *Journal of Environmental Management*, *287*, 112307.

Silambarasan, S., Logeswari, P., Sivaramakrishnan, R., Incharoensakdi, A., Kamaraj, B., & Cornejo, P. (2023). Scenedesmus sp. strain SD07 cultivation in municipal wastewater for pollutant removal and production of lipid and exopolysaccharides. *Environmental Research*, *218*, 115051.

Sindu, P. A., & Gautam, P. (2017). Studies on the biofilm produced by Pseudomonas aeruginosa grown in different metal fatty acid salt media and its application in biodegradation of fatty acids and bioremediation of heavy metal ions. *Canadian Journal of Microbiology*, *63*(1), 61–74.

Soanen, N., Da Silva, E., Gardarin, C., Michaud, P., & Laroche, C. (2016). Improvement of exopolysaccharide production by *Porphyridium marinum*. *Bioresource Technology*, *213*, 231–238

Tang, Wei, Song, Liyan, Dou, Li, Qiao, Jing, Zhao, Tiantao, & Zhao, Heping (2014). 'Production, characterization, and flocculation mechanism of cation independent, pH tolerant, and thermally stable bioflocculant from Enterobacter sp. ETH-2. *PloS One*, *9*, e114591.

Tang, Y. J., Zhang, W., & Zhong, J. J. (2009). Performance analyses of a pH-shift and DOT-shift integrated fed-batch fermentation process for the production of ganoderic acid and Ganoderma polysaccharides by medicinal mushroom Ganoderma lucidum. *Bioresource Technology*, *100*(5), 1852–1859.

Taskin, M., Erdal, S., & Canli, O. (2010). Utilization of waste loquat (*Eriobotrya Japonica* Lindley) kernels as substrate for scleroglucan production by locally isolated *Sclerotium rolfsii*. *Food Science and Biotechnology*, *19*, 1069–1075.

Topal, M., & Arslan Topal, E.I. (2022). Extracellular polymeric substances in textile industry. In: Muthu, S. S. (ed) *Sustainable Approaches in Textiles and Fashion. Sustainable Textiles: Production, Processing, Manufacturing & Chemistry* (pp. 23–40). Springer, Singapore.,

Torres, C. A., Antunes, S., Ricardo, A. R., Grandfils, C., Alves, V. D., Freitas, F., & Reis, M. A. (2012). Study of the interactive effect of temperature and pH on exopolysaccharide production by Enterobacter A47 using multivariate statistical analysis. *Bioresource Technology*, *119*, 148–156.

Trivedi, R. (2020). Exopolysaccharides: Production and application in industrial wastewater treatment. In Shah, M., & Banerjee, A. (eds) *Combined Application of Physico-Chemical & Microbiological Processes for Industrial Effluent Treatment Plant* (pp. 15–27). Springer, Singapore.

Tyagi, B., Gupta, B. and Thakur, I. S. (2020). Biosorption of Cr (VI) fromnaqueous solution by extracellular polymeric substances (EPS) produced by Parapedobacter sp. ISTM3 strain isolated from Mawsmai cave, Meghalaya, India. *Environmental Research*, *191*, 110064.

Wan, C., Zhao, X. Q., Guo, S. L., Alam, M. A., & Bai, F. W. (2013). Bioflocculant production from Solibacillus silvestris W01 and its application in cost-effective harvest of marine microalga Nannochloropsis oceanica by flocculation. *Bioresource Technology*, *135*, 207–212.

Wan, W. A. A. Q. I., Latif, N. A., Harvey, L. M., & McNeil, B. (2016). Production of exopolysaccharide by *Ganoderma lucidum* in a repeated-batch fermentation. *Biocatalysis and Agricultural Biotechnology*, *6*, 91–101.

Wang, C., Fan, Q., Zhang, X., Lu, X., Xu, Y., Zhu, W., ... Hao, L. (2018). Isolation, characterization, and pharmaceutical applications of an exopolysaccharide from Aerococcus Uriaeequi. *Marine Drugs*, *16*(9), 337.

Wang, D., Chen, F., Wei, G., Jiang, M., & Dong, M. (2015). The mechanism of improved pullulan production by nitrogen limitation in batch culture of Aureobasidium pullulans. *Carbohydrate Polymers*, *127*, 325–331.

Wang, Z., Hessler, C. M., Xue, Z., & Seo, Y. (2012). The role of extracellular polymeric substances on the sorption of natural organic matter. *Water Research*, *46*(4), 1052–1060.

Wanner, J. (2021). The development in biological wastewater treatment over the last 50 years. *Water Science and Technology*, *84*(2), 274–283.

Wu, S., Lu, M., Chen, J., Fang, Y., Wu, L., Xu, Y., & Wang, S. (2016). Production of pullulan from raw potato starch hydrolysates by a new strain of Auerobasidium pullulans. *International Journal of Biological Macromolecules*, *82*, 740–743.

Yang, M., Zhu, Y., Li, Y., Bao, J., Fan, X., Qu, Y., Wang, Y., Hu, Z., & Li, Q. (2016). Production and optimization of curdlan produced by *Pseudomonas* sp. QL212. *The International Journal of Biological Macromolecules*, *89*, 25–34.

Zanna, S., Mercier, D., Gardin, E., Allion-Maurer, A., & Marcus, P. (2022). EPS for bacterial anti-adhesive properties investigated on a model metal surface. *Colloids and Surfaces B: Biointerfaces*, *213*, 112413.

Zhang, B. B., & Cheung, C. K. (2011). Use of stimulatory agents to enhance the production of bioactive exopolysaccharide from Pleurotus tuber-regium by submerged fermentation. *Journal of Agricultural and Food Chemistry*, *59*(4), 1210–1216.

Zhao, G., Ji, S., Sun, T., Ma, F., & Chen, Z. (2017). Production of bioflocculants prepared from wastewater supernatant of anaerobic co-digestion of corn straw and molasses wastewater treatment. *BioResources*, *12*(1), 1991–2003.

Zhao, H., Zhong, C., Chen, H., Yao, J., Tan, L., Zhang, Y., & Zhou, J. (2016). Production of bioflocculants prepared from formaldehyde wastewater for the potential removal of arsenic. *Journal of Environmental Management*, *172*, 71–76.

Zhao, J., Liu, S., Liu, N., Zhang, H., Zhou, Q., & Ge, F. (2019). Accelerated productions and physicochemical characterizations of different extracellular polymeric substances from Chlorella vulgaris with nano-ZnO. *Science of the Total Environment*, *658*, 582–589.

Zhao, Y., Liu, D., Huang, W., Yang, Y., Ji, M., Nghiem, L. D., & Tran, N. H. (2019). Insights into biofilm carriers for biological wastewater treatment processes: current state-of-the-art, challenges, and opportunities. *Bioresource Technology*, *288*, 121619.

Zhou, W., Shen, B., Meng, F., Liu, S., & Zhang, Y. (2010). Coagulation enhancement of exopolysaccharide secreted by an Antarctic sea-ice bacterium on dye wastewater. *Separation and Purification Technology*, *76*(2), 215–221.

15 Agro-industrial Waste-based Adsorbent for Wastewater Treatment
Current Pilot and Industrial-scale Applications

E. Parameswari, J. Dharani, T. Ilakiya, V. Davamani, P. Kalaiselvi, and S. Paul Sebastian

15.1 INTRODUCTION

Water contamination brought on by commercial, industrial, and municipal activity has become a serious issue for both people and the ecosystem (Chen et al., 2016). Over the years, numerous water purification methods have been suggested and put into practice (Chang et al., 2018). Because of its low cost, simplicity of operation, and ability to remove contaminants from wastewater and water, adsorption is currently the most effective method. Long-term use of activated carbon for water filtration has given way to the development of low-cost AIW (Agro-Industrial Waste)adsorbents such as fruits, rice husk, straw, coffee grounds, coconut trash, vegetable peels, sludges, bagasse, steel slag, etc (Bhatnagar and Anastopoulos, 2017). Numerous businesses and forms of agriculture generate a lot of garbage, which results in a variety of environmental problems (air, water, and soil) (Deng et al., 2015). Industrial wastes were waste residue, dust, and other waste discharged into the environment during the course of industrial production, whereas agricultural wastes were from agricultural production, agricultural product processing, livestock breedingand rural household life (Anastopoulos and Kyzas, 2014). If these wastes aren't treated properly, they could seriously harm the ecosystem. The majority of treatment techniques were pricy, complicated, and prone to secondary contamination. The pollution would be significantly reduced if these leftovers could be converted into useful materials (Li et al., 2018). AIWs are a resource for treating water and waste in the environment. It is crucial to know how to recycle and use them as resources.

15.2 INDUSTRIAL WASTE-DERIVED ADSORBENTS

Industrial waste, such as metal hydroxide sludge, red mud, and fly ash, has been prepared as adsorbents due to its low cost, wide distribution, and abundant supplies. The metal hydroxide sludge from industrial effluent from electroplating was used by Netpraditet al. (2004) to create adsorbents for the removal of azo-reactive colors. The greatest adsorption capacity for azo-reactive anionic dye was 48–62 mg/g for these adsorbents with a lot of metal hydroxides positive charge. According to Da Silva et al. (2018) the acetosolv treatment could transform discarded black acacia bark into a long-lasting adsorbent for color removal. Fly ash was another industrial waste with greater adsorption capabilities.

DOI: 10.1201/9781003441144-15

15.3 INDUSTRIAL AND AGRICULTURAL WASTES OR BY-PRODUCTS

Utilizing industrial and agricultural wastes and by-products is important for wastewater treatment as well as for the economy of the country. Despite the fact that they have little commercial worth and have disposal issues. In this situation, waste from the agricultural and industrial sectors can be employed as an inexpensive adsorbent that is readily available in huge numbers. They can be considered very good adsorbents and have the ability to remove dye from industrial effluent due to their physicochemical properties. These can also take the place of commercial activated carbon. Red mud, fly ash, and metal hydroxide sludge are examples of industrial waste products that are categorized as low-cost adsorbents for the remediation of pollutants like dyes (Ravalet al., 2017). The removal of dye from various agricultural wastes has been investigated under various operating circumstances. There are many types of agricultural waste-based adsorbents for eliminating water contaminants from industrial wastewater (Figure 15.1).

Aramiet al. (2005), Ahmad (2009), Gupta et al.(2007a), activated palm ash (Hameed et al., 2007), cattail root (Hu et al., 2010), neem sawdust (Khattri and Singh, 2009), peanut husk (Khalequeet al., 2018), Jujuba seeds and coir pith activated carbon are only (Santhy and Selvapathy, 2006). For the purpose of removing dye from industrial effluent, activated carbon can also be made from a variety of other agricultural wastes, including coconut tree sawdust, banana pith, silk cotton hull, maize cob, and sago waste (Kavipriyaet al., 2002). In Table 15.1, various adsorbent types utilizing agricultural wastes are covered.

Red mud, a by-product of the aluminum industry, is employed as a dye-removing adsorbent (Sharma and Kaur, 2018). Red mud is created during the caustic digestion of bauxite during the manufacturing of alumina. Rhodamine B, methylene blue, and fast green were successfully removed from industrial effluent by Gupta et al. (2007a, 2007b) using red mud under various pH, initial adsorbate concentration, time, adsorbent dose, temperature, and adsorbent particle size conditions. Red mud was also discovered by Bhatnagar et al. (2011) to be an efficient adsorbent with significant potential for adsorption for dye removal.

The other industrial by-product, fly ash, is a highly well-liked and successful adsorbent for dye removal. Bagasse-y ash from the sugar industry is extremely affordable and free of harmful metals (Sharma and Kaur, 2018). Gupta et al. (2000), used bagasse y ash as an adsorbent to remove basic dyes like rhodamine B and methylene blue. According to Gupta et al. (2000), the adsorption capacity was quite high and corresponded with both the Freundlich and Langmuir models. Ashes from thermal power plants were used to create activated carbon, which Mohan et al. (2002)

FIGURE 15.1 Different kinds of agricultural waste-based adsorbents for removing water pollutants from industrial wastewater.

TABLE 15.1
Different Types of Adsorbents Using Agricultural Wastes

S.No.	Adsorbent	Dye	Remarks	References
1.	Coniferous pinus bark powder	Crystal violet (basic dye)	Adsorption capacity increases from 80% to 99.5% at pH 2–8.	Ahmad (2009)
2.	Orange peel	Direct red 23 and direct red 80	Adsorption efficiency was 92% for direct red 23 and 91% for direct red 80 at 8 and 4 g/L concentration and maximum desorption at pH 2 was 97.7% for direct red 23% and 93% for direct red 80.	Arami et al. (2005)
3.	Rice husk	Methylene blue and Congo red	Removal efficiency was 99.939% for methylene blue and 98.835% for Congo red dye.	Chowdhury et al. (2009)
4.	Peanut hull particle	Methylene blue, brilliant cresyl blue, and neutral red	Removal efficiency increases for methylene blue from 34.75% to 89.51%, 36.82% to 96.62% for neutral red, and 79.14% to 96.47% for brilliant cresyl blue at 0.5–2.0 g/L dose	Gong et al. (2005)
5.	Activated palm ash	Acid green 25	Based on Langmuir model, the monolayer adsorption capacity around 123.4, 156.3, and 181.8 mg/g at temperature 30°C, 40°C, and 50°C.	Hameed et al. (2007)
6.	Cattail root	Congo red	The maximum adsorption capacities were 38.79, 34.59, and 30.61mg/g at temperature of 20°C, 30°C, and 40°C.	Hu et al. (2010)
7.	Peanut husk	Sunfix red (reactive dye)	At pH 7, the removal efficiency was 98% in 150 minutes	Khaleque et al. (2018)
8.	Neem sawdust	Malachite green	Abundantly available and also inexpensive. Its binding capacity for basic dye is also high. The maximum capacity for monolayer adsorption was 4.354mg/g	Khattri and Singh (2009)
9.	Coir pith	Reactive dyes	Maximum amount of dye is removed at acidic pH. Highest was reactive orange followed by reactive red and reactive blue was lowest.	Santhy and Selvapathy (2006)
10.	Jujuba seeds	Congo red	Maximum color removal at pH2. Fitted with the Langmuir model with maximum adsorption capacity of 55.56 mg/g	Somasekhara Reddy et al. (2012)

utilized to remove dyes from synthetic wastewater. Crystal violet and rosaniline hydrochloride were successfully eliminated by the product from the effluent. Both dyes have excellent adsorption capabilities (Mohan et al., 2002).

To remove dye from industrial wastewater, metal hydroxide sludge from various industries is employed as an inexpensive adsorbent. The primary components of metal hydroxide sludge aremetal hydroxide ions and its salts (Sharma and Kaur, 2018). In their 2003 study, Netpraditet al. tested the effectiveness of adsorption in removing azo dyes from industrial effluent. They performed experiments on anionic dyes and discovered that for colors with higher charge values, adsorption is greater than 90%. Smaller particle size and quantity speed up the treatment procedure (Netpradit et al., 2004).

Another waste product from the timber sector, saw dust, can be utilized as a cheap adsorbent for the removal of dyes. Garg et al. (2003) conducted an experiment to remove malachite green using sawdust,which has been formaldehyde and sulfuric acid treatment. The product is quite successful, and they also discovered that sawdust treated with sulfuric acid has a substantially greater adsorption efficiency than sawdust treated with formaldehyde (Garg et al., 2003).

15.3.1 Peat

Peats with a high porosity were mostly made up of organic matter from the soil. Peats are classified into four types, according to various sources: moss peat, woody peat, herbaceous peat, and sedimentary peat (Lin et al., 2018). It was abundant, inexpensive, and frequently utilized, and it had a high adsorption capacity for a wide range of contaminants, including heavy metals and organic pollutants (Chwastowskiet al., 2017). Unprocessed peat included lignin, cellulose, fulvic acid, and humic acid, as well as polar functional groups such as alcohols, aldehydes, ketones, carboxylic acids, and phenolic hydroxides (Zehra et al., 2016). Peat also has a high cation exchange capacity. The removal effectiveness of acidic and basic dyes was compared to that of activated carbon, and the behavior of adsorptive dyes was better.

15.3.2 Biosorbents

Biosorption is a technique that uses biological materials such as chitin, chitosan, yeasts, fungus, and bacteria to remove contaminants from solutions by chelation and complexation. These biosorption materials and their derivatives, which include diverse functional groups, have the potential to mix dye. Biosorption materials outperformed standard ion exchange resins and commercial activated carbon in terms of selectivity, with pollutant concentrations lowered to ppb levels. Biosorption was a low-cost, effective method of eliminating contaminants.

15.3.3 Biomass

Researchers have long been interested in using white rot fungi, biomass (dead or living body), and other microbes to remove and adsorb contaminants from dye effluent. Some biomasses have a high affinity for specific contaminants. Because of the cost savings and adsorption capacity, there was an increase in the number of applications for treating wastewater using biomass (Salvi and Chattopadhyay, 2017). A vast range of by-products of microbial fermentation were used to absorb contaminants in the synthesis of antibiotics, enzymes, and other biological products (Da Fontouraet al., 2017). Algae, yeast, bacteria, and fungi, for example, possessed great adsorption abilities(Geethakarthi and Phanikumar, 2012). Furthermore, laccase was employed to immobilize on AIW (Agro-Industrial Waste)to increase pollutant removal. Laccase was immobilized on biochar microparticles made from pine wood (BC-PW), pig manure (BC-PM), and almond shell (BC-AS). At an environmentally relevant concentration (500 g/L), the micropollutant diclofenac was degraded and nearly completely removed in 5 hours (Dhillon et al., 2013).

The primary benefits of bio-adsorption are low cost and high removal rate, particularly for extremely hazardous wastewater. Due to the low cost of producing fungal biomass, it is also viable to use a relatively basic fermentation method and an inexpensive culture substrate. Furthermore, biosorption is a developing method that has overcome the limitations of selectivity in classic adsorption methods. In fact, there are some disadvantages to bio-adsorption. For starters, the adsorption process is slow. Second, the adsorption impact was clearly influenced by pH. Third, fungal organisms' functional groups and surface characteristics influenced adsorption. Indeed, extrinsic factors such as competing salt and ion concentration have an impact on its adsorption performance. Furthermore, biomasses can quickly clog the adsorption column.

15.4 MECHANISM OF BIOSORPTION

The mechanism by which adsorbents remove the pollutant is vital as it decides the removal and recovery of pollutants from the adsorbent(Mishra and Doble, 2008). There are many ways by which adsorbents uptake the pollutant, but they are not fully known. Several Factors like biomass nature (living/dead), adsorbate chemistry and properties and environmental conditions like pH, temperature etc.

From the aforesaid figure, it is clear that biosorption by dead biomass takes place throughthe process of physical adsorption, complexation, precipitation, and ion exchange. These mechanisms can take place simultaneously. The adsorption by physical method occurs with the aid of weaker Vanderwaal's force of attraction, and electrostatic interaction whereas the ion exchange involves the exchange of pollutant ions with counter ions (Muraleedharan and Venkobachar, 1990). The process of precipitation occurs as the outcome of the chemical interaction between the pollutant ion and the adsorbent surface. Aksuet al.(1992) reported that complexation takes place by the formation of coordination bonds between the pollutant and functional groups like amino groups of adsorbent.

15.5 FACTORS AFFECTING METAL BIOSORPTION

The study on the efficiency of biosorbent on metal sorption is absolutely essential for applicability in industrial aspects i.e., design of the equipment. Some of the elements thataffect the process of biosorption include temperature, pH, biosorbent dosage, and metal ions for active sites available on the surface of biosorbent. To tap the fullest potential of the biosorbent, it is essential to determine the optimum process parameters for metal adsorption which is better initiated through the present investigation.

15.5.1 Solution pH

In the process of biosorption, the pH of the solution has its impact in two ways: (i) metal insolubility, (ii) charge of biosorbent as the protons can be adsorbed or released (Abdel-Ghani et al., 2015). According to Romeraet al.(2007), the state of system's equilibrium depends on the solution pH and it is represented by the equation:

$$pK_a - pH = \log\left(\frac{[A-H]}{[A-]}\right)$$

If the pK_a value is greater than pH, the equilibrium shifts to the left with the consumption of protons till the value of pH equals pK_a. For pH values higher than pK_a, the contrary will occur (Guptaet al., 2009).In the medium with extremely high acidity ($pH \approx 2$), thereis a negligible removal of metal ions. This is due to the fact that H^+ ions compete with metal ions for active sites available on the surface of biosorbent. With the improvement in solution's pH upto certain level, the removal of heavy metal increases. Minimum biosorption at low solution pH is due to higherH^+ ion mobility, concentration, and its preferential sorption than other metal ions.When the pH is higher, the amount of H^+ ions will be reduced and negatively charged ligands will be increased resulting in a greater possibility of biosorption.

With reference to the role of solution's pH on the charge of the biosorbent surface, the different functional groups present in the adsorbent should be considered. The main functional groups usually found on thesurface of adsorbent include carbonyl, sulfhydryl, hydroxyl, and amino groups (Sonawane and Nimse, 2016). On increasing the solution's pH, deprotonation of functional groups occurs making the surface of the biosorbent a negatively charged moiety thatattracts the positively charged metal ion (Farooqet al., 2010). On the other hand, with the reduction in pH, the charge of the biosorbent surface will become positive making the positively charged metal ion to repel and favors the attraction of negatively charged ion (Guptaet al., 2009).

15.5.2 Adsorbate Concentration

The initial concentration of adsorbate ions in the solution helps in overcoming the mass transfer resistance between the two phases (solid and solution) (Danget al., 2009). It is generally agreed that biosorption capacity increases as the concentration of adsorbate ions increases (Pahlavanzadeh et al., 2010) but the removal efficiency decreases on increasing the initial metal ion concentration. When the concentration of sorbate is less in the solution, there will be 100% adsorption whereas, at higher concentrations, more ions are left unadsorbed (Naiyaet al., 2009).

15.5.3 Dosage of Adsorbent

The sorption of adsorbate ions on the adsorbent depends on the binding sites present and hence its dosage strongly affects the metal ions sorption(Nethajiet al., 2013). With the fixed adsorbate ion concentration, increase in adsorbent dosage contributes to a greater surface area and increases the binding sites available thereby enhancing the ion uptake compared to a lower dosage which provides lower active binding sites to adsorb a similar concentration of adsorbate ions (Chonget al., 2013). The adsorption capacity is reduced even in the increased adsorbent dosage which might be due to lower adsorbate to binding site ratio.According to Veneuet al.(2013), the optimum adsorbent dosage is defined as the lowest quantity of adsorbent that gives reasonable removal efficiency.

15.5.4 Adsorbent Size

With reduced adsorbent size, the surface area gets increased and greater numbers of active binding sites on the adsorbent are better exposed to adsorbate (Ahmadet al., 2005). The reduction in removal capacity with the increase in adsorbent size gives an idea about the porosity of the sorbent (Khazaeiet al., 2011).

15.5.5 Contact Time

Determination of optimum contact time is a key factor in biosorption experiments. The removal percent of the metal is higher in the beginning due to the greater adsorbentsurface area available for metal sorption which slows down gradually with time until it reaches equilibrium (Abdel-Ghaniet al., 2009). This equilibrium time is one of the important parameters in selecting a water treatment system.

Apart from these above-mentioned factors, parameters like temperature and agitation speed also affect the efficiency of biosorption process.

15.6 ISOTHERMS FOR ADSORPTION

To enhance the adsorption process, isotherms must be studied well. These equilibrium relationships are essential for the assessment of effective adsorption system and how the adsorbate interacts with the surface of adsorbents. As a result, isotherm explains the mobility of heavy metal ions from aqueous solution into the porous solid adsorbent. It also describes the degree of affinity, surface properties, and capacity of the adsorbent used. The isotherms are classified as one, two, three, four, and five parameter isotherms. One parameter isotherm is the simplest isotherm model where the adsorbate is directly proportional to the partial pressure of the gas. The most commonly used isotherm models are Langmuir and Freundlich isotherms (Dula et al., 2014).

Langmuir isotherm is used to quantify the adsorption potential of various adsorbents though it was initiated for gas-solid phase. It is a monolayer adsorption that occurs at a fixed number of sites. It follows a dynamic equilibrium thatbalances the rate of adsorption and desorption. This theory has arrived that the rise of distance results in a rapid decrease of the intermolecular forces. In contrast, Freundlich isotherm is based on the multilayer adsorption between the liquid and solid phase. It occurs on a heterogeneous. In this model, stronger binding sites are filled first and then the energy of adsorption decreases exponentially. Temkin isotherm takes into account the indirect interactions of the adsorbate and adsorbent. Increase in surface coverage leads to decrease in the heat of the adsorption of molecules. Though nonlinear regression analysis is used by the researchers, linear analysis is frequently used for analyzing adsorption performance. Therefore, the successful adsorption isotherm modeling influences the accuracy of the sorption process (Ayaweiet al., 2017).

15.7 KINETIC MODELS FOR ADSORPTION

Kinetic study is used to express the rate of chemical processes that govern the residence time in the adsorption process. It is used to determine the performance and mechanism of adsorption. The rate of kinetics depends on the species concentration and the law is expressed as follows (Gupta and Bhattacharyya, 2011)

$$R = K\,[A]^a\,[B]^b \ldots$$

where K is the kinetics coefficient, A, B.... etc. are the species involved in adsorption which have the order a,b...etc.

The rate-limiting step and kinetics have been portrayed by different kinetic models. They are pseudo first-order, pseudo second-order, Adam-Bohart-Thomas model, Elovich's model, Ritchie's equation, intra particle diffusion, and external mass transfer model (Febriantoet al., 2009). Pseudo first-order rate equation was first presented by Lagergren in 1898. He explained the oxalic and malonic acid adsorption by the charcoal. i.e., liquid-solid phase adsorption. Though it is the earliest model it was found that it is not suitable for whole contact time whereas appreciable only for about initial 20–30 minutes. Pseudo second-order kinetic model was introduced by Ho which involves the divalent metal ion adsorption on peat. This model is interpreted as a special kind of Langmuir kinetics that gives two assumptions. They are: (i) amount of ion species determines the total number of sorption sites at equilibrium; (ii) ion concentration is constant with time. Intraparticle diffusion model is found to be a suitable model in case of diffusion of ions into the pores for porous substance. It is a rate-limiting factor but not for the whole process where other models control the rate simultaneously. Homogeneous Solid Diffusion Model (HSDM) is the typical intraparticle diffusion model. Elovich rate equation was initially introduced by Zledowitsch for chemisorption of gases onto solids whereas it has expanded to the aqueous phase adsorption. Elovich model states that adsorption takes place on particular sites and the energy is directly proportional to surface coverage in adsorption.

15.8 REMEDIATION OF DYES FROM INDUSTRIAL WASTEWATER

Due to the extensive use of dye to color products in numerous industries, including pigments, leather, and textiles, a significant quantity of dyes are released through the effluent, which is intrinsically hazardous to both human health and the environment. As a result, dye removal from industrial wastewater has recently attracted a lot of attention. There are several conventional therapeutic modalities in use, including physical, chemical, and biological ones. Adsorption holds a prominent position among treatment technologies due to its low cost and efficiency in removing dyes. Therefore, the growing need for affordable and efficient treatment methods drives research and development of more cost-effective, high-efficiency adsorbents for dye removal from industrial effluent. The various low-cost adsorbents, including naturally occurring materials, industrial and agricultural waste or by-products, as well as synthetic products made from low-cost materials, and their uses in the treatment of water, have therefore been compiled here. There have been studies done on these adsorbents to remove different kinds of dyes, ranging from 80% to 99.9%.

15.9 CHALLENGES

Because of the inherent deficiency in raw AIWs(Agro-Industrial Waste), waste-derived adsorbents must be modified via physical or chemical procedures. These changes may result in an increase in the cost of preparing for their environmental application. The functional groups and fractions influence whether AIW may be employed to make a suitable adsorbent (Li et al., 2018). Because cellulose, lignin, and hemicelluloses offer the primary adsorption area for pollutants such as organic and inorganic pollutants, their quantities in AIW determine the adsorption rate.

15.10 PERSPECTIVES FOR FUTURE RESEARCH

AIWs (Agro-Industrial Waste) have been frequently used to create waste-derived adsorbents. These waste-derived adsorbents have produced good results for the removal of both heavy metals and organic contaminants. The majority of these studies concentrate on the structure and composition of waste-derived adsorbents, as well as the adsorption mechanism and surface modification by physical or chemical techniques. However, utilizing these works to promote its further development is problematic. The intended research keystones are the practicability and industrialization of waste-derived adsorbents. To determine the application potential of AIWs-derived adsorbents, actual wastewaters should be done. Apart from conventional organic pollutants, more AIW adsorbents are being investigated for the removal of other developing contaminants such as endocrine disruptors, medicines, and radionuclides. Furthermore, the reusability of AIWs-derived adsorbents is a hurdle for their application value. Adsorption capacity recovery for saturated AIW adsorbents could increase their economic feasibility. Some pollutant-laden wastes are produced during the adsorption process, and their subsequent treatment is a major environmental concern. Magnetic adsorbents are extremely beneficial for recovering adsorbents and controlling the adsorption process. Magnetic and AIW adsorbents can be integrated in future study, making AIW adsorbents more controlled and practical. Furthermore, to improve treatment efficiency, combining techniques for adsorption and other water treatment technologies should be researched.

15.11 CONCLUSION

Therefore, a wide range of affordable adsorbents for dye extraction from industrial effluent have been provided in the book chapter. They can be obtained from a variety of sources, including abundant natural resources, the use of diverse industrial and agricultural waste products and biosorbents, and the synthesis of novel goods employing reasonably priced dye removal ingredients. The low-cost adsorbents have various benefits, including the fact that they are made from waste materials with no commercial value and no disposal issues. The usage of natural resources has also been done without or with minimal pretreatment. The inexpensive adsorbents are shown to be very promising and highly efficient at removing different kinds of dye. Certain synthetic adsorbers that are inexpensive do better than commercial activated carbon at removing dye. At pilot and industrial stages, the adsorption process is still not fully expanded. Therefore, much work is needed to (i) predict the performance of adsorption under certain operating conditions, (ii) gain a better understanding of the mechanism of adsorption, (iii) develop more low-cost adsorbents with high efficiency, and (iv) use low-cost adsorbents in dye removal at pilot and industrial scales.

REFERENCES

Abdel-Ghani, N.T., El-Chaghaby, G.A., & Helal, F.S. (2015). Individual and competitive adsorption of phenol and nickel onto multiwalled carbon nanotubes. *Journal of Advanced Research, 6*(3), 405–415

Abdel-Ghani, N., Hegazy, A., & El-Chaghaby, G. (2009). *Typha domingensis* leaf powder for decontamination of aluminium, iron, zinc and lead: biosorption kinetics and equilibrium modeling. *International Journal of Environmental Science & Technology, 6*(2), 243–248.

Ahmad, R. (2009). Studies on adsorption of crystal violet dye from aqueous solution onto coniferous pinus bark powder (CPBP). *Journal of Hazardous Materials, 171*, 767–773. https://doi.org/10.1016/j.jhazmat.2009.06.060

Ahmad, A., Sumathi, S., & Hameed, B. (2005). Residual oil and suspended solid removal using natural adsorbents chitosan, bentonite and activated carbon: a comparative study. *Chemical Engineering Journal, 108*(1–2), 179–185.

Aksu, Z., Sag, Y., & Kutsal, T. (1992). The biosorption of copper by C. vulgaris and Z. ramigera. *Environmental Technology, 13*(6), 579–586.

Anastopoulos, I., & Kyzas, G.Z. (2014). Agricultural peels for dye adsorption: a review of recent literature. *Journal of Molecular Liquids, 200*, 381–389.

Arami, M., Limaee, N.Y., Mahmoodi, N.M., & Tabrizi, N.S. (2005). Removal of dyes from colored textile wastewater by orange peel adsorbent: equilibrium and kinetic studies. *Journal of Colloid and Interface Science, 288*, 371–376. https://doi.org/10.1016/j.jcis.2005.03.020

Ayawei, N., Ebelegi, A.N.,& Wankasi, D. (2017). Modelling and interpretation of adsorption isotherms. *Journal of Chemistry*, Article ID 3039817, Page 11. doi.org/10.1155/2017/3039817

Bhatnagar, A., & Anastopoulos, I. (2017). Adsorptive removal of bisphenol A (BPA) from aqueous solution: a review. *Chemosphere, 168*, 885–902.

Bhatnagar, A., Vilar, V.J.P., Botelho, C.M.S., & Boaventura, R.A.R. (2011). A review of the use of red mud as adsorbent for the removal of toxic pollutants from water and wastewater. *Environmental Technology, 32*, 231–249. https://doi.org/10.1080/09593330.2011.560615

Chang, H., Quan, X., Zhong, N., Zhang, Z., Lu, Z., Li, C., Cheng, Z., & Yang, L. (2018). Highefficiency nutrients reclamation from landfill leachate by microalgae *Chlorella vulgaris* in membrane photobioreactor for bio-lipid production. *Bioresource Technology, 266*, 374–381.

Chen, Y.J., Wang, F.H., Duan, L.C., Yang, H., & Gao, J. (2016). Tetracycline adsorption onto rice husk ash, an agricultural waste: its kinetic and thermodynamic studies. *Journal of Molecular Liquids, 222*, 487–494.

Chong, H., Chia, P., & Ahmad, M. (2013). The adsorption of heavy metal by Bornean oil palm shell and its potential application as constructed wetland media. *Bioresource Technology, 130*, 181–186.

Chowdhury, A.K., Sarkar, A.D., & Bandyopadhyay, A. (2009). Rice husk ash as a low cost adsorbent for the removal of methylene blue and Congo red in aqueous phases. *Clean (Weinh), 37*(7), 581–591. https://doi.org/10.1002/clen.200900051

Chwastowski, J., Staroń, P., Kołoczek, H., & Banach, M. (2017). Adsorption of hexavalent chromium from aqueous solutions using Canadian peat and coconut fiber. *Journal of Molecular Liquids, 248*, 981–989.

Da Fontoura, J.T., Rolim, G.S., Mella, B., Farenzena, M., & Gutterres, M. (2017). Defatted microalgal biomass as biosorbent for the removal of Acid Blue 161 dye from tannery effluent. *Journal of Environmental Chemical Engineering, 5*, 5076–5084.

Da Silva, J.S., Da Rosa, M.P., Beck, P.H., Peres, E.C., Dotto, G.L., Kessler, F., & Grasel, F.S. (2018). Preparation of an alternative adsorbent from *Acacia mearnsii* wastes through acetosolv method and its application for dye removal. *Journal of Cleaner Production, 180*, 386–394.

Dang, V., Doan, H., Dang-Vu, T., & Lohi, A. (2009). Equilibrium and kinetics of biosorption of cadmium (II) and copper (II) ions by wheat straw. *Bioresource Technology, 100*(1), 211–219.

Deng, S.B., Yao, N., Du, Z.W., Qian, H., Meng, P.P., Wang, B., Huang, J., & Gang, Y. (2015). Enhanced adsorption of perfluorooctane sulfonate and perfluorooctanoate by bamboo-derived granular activated carbon. *Journal of Hazardous Materials, 282*, 150–157.

Dhillon, G., Dhillon, S., Brar, SK. (2013). Perspective of apple processing wastes as low-cost substrates for bioproduction of high value products: A review. *Renewable and Sustainable Energy Reviews, 27*. 10.1016/j.rser.2013.06.046.

Dula, T., Siraj, K., & Kitte, S.A. (2014). Adsorption of hexavalent chromium from aqueous solution using chemically activated carbon prepared from locally available waste of bamboo (*Oxytenanthera abyssinica*). *International Scholarly Research Notices*, Article ID 438245, page 9. doi.org/10.1155/2014/438245

Farooq, U., Kozinski, J.A., Khan, M.A., & Athar, M. (2010). Biosorption of heavy metal ions using wheat based biosorbents-A review of the recent literature. *Bioresource Technology, 101*(14), 5043–5053.

Febrianto, J., Kosasih, A.N., Sunarso, J., Ju, Y.-H.,Indraswati, N.,& Ismadji, S. (2009). Equilibrium and kinetic studies in adsorption of heavy metals using biosorbent: a summary of recent studies. *Journal of Hazardous Materials, 162*(2-3), 616–645.

Garg, V.K., Gupta, R., Yadav, A.B., & Kumar, R. (2003). Dye removal from aqueous solution by adsorption on treated sawdust. *Bioresource Technology, 89*, (2), 121–124. https://doi.org/10.1016/S0960-8524(03)00058-0

Geethakarthi, A., & Phanikumar, B.R. (2012). Characterization of tannery sludge activated carbon and its utilization in the removal of azo reactive dye. *Environmental Science and Pollution Research,19*, 656–665.

Gong, R., Li, M., Yang, C., Sun, Y., & Chen, J. (2005). Removal of cationic dyes from aqueous solution by adsorption on peanut hull. *Journal of Hazardous Materials, 121*, 247–250. https://doi.org/10.1016/j.jhazmat.2005.01.029

Gupta, M., Singh, A., & Srivastava, R. (2009). Kinetic sorption studies of heavy metal contamination on Indian expansive soil. *Journal of Chemistry, 6*(4), 1125–1132.

Gupta, S.S., & Bhattacharyya, K.G. (2011). Kinetics of adsorption of metal ions on inorganic materials: a review. *Advances in Colloid and Interface Science*, *162*(1–2), 39–58.

Gupta, V.K., Jain, R., & Varshney, S. (2007a). Removal of Reactofix golden yellow 3 RFN from aqueous solution using wheat husk-an agricultural waste. *Journal of Hazardous Materials, 142*, 443–448. https://doi.org/10.1016/j.jhazmat.2006.08.048

Gupta, V.K., Mohan, D., Sharma, S., & Sharma, M. (2000). Removal of basic dyes (rhodamine B and methylene blue) from aqueous solutions using bagasse fly ash. *Separation Science and Technology, 35*, 2097–2113. https://doi.org/10.1081/SS-100102091

Gupta, V.K., Suhas, Ali, I., Saini, V.K. (2007b). Removal of Rhodamine B, fast green, and methylene blue from wastewater using red mud, an aluminum industry waste. *Industrial & Engineering Chemistry Research, 43*, 1740–1747. https://doi.org/10.1021/ie034218g

Hameed, B.H., Ahmad, A.A., & Aziz, N. (2007). Isotherms, kinetics and thermodynamics of acid dye adsorption on activated palm ash. *Chemical Engineering Journal*, *133*, 195–203. https://doi.org/10.1016/j.cej.2007.01.032

Hu, Z., Chen, H., Ji, F., & Yuan, S. (2010). Removal of Congo red from aqueous solution by cattail root. *Journal of Hazardous Materials, 173*, 292–297. https://doi.org/10.1016/j.jhazmat.2009.08.082

Kadirvelu, K., Kavipriya, M., Karthika, C., Radhika, M., Vennilamani, N., & Pattabhi, S. (2002). Utilization of various agricultural wastes for activated carbon preparation and application for the removal of dyes and metal ions from aqueous solutions. *Bioresource Technology, 87*, 129–132. https://doi.org/10.1016/S0960-8524(02)00201-8

Khaleque, M.A., Ahammed, S.S., Khan, S., Islam, M.E., & Alam, M.S. (2018). Efficient removal of reactive dyes from industrial wastewater by peanut husk. *IOSR Journal of Environmental Science, Toxicology and Food Technology, 12*, 24–28. https://DOI: 10.9790/2402-1209012428

Khattri, S.D., & Singh, M.K. (2009). Removal of malachite green from dye wastewater using neem sawdust by adsorption. *Journal of Hazardous Materials*, *167*, 1089–1094. https://doi.org/10.1016/j.jhazmat.2009.01.101

Khazaei, Y., Faghihian, H., & Kamali, M. (2011). Removal of thorium from aqueous solutions by sodium clinoptilolite. *Journal of Radioanalytical and Nuclear Chemistry, 289*(2), 529–536.

Li, Y.F., Wang, L.A., Zhang, Z.E., Hu, X.Y., Cheng, Y., & Zhong, C. (2018). Carbon dioxide absorption from biogas by amino acid salt promoted potassium carbonate solutions in a hollow fiber membrane contactor: a numerical study. *Energy Fuel, 32*(3), 3637–3646.

Lin, J., Chen, X., Chen, C., Hu, J., Zhou, C., Cai, X., Wang, W., Zheng, C., Zhang, P., & Cheng, J. (2018). Durably antibacterial and bacterially anti-adhesive cotton fabrics coated by cationic fluorinated polymers. *ACS Applied Materials & Interfaces,10*, 6124–6136.

Mishra, S., & Doble, M. (2008). Novel chromium tolerant microorganisms: isolation, characterization and their biosorption capacity. *Ecotoxicology and Environmental Safety, 71*(3), 874–879.

Mohan, D., Singh, K.P., Singh, G., & Kumar, K. (2002). Removal of dyes from wastewater using Flyash, a low-cost adsorbent. *Industrial & Engineering Chemistry Research, 41*, 3688–3695. https://doi.org/10.1021/ie010667+

Muraleedharan, T., & Venkobachar, C. (1990). Mechanism of biosorption of Copper (Ii) by ganoderma iucidum. *Biotechnology and Bioengineering, 35*(3), 320–325.

Naiya, T.K., Bhattacharya, A.K., & Das, S.K. (2009). Adsorption of Cd (II) and Pb (II) from aqueous solutions on activated alumina. *Journal of Colloid and Interface Science, 333*(1), 14–26

Nethaji, S., Sivasamy, A., & Mandal, A. (2013). Adsorption isotherms, kinetics and mechanism for the adsorption of cationic and anionic dyes onto carbonaceous particles prepared from Juglans regia shell biomass. *International Journal of Environmental Science and Technology, 10*(2), 231–242.

Netpradit, S., Thiravetyan, P., Towprayoon, S. (2004). Adsorption of three azo reactive dyes by metal hydroxide sludge: effect of temperature, pH, and electrolytes. *Journal of Colloid and Interface Science*, *270*, 255–261.

Pahlavanzadeh, H., Keshtkar, A., Safdari, J., & Abadi, Z. (2010). Biosorption of nickel (II) from aqueous solution by brown algae: equilibrium, dynamic and thermodynamic studies. *Journal of Hazardous Materials, 175*(1-3), 304–310.

Raval, N.P., Shah, P.U., & Shah, N.K. (2017). Malachite green "a cationic dye" and its removal from aqueous solution by adsorption. *Applied Water Science*, *7*, 3407–3445. https://doi.org/10.1007/s13201-016-0512-2

Romera, E., González, F., Ballester, A., Blázquez, M., & Munoz, J. (2007). Comparative study of biosorption of heavy metals using different types of algae. *Bioresource Technology, 98*(17), 3344–3353.

Salvi, N.A., & Chattopadhyay, S. (2017). Biosorption of Azo dyes by spent *Rhizopus arrhizus* biomass. *Applied Water Science, 7*, 3041–3054.

Santhy, K., & Selvapathy, P. (2006). Removal of reactive dyes from wastewater by adsorption on coir pith activated carbon. *Bioresource Technology, 97*, 1329–1336. https://doi.org/10.1016/j.biortech.2005.05.016

Sharma, S., & Kaur, A. (2018). Various methods for removal of dyes from industrial effluents - a review. *Indian Journal of Science and Technology, 11*, 1–21. https://doi.org/10.17485/ijst/2018/v11i12/120847

Somasekhara Reddy, M.C., Sivaramakrishna, L., & Varada Reddy, A. (2012). The use of an agricultural waste material, Jujuba seeds for the removal of anionic dye (Congo red) from aqueous medium. *Journal of Hazardous Materials, 203–204*, 118–127. https://doi.org/10.1016/j.jhazmat.2011.11.083

Sonawane, M.D., & Nimse, S.B. (2016). Surface modification chemistries of materials used in diagnostic platforms with biomolecules. *Journal of Chemistry,* Article id 9241378. doi.org/10.1155/2016/9241378.

Veneu, D.M., Torem, M.L., & Pino, G.A. (2013). Fundamental aspects of copper and zinc removal from aqueous solutions using a Streptomyces lunalinharesii strain. *Minerals Engineering, 48*, 44–50.

Zehra, T., Priyantha, N., & Lim, L.B.L. (2016). Removal of crystal violet dye from aqueous solution using yeast-treated peat as adsorbent: thermodynamics, kinetics, and equilibrium studies. *Environmental Earth Sciences, 75*, 357.

16 Bacterial Ligninolytic Enzymes as Green and Versatile Agents for Wastewater Treatment

Adarsh Kumar, Pawan Kumar Bhargawa, and Subhash Chandra

16.1 INTRODUCTION

The World-Wide Fund (WWF) for Nature claims that harmful compounds that enter water bodies, including lakes, rivers, and the ocean, dissolve in the water, remain suspended in the water, or deposit at the bottom, causing water pollution (Sook et al., 2022). These hazardous materials, which contain nitrogen, organic matter, and trace inorganic elements such as heavy metals, can have negative effects on ecosystems and populations of living things (Kumar and Chandra, 2021a). The industries that use the most water pollute the environment the most. For example, traditional paper manufacturing often produces large amounts of harmful wastewater and requires 300 cubic meters of water per ton of product (Kumar et al., 2020c; Singh et al., 2021). Similar to this, the oil industries require10–300 m^3 of water for every tonne of output, while the mining industry needs 40 m^3/tonne of ore (Kumar et al., 2020c, 2021c). The iron and steel sector needs 20–60 m^3 of water for every tonne of product, and the synthetic fertilizer industry significantly contributes up to 270 m^3 of water for every tonne of product (Kumar et al., 2021c). However, the distillery industry's traditional methods for producing one litre of alcohol produce 15 L of wastewater (Pant and Adholeya, 2007). Additionally, food firms usually produce between 0.6 and 20 m^3 of wastewater per tonne of items including fruit juice, bread, butter, and milk (Kumar et al., 2021c). As a result, the following industries are on the list of industries that produce waste: mining, pulp and paper mills, refineries, sugar production/distilleries, tanneries, chemicals, textiles, pharmaceuticals, and tanneries (Kumar et al., 2021a, 2021b).

Microorganisms (often bacteria or fungi) produce ligninolytic enzymes for bioremediation of pollutants into less or non-hazardous compounds (Kumar and Chandra, 2020). Bacterial ligninolytic enzymes are efficient and eco-friendly tools for the degradation of polluted industrial wastewater. Through the efficient release of their enzymes to break down environmental pollutants, bacteria can contribute to the remediation process (Kumar and Chandra, 2021a, b). Four oxidative enzymes make up the complicated ligninolytic enzyme system: Lac, MnP, LiP, and VP (Kumar et al., 2021b, 2022a). These enzymes have their own unique mechanism to biologically remediate industrial refractory pollutants found in wastewater to reduce the pollution load (Kumar and Chandra, 2021a, 2021b). Lignocellulosic materials make up most polluted industries such as food industry, paper and pulp industry, distilleries, etc. Lignocellulose, which includes cellulose, hemicellulose, and lignin, is the primary structural chemicalof both non-woody and woody plants (Kumar et al., 2021a). Lignin is residual in nature because it degrades more slowly, harms the environment, and has an indirect impact on people due to the food chain. Lignocellulosic material is broken down into small molecules part by ligninolytic enzymes. These enzymes break down lignocellulose as well as stubborn environmental pollutants such as agrochemicals containing organochlorines (OCs), in paper mills wastewater, tanneries wastewater, distilleries wastewater, and textile mills wastewater (Kumar and Chandra, 2021a). They can also be bioremediate polluted soils (Kumar and Chandra, 2021b). The

DOI: 10.1201/9781003441144-16

important role that ligninolytic enzymes play in the breakdown of many xenobiotic and refractory polymers has made them very favorable in nature (Kumar and Chandra, 2021a). Many bacterial species are potential to produce ligninolytic enzymes, but some higher fungi are more capable reported by several researchers. Numerous researchers have documented the use of lignocellulosic wastes as a substrate for ligninolytic enzyme synthesis in response to the rising demand for these enzymes on an industrial scale. A cost-effective substrate for bacterial synthesis of ligninolytic enzymes on a commercial scale has been agricultural waste (Kumar et al., 2020b). The chapter discusses in detail bacterial ligninolytic enzymes as an eco-friendly and adaptable tool for the breakdown and detoxification of wastewater and industrial wastes.

16.2 BACTERIAL LIGNINOLYTIC ENZYMES

Bacteria, fungi, plants, and even insects have been reported by many researchers for ligninolytic enzyme production. Lac (EC 1.10.3.2), LiP (EC 1.11.1.14), MnP (EC 1.11.1.13), and VP (EC 1.11.1.16) are the most frequent ligninolytic enzymes. Applications for bacterial ligninolytic enzymes in industry are currently limited, although they might be used in bioremediation of various industrial wastes. Ligninolytic enzyme producing major bacterial strains such as *M. tuberculosum B. subtilis, Caulobacter crescentus, Yersinia pestis, Stenotrophomonas maltophilia, E. coli, Bordetella compestris, P. syringae, P. aeruginosa*, etc (Kumar and Chandra, 2020).

16.2.1 LACCASE

A copper-containing extracellular enzyme such as laccase is composed of monomeric, dimeric, and tetrameric glycoproteins (Figure 16.1). It is mostly reported in bacteria, actinomycetes, and fungi (Janusz et al., 2017). Lac was first reported from exudates of the Japanese tree *Rhus vernicifera* by Yoshida (1883). In microbes, insects, and plants, laccases are widely reported and have different roles. Laccases can cleave a broad range of compounds with phenolic structures because of their limited substrate specificity (Kumar and Chandra, 2020). Because of this, laccase has been used in several processes i.e. bioremediation of residual pollutants, textile dyes, chemical synthesis, lignin-containing derivatives, and biomass pretreatment for biofuel production (Kumar and Chandra, 2020; Kumar et al., 2020a, 2021b). Laccase is regarded as the ideal "green catalyst" because it

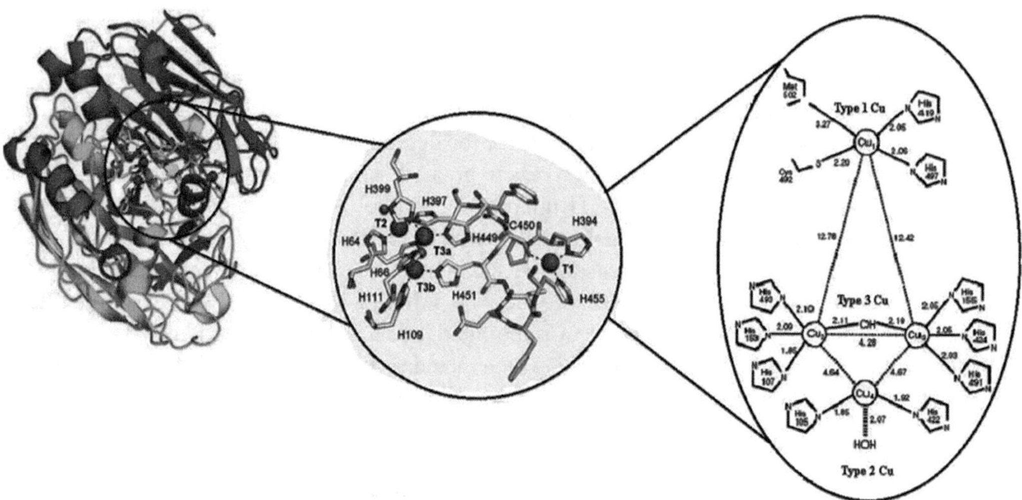

FIGURE 16.1 Schematic structure and active site of laccase.

Source: Adopted from Kumar and Chandra (2020).

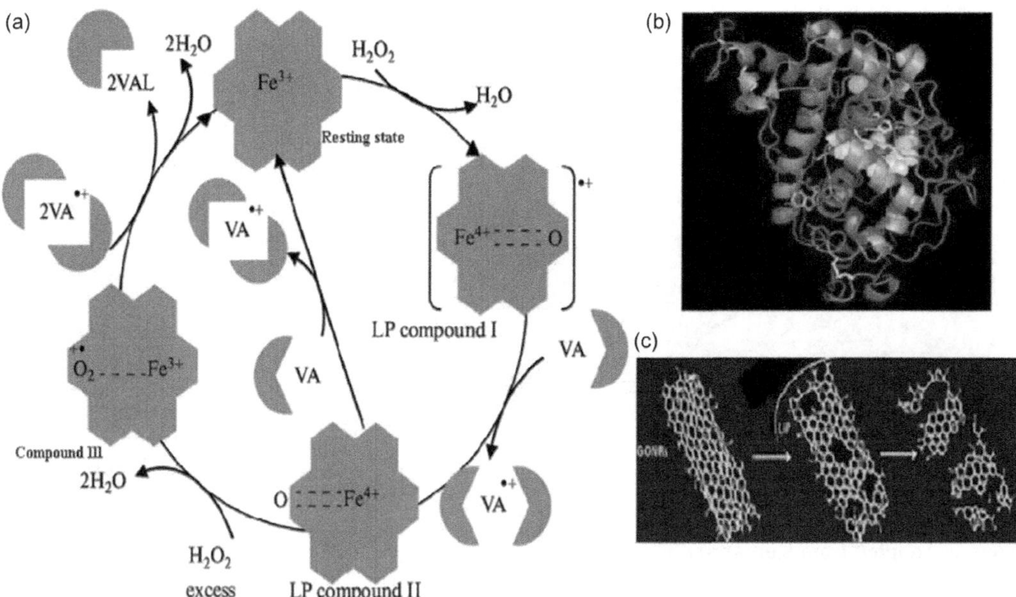

FIGURE 16.2 (a) Schematic catalytic mechanisms of lignin peroxidase, (b) Structure (c) Degraded pollutants structure.

Source: Adopted from Kumar and Chandra (2020).

can oxidize various typesof pollutants with O_2 and only produce H_2O as a by-product (Kumar and Chandra, 2020; Kumar et al., 2021b).

16.2.2 Lignin Peroxidase

LiP is a member of an oxidoreductase family that breaks down lignin and its derivatives when H_2O_2 is present (Figure 16.2) (Edwards et al., 1993). LiPsareheme-containing enzymes that break down polymers oxidative and are released mostly by some higher fungi and bacteria (Pothiraj et al., 2006). Additionally, several insects, including *R. flavipes* and the eastern subterranean termite, have employed this enzyme to decompose higher woody waste (Kumar and Chandra, 2020). LiPs are able to breakdown of a variety of residual pollutants including methoxybenzene and veratryl alcohol (3,4-dimethoxybenzyl) due to their strong redox potential (–1,400 mV), low pH range (3.0–4.5) (Dias et al., 2007). Instead of highly activated aromatic substrates catalyzed by common peroxidases, LiP oxidizes phenolic rings that are fairly activated by electron-donor substituents (amines, hydroxyls, etc.). The generation of veratyl alcohol radicals can be used to explain this type of catalysis. For the reason that they have a greater redox potential than LiP's parts I and II. These radicals can aid in the oxidation of substances with a high redox potential (Khindaria et al., 1996).

16.2.3 Manganese Peroxidase

MnPare heme-containing enzymes belonging to the oxidoreductase family (Figure 16.3). Lignocellulolytic bacteria produce MnP enzymes in their microenvironments in both solid and liquid states (Hatakka et al., 2003). In batch cultures of *P. chrysosporium* discovered the first MnP (Kuwahara et al., 1984). They are 350–350 amino acid residues glycoproteins with a molecular weight betweenthe range of 38 and 62.5 kDa, with 43% similarity to the LiP sequence (Martin, 2002). The MnP structure includes ten major helix, one minor helix, five disulfide bridges, and two domains with a heminic group in the middle (Kumar and Chandra, 2020; Kumar et al., 2022c). One

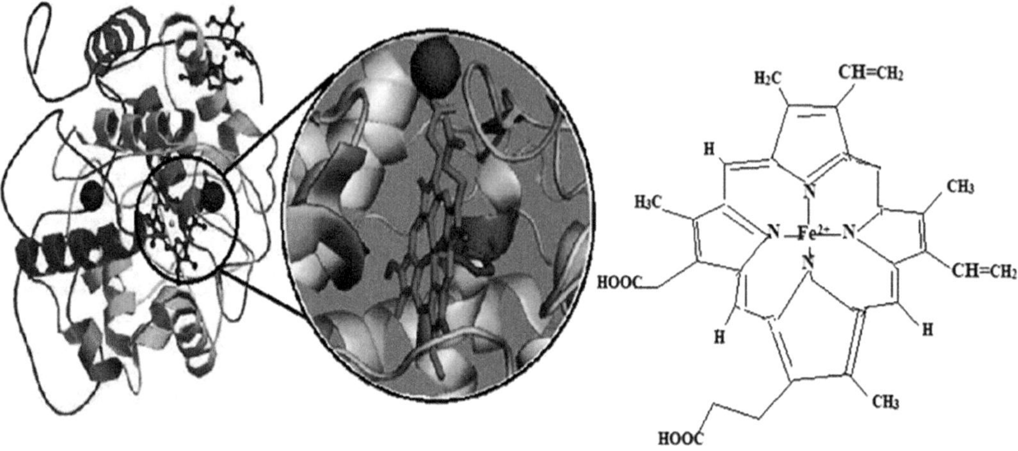

FIGURE 16.3 Schematic structure and active site of MnP.

Source: Adopted from Kumar and Chandra (2020).

of the bridges gets involved in manganese's binding site (Mn)because this is a feature of MnP that differentiates from other peroxidases (Sundaramoorthy et al., 1994). The two electrons transfer from the heminic group to H_2O_2 is the first step in the catalytic cycle of the enzyme and generated compound I along with water (Kumar and Chandra, 2020). After that oxidation of a substrate molecule is then catalyzed by compound I and generatedcompound II and free radical (Kumar and Chandra, 2020). The cation needed to oxidize phenolic compounds, Mn^{3+} is generated when compound II reacts with Mn^{2+}. It is important to keep in mind that compound II can only react with $Mn^{2+,}$ whereas compound I can oxidize both Mn^{2+} and the substrate. Mn^{3+} interacts non-specifically with organic molecules after being stabilized by organic acids by withdrawing an electron and a proton from the substrate (Martin, 2002; Dias et al., 2007). The redox-potential-rich small molecule Mn^{3+} diffuses easily through the lignified cell wall (Kumar et al., 2022c). To facilitate the entry and action of other enzymes, Mn^{3+} initiates invasion of the interior of the residual pollutants (Angel, 2002).

16.2.4 Versatile Peroxidase

VP, a member of the oxidoreductase family and also known as hybrid peroxidase (manganese-lignin peroxidase), is made up of glycoproteins (Kumar and Chandra, 2020). VP as a model biocatalyst, the relatively recent heme peroxidase has generated some interest in recent years (Figure 16.4). Because of their ability to break down various resistant substrates without the need for mediators, VPs are interesting for effective denaturation processes (Yadav et al., 2014).

VP is made up of secreted glycoproteins in a variety of isoenzymes with pHranging from 3.4 to 3.9 and molecular weights between 40 and 45 kDa (Mester and Field, 1998). VPs havea broad range of substrate-specific featuresto catalyze molecules with moderate and high redox potential. Moreover, VPshave high redox potential and can oxidize phenolic and non-phenolic pollutants, azo dyes, and lignin derivatives in the absence of mediators (Kumar et al., 2021a). The molecular structure of VP contains aheme pocket containing His residues distal and proximal, two calcium binding sites, four disulfide bridges, as well as Mn^{2+} binding site (Kumar et al., 2021a). Similar to LiP and other heme peroxidases, VP's catalytic cycle involves the transit of electrons from the oxidation of the substrate, which results in the synthesis and reduction of compounds I and II (Kumar and Chandra, 2020). A number of biotechnological applications could benefit from the use of VP, a potential biocatalyst. The use of VP as a model enzyme has attracted much attention in recent years.

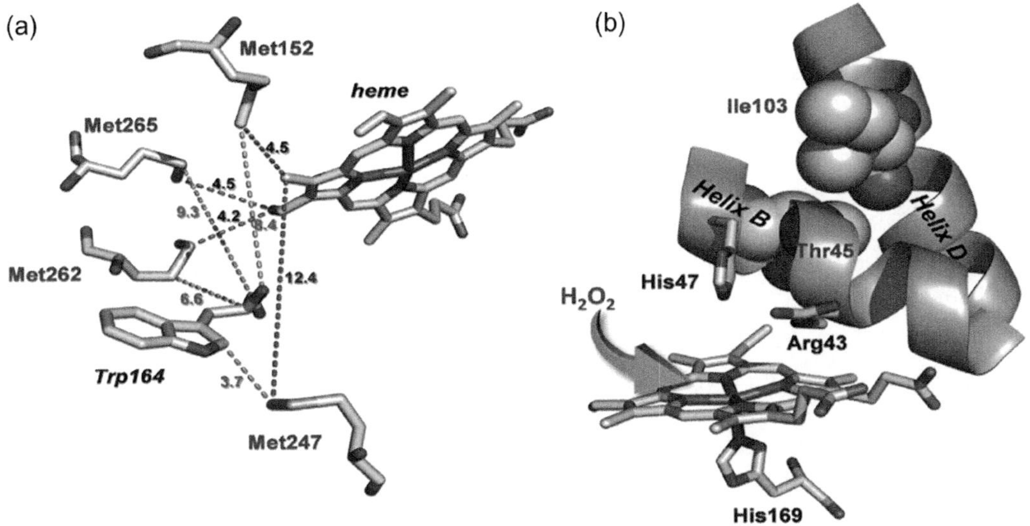

FIGURE 16.4 (a and b) Schematic molecular properties of hybrid VP enzyme.

Source: Adopted from Kumar and Chandra (2020).

16.3 EDCS DETECTED IN WATER AND REMOVED BY LIGNINOLYTIC ENZYMES

This concern has prompted research into identification of EDCs (endocrine-disrupting chemicals) present in the wastewater and developed efficient techniques for treatment. Recent investigations have shown EDCs in wastewater and receiving water bodies, among many researchers (Falade et al., 2018). The ligninolytic enzymes are potential biocatalytic tools for the degradation of EDCs existence in wastewater. Ligninolytic enzymes are effective in eliminating EDCs due to their high redox potential for the oxidation of various types of chemical pollutants. Additionally, they have demonstrated excellent potential for the transformation of many pollutants known to have endocrine-disrupting abilities, such as pesticides, PCBs, APs, and PAHs.

It has been extensively researched how well Lac and MnP can remove EDCs from the wastewater (Lloret et al., 2013). The use of VP for EDC elimination is not well understood (Cabana et al., 2007a). Given its unusual property of hybrid molecular structures, VP appears to be the most promising enzyme for the degradation of EDCs present in wastewater (Cabana et al., 2007b). Moreover, VP does not require redox mediators to breakdown micropollutants (Table 16.1) (Ravichandran and Sridhar, 2016).

16.4 GREEN AND VERSATILE AGENTS FOR WASTEWATER TREATMENT

The natural recycling of harmful substances into benign or harmless products is known as micro-organisms. Industrial wastewater contains a variety of organisms that break down contaminants, including bacteria, fungi, and actinomycetes. As a result of their complexities, which enable them to act on various harmful substances, several studies have shown that combinations of micro-organisms (consortia) destroy pollutants better than individual (axenic) organisms (Kumar et al., 2022b). An effective alternative method to deal with the cleaning of an environment contaminated with a variety of contaminants may be provided by ligninolytic enzymes (Kumar et al., 2022b). Ligninolytic enzymes have the ability to break down many residual organic pollutantsand polycyclic aromatic hydrocarbons. Phenols exist in agricultural activities and industrial effluents are harmful to living beings even in extremely small amounts (Kumar and Chandra, 2021a). Various phenolic chemicals such as hydroquinone, catechol, and resorcinol were produced as byproducts in wastes from several

TABLE 16.1
Ligninolytic Enzymes Were Used to Degrade EDCs in the Wastewater

Enzyme used	Reaction Matrix	Types of EDC	Removal Efficiency (%)	References
Immobilized Lac	AS	Bisphenol A bisphenol F bisphenol S	≈100 ≈100 >40	Zdarta et al. (2018)
Immobilized Lac	AS	Acetaminophen diclofenac	68 90	Garcia-Morales et al. (2018)
Immobilized Lac	AS	Bisphenol A carbamazepine	90 40	Ji et al. (2017)
Crude Lac	AS	Bisphenol A	100	de Freitas et al. (2017)
Lac with mediator (Hydroxybenzotriazole)	AS	Bisphenol A	100	Daasi et al. (2016)
Free Lac cocktail	Synthetic and groundwater	Bisphenol A 4-nonylphenol 17α-ethinylestradiol triclosan	89 93 100 90	Garcia-Morales et al. (2015)
Lac	AS	Nonylphenol and triclosan	>95%	Ramírez-Cavazos et al. (2014)
Cross-linked Lac aggregates and polysulfone hollow fiber microfilter membrane	Wastewater	Acetaminophen	93	Ba et al. (2014)
Immobilized lac	AS	Bisphenol A	100	Debaste et al. (2014)
Lac using enzymatic membrane reactor (EMR)	Wastewater	Estrone 17β-estradiol (E2) 17α-ethinylestradiol (EE2)	83.6 94 93.6	Lloret et al. (2013)
Immobilized lac	AS	Bisphenol A nonylphenol triclosan	80 40 60	Songulashvili et al. (2012)
Lac	Simulated wastewater	Nonylphenol bisphenol A triclosan	100 100 65	Cabana et al. (2007a)
Lac	Wastewater	Estrone	100	Auriol et al. (2008)
Immobilized Lac in fluidized bed reactor	Simulated wastewater	Nonylphenol	100	Cabana et al. (2007b)
Immobilized Lac	Wastewater	Bisphenol A, B, F	100	Diano and Mita (2011)
Encapsulated ligninolytic enzymes (Lac, LiP, and MnP)	Water	Bisphenol A	90	Gassara et al. (2013)
MnP	AS	Triclosan	99.4	Inoue et al. (2010)
MnP and Lac	AS	Natural steroidal hormone, estrone	98	Tamagawa et al. (2006)

(*Continued*)

TABLE 16.1 (*Continued*)
Ligninolytic Enzymes Were Used to Degrade EDCs in the Wastewater

Enzyme used	Reaction Matrix	Types of EDC	Removal Efficiency (%)	References
MnP and Lac	AS	Genistein	93	Tamagawa et al. (2005)
MnP and Lac - 1-hydroxybenzotriazole (lac-HBT) system	AS	Bisphenol A Nonylphenol	100	Tsutsumi et al. (2001)
VP	AS	Diclofenac and estrogen hormones Sulfamethoxazole and naproxen	100 80	Eibes et al. (2011)
VP using TSS	Wastewater	Nonylphenol	99.2	Mendez-Hernandez et al. (2015)

Lac: Laccase, LiP: Lignin peroxidase MnP: Manganese peroxidase, VP: Versatile peroxidase, AS: Aqueous system

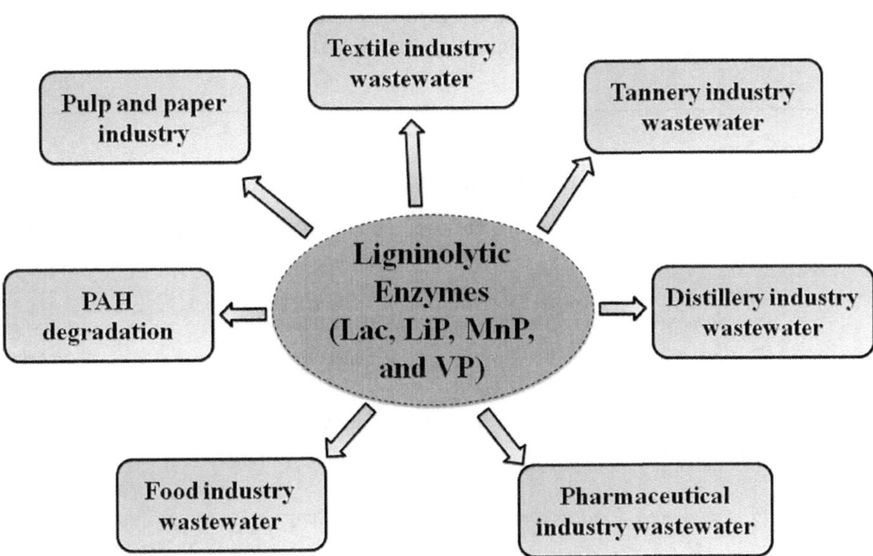

FIGURE 16.5 Ligninolytic enzymes as green and versatile agents for wastewater treatment.

industries (Schweigert et al., 2001). These residual pollutants dissipate into the aquatic resources, where they build up in the groundwater and soil, posing a major threat to the ecosystem. Figure 16.5 and Table 16.2 depict ligninolytic enzymes and how they are used in the bioremediation of industrial wastewater.

Ligninolytic enzymes have made important contributions to the treatment of several residual pollutants (Husain, 2006). The use of peroxidases as biocatalysts by biotechnological industries for environmental goals embodies and assumes this function. These protein structures facilitate the biodegradation of harmful contaminants, assist in precipitation for their oxidation or removal, and allow their transformation into value-added products with particular applications (Chen & Wan, 2017). The main technical barrier to enzymatic bioremediation is the efficient instability of free enzymes in challenging environments (Kumar et al., 2022a). The problem with using native enzymes for remediation is that the biological catalyst cannot be recovered and reused. Thus, the construction and intentof enzyme structures with optimized and enhanced performance greatly benefited from enzyme engineering (Kumar et al., 2022a,2022c). Researchers have studied immobilization using a variety of techniques, including covalent attachment to a support, adsorption on a carrier (support), encapsulation, and cross-linking (carrier-free) (Quilles Junior et al., 2016; Besharati Vineh et al., 2018).

Environmental pollution caused by wastewater from paper and pulp mills has significant effects in aquatic and terrestrial ecosystems (Kumar et al., 2020a). Researchers around the world have recently focused heavily on microbial techniques to optimize various parameters under laboratory conditions for sustainable wastewater treatment technology (Kumar and Chandra, 2021a, 2021b). The textile industryuses various types of synthetic dyes, which are structurally classified as azo, anthraquinone, sulfur, heterocyclic, triarylmethane, and phthalocyanine (Devi Saini, 2017). Depending on how they are applied, dyes are classified as dispersion, direct, reactive, basic, or vat (Popli and Patel, 2015). Currently, 43 million tons of various dyes and chemicals are required annually to produce 110 million tons of textile fibers and clothing produced worldwide (Ütebay et al., 2020). Around 79–93 billion cubic meters of water is used during the textile production process all over the world, with a significant portion of this water being released as colored waste (Ütebay et al., 2020). It pollutes the ecosystem severely (Yaseen and Scholz, 2019). Waste or waste disposal sites are often associated with dye-degrading bacteria. It has been demonstrated that *Bacillus*species isolated from dye wastewater sites have the ability to bleach the dye. Many bacterial strains such as *B. Altitudinis* AD14, *B. flexus*, *E. homiens*, *S. aureus*, *P. aeruginosa*, and *B. Amyloliquefications* AD20 have been performed for the decolonization and degradation of tannery wastewater. In batch

TABLE 16.2
Applications of Ligninolytic Enzymes in the Treatment of Various Industrial Wastewater

Industries	Bacteria	Enzyme	Bioremediation Application	References
Pulp and paper industry	*Paenibacillus* sp.	Lac	Effluent treatment	Raj et al. (2014)
	Pseudomonas sp.		Effluent treatment	Raj et al. (2014)
	Bacillus sp.		Effluent treatment	Jeenathunisa et al. (2017)
	S. liquefaciens	LiP	Wastewater treatment	Haq et al. (2016)
	B. aryabhattai	MnP	Wastewater treatment	Zainith et al. (2019b)
	P. aeruginosa & S. marcescens	LiP	Biobleaching	Bholay et al. (2012)
	K. pneumoniae	Lac, MnP	Effluent treatment	Gaur et al., (2018)
	Bacillus cereus AKRC03	Lac	Effluent treatment	Kumar and Chandra (2021a,b)
	A. aneurinilyticus	Lac	Decolorize kraft lignin	Chandra et al., (2007)
	Azotobacter	Lac	Decolorize and solubilize lignin	Morii et al. (1995)
	Pantoea sp.	Lac, MnP	Decolonization, reduce BOD, COD and degrade chlorophenol and lignin	Chandra et al. (2012)
Distillery industry	*P. arruginosa*	Lac	Bioremediation and decolorization of effluent	Pal and Vimala (2012)

(Continued)

TABLE 16.2 (*Continued*)
Applications of Ligninolytic Enzymes in the Treatment of Various Industrial Wastewater

Industries	Bacteria	Enzyme	Bioremediation Application	References
	Raoultella planticola, Enterobacter sakazaki and *Bacillus* sp.,	MnP	Degradation and decolorization of melanoidin	Yadav et al. (2014)
	Stenotrophomonas sp.	Lac, MnP	Degradation of sugarcane molasses-based distillery waste	Chandra and Kumar (2017)
	Bacillus sp.	Lac, MnP	Degradation of sugarcane molasses-based distillery waste	Chandra and Kumar (2017)
Pharmaceutical industry wastewater	*Bacillus subtilis*	Lac	Biodegradation and detoxification	Gangola et al. (2018)
	Marinomonas mediterranea	Lac, MnP, LiP	Used in diagnostic kits	Plácido and Capareda (2015)
	–	Lac, MnP, LiP	Effluent treatment	Kumar and Chandra (2020)
	–	Lac, VP	Coupling of phenols and steroids Medical agents	Kumar and Chandra (2020)
	Bacillus subtilis	Lac	Effluent treatment	Lawal et al. (2022)
	–	VP	Industrial biocatalyst, bioremediation and so forth.	Kumar and Chandra (2020)
Food industry wastewater	–	Lac	Food and beverage	Minussi et al. (2007)
	–	MnP	Natural aromatic flavours	Barbosa et al. (2008)
	–	LiP	Natural aromatics production of vanillin	Barbosa et al. (2008)
	–	Lac	Sugar beet pectin gelation	Minussi et al. (2007)
Textile industry wastewater	*Bacillus amyloliquefaciens* AD 20 and	Lac	Decolorization and degradation of remazol brilliant blue R (RBFR) dye	Odunsi et al. (2022)
	B. altitudinis AD 14	Lac	Decolorization and degradation RBFR dye	Odunsi et al. (2022)
	Bacillus subtilis WD23	Lac	decolorization of congo Red	Wang et al., (2010)
	Bacillus subtilis	Lac	Decolorization and degradation of eriochrome black T	Zaid et al. (2015)
	Bacillus licheniformis	LiP, MnP	Decolorization and degradation of alphazurine A	Bu et al. (2020)
	B. amyloliquefaciens	Lac	Decolorization and degradation of reactive blue 19 dyes	Wang et al. (2017)
	Bacillus sp.	LiP, MnP	Decolorization and degradation of azo dye	Horitsu et al. (1977)
Tannery industry wastewater	*Pseudomonas* sp. strain DY1	Lac, MnP	Biodegradation of tannery wastewater	Du et al. (2011)
	P. aeruginosa	MnP	Biodegradation of tannery wastewater	Kimata et al. (2003)

(*Continued*)

TABLE 16.2 (*Continued*)
Applications of Ligninolytic Enzymes in the Treatment of Various Industrial Wastewater

Industries	Bacteria	Enzyme	Bioremediation Application	References
	Bacillus flexus	Lac	Biodegradation of tannery wastewater	Sivaprakasam et al. (2008)
	Exiguobacterium homiense	LiP	Biodegradation of tannery wastewater	Sivaprakasam et al. (2008)
	Staphylococcus aureus	MnP	Biodegradation of tannery wastewater	Sivaprakasam et al. (2008)

studies, salinity, growth factor media, pH, temperature, agitation rate, and inoculum concentration were tuned for the reported bacterial strains. The commercial tannery brackish wastewater was biologically treated using the single and consortium of halotolerant bacterial isolates.

Bioactive substances, which also have positive effects on animal and human health and nutritional benefits, influence food quality. The applications of ligninolytic enzymes in the food industry include beer, wine, haze formation, removing unfavorable phenolics, changing or improving the appearance of food or beverage turbidity and color in clarified fruit juices as well as treatingfood industry wastewater treatment (Kumar et al., 2021b). Bilal et al. (2016) reported that immobilizing the MnP enzymes significantly reduced the turbidity and colour of apple juice by 36.3% and 42.7%, respectively. According to a different investigation, there was a 51.5% and 43.6% reduction in colour turbidity, respectively. Furthermore, Bilal et al. (2016) reported that stabilized ligninolytic enzyme treatment could reduce turbidity in apple, grape, orange, and pomegranate juices by 84.02%, 57.84%, 86.14%, and 82.13%, respectively. MnP has seen an increase in demand in recent years for various uses in the food industry as well as wastewater treatment.

16.5 CONCLUSION

Research on the use of bacterial ligninolytic enzymes in different types of industrial wastewater treatment is being addressed in a demanding, targeted, and challenging approach. Currently, various studies are being conducted on different aspects of the biodegradation of industrial wastewater employing different microorganisms. Ligninolytic enzymes are potentially capable of treating the xenobiotic or residual organic pollutants existingin the wastewater produced by several industries. Ligninolytic enzymes are a potential replacement for the traditional chemical processes now used in industry. Several countries have given substantial attention to the biodegradation of wastewater as well as the reductionof contaminants that pose a risk to aquatic and terrestrial ecosystem. In this chapter, important information regarding the potential application of ligninolytic enzymes in treated wastewater and being environmentally beneficial is evaluated.

REFERENCES

Angel, T.M. (2002). Molecular biology and structure-function of lignindegrading heme peroxidases. *Enzyme and Microbial Technology*, 30, 425–444.

Auriol, M., Filali Meknassi, Y., Adams, C.D., Tyagi, R.D., Noguerol, T.N., & Pina, B. (2008). Removal of estrogenic activity of natural and synthetic hormones from a municipal wastewater: Efficiency of horseradish peroxidase and laccase from *Trametes versicolor*. *Chemosphere*, 70, 445–452.

Ba, S., Jones, J.P., & Cabana, H. (2014). Hybrid bioreactor (HBR) of hollow fiber microfilter membrane and cross linked laccase aggregates eliminate aromatic pharmaceuticals in wastewaters. *Journal of Hazardous Materials*, 280, 662–670.

Barbosa, E.S., Perrone, D., Vendramini, A.L.A., & Leite, S.G.F. (2008). Vanillin production by *Phanerochaete chrysosporium* grown on green coconut agro-industrial husk in solid state fermentation. *BioResources*, 3(4), 1042–1050.

Besharati Vineh, M., Saboury, A.A., Poostchi, A.A., Rashidi, A.M., & Parivar, K. (2018). Stability and activity improvement of horseradish peroxidase by covalent immobilization on functionalized reduced graphene oxide and biodegradation of high phenol concentration. *International Journal of Biological Macromolecules*,106, 1314–1322.

Bholay, A. D., Borkhataria, B. V., Jadhav, P. U., Palekar, K. S., Dhalkari, M. V., & Nalawade, P. M. (2012). Bacterial Lignin Peroxidase: A Tool for Biobleaching and Biodegradation of Industrial Effluents. *Universal Journal of Environmental Research and Technology*, 2, 58–64.

Bilal, M., Asgher, M., Iqbal, H., Hu, H., & Zhang, X. (2016). Gelatin-immobilized manganese peroxidase with novel catalytic characteristics and its industrial exploitation for fruit juice clarification purposes. *Catalysis Letters*, 146. https://doi.org/10.1007/s10562-016-1848-9.

Bu, T., Yang, R., Zhang, Y., Cai, Y., Tang, Z., Li, C., Wu, Q., & Chen, H. (2020). Improving decolorization of dyes by laccase from Bacillus licheniformis by random and site-directed mutagenesis. *PeerJ*, 8, e10267. https://doi.org/10.7717/peerj.10267.

Cabana, H., Jiwan, J.L.H., Rozenberg, R., Elisashvili, V., Penninckx, M, Agathos, S.N., & Jones, J.P. (2007a). Elimination of endocrine disrupting chemicals nonylphenol and bisphenol and personal care product ingredient triclosan using enzyme preparation from the white rot fungus, *Coriolopsis polyzona*. *Chemosphere*, 67, 770–778.

Cabana, H., Jones, J.P., & Agathos, S.N. (2007b). Preparation and characterization of cross linked laccase aggregates and their application to the elimination of endocrine disrupting chemicals. *Journal of Biotechnology*, 132, 23–31.

Chandra, R., & Kumar, V. (2017). Detection of Bacillus and Stenotrophomonas species growing in an organic acid and endocrine-disrupting chemical-rich environment of distillery spent wash and its phytotoxicity. *Environmental Monitoring and Assessment*, 189, 26. https://doi.org/10.1007/s10661-016-5746-9

Chandra, R., Raj, A., Purohit, H.J., & Kapley, A. (2007). Characterization and optimization of three potential aerobic bacterial strains for kraft lignin degradation from pulp paper waste. *Chemosphere*, 67, 839–846.

Chandra, R., Singh, R., & Yadav, S. (2012). Effect of bacterial inoculum ratio in mixed culture for decolourization and detoxification of pulp paper mill effluent. *Journal of Chemical Technology & Biotechnology*, 87, 436–444.

Chen, Z., & Wan, C. (2017). Biological valorization strategies for converting lignin into fuels and chemicals. *Renewable and Sustainable Energy Reviews*, 73, 610–621.

Daasi, D., Prieto, A., Zouari Mechichi, H., Martinez, M.J., Nasri, M., & Mechichi, T. (2016). Degradation of bisphenol A by different fungal laccases and identification of its degradation products. *International Biodeterioration and Biodegradation*, 110, 181–188.

de Freitas, E.N., Bubna, G.A., Brugnari, T., Kato, C.G., Nolli, M., Rauen, T.G., … Peralta, R.M. (2017). Removal of bisphenol A by laccase from *Pleurotus ostreatus* and *Pleurotus pulmonarius* and evaluation of ecotoxicity of degradation products. *Chemical Engineering Journal*, 330, 1361–1369.

Debaste, F., Songulashvili, G., & Penninckx, M.J. (2014). The potential of *Cerrena unicolor* laccase immobilized on mesoporous silica beads for removal of organic micropollutants in wastewaters. *Desalination and Water Treatment*, 52, 2344–2347.

Devi Saini, R. (2017). Textile organic dyes: Polluting effects and elimination methods from textile waste water. *International Journal of Chemical Engineering Research*, 9(1), 121–136.

Diano, N., & Mita, D.G. (2011). Removal of endocrine disruptors in waste waters by means of bioreactors. In F.S.G.Einschlag (Ed.) *Waste Water - Treatment and Reutilization* (pp. 29–48). Croatia: InTech. https://www.intechopen.com/books/waste-water-treatmentand-reutilization/removal-of-endocrine-disruptors-in-waste-waters-by-means-of-bioreactors

Dias, A., Sampaio, A., & Bezerra, R. (2007). Environmental applications of fungal and plant systems: Decolourisation of textile wastewater and related dyestuffs. In S. Singh and R. Tripathi (Eds.) *Environmental Bioremediation Technologies* (pp. 445–463). Berlin and Heidelberg: Springer.

Du, L.N., Wang, S., Li, G., Wang, B., Jia, X.M., Zhao, Y.H., Chen Y.L. (2011). Biodegradation of malachite green by Pseudomonas sp. strain DY1 under aerobic condition: characteristics, degradation products, enzyme analysis and phytotoxicity. *Ecotoxicology*, 20(2011), 438–446.

Edwards, S.L., Raag, R., Wariishi, H., Gold, M.H., & Poulos, T.L. (1993). Crystal structure of lignin peroxidase. *Proceedings of the National Academy of Sciences*, 90, 750–754.

Eibes, G., Debernardi, G., Feijoo, G., Moreira, M.T., & Lema, J.M. (2011). Oxidation of pharmaceutically active compounds by a ligninolytic fungal peroxidase. *Biodegradation*, 22, 539–550.

Falade, A.O., Mabinya, L.V., Okoh, A.I., & Nwodo, U.U. (2018). Ligninolytic enzymes: Versatile biocatalysts for the elimination of endocrine-disrupting chemicals in wastewater. *Microbiologyopen*, 7(6), e00722. https://doi.org/10.1002/mbo3.722

Gangola, S., Sharma, A., Bhatt, P., et al. (2018). Presence of esterase and laccase in Bacillus subtilis facilitates biodegradation and detoxification of cypermethrin. *Scientific Reports*, 8, 12755. https://doi.org/10.1038/s41598-018-31082-5

Garcia Morales, R., Garcia Garcia, A., Orona Navar, C., Osma, J.F., Nigam, K.D.P., & Ornelas Soto, N. (2018). Biotransformation of emerging pollutants in groundwater by laccase from *P. sanguineus* CS43 immobilized onto Titania nanoparticles. *Journal of Environmental Chemical Engineering*, 6, 710–717. https://doi.org/10.1016/j.jece.2017.12.006.

Garcia Morales, R., Rodríguez Delgado, M., Gomez Mariscal, K., Orona Navar, C., Hernandez Luna, C., Torres, E., … Ornelas Soto, N. (2015). Biotransformation of endocrine disrupting compounds in groundwater: Bisphenol A, nonylphenol, ethynylestradiol and triclosan by a laccase cocktail from *Pycnoporus sanguineus* CS43. *Water, Air, and Soil Pollution*, 226, 251. https://doi.org/10.1007/s11270-015-2514-3

Gassara, F., Brar, S.K., Verma, M., & Tyagi, R.D. (2013). Bisphenol A degradation in water by ligninolytic enzymes. *Chemosphere*, 92(10), 1356–1360.

Gaur, N., Narasimhulu, K., & Pydi Setty, Y. (2018). Extraction of ligninolytic enzymes from novel *Klebsiella pneumoniae* strains and its application in wastewater treatment. *Applied Water Science*, 8, 111. https://doi.org/10.1007/s13201-018-0758-y

Haq, I., Raj, A., & Markandeya (2018). Biodegradation of Azure-B dye by Serratia liquefaciens and its validation by phytotoxicity, genotoxicity and cytotoxicity studies. *Chemosphere*, 196, 58–68. https://doi.org/10.1016/j.chemosphere.2017.12.153.

Hatakka, A., Lundell, T., Hofrichter, M., & Maijala, P. (2003). Manganese peroxidase and its role in the degradation of wood lignin. In S.D. Mansfield & J.N. Saddler (Eds.) *Applications of Enzymes to Lignocellulosic*, ACS Symposium Series (vol. 855, pp. 230–243). Washington, DC: American Chemical Society.

Horitsu, H., Takada, M, Idaka, E., Tomoyeda, M., & Ogawa, T. (1977) Degradation of aminoazobenzene by *Bacillus subtilis. European Journal of Applied Microbiology and Biotechnology*, 4, 217–224.

Husain, Q. (2006). Potential applications of the oxidoreductive enzymes in the decolorization and detoxification of textile and other synthetic dyes from polluted water: A review. *Critical Reviews in Biotechnology*, 26(4), 201–221.

Inoue, Y., Hata, T., Kawai, S., Okamura, H., & Nishida, T. (2010). Elimination and detoxification of triclosan by manganese peroxidase from white rot fungus. *Journal of Hazardous Materials*, 180, 764–767.

Janusz, G., Pawlik, A., Sulej, J., widerska-Burek, U., Jarosz-Wilkołazka, A., & Paszczy´nski, A. (2017). Lignin degradation: Microorganisms, enzymes involved, genomes analysis and evolution. *FEMS Microbiology Reviews*, 41(6), 941–962.

Jeenathunisa, N. Jeyabharathi, S. & Arthi J. (2017). Bioremediation of paper and pulp industrial effluent using bacterial isolates. *International Journal for Research in Applied Science and Engineering Technology*, 5, (10), 692–698.

Ji, C., Nguyen, L.N., Hou, J., Hai, F.I., & Chen, V. (2017). Direct immobilization of laccase on Titania nanoparticles from crude enzyme extracts of *P. ostreatus* culture for micro pollutant degradation. *Separation and Purification Technology*, 178, 215–223.

Khindaria, A., Yamazaki, I., & Aust, S.D. (1996). Stabilization of the veratryl alcohol cation radical by lignin peroxidase. *Biochemistry (N Y)*, 35, 6418–6424.

Kimata, N., Nishino, T., Suzuki, S., & Kogure, K. (2003). *Pseudomonas aeruginosa* isolated from marine environments in Tokyo Bay. *Microbial Ecology*, 47, 41–47. https://doi.org/10.1007/s00248-003-1032-9.

Kumar, A., Bilal, M., Singh, A.K., Ratna, S., & Rameshwari, K.R.T. (2022a). Enzyme cocktail: A greener approach for biobleaching in paper and pulp industry. In *Nanotechnology in Paper and Wood Engineering, Fundamentals, Challenges and Applications, Micro and Nano Technologies* (pp. 303–328). Elsevier. https://doi.org/10.1016/B978-0-323-85835-9.00007-6.

Kumar, A., & Chandra, R. (2020). Ligninolytic enzymes and its mechanisms for degradation of lignocellulosic waste in environment. *Heliyon*, 6, e03170.

Kumar, A., & Chandra, R. (2021a). Biodegradation and toxicity reduction of pulp paper mill wastewater by isolated laccase producing *Bacillus cereus* AKRC03. *Cleaner Engineering and Technology*, 4, 100193. https://doi.org/10.1016/j.clet.2021.100193

Kumar, A., & Chandra, R. (2021b). Extremophilic nature of microbial ligninolytic enzymes and their role in biodegradation. In S. Das & R. Dash (Eds.) *Microbial Biodegradation and Bioremediation*. https://doi.org/10.1016/B978-0-323-85455-9.00012-6.

Kumar, A., Saxcena, G., Kumar, V., & Chandra, R. (2020a). Environmental contamination, toxicity profile and bioremediation approaches for treatment and detoxification of pulp paper industry effluent. In G. Saxcena, V. Kumar, & M. Shah (Eds.) *Bioremediation for Environmental Sustainability Toxicity, Mechanisms of Contaminants Degradation, Detoxification, and Challenges*. https://doi.org/10.1016/B978-0-12-820524-2.00015-8.

Kumar, A., Singh, A.K., Ahmad, S., & Chandra, R. (2020b). Optimization of laccase production by *Bacillus* sp. Strain AKRC01 in presence of agro-waste as effective substrate using response surface methodology. *Journal of Pure and Applied Microbiology*, 14(1), 351–362.

Kumar, A., Singh, A.K., & Chandra, R. (2020c). Comparative analysis of residual organic pollutants from bleached and unbleached paper mill wastewater and their toxicity on *Phaseolus aureus* and *Tubifex tubifex*. *Urban Water Journal*, 17(10), 860–870.

Kumar, A., Singh, A.K., & Chandra, R. (2021c). Recent advances in physicochemical and biological approaches for degradation and detoxification of industrial wastewater. In I. Haq & A. Kalamdhad (Eds.) *Emerging Treatment Technologies for Waste Management.* https://doi.org/10.1007/978-981-16-2015-7-1.

Kumar, A., Singh, A.K., & Kumar, S. (2022c). MnP enzyme: Structure, mechanisms, distributions and its ample opportunities in biotechnological application. In P. Verma & M.P. Shah (Eds.) *Bioprospecting of Microbial Diversity Challenges and Applications in Bio-chemical Industry, Agriculture and Environment Protection* (pp. 265–283). https://doi.org/10.1016/B978-0-323-90958-7.00014-5.

Kumar, A., Singh, A.K., Bilal, M., & Chandra, R. (2021a). Sustainable production of thermostable laccase from agro-residues waste by Bacillus aquimaris AKRC02. *Catalysis Letters.* https://doi.org/10.1007/s10562-021-03753-y

Kumar, A., Singh, A.K., Bilal, M., & Chandra, R. (2021b). Extremophilic ligninolytic enzymes- a versatile tool with huge biotechnological promise. *Catalysis Letters.* https://doi.org/10.1007/s10562-021-03800-8

Kumar, A., Singh, A.K., & Sonal, P. (2022b). Chapter 17: Paper and pulp mill wastewater: Characterization, Microbial mediated degradation and challenges. In *Nanotechnology in Paper and Wood Engineering, Fundamentals, Challenges and Applications, Micro and Nano Technologies* (pp. 371–387). Elsevier. https://doi.org/10.1016/B978-0-323-85835-9.00011-8.

Kuwahara, M., Glenn, J.K., Morgan, M.A., & Gold, M.H. (1984). Separation and characterization of two extracellular H2O2 dependent oxidases from ligninolytic cultures of *Phanerochaete chrysosporium. FEBS (Federation of European Biochemical Societies) Letters*, 169, 247–250.

Lawal, I.K., Adekanmi, A.A., Ahmad, K, Akinkunmi, O.O., Busayo, O., & Micheal Ajewole, O. (2022). Bioremediation of waste water from pharmaceutical industry by bacteria (*Bacillus subtilis*). *Journal of Environmental Issues and Climate Change*, 1(1), 38–50.

Lloret, L., Eibes, G., Moreira, M.T., Feijoo, G., & Lema, J.M. (2013). Removal of estrogenic compounds from filtered secondary wastewater effluent in a continuous Enzymatic Membrane Reactor. Identification of biotransformation products. *Environmental Science and Technology*, 47, 4536–4543.

Martin, H. (2002). Review: Lignin conversion by manganese peroxidase (MnP). *Enzyme Microb Technol*, 30, 454–466.

Mendez Hernandez, J.E., Eibes, G., Arca Ramos, A., Lu chau, T.A., Feijoo, G., Moreira, M.T., & Lema, J.M. (2015). Continuous removal of nonylphenol by versatile peroxidase in a two stage membrane bioreactor. *Applied Biochemistry and Biotechnology*, 175(6), 3038–3047.

Mester, T., & Field, J.A. (1998). Characterization of a novel manganese peroxidase-lignin peroxidase hybrid isozyme produced by *Bjerkandera species* strain BOS55 in the absence of manganese. *Journal of Biological Chemistry*, 273(25), 15412–15417.

Minussi, R.C., Miranda, M.A., Silva, J.A., Ferreira, C.V., Aoyama, H., Marangoni, S., Rotilio, D., Pastore, G.M., & Duran, N. (2007). Purification, characterization and application of laccase from *Trametes versicolor* for colour and phenolic removal of olive mill wastewater in the presence of 1-hidroxybenzotriazole. *African Journal of Biotechnology*, 6(10), 1248–1254.

Morii, H., Nakamiya, K., & Kinoshita, S. (1995). Isolation of a lignin decolorizing bacterium. *Journal of Fermentation and Bioengineering*, 80(3), 296–299.

Odunsi, A.A., Nsa, I.Y., Omolere, B.M., Eze, C.O., Omonigbehin, E.A., Nwinyi, O.C., & Amund, O.O. (2022). *Bacillus amyloliquefaciens* AD 20 and *Bacillus altitudinis* AD 14isolated from a dye pond decolorize synthetic textile reactive dyes. *Covenant Journal of Physical and Life Sciences*,10, 2354–3485.

Pal. S., & Vimala, Y. (2012). Bioremediation and decolorization of Distillery effluent by novel Microbial Consortium. *European Journal of Experimental Biology*, 2(3),496–504.

Pant, D., & Adholeya, A. (2007). Biological approaches for treatment of distillery wastewater: A review. *Bioresource Technology*, 98(12), 2321–2334

Plácido, J., & Capareda, S. (2015). Ligninolytic enzymes: A biotechnological alternative for bioethanol production. *Bioresources and Bioprocessing*, 2(1), 1–12.

Popli, S., & Patel, U.D. (2015). Destruction of azo dyes by anaerobic-aerobic sequential biological treatment: A review. *International Journal of Environmental Science and Technology*, 12, 405–420.

Pothiraj, C., Kanmani, P., & Balaji, P. (2006). Bioconversion of lignocellulose materials. *Mycobiology*, 34(4), 159–165.

Quilles Junior, J.C., Ferrarezi, A.L., Borges, J.P., Brito, R.R., Gomes, E., da Silva, R., Guisán, J.M., & Boscolo, M. (2016). Hydrophobic adsorption in ionic medium improves the catalytic properties of lipases applied in the triacylglycerol hydrolysis by synergism. *Bioprocess and Biosystems Engineering*, 39(12), 1933–1943.

Raj, A., Kumar, S., Haq, I., & Singh, S.K. (2014). Bioremediation and toxicity reduction in pulp and paper mill effluent by newly isolated ligninolytic Paenibacillus sp., *Ecological Engineering*, 71, 355–362.

Ramírez Cavazos, L.I., Junghanns, C., Ornelas Soto, N., Cárdenas Chávez, D.L., Hernández Luna, C., Demarche, P., … Parra, R. (2014). Purification and characterization of two thermostable laccases from *Pycnoporus sanguineus* and potential role in degradation of endocrine disrupting chemicals. *Journal of Molecular Catalysis B: Enzymatic*, 108, 32–42.

Ravichandran, A., & Sridhar, M. (2016). Versatile peroxidase: Super peroxidases with potential biotechnological applications. *Journal of Diary, Veterinary & Animal Research*, 4(2), 00116.

Schweigert, N., Alexander, J.B., Zehnder, R., & Eggen I.L. (2001). Chemical properties of catechols and their molecular modes of toxic action in cells, from microorganisms to mammals. *Environmental Microbiology*, 3, 81–91. https://doi.org/10.1046/j.1462-2920.2001.00176.x

Singh, A.K., Kumar, A., Bilal, M., & Chandra, R. (2021). Organometallic pollutants of paper mill wastewater and their toxicity assessment on Stinging catfish and Sludge Worm. *Environmental Technology and Innovation*, 24, 101831. https://doi.org/10.1016/j.eti.2021.101831

Sivaprakasam, S., Mahadevan, S., & Sekar, S., et al. (2008). Biological treatment of tannery wastewater by using salt-tolerant bacterial strains. *Microbial Cell Factories*, 7, 15. https://doi.org/10.1186/1475-2859-7-15

Songulashvili, G., Jimenez tobon, G.A., Jaspers, C., & Penninckx, M.J. (2012). Immobilized laccase of *Cerrena unicolor* for elimination of endocrine disruptor micropollutants. *Fungal Biology*, 116, 883–889.

Sook Sin Chan, Kuan Shiong Khoo, Kit Wayne Chew, Tau Chuan Ling, & Pau Loke Show, (2022). Recent advances biodegradation and biosorption of organic compounds from wastewater: Microalgae-bacteria consortium –A review. *Bioresource Technology,* 344, Part A, 126159, doi.org/10.1016/j.biortech.2021.126159.

Sundaramoorthy, M., Kishi, K., Gold, M.H., & Poulos, T.L. (1994). The crystal structure of manganese peroxidase from *Phanerochaete chrysosporium* at 2.06-A resolution. *Journal of Biological Chemistry*, 269, 32759–32767.

Tamagawa, Y., Hirai, H., Kawai, S., & Nishida, T. (2005). Removal of estrogenic activity of endocrine disrupting genistein by ligninolytic enzymes from white rot fungi. *FEMS Microbiology Letters*, 244, 93–98.

Tamagawa, Y., Yamaki, R., Hirai, H., Kawai, S., & Nishida, T. (2006). Removal of estrogenic activity of natural steroidal hormone estrone by ligninolytic enzymes from white rot fungi. *Chemosphere*, 65, 97–101.

Tsutsumi, Y., Haneda, T., & Nishida, T. (2001). Removal of estrogenic activities of bisphenol A and nonylphenol by oxidative enzymes from lignin degrading basidiomycetes. *Chemosphere*, 42, 271–276.

Ütebay, B., Çelik, P., & Çay, A., (2020). Textile wastes: Status and perspectives. In A. Körlü (ed.), *Waste in Textile and Leather Sectors*. London: IntechOpen. https://doi.org/10.5772/intechopen.92234

Wang, C., Zhao, M., Wei, X., Li, T., & Lu, L. (2010) Characteristics of spore-bound laccase from *Bacillus subtilis* WD23 and its use in dye decolorization. *Advanced Materials Research*, 113–116, 226–230.

Wang, J., Lu, L. & Feng, F. (2017). Improving the indigo carmine decolorization ability of a *Bacillus amyloliquefaciens*laccase by site-directed mutagenesis. *Catalysts*,7, 275.

Yadav, S., Gangwar, N., Mittal, P., Sharama, S., & Bhatnagar, T. (2014). Isolation, screening and biochemical characterization of laccase producing bacteria for degradation of lignin. *International Journal of Science Education*, 1, 17–21.

Yaseen, D., & Scholz, M. (2019) Textile dye wastewater characteristics and constituents of synthetic effluents: A critical review. *International Journal of Environmental Science and Technology*, 16, 1193–1226.

Yoshida, H. (1883). LXIII.-chemistry of lacquer (Urushi). Part I. communication from the chemical society of Tokio. *Journal of the Chemical Society, Transactions*,43, 472–486.

Zaid, A., Hashim, A., & Alkhozai, Z. (2015). Dyes decolorization by spore-bound laccase from local isolate of *Bacillus subtilis. International Journal of Advanced Research*, 3(10), 1417–1424.

Zainith, S., Purchase, D., Saratale, S.D., Ferreira, L.F.R., Bilal, M., & Bharagava, R.N. (2019). Isolation and characterization of lignin-degrading bacterium Bacillus aryabhattai from pulp and paper mill wastewater and evaluation of its lignin- degrading potential. *3 Biotech.*, 9(3), 1–11. https://doi.org/10.1007/s13205-019-1631-x.

Zdarta, J., Antecka, K., Frankowski, R., Zgola Grzeskowiak, A., Ehrlich, H., & Jesionowski, T. (2018). The effects of operational parameters on the biodegradation of bisphenols by *Trametes versicolor* laccase immobilized on *Hippospongia communis* spongin scaffolds. *Science of the Total Environment*, 615, 784–795.

17 Application of Water Hyacinth in Toxic Metal Phytoremediation with Special Reference to Arsenic

Kelvin Mutugi Kithaka, Kumar Abhishek, Nidhi Mourya, and Sayan Bhattacharya

17.1 INTRODUCTION

The modern world is experiencing rapid human population growth, and, in turn, demand for goods and services is increasing in order to satisfy the needs of the people. On the other hand, excessive production and consumption increase the generation of diverse types of hazardous wastes. This calls for urgent action to control the generation and effects of hazardous wastes and to reduce their detrimental impacts on ecosystems and human health.

The term 'heavy metal' includes all the metals and metalloids which have gravity equal and greater than 4,000 kg/m^3 (Nammalwar, 1983; Nnaji and Okoye, 2007; Vardhan et al., 2019). Several heavy metals are considered to be toxic; however, as per the guidelines of US FDA, heavy metals are divided into two broad categories: essential and non-essential. Metals like manganese, iron, nickel, and copper are important for marine organisms to carry out their metabolic activities, whereas arsenic, cadmium, lead, and mercury exhibit toxicity and are categorized as non-essential hazardous metals (Kumar et al., 2012; Vardhan et al., 2019).

Hazardous heavy metals can impact different environmental conditions and biological organisms by the processes of accumulation, mobilization, chelation, bioaccumulation, biomagnification, etc. Heavy metals can exert toxicity on living organisms even in extremely low concentrations (Mohan and Singh, 2002; Vardhan et al., 2019).

Arsenic is a metalloid with detrimental toxicity and has been identified as carcinogenic to humans. Arsenic (As) has been known and used since ancient times in different parts of the world (Bhattacharya et al., 2012; Bhattacharya et al., 2021; Ivy et al., 2022). Arsenic exists in several oxidation states, –3, 0, +3, and +5 (Smedley and Kinniburgh, 2002); different compounds of arsenic present on Earth include arsenious acids (H_3AsO_3, H_3AsO_3, $H_3AsO_2^{-3}$), arsenic acids (H_3AsO_4, H_3AsO^{-4}, $H_3AsO_2^{-4}$), arsenites, arsenates, methylarsenic acid, dimethylarsinic acid, arsine, etc. Arsenic is primarily found in the environment due to geological processes and mineral deposits (Bissen and Frimmel, 2003). Anthropogenic activities like mining, burning fossil fuels and using chemicals containing arsenic in agriculture also contribute to the occurrence and deposition of arsenic in the environment. Arsenic can affect human health when exposed to a specific dose at a given period. It is now widely accepted that arsenic causes cancer, even at low doses (Mandal and Suzuki, 2002).

According to the provisional guideline given by the World Health Organization (WHO), 10 ppb (0.01 mg/L) is considered as the permissible limit of arsenic in drinking water, however, some countries still retain the previous WHO guideline of 50 ppb (0.05 mg/L), including Bangladesh and China (Raju, 2022). Arsenic pollution has been found in several countries such as the USA, Japan, China, Chile, Mexico, Taiwan, Poland, Hungary, Argentina, New Zealand, Canada, Bangladesh, and India. Bengal basin, encompassing parts of both India and Bangladesh, is considered to be the

DOI: 10.1201/9781003441144-17

largest geological area affected by arsenic pollution in groundwater, where more than 120 million people use arsenic-contaminated water for drinking and irrigation purposes (Bhattacharya et al., 2012; Bhattacharya et al., 2021; Ivy et al., 2022). The use of arsenic-contaminated irrigation water elevates arsenic concentration in the agricultural fields. Arsenic transmits through food chain in the Bengal basin and considerable deposition of arsenic was found in edible crops including rice, vegetables, meat, milk, and other foodstuffs (Bhattacharya et al., 2012; 2021).

In the last three decades, several methods are developed and become available for arsenic removal and treating contaminated water which is largely dependent on the chemical state in which arsenic is present in a particular area (Bhattacharya et al., 2013; Ray and Shipley, 2015; Alka et al., 2021). The chemical composition of arsenic-contaminated water is a significant factor in determining the extent of arsenic removal from water (Bhattacharya et al., 2023). Most of the removal methods available are more effective in removing arsenate than arsenite Bhattacharya et al., 2023). Because of this reason, trivalent arsenic is less available for adsorption, precipitation, or ion exchange. Accordingly, treatment technologies can be more effective by applying a two-step approach: oxidation process which converts arsenite to arsenate, followed by a suitable method to remove arsenate.

Some commonly used techniques for arsenic removal are (i) oxidation followed by precipitation, (ii) adsorption onto sorptive media, (iii) coagulation/electrocoagulation/co-precipitation, (iv) membrane technique, and (v) ion exchange (Nicomel et al., 2016; Ghosh (Nath) et al., 2019; Yadav et al., 2021). Metals, metal oxides, and nanoparticle-based materials are commercially available for arsenic removal. Some low-cost nanotubes, nanoparticle-impregnated adsorbents, and other nanocomposites are also available commercially for use (Bhattacharya et al., 2012; Bhattacharya et al., 2021; Alka et al., 2021; Mohan and Pittman, 2006; Mohmood et al., 2013). Advantages of some of the arsenic removal technologies are summarized in Table 17.1.

Several plants can uptake and accumulate contaminants from the environment and can detoxify the pollutants by various molecular and biochemical mechanisms. Phytoremediation is a remediation technique by which living plants and associated microorganisms present in the soil are applied

TABLE 17.1
Summary of Available Technologies Used in Arsenic Remediation

Removal Technologies	Advantages	References
Coagulation • Alum coagulation (aluminium sulphate) • Iron coagulation (Ferric chloride/ ferric sulphate) • Lime Softening	• Applicable in neutral pH levels • Chemicals are available commercially • Coagulation-filtration process is effective	Song and Gallegos-Garcia (2014); Pramanik et al. (2016)
Adsorption • Activated Alumina • Iron coated sand • Ion exchange resin • Biochar • Mixed oxide nanoparticles	• Low cost, secure operation, and handling • Suitable for removing both As(V) and As(III) • Well-defined medium and capacity; pH independent; high removal efficiency; sludge-free	Mohan and Pittman (2007); Alka et al. (2021)
Membrane techniques • Nanofiltration • Reverse osmosis Electrolysis	• Efficiently remove both inorganic arsenic forms from water • Well-defined and high-removal efficiency	Bhattacharya et al. (2023); Mohan and Pittman (2007).
Chemical/Air oxidation Precipitation	• Removes dissolved metal ions from solutions • Cost-effective method • Removal of specific contaminants	Alka et al. (2021)

TABLE 17.2
Different Types of Phytoremediation Mechanisms

Phytoextraction	In this process, plants uptake pollutants from soil and water, and translocate to and store in the harvestable biomass of the plants. Phytoextraction aims to remove pollutants form the contaminated sites. This process is usually observed in hyperaccumulating plants resistant to the pollutants
Phytostabilization	Plants reduce mobility and phytoavailibility of contaminants in the environment. This process does not remove pollutants from contaminated sites but reduces mobility and excludes metals from plant uptake
Phytovolatilization	Hyperaccumulating plants uptake pollutants from soil and water, and translocate to the aerial parts of the plants, and volatilize the pollutants in the air
Phytotransformation	This process is one kind of plant's defense mechanism to the environmental pollutants. The hyperaccumulating plants modify, inactivate, degrade (phytodegradation), or immobilize (phytostabilization) the pollutants through their metabolism
Rhizofiltration	Usually aquatic plants perform this process. The hyperaccumulating aquatic plants adsorb and absorb pollutants from aquatic environments (water and wastewater)

Source: Reprinted with Permission from Elsevier (Rahman and Hasegawa, 2011), License Number 5460671351860

to reduce the concentrations or toxic impacts of contaminants in the environment (Bhattacharya et al., 2021). Identifying and exploring effective hyperaccumulating plants is a key step in the process of phytoremediation. It is a cost-effective, eco-friendly method to remove heavy metals and metalloids, radionuclides, organic pollutants, etc. from contaminated water and soil. Phytoremediation has much lower cost of implementation and maintenance when compared to other available remediation methods (Van Aken, 2009). Different types of phytoremediation are shown in Table 17.2 (Rahman and Hasegawa, 2011).

Aquatic macrophytes are useful in phytoremediation of contaminated water due to their fast growth rate, production of considerable biomass, and high capacity of heavy metals and pollutant uptake. They perform better in water purification because of direct contact with polluted water (Sood et al., 2012).

Water hyacinth can be considered as a sustainable alternative for heavy metal bioremediation in aquatic systems. However, the plant can cause serious water management problems because of its high growth rate and vegetative reproduction. Proper management and disposal of phytoremediating water hyacinth are also a major constrain in the way of successful implementation of phytoremediation technology. This present work focuses on the attainment of ecological balance by the removal of arsenic by water hyacinth (an invasive species), which helps to integrate ecological balance with hazardous waste management.

17.2 USE OF INVASIVE PLANTS IN PHYTOREMEDIATION: A NEW PATH TO ESTABLISH ECOLOGICAL BALANCE WITH HAZARDOUS WASTE MANAGEMENT

Phytoremediation has been experimented and applied globally for cleaning of environmental wastes and is considered as an innovative method in sustainable waste management. The idea of using hyperaccumulating plants to reduce toxic metals and other hazardous compounds was first proposed in 1983, but the concept has been applied to wastewater discharges for long time (Blaylock, 2008). Phytoremediation is a novel, cost-effective, efficient, environment, and eco-friendly remediation strategy that is accepted widely (Turan and Esringu, 2007; Singh et al., 2009; Bhattacharya et al., 2021). Phytoremediation can be extremely useful in removing heavy metals such as lead, chromium, arsenic, zinc, cadmium, copper, mercury, etc. from soil, thereby reducing their intake by cultivable plants (Ali et al., 2013).

Water hyacinth has been applied in several cases for phytoremediation of toxic metals. It is a fast growing, floating hydrophyte with a well-developed fibrous root system and is capable of producing high biomass. It also has strong adaptive capacity in various environmental conditions and is able to remove heavy metals (ex. Cr, Hg, As, Cd, etc.) from water (Liao and Chang, 2004). Arsenic is translocated from the root to the shoot in water hyacinth plants, and the mechanism can be effective in removing the contaminant from the contaminated sites. Water hyacinth plants can perform phytoremediation by translocation, eliminating arsenic contaminants from different sites. Translocation ability is a crucial factor for plant species since it affects their capacity for phytoremediation (Xie et al., 2009; Bhattacharya et al., 2021). As(V) can be transformed to As(III) by the action of arsenic reductase enzyme, resulting in change of inorganic arsenic species in the plant system (Bhattacharya et al., 2012). It is noticed that arsenic generally has low translocation mobility from roots to shoots (Zhao et al., 2003; Bhattacharya et al., 2012).

Taking an explicit example, the ability of water hyacinth to purify pond water tainted with hazardous sewage has been established (Qin et al., 2016). According to a study by Sayago (2019), water hyacinth is efficient in phytoremediation of heavy metals and metalloids from wastewater and subsequently can be used for manufacturing bioethanol, a step towards sustainability. The plant also has been used to treat textile wastewater as an adsorbent (Priya and Selvan, 2017). Water hyacinth has diversified use, whereby alternative uses such as biofuel production have dominated the current market, focusing on maximizing the use of renewable energy. However, water hyacinth has been considered as a potential agent of phytoremediation among invasive species, therefore its use in phytoremediation can help in pollutant remediation and invasive species management in ecosystems (Singh et al., 2021). In terms of its efficiency in combustion and fewer carbon dioxide emissions and biodiesel-producing capacity, water hyacinth has been considered advantageous in diesel engines (Alagu et al., 2019).

17.3 APPLICATION OF WATER HYACINTH IN HEAVY METAL PHYTOREMEDIATION

17.3.1 Diverse Applications of Water Hyacinth in Environmental Remediation

Water hyacinth (*Eichhornia crassipes*) is an aquatic plant or weed with aggressive growth and is free-floating on the surface of water bodies (free-floating aquatic hydrophyte) (Rezania et al., 2015b). It is a short-height plant with medium-sized light-dark green leaves, rhizomes, beautiful purple-coloured flowers, and pendant roots. Stems and leaves of water hyacinth plants contain air-filled tissues that help the plant to float on the water surface. Being a member of the Pontederiaceae family, water hyacinth is indigenous to Brazil, the Amazon basin, and the Ecuador regions, concentrating mostly in the tropical and sub-tropical regions of the world (Téllez et al., 2008). Water hyacinth is able to tolerate prolonged dry condition and is able to survive in moist sediments for a long period of several months (Rezania et al., 2015b). Several initiatives have been taken in order to eradicate water hyacinth from different water bodies, however, this weed manages to propagate successfully in wide geographical locations.

Water hyacinth is considered as the fast growing and perishable plant; at the same time, it would have been considered a cheap and effective living agent in water treatment and remediation (Rezania et al., 2015b). It helps in removing detrimental pollutants, toxins, and heavy metals present in the water very quickly by absorbing them from water (Rezania et al., 2015b; Ajayi and Ogunbayo, 2012). Moreover, several studies have pointed out its diverse applications, which are classified and discussed as follows:

17.3.1.1 Animal Feed and Production

Water hyacinth spreads vigorously in the water bodies. To control their growth and because of the increase in the price of animal feed products available in the market, water hyacinth is used effectively as animal feed. The plant also has more protein and other nutrients than the available feed products in the market. Its leaves and roots are rich in protein and nutrients, which are mainly useful in preparing animal feeds. It is used primarily to provide essential nutrients to animals such as cattle, pigs, horses, and poultry (Rezania et al., 2015a, 2015b).

17.3.1.2 Organic Fertilizer Production

Water hyacinth has a very high amount of protein and plant nutrients like N, Ca, and K, therefore, can be a highly effective precursor for the production of composts. Water hyacinth is considered more beneficial than other plant residues as it absorbs more nutrients and releases them easily in manure (Abdel-Sabour, 2010).

17.3.1.3 Bioenergy Production

Water hyacinth is effectively used in bioenergy production. Biogas and bioethanol can be produced by using water hyacinth biomass. A mixture of dry water hyacinth and cow dung in a ratio of 70: 30% is considered perfect for methane production (Abdel-Sabour, 2010).

17.3.1.4 Production of Medicines

Due to the high nutritional value of water hyacinth, research has also found that it is used in many ways for medicinal purposes. Also, this plant is used for curing skin diseases in animals. (Rezania et al., 2015b; Guna et al., 2017).

17.3.1.5 Wastewater Treatment

Wastewater is water contaminated with many types of pollutants coming from domestic, agricultural, industrial, and other commercial sources, in which several types of chemicals and heavy metals are present. Several technologies and methods are used for the removal and treatment of wastewater contaminants. However, many of them are often considered as expensive and inconvenient. However, phytoremediation by using aquatic plants is considered as a low-cost and efficient technique of contaminated water and soil treatment (Bhattacharya et al., 2021; Roongtanakiat et al., 2007). Water hyacinth plant is extensively used in the remediation of contaminated wastewater, especially for heavy metal removal (Rezania et al., 2015b). Recent studies have also found that water hyacinth is a good source of bio-sorption and a biological indicator of heavy metals in contaminated water (Priya and Selvan, 2017).

17.3.2 Use of Water Hyacinth Whole Plant

Several parts of water hyacinth plants, such as leaves, rhizomes, flowers, stems and young whole plants are used in contaminated water treatment. Water hyacinth roots are useful in purifying water by absorbing pollutants and heavy metals present in water. Previous research showed that the heavy metals can be absorbed into the plant through its roots and can be transferred to upper parts of the plant, such as stems, shoots, and leaves, where they can bioaccumulate subsequently. Arsenic can be removed by the whole plant of water hyacinth; the outline of the process is shown in Figure 17.1.

In constructed wetlands, different types of aquatic species are used for water remediation and purification. The floating hydrophytes can accumulate different heavy metals by their root systems while in case of submerged plants, accumulation of heavy metals can occur by the entire plant body (Rahman and Hasegawa, 2011). Water hyacinth (*Eichhornia crassipes*) has been applied for heavy metal phytoremediation in constructed wetlands. The plant also adapts comfortably in various aquatic ecosystems and conditions and plays an important role in bioaccumulating heavy metals from water (Liao and Chang, 2004).

In one of the early experiments, water hyacinth plant was applied to toxic metal removal from water. Batch studies were conducted and the uptake of heavy metals (As, Cr, Hg, Ni, Pb, and Zn) from the aqueous solution with 6 different concentrations (5 mg/L to 50 mg/L) was studied. It was

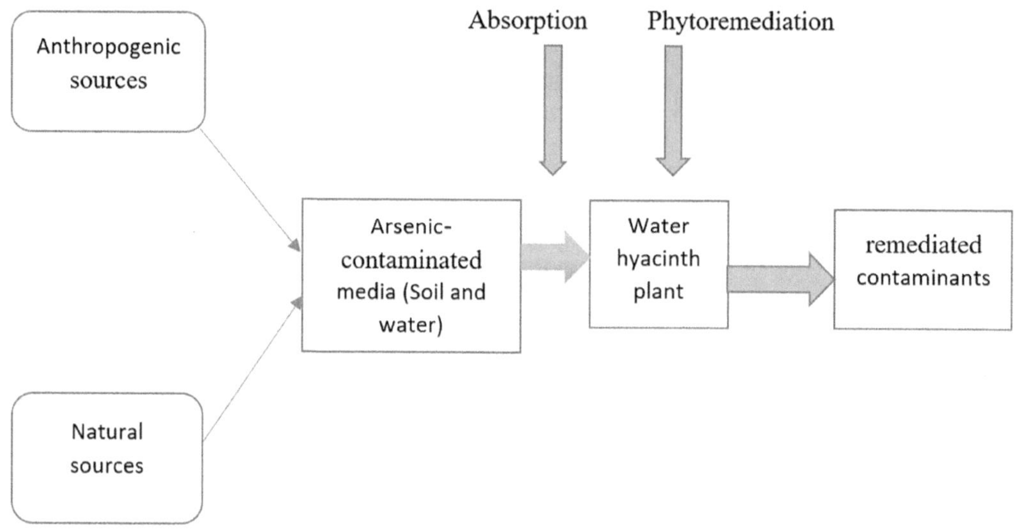

FIGURE 17.1 Phytoremediation of arsenic-contaminated water by water hyacinth plants.

seen that the efficiency of heavy metal removal by water hyacinth decreased with increase in the concentration of the metals. Results also showed that at lower concentrations (less than 5 mg/L of heavy metals), normal plant growth was recorded and removal efficiency was higher. At increased concentrations (more than 10 mg/L), wilting of the plant was initiated and removal efficiency was reduced (Ingole and Bhole, 2003). Another recent study showed that water hyacinth can reduce cadmium concentration by 59.4%, arsenic concentration by 60.8%, lead concentration by 92.4%, zinc concentration by 60.22%, and copper concentration by 60.74% from contaminated wastewater streams (Huynh et al., 2021).

Arsenic removal potential of water hyacinth (*Eichhornia crassipes*) and lesser duckweed (*Lemna minor*) was studied in another experiment. The removal efficiency of water hyacinth was found to be higher (18% compared to 5% in *Lemna*) due to the production of biomass and the more favourable climatic conditions of the experimental area for water hyacinth (Alvarado et al., 2008).

Plant biomass containing bioaccumulated heavy metals can be burned under appropriate conditions to produce energy and the ash produced (which is called bio-ore) can be processed further for the recovery or extraction of the heavy metals. This process of heavy metal extraction is called phytomining (Rezania et al., 2015b; Ali et al., 2013). The collected metals can be reused and the problem of disposal of HM contaminated plants can be minimized.

17.3.3 Use of Water Hyacinth Plant Biomass Powder

Biomass of water hyacinth has a considerable capacity for adsorption because of the presence of high amount of cellulose and functional groups like carboxyl (–COO-) and hydroxyl (–OH) (Patel, 2012). The biochar materials derived from water hyacinth are endowed with many properties, such as being highly carbon-rich and having porous structures. Application of water hyacinth biomass in biochar production can convert this invasive plant into a valuable resource and a substantial method of management and utilization of this species can be established. Several researchers have developed biochar from water hyacinth plant biomass. It has wide applications as soil fertilizer, biofuel, and bio-manure. Additionally, it is used as a sustainable purification agent in water purification and treatment (Zhang et al., 2016; Chen et al., 2018; Yu et al., 2018).

It was demonstrated that the dead and dried roots of *Eichhornia crassipes* plants are capable of quick and effective removal of arsenic from contaminated water; more than 93% of arsenite (As(III)) and 95% of arsenate (As(V)) were removed from a 200 ppb As solution by giving 60 minutes exposure to dried water hyacinth roots powder (Al Rmalli et al., 2005). In another study, phytoremediation in arsenic removal using dried roots of *Eichhornia crassipes* was studied (Govindaswamy et al., 2011). Water hyacinth roots were washed, dried, and powdered to develop dried hyacinth roots (DHR). Batch and continuous column experiments showed effective removal of arsenic, thereby further confirming the results shown by Al Rmalli et al. (2005). Citric acid (CA) cross-linked water hyacinth root powder (RP) was applied to remove arsenic from water (Gogoi et al., 2017). The material can reduce arsenic concentration within the permissible limit recommended by the World Health Organization.

17.3.4 Genetic Perspectives of Heavy Metal Hyperaccumulation in Water Hyacinth: Potential and Prospect

Water hyacinth plants may contain certain genes that develop traits to absorb, hyper-accumulate, assimilate, and convert harmful toxicants into innocuous substances. However, no concrete study has been done so far to explore the genes present in water hyacinth plants to facilitate the process of phytoremediation. Plasmalemma of a water hyacinth plant absorbs contaminants through the root, which are then stored in the root cells and stem of the plant.

Genetic resistance to toxic substances in water hyacinths is a promising aspect of future solutions for waste management in contaminated ecosystems. Experimental evidence showed when the plant absorbs heavy metals through its roots, those can subsequently spread into the upper parts through translocation. Several heavy metals can be collected and stored in the water hyacinth roots, presumably in multiple vacuoles in the root tissues. At the same time, some are transferred to other aerial portions of the plant via the xylem (Abbas et al., 2019).

Phytoremediation potential of water hyacinth can lead to opportunities for industries to start investing in local areas by buying water hyacinths and using them in remediation of industrial hazardous contaminated sites. Additionally, the living standards of the local people can be improved due to economic gains by selling water hyacinth plants to industrial sectors and the process can help to reduce environmental contaminants in water bodies at local levels.

17.4 APPLICATION OF WATER HYACINTH BIOCHAR IN ARSENIC REMOVAL

Biochar is considered as low-cost, sustainable, and efficient biomaterial for remediation of arsenic from contaminated water. Biochar is a carbon-based stable product, which is generally synthesized by pyrolysis technique (250 to 750°C), using solid wastes as feedstock materials with little or no oxygen supply (Sohi, 2012; Afshar and Mofatteh, 2024). The application of biochar/nano char helps to improve soil organic matter, soil nutrient capacity, and carbon sequestration and can immobilize heavy metals in soil (Mohan et al., 2018; Sohi, 2012; Weber and Quicker, 2018). The development of nanomaterials, especially from invasive plant species is a major step toward sustainable remediation of arsenic from wastewater. Nanomaterials resemble small atomic clusters with less than 100 nm size (Mekuye & Abera, 2023). The selection of nanomaterial/nano adsorbent needs to be done carefully to achieve maximum efficiency in arsenic removal from drinking water. Various types of nanocomposites and nanomaterials have been used previously for the removal of arsenic from wastewater because these materials can absorb contaminants by ion exchange, ion precipitation, and adsorption. Different types of biochar are developed from wood and bark, oak wood and bark, solid waste, rice husks, biosolids, and animal products, and are used to remove arsenic from contaminated water. Biochar particles are developed from diverse types of biomaterials, and their applications in arsenic removal from water have been effectively performed (Sadegh-Zadeh and Seh-Bardan, 2013; Bhattacharya et al., 2023). For example, oakwood, and pinewood empty fruit brunch biochar, *Tectona* and *Lagerstroemia* leaves biochar, perilla leaves and Japanese oak wood biochar, paper mill sludge derived biochar, tea waste, almond shell biochar, wheat straw derived biochar, potato peel and rice husk ash, and bagasse fly ash are used in arsenic removal from aqueous solutions (Mohan and Pittman, 2007; Sinha et al., 2022). Van Vinh et al. (2015) showed 66% removal efficiencies in case of As(III) by biochar derived from raw cone pine wood. Moreover, sewage sludge-derived biochar showed efficiency in removing arsenic from water (Khan et al., 2023). Removal of arsenic in both the forms, As(III) and As(V) in aqueous solution has been highly successful by using biochar materials (Bhattacharya et al., 2023; Sadegh-Zadeh and Seh-Bardan, 2013). The presence of the functional group in biochar plays a key role in arsenic removal from contaminated solution (Sharma et al., 2022).

Several types of biochars have been developed from water hyacinth plants by researchers. Many of them showed promising results in the removal of heavy metals including arsenic from contaminated water. The full body of *Eichhornia crassipes* was applied to produce biochar in one of the recent studies. The biochar showed 99% efficiency in chromium removal from tannery wastewater under optimum conditions (Hashem et al., 2020). Another study successfully explored the capacity of water hyacinth biochar-alginate capsules to remove Cd^{+2} from aqueous medium. The materials showed excellent adsorption capacity for cadmium, with maximum adsorption of 46 mg Cd^{+2g_1} (Liu et al., 2020). Water hyacinth biochar was also utilized in the treatment of U(VI)-containing aqueous solutions; the solution pH has considerable influence on U(VI) adsorption, while the ionic strength has no significant effect (Xu et al., 2020). Yu et al., (2018) prepared water hyacinth biochar, modified the material with ZnO nanoparticles and subsequently applied it for Cr(VI) removal from an aqueous solution; more than 95% removal efficiency of Cr(VI) was shown by the material, even at natural pH. In another recent research work, response surface method (RSM) was found to be capable of optimizing the conditions of water hyacinth biochar preparation and was able to improve the adsorption ability of lead (Pb^{2+}) (Zhou et al., 2021). Similar materials were also successful in cadmium removal from contaminated water (Zhou et al., 2019). Results from another study highlight that the production of biochar from water hyacinth could become a "win-win" strategy for both the removal of cadmium from contaminated water and better management of this invasive species. This study explored the mechanisms and effectiveness of Cd removal by biochar pyrolyzed from water and found it satisfactory (Liu et al., 2020). Recently, a novel nanoscale zero-valent iron (nZVI) coated *E. crassipes* biochar was developed by a group of researchers, and the material was applied successfully to remove Cd(II) from aqueous solutions (Chen et al., 2018). Ding et al. (2014) studied the competitive removal of Cd (II) and Pb (II) by water hyacinth biochar. Pb^{2+} was found to have a higher amount of adsorption on water hyacinth biochar compared to Cd^{2+}, and the coexistence of Pb^{2+} and Cd^{2+} could affect the adsorption capacity of each other. The desorption studies showed that water hyacinth biochar had the potential of regeneration for Cd^{2+} and Pb^{2+} adsorption.

Magnetic biochars are gaining attention in wastewater treatment nowadays because of their high adsorption potential. Magnetic biochars synthesized by chemical co-precipitation of ferrous or ferric ions on water hyacinth biomass followed by pyrolysis showed promising results in As(V) removal from contaminated water. This work shows an effective way to manage and make profit from water hyacinth (Zhang et al., 2016).

17.5 ADVANTAGES AND LIMITATIONS OF HEAVY METAL REMEDIATION BY WATER HYACINTH

There are many uses and benefits of water hyacinth, such as for water treatment, animal feeds, making organic fertilizers, bioenergy, and medicinal purposes. Phytoremediation of heavy metals and arsenic by using water hyacinth is an environment-friendly, cost-effective, and sustainable method. Besides, the synthesis of biochar from water hyacinth and its application in arsenic and other heavy metal adsorption showed immense potential while keeping sustainability in focus. However, harvesting of water hyacinth is also very difficult; as several physical and mechanical methods need to be resorted in order to maximize water hyacinth harvesting, which is very expensive and inconvenient.

Nano-biochar derived from water hyacinth plants can also pose health risks after being released into the environment. Suitable risk assessment and safety measures need to be implemented before synthesizing and applying these nanomaterials in a large scale.

17.6 RESEARCH POTENTIAL AND FUTURE DIRECTION

Several researches have been conducted to determine how different invasive species can be used to remove hazardous waste from the environment. Phytoremediation cleanup techniques using such plants are proved to be helpful, especially in the areas where pollution in water and soil is reported. The specific percentage of efficiency to how phytoremediation can reliably remove contaminants is yet to be stated clearly. Exploring suitable genes present in the aquatic macrophytes (like water hyacinth) which are responsible for heavy metal phytoremediation can be useful in implementing phytoremediation strategies. Isolation of appropriate genes present in water hyacinths and other aquatic macrophytes and the insertion of those genes in other plants can open up new dimensions of phytoremediation by the application of genetic engineering.

REFERENCES

Abbas, Z., Arooj, F., Ali, S., Zaheer, I. E., Rizwan, M., & Riaz, M. A. (2019). Phytoremediation of landfill leachate waste contaminants through floating bed technique using water hyacinth and water lettuce. *International Journal of Phytoremediation*, *21*(13), 1356–1367.

Abdel-Sabour, M. F. (2010). Water hyacinth: available and renewable resource. *Electronic Journal of Environmental, Agricultural & Food Chemistry*, *9*(11), 1746–1759.

Afshar, M., & Mofatteh, S. 2024. Biochar for a sustainable future: environmentally friendly production and diverse applications. *Results in Engineering, 23,* 102433.

Ajayi, T. O., & Ogunbayo, A. O. (2012). Achieving environmental sustainability in wastewater treatment by phytoremediation with water hyacinth (Eichhornia crassipes). *Journal of Sustainable Development*, *5*(7), 80.

Al Rmalli, S. W., Harrington, C. F., Ayub, M., & Haris, P. I. 2005. A biomaterial based approach for arsenic removal from water. *Journal of Environmental Monitoring, 7(4)*, 279–282.

Alagu, K., Venu, H., Jayaraman, J., Raju, V. D., Subramani, L., Appavu, P., & Dhanasekar, S. (2019). Novel water hyacinth biodiesel as a potential alternative fuel for existing unmodified diesel engine: Performance, combustion and emission characteristics. *Energy*, *179*, 295–305.

Ali, H., Khan, E., & Sajad, M. A. (2013). Phytoremediation of heavy metals--concepts and applications. *Chemosphere, 91(7)*, 869–881.

Alka, S., Shahir, S., Ibrahim, N., Ndejiko, M. J., Vo, D.-V. N., & Abd Manan, F. (2021). Arsenic removal technologies and future trends: a mini review. *Journal of Cleaner Production*, *278*, 123805.

Alvarado, S., Guédez, M., Lué-Merú, M. P., Nelson, G., Alvaro, A., Jesús, A. C., & Gyula, Z. (2008). Arsenic removal from waters by bioremediation with the aquatic plants Water Hyacinth (Eichhornia crassipes) and Lesser Duckweed (Lemna minor). *Bioresource Technology*, *99*(17), 8436–8440.

Bhattacharya, S., Gupta, K., Debnath, S., Ghosh, U. C., Chattopadhyay, D., & Mukhopadhyay, A. (2012). Arsenic bioaccumulation in rice and edible plants and subsequent transmission through food chain in Bengal basin: a review of the perspectives for environmental health. *Toxicological & Environmental Chemistry*, *94*(3), 429–441.

Bhattacharya, S., Sharma, P., Mitra, S., Mallick, I., & Ghosh, A. (2021). Arsenic uptake and bioaccumulation in plants: A review on remediation and socio-economic perspective in Southeast Asia. *Environmental Nanotechnology, Monitoring & Management*, *15*, 100430.

Bhattacharya, S., Talukdar, A., Sengupta, S., Das, T., Dey, A., Gupta, K., & Dutta, N. (2023). Arsenic contaminated water remediation: A state-of-the-art review in synchrony with sustainable development goals. *Groundwater for Sustainable Development, 23,* 101100.

Bissen, M., & Frimmel, F. H. (2003). Arsenic-a review. Part I: occurrence, toxicity, speciation, mobility. *Acta Hydrochimica et Hydrobiologica*, *31*(1), 9–18.

Blaylock, M. (2008). *Phytoremediation of contaminated soil and water: field demonstration of phytoremediation of lead contaminated soils*. Boca Raton, FL: Lewis Publishers, .

Chen, L., Li, F., Wei, Y., Li, G., Shen, K., & He, H. 2018. High cadmium adsorption on nanoscale zero-valent iron coated Eichhornia crassipes biochar. *Environmental Chemistry Letters, 17*, 589–594.

Ding, W., Dong, X., Ime, I. M., Gao, B., & Ma, L. Q. (2014). Pyrolytic temperatures impact lead sorption mechanisms by bagasse biochars. *Chemosphere, 105*, 68–74.

Ghosh (Nath), S., Debsarkar, A., & Dutta, A. (2019). Technology alternatives for descontamination of arsenic-rich groundwater - a critical review. *Environmental Technology and Innovation, 13*, 277–303.

Gogoi, P.J., Adhikari, P., & Maji, T.K. 2017. Bioremediation of arsenic from water with citric acid cross-linked water hyacinth (E. crassipes) root powder. *Environmental Monitoring and Assessment, 189*, 1–11.

Govindaswamy, S., Schupp, D. A., & Rock, S. A. (2011). Batch and continuous removal of arsenic using hyacinth roots. *International Journal of Phytoremediation, 13*(6), 513–527.

Guna, V., Ilangovan, M., Prasad, M., & Reddy, N. 2017. Water Hyacinth: A Unique Source for Sustainable Materials and Products. *ACS Sustainable Chemistry & Engineering, 5,* 4478–4490.

Hashem, M. A., Hasan, M., Momen, M. A., Payel, S., & Nur-A-Tomal, M. S. (2020). Water hyacinth biochar for trivalent chromium adsorption from tannery wastewater. *Environmental and Sustainability Indicators, 5*, 100022.

Huynh, A. T., Chen, Y.-C., & Tran, B. N. T. (2021). A small-scale study on removal of heavy metals from contaminated water using water hyacinth. *Processes, 9*(10), 1802.

Ingole, N. W., & Bhole, A. G. (2003). Removal of heavy metals from aqueous solution by water hyacinth (Eichhornia crassipes). *Journal of Water Supply: Research and Technology-AQUA, 52*(2), 119–128.

Ivy, N., Mukherjee, T., Bhattacharya, S., Ghosh, A., & Sharma, P. (2022). Arsenic contamination in groundwater and food chain with mitigation options in Bengal delta with special reference to Bangladesh. *Environmental Geochemistry and Health*, 1–27.

Jia, R., Qu, Z., You, P., & Qu, D. (2018). Effect of biochar on photosynthetic microorganism growth and iron cycling in paddy soil under different phosphate levels. *Science of the Total Environment, 612*, 223–230. https://doi.org/https://doi.org/10.1016/j.scitotenv.2017.08.126

Khan, R., Shukla, S., Kumar, M., Zuorro, A., & Pandey, A. 2023. Sewage sludge derived biochar and its potential for sustainable environment in circular economy: Advantages and challenges. *Chemical Engineering Journal, 471,* 144495.

Kumar, B., Sajwan, K. S., & Mukherjee, D. P. (2012). Distribution of heavy metals in valuable coastal fishes from North East Coast of India. *Turkish Journal of Fisheries and Aquatic Sciences, 12*, 917–924. DOI : 10.4194/1303-2712-v12-1-10

Liao, S. W., & Chang, W. L. (2004). Heavy metal phytoremediation by water hyacinth at constructed wetlands in Taiwan. *Journal of Aquatic Plant Management, 42,* 60–68.

Liu, X., Sun, J., Duan, S., Wang, Y., Hayat, T., Alsaedi, A., Wang, C., & Li, J. (2017). A valuable biochar from poplar catkins with high adsorption capacity for both organic pollutants and inorganic heavy metal ions. *Scientific Reports, 7*(1), 1–12.

Liu, C., Ye, J., Lin, Y., Wu, J., Price, G. W., Burton, D. L., & Wang, Y. 2020. Removal of Cadmium (II) using water hyacinth (Eichhornia crassipes) biochar alginate beads in aqueous solutions. *Environmental Pollution, 264*, 114785.

Mandal, B. K., & Suzuki, K. T. (2002). Arsenic round the world: a review. *Talanta, 58*(1), 201–235.

Mekuye, B., & Abera, B. H. 2023. Nanomaterials: An overview of synthesis, classification, characterization, and applications. *Nano Select, 4, 486–501.*

Mohan, D., & Pittman, C. U. 2007. Arsenic removal from water/wastewater using adsorbents--A critical review. *Journal of Hazardous Materials, 142 1–2*, 1–53.

Mohan, D., Abhishek, K., Sarswat, A., Patel, M., Singh, P., & Pittman, C. U. (2018). Biochar production and applications in soil fertility and carbon sequestration-a sustainable solution to crop-residue burning in India. *RSC Advances, 8*(1), 508–520.

Mohan, D., & Pittman Jr, C. U. (2006). Activated carbons and low cost adsorbents for remediation of tri-and hexavalent chromium from water. *Journal of Hazardous Materials, 137*(2), 762–811.

Mohan, D., & Singh, K. P. (2002). Single-and multi-component adsorption of cadmium and zinc using activated carbon derived from bagasse-an agricultural waste. *Water Research, 36*(9), 2304–2318.

Mohmood, I., Lopes, C. B., Lopes, I., Ahmad, I., Duarte, A. C., & Pereira, E. (2013). Nanoscale materials and their use in water contaminants removal-a review. *Environmental Science and Pollution Research, 20*(3), 1239–1260.

Nammalwar, P. (1983). Heavy metal pollution in the marine environment. *Science Reporter*, (March), 158–160.

Nicomel, N. R., Leus, K., Folens, K., Van Der Voort, P., & Du Laing, G. (2016). Technologies for arsenic removal from water: current status and future perspectives. *International Journal of Environmental Research and Public Health, 13*, 62.

Nnaji, J., & Okoye, F. C. (2007). *Toxicity of heavy metals to fish: An important consideration for succesful aquaculture* (21st Annual). Fisheries Society of Nigeria (FISON).

Patel, S. (2012). Threats, management and envisaged utilizations of aquatic weed Eichhornia crassipes: an overview. *Reviews in Environmental Science and Bio/Technology*, *11*(3), 249–259.

Priya, E. S., & Selvan, P. S. (2017). Water hyacinth (*Eichhornia crassipes*) – An efficient and economic adsorbent for textile effluent treatment – A review. *Arabian Journal of Chemistry,* 10(2), S3548–S3558.

Priya, E. S., & Selvan, P. S. (2017). Water hyacinth (Eichhornia crassipes)-An efficient and economic adsorbent for textile effluent treatment-A review. *Arabian Journal of Chemistry*, *10*, S3548–S3558.

Qin, H., Zhang, Z., Liu, M., Liu, H., Wang, Y., Wen, X., Zhang, Y., & Yan, S. (2016). Site test of phytoremediation of an open pond contaminated with domestic sewage using water hyacinth and water lettuce. *Ecological Engineering*, *95*, 753–762.

Rahman, M. A., & Hasegawa, H. (2011). Aquatic arsenic: phytoremediation using floating macrophytes. *Chemosphere*, *83*(5), 633–646.

Raju, N. J. (2021). Arsenic in the geo-environment: A review of sources, geochemical processes, toxicity and removal technologies. *Environmental Research*, *203,* 111782.

Rezania, S., Ponraj, M., Din, M. F., Songip, A. R., Sairan, F. M., & Chelliapan, S. (2015a). The diverse applications of water hyacinth with main focus on sustainable energy and production for new era: an overview. *Renewable and Sustainable Energy Reviews*, *41,* 943–954.

Rezania, S., Ponraj, M., Talaiekhozani, A., Mohamad, S. E., Din, M. F. M., Taib, S. M., Sabbagh, F., & Sairan, F. M. (2015b). Perspectives of phytoremediation using water hyacinth for removal of heavy metals, organic and inorganic pollutants in wastewater. *Journal of Environmental Management*, *163*, 125–133.

Roongtanakiat, N., Tangruangkiat, S., & Meesat, R. (2007). Utilization of vetiver grass (Vetiveria zizanioides) for removal of heavy metals from industrial wastewaters. *Science Asia*, *33*(4), 397–403.

Sadegh-Zadeh, F., & Seh-Bardan, B. J. (2013). Adsorption of As (III) and As (V) by Fe coated biochars and biochars produced from empty fruit bunch and rice husk. *Journal of Environmental Chemical Engineering*, *1*(4), 981–988.

Sayago, U. F. C. (2019). Design of a sustainable development process between phytoremediation and production of bioethanol with Eichhornia crassipes. *Environmental Monitoring and Assessment*, *191*(4), 1–8.

Sharma, P. K., Kumar, R., Singh, R. K., Sharma, P., & Ghosh, A. K. 2022. Review on arsenic removal using biochar-based materials. *Groundwater for Sustainable Development*, *17*, 100740.

Singh, A., Eapen, S., & Fulekar, M. H. (2009). Potential of Medicago sativa for uptake of cadmium from contaminated environment. *Romanian Biotechnological Letters*, *14*(1), 4164–4169.

Singh, S., Saha, L., Kumar, M., & Bauddh, K. (2021). Phytoremediation potential of invasive species growing in mining dumpsite. In *Phytorestoration of Abandoned Mining and Oil Drilling Sites* (pp. 287–305). Elsevier.

Smedley, P. L., & Kinniburgh, D. G. (2002). A review of the source, behaviour and distribution of arsenic in natural waters. *Applied Geochemistry,* *17*, 517–568.

Sohi, S. P. 2012. Carbon storage with benefits. *Science, 338*, 1034–1035.

Song, S., & Gallegos-Garcia, M., 2014. *Arsenic Removal from Water by the Coagulation Process. The Role of Colloidal Systems in Environmental Protection*. Elsevier, pp. 261–277.

Sood, A., Uniyal, P. L., Prasanna, R., & Ahluwalia, A. S. (2012). Phytoremediation potential of aquatic macrophyte, Azolla. *Ambio*, *41*(2), 122–137.

Téllez, T. R., López, E., Granado, G. L., Pérez, E. A., López, R. M., & Guzmán, J. M. S. (2008). The water hyacinth, Eichhornia crassipes: an invasive plant in the Guadiana River Basin (Spain). *Aquatic Invasions*, *3*(1), 42–53.

Turan, M., & Esringu, A. (2007). Phytoremediation based on canola (Brassica napus L.) and Indian mustard (Brassica juncea L.) planted on spiked soil by aliquot amount of Cd, Cu, Pb, and Zn. *Plant Soil and Environment*, *53*(1), 7.

Van Aken, B. (2009). Transgenic plants for enhanced phytoremediation of toxic explosives. *Current Opinion in Biotechnology*, *20*(2), 231–236.

Van Vinh, N., Zafar, M., Behera, S. K., & Park, H.-S. (2015). Arsenic (III) removal from aqueous solution by raw and zinc-loaded pine cone biochar: equilibrium, kinetics, and thermodynamics studies. *International Journal of Environmental Science and Technology*, *12*(4), 1283–1294.

Wang, X.-W., Li, J.-S., Guo, T.-K., Zhen, B., Kong, Q.-X., Yi, B., Li, Z., Song, N., Jin, M., Xiao, W.-J., Zhu, X.-M., Gu, C.-Q., Yin, J., Wei, W., Yao, W., Liu, C., Li, J.-F., Ou, G.-R., Wang, M.-N., … Li, J.-W. (2005). Concentration and detection of SARS coronavirus in sewage from Xiao Tang Shan Hospital and the 309th Hospital. *Journal of Virological Methods*, *128*(1), 156–161. https://doi.org/https://doi.org/10.1016/j.jviromet.2005.03.022

Woolf, D., Amonette, J. E., Street-Perrott, F. A., Lehmann, J., & Joseph, S. (2010). Sustainable biochar to mitigate global climate change. *Nature Communications*, *1*(1), 1–9.

Xie, Q.-E., Yan, X.-L., Liao, X.-Y., & Li, X. (2009). The arsenic hyperaccumulator fern Pteris vittata L. *Environmental Science & Technology*, *43*(22), 8488–8495.

Xu, Z., Xing, Y., Ren, A., Ma, D., Li, Y., & Hu, S. (2020). Study on adsorption properties of water hyacinth-derived biochar for uranium (VI). *Journal of Radioanalytical and Nuclear Chemistry*, *324*(3), 1317–1327.

Vardhan, K. H., Kumar, P. S., & Panda, R. C. (2019). A review on heavy metal pollution, toxicity and remedial measures: Current trends and future perspectives. *Journal of Molecular Liquids*, *290*, 111197.

Yadav, M. K., Saidulu, D., Gupta, A. K., Ghosal, P. S., & Mukherjee, A. (2021). Status and management of arsenic pollution in groundwater: A comprehensive appraisal of recent global scenario, human health impacts, sustainable field-scale treatment technologies. *Journal of Environmental Chemical Engineering*, *9*, 105203.

Yu, J., Jiang, C., Guan, Q., Ning, P., Gu, J., Chen, Q., Zhang, J., & Miao, R. (2018). Enhanced removal of Cr (VI) from aqueous solution by supported ZnO nanoparticles on biochar derived from waste water hyacinth. *Chemosphere*, *195*, 632–640.

Zhang, F., Wang, X., Xionghui, J., & Ma, L. (2016). Efficient arsenate removal by magnetite-modified water hyacinth biochar. *Environmental Pollution*, *216*, 575–583.

Zhao, F. J., Wang, J. R., Barker, J. H. A., Schat, H., Bleeker, P. M., & McGrath, S. P. (2003). The role of phytochelatins in arsenic tolerance in the hyperaccumulator Pteris vittata. *New Phytologist*, *159*(2), 403–410.

Zhou, R., Zhang, M., Zhou, J., & Wang, J. (2019). Optimization of biochar preparation from the stem of Eichhornia crassipes using response surface methodology on adsorption of Cd2+. *Scientific Reports*, *9*(1), 1–17.

Zhou, Y., Qin, S., Verma, S., Sar, T., Sarsaiya, S., Ravindran, B., Liu, T., Sindhu, R., Patel, A. K., & Binod, P. (2021). Production and beneficial impact of biochar for environmental application: A comprehensive review. *Bioresource Technology*, *337*, 125451.

18 Wastewater Remediation via Actinomycetes

Current Trends and Future Prospects

C. Poornachandhra, Thangaraj Gokul Kannan, M. Saratha, J. Ezra John, and K. Angappan

18.1 INTRODUCTION

Revamping economies, rapid industrialisation, urban growth, and tremendous population inflation are the main contributors to water quality deterioration. Industrial effluents contain harmful inorganic and organic contaminants that severely pollute nearby soil and water streams, exerting an adverse effect on all life forms (Chandra et al., 2015). Recently developed microbe-based bioremediation techniques have made it possible to successfully remediate wastewater in a sustainable way. In the bioremediation technique, harmful toxins and compounds in wastewater could transform into harmless end products like CO_2 and H_2O via cellular metabolisms and also favours the recovery of valuable metals, nutrients, energy, and certain organic substances from wastewater. For millennia, microorganisms have been used to develop an array of products, inclusive of probiotics, enzymes, and biofuels like bioethanol, hydrogen gas, etc. These microbes are forthwith playing significant roles in the industrial scale clean-up of hazardous wastes and are the primary engineers in controlling all ecological processes. Numerous microorganisms, including bacteria, fungi, yeasts, and algae, have the ability to operate as biologically active methylators, which can at least change dangerous species. In this line, Actinomycetes, a broad group of bacteria, is one such great example to remediate wastewater (Hamedi and Mohammadipanah, 2015). They are interesting candidates for wastewater treatment because of their high catabolic rate, capacity to produce a vast range of bioactive chemicals, and ubiquity in the environment. Likely, Actinomycetes can use a variety of growth substrates in wastewater treatment plants, including sugars, high molecular weight polysaccharides, proteins, and aromatic molecules. Actinomycetes have several advantages over other wastewater bacteria, including better resilience to ultraviolet irradiation and desiccation, as well as the capacity to store polyphosphates and poly-hydroxybutyric acid. Actinomycetes have been holding active records in the degradation of organic components inside wastewater bioreactors. Some actinomycetes can also create biofilms, which they can use to break down complex polymers in the wastewater (Timková et al., 2018). This property makes them useful for composting or bioremediation activities, especially those involving harmful pesticides (Jain et al., 2022). Accordingly, the precise function of actinomycetes in wastewater remediation is covered in this chapter.

18.2 BIOREMEDIATION OF WASTEWATER

18.2.1 Concept of Bioremediation: A Viable Solution to Pollution

In accordance with the US EPA, United States Environmental Protection Agency, bioremediation is the "use of living organisms to clean up or remove pollutants from soil, water, or wastewater; use of

DOI: 10.1201/9781003441144-18

organisms such as non-harmful insects to remove agricultural pests or counteract diseases of trees, plants, and garden soil" (Cristaldi et al., 2017). This method can be used to remove heavy metals, metalloids, and other inorganic pollutants from soil or water. Effective bioremediation of eliminating inorganic pollutants depends on the physicochemical properties of the soil, water, microbial, and plant exudates, and the capacity of living organisms to accumulate, sequester, assimilate, and detoxify the pollutants (Khalid et al., 2017). Bioremediation process revealed to be more inexpensive, effective, distinctive, ecologically friendly, and solar-powered than mechanical treatments like incineration, excavation, solidification, soil washing, and flushing (Sarwar et al., 2017).

The concept of bioremediation is based on biodegradation. Bioremediation involves living microorganisms, usually naturally occurring bacteria, to break down toxic compounds found in wastewater into simpler less or non-toxic forms. Although it is green and commercially feasible technology, its potency can be variable depending on the region (Wang and Tam, 2019). During the process of bioremediation, microbes utilise these pollutants as nutrients or energy sources. Some native microorganisms may already be on the site, while others may be isolated from other sources and augmented into the polluted area. Even though the efficacy of bioremediation depends on the growth rate and degradation activity of microbes, it is also influenced by environmental conditions (Ojha et al., 2021).

18.2.2 Mechanisms Associated with Microbial Bioremediation

Immobilising and reducing the bioavailability of contaminants is the main mechanism of bioremediation. Heavy metals are an example of an inorganic contaminant that can only be transformed into other forms by microbes (Ashraf et al., 2019). For instance, to form biofilms, microbes have been supported by solid materials (Agro-residue, polypropylene, etc.) which in turn improves the dissolution and decomposition of pollutants (Mahdi et al., 2017). The primary bioremediation mechanisms are intracellular accumulation, extracellular complexation, precipitation, and oxidation-reduction processes. Bioremediation involves two mechanisms, such as biosorption and bioaccumulation. By absorbing toxins from the environment, a process known as bioaccumulation causes their concentration or accumulation to increase (Wang et al., 2016). There are both intracellular and external processes. The only kind of biomass that is involved in bioaccumulation is living biomass, which makes this approach expensive. A rapid and reversible passive adsorption mechanism is biosorption. Both living and dead biomass are involved in biosorption. Since the biomass obtained from industrial waste can be repeatedly regenerated and used, this method is inexpensive (Azubuike et al., 2020).

18.3 GENERAL CHARACTERISTICS OF ACTINOMYCETES

"Actinomycetes" belongs to the prokaryotic group as they lack nucleus and organelles in a subclass of Gram-positive and phylum Bacteria. The term "actinomycetes" is derived from the Greek word "actys" meaning ray and "mykes" meaning fungus, referring to their mycelium-like growth usually seen in fungi morphology (Villegas et al., 2013). Moreover, their cell wall is rigid and made up of muramic acid. Free-living species of the phylum, the majority of which are chemo-organotrophs, have long been thought of as having large G+C contents in their genomes. However, the cosmopolitan and many fresh water Actinomycetes have recently challenged this perception (Ghai et al., 2012). Actinomycetes phylum contains members with variable morphologies, from cocci-shaped cells to distinct mycelia and their afar dissemination is possible with spore production. This taxonomic group has widely distributed population where individuals are found in aquatic, terrestrial, and extreme habitats. They are significant organisms in the formation of organic matter, particularly in soil. Actinomycetes exhibit a variety of life patterns including soil actinomycetes, nitrogen-fixing symbionts, free-living saprophytes, plant commensals, and inhabitants of other organisms as beneficial and also as harmful ones (Goodfellow et al., 2018). They are also crucial in recycling materials because they can metabolise complex organic compounds. Actinomycetes have a vital ecological role, as evidenced by their capacity to detoxify pollutants inclusive of agrochemicals and heavy metal pollutants from the environment (Nazari et al., 2022). Due to their versatile metabolism, actinomycetes have garnered a lot of interest from scientists all over the world for a number of environmental and biotechnological applications.

18.4 HABITAT AND CLASSIFICATION OF ACTINOMYCETES

Actinobacteria embodies the largest and most diversified phyla within the bacterial kingdom including 6 classes, 19 orders, 50 families, and 221 genera while new taxa are constantly being discovered (Goodfellow et al., 2018). The six classes include *Thermoleophilia*, *Rubrobacteria*, *Acidimicrobiia*, *Coriobacteriia*, *Nitriliruptoria,* and *Actinomycetes.* The order Actinomycetales belongs to the class Actinomycetes, which is now curbed to individuals of the family Actinomycetaceae (Ludwig et al., 2012). *Catenuloplanes*, *Kineospora*, *Microbispora*, *Nonomuraea*, *Dactylosporangium*, *Actinoplanes*, *Amycolatopsis,* and *Micromonospora* are some of the actinomycetes genera that are classified as rare actinomycetes because of their slow development and difficulty in isolation and cultivation (Hayakawa, 2008). A vast variety of habitats including terrestrial, wide range of soils, freshwater, marine habitats, tidal flat ecosystems, organic matter, plants, and extreme environments are home to the majority of actinomycetales (Bull et al., 2005). Additionally, they have symbiotic interactions with a variety of macro-organisms, such as termites, ants, marine sponges, and tunicates (Kurtböke et al., 2015). *Streptomyces* is a widespread species that typically makes up a significant proportion of the actinomycetes population in natural environments. The distribution of Actinomycetes especially *Streptomyces* sp. has been documented in different environments, like sandy soil, black alkaline soil, sandy loam soil, alkaline desert soil, and subtropical desert soil. The pH range between 6 and 9 is ideal for actinomycetes, often exhibiting their optimum growth around neutral pH. However, some members have been identified from both acidic and alkaline soil. The acidophilic actinomycetes can either be strict acidophiles (pH range – 3.5–6.5) or neutrotolerant acidophiles (pH range–4.5–7.5) (Hazarika and Thakur, 2020). The other acidophilic genera such as *Acidimicrobium*, *Micromonospora*, *Acidothermus*, *Aciditerrimonas*, and *Ferrimicrobium*, were reported to grow best in the pH range of 1.8–4 (Zenova et al., 2011). Alkaline soil and lakes all support the growth of alkaliphilic actinomycetes throughout a wider pH range, from neutral to alkaline. Three main categories of alkaliphilic actinomycetes prevail: alkalitolerant actinomycetes (pH range 6–11), moderately alkaliphilic actinomycetes (pH range 7–10) and alkaliphilic actinomycetes (pH range 10–11). *Streptomyces*, *Cornyebacterium*, *Arthrobacter*, *Kocuria*, *Cellulomonas*, *Nocardioides*, *Microellobosporia*, *Elytrosporangium*, *Saccharothrix*, *Nocardiopsis Micromonospora*, and *Streptoverticillium* are some of the genera that have alkaliphilic forms (Shivlata and Satyanarayana, 2015; Zenova et al., 2011). Actinomycetes that are truly marine-adapted have also been discovered using sophisticated sampling and molecular techniques (Hall et al., 2007). Examples include the new genus *Salinispora* (Ahmed et al., 2014; Jensen et al., 2015) and *Rhodococcus marinonascens* (Wink et al., 2017), both have shown extensive spreading of their species in marine habitats (Jensen and Mafnas, 2006; Freel et al., 2012). Actinomycetes are no longer thought of as indigenous soil and freshwater species, but rather as some of the most successful colonisers of all conditions in the extreme biosphere. Further, they play a significant part in the natural biogeochemical cycles of metals/metalloids and few of the processes they involve could be used to treat contaminated environments.

Actinomycetes spread via spores that develop on specialised reproductive structures known as aerial mycelia (Figure 18.1). Actinomycetes perform their growth by forming branching hyphae that eventually form a vegetative mycelium. As a result, their life cycle closely aligns with that of filamentous fungi. The life cycle of the species *Streptomyces* has received the most attention due to its commercial significance in the production of secondary metabolites like antibiotics. A *Streptomyces* spore will germinate that tend to generate hyphae and will eventually develop by

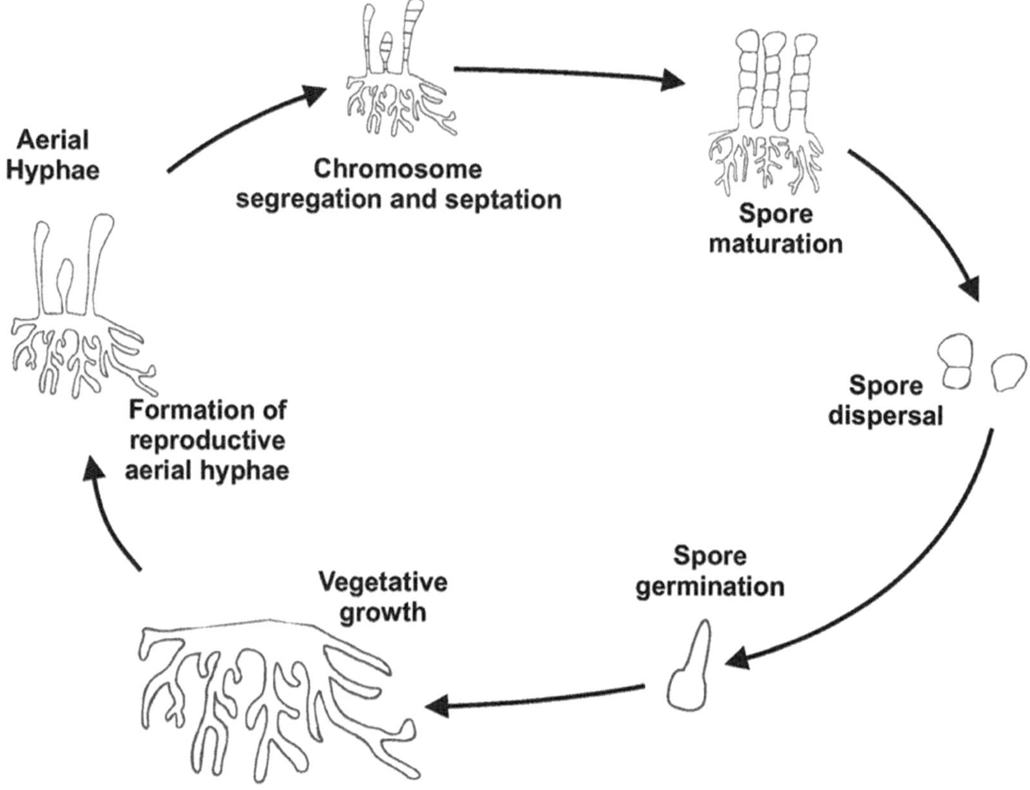

FIGURE 18.1 Life cycle of actinomycetes.

extension of tip and in optimum condition form substrate mycelia. Numerous morphological differentiations will take place when the cells are not provided with enough nutrients or when other signals are detected. Different actinomycetes may have some stage extensions in their life cycles. For instance, it has been observed that biofilm formation appears to be a significant part in the life cycle of cellulose-degrading *Thermobifida* (Alonso Gutiérrez et al., 2011).

18.5 ACTINOMYCETES – A BIOFACTORY AND ENVIRONMENTAL CLEANER

Compared to other physiochemical methods, actinomycetes offer more trustworthy and secured biological assistance for the degradation or removal of environmental contaminants. The pollutants can be metabolised by them to aid in growth and reproduction. They secrete extracellular bio-enzymes which break down a number of substances and drought spores. Additionally, filamentous growth favors soil particle colonisation (Amoroso et al., 2013). These distinct features of Actinomycetes make them an efficient candidate for the remediation of environmental pollutants through several processes like biosorption, biomineralisation, biostimulation, cell immobilisation, synthesis of biosurfactants, and plant-microbe interaction (Alvarez et al., 2017).

18.5.1 Organic Pollutant Removal

Organic pollutants such as agricultural chemicals and other xenobiotics in the environment are effectively broken down and can be detoxified by a significant group of Actinomycetes. *Actinomycete* strains survive in extreme habitats and secrete some extracellular enzymes which were found to be highly stable and had remarkable substrate selectivity. Actinomycetes, which are organisms that live in harsh environments and have a great deal of commercial potential, can be used to produce novel enzymes. The most significant lignin-degrading enzyme, lignin peroxidase, is secreted by actinomycetes in various forms. By breaking down lignin and lignocelluloses, lignin peroxidases produce H_2O_2, as well as other easily dissolvable activated oxygen species like singlet oxygen (O), superoxide anion radicals (O_{2}_), and hydroxyl radicals (OH_). *Streptomyces* are well-known lignin degraders that can solubilise certain lignin, resulting in polymeric lignin that is water-soluble and acid-perceptible. The processes in the breakdown of lignin compounds include demethylation of aromatic ring structures, oxidation of C-α for carbonyl group introduction, and α-β cleavage to monomeric products. Actinomycetes that produce cellulase include *Thermobifida fuscahave, Cellulolomonas fimi, and Microbispora bispora* (Gomez del Pulgar and Saadeddin, 2014) are involved in cellulolytic activities. The rate of bioremediation/biodegradation of organic pollutants is accelerated by incorporating bioemulsifiers and biosurfactants into the environment (Sánchez-Andrea et al., 2011). Immiscible hydrocarbons and microbial cells interact through a variety of methods, including micellisation, emulsification, adhesion/de-adhesion of microbes to and from hydrocarbons and desorption of pollutants. A possible biosurfactant-producing strain known as *Nocardiopsis* B4 actinomycetes was discovered using phenylalanine as the nitrogen source and olive oil as the carbon source. *Nocardiopsis* produced the most biosurfactant when the C/N ratio, temperature, pH, and salt concentration were 2:1, 30°C, 7.0, and 3%, respectively. The degradation of PAHs in soil depends on this strain.

Table 18.1 shows some of the compounds degraded by *Actinomycete* species. Cells after immobilisation have multiple advantages over free suspended cells, inclusion of improved cellular viability, microbial retention, and toxin protection. Saez et al. (2012), recorded the effect of immobilised *Streptomyces* cultures as individual and consortiums in distinct media including silicone tubes, agar, cloth sachets, and polyvinyl alcohol. Compared to free cells, immobilised microbes considerably removed more lindane and they could be employed again, thus, reducing the cost involved in the biotechnology method. Furthermore, actinomycetes might degrade pollutants along with plants through co-metabolism. (Alvarez et al., 2017) used the root of maize as the main source of carbon for *Streptomyces sp.* M7 and *Streptomyces sp.* A5 to degrade an organophosphate pesticide lindane. The maize root exudates have improved the microorganisms ability to degrade lindane in contrast to the minimal synthetic media with glucose as the carbon source. A recent publication by Mesquini et al. (2015), identified the *Streptomyces* sp. atz2, endophytic actinobacterium from

TABLE 18.1
Actinomycetes for Organic Pollutant Degradation

Compound	Species	References
Terephthalic acid	*Thermobifida alba*	Hu et al. (2010)
Prestige oil spill and Long chain *n*-alkane	*Dietzia* sp.	Alonso Gutiérrez et al. (2011)
PHB	*Streptomyces bangladeshensis*	Hsu et al. (2012)
Aniline	*Dietzia natronolimnaea*	Jin et al. (2012)
4- Chlorophenol	*Georgenia daeguensis*	Woo et al. (2012)
Naphthalene, alkylbenzenes, phenoxyacetate	*Nocardia* sp.	Das and Adholeya (2012)
PCL	*Streptomyces thermoviolaceus*	Chua et al. (2013)
DDT	*Kocuria rosea*	Wu et al. (2014)
Pentachlorophenol	*Janibacter* sp.	Khessairi et al. (2014)
Hydrocarbons	*Nocardia* sp., *Rhodococcus* sp., and *Streptomyces* sp.	Shekhar et al. (2014)
S-triazine (atrazine)	*Arthrobacter* sp.	Sagarkar et al. (2016)

sugarcane leaves, which showed atrazine content reduction of 98% under aerobic and microaerophilic conditions. Potential microbial groups including *Micromonospora, Micrococcus, Gordonia, Rhodococcus, Nocardia* of actinomycetes, and bacterial species were found for decomposing lubricating oil, according to a culture-dependent microbial diversity analysis (Idemudia et al., 2014). Oil degradation capacity of actinomycetes species ranged from 1.035% to 7.53%. Petroleum compounds containing both long chain n-alkanes and c-alkanes could be broken down by actinomycetes isolates from the genera *Rhodococcus* and *Gordonia* (Kubota et al., 2008; Kuyukina & Ivshina, 2019).

18.5.2 Inorganic Pollutant Removal

With diversity of physicochemical and biological mechanisms which influence changes between the soluble and insoluble phases, actinomycetes play key roles in the environmental fate of hazardous metals. The detoxifying mechanisms of Actinomycetes include volatilisation, extracellular chemical precipitation, and changing the valence state of metals. This could make the metals immobile or convert them into less toxic forms. Actinomycetes proteinaceous layer called (S-layer) on its surface serves as a barrier between their cell contents and environment. Numerous functional groups, including COOH, NH_2, OH, PO_4, and SO_4, that interact with metals are present in the protein layers. Additionally, the S layer traps ions and/or inhibits harmful metals from penetrating the bacterial cell. Additionally, it serves as the best protection against harmful dissolved metal ions. Binding forms layered metal complexes as the first stage of biomineralisation (Phoenix et al., 2005). Gremion et al. (2003) assessed the rhizospheric and bulk soil bacterial diversity of *Thlaspi caerulescens,* metal accumulating plant, and revealed Actinomycetes was the dominant group of active microbial population. Several Ni-resistant bacteria were also identified in the rhizospheric soil of the Ni-hyper accumulator *T. goesingense* (Idris et al., 2004). They also observed the bacteria mostly isolated from Ni-contaminated medium were *Rhodococcus* sp. and *Okibacterium* sp. belonged to phyla Actinomycetes and that the Streptomyces and Arthrobacter genera constitute most of the Ni-resistant bacteria. Likewise, *Arthrobacter crystallopoites, Pseudomonas* sp., *Corynebacterium hoagie, Bacillus maroccanus,* and *Bacillus cereus* have the ability to tolerate, resist, and reduce chromium (Viti et al., 2003). Accordingly, Dávila Costa et al. (2011), *Amycolatopsis tucumanensis* showed significant cupric reductase activity and also markedly increased the rate of Cu (II) reduction. Actinomycetes from marine areas were also reported to adsorb radioactive elements in their cell walls. Kamala et al. (2020), obtained *Streptomyces* sp. from marine sediments that aided in the biosorption of Sr^{2+} by secreting extracellular polymeric substance (EPS). Another species namely *Streptomyces sporoverrucosus* was found to accumulate high amounts of uranium in its cell walls (Li et al., 2016). Likely, Sivaperumal et al. (2018), obtained *Nocardiopsis* sp. showing the ability to adsorb radioactive cesium. Several actinomycete strains used in the removal of inorganic pollutants are given in Table 18.2.

TABLE 18.2
Actinomycetes for Inorganic Pollutant Degradation

Compound	Species	References
CrO_4^{2-}	*Arthrobacter rhombi*	Elangovan et al. (2010)
Cadmium, chromium, cobalt, zinc, mercury	*Arthrobacter ramosus*	Bafana et al. (2010)
Cadmium, lead, arsenic	*Bifidobacterium angulatum*	Elsanhoty et al. (2016)
Chromium, zinc, nickel	*Nocardia sp.*	El-Gendy & El-Bondkly (2016)
Chromium, cadmium, lead	*Streptomyces rochei*	Hamdan et al. (2021)
Ag^+	*Corynebacterium glutamicum and Rhodococcus sp.*	Otari et al. (2015)
Zn^{2+}	*Rhodococcus pyridinivorans*	Kundu et al. (2014)

18.6 ROLE OF ACTINOMYCETES IN WASTEWATER REMEDIATION

Actinomycetes with their tremendous potential for industrial wastewater remediation attracted the huge attention of scientific community (Gousterova et al., 2011). Because of their various metabolic pathways that may function in a variety of environmental conditions and their ability to survive in extreme environments, actinomycetes are particularly well suited for wastewater treatment. Due to the secretion of numerous substances including flocculant molecules, actinomycetes represent its significant role in the wastewater treatment. Elimination of sludge sedimentation was made possible by rapid colonisation that effectively involved the formation of sludge flocs and will be settled at the bottom of the tank. Additionally, actinomycetes best suited for removing heavy metals as they capacity to store significant levels of polyphosphate and poly-hydroxybutyric acid (Theerthagiri et al., 2019). Besides heavy metal reduction, Actinomycete also has resistance against desiccation and UV-irradiation rather than other microbes used in wastewater treatment.

Several Actinomycete species secrete polysaccharides which act as bioflocculants having unique functional structure and properties (More et al., 2014). Active functional groups of bioflocculants such as acetyl, carboxyl, hydroxyl, and carbonyl. These groups binds well with variety of pollutants like metals, dyes, pathogens, emerging contaminants, and even suspended solids. Among all the actinomycete species, *Kocuria rosea* was reported as an excellent bioflocculant producing species (Chouchane et al., 2018). Because of this unique feature, they could be used in treating poultry wastewater (Ghosh et al., 2009), brewery (Ugbenyen and Okoh, 2014), textiles (Li et al., 2013), swine (Guo and Ma, 2015), pulp and paper mill (Hayakawa, 2008), and whey effluent (Patil et al., 2011). Unlike chemical-based flocculating agents, bioflocculants are non-toxic and more eco-friendly as they are from biological origin (Okaiyeto et al., 2014).

18.6.1 A Dye Remover

Textile, food, pharmaceutical, biochemical, cosmetic, paper, graphics, and other industries all utilise dyes. As a result, substantial amounts of effluents with high levels of toxicity, colour, turbidity, and oxygen demand (chemical and biological) are produced. These effluents can contaminate surface and groundwater if they are not properly disposed of or treated, which causes a series of problems for the ecosystem (Al-Tohamy et al., 2022). Colour from the pollutants limits the amount of sunlight that entering aquatic environment resulting the reduction of dissolved oxygen and photosynthetic activity. Additionally, the by-products of dye breakdown might be mutagenic and harmful to aquatic life, harming both flora and fauna (Varjani et al., 2020). The two main types of dyes including azo and triphenyl methane have garnered the most research. Azo dyes with azo-chromophore groups (–N– –N–) are the broadest and most diverse class of synthetic dyes used commercially. (Maniyam et al., 2020). The central carbon atom of chromogen in triphenylmethane dyes is linked with three phenyl groups. Both classes contain poisonous, mutagenic, and recalcitrant characteristics pose a serious threat to the environment when present in effluents (Adenan et al., 2020). The dye biodegradation experiments were conducted in a variety of ways, including: I. Microbial culture medium should be added with dye compound to assess its effects on microbial development and dye removal; II. Using live/ dormant and/or autoclaved microbial cell biomass; III. Using immobilised cell matrices.

The actinomycetes are capable of secreting enzymes such as peroxidase, laccase, and azo-reductase enzymes which in turn degrade various synthetic organic dyes (Sahasrabudhe et al., 2014; Ting, 2020). The majority of studies mentioned dye degrading actinomycetes species belong to the genera *Streptomyces, Micromonospora, Micropolyspora* (Raja et al., 2016), *Nocardiopsis* (Chittal et al., 2019), *Kitasatospora* (Adenan et al., 2020) and *Rhodococcus* (Maniyam et al., 2020) were reported as dye degraders by majority of the studies. In a study by Chengalroyen (2011), *Streptomyces coelicolor* produced azo-reductase enzyme to degrade the complex dye Congo red. Similarly, laccase producing *Georgenia* sp. was reported to degrade reactive orange dye with 94.2% efficiency at a wide range of temperatures (28–45) within 8 hours (Sahasrabudhe et al., 2014). *Rhodococcus* sp. and *Kocuria rosea* showed degrading ability against a wide range of dyes including reactive black 5 (Martorell et al., 2012), methyl red (Maniyam et al., 2018), reactive blue 4, Acid Yellow, basic blue 3, basic red (Chouchane et al., 2018) as revealed by many researchers. Azo-reductase enzyme producing *Rhodococcus opacus* found to degrade methyl red and other complex dyes (Qi et al., 2017). *Rhodococcus* sp. is more tolerant of environmental perturbations that interfere with the biodegradation process when it is immobilised in alginate, making it more stable (Sundarajoo and Maniyam, 2019). *Rhodococcus* has been successfully used as a biocatalyst multiple times, withstanding multiple cycles (9–15) (Maniyam et al., 2020), making its application even more alluring and less expensive.

The enzymes created by the bacterium drive the biodegradation process. The dye molecules are broken down by these enzymes, producing intermediates that may or may not be utilised as nutrition to maintain the viability of cells (Al-Tohamy et al., 2022). Biosorption involves the interaction of functional group with dye molecule on the surface of microbial cell wall either in alive or dead condition irrespective of their age and enzyme production. Due to the continuous process of binding of dye molecules, the microbial cells typically exhibit dye staining (depending on the amount) (Roy et al., 2018). However, immobilised cell matrices would also involve biosorption along with biodegradation of dye molecules. Dye molecules' interaction with the structure of matrix material defines adsorption (Selvaraj et al., 2021). This method has a high cell density and vitality, improves chemical stability and operational flexibility, and makes it easier to separate cells from the medium after usage (Bouabidi et al., 2019; Wu et al., 2021). A readily available and moderately priced polymer is alginate. Its physicochemical characteristics make it a good choice for use as a support. Additionally, the product produced is non-toxic and biodegradable, which benefits the environment (Maniyam et al., 2020). Utilising immobilised cells improves this reuse since it boosts productivity and makes it easier to separate the material from the medium.

18.6.2 Removal of Heavy Metals

Heavy metals are extremely dense and deadly, even in lower concentrations. Therefore, it is alarming that the usage of heavy metals like nickel, copper, cobalt, and zinc in a variety of businesses, from basic industries to high-tech products (Baltazar et al., 2019). Additionally, leftovers from the leather tanning, metal finishing, fertiliser, electroplating, battery, textile, and nuclear power sectors might contain heavy metals (Dobrowolski et al., 2017; Goswami et al., 2017). The widespread metal usage inclined their availability in the environment, exposing different stages of the food chain to high concentrations of this contaminant, which can restrict biodiversity and cause diseases in humans (Baltazar et al., 2019). Moreover, heavy metals contribute to the aetiology of neurological conditions such as multiple sclerosis, amyotrophic lateral sclerosis, Alzheimer's disease, etc. (Kiray et al., 2017). Heavy metal deposition in water is seen as a severe environmental issue because these metals are resistant and harmful to life (Baltazar et al., 2019). There are methods that are commonly used to remove heavy metals from aqueous matrices, including microfiltration, ion exchange, reverse osmosis, precipitation, and flocculation. However, these methods produce more sludge, thus, removing it frequently raises the involvement of cost and energy (Yaashikaa et al., 2022). Bioremediation of heavy metals was found to be the best option in this hour due to its affordability and attention to the environment (Dobrowolski et al., 2017; Ali et al., 2021).

Typically, actinomycetes have polyphosphate reserves, which act both as phosphorus reserves and as metal chelating spaces, which can reduce the toxicity of metals in industrial wastewater. Actinomycete cells can tolerate or alter heavy metal toxicity by a variety of metal-independent mechanisms like siderophore-metal complex formation, exclusion through permeability barriers, intracellular meta-sequestration and reduction of cellular sensitivity (Bankar and Nagaraja, 2018). Various mechanisms adapted actinomycetes for removing and tolerating heavy metals were given in Figure 18.2. Likewise, actinomycetes synthesise proteins binding on metals including

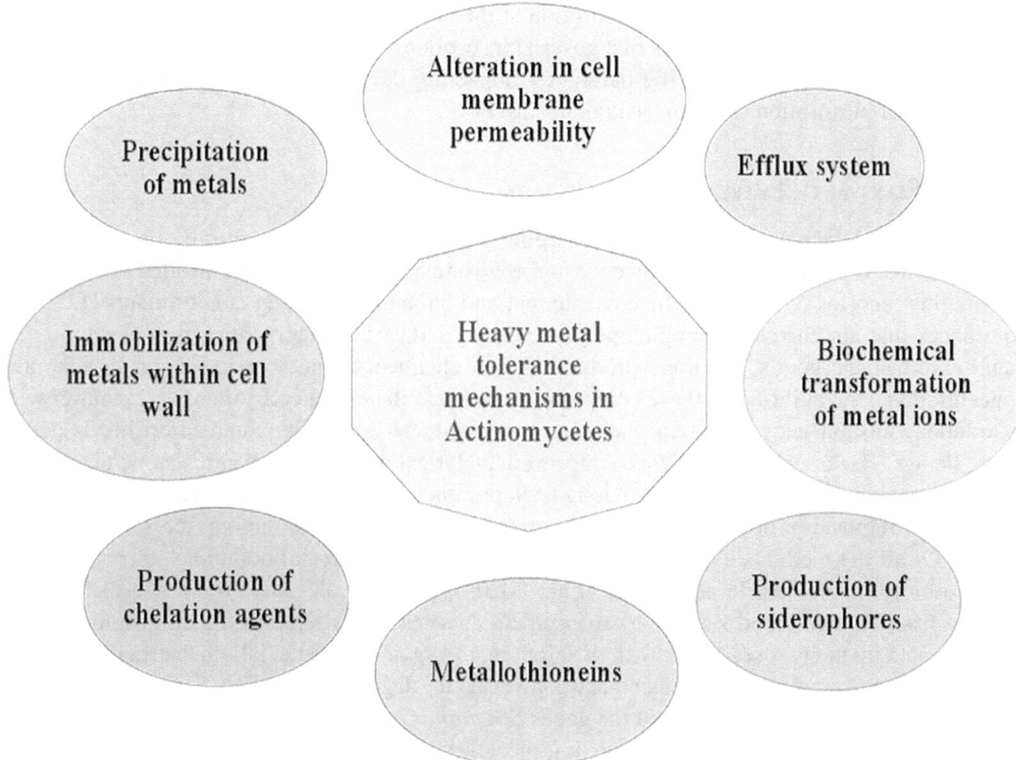

FIGURE 18.2 Heavy metal tolerance mechanisms in Actinobacteria.

phytochelatins and metallothioneins as defense systems under heavy metal stress. These proteins contain more cycteine and histidine peptides as they play remarkable role in binding and sorption inside the cells (Schmidt et al., 2010). Similarly, reduction of metal toxicity related to siderophore production and binding to heavy metals. Actinomycetes were reported to secrete various kinds of siderophores compounds including desferrioxamine (B, E, G) and rhodotorulic acid (Sessitsch et al., 2013). Efflux system involving chemiosmotic ion/proton pumps and ATPases are the prime mechanism of actinomycetes in tolerance of heavy metal. The efflux system pumps back the toxic metal ions which have infiltrated into the cells. Other mechanisms followed by Actinomycete such as bioaccumulation, enzymatic reduction, and biosorption were also reported. Actinomycete exhibits an amphoteric feature by the presence of polysaccharides, lipids, carboxylic groups, and mycolic acids which aids in biosorption (Nazari et al., 2022). Albarracín et al. (2010) reported 71.2% reduction of copper ions (Cu II) in six days of incubation in copper solution (39 ppm) by an indigenous copper-resistant actinomycete. A novel species, *Streptomyces zinciresistens* in their cell wall accumulated 85.0 and 26.81 mg/g of Cd (II) and Zn (II) respectively (Lin et al., 2012). Jastaniah and Aburas (2016), isolated *Streptomyces* sp. from wastewater remediation plant in Saudi Arabia which was shown to bioabsorb cadmium (12%), chromium (22%), copper (16%), zinc (11%), nickel (12%), lead (32%), and manganese (77%).

According to Hozzein et al. (2012), El Baz et al. (2015), and Theerthagiri et al. (2021), for treating industrial contaminants and the eradication of dangerous metals from wastewater, actinomycete has excellent biosorbent efficiency, thus economically advantageous. In most investigations using liquid matrices, metal biosorption involves the interaction with active sites of metal ions on the surfaces of microorganisms. Electrostatic interactions, precipitation, ion exchange, chelation, and reduction of metal ions are some of these interactions (Kiray et al., 2017). pH of the medium is typically the primary factor that controls metal ions sorption at the cell surface and also the number of species availability, among the major factors that govern biosorption (Vásquez et al., 2007). Following temperature is another crucial operating parameter impacting the metal-biosorbent complex stability and the rate of biosorption (Dobrowolski et al., 2017).

18.6.3 Removal of Emerging Contaminants

A new global environmental concern is emerging pollutants that occur naturally or synthetically in ecosystems. In contrast, their occurrence and environmental fate are not controlled or regulated, despite their eco-toxic effects on the environment and humans. Emerging contaminants (ECs) are substances that are currently being employed in many industrial sectors, agriculture sectors, personal care/hygiene goods, hormone production, and chemical manufacturing. Even in very low concentrations, the build-up of these compounds can harm the entire ecosystem, *viz*., plant growth retardation, soil nutrients depletion, and also cause endocrine disorders, and infertility issues in living beings (Baek et al., 2022). Due to improved analytical methods, they can now be detected in the environment, even though they have long been present in the environment. With this knowledge on ECs, the removal or degradation of those toxins is a rising concern among the scientific community. Of all the methods used to detoxify/remove the environmental pollutants, bioremediation was proven to be efficient in action (He et al., 2020; Egorova et al., 2017). Removing ECs either directly from contaminated sites or by using them as a nutrient source in the culture media was demonstrated by many researchers with different microorganisms, especially bacteria (Song et al., 2022). Table 18.3 presents reports using actinomycetes for degradation of ECs.

Ungureanu et al. (2015) mentioned the genus *Streptomyces* secrete ligninolytic enzymes such as laccase which could be involved in removing recalcitrant ECs. They also reported 30% degradation of carmabazepine (CBZ) by a *Streptomyces* strain MIUG 4.89. The genus *Rhodococcus* was known to degrade a variety of pharmaceutical compounds. Biodegradation of nitrogen-containing sulfamethisole (14%), sulfamethoxazole (20%) and CBZ (15%) within 36 days by *Rhodococcus rhodochrous* was shown by Gauthier et al. (2010). Similarly, Chuang et al. (2021), *Rhodococcus qingshengii* degraded a pharma compound carbedazim by 93% in just 14 days. *Rhodococcus rhodochrous* was also found to deplete 17a-ethinylestradiol by utilising it as the only carbon source (Larcher and Yargeau, 2013). Xu et al. (2018) studied the degradation of brominated flame retardant, Tetrabromobisphenol A (TBBPA) with *Rhodococcus jostii* species. TBBPA was transferred to monomethyl ether and this was accelerated by the addition of heavy metals Cu and Fe. *Rhodococcus erythropolis* encapsulated on alginate beads reduced the levels of atrazine in aqueous solution as

TABLE 18.3
Emerging Contaminants Bioremediation Using Actinomycetes

Species	Pollutant	Mechanism	Removal Efficiency	References
Rhodococcus sp.	Carbendazin	Enzyme production	98.2%	Bai et al. (2017)
Rhodococcus zopfii	17α-Ethynylestradiol		80%	Menashe et al. (2020)
Rhodococcus sp.	benzophenone-3 (Personal care product)	Maleylacetate reductase, flavin-dependent oxidoreductase, hydroxyquinol 1,2-dioxygenase, and α/β hydrolase	–	Baek et al. (2022)
Rhodococcus sp.	Steroids	17β-HSD dehydrogenase	–	Ye et al. (2020)

reported by Vancov et al. (2005). There are a few challenges in the bioremediation of ECs that need to be resolved because of the low polarity, complexity, and stability of the major compounds (Song et al., 2022). Along with how pollutants interact with microorganisms present in soil and aquatic habitats when they are used singly, in complex, it is also important to understand the intermediates production involved in the biodegradation of ECs (Baek et al., 2022).

18.7 UTILISATION OF ACTINOMYCETE GENETIC RESOURCES FOR WASTEWATER REMEDIATION

The actinomycetes is a unique group of organisms that could be used for the benefit of mankind. The discovery of the antibiotics production potential of actinomycetes through metabolic activity paved a way of its application in medicine. The renowned actinomycetes, *Streptomyces* sp. has revolutionised the biogenic medicine field. Production of bioactive agents by these groups of microbes has only been partially exploited in the last four decades. This is due to that the Actinomycetes species could not be reproduced efficiently under laboratory conditions. For which molecular tools such as genome mining, sequencing, and bioactivity screening could be helpful in disclosing many of biosynthetic gene clusters present but not expressed in genome of actinomycetes. Consequently, genetic technologies are being developed to activate the biosynthetic pathways in the genome of Actinomycetes. Actinomycetes are currently being genetically engineered to activate biosynthetic gene clusters (BGC) expression and produce the appropriate compounds. These methods consist of expressing whole BGC in multiple copies or components that restrict production, activator genes, the removal of genes encoding BGC repressors, BGC refactoring in native, augmented, or heterologous hosts. Actinomycetes are difficult to genetically modify due to their high guanine-cytosine (GC-content), particular codon usage, lack of suitable molecular tools, DNA degradation, instability in their genes, limited cloning, and DNA transfer techniques. Each actinomycete has a specific vector that can be used to create genetic constructs and introduce them into the host through protoplast transformation or conjugation. This significant development aided in the progress of metabolic engineering for producing bioactive natural compounds like antibiotics. The application of these gene-constructed actinomycetes starts from biotransformation of the compound of interest. The production of antibiotics by actinomycetes can be applied to remove the pathogenic microbes from the effluent. This could substitute chemical disinfection and cut the cost down for treatment. However, the remediation of wastewater through actinomycetes remains dull due to above-said barriers. Nevertheless, laboratory scale remediation and treatment studies are under recent research. Removing polyaromatic sulphur heterocycle is particularly possible in *Rhodococcus erthropolis* I-19, by the multiple expression of key dsz genes which convert dibenzothiophene (DBT) to 2-hydroxybiphenyl and sulphite. *Streptoverticillium kashmirense* AF1 secreted the enzyme called PHBV depolymerases was purified for use in PHBV film degradation. Similarly, thermophilic actinomycetes *viz., Streptomyces, Thermoactinomyces, Actinomadura, Saccharomonospora,* and *Microbispora* showed their degrading ability of poly ethylene succinate (PES), poly ε-caprolactone (PCL) and poly β-hydroxybutyrate (PHB). Polychlorinated biphenyl (PCB) degrading *Rhodococcus* sp. RHAI has PCB degrading genes and linear plasmids including *pRHL* (1, 2, and 3). Further linear plasmids such as bphB2, etbD2, etbC, bphDEF, bphC2, and bphC4 will be studied for bioremediation in the coming years. The Recipient Ratio Conjugal transfer method was adapted for transferring foreign DNAs into rare Antinomycetes (Rebets et al., 2017; Zhang et al., 2019; Song et al., 2022). *Streptomyces coelicolor* M145 showed great efficiency in digesting n-hexadecane when the alkB gene was reprogrammed to produce alkane mono-oxygenase. The structured DNA encoded with the *xylE* gene has been shown to be effective for degradation and transformation of aromatic hydrocarbons than the original strain with the gene (Ma et al., 2021). Further, it is unavoidable to explain the use of specific bioremediation procedure using the novel biomarkers exploration. Developing entire picture remedial processes by merging the available genomic data via molecular tools could brighten the future for wastewater treatment using actinomycetes.

18.8 CHALLENGES IN APPLICATION OF ACTINOMYCETES FOR WASTEWATER REMEDIATION

Actinomycetes have very promising results across various wastewater from industries. The search for efficient actinomycetes strains is still in its early stages. This is due to the difficulty in identifying the scope of application with conventional tools and inherent limitations in using a single strain in multi-pollutant decontamination. Moreover, several strains with compatibility would work together effectively in removing the pollutants from wastewater. The genetic tools from actinomycetes that could help in degradation/breakdown of new environmental pollutants are a top priority and are yet to be discovered. Genetically modified organisms could provide improved adaptability in removing pollutants that emerge on a regular basis from growing industries in a sustainable manner. In a complex environment with diverse substrates and many microbial interactions, only a few engineered bacteria are currently engaged in treating and removing toxins. As a result, constructed plasmids can slow down the growth and reproduction of a strain, reducing its ability to treat contaminants in the long run. When plasmids are lost during microbial reproduction, microbial breakdown, and wastewater treatment are negatively affected. Hence, multi-plasmid bacteria with several desirable features are challenging to engineer. Despite this, genetically modified bacteria's widespread use in the natural environment is still controversial, given their potential negative impacts on human health and the delicate balance of biological ecosystems (Mukherjee et al., 2017). However, protocols for constructing the perfect plasmids for remediating industrial wastewater should be carried out using the multi-omics methods. To understand the current importance of "omics" in the microbial remediation, extensive research is required. Mining for bioremediation data is required for understanding the pathways of degradation which could help in developing novel algorithms thereby facilitating modelling. Striving to reduce these challenges could reduce the time take to treat industrial wastewater exclusively with microbes especially actinomycetes.

18.9 CONCLUSION

Millions of people's lives have been completely transformed by industrial technologies around the globe. The quantity and quality of wastewater produced by several industries is the main cause of concern as it is discharged untreated into the environment ultimately harming the ecosystem. As a result, treating wastewater through bioremediation using Actinomycetes is advantageous and environmentally beneficial. Numerous research have shown that Actinomycetes have the potential to degrade pesticides, remove heavy metals, and emerging pollutants, highlighting their potential capacities as tool for wastewater treatment. Furthermore, it has been demonstrated that actinomycetes are widespread in contaminated soils and that they may metabolise pollutants in industrial effluents to support their growth. Additionally, to increase the effectiveness of bioremediation, consortia of Actinomycetes strains with other microorganisms, such as bacteria, fungi, and microalgae, could be used. The use of genetic engineering techniques to increase Actinomycetes' potential for bioremediation represents another opportunity in this scenario. However, most of the research with Actinomycetes was done in laboratory conditions. The potential of Actinomycetes in the remediation of wastewater under the influence of biotic and abiotic factors in pilot and field scale studies should be carried out to promote large-scale application. Likewise, only few genera in Actinomycetes phyla were isolated and characterised for its potential (e.g. *Streptomyces*, *Rhodococcus*). Many genera with numerous potentials were yet to be identified. Future research should focus on the isolation and characterisation of new Actinomycete strains from extreme habitats. Also, the secondary metabolites such as enzymes, biosurfactants produced by Actinomycetes genera should be purified and tested for its efficacy in removal of pollutants from wastewater. As more becomes known about the unique properties of Actinomycetes, it will become ubiquitous in sustainable applied research in decontamination of environment.

REFERENCES

Adenan, N. H., Lim, Y. Y., & Ting, A. S. Y. (2020). Discovering decolorization potential of triphenylmethane dyes by actinobacteria from soil. *Water, Air, & Soil Pollution*, *231*(12), 1–13.

Ahmed, I., Kudo, T., Abbas, S., Ehsan, M., Iino, T., Fujiwara, T., & Ohkuma, M. (2014). *Cellulomonas pakistanensis* sp. nov., a moderately halotolerant Actinobacteria. *International Journal of Systematic and Evolutionary Microbiology*, *64*(7), 2305–2311.

Albarracín, V. H., Amoroso, M. J., & Abate, C. M. (2010). Bioaugmentation of copper polluted soil microcosms with *Amycolatopsis tucumanensis* to diminish phytoavailable copper for Zea mays plants. *Chemosphere*, *79*(2), 131–137.

Ali, M., Song, X., Ding, D., Wang, Q., Zhang, Z., & Tang, Z. (2021). Bioremediation of PAHs and heavy metals co-contaminated soils: Challenges and enhancement strategies. *Environmental Pollution*, *118686*.

Alonso Gutiérrez, J., Teramoto, M., Yamazoe, A., Harayama, S., Figueras, A., & Novoa, B. (2011). Alkane degrading properties of *Dietzia* sp. H0B, a key player in the Prestige oil spill biodegradation (NW Spain). *Journal of Applied Microbiology*, *111*(4), 800–810.

Al-Tohamy, R., Ali, S.S., Li, F., Okasha, K.M., Mahmoud, Y. A.-G., Elsamahy, T., Jiao, H., Fu, Y., & Sun, J. (2022). A critical review on the treatment of dye-containing wastewater: Ecotoxicological and health concerns of textile dyes and possible remediation approaches for environmental safety. *Ecotoxicology and Environmental Safety*, *231*, 113160.

Alvarez, A., Saez, J. M., Costa, J. S. D., Colin, V. L., Fuentes, M. S., Cuozzo, S. A., Benimeli, C. S., Polti, M. A., & Amoroso, M. J. (2017). Actinobacteria: current research and perspectives for bioremediation of pesticides and heavy metals. *Chemosphere*, *166*, 41–62.

Amoroso, M. J., Benimeli, C. S., & Cuozzo, S. A. (2013). Actinobacteria: application in bioremediation and production of industrial enzymes. CRC Press.

Ashraf, S., Ali, Q., Zahir, Z. A., Ashraf, S., & Asghar, H. N. (2019). Phytoremediation: Environmentally sustainable way for reclamation of heavy metal polluted soils. *Ecotoxicology and Environmental Safety*, *174*, 714–727.

Azubuike, C. C., Chikere, C. B., & Okpokwasili, G. C. (2020). Bioremediation: An eco-friendly sustainable technology for environmental management. In: Saxena, G., & Bharagava, R. (eds) *Bioremediation of industrial waste for environmental safety* (pp. 19–39). Springer.

Baek, J. H., Kim, K. H., Lee, Y., Jeong, S. E., Jin, H. M., Jia, B., & Jeon, C. O. (2022). Elucidating the biodegradation pathway and catabolic genes of benzophenone-3 in Rhodococcus sp. S2-17. *Environmental Pollution*, 299, 118890.

Bafana, A., Krishnamurthi, K., Patil, M., & Chakrabarti, T. (2010). Heavy metal resistance in *Arthrobacter ramosus* strain G2 isolated from mercuric salt-contaminated soil. *Journal of Hazardous Materials*, *177*(1-3), 481–486.

Bai, N., Wang, S., Abuduaini, R., Zhang, M., Zhu, X., & Zhao, Y. (2017). Rhamnolipid-aided biodegradation of carbendazim by *Rhodococcus* sp. D-1: Characteristics, products, and phytotoxicity. *Science of the Total Environment*, *590*, 343–351.

Baltazar, M. dos P. G., Gracioso, L. H., Avanzi, I. R., Karolski, B., Tenório, J. A. S., do Nascimento, C. A. O., & Perpetuo, E. A. (2019). Copper biosorption by *Rhodococcus erythropolis* isolated from the Sossego Mine-PA-Brazil. *Journal of Materials Research and Technology*, *8*(1), 475–483.

Bankar, A., & Nagaraja, G. (2018). Recent trends in biosorption of heavy metals by Actinobacteria. In: Singh, B. P., Gupta, V. K., & Passari, A. K. (eds) *New and future developments in microbial biotechnology and bioengineering* (pp. 257–275). Elsevier.

Bouabidi, Z. B., El-Naas, M. H., & Zhang, Z. (2019). Immobilization of microbial cells for the biotreatment of wastewater: a review. *Environmental Chemistry Letters*, *17*(1), 241–257.

Bull, A. T., Stach, J. E. M., Ward, A. C., & Goodfellow, M. (2005). Marine actinobacteria: perspectives, challenges, future directions. *Antonie Van Leeuwenhoek*, *87*(1), 65–79.

Chandra, R., Saxena, G., & Kumar, V. (2015). Phytoremediation of environmental pollutants: an eco-sustainable green technology to environmental management. In: Chandra, R. (ed) *Advances in biodegradation and bioremediation of industrial waste* (pp. 1–30). Boca Raton: CRC Press.

Chengalroyen, M. D. (2011). Studies on triphenylmethane, azo dye and latex rubber biodegradation by actinomycetes (Doctoral dissertation).

Chittal, V., Gracias, M., Anu, A., Saha, P., & Rao, K. V. B. (2019). Biodecolorization and biodegradation of azo dye reactive orange-16 by marine *Nocardiopsis* sp. *Iranian Journal of Biotechnology*, *17*(3), e1551.

Chouchane, H., Mahjoubi, M., Ettoumi, B., Neifar, M., & Cherif, A. (2018). A novel thermally stable heteropolysaccharide-based bioflocculant from hydrocarbonoclastic strain Kocuria rosea BU22S and its application in dye removal. *Environmental Technology*, *39*(7), 859–872.

Chua, T.-K., Tseng, M., & Yang, M.-K. (2013). Degradation of Poly (ε-caprolactone) by thermophilic *Streptomyces thermoviolaceus* subsp. thermoviolaceus 76T-2. *Amb Express*, *3*(1), 1–7.

Chuang, S., Yang, H., Wang, X., Xue, C., Jiang, J., & Hong, Q. (2021). Potential effects of *Rhodococcus qingshengii* strain djl-6 on the bioremediation of carbendazim-contaminated soil and the assembly of its microbiome. *Journal of Hazardous Materials*, *414*, 125496.

Cristaldi, A., Conti, G. O., Jho, E. H., Zuccarello, P., Grasso, A., Copat, C., & Ferrante, M. (2017). Phytoremediation of contaminated soils by heavy metals and PAHs. A brief review. *Environmental Technology & Innovation*, *8*, 309–326.

Das, M., & Adholeya, A. (2012). Role of microorganisms in remediation of contaminated soil. In: Satyanarayana, T., & Johri, B. (eds) *Microorganisms in environmental management* (pp. 81–111). Springer.

Dávila Costa, J. S., Albarracín, V. H., & Abate, C. M. (2011). Cupric reductase activity in copper-resistant *Amycolatopsis tucumanensis*. *Water, Air, & Soil Pollution*, *216*(1), 527–535.

Dobrowolski, R., Szcześ, A., Czemierska, M., & Jarosz-Wikołazka, A. (2017). Studies of cadmium (II), lead (II), nickel (II), cobalt (II) and chromium (VI) sorption on extracellular polymeric substances produced by Rhodococcus opacus and Rhodococcus rhodochrous. *Bioresource Technology*, *225*, 113–120.

Egorova, D. O., Buzmakov, S. A., Nazarova, E. A., Andreev, D. N., Demakov, V. A., & Plotnikova, E. G. (2017). Bioremediation of hexachlorocyclohexane-contaminated soil by the new *Rhodococcus wratislaviensis* strain Ch628. *Water, Air, & Soil Pollution*, *228*(5), 1–16.

El Baz, S., Baz, M., Barakate, M., Hassani, L., el Gharmali, A., & Imziln, B. (2015). Resistance to and accumulation of heavy metals by actinobacteria isolated from abandoned mining areas. *The Scientific World Journal*, 2015.

Elangovan, R., Philip, L., & Chandraraj, K. (2010). Hexavalent chromium reduction by free and immobilized cell-free extract of *Arthrobacter rhombi*. *Applied Biochemistry and Biotechnology*, *160*(1), 81–97.

El-Gendy, M. M. A. A., & El-Bondkly, A. M. A. (2016). Evaluation and enhancement of heavy metals bioremediation in aqueous solutions by *Nocardiopsis* sp. MORSY1948, and Nocardia sp. MORSY2014. *Brazilian Journal of Microbiology*, *47*, 571–586.

Elsanhoty, R. M., Al-Turki, I. A., & Ramadan, M. F. (2016). Application of lactic acid bacteria in removing heavy metals and aflatoxin B1 from contaminated water. *Water Science and Technology*, *74*(3), 625–638.

Freel, K. C., Edlund, A., & Jensen, P. R. (2012). Microdiversity and evidence for high dispersal rates in the marine actinomycete '*Salinispora pacifica*'. *Environmental Microbiology*, *14*(2), 480–493.

Gauthier, H., Yargeau, V., & Cooper, D. G. (2010). Biodegradation of pharmaceuticals by *Rhodococcus rhodochrous* and *Aspergillus niger* by co-metabolism. *Science of the Total Environment*, *408*(7), 1701–1706.
Ghai, R., McMahon, K. D., & Rodriguez-Valera, F. (2012). Breaking a paradigm: cosmopolitan and abundant freshwater actinobacteria are low GC. *Environmental Microbiology Reports*, *4*(1), 29–35.

Ghosh, M., Ganguli, A., & Pathak, S. (2009). Application of a novel biopolymer for removal of *Salmonella* from poultry wastewater. *Environmental Technology*, *30*(4), 337–344.

Gomez del Pulgar, E. M., & Saadeddin, A. (2014). The cellulolytic system of *Thermobifida fusca*. *Critical Reviews in Microbiology*, *40*(3), 236–247.

Goodfellow, M., Nouioui, I., Sanderson, R., Xie, F., & Bull, A. T. (2018). Rare taxa and dark microbial matter: novel bioactive actinobacteria abound in Atacama Desert soils. *Antonie van Leeuwenhoek*, *111*(8), 1315–1332.

Goswami, L., Arul Manikandan, N., Pakshirajan, K., & Pugazhenthi, G. (2017). Simultaneous heavy metal removal and anthracene biodegradation by the oleaginous bacteria Rhodococcus opacus. *3 Biotech*, *7*(1), 1–9.

Gousterova, A., Nustorova, M., Paskaleva, D., Naydenov, M., Neshev, G., & Vasileva-Tonkova, E. (2011). Assessment of feather hydrolysate from thermophilic actinomycetes for soil amendment and biological control application. *International Journal of Environmental Research*, *5*(4), 1065–1070.

Gremion, F., Chatzinotas, A., & Harms, H. (2003). Comparative 16S rDNA and 16S rRNA sequence analysis indicates that Actinobacteria might be a dominant part of the metabolically active bacteria in heavy metal-contaminated bulk and rhizosphere soil. *Environmental Microbiology*, *5*(10), 896–907.

Guo, J., & Ma, J. (2015). Bioflocculant from pre-treated sludge and its applications in sludge dewatering and swine wastewater pretreatment. *Bioresource Technology*, *196*, 736–740.

Hall, N. (2007). Advanced sequencing technologies and their wider impact in microbiology. *Journal of Experimental Biology*, *210*(9), 1518–1525.

Hamdan, A. M., Abd-El-Mageed, H., & Ghanem, N. (2021). Biological treatment of hazardous heavy metals by Streptomyces rochei ANH for sustainable water management in agriculture. *Scientific Reports*, *11*(1), 1–12.

Hamedi, J., & Mohammadipanah, F. (2015). Biotechnological application and taxonomical distribution of plant growth promoting actinobacteria. *Journal of Industrial Microbiology and Biotechnology*, *42*(2), 157–171.

Hayakawa, M. (2008). Studies on the isolation and distribution of rare actinomycetes in soil. *Actinomycetologica*, *22*(1), 12–19.

Hazarika, S. N., & Thakur, D. (2020). Actinobacteria. In: Amaresan, N., Senthil Kumar, M., Annapurna, K., Krishna Kumar, & Sankaranarayanan, A. (eds) *Beneficial Microbes in Agro-Ecology* (pp. 443–476). Elsevier.

He, J., Chen, Y., Dai, L., Yao, J., Mei, Y., Hrynsphan, D., Tatsiana, S., & Chen, J. (2020). Rapid and complete biodegradation of acrylic acid by a novel strain *Rhodococcus ruber* JJ-3: Kinetics, carbon balance, and degradation pathways. *Biotechnology and Bioprocess Engineering*, *25*(4), 589–598.

Hozzein, W. N., Ahmed, M. B., & Tawab, M. S. A. (2012). Efficiency of some actinomycete isolates in biological treatment and removal of heavy metals from wastewater. *African Journal of Biotechnology*, *11*(5), 1163–1168.

Hsu, K.-J., Tseng, M., Don, T.-M., & Yang, M.-K. (2012). Biodegradation of poly (β-hydroxybutyrate) by a novel isolate of Streptomyces bangladeshensis 77T-4. *Botanical Studies*, 53(3).

Hu, X., Thumarat, U., Zhang, X., Tang, M., & Kawai, F. (2010). Diversity of polyester-degrading bacteria in compost and molecular analysis of a thermoactive esterase from *Thermobifida alba* AHK119. *Applied Microbiology and Biotechnology*, *87*(2), 771–779.

Idemudia, M. I., Nosagie, O. A., & Omorede, O. (2014). Comparative assessment of degradation potentials of bacteria and actinomycetes in soil contaminated with motorcycle spent oil. *Asian Journal of Science and Technology*, *5*(8), 482–487.

Idris, R., Trifonova, R., Puschenreiter, M., Wenzel, W. W., & Sessitsch, A. (2004). Bacterial communities associated with flowering plants of the Ni hyperaccumulator *Thlaspi goesingense*. *Applied and Environmental Microbiology*, *70*(5), 2667–2677.

Jain, S., Gupta, I., Walia, P., & Swami, S. (2022). Application of actinobacteria in agriculture, nanotechnology, and bioremediation. In: Hozzein, W. H. (eds) *Actinobacteria-Diversity, Applications and Medical Aspects*. IntechOpen.

Jastaniah, S. D., & Aburas, M. M. A. (2016). Effects of Some Heavy Metals on Growth, Protein Content and Pigment Production by *Streptomyces Coelicolor* SM1. *Biosciences Biotechnology Research Asia*, *13*(4), 1975–1981.

Jensen, P. R., & Mafnas, C. (2006). Biogeography of the marine actinomycete *Salinispora*. *Environmental Microbiology*, *8*(11), 1881–1888.

Jensen, P. R., Moore, B. S., & Fenical, W. (2015). The marine actinomycete genus Salinispora: a model organism for secondary metabolite discovery. *Natural Product Reports*, *32*(5), 738–751.

Jin, Q., Hu, Z., Jin, Z., Qiu, L., Zhong, W., & Pan, Z. (2012). Biodegradation of aniline in an alkaline environment by a novel strain of the halophilic bacterium, Dietzia natronolimnaea JQ-AN. *Bioresource Technology*, *117*, 148–154.

Kamala, K., Sivaperumal, P., Thilagaraj, R., & Natarajan, E. (2020). Bioremediation of Sr2+ ion radionuclide by using marine *Streptomyces* sp. CuOff24 extracellular polymeric substances. *Journal of Chemical Technology & Biotechnology*, *95*(4), 893–903.

Khalid, S., Shahid, M., Niazi, N. K., Murtaza, B., Bibi, I., & Dumat, C. (2017). A comparison of technologies for remediation of heavy metal contaminated soils. *Journal of Geochemical Exploration*, *182*, 247–268.

Khessairi, A., Fhoula, I., Jaouani, A., Turki, Y., Cherif, A., Boudabous, A., Hassen, A., & Ouzari, H. (2014). Pentachlorophenol degradation by Janibacter sp., a new actinobacterium isolated from saline sediment of arid land. *BioMed Research International*, 2014, 1–9

Kiray, E., Er, C., Kariptas, E., & Ciftci, H. (2017). Biosorption and separation/preconcentration of lead and nickel on *Rhodococcus Ruber* biomass. *Fresenius Environmental Bulletin, 7*(8), 7740–7749.

Kubota, K., Koma, D., Matsumiya, Y., Chung, S.-Y., & Kubo, M. (2008). Phylogenetic analysis of long-chain hydrocarbon-degrading bacteria and evaluation of their hydrocarbon-degradation by the 2, 6-DCPIP assay. *Biodegradation*, *19*(5), 749–757.

Kundu, D., Hazra, C., Chatterjee, A., Chaudhari, A., & Mishra, S. (2014). Extracellular biosynthesis of zinc oxide nanoparticles using *Rhodococcus pyridinivorans* NT2: multifunctional textile finishing, biosafety evaluation and in vitro drug delivery in colon carcinoma. *Journal of Photochemistry and Photobiology B: Biology*, *140*, 194–204.

Kurtböke, D. İ., Grkovic, T., & Quinn, R. J. (2015). Marine actinomycetes in biodiscovery. In : Kim, S.K. (ed) *Springer Handbook of Marine Biotechnology* (pp. 663–676). Springer.

Kuyukina, M. S., & Ivshina, I. B. (2019). Bioremediation of contaminated environments using *Rhodococcus*. In: Alvarez, H. (ed) *Biology of Rhodococcus* (pp. 231–270). Springer.

Larcher, S., & Yargeau, V. (2013). Biodegradation of 17α-ethinylestradiol by heterotrophic bacteria. *Environmental Pollution*, 173, 17–22.

Li, O., Lu, C., Liu, A., Zhu, L., Wang, P.-M., Qian, C.-D., Jiang, X.-H., & Wu, X.-C. (2013). Optimization and characterization of polysaccharide-based bioflocculant produced by *Paenibacillus elgii* B69 and its application in wastewater treatment. *Bioresource Technology*, *134*, 87–93.

Li, X., Ding, C., Liao, J., Du, L., Sun, Q., Yang, J., Yang, Y., Zhang, D., Tang, J., & Liu, N. (2016). Bioaccumulation characterization of uranium by a novel *Streptomyces sporoverrucosus* dwc-3. *Journal of Environmental Sciences*, *41*, 162–171.

Lin, Y., Wang, X., Wang, B., Mohamad, O., & Wei, G. (2012). Bioaccumulation characterization of zinc and cadmium by *Streptomyces zinciresistens*, a novel actinomycete. *Ecotoxicology and Environmental Safety*, *77*, 7–17.

Ludwig, W., Euzéby, J., Schumann, P., Busse, H. J., Trujillo, M. E., Kämpfer, P., & Whitman, W. B. (2012). Road map of the phylum Actinobacteria. In: Goodfellow, M., et al. (eds) *Bergey's Manual(r) of Systematic Bacteriology* (pp. 1–28). Springer.

Ma, L., Li, Y., Yao, L., & Du, H. (2021). Polycyclic aromatic hydrocarbons in soil-turfgrass systems in urban Shanghai: Contamination profiles, in situ bioconcentration and potential health risks. *Journal of Cleaner Production*, 289, 125833.

Mahdi, A. M. el, Aziz, H. A., & Eqab, E. S. (2017). Review on innovative techniques in oil sludge bioremediation. *AIP Conference Proceedings*, 1892(1), 040026.

Maniyam, M. N., Hari, M., & Yaacob, N. S. (2020). Enhanced methylene blue decolourization by *Rhodococcus* strain UCC 0003 grown in banana peel agricultural waste through response surface methodology. *Biocatalysis and Agricultural Biotechnology*, *23*, 101486.

Maniyam, M. N., Ibrahim, A. L., & Cass, A. E. G. (2018). Decolourization and biodegradation of azo dye methyl red by Rhodococcus strain UCC 0016. *Environmental Technology*, 41(1), 71–85.

Martorell, M. M., Pajot, H. F., Rovati, J. I., & Figueroa, L. I. C. (2012). Optimization of culture medium composition for manganese peroxidase and tyrosinase production during Reactive Black 5 decolourization by the yeast *Trichosporon akiyoshidainum*. *Yeast*, *29*(3-4), 137–144.

Menashe, O., Raizner, Y., Kuc, M. E., Cohen-Yaniv, V., Kaplan, A., Mamane, H., Avisar, D., & Kurzbaum, E. (2020). Biodegradation of the endocrine-disrupting chemical 17α-ethynylestradiol (EE2) by Rhodococcus zopfii and Pseudomonas putida encapsulated in small bioreactor platform (SBP) capsules. *Applied Sciences*, 10(1), 336.

Mesquini, J. A., Sawaya, A. C. H. F., López, B. G. C., Oliveira, V. M., & Miyasaka, N. R. S. (2015). Detoxification of atrazine by endophytic *Streptomyces* sp. isolated from sugarcane and detection of nontoxic metabolite. *Bulletin of Environmental Contamination and Toxicology*, *95*(6), 803–809.

More, T. T., Yadav, J. S. S., Yan, S., Tyagi, R. D., & Surampalli, R. Y. (2014). Extracellular polymeric substances of bacteria and their potential environmental applications. *Journal of Environmental Management*, *144*, 1–25.

Mukherjee, A., & Chattopadhyay, D. (2017). Exploring environmental systems and processes through next-generation sequencing technologies: insights into microbial response to petroleum contamination in key environments. *The Nucleus*, *60*, 175–186.

Nazari, M. T., Machado, B. S., Marchezi, G., Crestani, L., Ferrari, V., Colla, L. M., & Piccin, J. S. (2022). Use of soil actinomycetes for pharmaceutical, food, agricultural, and environmental purposes. *3 Biotech*, *12*(9), 1–26.

Ojha, N., Karn, R., Abbas, S., & Bhugra, S. (2021). Bioremediation of industrial wastewater: A review. *IOP Conference Series: Earth and Environmental Science*, *796*(1), 012012.

Okaiyeto, K., Nwodo, U. U., Mabinya, L. v, & Okoh, A. I. (2014). Evaluation of the flocculation potential and characterization of bioflocculant produced by *Micrococcus* sp. Leo. *Applied Biochemistry and Microbiology*, *50*(6), 601–608.

Otari, S. v, Patil, R. M., Ghosh, S. J., Thorat, N. D., & Pawar, S. H. (2015). Intracellular synthesis of silver nanoparticle by actinobacteria and its antimicrobial activity. *Spectrochimica Acta Part A: Molecular and Biomolecular Spectroscopy*, *136*, 1175–1180.

Patil, S. V., Patil, C. D., Salunke, B. K., Salunkhe, R. B., Bathe, G. A., & Patil, D. M. (2011). Studies on characterization of bioflocculant exopolysaccharide of *Azotobacter indicus* and its potential for wastewater treatment. *Applied Biochemistry and Biotechnology*, *163*, 463–472.

Phoenix, V. R., Renaut, R. W., Jones, B., & Ferris, F. G. (2005). Bacterial S-layer preservation and rare arsenic-antimony-sulphide bioimmobilization in siliceous sediments from Champagne Pool hot spring, Waiotapu, New Zealand. *Journal of the Geological Society*, *162*(2), 323–331.

Qi, J., Anke, M. K., Szymańska, K., & Tischler, D. (2017). Immobilization of *Rhodococcus opacus* 1CP azoreductase to obtain azo dye degrading biocatalysts operative at acidic pH. *International Biodeterioration & Biodegradation*, *118*, 89–94.

Raja, M. M. M., Raja, A., Salique, S. M., & Gajalakshmi, P. (2016). Studies on effect of marine actinomycetes on amido black (azo dye) decolorization. *Journal of Chemical and Pharmaceutical Research*, *8*(8), 640–644.

Rebets, Y., Kormanec, J., Luzhetskyy, A., Bernaerts, K., & Anné, J. (2017). Cloning and expression of metagenomic DNA in Streptomyces lividans and subsequent fermentation for optimized production. *Metagenomics: Methods and Protocols*, 99–144.

Roy, U., Manna, S., Sengupta, S., Das, P., Datta, S., Mukhopadhyay, A., & Bhowal, A. (2018). Dye removal using microbial biosorbents. *Green Adsorbents for Pollutant Removal*, *19*, 253–280.

Saez, J. M., Benimeli, C. S., & Amoroso, M. J. (2012). Lindane removal by pure and mixed cultures of immobilized actinobacteria. *Chemosphere*, *89*(8), 982–987.

Sagarkar, S., Bhardwaj, P., Storck, V., Devers-Lamrani, M., Martin-Laurent, F., & Kapley, A. (2016). s-triazine degrading bacterial isolate *Arthrobacter* sp. AK-YN10, a candidate for bioaugmentation of atrazine contaminated soil. *Applied Microbiology and Biotechnology*, *100*(2), 903–913.

Sahasrabudhe, M. M., Saratale, R. G., Saratale, G. D., & Pathade, G. R. (2014). Decolorization and detoxification of sulfonated toxic diazo dye CI Direct Red 81 by *Enterococcus faecalis* YZ 66. *Journal of Environmental Health Science and Engineering*, *12*(1), 1–13.

Sánchez-Andrea, I., Rodríguez, N., Amils, R., & Sanz, J. L. (2011). Microbial diversity in anaerobic sediments at Rio Tinto, a naturally acidic environment with a high heavy metal content. *Applied and Environmental Microbiology*, *77*(17), 6085–6093.

Sarwar, N., Imran, M., Shaheen, M. R., Ishaque, W., Kamran, M. A., Matloob, A., Rehim, A., & Hussain, S. (2017). Phytoremediation strategies for soils contaminated with heavy metals: modifications and future perspectives. *Chemosphere*, *171*, 710–721.

Schmidt-Wygasch, C., Schamuhn, S., Meurers-Balke, J., Lehmkuhl, F., & Gerlach, R. (2010). Indirect dating of historical land use through mining: linking heavy metal analyses of fluvial deposits to archaeobotanical data and written accounts. *Geoarchaeology*, *25*(6), 837–856.

Selvaraj, V., Karthika, T. S., Mansiya, C., & Alagar, M. (2021). An over review on recently developed techniques, mechanisms and intermediate involved in the advanced azo dye degradation for industrial applications. *Journal of Molecular Structure*, 1224, 129195.

Sessitsch, A., Kuffner, M., Kidd, P., Vangronsveld, J., Wenzel, W. W., Fallmann, K., & Puschenreiter, M. (2013). The role of plant-associated bacteria in the mobilization and phytoextraction of trace elements in contaminated soils. *Soil Biology and Biochemistry*, *60*, 182–194.

Shekhar, S. K., Godheja, J., Modi, D. R., & Peter, J. K. (2014). Growth potential assessment of Actinomycetes isolated from petroleum contaminated soil. *Journal of Bioremediation & Biodegredation*, *5*(7), 1.

Shivlata, L., & Satyanarayana, T. (2015). Thermophilic and alkaliphilic Actinobacteria: biology and potential applications. *Frontiers in Microbiology*, *6*, 1014.

Sivaperumal, P., Kamala, K., & Rajaram, R. (2018). Adsorption of cesium ion by marine actinobacterium *Nocardiopsis* sp. 13H and their extracellular polymeric substances (EPS) role in bioremediation. *Environmental Science and Pollution Research*, *25*(5), 4254–4267.

Song, X., Zhang, Z., Dai, Y., Cun, D., Cui, B., Wang, Y., Fan, Y., Tang, H., Qiu, L., & Wang, F. (2022). Biodegradation of phthalate acid esters by a versatile PAE-degrading strain *Rhodococcus* sp. LW-XY12 and associated genomic analysis. *International Biodeterioration & Biodegradation*, *170*, 105399.

Sundarajoo, A., & Maniyam, M. N. (2019). Enhanced decolourization of congo red dye by *Malaysian rhodococcus* UCC 0010 immobilized in calcium alginate. *Journal of Advanced Research Design*, *62*(1), 1–9.

Theerthagiri, J., Lee, S. J., Karuppasamy, K., Arulmani, S., Veeralakshmi, S., Ashokkumar, M., & Choi, M. Y. (2021). Application of advanced materials in sonophotocatalytic processes for the remediation of environmental pollutants. *Journal of Hazardous Materials*, 412, 125245.

Theerthagiri, J., Salla, S., Senthil, R. A., Nithyadharseni, P., Madankumar, A., Arunachalam, P., Maiyalagan, T., & Kim, H.-S. (2019). A review on ZnO nanostructured materials: energy, environmental and biological applications. *Nanotechnology*, *30*(39), 392001.

Timková, I., Sedláková-Kaduková, J., & Pristaš, P. (2018). Biosorption and bioaccumulation abilities of actinomycetes/streptomycetes isolated from metal contaminated sites. *Separations*, *5*(4), 54.

Ting, A. S. Y. (2020). Actinobacteria for the effective removal of toxic dyes. In: Chowdhary, P., Abhay, R., Verma, D., & Akhter, Y. (eds) *Microorganisms for Sustainable Environment and Health* (pp. 37–52). Elsevier.

Ugbenyen, A. M., & Okoh, A. I. (2014). Characteristics of a bioflocculant produced by a consortium of *Cobetia* and *Bacillus* species and its application in the treatment of wastewaters. *Water SA*, *40*(1), 140–144.

Ungureanu, C. P., Favier, L., Bahrim, G., & Amrane, A. (2015). Response surface optimization of experimental conditions for carbamazepine biodegradation by Streptomyces MIUG 4.89. *New Biotechnology*, *32*(3), 347–357.

Vancov, T., Jury, K., & Van Zwieten, L. (2005). Atrazine degradation by encapsulated *Rhodococcus erythropolis* NI86/21. *Journal of Applied Microbiology*, *99*(4), 767–775.

Varjani, S., Rakholiya, P., Ng, H. Y., You, S., & Teixeira, J. A. (2020). Microbial degradation of dyes: an overview. *Bioresource Technology*, *314*, 123728.

Vásquez, T. G. P., Botero, A. E. C., de Mesquita, L. M. S., & Torem, M. L. (2007). Biosorptive removal of Cd and Zn from liquid streams with a *Rhodococcus opacus* strain. *Minerals Engineering*, *20*(9), 939–944.

Villegas, L. B., Rodríguez, A., Pereira, C. E., & Abate, C. M. (2013). Cultural factors affecting heavy metals removal by actinobacteria (pp. 26–43). CRC Press.

Viti, C., Pace, A., & Giovannetti, L. (2003). Characterization of Cr (VI)-resistant bacteria isolated from chromium-contaminated soil by tannery activity. *Current Microbiology*, *46*(1), 1–5.

Wang, S.-S., Ye, S.-L., Han, Y.-H., Shi, X.-X., Chen, D.-L., & Li, M. (2016). Biosorption and bioaccumulation of chromate from aqueous solution by a newly isolated *Bacillus mycoides* strain 200AsB1. *RSC Advances*, *6*(103), 101153–101161.

Wang, Y., & Tam, N. F. Y. (2019). Microbial remediation of organic pollutants. In *World Seas: An Environmental Evaluation* (pp. 283–303). Elsevier.

Wink, J., Mohammadipanah, F., & Hamedi, J. (2017). Biology and biotechnology of actinobacteria. Springer.

Woo, S.-G., Cui, Y., Kang, M.-S., Jin, L., Kim, K. K., Lee, S.-T., Lee, M., & Park, J. (2012). *Georgenia daeguensis* sp. nov., isolated from 4-chlorophenol enrichment culture. *International Journal of Systematic and Evolutionary Microbiology*, *62*(Pt_7), 1703–1709.

Wu, C.-Y., Chen, N., Li, H., & Li, Q.-F. (2014). *Kocuria rosea* HN01, a newly alkaliphilic humus-reducing bacterium isolated from cassava dreg compost. *Journal of Soils and Sediments*, *14*(2), 423–431.

Wu, P., Wang, Z., Bhatnagar, A., Jeyakumar, P., Wang, H., Wang, Y., & Li, X. (2021). Microorganisms-carbonaceous materials immobilized complexes: Synthesis, adaptability and environmental applications. *Journal of Hazardous Materials*, *416*, 125915.

Xu, S., Wang, Y. F., Yang, L. Y., Ji, R., & Miao, A. J. (2018). Transformation of tetrabromobisphenol A by *Rhodococcus jostii* RHA1: Effects of heavy metals. *Chemosphere*, *196*, 206–213.

Yaashikaa, P. R., Kumar, P. S., Jeevanantham, S., & Saravanan, R. (2022). A review on bioremediation approach for heavy metal detoxification and accumulation in plants. *Environmental Pollution*, *119035*.

Ye, Z., Li, H., Jia, Y., Fan, J., Wan, J., Guo, L., ... Shen, C. (2020). Supplementing resuscitation-promoting factor (Rpf) enhanced biodegradation of polychlorinated biphenyls (PCBs) by *Rhodococcus biphenylivorans* strain TG9T. *Environmental Pollution*, *263*, 114488.

Zenova, G. M., Manucharova, N. A., & Zvyagintsev, D. G. (2011). Extremophilic and extremotolerant actinomycetes in different soil types. *Eurasian Soil Science*, *44*(4), 417–436.

Zhang, Y., Ji, J., Xu, S., Wang, H., Shen, B., He, J., Qiu, J., & Chen, Q. (2019). Biodegradation of Picolinic Acid by Rhodococcus sp. PA18. *Applied Sciences*, *9*(5), 1006.

19 Integration of Natural Coagulants with Heterogeneous Photo-Fenton Processes for an Agro-industrial Wastewater Treatment

Nuno Jorge, Ana R. Teixeira, José A. Peres, and Marco S. Lucas

19.1 INTRODUCTION

In wine countries, the liquid effluents generated during the production of wine are an environmental issue with difficult management due to their reduced biodegradability. These wastewaters can be treated in dedicated wastewater treatment plants (WWTP) nearby the cellar or sent through the sewage network to a municipal WWTP, therefore, the WW requires proper pre-treatment (Johnson and Mehrvar, 2022; Jorge et al., 2022) to avoid the collapse of the biological reactors of the municipal WWTP. In this work, a coagulation-flocculation-decantation (CFD) pre-treatment is proposed followed by a heterogeneous photocatalytic process.

CFD is a common physical-chemical process used in water and wastewater treatment plants. The efficiency of the CFD process was mentioned in various studies on the treatment of wastewaters, focused on optimization, throughout selecting the best type of coagulant and dosage, evaluation of pH effect, stirring conditions, and dosage of flocculant (Tatsi et al., 2003). Considering the need to obtain more sustainable products which are cheaper and environmentally friendly, natural coagulants originated from plants (D.g. seeds, F.a. seeds and V.v. rachis) (Prozil et al., 2014), and conventional coagulants often used in wine treatment such as activated sodium bentonite (Horvat et al., 2019), potassium caseinate (Cosme et al., 2008) and PVPP (Laborde et al., 2006) were investigated.

Afterwards, to complement the CFD process, Advanced Oxidation Processes (AOP) were studied due to the capacity to enhance the degradation of organic pollutants from WW. In AOPs, the photo-Fenton process has the capacity to generate hydroxyl radicals ($HO^{\bullet}$) by the mixture of iron ions, and hydrogen peroxide under UV radiation (Pignatello et al., 2006). These powerful radicals degrade organic contaminants very quickly and non-selectively, in accordance with Equation 19.1 (Tokumura et al., 2011):

$$RH\,(\text{organic compound}) + HO^{\bullet} \rightarrow \text{intermediates} + HO^{\bullet} \rightarrow CO_2 + H_2O \qquad (19.1)$$

Although the photo-Fenton process has shown often good results, there are some limitations that should be addressed, such as: (i) prevention of metal precipitation by low pH; (ii) recovery of the

DOI: 10.1201/9781003441144-19

catalyst; (iii) neutralization of treated wastewater according to legal limits (pH 6.0–9.0) (Guimarães et al., 2019b; Jorge, Teixeira, Lucas et al., 2021). In addition, to make more attractive the application of homogeneous photo-Fenton, the use of heterogeneous catalysts can reduce the costs related to sludge treatment, allowing catalyst recovery, and its reuse for application in multiple reaction cycles (Guimarães et al., 2019b; Loddo et al., 2018). In this work, it was explored the use of ferrocene (Fc) as a heterogeneous catalyst. Ferrocene is an organometallic compound with the formula $Fe(C_5H_5)_2$. The molecule is a complex consisting of two cyclopentadienyl rings sandwiching a central iron atom. The high effect and stability of ferrocene in methyl orange degradation by heterogeneous Fenton was observed by Jia et al. (2019).

The aim of this work is (i) apply conventional and plant natural-based coagulants powder from D.g. seeds, F.a. seeds, and V.v. rachis in coagulation process, to produce a potentially reusable organic sludge, (ii) optimize heterogeneous photo-Fenton and (iii) study the effect of combined CFD-heterogeneous photo-Fenton.

19.2 MATERIAL AND METHODS

19.2.1 Reagents

Bentonite was acquired from Angelo Coimbra & Ca., Lda (Portugal), potassium caseinate and polyvinylpyrrolidone (PVPP) from A. Freitas Vilar (Portugal), ferric chloride hexahydrate from Merck (Germany), hydrogen peroxide (30% w/w) from Sigma-Aldrich (USA) and ferrocene from Alfa Aesar (USA). Sodium hydroxide (Labkem, Spain) and sulphuric acid (Scharlau, Spain) were applied in pH adjustment and deionised water was used to prepare work solutions.

19.2.2 Analytical Technics

Total polyphenols were determined by the Folin-Ciocalteau method (Singleton and Rossi, 1965), chemical oxygen demand (COD) was digested in a COD reactor and data was recorded by spectrophotometer (at $\lambda = 600\,nm$) DR2100 (Hach, USA), biochemical oxygen demand (BOD_5) was analyzed with OxiTop system (WTW, USA). Dissolved organic carbon (DOC) was determined by a Shimadzu TOC-L_{CSH} analyzer (Japan). The turbidity, total suspended solids and pH (Jenway, UK) were calculated following the Standard Methods (APHA et al., 1999). Atomic absorption spectroscopy (Thermo Fisher Scientific, USA) was used to determine the iron remaining in solution. Table 19.1 shows the main physicochemical parameters of WW.

19.2.3 Plant Natural-Based Coagulants Preparation

Plant natural-based coagulants were prepared using the methodology proposed by Jorge et al., (2023). To study the composition of the plant natural-based coagulants, it was employed Fourier Transform Infrared spectrometer (FTIR) (Shimadzu, Japan), the inner regions of the coagulants were studied by scanning electron microscopy (SEM, FEI Quanta, USA) and the textural parameters were obtained according to Brunauer, Emmett, Teller method.

TABLE 19.1
Physicochemical Characterization of WW

pH	COD mg O_2/L	BOD_5 mg O_2/L	BOD_5/COD	Turbidity NTU	TSS mg/L	Total Polyphenols mg gallic acid/L
4.0	2145	550	0.26	296	750	23

19.2.4 CFD Experimental Organization

The CFD experiments were executed with 500mL of WW in a Jar-test (ISCO JF-4, USA), with a sedimentation period of 2hours. Two oenological coagulants (bentonite and potassium caseinate), three plant natural-based coagulants, and one common coagulant (ferric chloride) were tested, carrying out the following steps:

1. The coagulant concentrations used were within the range of 0.1–2.0g/L and the pH ranged from 3.0 to 11.0, with a fast mixing of 150rpm/3min, a slow mixing of 20rpm/20min, and a sedimentation period of 2hours;
2. In the second step, the mixing process (rpm/min) consisted in the variation of fast mix (rpm/min) – slow mix (rpm/min) within the range 120/1–20/30 to 200/2–60/30);
3. As a final step in CFD optimization, different concentrations of PVPP range from 5 to 100 mg/L.

19.2.5 Heterogeneous Photo-Fenton Setup

A cylindrical photoreactor, carrying a UV-C low pressure mercury vapour lamp (TNN 15/32) - working power=15 W and λ_{max}=254nm (*Heraeus*, Germany) was used. The experiments were performed with 500mL volume for 240min. For the optimization of the oxidation process, initially different AOPs were tested. Then, the pH of the WW was varied (3.0–6.0). After, the addition of H_2O_2 (single and multiple dosing) was tested. In the best manner, the H_2O_2 concentration was varied (97–291mM). Finally, the concentration of ferrocene was investigated (0.25–1.0g/L) as a following step.

Equation 19.2 was employed to calculate the parameters removal (C_0 and C_f represent concentrations at time 0 and t) (Jorge, Teixeira, & Matos et al., 2021).

$$\text{Removal percentage (\%)} = \frac{C_0 - C_f}{C_0} \times 100 \quad (19.2)$$

Triplicate analyses were carried out and less than 5% standard deviation was shown, using OriginLab 2019 software (USA) for statistical analysis.

19.3 RESULTS AND DISCUSSION

19.3.1 CFD Experiments

19.3.1.1 Effect of pH and Coagulant Dosage

Bentonite, potassium caseinate, D.g. seeds, F.a. seeds, V.v. rachis and ferric chloride were tested by dosing four distinct concentrations at pH 3.0 (value obtained in preliminary studies). Under the best pH conditions, it was analyzed the effect of different concentrations of coagulants. The results in Figure 19.1 showed that coagulant concentration had a considerable outcome in CFD efficiency. For the above-mentioned six coagulants, the optimal concentrations obtained were 0.1, 0.1, 2.0, 2.0, 0.1 and 0.5g/L, respectively. These concentrations allowed to achieve 4.6%, 11.4%, 15.5%, 16.3%, 44.9% and 5.1% of DOC removal, 96.7%, 96.9%, 97.5%, 92.8%, 93.0% and 91.4% of turbidity removal and 92.8%, 93.5%, 93.6%, 90.1%, 90.8% and 81.1% of TSS removal, respectively.

The actions of natural coagulants can be related to the existence of cationic proteins, that in water attracts the electrons (in the surface of the colloidal particles), destabilizing these particles and then, by slow mixing, promotes the generation of flocs that will precipitates. This is in agreement with the results shown by Vunain et al. (2019). The FTIR analysis (Figure 19.2) indicated that the plant parts used had proteins in their constitution such as amide II band of proteins at 1,543 per cm (Kong and Yu, 2007), amide I band of proteins at 1,651 per cm (Kong and Yu, 2007) and protein band at 2,850 per cm (Meade et al., 2010). This mechanism is involved in the combination of

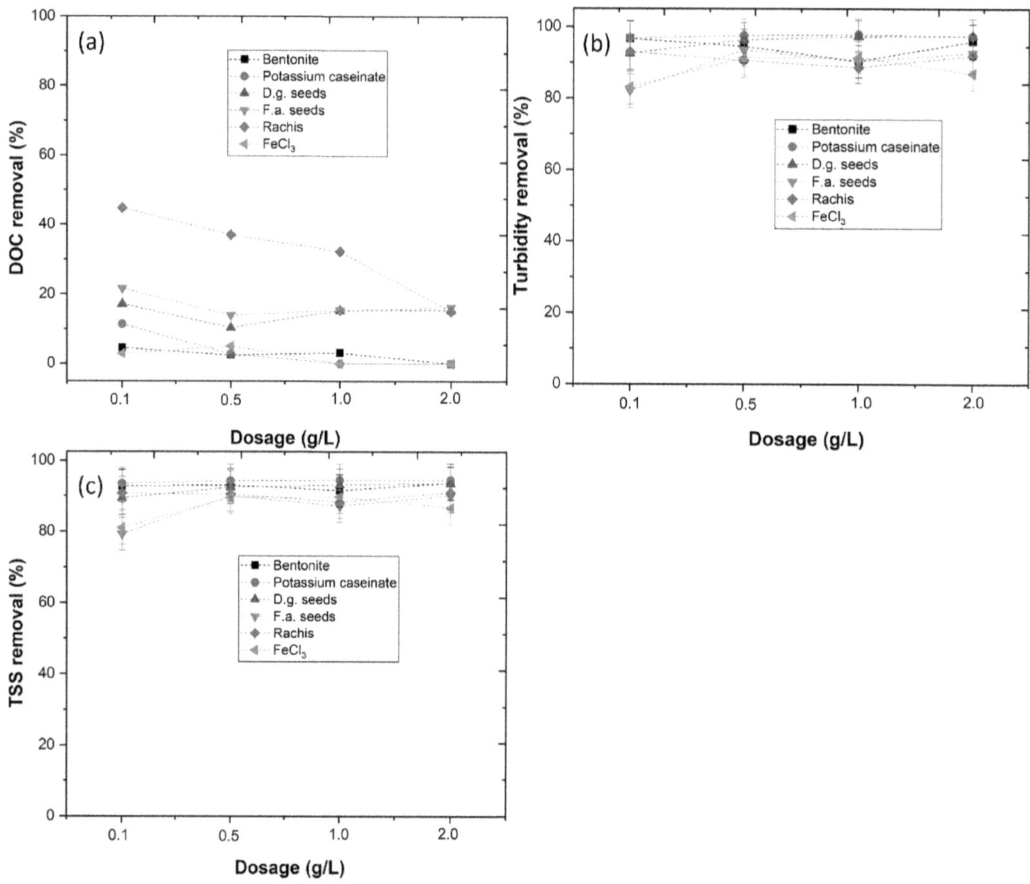

FIGURE 19.1 Application and evaluation of different coagulant dosages (0.1–2.0 g/L) at pH 3.0 in the removal of DOC (a), turbidity (b), TSS (c). (pH = 3.0, rapid mix = 150 rpm/3 min, slow mix = 20 rpm/20 min, sedimentation = 2 hours).

adsorption, charge neutralization and particle bridging, occurring simultaneously (Camacho et al., 2017; Vunain et al., 2019).

The bentonite has a pH of 7.4 in water (isoelectric point 7.0) (Jorge, Teixeira, & Matos et al., 2021), becoming electropositive at pH 3.6–4.0 (Benna et al., 1999). Considering the high efficiency of bentonite at pH 3.0, after its addition, the pH of the WW was corrected to 3.0. At pH 3.0, occurs an electrostatic interaction between bentonite and colloidal particles such as proteins and polyphenols present in the WW, due to the considerable concentration of hydrogen ions that protonates the bentonite's negative surface becoming extra positively charged and an adsorption mechanism occurs, as shown by Guimarães et al. (2019a).

Potassium caseinate's higher effect at pH 3.0 can be explained by its chemical composition. Caseins are formed by sequences of amino acids with hydrophobic and hydrophilic characteristics, generating ambiphilic character. Under pH 3.0, it was observed a mechanism of (i) compression of the electrical double layer of the colloidal particles, (ii) adsorption and charge neutralization, (iii) adsorption and interparticle bridging and (iv) enmeshment in a precipitate or "sweep floc" (Howe et al., 2012). The potassium caseinate, an oenological coagulant, can be employed in wines (within the legal dosage range of 10–50 g/hL (Cardoso, 2007; OIV, 2016)), therefore, the dosage employed should pose no threat to the environment.

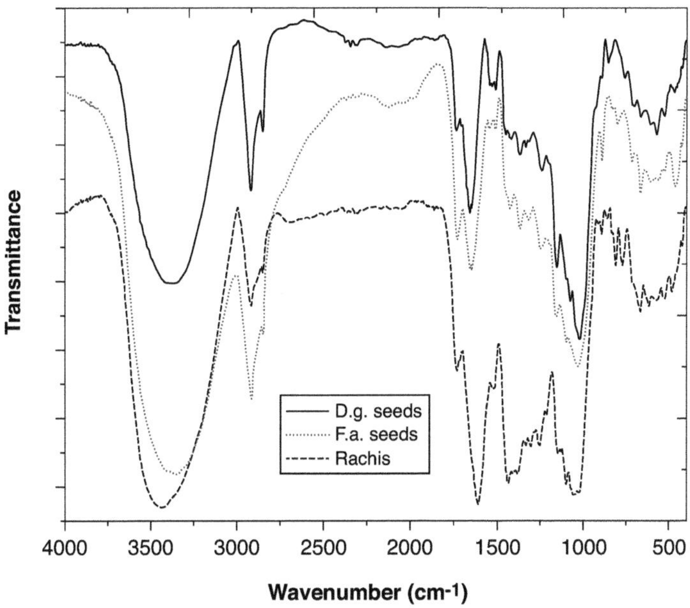

FIGURE 19.2 FTIR spectra of plant natural-based coagulants.

19.3.1.2 Effect of Agitation

To achieve maximum performance from CFD process, it was necessary to study how agitation affected capability of the coagulants to promote flocculation, and so, the mixing process was optimized by varying fast mix and slow mix, as shown in Figure 19.3. The results showed that the mixing conditions affected the removal of DOC, turbidity and TSS in different manners. Although no significant changes were observed in DOC removal, the turbidity and TSS were affected. Based in the results it were selected the following conditions (rpm/min): 150/3, 150/2, 200/2, 200/2, 150/3 and 120/1, for bentonite, potassium caseinate, D.g. seeds, F.a. seeds, V.v. rachis and ferric chloride, respectively. Although D.g. and F.a. seeds require higher agitation speeds, the remaining coagulants achieved better results with lower agitation speeds. This could be explained by the disruptive forces caused by the fast mixing, which broke the recently generated flocs, enabling the formation of large particles and reducing the sedimentation of colloids (Jorge et al., 2023).

19.3.1.3 Effect of PVPP

Previous sections showed that the coagulants were able to remove >95% of turbidity and TSS, however, DOC removal remained low. To improve this removal, a polyelectrolyte often used in oenological processes (polyvinylpyrrolidone, PVPP) was added as a flocculant. Polyelectrolytes have been reported to enhance coagulation (Amuda and Amoo, 2007), and considering that they are safe to work and biodegradable, they are advantageous (Zhu et al., 2004). In this work PVPP was added in small concentrations (5, 45, 85 and 100 mg/L), below the legal limit allowed for wine treatment (500 mg/L), which is in accordance with the International Code of Oenological Practices of the OIV (OIV, 2016).

Results showed that for the six coagulants under study (bentonite, potassium caseinate, plant natural-based coagulants (D.g. and F.a. seeds and V.v. rachis) and $FeCl_3$), the addition of 5, 5, 85, 5, 85 and 5 mg/L of PVPP, respectively, achieve DOC (25.9%, 28.9%, 20.3%, 23.5%, 48.2% and 30.4%) (Figure 19.4a), turbidity (97.3%, 99.6%, 98.7%, 98.8%, 98.4% and 95.7%) (Figure 19.4b) and TSS removals (94.3%, 95.3%, 93.5%, 93.7%, 94.5% and 93.5%) (Figure 19.4c), respectively.

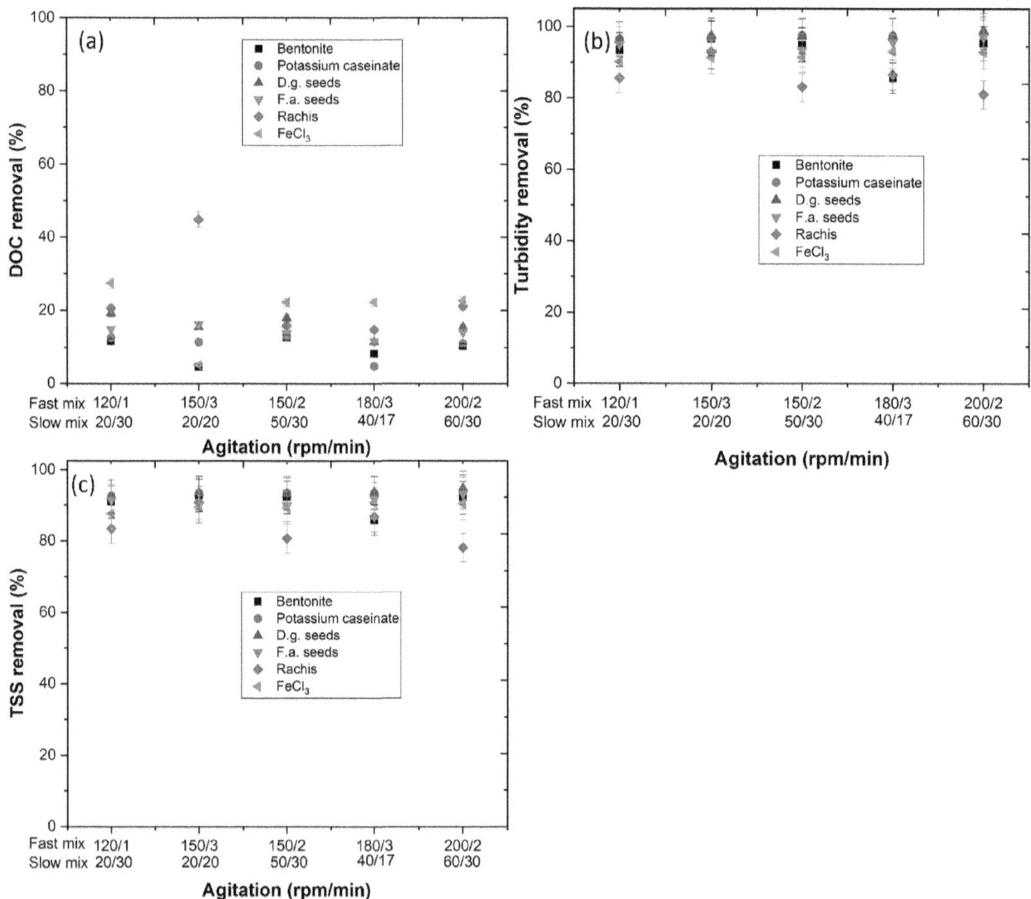

FIGURE 19.3 Assessment of agitation conditions in the removal of DOC (a), turbidity (b), TSS (c). (pH=3.0, [coagulant]=0.1, 0.1, 2.0, 2.0, 0.1 and 0.5 g/L, respectively for bentonite, potassium caseinate, D.g. seeds, F.a. seeds and ferric chloride; sedimentation=2 hours).

The coagulants and PVPP have many sites that can absorb/adsorb the organic matter, as observed by the SEM images (Figure 19.5), due to their porous nature (Table 19.2). These results agree with Teixeira et al., (2022), in which plant-based products can absorb large amounts of contaminant. With the addition of PVPP is generated a sheet which spreads throughout the wastewater, increasing the size and weight of the flocs, as well as the enhancement of small particle sedimentation, improving the organic matter removal. The mechanism involved in the flocculation by PVPP, comprehends adsorption and interparticle bridging, due to van der Walls forces, dipole interaction, coulombic (charge–charge) interactions and hydrogen bounds (Hunter, 2001; Jorge, Teixeira, & Matos et al., 2021).

19.3.2 Heterogeneous Catalysis

19.3.2.1 Assessment of AOPs in DOC Removal

The CFD process was revealed to be capable of removing the excessive sediments, but only limited DOC removal was achieved. In this way, a complementary system should be studied to achieve a higher mineralization of the WW. In this section, different chemical oxidation processes were tested (Figure 19.6), using ferrocene (Fc) as a catalyst and H_2O_2 as an oxidant agent.

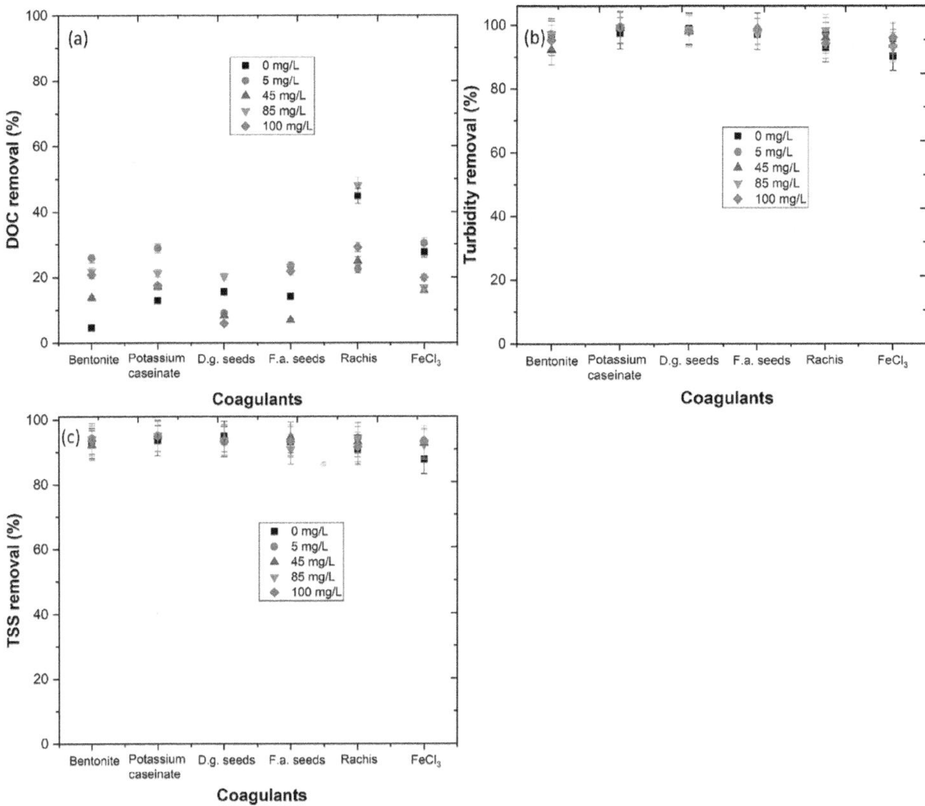

FIGURE 19.4 Appraisal of the concentration of PVPP (0–100 mg/L) in the removal of DOC (a), turbidity (b) TSS (c) TSS. (pH = 3.0, [coagulant] = 0.1, 0.1, 2.0, 2.0, 0.1 and 0.5 g/L, respectively for bentonite, potassium caseinate, D.g. seeds, F.a. seeds, V.v. rachis and ferric chloride, sedimentation = 2 hours).

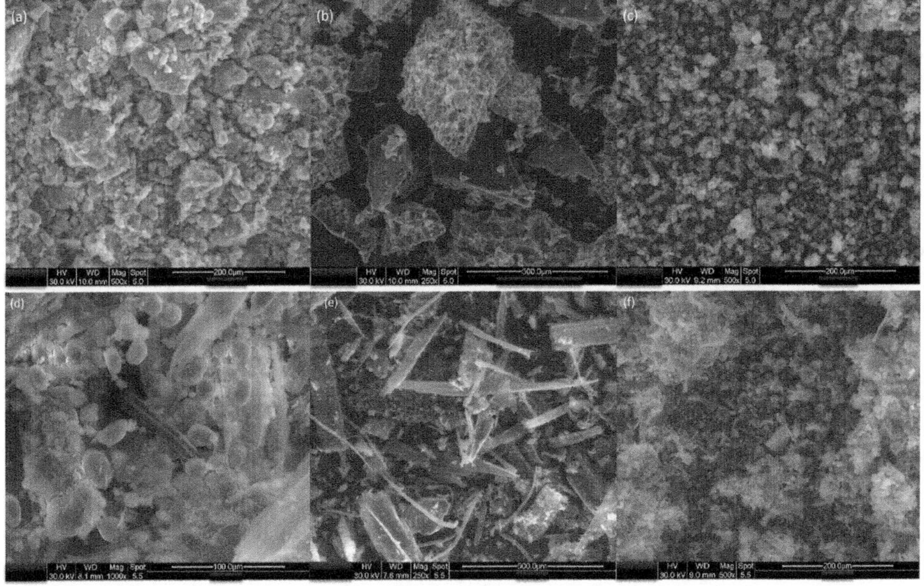

FIGURE 19.5 SEM images of bentonite (a), potassium caseinate (b), PVPP (c), D.g. seeds (d) F.a. seeds, (e) and V.v. rachis (f).

TABLE 19.2
B.E.T. Analysis of Coagulants

Coagulants	S_{BET} (m²/g)	$V_{total\ pore}$ (cm³/g)	Particle Size (nm)
Bentonite		0.045	4.0
Potassium caseinate	1.0	n.q.	n.q.
Dactylis glomerata L. seeds	0.04	n.q.	n.q.
Festuca ampla Hack. seeds	0.16	n.q.	n.q.
Vitis vinifera L. rachis	0.45	n.q.	n.q.

n.q. – not quantifiable.

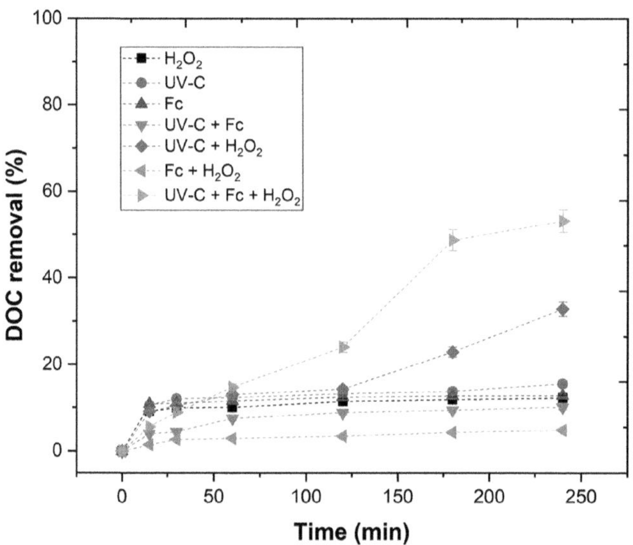

FIGURE 19.6 Removal of DOC from WW by different oxidants. Experimental conditions: $[H_2O_2]=194\,mM$, $[Fc]=0.50\,g/L$, $pH=3.0$, UV-C radiation (254 nm).

From the results, the organic matter dissolved in the WW is resistant to oxidation, with 12.3%, 15.6%, 10.2% and 4.9% DOC removal for H_2O_2, UV-C, UV-C+Fc and Fc+H_2O_2, respectively. Clearly, when dealing with wastewaters with high content of impurities, these chemical oxidants are not sufficient to achieve satisfactory removals of DOC, a fact observed in previous works dealing with mineralization of organic matter from the WW by heterogeneous catalysis (Guimarães et al., 2020). With the implementation of UV-C+H_2O_2, soaring DOC degradation was achieved (32.9%), which could be attributed to a higher production of $HO^{\bullet}$ radicals, surpassed only by UV-C+Fc+H_2O_2 (53.3%). With these results, UV-C+Fc+H_2O_2 was selected as the most efficient AOP.

19.3.2.2 Heterogeneous Photocatalysis Enhancement

Heterogeneous photocatalysis was shown to be more efficient than other AOPs, however, it is not sufficient to degrade the organic matter from the wastewater. In this way, a model was used to predict the most probable outcome of oxidation processes. The results obtained fitted the Fermi's kinetic model. Fermi's model was used considering the good fitting of the results, being the best model to explain the results, as shown in Equation 19.3 (Ghime and Ghosh, 2018; Herney-Ramirez et al., 2011; Rache et al., 2014):

TABLE 19.3
Experimental Data Fitting to the Fermi's Kinetic Model

		Kinetic Parameters			
Experiment	**Parameters**	k_{DOC} **(min⁻¹)**	t^* **(min⁻¹)**	x_{DOC}	r^2
pH variation	pH 3.0	0.0226	126	58.75	0.978
	pH 4.0	0.0207	146	48.96	0.962
	pH 6.0	0.0212	107	35.83	0.920
$[H_2O_2]$ addition	One addition	0.0226	126	58.75	0.978
	Six additions	0.0432	155	86.81	0.990
$[H_2O_2]$ variation	97 mM	0.0153	187	71.19	0.937
	194 mM	0.0432	155	86.81	0.990
	291 mM	0.0156	196	118.2	0.987
Fc dosage	0.25 g/L	0.0262	138	82.65	0.996
	0.50 g/L	0.0432	155	86.81	0.990
	1.0 g/L	0.0140	329	240.6	0.992

$$\frac{DOC}{DOC_0} = \frac{1 - x_{DOC}}{1 + \exp[k_{DOC}(t - t^*_{DOC})]} + x_{DOC} \tag{19.3}$$

where k_{DOC} is kinetic constant, x_{DOC} is fraction of non-oxidizable organic molecules and t^*_{DOC} is transition time.

Initially, it was applied a pH within the range of 3.0–6.0 (Figure 19.7a), keeping the conditions previously established. The results obtained fitted the Fermi's kinetic model as observed in Table 19.3. The pH 3.0 showed the best kinetic rate (0.0226 per min) decreasing as the pH increased further to pH 4.0 and 6.0. This behavior could be explained by two different reasons: (i) iron precipitation at pH 6.0, decreasing its accessibility in the WW (Table 19.5), decreasing the H_2O_2 consumption and consequently production of $HO^{\bullet}$ radicals (Table 19.4); (ii) authors, such as Li et al. (2018) and Wang et al. (2013), showed that ferrocene has its highest performance at pH 3.0, and an increase above pH 5.0 reduced significantly ferrocene's degradation capability. In fact, Li et al. (2018) observed on the treatment of sulfamethoxazole, that ferrocene exhibited a platform at pH>6. The reaction involved in the Fc-catalyzed photo-Fenton system were as follows in Equations 19.4–19.10 (Li et al., 2018):

$$Fc + h\nu(\lambda > 290\ nm) \rightarrow Fc^+ + e^- \tag{19.4}$$

$$Fc + H_2O_2 \rightarrow Fc^+ + HO^{\bullet} + HO^- \tag{19.5}$$

$$e^- + O_2 \rightarrow O_2^{\bullet -} \tag{19.6}$$

$$2O_2^{\bullet -} + 2H^+ \rightarrow H_2O_2 + O_2 \tag{19.7}$$

$$H_2O_2 + e^- \rightarrow +HO^{\bullet} + HO^- \tag{19.8}$$

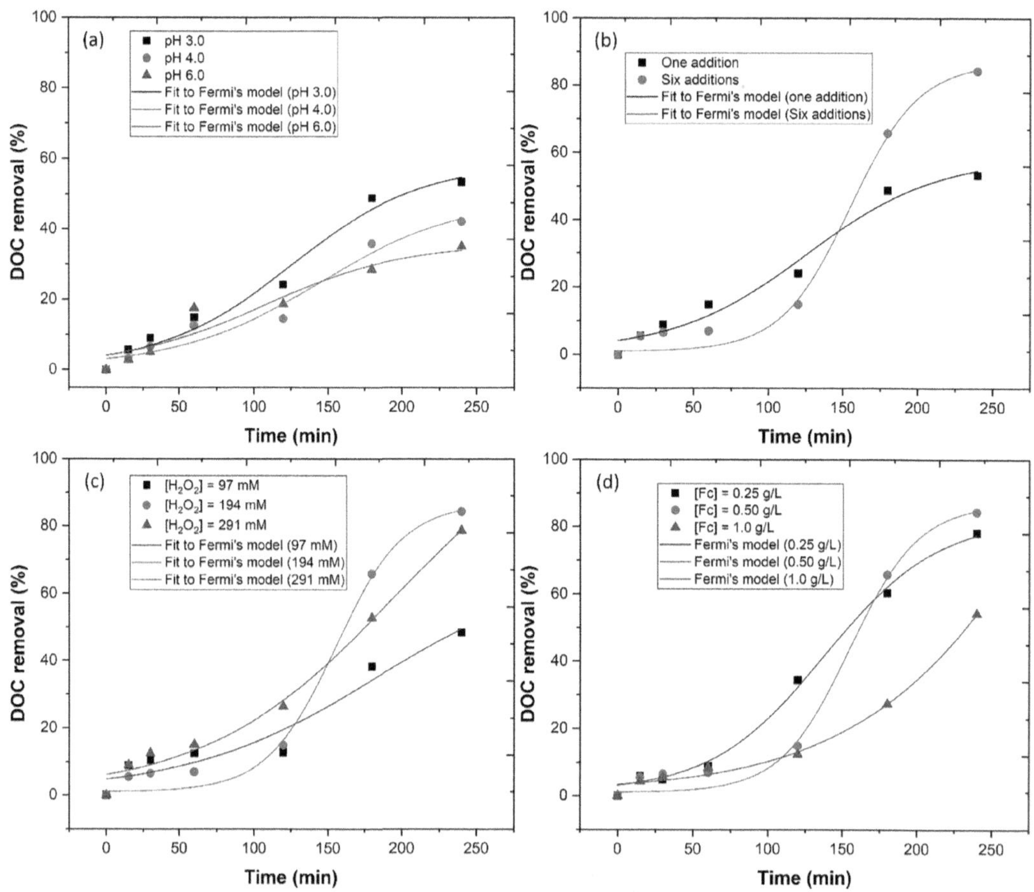

FIGURE 19.7 DOC removal fitting to Fermi's model by variation of (a) pH from 3.0 to 6.0 (b) H_2O_2 addition manner; (c) $[H_2O_2]$ from 97 to 291 mM and (d) [Fc] from 0.25 to 1.0 g/L.

TABLE 19.4

H_2O_2 Consumption (%) Evolution after Different Optimization Steps

Time	pH			H_2O_2 Addition		$[H_2O_2]$ (mM)			[Fc] (g/L)		
(min)	3.0	4.0	6.0	One	Six	97	194	291	0.25	0.50	1.0
0	0.0	0.0	0.0	0.0	0.0	0.0	0.0	0.0	0.0	0.0	0.0
15	23.9	13.1	11.7	23.9	4.4	2.8	4.4	1.3	2.6	4.4	1.4
30	23.9	13.1	11.7	23.9	4.4	4.7	4.4	1.3	3.4	4.4	2.2
60	23.9	13.1	11.7	23.9	7.5	16.8	7.5	6.9	8.2	7.5	8.2
120	24.2	21.9	34.8	24.2	28.9	38.6	28.9	14.5	21.6	28.9	29.7
180	88.2	89.4	53.2	88.2	79.2	81.1	79.2	57.8	72.5	79.2	72.2
240	99.6	99.4	79.6	99.6	99.9	99.9	99.9	91.3	97.5	99.9	99.9

$$H_2O_2 + hv \rightarrow 2HO^{\bullet} \tag{19.9}$$

$$Fc^{+} + H_2O_2 \rightarrow Fc + H^{+} + HO_2^{\bullet} \tag{19.10}$$

Following the optimization process, it should be understood that the manner of addition could influence the DOC removal. Thus, the next step of optimization consisted in the variation of the addition manner of H_2O_2, which was added in a single or in six steps. The results showed a good fitting to Fermi's kinetic model, exhibiting an enhancement of k_{DOC} from 0.0226 to 0.0432 per min (Figure 19.7b). With the single dosing step of H_2O_2, the generation of radicals is very quick in the beginning of the reaction, considering the high concentration of H_2O_2 available, resulting in rapid degradation of DOC, but, scavenging reactions also occurs very quickly, as observed in Equation 19.11 (Lucas and Peres, 2009; Rodríguez-Chueca et al., 2017). In multiple dosing steps of H_2O_2, the concentration is kept low in the reactor (Table 19.4), decreasing reactions between H_2O_2 and $HO^{\bullet}$ radicals, increasing the degradation of DOC.

$$H_2O_2 + HO^{\bullet} \rightarrow H_2O + HO_2^{\bullet} \tag{19.11}$$

Following the manner of H_2O_2 addition, the concentration of H_2O_2 was studied, varying from 97 to 194 to 291 mM. In Figure 19.7(c), it was observed a DOC removal, up to 120 min of 12.8%, 14.9% and 26.4%, respectively. These results were in agreement with other authors such as Wang et al. (2013), who observed that ferrocene has an induction period of 120 min. The results were fitted into Fermi's kinetic model, and it was shown a k_{DOC} = 0.0432 per min (highest) with the use of 194 mM H_2O_2. The lower DOC removal observed with 97 mM, could be attributed to the lack of H_2O_2 necessary to complete the catalytic process. The increase of H_2O_2 concentration (291 mM) wasn't favorable for heterogeneous photo-Fenton process. In Table 19.4, it is shown a 91.3% H_2O_2 consumption, thus $HO^{\bullet}$ radicals were produced. This suggested that the overabundance of H_2O_2 had a scavenging effect, which hampered the production of $HO^{\bullet}$ species (Equation 19.11).

Finally, catalyst concentration varied from 0.25 to 1.0 g/L, and Figure 19.7(d) shows the results after 240 min of reaction, in which the highest DOC degradation was reached with 0.50 g/L Fc. The results fitted the Fermi's kinetic model, showing a k=0.0432 per min. Below this concentration, the kinetic rate dropped to 0.0262 per min, most likely, the concentration of iron made available was much lower (Table 19.5), leading to lower production of $HO^{\bullet}$ radicals. With the addition of 1.0 g/L Fc, it was observed a gradual increase in Fe^{2+} leaching, until reached 240.8 mg/L after 240 min, which could have originated scavenger reactions by the iron onto $HO^{\bullet}$ radicals, thus decreasing the kinetic rate.

19.3.3 Combined CFD + Heterogeneous Photo-Fenton

Previous sections showed that CFD process achieved a near complete removal of turbidity and TSS, thus allowing the WW to become clearer for the UV-C radiation to pull through. Therefore,

TABLE 19.5
Fe^{2+} Concentration (mg/L) Available in Solution after Different Optimization Steps

Time	pH			H_2O_2 Addition		$[H_2O_2]_0$ (mM)			[Fc] (g/L)		
(min)	3.0	4.0	6.0	One	Six	97	194	291	0.25	0.50	1.0
0	0.1	0.1	0.1	0.1	0.1	0.1	0.1	0.1	0.1	0.1	0.1
15	2.1	2.1	0.0	2.1	6.0	1.8	6.0	0.5	2.3	6.0	1.4
30	5.3	7.1	0.8	5.3	8.6	7.1	8.6	5.8	5.3	8.6	11.1
60	9.9	10.3	3.8	9.9	14.7	16.8	14.7	12.8	11.9	14.7	34.4
120	15.8	18.5	1.2	15.8	27.4	36.6	27.4	33.5	13.7	27.4	99.5
180	18.1	35.9	9.1	14.9	40.4	94.7	40.4	37.0	22.3	40.4	240.4
240	42.2	58.2	5.1	42.2	19.5	143.5	19.5	52.0	7.5	19.5	240.8

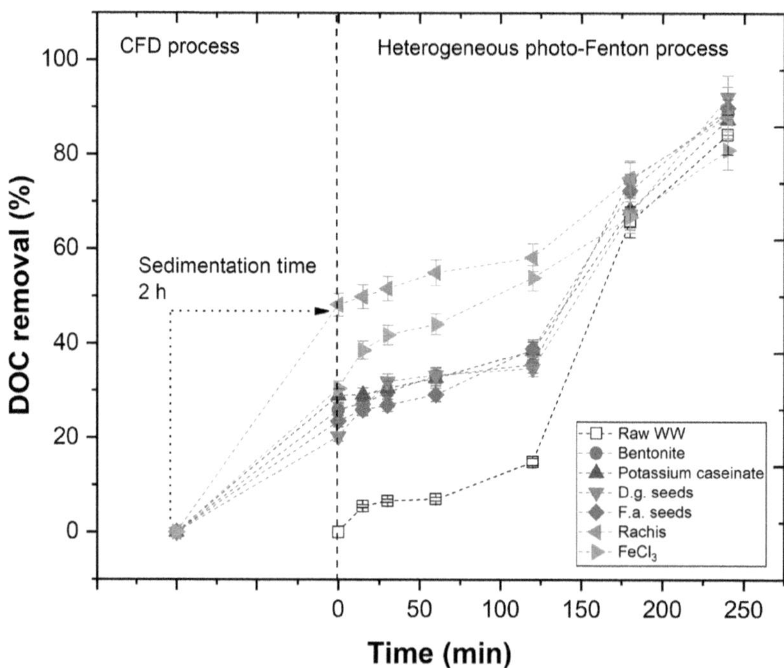

FIGURE 19.8 Degradation of DOC after merging of CFD and heterogeneous photo-Fenton. CFD conditions: pH=3.0, [coagulant]=0.1, 0.1, 2.0, 2.0, 0.1 and 0.5 g/L, respectively for bentonite, potassium caseinate, D.g. seeds, F.a. seeds, V.v. rachis and ferric chloride, sedimentation=2 hours. Heterogeneous photo-Fenton conditions: [Fc]=0.50 g/L, [H_2O_2]=194 mM, pH=3.0, UV-C radiation (254 nm) and reaction time=240 minutes.

after the decantation of the wastewater, it was performed the heterogeneous photo-Fenton process. Figure 19.8 presents the progression of DOC removal with the combination of both processes and it was compared with the performance of heterogeneous photo-Fenton process in raw WW. The results presented a DOC removal of 88.7%, 87.4%, 92.2%, 89.9%, 87.9% and 81.0%, respectively, for bentonite, potassium caseinate, D.g. seeds, F.a. seeds, V.v. rachis and ferric chloride, against 82.7% DOC removal (raw WW). The results showed that a combination of CFD (using oenological and plant natural-based coagulants) with heterogeneous photo-Fenton enhanced DOC removal, higher than application to raw WW, agreeing with Amor et al. (2015), who showed that the oxidation process was enhanced by the removal of turbidity from wastewater. The use of ferric chloride as a pre-treatment, showed a reduction in DOC removal after the 240 min of oxidation. Clearly, the iron present in excess reacted as a $HO^{\bullet}$ radical scavenger (Equation 19.12) (Martins et al., 2022), decreasing DOC degradation by heterogeneous photo-Fenton.

$$Fe^{2+} + HO^{\bullet} \rightarrow Fe^{3+} + HO^{-} \quad (19.12)$$

Considering the good results reached by the combination of both processes in DOC reduction (Figure 19.8), it was also studied the evolution of COD removal. Results showed significant differences between each treated wastewater, after both CFD and heterogeneous photo-Fenton. Combining both processes, it was observed that pre-treatment with bentonite, *Festuca ampla Hack.* seeds and *Vitis vinifera L.* rachis allowed to reach the Portuguese legal values (COD ≤ 150 mg O_2/L) for wastewater discharge (Figure 19.9a). As the raw WW showed a low biodegradability (BOD_5/COD < 0.30), this ratio was checked at the end of each treatment. Results have shown that all treated wastewaters achieved BOD_5/COD > 0.30, with exception of ferric chloride pre-treatment (BOD_5/COD = 0.23) (Figure 19.9b). Thus, all treated wastewaters with oenological and plant natural-based coagulants, that did not reach a COD < 150 mg O_2/L, could follow through a biologic process.

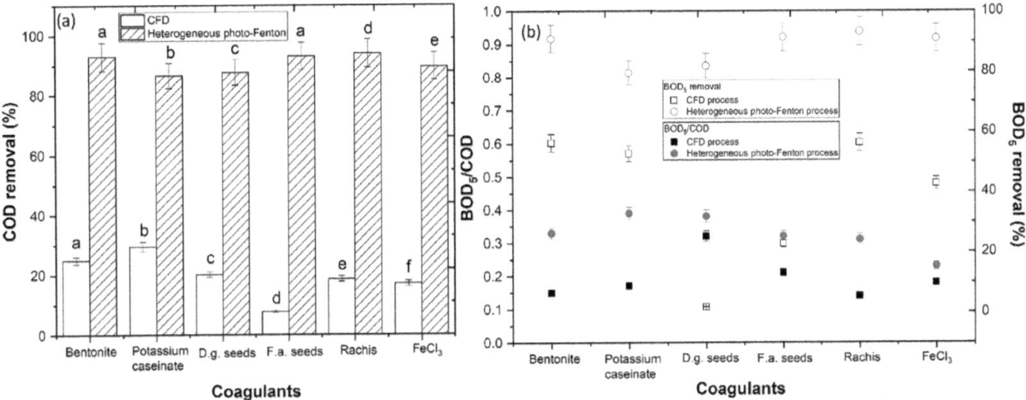

FIGURE 19.9 Appraisal of CFD and CFD/heterogeneous photo-Fenton in degradation of COD (a), biodegradability index and BOD_5 removal (b). Different letters in bars act for means significant differences ($p<0.05$) in COD removal under different treatment processes.

19.4 CONCLUSIONS

A real WW was pre-treated by the CFD process, employing conventional (bentonite, potassium caseinate, $FeCl_3$) and plant natural-based coagulants (*Dactylis glomerata L.* (D.g.) seeds, *Festuca ampla Hack.* (F.a.) seeds and *Vitis vinifera L.* (V.v.) rachis) followed by a flocculant (PVPP). These experiments allowed to achieve good DOC (25.9%, 28.9%, 20.3%, 23.5%, 48.2% and 30.4%), high turbidity (97.3%, 99.6%, 98.7%, 98.8%, 98.4% and 95.7%) and high TSS removals (94.3%, 95.3%, 93.5%, 93.7%, 94.5% and 93.5%), respectively.

Afterwards, to complement the CFD process, a heterogeneous photo-Fenton process was used with ferrocene (Fc) as a source of iron. The results obtained for the heterogeneous photo-Fenton fitted into a non-linear Fermi's kinetic model and under the top conditions it was achieved a $k_{DOC}=0.0432$ per min. Combination of both processes allowed DOC removals of 88.7%, 87.4%, 92.2%, 89.9%, 87.9% and 81.0%, respectively, for bentonite, potassium caseinate, D. g. seeds, F. a. seeds, V.v. rachis and ferric chloride, in comparison to 82.7% DOC removal achieved by heterogeneous photo-Fenton with raw WW.

Finally, based on the BOD_5/COD index, it is concluded that the combination of treatments increases the biodegradability, and the WW can be further treated in a biologic reactor. Based on the results, the application of CFD with heterogeneous-photo-Fenton is a sustainable and feasible treatment for WW.

ACKNOWLEDGMENT

The authors are grateful for the financial support of the Project AgriFood XXI, operation nº NORTE-01–0145-FEDER-000041, and to the Fundação para a Ciência e a Tecnologia (FCT) for the financial support provided to CQVR through UIDB/00616/2020. Ana R. Teixeira also thanks the FCT for the financial support provided through the doctoral scholarship UI/BD/150847/2020.

REFERENCES

Amor, C., Torres-Socías, E. De, Peres, J.A., Maldonado, M.I., Oller, I., Malato, S., Lucas, M.S. (2015). Mature Landfill Leachate Treatment by Coagulation/Flocculation Combined with Fenton and Solar Photo-Fenton Processes. *J. Hazard. Mater.*, 286, 261–268. https://doi.org/10.1016/j.jhazmat.2014.12.036.

Amuda, O.S., Amoo, I.A. (2007). Coagulation/Flocculation Process and Sludge Conditioning in Beverage Industrial Wastewater Treatment. *J. Hazard. Mater.*, 141, 778–783. https://doi.org/10.1016/j.jhazmat.2006.07.044.

APHA, AWWA, WEF. (1999). *Standard Methods for the Examination of Water and Wastewater*, 20th ed. American Public Health Association, American Water Works Association, Water Environment Federation.

Benna, M., Magnin, A., Bergaya, F. (1999). Effect of pH on Rheological Properties of Purified Sodium Bentonite Suspensions. *J. Colloid Interface Sci.*, 218, 442–455.

Camacho, F.P., Sousa, V.S., Bergamasco, R., Teixeira, M.R. (2017). The Use of Moringa Oleifera as a Natural Coagulant in Surface Water Treatment. *Chem. Eng. J.*, 313, 226–237. https://doi.org/10.1016/j.cej.2016.12.031.

Cardoso, A.D. (2007). *O vinho - da uva à garrafa*, 1a Edição. ed. Barcelos: Âncora Editora.

Cosme, F., Ricardo-da-Silva, J.M., Laureano, O. (2008). Interactions between Protein Fining Agents and Proanthocyanidins in White Wine. *Food Chem.*, 106, 536–544. https://doi.org/10.1016/j.foodchem.2007.06.038.

Ghime, D., Ghosh, P. (2018). Decolorization of Diazo Dye Trypan Blue by Electrochemical Oxidation: Kinetics with a Model Based on the Fermi's Equation. *J. Environ. Chem. Eng.* https://doi.org/10.1016/j.jece.2018.11.037.

Guimarães, V., Lucas, M.S., Peres, J.A. (2019a). Combination of Adsorption and Heterogeneous photo-Fenton Processes for the Treatment of Winery Wastewater. *Environ. Sci. Pollut. Res.*, 26, 31000–31013. https://doi.org/10.1007/s11356-019-06207-6.

Guimarães, V., Teixeira, A.R., Lucas, M.S., Peres, J.A. (2020). Effect of Zr Impregnation on Clay-Based Materials for H2O2-Assisted Photocatalytic Wet Oxidation of Winery Wastewater. *Water*, 12, 3387. https://doi.org/10.3390/w12123387.

Guimarães, V., Teixeira, A.R., Lucas, M.S., Silva, A.M.T., Peres, J.A. (2019b). Pillared Interlayered Natural Clays as Heterogeneous Photocatalysts for H2O2-assisted Treatment of a Winery Wastewater. *Sep. Purif. Technol.*, 228, 115768. https://doi.org/10.1016/j.seppur.2019.115768.

Herney-Ramirez, J., Silva, A.M.T., Vicente, M.A., Costa, C.A., Madeira, L.M. (2011). Degradation of Acid Orange 7 Using a Saponite-based Catalyst in Wet Hydrogen Peroxide Oxidation: Kinetic Study with the Fermi's Equation. *Applied Catal. B, Environ.*, 101, 197–205. https://doi.org/10.1016/j.apcatb.2010.09.020.

Horvat, I., Radeka, S., Plav, T., Luki, I. (2019). Bentonite Fining During Fermentation Reduces the Dosage Required and Exhibits Significant Side-effects on Phenols, Free and Bound Aromas, and Sensory Quality of White Wine. *Food Chem.*, 285, 305–315. https://doi.org/10.1016/j.foodchem.2019.01.172.

Howe, K.J., Hand, D.W., Crittenden, J.C., Trussell, R.R., Tchobanoglous, G. (2012). *Principles of Water Treatment.* Hoboken, NJ: John Wiley & Sons, Inc.

Hunter, R.J. (2001). *Foundations of Colloid Science*, 2nd ed. Oxford University Press, United Kingdom.

Jia, X., Chen, X., Liu, Y., Zhang, B., Zhang, H., Zhang, Q. (2019). Hydrophilic Fe3O4 Nanoparticles Prepared by Ferrocene as High Efficiency Heterogeneous Fenton Catalyst for the Degradation of Methyl Orange. *Appl. Organomet. Chem.*, 33, 1–12. https://doi.org/10.1002/aoc.4826.

Johnson, M.B., Mehrvar, M. (2022). Treatment of Actual Winery Wastewater by Fenton-like Process: Optimization to Improve Organic Removal, Reduce Inorganic Sludge Production and Enhance Co-Treatment at Municipal Wastewater Treatment Facilities. *Water*, 14, 39. https://doi.org/10.3390/w14010039.

Jorge, N., Teixeira, A.R., Lucas, M.S., Peres, J.A. (2023). Combined Organic Coagulants and Photocatalytic Processes for Winery Wastewater Treatment. *J. Environ. Manage.*, 326, 116819. https://doi.org/10.1016/j.jenvman.2022.116819.

Jorge, N., Teixeira, A.R., Lucas, M.S., Peres, J.A. (2022). Agro-Industrial Wastewater Treatment with Acacia dealbata Coagulation/Flocculation and Photo-Fenton-Based Processes. *Recycling,* 7, 54. https://doi.org/10.3390/recycling7040054.

Jorge, N., Teixeira, A.R., Lucas, M.S., Peres, J.A. (2021). Combination of Adsorption in Natural Clays and Photo-Catalytic Processes for Winery Wastewater Treatment. In: M. Abrunhosa, A. Chambel, S. Peppoloni, and H. I. Chaminé (Eds.), *Advances in Geoethics and Groundwater Management : Theory and Practice for a Sustainable Development.* Cham: Springer, pp. 291–294. https://doi.org/10.1007/978-3-030-59320-9-60.

Jorge, N., Teixeira, A.R., Matos, C.C., Lucas, M.S., Peres, J.A. (2021). Combination of Coagulation-Flocculation-Decantation and Ozonation Processes for Winery Wastewater Treatment. *Int. J. Environ. Res. Public Health,* 18, 8882. https://doi.org/10.3390/ijerph18168882.

Kong, J., Yu, S. (2007). Fourier Transform Infrared Spectroscopic Analysis of Protein Secondary Structures. *Acta Biochim. Biophys. Sin. (Shanghai)*, 39, 549–559. https://doi.org/10.1111/j.1745-7270.2007.00320.x.

Laborde, B., Moine-Ledoux, V., Richard, T., Saucier, C., Dubourdieu, D., Monti, J.-P. (2006). PVPP–Polyphenol Complexes: A Molecular Approach. *J. Agric. Food Chem.,* 54, 4383–4389. https://doi.org/10.1021/jf060427a.

Li, Y., Zhang, B., Liu, X., Zhao, Q., Zhang, H., Zhang, Y. (2018). Ferrocene-catalyzed Heterogeneous Fenton-like Degradation Mechanisms and Pathways of Antibiotics under Simulated Sunlight: A Case Study of Sulfamethoxazole. *J. Hazard. Mater.*, 353, 26–34. https://doi.org/10.1016/j.jhazmat.2018.02.034.

Loddo, V., Bellardita, M., Camera-Roda, G., Parrino, F., Palmisano, L. (2018). Heterogeneous Photocatalysis: A Promising Advanced Oxidation Process, In (Eds): Angelo Basile, Sylwia Mozia, Raffaele Molinari, *Current Trends and Future Developments on (Bio-) Membranes, Elsevier*, 1–43. https://doi.org/10.1016/B978-0-12-813549-5.00001-3

Lucas, M.S., Peres, J.A. (2009). Removal of COD from Olive Mill Wastewater by Fenton's Reagent: Kinetic Study. *J. Hazard. Mater.*, 168, 1253–1259. https://doi.org/10.1016/j.jhazmat.2009.03.002.

Martins, R.B., Jorge, N., Lucas, M.S., Raymundo, A., Barros, A.I., Peres, J.A. (2022). Food By-Product Valorization by Using Plant-Based Coagulants Combined with AOPs for Agro-Industrial Wastewater Treatment. *Int. J. Environ. Res. Public Health*, 19, 4134. https://doi.org/10.3390/ijerph19074134.

Meade, A.D., Clarke, C., Byrne, H.J., Lyng, F.M. (2010). Fourier Transform Infrared Microspectroscopy and Multivariate Methods for Radiobiological Dosimetry. *Radiat. Res.*, 173, 225–237. https://doi.org/10.1667/RR1836.1.

Organisation Internationale de la Vigne et du Vin (OIV). (2016). International Code of Oenological Practices. Paris.

Pignatello, J.J., Oliveros, E., MacKay, A. (2006). Advanced Oxidation Processes for Organic Contaminant Destruction Based on the Fenton Reaction and Related Chemistry. *Crit. Rev. Environ. Sci. Technol.*, 36, 1–84. https://doi.org/10.1080/10643380500326564.

Prozil, S.O., Evtuguin, D. V., Silva, A.M., Lopes, L.P. (2014). Structural Characterization of Lignin from Grape Stalks (*Vitis vinifera* L.). *J. Agric. Food Chem.*, 62, 5420–5428. https://doi.org/10.1021/jf502267s.

Rache, M.L., García, A.R., Zea, H.R., Silva, A.M.T., Madeira, L.M., Ramírez, J.H. (2014). Azo-dye Orange II Degradation by the Heterogeneous Fenton-like Process Using a Zeolite Y-Fe Catalyst - Kinetics with a Model Based on the Fermi's Equation. *Applied Catal. B, Environ.*, 146, 192–200. https://doi.org/10.1016/j.apcatb.2013.04.028.

Rodríguez-Chueca, J., Amor, C., Silva, T., Dionysiou, D.D., Li, G., Lucas, M.S., Peres, J.A. (2017). Treatment of Winery Wastewater by Sulphate Radicals: HSO5-/transition metal/UV-A LEDs. *Chem. Eng. J.*, 310, 473–483. https://doi.org/10.1016/j.cej.2016.04.135.

Singleton, V.L., Rossi, J.A. (1965). Colorimetry of Total Phenolics with Phosphomolybdic-Phosphotungstic Acid Reagents. *Am. J. Enol. Vitic.*, 16, 144–158.

Tatsi, A.A., Zouboulis, A.I., Matis, K.A., Samaras, P. (2003). Coagulation - Flocculation Pretreatment of Sanitary Landfill Leachates. *Chemosphere*, 53, 737–744. https://doi.org/10.1016/S0045-6535(03)00513-7.

Teixeira, A.R., Jorge, N., Fernandes, J.R., Lucas, M.S., Peres, J.A. (2022). Textile Dye Removal by Acacia dealbata Link. Pollen Adsorption Combined with UV-A/NTA/Fenton Process. *Top. Catal.*, 65, 1–17. https://doi.org/10.1007/s11244-022-01655-w.

Tokumura, M., Morito, R., Hatayama, R., Kawase, Y. (2011). Iron Redox Cycling in Hydroxyl Radical Generation During the Photo-Fenton Oxidative Degradation: Dynamic Change of Hydroxyl Radical Concentration Fe (II) Compounds Fe (III) Compounds. *Applied Catal. B, Environ.*, 106, 565–576. https://doi.org/10.1016/j.apcatb.2011.06.017.

Vunain, E., Mike, P., Mpeketula, G., Monjerezi, M., Etale, A. (2019). Evaluation of Coagulating Efficiency and Water Borne Pathogens Reduction Capacity of Moringa Oleifera Seed Powder for Treatment of Domestic Wastewater from Zomba, Malawi. *J. Environ. Chem. Eng.*, 7, 103118. https://doi.org/10.1016/j.jece.2019.103118.

Wang, Q., Tian, S., Cun, J., Ning, P. (2013). Degradation of Methylene Blue Using a Heterogeneous Fenton Process Catalyzed by Ferrocene. *Desalin. Water Treat.*, 28, 5821–5830. https://doi.org/10.1080/19443994.2012.763047.

Zhu, K., El-din, M.G., Moawad, A.K., Bromley, D. (2004). Physical and Chemical Processes for Removing Suspended Solids and Phosphorus from Liquid Swine Manure. *Environ. Technol.*, 25, 1177–1187. https://doi.org/10.1080/09593332508618385.

20 Transitioning Existing Wastewater Treatment Systems Toward a Circular Economy by Increasing Resource Circularity

Ganesan Karthikeyan, Thangaraj Gokul Kannan, Raveendra Gnana Keerthi Sahasa, Ramesh Poornima, Sundarajayanthan Ramakrishnan, and Periyasamy Dhevagi

20.1 INTRODUCTION

Increasing population, urbanisation, climate change, and agricultural cultivation have all had an influence on the quality and amount of our water resources, placing pressure on water which as we know it is essential to life in an unprecedented way. The issue for water utilities (WUs) is to deliver safe and clean drinking water in the face of old, expensive infrastructure, a price for water that does not represent its actual worth, and the requirement for infrastructure to be robust in an era where dangers of floods and droughts are widespread (Mbavarira and Grimm, 2021). The need for modern wastewater treatment technology is being stifled globally due to ageing infrastructure, strict environmental regulations, and problems with the aquatic environment. In addition to removing organic carbon, modern wastewater treatment systems also aim to recover useful products. Despite the fact that removing the three main nutrients (nitrogen, carbon, and phosphorus) is essential for environmental preservation, the industry's current practices require a lot of energy, chemicals, and resources, which is an unsustainable trend considering the world's population growth and the rapid urbanisation (Ghimire et al., 2021). The creation of hazardously polluted wastewater that also contains unidentified chemicals. Additionally, if wastewater is not adequately treated, hazardous contaminants will eventually drain back into the earth, damaging groundwater resources. Treatment, reuse, and safe disposal of wastewater are now essential for a sustainable life (Yenkie, 2019).

According to the circular economy (CE), waste, materials (including unprocessed material), and goods should all be considered secondary raw resources that can be processed and used again (Ghisellini et al., 2016; Neczaj and Grosser, 2018). This differentiates it from a linear economy founded on the "take-make-use-dispose" paradigm, where waste is frequently the last phase of the product's life cycle. By decreasing waste generation and repurposing it as a secondary resource, CE is a concept that aids sustainable material and energy management. Reclaimed water is effluent from industrial or municipal sources that has been transformed into usable water. Since reclaimed water contains pathogens, heavy metals, and other potentially harmful contaminants like active pharmaceutical chemicals and endocrine disruptors, its water quality is fundamentally distinct when compared to potable water supplies (Chen et al., 2013). The utilisation of reclaimed wastewater to irrigate vegetables and cereal crops was studied for potential negative impacts on people, animals, and the environment. The ideal reclamation techniques needed to meet water quality criteria were researched, as well as the operating features (Jang et al., 2010). Around the world, several industrial or governmental sectors produce a sizable volume of wastewater. Conventional wastewater treatment (WWT) plants have primarily focused on treating wastewater rather than recovering useful

DOI: 10.1201/9781003441144-20

materials. The shift from a linear to a circular economy could offer a unique basis for salvaging valuable resources from wastewater, including nutrients, energy, and high-value goods. However, transitioning from traditional framework to sustainable WWT systems is still fairly challenging (Yadav et al., 2021).

Urban wastewater treatment plants (WWTPs) have the potential to significantly contribute to circular sustainability since they combine energy generation and resource recovery during the process of producing clean water (Mo and Zhang, 2013; Neczaj and Grosser, 2018). In the foreseeable future, WWTPs will be "ecologically sustainable" technical systems. However, during the last 25 years, a large number of key factors have emphasised the need of recovering the resources present in wastewater. This chapter addresses the types of wastewater, the status of reclaimed water, current status and challenges, utilisation and resource recovery of wastewater and ways to achieve circular economy.

20.2 TYPES OF WASTE WATER TREATMENT SYSTEMS

A wastewater treatment system is an amalgamation of numerous technologies that act together to treat wastewater. The process of treating wastewater is rarely static; thus, a wastewater treatment system should be designed to adjust to various treatment requirements to avoid expensive replacements and upgrades in the future. An effective and well-designed water treatment system must be able to handle: (i) changes in flow and contaminant levels, (ii) fluctuations in water chemistry demands and necessary chemical volume adjustments, and (iii) potential changes in wastewater treatment requirements. The various technologies used in wastewater treatment systems are given in Table 20.1. Waste water treatment system removes organic matter (BOD and COD), phosphates, nitrates, metals, pathogens, total dissolved solids (TDS), total suspended solids (TDS) and synthetic chemicals (EPA, 1998). The basic components of wastewater treatment plant are given in Figure 20.1.

TABLE 20.1
Types of Waste Water Treatment Systems

Type	Conventional Method	Existing Methods	Emerging Systems	References
Physical	Filtration Flotation	Evaporation Reverse osmosis	Nano filtration Ultrafiltration Adsorption Nanomaterials	Crini and Lichtfouse (2019); Manchisi et al. (2020); Liu et al. (2020); Rashid et al. (2021); Yang et al. (2020a); Zhu et al. (2020)
Chemical	Precipitation Flocculation/ Coagulation	Solvent extraction Oxidation Electrochemical treatment Ion-exchange Incineration Membrane separation Ozonation	Advanced oxidation or Fenton's oxidation Electrochemical oxidation Photo-electrochemical oxidation	Crini and Lichtfouse (2019); Yogalakshmi et al. (2020); Rashid et al. (2021)
Biological	Biodegradation Oxidation ponds	Biofilters/trickling filters Activated sludge processes Rotating biological contactors	Biosorption Biomass	Duan et al. (2016); Crini and Lichtfouse (2019); Rashid et al. (2021)

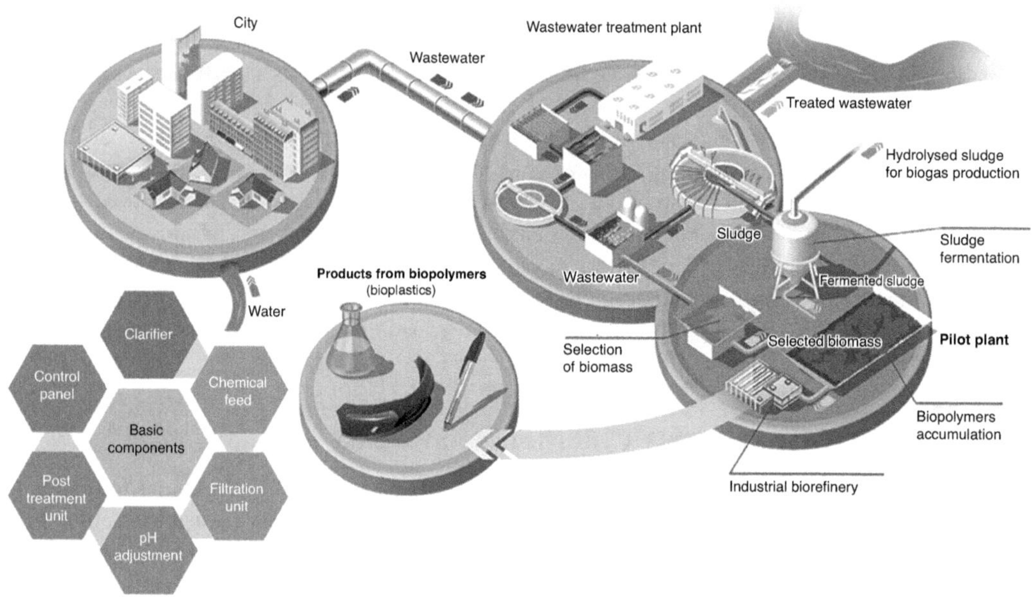

FIGURE 20.1 Basic components of wastewater treatment plant.

20.2.1 Working Mechanism of Waste Water Treatment System

Specific treatment methods may differ, but a conventional wastewater treatment plant will usually contain the following steps (EPA, 1998):

Coagulation → Flocculation → Sedimentation → Filtration → Disinfection → Distribution/Discharge

20.2.2 Stages of Waste Water Treatment

Water treatment is categorised into three stages/phases: primary, secondary, and tertiary (or advanced). The purpose of preliminary treatment is to remove coarse particles and other larger particles commonly found in raw wastewater. The usual preliminary treatment procedures involve coarse screening, grit removal, and occasionally, the milling of large particles (Zagklis and Bampos, 2022).

20.2.2.1 Primary Treatment

The purpose of primary treatment is to separate sum (floatable materials) by skimming and settleable inorganic and organic solids through sedimentation. In primary treatment, between 50%–70% of the total suspended solids (SS), 65% of the oil and grease, and 25%–50% of the total biological oxygen demand (BOD_5) are removed (Hussein et al., 2019). In addition to removing considerable organic phosphorus, organic nitrogen, and heavy metals that are connected to solids, primary sedimentation has little effect on colloidal or dissolved substances. The waste discharged by primary sedimentation unit is referred to as primary effluent. Primary sedimentation tanks or clarifiers can be circular or rectangular basins which are 2–3 hours hydraulically retained and typically 3–5 m deep (Harb et al., 2019; Hand and Cusick, 2021).

20.2.2.2 Secondary Treatment

Removing any leftover organic debris and suspended particles from the primary treatment output is the major objective of secondary treatment. Most of the time, secondary treatment is done following primary treatment and entails the utilisation of aerobic biological treatment methods to separate bio-degradable colloidal and dissolved organic waste. By utilising the organic matter in the wastewater, aerobic microbes (mostly bacteria) performing aerobic biological treatment create additional inorganic end products (mainly CO_2, H_2O and NH_3) and microorganisms. Numerous aerobic biological systems are employed for secondary treatment. These processes vary largely in how oxygen is given to the microorganisms and in how rapidly the organisms metabolise these organic materials. When compared to low-rate processes, high-rate biological processes exhibit relatively small reactors in sizes and greater microbe populations. Because of the tightly controlled environment, new organisms grow at a significantly faster rate in high-rate systems. To create clarified secondary effluent, the micro-organisms have to be removed by sedimentation from the previously treated wastewater. The main clarifiers previously mentioned act similarly to the secondary clarifiers, which are sedimentation tanks employed in secondary treatment and are frequently called secondary clarifiers (Topare et al., 2011).

The biofilters or trickling filters, activated sludge processes, rotating biological contactors and oxidation ponds are examples of common high-rate processes (RBC). Municipal wastewater which has a significant amount of organic matters and is from industrial sources is often treated using a mixture of two of the RBC procedures run in succession (for example, biofilter succeeded by activated sludge) (Krzeminski et al., 2019).

20.2.2.3 Tertiary Treatment

Whenever certain wastewater elements that cannot be eliminated by secondary treatment need to be removed, tertiary wastewater treatment can be used. Separate treatment procedures are required to remove heavy metals, dissolved solids, refractory organics, extra suspended particles, nitrogen, and phosphorus. Advanced treatment is occasionally mentioned as tertiary treatment since it is frequently employed after high-rate secondary treatment. To remove phosphorus, chemical addition to aeration basins or primary clarifiers or is one example of an advanced treatment procedure that is sometimes employed in conjunction with primary or secondary treatment (Zagklis and Bampos, 2022).

20.3 STATUS OF RECLAIMED WATER

Even though the world is covered in water to a depth of more than 70%, only 3% is fit for human consumption, with the remaining 97% being salt water (Ahmad et al., 2019). Wastewater recovery or reuse is among the most essential requirements of the current situation as freshwater supplies are rapidly diminishing and demand is rising. About 92% of the world's total water consumption is used for agriculture (Hoekstra and Mekonnen, 2012), in which 70% of freshwater is supplied by subterranean water sources and rivers, and is utilised for irrigation (WRI, 2020). The data indicate grave concern for the nations experiencing a water shortage. The status of reclaimed water is given in Table 20.2. According to Shen et al. (2014), nearly 40% of the population in the world live in basins with high water stress, which highlights the water issue.

Consequently, using reclaimed water for agriculture instead of freshwater is a great tool (Contreras et al., 2005). The majority of the time, treated wastewater is used for nonpotable activities like firefighting, building construction, vehicle washing, irrigation, groundwater replenishment, irrigation for golf courses, and agriculture. It can also be utilised in thermal power plants for cooling purposes (Yang et al., 2017; Katsoyiannis et al., 2017). Irrigation of treated wastewater helps millions of small-scale farmers throughout the world support their livelihoods and increase agricultural output (Sato et al., 2013). Reusing treated wastewater for agriculture around the world varies greatly, from 1.5% to 6.6% (Ungureanu et al., 2018).

Above 10% of the world's population utilises agricultural goods that are grown using wastewater irrigation (WHO, 2006). In China, the USA, and Europe, volumes of reused treated wastewater have grown up to 10%–29% annually, and by 41% in Australia (Aziz and Farissi, 2014). China leads as the top Asian nation for wastewater reuse, with an estimated 1.3 million hectares (ha) of

TABLE 20.2
Status of use of Reclaimed Water

Country	Water Consuming Sectors	Water Use (%)	Reclaimed Water Use (%)	Sectors that Use Reclaimed Water	References
India	Agriculture	87	78	Irrigation	Jindal and Kamat (2011)
	Industry	7	12	Industrial reuse	
	Domestic	4	4	Thermal plant	
	Electricity	2	6	Recharging groundwater and man-made lakes	
Europe	Agriculture	44	2.2	Groundwater recharge	EEA CSI (2018)
	Energy and industry	40	20	Landscape irrigation	
	Public supply	16	6.8	Recreational uses	
			32	Irrigation	
			8.3	Urban non-potable uses	
			2.3	Potable uses (indirect)	
			19	Industrial uses	
			8	Environmental restoration	
			1.5	Other uses	
USA	Thermal plants	41	37	Irritation	SWRCB (2011)
	Irrigation	37	2	Geothermal energy	
	Domestic use	14	17	Landscape irrigation	
	Industries	6	7	Golf court irrigation	
	Aquaculture and livestock	3	12	Recharging groundwater	
			8	Commercial and industrial	
			7	Seawater barrier	
			4	Wildlife habitat, wetlands	
			4	Recreational uses	
			2	Other uses	
South Africa	Agriculture	60	9	Sports field and landscape irrigation	Adewumi et al. (2010)
	Domestic use	27	na		
	Power	4	na	Industrial reuse	
	Industrial	3	48	Agriculture	
	Mining	3	43		
	Others	3	na		
Greece	Irrigation	83	58	Irrigation	Frontistis et al. (2011)
	Public use	13	na	Landscape irrigation	
	Industry	2.2	23	Forest irrigation and fire department	
	Animal husbandry	1.3	na		
	Others	1.2	na		

na – not available

land, and other countries include India, Pakistan and Vietnam (Zhang and Shen, 2017). According to estimates, only 37.6% of India's urban wastewater is currently being treated (Singh et al., 2019). Israel is the highest user of reclaimed water for agriculture irrigation, using 90% of reclaimed water (Angelakis and Snyder, 2015). Untreated wastewater is a common source of irrigation in low-income nations throughout Asia, Africa and Latin America. However, middle-income nations like Saudi Arabia, Jordan and Tunisia use reclaimed water for agriculture land irrigation (Balkhair, 2016a, 2016b).

Treated wastewater and domestic water contain a variety of nutrients, including phosphate, sulfur, nitrogen, and potassium, and the majority of these nutrients are readily absorbed by plants in wastewater, which is why it is frequently used for irrigation (Sengupta et al., 2015; Poustie et al.,

2020). Reclaimed wastewater has a high nutritional content, which decreases the need for fertilisers, boosts agricultural output, improves soil fertility, and may even lower crop production costs (Jeong et al., 2016).

20.4 CURRENT STATUS AND CHALLENGES IN WASTEWATER TREATMENT

In comparison to other reuse options around the world (industrial, fire protection, toilet wetlands, flushing, groundwater recharge, recreation and drinkable), reutilising the treated wastewater for agriculture and landscape irrigation has the greatest rate (52%) (Yang et al., 2020b). Reusing treated wastewater is favoured over other options since it is more practical, especially in areas with limited access to water (such as seawater). The majority (44%) of the recovered water in southern Europe is used for irrigation, whereas 51% of it is utilised for environmental restoration purposes in northern Europe (such as expanding existing water sources and creating wetland areas). Pioneers in the utilisation of reclaimed water in agriculture include Italy, Spain, Cyprus, and Greece (Voulvoulis, 2018). 76% of treated wastewater was used for irrigation in Cyprus in 2013 (Kirhensteine et al., 2016). In 2016, 61% of the wastewater in Spain was used in agriculture, with 10% of the wastewater being recovered (SUWANU, 2019). According to Hristov et al. (2021), despite a projected reduction in water stress of 14%, the EU's present water reuse applications are expected to reduce water stress by only 1% by 2030.

The Europe Commission issued a regulation requiring a minimum standard of reclaimed water quality for the European Union, but each country has its own regulatory structure, and strict regulations discourage the use of reclaimed water. For instance, Israel uses water reuse to meet half of its requirement for irrigation (Tal, 2016), monitors fewer than a dozen variables, but specifies the crop/produce type, required irrigation technique and treatment level. Similarly, Italy establishes limits for indicators, some of which are not taken into account when analysing drinking water (Ait-Mouheb et al., 2020).

The objective of achieving global carbon neutrality is greatly advanced by the reuse of water from treated wastewater. Using a circular economy strategy, resources like nutrients, carbon, and bioplastics, could also be salvaged from wastewater. According to the literature, barriers—technological, societal, legal, and organisational—are the primary culprits. A robust and detailed roadmap that prepares the path for the transformation into a circular economy model in the water sector may be one method to get over these difficulties. Although there have been some published initiatives (such as the wider-uptake EU project), still there is a need to solidify this path's dissemination across the associated interdisciplinary stakeholders (such as business, law, researchers, and government agencies, among others). Even though they play a significant role; policymakers must also educate the next generation about the value of resource recovery and water reuse if the shift is to be successful. Regarding this, significant attempts have been made as part of the Wider-Uptake project to establish the first WRRF on a university campus (Palermo University in Italy). A successful public engagement approach like this one could give the university a standout role. There are currently no other case studies, hence research should focus on encouraging the development of such models to overcome the aforementioned obstacles. To enable a genuine transition to the circular economy in the water sector, it is vital to develop a better focused solution (water smart).

20.5 UTILISATION OF RECLAIMED WATER

The transformation to a circular from a linear economy in the water sector is being driven by the urge to preserve the environment and improve public health without sacrificing economic progress (Smol et al., 2020). Resource utilisation and waste management are unrelated in the linear economy model. In a linear economy, extract-produce-use-dispose is the primary strategy (Korhonen et al., 2018). In opposition, the circular economy model emphasises the close connection between resource usage and waste management. By reusing and recycling materials and recovering resources, the circular economy strives to maintain resources and products in use for as long as plausible. It reduces trash production as a result and restores natural systems (Fitch-Roy et al., 2021). The goal of the circular economy is to promote economic growth while safeguarding social justice and the environment (Kirchher et al., 2017).

The first circular economy action plan was modified by the European Commission (EC) in 2015, and "A new Circular Economy Action Plan" was issued in 2020 (EC, 2020). Agriculture, industry, and municipal systems are all connected by water. Water is everywhere, which makes it crucial for implementing the circular economy concept. For urban, industrial, and agricultural purposes (such as street cleaning, landscape irrigation, fire prevention, flushing, toilet, etc.), water reclamation techniques are used to create or recirculate water (Lazarova, 2022). The revised plan's "Food, water, and nutrient" product value chain includes water reuse in agriculture. The EC brought attention to the possibility of reducing the usage of industrial fertilisers by reusing nutrients through water reuse (EC, 2015). By using recovered water in irrigation, the circular economy in the water sector can also be achieved while maintaining freshwater resources and extending the life cycle of water (Smol et al., 2020). Even though the EC's action plan assures the quality of recycled water for the production of safe food, there are still obstacles in the way of the switch to a circular economic model.

20.5.1 Barriers in the Applications of Reclaimed Water

It is feasible to recover irrigation water acceptable for agricultural use using the present procedures applied in the treatment of household wastewater. However, the following barriers reduce the use of reclaimed water:

a. Economic and technical barriers
b. Regulatory/institutional barriers
c. Social barriers

20.5.1.1 Economical and Technical Barriers

Irrigation water shouldn't contain particulate matter, pathogens, emerging pollutants, salt, heavy metals, or organic matter found in domestic wastewater. However, they do not create effluent water suitable for irrigation. Organics and nutrients are removed from wastewater by secondary wastewater treatment plants. Even potable water can be made from residential wastewater due to the use of cutting-edge treatment technologies. Examples include membrane filtration, sophisticated oxidation techniques, and adsorption (Rubiano et al., 2012). The membrane filtration approach relies on the filtration medium's gap size and is dependent on the physical retention of contaminants in the filtered stream (Davis and Cornwell, 2019). The membrane technique is both environmentally friendly and highly effective. Membranes used for ultrafiltration (UF) can hold on to bacteria, viruses, and other particulate and colloid material. Due to this, UF membranes are preferred in the majority of water reuse projects following secondary treatment (Kehrein et al., 2021). Membranes are employed in membrane bioreactors (MBRs), which combine the activated sludge processes and membrane, to exclude treated waste water from biomass instead of traditional settling tanks. In MBR systems, hydraulic and sludge retention time are separated, and increased biomass concentrations can result in better removal efficiencies. However, the high operating and investment expenses associated with membrane processes are a drawback (Yang et al., 2020a).

Advanced oxidation processes (AOP) are notable, exclusively when it comes to the elimination of new contaminants. This approach guarantees the oxidation of non-biodegradable organic contaminants by using highly reactive oxidant agents (Silva et al., 2017). The proper AOP can be determined depending on the emerging pollution that has to be eliminated. The oxidation process of chlorination is utilised to disinfect WWTP effluents. Organic debris in the effluent causes the synthesis of disinfection byproducts which are hazardous to human health. Hence, it is crucial for reuse applications to have a high organic matter removal efficiency. Ozonation is yet another additional AOP that employs the ozonation process, organic debris, pathogens and emerging pollutants may be eliminated from wastewater; however, if bromide is present, there is a potential that bromated organic (carcinogenic) compounds will emerge (Kehrein et al., 2020). AOP systems' high energy requirements are another drawback of this particular technology (Li et al., 2022).

The raw carbonaceous substances/materials possess a very high adsorption capacity after activation. Activated carbon may be resurrected and used as a filter substratum/medium to absorb fresh impurities from wastewater after it has been used (Alvarino et al., 2018). Additionally, in reactors using granular activated carbon (GAC), dissolved organic matter (DOM) from treated wastewater might compete with specific emergent contaminants. Since the pollutants are still not eliminated by adsorption, the growth in biomass brought on by absorbed DOM in the GAC reactor may speed up the breakdown of emerging contaminants and minimise the need for further processes. The pathogens in the treated wastewater must be eliminated in accordance with the intended purpose for it to be reused in agriculture.

The most common disinfection technique is chlorination. Despite being effective in killing microorganisms, large concentrations of chlorine could interact with organic materials in treated wastewater to create chlorination byproducts that may be carcinogenic (Furst et al., 2018). UV (ultraviolet) radiation may disinfect objects rapidly and affordably. However, it has been observed that as little as five days following UV disinfection, the total quantity of bacteria in the water may approach the same level as untreated wastewater (Banach et al., 2021).

The transfer from current wastewater treatment plants into water recovery plants still faces gaps in technical knowledge, skills, and training in spite of the scientific basis for effective and efficient wastewater treatment (Yadav et al., 2021). Additionally, after adopting cutting-edge technologies to create irrigation water from wastewater, it was necessary to build a separate distribution network to deliver the reclaimed water to the customers (Guo et al., 2014). In typical water systems, wastewater is collected by the collection network and sent to the wastewater treatment facilities while the water that is drawn from the source is distributed to the customers. Treatment facilities are typically situated farther away from the wastewater basin than wastewater basins themselves in order to transport collected wastewater to the WWTP by gravity and save energy expenses related to pumping. The treatment facility's low elevation could make distributing reclaimed water more expensive because pu\mping is necessary (Kehrein et al., 2020). Additionally, a sophisticated reclaimed water distribution network is needed to transfer the produced reclaimed water to the farms that are situated far from one another. The cost of the methods rises as treatment performance does as well. The costs of the reclamation project are further increased by the distribution of recycled water. Because of this, it is difficult to produce irrigation water from residential wastewater in places with easy availability to irrigation water.

Producing irrigation water from wastewater, as opposed to other options like desalination or water import, is less expensive and has a reduced carbon impact in areas/regions where access to water is problematic. Making the use of recovered water economically viable is necessary to advance the circular water economy (Salminen et al., 2022). However, it is challenging to accomplish this without government assistance (economic tools) using the free market process (Lyu et al., 2016).

20.5.1.2 Regulatory Barriers

According to Lyu et al. (2016), policy factors influence recovered water uses more so than technical or financial factors. Politicians and decision-makers are sometimes hesitant to take action to promote water reclamation because they lack sufficient knowledge regarding the reuse of wastewater in agriculture (Mannina et al., 2021). Some institutional obstacles to water reuse include the absence of standard rules, a lack of economic incentives (such as tax breaks and intensives), a lack of effective communication between decision-makers and farmers, and a lack of public awareness of water reuse (Saliba et al., 2018). The calibre of recycled water used in agriculture has a significant impact on the quality of the crop and, consequently, on the public health. Therefore, the quality of recycled water needs to be tight enough to reduce threats to human health. On the other side, it's crucial to establish reclaimed water quality criteria that don't restrict applications for water reuse (Mizyed, 2013). In fact, the strict recovered water quality regulations prevent more people from using water reuse techniques.

For instance, technical practices (appropriate crop type, treatment technology, adept irrigation technique, etc.) that should be used for water reclamation are shown to be necessary, rather than exceedingly limiting reclaimed water's parameters. However, there are only little water reclamation projects in Italy due to the country's stringent water reuse laws (Ait-Mouheb et al., 2020a). Additionally, standards shouldn't lag behind advances in science. The inconsistent laws among various administrations (like agricultural administrations and water treatment) are another problem with regard to water reuse requirements. Since there are no uniform regulations for water re-use in agriculture, projects including this practice are unpredictable, which discourages decision-makers from investing in them (Kehrein et al., 2020). The success of reclamation operations is additionally increased by the existence of a general circular market among numerous reclamation applications. In fact, the goal of unifying European members is to advance shared social and economic advancement. The EU is currently transitioning its water reuse policy from a linear economy paradigm towards a circular economy model (Mannina et al., 2021). One of the most significant improvements in the EU's water reuse policy is Regulation 2020/741 on minimum recovered water quality criteria, which limits turbidity, total suspended particles, *Escherichia coli*, biological oxygen demand (BOD) and in reclaimed water.

This act intends to stimulate water reuse techniques by informing the public that the re-use of treated water in agriculture is safe for human and environmental health. The potential to reduce the cost of commercial fertiliser by allowing phosphorus and nitrogen in treated wastewater is also underlined in Regulation 2020/741, along with the safety of reclaimed water quality. The circularity of resources (water, nitrogen, and phosphorus) is thus intended to be facilitated. Despite the EU's progressive actions, some of its members may still believe that reusing water for agriculture is unacceptable. The designing of a common circular economy market and the unification of European law are both hampered by this predicament (Mannina et al., 2021).

20.5.1.3 Societal Barriers

Water reclamation programmes cannot succeed without support from the general population. A high level of pathogen revulsion is associated with an unwillingness to use reclaimed water in agriculture and it is the social and psychological barriers in the reuse of reclaimed water. This results from the public's lack of knowledge regarding the quality and advantages of recovered water (Saliba et al., 2018). Another cause for not gaining public approval is the public's lack of faith in the authorities (Beveridge et al., 2017). However, it was discovered that the acceptability of treated water use in agriculture was closely correlated with growing water scarcity and water consumption limitations (Wester et al., 2015). Farmers' worries regarding the usage of reclaimed water include dangers associated with soil and crop quality as well as the crops' market value. According to Frijns et al. (2016), farmers think that consumers would not appreciate items that were watered with recycled water. Farmers are more inclined to reuse water because the nitrogen along with phosphate concentration of reclaimed water lowers the cost/expenditure of commercial fertilisers (Ricart and Rico, 2019).

20.6 RESOURCE RECOVERY TECHNOLOGIES AND LIFE CYCLE ASSESSMENT

20.6.1 Nutrient Recovery

Nutrient recycling from WWTPs benefits the environment by lowering the demand for traditional fossil-based fertilisers and, as a result, lowering energy and water consumption. Sewage sludge (biosolids), raw, and semi-treated wastewater can all be used to recover nutrients (Zhang et al., 2017). The first method uses the land application of biosolids, wherein wastewater treatment byproducts are applied to the soil's surface or injected into the ground as fertiliser. Biosolids can be treated before application using aerobic or anaerobic digestion, chemical treatment, drying, and composting

(Poornima et al., 2021). However, the main issues with this method of biosolids removal are the risks to public health and safety, odor, disturbance, and acceptability. In addition to direct land application, technological recycling from sewage sludge or wastewater, as well as ashes of burned sludge, may be used to recycle phosphorus from wastewater. At present, phosphorus is recycled in WWTPs largely through struvite crystallisation method (Neczaj and Grosser, 2018). The primary issues of struvite crystallisation are high costs of chemicals and inadvertent struvite production, which causes valves, pipelines, and pumps to get clogged. Another strategy for nutrient recovery is the isolation of urine from the primary wastewater stream. It is believed that urine constitutes 50% phosphorus and 70%–80% nitrogen, which may potentially be recovered at a level of 70% utilising the urine-collecting mechanism in toilets (Batstone et al., 2015). When urine is utilised for land application, urine collection systems are frequently employed in industrialised countries. However, because of major technological issues and a lack of public acceptability, this technology has not been extensively used in industrialised nations.

Aqua-species such as microalgae, vegetables, wetland plants, and duckweed may also recover nutrients from wastewater, which can subsequently be utilised as fertilisers or animal feeds (El-Shafai et al., 2007). Aqua-species nutrient recovery is being recognised as an ecologically benign method, owing to the lower energy requirement and synergistic effects between nutrient recycling and wastewater treatment. Nonetheless, this technique is not frequently employed. The only method employed on a technical or pilot scale is manmade wetlands, although its use does not include nutrient recycling for secondary use.

20.6.2 Energy Recovery

To achieve sustainability, energy recovery at wastewater treatment facilities is a key policy instrument. This can be accomplished by producing biogas, using heating systems in treatment plant effluents, and recovering heat exchanger energy from a variety of high-temperature streams (Bertanza et al., 2018). The principal energy source in a WWTP is biogas produced by anaerobic digestion (AD) in a digester, which has a generation capacity of 6.5 kWh/m^3 (65% methane content). The anticipated net energy consumption of WWTPs including sludge digestion is 40% lower than that of WWTPs without AD digestion. Depending on the amount of volatile solids removed, gas output can range between 0.75 and 1.12 m^3/kg, with biogas having a low heating value of 22.4 kJ/m^3. The biogas has the ability to produce heat and/or electricity.

Improving AD efficiency is a frequent strategy for improving WWTP energy self-sufficiency. AD optimisations encompass several sewage sludge pretreatment procedures aimed at increasing sludge biodegradability. These approaches can be classified as mechanical, thermal, chemical, or biological, or various combinations of these (Zhen et al., 2017) in the past, spelled in the past. Another method that gives a variety of environmental and economic benefits is co-digestion of sewage sludge with other biodegradable trash (Hagos et al., 2017). In addition to making WWTPs energy-neutral, co-digestion of organic waste material with sewage sludge significantly lowers the cost of treating industrial and municipal organic waste. Municipal WWTP plant effluents are dependable and cost-effective sources for heat to use in heat pumps (HP) (Culha et al., 2015). The heat from HPs can be utilised to heat and cool the plant's residential, social, and administrative facilities, as well as nearby infrastructure. It must be highlighted that WWTPs that use high-temperature sludge treatment procedures (e.g., thermal drying, THP, AD digestion) need to consider installing heat exchangers in order to recover energy from all the high-temperature streams (reject water, sludge, condensate, etc.). The energy recovered may be used for a variety of purposes, including water heating, sludge heating, and so on.

20.6.3 Water Reuse

The reuse of treated wastewater from WWTPs for land and agriculture irrigation, toilet flushing, groundwater replenishment and industrial purposes is a key component of the currently employed strategy aimed at supplying freshwater for domestic usage, improving the quality of WWTPs effluent, and, as a result, higher quality river waters used for drinking water abstraction (Becerra-Castro et al., 2015). The re-use of treated wastewater for agricultural irrigation have been recognised for many decades and has the potential to meet agricultural demand while also reducing regional water stress. Furthermore, the nutrient content in the wastewater lessens the demand for expensive fertilisers. It is advised to utilise secondary treatment effluent for irrigating non-food crops and to use tertiary effluent for food crops.

20.6.4 Other Resources

Sewage sludge's application in the construction industry is entirely compliant with the CE presumptions. The ash from sewage sludge can be used to create building materials like bricks and tiles. In addition, it can be used as a raw ingredient in the production of lightweight materials like cement, concrete, and mortal. Another possibility is to inexpensively extract precious metals like copper, silver, or gold from the ashes produced by burning sewage sludge (Frijns, 2014). To create biodegradable polymers from polyhydroxyalkanoates (PHA) gathered in biomass growing in wastewater treatment reactors, researchers from the world's finest technical institutions are also working on biotechnology for treating wastewater (Bengtsson et al., 2017). Similarly, Biological Fuel Cells (BFCs) are being used to directly create power while eliminating impurities from waste water (Pandey et al., 2016).

20.7 ROADMAPS TO ACHIEVE CIRCULAR ECONOMY IN WATER SECTOR

Multi-level policy packages are required for the shift towards a circular economy from a linear economy. Policy packages for the circular economy comprise multi-tool solutions/strategies, framework laws, plans, and road maps (Benson and Monciardini, 2018). Water is a valuable resource that is used in various industries, therefore using circular water helps to increase the total circularity of resources across all industries. According to Salminen et al. (2022), the grounds for disregarding water as well as ecosystems associated with it in a circular economy policy instruments are that water is viewed as a danger that may jeopardise world peace and is priceless in comparison to the rest of the resources (such as energy and materials). Water must be incorporated in roadmaps for the circular economy and managed in accordance with this strategy, though, as it is essential for all sectors and creates a connection between them. As a result, it is vital that these present roadmaps be amended to take water consumption, particularly for the agriculture sector, into consideration.

The use of grey water, treated wastewater, industrial operations, groundwater recharging, recreational usage, and non-potable (urban) wastewater is referred to as water reuse or recycling (i.e., fire protection and toilet flushing). One of the most important tactics to lessen water shortage and prevent the depletion of water resources is the treatment of wastewater to make it useable (Figure 20.2). Modern technology is divided into two categories: intense, or requiring large quantities of energy, and vast, or requiring a lot of space. When compared to extensive technologies, which include constructed wetlands, waste stabilisation ponds, infiltration percolation systems, and organic removal processes, intensive technologies include physicochemical and biological systems, membrane technologies, disinfection technologies, and radionuclide removal. Membrane filtration is an example of an intensive technology that often consumes a lot of energy, whereas extensive processes like constructed wetlands and floating wetlands demand a lot of space for the plants that are used to treat wastewater.

As a result of "zero waste" programmes and fashion, society is becoming more and more environmentally conscious, and it can be predicted that they will strive to implement some modifications in their homes and workplaces in the near future. The introduction of strategies to minimise water use in businesses, which is crucial in industrial facilities that depend on water usage, is made possible by the behaviour change.

The European Union's "Achieving wider uptake of water smart solutions - Wider-Uptake" project is part of the Horizon 2020 initiative and seeks to remove obstacles that stand in the way of

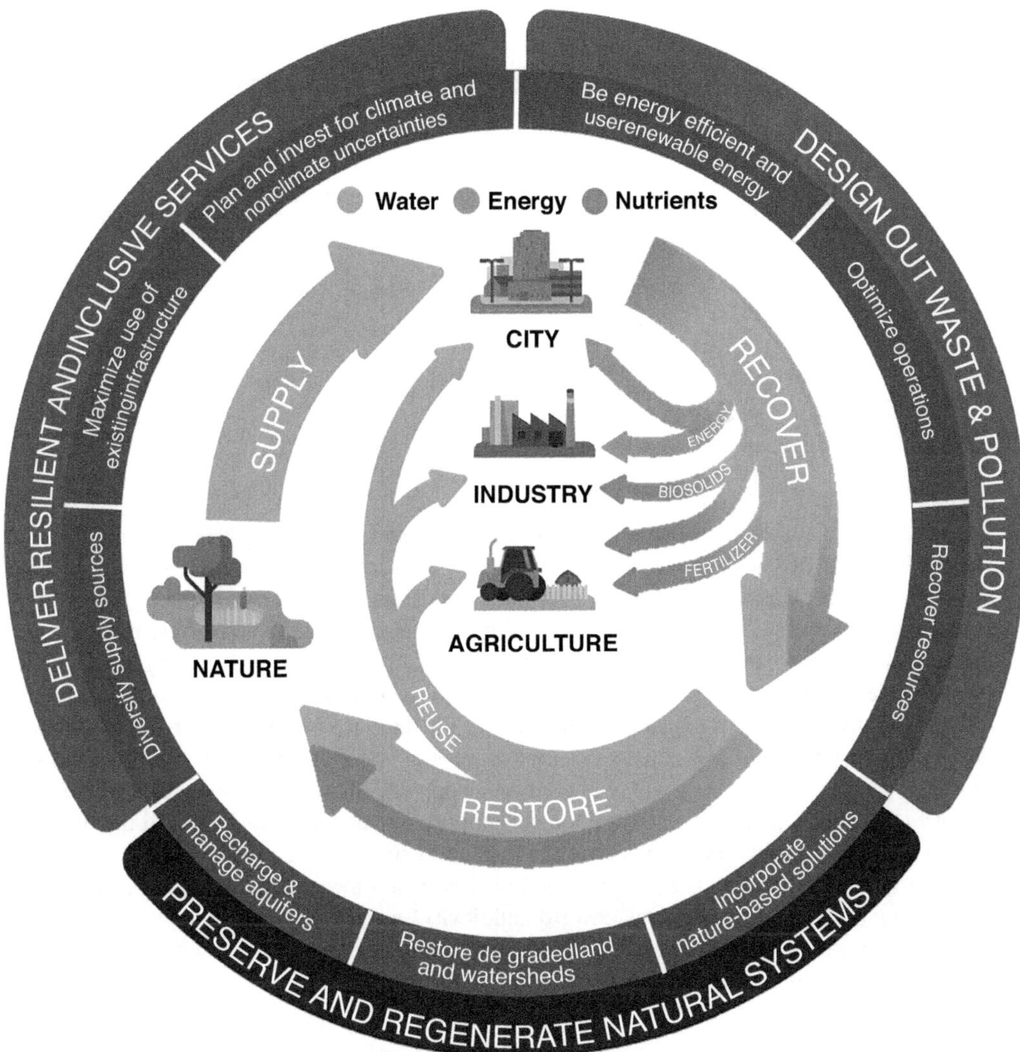

FIGURE 20.2 Water in circular economy and resilience framework.

the wastewater industry's shift from a linear towards a circular economy strategy (Mannina et al., 2021). As alternatives to traditional water-saving measures, the production of bioplastic, fertilisers, soil conditioner, and irrigation water has been proven in both lab and pilot-scale settings. Because resource circularity happens across industries, roadmaps for the circular economy must be designed jointly. As a result, networking and establishing responsibilities are essential from the start of the roadmap creation process. The existing situation should then be carefully studied to reveal potential transitional futures, hurdles, and strategies for overcoming these barriers. The plan for this transition process should then be established, with the milestones clearly marked. To put the defined roadmap into action, a strategy has to be prepared. The outcomes of the process have to be continuously monitored as the roadmap is being worked on during the transition, and it should be evaluated and revised as appropriate (McDowall et al., 2017).

Lessons learned from Communities of Practice (CoPs) on the potential for and challenges to water reuse and how to overcome them include: For water reclamation initiatives to be effective, a combination of policies, market accessibility, teamwork, openness, ongoing learning, and innovations are

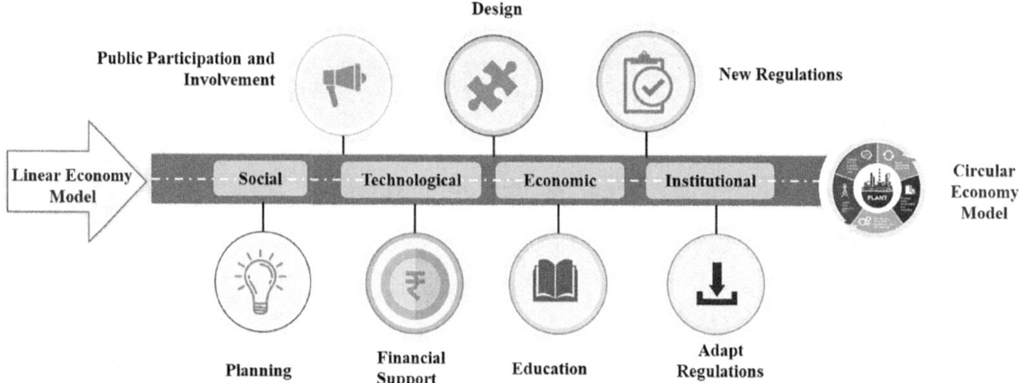

FIGURE 20.3 Transition towards circular economy in the water sector.

essential. Multilevel policy instruments must be flexible enough to accommodate regional demands while also being unified across regions. Because the region's infrastructure, agricultural traits, and water availability all have a significant impact on the projects' performance. In this context, policymakers must be open to new ideas and work with water businesses. To effectively manage water resources, water utilities and authorities must communicate openly about their underlying presumptions, principles, and objectives. Water managers have a responsibility to locate new consumers, develop circular economy strategies that are suitable for the purpose and implement those plans. Lessons from best practices and CoPs show that successful water reuse projects require educating the end users and next generation and end to raise awareness, showcasing water smart alternative solutions to the stakeholders to win their financial support, and updating and streamlining the regulations to get rid of the factors (complex regulations and extremely rigid water quality limits) that cause uncertainty. Additionally, wastewater authorities must adhere to the circular economy base from planning stage of new initiatives onward and devise the treatment technology in accordance with the reuse purpose (fit-for-purpose) to lower the treatment expenditure/cost and maximise the benefits from the obtained irrigation water (Figure 20.3).

20.8 ENERGY PATHWAY FOR TRANSITIONING TO A CIRCULAR ECONOMY WITHIN WASTEWATER SERVICES

Each year, humanity produces roughly 380 billion litres of wastewater, with Asia accounting for the majority of this production, followed by North America and Europe. That is five times as much water as flows over Niagara Falls annually, and it is anticipated that this volume will rise by 24% by 2030 and by 51% by 2050 (Fluence, 2020). Even though conventional wastewater treatment uses a lot of energy, wastewater really contains around five times as much energy as is needed to treat it, and the technology to do so is easily available. Approximately 80% of the untapped energy in wastewater, according to the American Biogas Council, is chemical, 20% is thermal, and less than 1% of the potential is present in hydraulic generation. One of the most significant and powerful resource streams that regularly goes untapped by business in the United States is the energy salvaged from wastewater as biogas generated by anaerobic digestion. The recovery of energy from waste water is enough to produce power for 158 million homes, which is the total number of homes in the US and Mexico (Fluence, 2020). About 70% of biogas is methane, although yields are based on how much putrescible (digestible) feedstock is present in the source water. 2,200 locations in the US now employ anaerobic digesters to generate biogas, including 1,269 wastewater treatment facilities, 250 farms, 66 independent businesses, and 652 landfill projects. In Europe, there are more than 10,000 digesters, and some towns use them to essentially abandon fossil fuels (Fluence, 2020).

The need to switch from present linear models to a circular economy that allows for the longer-term and multipurpose use of water resources is driven by the escalating global water security issues. Over the past ten years, the circular economy (CE) idea has gained acceptance on a worldwide scale, mostly in reaction to resource exhaustion and linear models' failure to eliminate waste. The idea differs significantly from the conventional linear model in that it makes it possible to make the most of resources and extract considerable value from them for a longer period of time, recovering and renewing valuable goods and materials. Water, material, and energy have been identified as the three main paths enabling the adaptation of a CE for wastewater services, and each of these is supported by innovation. The 5R concept (Reduce, Reuse, Recycle, Recovery, and Restore) further supports the link between waste and energy. Traditional technical platforms including combustion, anaerobic digestion, gasification, and pyrolysis are proposed for converting sludge into useable energy. Because wastewater service activities consume a significant amount of energy and offer potential savings opportunities, the energy pathway can further reduce dependence on fossil-based sources of energy while increasing reliance on renewable energy sources. This could assist to reduce carbon emissions.

With untreated PS (Primary Sludge) energy content ranging from 17 to 24 MJ/kgDS and WAS (Waste Activated Sludge) energy content ranging from 14 to 20 MJ/kgDS (Kurniawan et al., 2018), investigations conducted globally have demonstrated that wastewater sludge has a high energy content comparable to that of low-grade coal. Based on a 65% TSS removal in PST and a sludge output of 0.25 kg TSS/kg COD applied for plants running at an average sludge age of 20 days, the ASPs are predicted to produce 1,960 tDS/d PS and 645 tDS/d WAS, respectively. The energy potential of wastewater sludge from the national ASPs is 355 MJ/s, or 355 MWe, assuming that the average energy content of PS and WAS is 15 MJ/kgDS (Van der Merwe-Botha et al., 2016). An economically viable alternative to energy generation from wastewater sludge can be provided by treating this potentially high energy content wastewater sludge utilising new technologies. The produced energy can be utilised locally to reduce the expensive aeration energy costs incurred by ASPs. Additionally, in accordance with the waste-to-energy 5R concept, this presents options for co-processing sludge with a variety of waste streams to assist the energy pathway to CE. The biological digestion of waste materials, such as agricultural and industrial wastes, can be used to produce biohydrogen. Hydrogen may be produced from wastewater using electrohydrolysis and biological processes like microbial fermentation in the dark or photofermentation. Heat pumps draw energy from a variety of sources, including the earth, water, and air. The effluents from WWTPs, however, offer a promising and cost-effective source of heat for heat pumps. However, the heat extracted from wastewater is referred to as being of inferior quality. The technology in question is referred to as a wastewater source heat pump. This heat may be used for infrastructure, water heating, sludge heating, industrial plants, public places, and commercial and residential buildings. The latent energy of wastewater can be recovered in its journey together with a WWTP by installing hydropower plants.

20.9 PERSPECTIVES AND PROSPECTS OF WATER IN CIRCULAR WATER ECONOMY

Discussions in the water business are more dominated than ever by sustainability and water management issues. Applying these economic ideas to the water industry has the potential to mitigate serious environmental issues. When it comes to resources, products, and services, societies have traditionally depended on the linear economic model of "extract, make, use, and discard" to accomplish growth. In such systems, resources are utilised to make goods, which are then consumed and finally thrown away as trash. An alternate strategy to address these issues and build a more sustainable system is the circular economy. In general, the circular economy produces changes that completely rebuild the system on all fronts—ecological, social, and financial (Del Borghi et al., 2020). Sustainability and reuse strategies are the main areas of distinction between linear and circular economies. The circular economy has the potential to improve water management, just like it can in many other industries. It may include water-related goods, facilities, technology, and services. Optimising water extraction from sources, enhancing water use habits, recycling used water, and recovering and reusing valuable components from wastewater are all examples of how to apply the fundamental principles of the circular economy to the water sectors and create a closed loop within the system. Every phase of the water cycle—from extraction through treatment, distribution, consumption, and waste generation—offers chances for a circular economy (Tarpeh and Chen, 2021).

Due to the fact that water is a resource, a good, and a service without an economic counterpart, a special definition of the Circular Economy of Water (CEW) was developed with nine strategies: reduce, rethink, replace, recycle, avoid, reuse, recover, cascade, and store. The CEW may be used as a foundation for sustainable water management through water conservation and efficient usage, which supports various SDGs (Sustainable Development Goals). The development and implementation of CE (circular economy) solutions for water in this context depend heavily on CE strategies. If CE ideas and solutions are to be successfully implemented, favourable circumstances must be present. As a result, the transition to a full CEW is made easier by addressing the three primary issues that have been identified: normative, governance, and implementation. Normative challenges are situations when new law or changes to current legislation are required (Mannina et al., 2021). Legal standards are effective vehicles for bringing about radical change, but water legislation is typically intricate, broad, and fragmented on both a national and international scale. The CE may be a mechanism to improve and update present water law such that it promotes more sectoral integration. To achieve this, it is necessary to deploy CEW principles and tactics based on a precise vocabulary, as this study suggests. The adoption of enabling mechanisms, for instance a system of rewards and incentives for adopting CEW initiatives, such as nutrient Recovery, or the use of disincentives, like taxation, to discourage wasteful linear operations, should be taken into consideration while establishing normative features. International agreements should also be included in new laws to support the development of transboundary CE plans and solutions (Mainardis et al., 2021).

Roles and duties for the creation and execution of CEW initiatives are among the governance problems. Expanding the CE's comprehension and acceptability across all water-using actors is essential, for instance when water is reused numerous times or used sequentially by various actors in cascading. For this aim to be accomplished, acceptance, education, and involvement with regard to CEW ideas and solutions are crucial (Nikolaou and Tsagarakis, 2021). Additionally, in light of regional specifics and contextual circumstances, water authorities and governments are responsible for establishing the most efficient water governance processes and arrangements (such as centralised/decentralised; public, private, or mixed solutions) for the CEW. All necessary parties must be involved for a smooth transition to the CEW, especially when rethinking a system. Barriers to and possibilities for a systematic use of the CEW are examples of implementation issues. These issues can arise when numerous techniques and solutions must be merged, as in cascading, or when conservative water management methods in use create a route dependence and lock-in that prohibit the adoption of new approaches. This is because many CEW solutions in the pilot phase need sufficient funding to be adopted widely (Salminen et al., 2022).

20.10 CONCLUSION

The circular economy, which aims to minimise waste and keep resources in the cycle for as long as possible, is gaining a foothold in the water sector. Nevertheless, the implementation of the CE concept with WTS is fairly challenging, partly because sludge from different WTPs and even from a single WTP might vary in physical and chemical qualities over the course of a year. Reusing water from treated wastewater significantly helps achieve the goal of global carbon neutrality. A circular economy strategy can also be used to recover resources that are obtained during wastewater treatment, including nutrients, carbon, bioplastics, and others. Despite these possibilities, research indicates there are not many real-world case studies. The main factors, according to the literature, are organisational, social, legal, and technological constraints. One way to get over these challenges would be to develop a solid and comprehensive roadmap that guides the water sector's transition to a circular economy model from a linear model. Although certain attempts have previously been made, it is necessary to strengthen this path so that it involves active, interdisciplinary stakeholders (i.e., industry, legislation, researchers, administrative, etc.). However, policymakers are not the only ones who can make a difference. Young people also need to be educated about the necessity of resource recovery and water reuse for an effective transformation to take place. Such a strategy encourages public participation, with the institution maybe taking the lead in this process. There are currently no more case studies, so in order to get beyond the aforementioned limitations, research should concentrate on stimulating the creation of such models. To enable a significant transition to the circular economy in the water sector, it is also necessary to develop more targeted solutions (water smart).

REFERENCES

Adewumi, J. R., Ilemobade, A. A., & Van Zyl, J. E. (2010). Treated wastewater reuse in South Africa: Overview, potential and challenges. *Resources, Conservation and Recycling*, *55*(2), 221–231.

Ahmad, M., Yousaf, M., Nasir, A., Bhatti, I. A., Mahmood, A., Fang, X., ... Mahmood, N. (2019). Porous eleocharis@ MnPE layered hybrid for synergistic adsorption and catalytic biodegradation of toxic Azo dyes from industrial wastewater. *Environmental science & technology*, *53*(4), 2161–2170.

Ait-Mouheb, N., Mayaux, P. L., Mateo-Sagasta, J., Hartani, T., & Molle, B. (2020a). Water reuse: A resource for Mediterranean agriculture. In *Water resources in the mediterranean region* (pp. 107–136). Elsevier.

Ait-Mouheb, N., Mayaux, P. L., Mateo-Sagasta, J., Hartani, T., & Molle, B. (2020b). Water reuse: A resource for Mediterranean agriculture. In *Water resources in the mediterranean region* (pp. 107–136). Elsevier.

Alvarino, T., Suarez, S., Lema, J., & Omil, F. (2018). Understanding the sorption and biotransformation of organic micropollutants in innovative biological wastewater treatment technologies. *Science of the Total Environment*, *615*, 297–306.

Angelakis, A. N., & Snyder, S. A. (2015). Wastewater treatment and reuse: Past, present, and future. *Water*, *7*(9), 4887–4895.

Aziz, F., & Farissi, M. (2014). Reuse of treated wastewater in agriculture: Solving water deficit problems in arid areas. *Annales of West University of Timisoara. Series of Biology*, *17*(2), 95.

Balkhair, K. S. (2016a). Impact of treated wastewater on soil hydraulic properties and vegetable crop under irrigation with treated wastewater, field study and statistical analysis. *Journal of Environmental Biology*, *37*(5), 1143.

Balkhair, K. S. (2016b). Microbial contamination of vegetable crop and soil profile in arid regions under controlled application of domestic wastewater. *Saudi Journal of Biological Sciences*, *23*(1), S83–S92.

Banach, J. L., Hoffmans, Y., Appelman, W. A. J., Van Bokhorst-van de Veen, H., & van Asselt, E. D. (2021). Application of water disinfection technologies for agricultural waters. *Agricultural Water Management*, *244*, 106527.

Batstone, D. J., Hülsen, T., Mehta, C. M., & Keller, J. (2015). Platforms for energy and nutrient recovery from domestic wastewater: A review. *Chemosphere*, *140*, 2–11.

Becerra-Castro, C., Lopes, A. R., Vaz-Moreira, I., Silva, E. F., Manaia, C. M., & Nunes, O. C. (2015). Wastewater reuse in irrigation: A microbiological perspective on implications in soil fertility and human and environmental health. *Environment International*, *75*, 117–135.

Bengtsson, S., Karlsson, A., Alexandersson, T., Quadri, L., Hjort, M., Johansson, P., ... Werker, A. (2017). A process for polyhydroxyalkanoate (PHA) production from municipal wastewater treatment with biological carbon and nitrogen removal demonstrated at pilot-scale. *New Biotechnology*, *35*, 42–53.

Benson, D., & Monciardini, D. (2018). Governing the circular economy: Multi-level comparative analysis. *Circular Economy Disruptions e Past, Present and Future, 17th–19th June. University of Exeter, UK.*

Bertanza, G., Canato, M., & Laera, G. (2018). Towards energy self-sufficiency and integral material recovery in waste water treatment plants: Assessment of upgrading options. *Journal of Cleaner Production*, *170*, 1206–1218.

Beveridge, R., Moss, T., & Naumann, M. (2017). Sociospatial understanding of water politics: Tracing the multidimensionality of water reuse. *Water Alternatives*, *10*(1), 22–40.

Chen, W., Lu, S., Jiao, W., Wang, M., & Chang, A. C. (2013). Reclaimed water: A safe irrigation water source?. *Environmental Development*, *8*, 74–83.

Contreras-Ramos, S. M., Escamilla-Silva, E. M., & Dendooven, L. (2005). Vermicomposting of biosolids with cow manure and oat straw. *Biology and Fertility of Soils*, *41*(3), 190–198.

Crini, G., & Lichtfouse, E. (2019). Advantages and disadvantages of techniques used for wastewater treatment. *Environmental Chemistry Letters*, *17*(1), 145–155.

Culha, O., Gunerhan, H., Biyik, E., Ekren, O., & Hepbasli, A. (2015). Heat exchanger applications in wastewater source heat pumps for buildings: A key review. *Energy and Buildings*, *104*, 215–232.

Del Borghi, A., Moreschi, L., & Gallo, M. (2020). Circular economy approach to reduce water-energy-food nexus. *Current Opinion in Environmental Science & Health*, *13*, 23–28.

Duan, P., Yan, C., Zhou, W., & Ren, D. (2016). Development of fly ash and iron ore tailing based porous geopolymer for removal of Cu (II) from wastewater. *Ceramics International*, *42*(12), 13507–13518.

EEA CSI. (2018). Available from https://www.eea.europa.eu/data-and-maps/daviz/annual-and-seasonal-water-abstraction-7#tab-dashboard-02 (Accessed 13 December 2022).

El-Shafai, S. A., El-Gohary, F. A., Nasr, F. A., Van Der Steen, N. P., & Gijzen, H. J. (2007). Nutrient recovery from domestic wastewater using a UASB-duckweed ponds system. *Bioresource Technology*, *98*(4), 798–807.

EPA. (1998). How wastewater treatment works...basics. Available online: https://www3.epa.gov/npdes/pubs/bastre.pdf/ (Accessed 13 December 2022).

European Council. (2020). Water reuse for agricultural irrigation: Council adopts new rules. Available online: https://www.consilium.europa.eu/en/press/ pressreleases/2020/04/07/water-reuse-for-agricultural-irrigation-council-ado pts-new-rules/

Fitch-Roy, O., Benson, D., & Monciardini, D. (2021). All around the world: Assessing optimality in comparative circular economy policy packages. *Journal of Cleaner Production*, *286*, 125493.

Fluence. (2020). How much energy exists in wastewater? Fluence. Available online: https://www.fluencecorp.com/how-much-energy-exists-in-wastewater/ (Accessed 10 December 2022).

Frijns, J. (2014). Intervention Concepts for energy saving, recovery and generation from the urban water sytems D45-1. KWR Watercycle Reasearch Instutute. Available online: https://www.researchgate.net/profile/Nelson_Carrico/publication/275213940_Intervention_concepts_for_energy_saving_recovery_and_generation_from_the_urban_water_system/links/553576530cf20ea35f10d908.pdf (Accessed 15 December 2022).

Frijns, J., Smith, H. M., Brouwer, S., Garnett, K., Elelman, R., & Jeffrey, P. (2016). How governance regimes shape the implementation of water reuse schemes. *Water*, *8*(12), 605.

Frontistis, Z., Xekoukoulotakis, N. P., Hapeshi, E., Venieri, D., Fatta-Kassinos, D., & Mantzavinos, D. (2011). Fast degradation of estrogen hormones in environmental matrices by photo-Fenton oxidation under simulated solar radiation. *Chemical Engineering Journal*, *178*, 175–182.

Furst, K. E., Pecson, B. M., Webber, B. D., & Mitch, W. A. (2018). Tradeoffs between pathogen inactivation and disinfection byproduct formation during sequential chlorine and chloramine disinfection for wastewater reuse. *Water Research*, *143*, 579–588.

Ghimire, U., Sarpong, G., & Gude, V. G. (2021). Transitioning wastewater treatment plants toward circular economy and energy sustainability. *ACS Omega*, *6*(18), 11794–11803.

Ghisellini, P., Cialani, C., & Ulgiati, S. (2016). A review on circular economy: The expected transition to a balanced interplay of environmental and economic systems. *Journal of Cleaner Production*, *114*, 11–32.

Guo, T., Englehardt, J., & Wu, T. (2014). Review of cost versus scale: Water and wastewater treatment and reuse processes. *Water Science and Technology*, *69*(2), 223–234.

Hagos, K., Zong, J., Li, D., Liu, C., & Lu, X. (2017). Anaerobic co-digestion process for biogas production: Progress, challenges and perspectives. *Renewable and Sustainable Energy Reviews*, *76*, 1485–1496.

Hand, S., & Cusick, R. D. (2021). Electrochemical disinfection in water and wastewater treatment: Identifying impacts of water quality and operating conditions on performance. *Environmental Science & Technology*, *55*(6), 3470–3482.

Harb, M., Lou, E., Smith, A. L., & Stadler, L. B. (2019). Perspectives on the fate of micropollutants in mainstream anaerobic wastewater treatment. *Current Opinion in Biotechnology*, *57*, 94–100.

Hoekstra, A. Y., & Mekonnen, M. M. (2012). Reply to Ridoutt and Huang: From water footprint assessment to policy. *Proceedings of the National Academy of Sciences*, *109*(22), E1425–E1425.

Hristov, J., Barreiro-Hurle, J., Salputra, G., Blanco, M., & Witzke, P. (2021). Reuse of treated water in European agriculture: Potential to address water scarcity under climate change. *Agricultural Water Management*, *251*, 106872.

Hussein, A. M., Jabbar, D. N., & Abdulridha, S. Q. (2019). Impact of primary sedimentation tank on wastewater treatment plant units using computer simulation program. DOI:10.46617/icbe6003

Jang, T., Lee, S. B., Sung, C. H., Lee, H. P., & Park, S. W. (2010). Safe application of reclaimed water reuse for agriculture in Korea. *Paddy and Water Environment*, *8*(3), 227–233.

Jeong, H., Kim, H., & Jang, T. (2016). Irrigation water quality standards for indirect wastewater reuse in agriculture: A contribution toward sustainable wastewater reuse in South Korea. *Water*, *8*(4), 169.

Jindal, A., & Kamat, S. (2011). Water recycling and reuse for domestic and industrial sectors. *Chemical Engineering World*, 52–62.

Katsoyiannis, I. A., Gkotsis, P., Castellana, M., Cartechini, F., & Zouboulis, A. I. (2017). Production of demineralized water for use in thermal power stations by advanced treatment of secondary wastewater effluent. *Journal of Environmental Management*, *190*, 132–139.

Kehrein, P., Jafari, M., Slagt, M., Cornelissen, E., Osseweijer, P., Posada, J., & van Loosdrecht, M. (2021). A techno-economic analysis of membrane-based advanced treatment processes for the reuse of municipal wastewater. *Water Reuse*, *11*(4), 705–725.

Korhonen, J., Honkasalo, A., & Seppälä, J. (2018). Circular economy: The concept and its limitations. *Ecological Economics*, *143*, 37–46.

Krzeminski, P., Tomei, M. C., Karaolia, P., Langenhoff, A., Almeida, C. M. R., Felis, E., … Fatta-Kassinos, D. (2019). Performance of secondary wastewater treatment methods for the removal of contaminants of emerging concern implicated in crop uptake and antibiotic resistance spread: A review. *Science of the Total Environment*, *648*, 1052–1081.

Kurniawan, T., Hakiki, R., & Sidjabat, F. M. (2018). Wastewater sludge as an alternative energy resource: A review. *Journal of Environmental Engineering and Waste Management*, *3*(1), 1–12.

Lazarova, V. (2022). Water reuse: A pillar of the circular water economy. In *Resource Recovery from Water* (pp. 61–98). Academia.

Li, Y., Yang, Z., Yang, K., Wei, J., Li, Z., Ma, C., … Zhang, C. (2022). Removal of chloride from water and wastewater: Removal mechanisms and recent trends. *Science of the Total Environment*, 153174. DOI: 10.1016/j.scitotenv.2022.153174

Liu, Q., Zhou, Y., Lu, J., & Zhou, Y. (2020). Novel cyclodextrin-based adsorbents for removing pollutants from wastewater: A critical review. *Chemosphere*, *241*, 125043.

Lyu, S., Chen, W., Zhang, W., Fan, Y., & Jiao, W. (2016). Wastewater reclamation and reuse in China: Opportunities and challenges. *Journal of Environmental Sciences*, *39*, 86–96.

Mainardis, M., Cecconet, D., Moretti, A., Callegari, A., Goi, D., Freguia, S., & Capodaglio, A. G. (2021). Wastewater fertigation in agriculture: Issues and opportunities for improved water management and circular economy. *Environmental Pollution*, *296*, 118755.

Manchisi, J., Matinde, E., Rowson, N. A., Simmons, M. J., Simate, G. S., Ndlovu, S., & Mwewa, B. (2020). Ironmaking and steelmaking slags as sustainable adsorbents for industrial effluents and wastewater treatment: A critical review of properties, performance, challenges and opportunities. *Sustainability*, *12*(5), 2118.

Mannina, G., Badalucco, L., Barbara, L., Cosenza, A., Di Trapani, D., Gallo, G., … Helness, H. (2021). Enhancing a transition to a circular economy in the water sector: The EU project wider uptake. *Water*, *13*(7), 946.

Mannina, G., Gulhan, H., & Ni, B. J. (2022). Water reuse from wastewater treatment: The transition towards circular economy in the water sector. *Bioresource Technology*, 363, 127951. DOI: 10.1016/j.biortech.2022.127951

Mbavarira, T. M., & Grimm, C. (2021). A systemic view on circular economy in the water industry: Learnings from a Belgian and Dutch case. *Sustainability*, *13*(6), 3313.

Mizyed, N. R. (2013). Challenges to treated wastewater reuse in arid and semi-arid areas. *Environmental Science & Policy*, *25*, 186–195.

Mo, W., & Zhang, Q. (2013). Energy-nutrients-water nexus: Integrated resource recovery in municipal wastewater treatment plants. *Journal of Environmental Management*, *127*, 255–267.

Neczaj, E., & Grosser, A. (2018). Circular economy in wastewater treatment plant-challenges and barriers. *Multidisciplinary Digital Publishing Institute Proceedings*, *2*(11), 614.

Nikolaou, I. E., & Tsagarakis, K. P. (2021). An introduction to circular economy and sustainability: Some existing lessons and future directions. *Sustainable Production and Consumption*, *28*, 600–609.

Pandey, P., Shinde, V. N., Deopurkar, R. L., Kale, S. P., Patil, S. A., & Pant, D. (2016). Recent advances in the use of different substrates in microbial fuel cells toward wastewater treatment and simultaneous energy recovery. *Applied Energy*, *168*, 706–723.

Poornima, R., Suganya, K., & Sebastian, S. P. (2021). Biosolids towards Back-To-Earth alternative concept (BEA) for environmental sustainability: A review. *Environmental Science and Pollution Research*, 1–42.

Poustie, A., Yang, Y., Verburg, P., Pagilla, K., & Hanigan, D. (2020). Reclaimed wastewater as a viable water source for agricultural irrigation: A review of food crop growth inhibition and promotion in the context of environmental change. *Science of the Total Environment, 739*, 139756.

Rashid, R., Shafiq, I., Akhter, P., Iqbal, M. J., & Hussain, M. (2021). A state-of-the-art review on wastewater treatment techniques: The effectiveness of adsorption method. *Environmental Science and Pollution Research, 28*(8), 9050–9066.

Ricart, S., & Rico, A. M. (2019). Assessing technical and social driving factors of water reuse in agriculture: A review on risks, regulation and the yuck factor. *Agricultural Water Management, 217*, 426–439.

Rubiano, M. E., Agulló-Barceló, M., Casas-Mangas, R., Jofre, J., & Lucena, F. (2012). Assessing the effects of tertiary treated wastewater reuse on a Mediterranean river (Llobregat, NE Spain), part III: Pathogens and indicators. *Environmental Science and Pollution Research, 19*(4), 1026–1032.

Saliba, R., Callieris, R., D'Agostino, D., Roma, R., & Scardigno, A. (2018). Stakeholders' attitude towards the reuse of treated wastewater for irrigation in Mediterranean agriculture. *Agricultural Water Management, 204*, 60–68.

Salminen, J., Määttä, K., Haimi, H., Maidell, M., Karjalainen, A., Noro, K., ... Pohjola, J. (2022). Water-smart circular economy-Conceptualisation, transitional policy instruments and stakeholder perception. *Journal of Cleaner Production, 334*, 130065.

Sato, T., Qadir, M., Yamamoto, S., Endo, T., & Zahoor, A. (2013). Global, regional, and country level need for data on wastewater generation, treatment, and use. *Agricultural Water Management, 130*, 1–13.

Sengupta, S., Nawaz, T., & Beaudry, J. (2015). Nitrogen and phosphorus recovery from wastewater. *Current Pollution Reports, 1*(3), 155–166.

Shen, Y., Oki, T., Kanae, S., Hanasaki, N., Utsumi, N.,&Kiguchi, M. (2014). Projection of future world water resources under SRES scenarios: An integrated assessment. *Hydrological Sciences Journal, 59*, 1775–1793.

Silva, L. L., Moreira, C. G., Curzio, B. A., & da Fonseca, F. V. (2017). Micropollutant removal from water by membrane and advanced oxidation processes-a review. *Journal of Water Resource and Protection, 9*(05), 411.

Singh, A., Sawant, M., Kamble, S. J., Herlekar, M., Starkl, M., Aymerich, E., & Kazmi, A. (2019). Performance evaluation of a decentralized wastewater treatment system in India. *Environmental Science and Pollution Research, 26*(21), 21172–21188.

Smol, M., Adam, C., & Preisner, M. (2020). Circular economy model framework in the European water and wastewater sector. *Journal of Material Cycles and Waste Management, 22*(3), 682–697.

SWRCB. (2011) Order No. R3-2011-0222: Waste discharge requirements NPDES general permit for discharges of highly treated groundwater to surface waters, NPDES NO. CAG993002. California State Water Quality Control Board.

Tal, A. (2016). Rethinking the sustainability of Israel's irrigation practices in the Drylands. *Water Research, 90*, 387–394.

Tarpeh, W. A., & Chen, X. (2021). Making wastewater obsolete: Selective separations to enable circular water treatment. *Environmental Science and Ecotechnology, 5*, 100078.

Topare, N. S., Attar, S. J., & Manfe, M. M. (2011). Sewage/wastewater treatment technologies: A review. *Scientific Reviews & Chemical Communications, 1*(1), 18–24.

Ungureanu, N., Vlăduț, V., Dincă, M., & Zăbavă, B. Ș. (2018, May). Reuse of wastewater for irrigation, a sustainable practice in arid and semi-arid regions. In *Proceedings of the 7th International Conference on Thermal Equipment, Renewable Energy and Rural Development (TE-RE-RD)*, Drobeta-Turnu Severin, Romania (Vol. 31, pp. 379–384).

Van der Merwe-Botha, M., Juncker, K., & Visser, A. (2016). *Guiding Principles in the Design and Operation of a Wastewater Sludge Digestion Plant with Biogas and Power Generation: Report to the Water Research Commission*. Water Research Commission.

Voulvoulis, N. (2018). Water reuse from a circular economy perspective and potential risks from an unregulated approach. *Current Opinion in Environmental Science & Health, 2*, 32–45.

Wester, J., Timpano, K. R., Çek, D., Lieberman, D., Fieldstone, S. C., & Broad, K. (2015). Psychological and social factors associated with wastewater reuse emotional discomfort. *Journal of Environmental Psychology, 42*, 16–23.

World Health Organization. (2006). *WHO Guidelines for the Safe Use of Wasterwater Excreta and Greywater* (Vol. 1). World Health Organization.

World Resources Institute (WRI). (2020). Aqueduct country rankings. Available online: https://www.wri.org/applications/aqueduct/country-rankings/ (Accessed 13 December 2022).

Yadav, G., Mishra, A., Ghosh, P., Sindhu, R., Vinayak, V., & Pugazhendhi, A. (2021). Technical, economic and environmental feasibility of resource recovery technologies from wastewater. *Science of the Total Environment, 796*, 149022.

Yang, C., Xu, W., Nan, Y., Wang, Y., Hu, Y., Gao, C., & Chen, X. (2020a). Fabrication and characterization of a high performance polyimide ultrafiltration membrane for dye removal. *Journal of Colloid and Interface Science*, *562*, 589–597.

Yang, J., Jia, R. S., Gao, Y. L., Wang, W. F., & Cao, P. Q. (2017, December). The reliability evaluation of reclaimed water reused in power plant project. *IOP Conference Series: Earth and Environmental Science*, *100*(1), 012189.

Yang, J., Monnot, M., Ercolei, L., & Moulin, P. (2020b). Membrane-based processes used in municipal wastewater treatment for water reuse: State-of-the-art and performance analysis. *Membranes*, *10*(6), 131.

Yenkie, K. M. (2019). Integrating the three E's in wastewater treatment: Efficient design, economic viability, and environmental sustainability. *Current Opinion in Chemical Engineering*, *26*, 131–138.

Yogalakshmi, K. N., Das, A., Rani, G., Jaswal, V., & Randhawa, J. S. (2020). Nano-bioremediation: A new age technology for the treatment of dyes in textile effluents. In *Bioremediation of Industrial Waste for Environmental Safety* (pp. 313–347). Singapore: Springer.

Zagklis, D. P., & Bampos, G. (2022). Tertiary wastewater treatment technologies: A review of technical, economic, and life cycle aspects. *Processes*, *10*(11), 2304.

Zhang, Q., Hu, J., Lee, D. J., Chang, Y., & Lee, Y. J. (2017). Sludge treatment: Current research trends. *Bioresource Technology*, *243*, 1159–1172.

Zhang, Y., & Shen, Y. (2017). Wastewater irrigation: Past, present, and future. Wastewater treatment: Aims and challenges. *Water*, *6*(3), e1234. https://doi.org/10.1002/wat2.1234.

Zhen, G., Lu, X., Kato, H., Zhao, Y., & Li, Y. Y. (2017). Overview of pretreatment strategies for enhancing sewage sludge disintegration and subsequent anaerobic digestion: Current advances, full-scale application and future perspectives. *Renewable and Sustainable Energy Reviews*, *69*, 559–577.

Zhu, Z., Liu, D., Cai, S., Tan, Y., Liao, J., & Fang, Y. (2020). Dyes removal by composite membrane of sepiolite impregnated polysulfone coated by chemical deposition of tea polyphenols. *Chemical Engineering Research and Design*, *156*, 289–299.

21 Bioconversion of Lignocellulosic Waste

Chemicals and Biofuels Biorefinery Concept

Bhoomika Yadav, Anusha Atmakuri, Shraddha Chavan, R.D. Tyagi, and Patrick Drogui

21.1 INTRODUCTION

The crises in energy and environment due to the increase in population have shifted the world's interest in search of renewable options for fuel, chemicals and materials. Fossil fuels are used to satisfy the energy demands and generate various products including fuels, polymers, platform chemicals, cosmetic and hygiene products, etc. (Okolie et al., 2021). However, the over-utilization of fossil resources has resulted in their reduction along with other environmental problems such as greenhouse gas emissions (Nanda et al., 2016). Therefore, now the interest is shifting towards the utilization of renewable resources for the production of fuels and chemicals.

Waste organic biomass is inexpensive, abundant, renewable and is a good substrate for production of biofuels, fine chemicals and biomaterials. It will not only decrease the burden on fossil fuels but also decrease the environmental impacts of the use of fossil resources and address the issues of waste management, energy security and rural development (Chavan et al., 2022). It is estimated that the research in biomass resource could cause more than 30% of fossil fuel replacement in the future for the production of biofuels, biochemicals and biomaterials (Okolie et al., 2021).

Lignocellulosic feedstocks refer to the non-edible parts of plants that are made of three main components i.e., cellulose, hemicellulose and lignin, and small amounts of extractives such as fats, lipids, resins, tannins, terpenes, steroids, phenols and flavonoids (Nanda et al., 2013). Lignocellulosic materials are promising feedstocks for the production of biofuels, biochemicals and biomaterials because of their socioeconomic advantages and non-seasonal availability. Also, these feedstocks are non-edible hence, they do not pose any competition with the food crops. Due to the high organic matter present, additional pre-treatment steps (physical, chemical, biological or their combinations) are required to enhance the depolymerization of the structural components, thus increasing the degradability of the biomass. After the pre-treatment, lignocellulosic biomass can be converted into biofuels and other biomaterials by various thermochemical and biochemical conversion technologies.

The first section of the chapter provides a comprehensive review on the types and origin of the lignocellulosic wastes along with the composition of the constituents of the biomass. Then, a review on the current pre-treatment technologies for lignocellulosic biomass is provided, followed by the industrial applications for the production of biofuels/bioenergy, biochemicals, enzymes, bioplastics, biosurfactants and bioactive compounds. Lastly, the technical challenges in the process with the future perspectives are briefly discussed.

DOI: 10.1201/9781003441144-21

21.2 LIGNOCELLULOSIC WASTES

21.2.1 Origin, Global Production and Types

Worldwide, a large amount of land is used for agricultural purposes; for example, east Asia, Europe and South America use around 648,396 464,405 and 620,418 ha, respectively for the production of common agricultural products such as crops, cereals, vegetable oil, roots, fruits and vegetables (Batista Meneses et al., 2020). As a result of intensive agriculture, large quantities of agro-industrial wastes are also produced worldwide that are lignocellulosic in nature. For example, around 740 million tonnes (MT)/year of rice straw is generated worldwide. The United States alone generates around 274 MT/year of corn bagasse. Brazil produces 166 MT/year of sugarcane bagasse. Out of all the agro-industrial wastes, rice straw, wheat stalk, corn bagasse, sugar beet, sugarcane bagasse, soy stalk, sawdust, cotton stalk, orange peel and barley straw have been produced in large quantities worldwide (Batista Meneses et al., 2020; Okolie et al., 2021).

The lignocellulosic biomass can be broadly categorized into crop residues (including rice straw, wheat straw, oat hall, rice hall, corn stover, wheat stalk and cotton stalk); forestry materials (wood chips, bark, sawdust and wood logs); and dedicated energy crops (switchgrass, elephant grass, timothy grass, willow, etc.). Each type of lignocellulosic residue has different proportions of the components depending on the source whether derived from hardwood, softwood or grasses. For example, canola straw consists of around 42.4 wt% cellulose, 16.4 wt% hemicellulose and 14.2 wt% lignin (Adapa et al., 2009), whereas rice straw consists of 32.1% cellulose, 24% hemicellulose and 18% lignin (Howard et al., 2003). The composition of common agro-industrial wastes is presented in Figure 21.1.

21.2.2 Composition and Structure

Generally, lignocellulosic biomass consists of 35–55 wt% cellulose, 20–40 wt% hemicellulose, 10%–25% lignin and traces of extractives (Okolie et al., 2021). Cellulose is the main structural

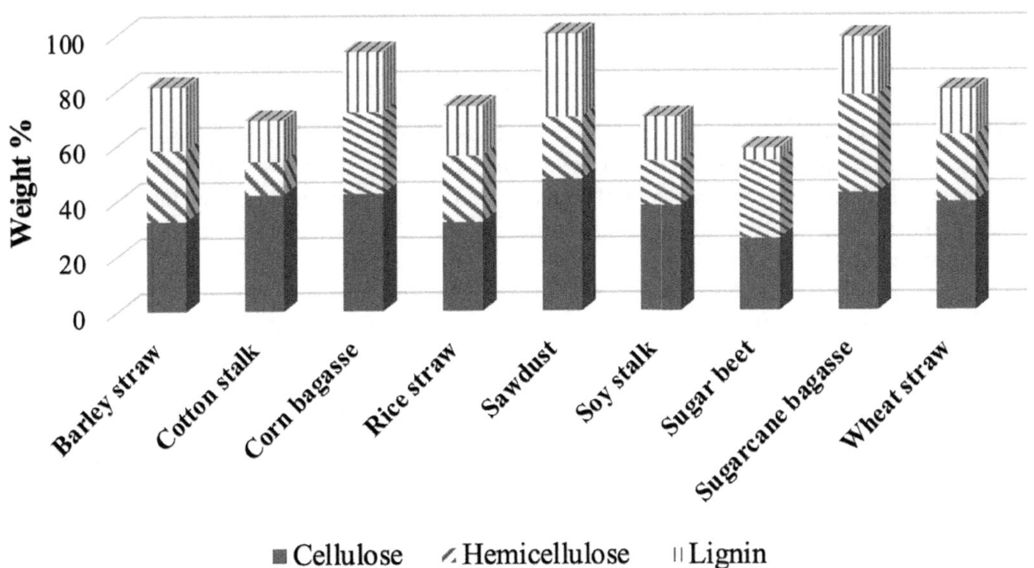

FIGURE 21.1 Composition of lignocellulosic components in common agro-industrial wastes.

component of plant cells responsible for strength, hemicellulose is composed of repeated polymers of pentoses and hexoses. Lignin consists of aromatic alcohols i.e., coniferyl alcohol, sinapyl alcohol and *p*-coumaryl alcohol forming a layer around cellulose and hemicellulose.

Cellulose is a very stable and abundant polymer that contains glucose attached with linear chains mainly composed of (1,4)-D-glucopyranose units which are attached by β-1,4 glycosidic linkages forming a dimer called cellobiose (fundamental unit of cellulose) (Anwar et al., 2014). The cellulose molecules are held together by the intermolecular hydrogen bonds in native state and have a strong tendency to form intra-molecular and intermolecular hydrogen bonds which makes cellulose insoluble and resistant to organic solvents. Common applications of cellulose include additive for adhesives, paper-based products, cement industries, food coating, biomedicine, etc. (Batista Meneses et al., 2020).

Hemicellulose is a second abundant polymer that consists of polysaccharide mixture of pentose sugars (xylose, arabinose), hexose sugars (glucose, mannose, galactose) and sugar acids (glucuronic acid, galacturonic acid) (Okolie et al., 2021). Sugars are linked mostly by β-1,4- and sometimes by β-1,3-glycosidic bonds (Sánchez, 2009). Hemicellulose can be degraded easily due to low degree of polymerization and amorphous structure as compared to cellulose. The composition of hemicellulose is dependent on the source of lignocellulosic biomass. Grasses and straws contain arabin, galactan and xylan whereas mannin is found as component in the hemicelluloses of hardwood and softwood (Anwar et al., 2014). Xylans are most abundant polysaccharide of hemicellulose consisting of 1,4-linked β-D-xylose units.

Lignin is the smallest fraction in lignocellulosic biomass but the most complex one. It is a long-chain, heterogenous polymer made largely of phenyl-propane units mostly linked with ether bonds. It consists of various types of phenols where coniferyl alcohols are predominated in softwoods, whereas p-cormaryl and sinapyl alcohols are present in high proportions in hardwoods. Lignin acts as a membrane around cellulose and hemicellulose components filling the gap between and around them (Anwar et al., 2014; Hafiz Muhammad Nasir et al., 2011). The presence of lignin in the lignocellulosic biomass makes it difficult to extract cellulose and hemicellulose for the production of fermentable sugars. Therefore, it is important to remove lignin to expose cellulose and hemicellulose sugars for hydrolysis and fermentation via microbes by pre-treatment of lignocellulosic materials (Okolie et al., 2021).

21.3 PRE-TREATMENT OF LIGNOCELLULOSIC MATERIALS

Due to the complex hierarchical structure and recalcitrant nature of lignocellulosic biomass, pretreatment steps render a crucial challenge to biomass utilization prior to conversion. Pre-treatment of lignocellulosic biomass breaks down the recalcitrant structure of the biomass, releasing the cellulose and hemicellulose from the matrix structure of lignin. This step enriches the yield of the final product in addition to facilitating the saccharification process (Ali et al., 2020).

During the chemical and biological conversion of lignocellulosic biomass, pre-treatment efficiently produces 2nd generation sugars (2G), the natural intermediates, also the chief ingredients in a biorefinery. However, pre-treatment is an energy and cost-intensive process and varies largely with the feedstock and the desired end product. Hence, it is essential to consider this step while evaluating the overall techno-economic viability and sustainability of the lignocellulosic biorefinery. In addition to high cost and energy consumption, most pre-treatment techniques have other drawbacks such as inhibitory intermediates and by-product formation diminishing the overall attractiveness of biomass valorization (Hassan et al., 2019).

The different kinds of pre-treatment methods include physical, chemical, biological and physiochemical methods. However, these methods have their own pros and cons. For instance, most physical pre-treatment processes need high energy and special equipment, raising the overall cost and limiting the industrial scalability from the lab-scale (Ali et al., 2020; Ana et al., 2010; Bhatia et al., 2020). Alternatively, chemical pre-treatment requires lower energy in comparison, yet uses digesters and chemicals which again escalates the cost and overall greenness of the process. Although these methods can achieve high solubilization of cellulose and hemicellulose, and removal of lignin, they cause a high environmental burden.

To speed up the efficiency of breakdown, these pre-treatments are performed in combination with thermal processes such as high pressure and temperature. In addition, these pre-treatments need the hydrolysate to be neutralized of inhibitors preceding the saccharification, becoming a reason for the increased cost of valorization (Adewuyi, 2022; Hassan et al., 2019). Even though the inhibitors are neutralized, there are some derivatives of the same being formed during the whole process, posing a risk to the environment, thereby urging the need for its treatment pre-disposal, hence adding to the total cost.

Consequently, there is a developing incline towards implementation of biological pre-treatments, deemed safe, low energy consuming, economic, and no by-product formation. It is believed that the microbes utilized in these treatments could also be used to produce enzymes, thus lowering the cost of enzyme retrieval.

Physical pre-treatment methods include mechanical (grinding, milling and chopping), sonication, mechanical extrusion, freezing, ozonolysis, pulsed-electric field and pyrolysis; chemical methods most commonly include acid and alkali pre-treatment methods, in addition to oxidative and organosolv pre-treatment, ozonolysis, the use of ionic liquids and novel natural deep eutectic solvents. A combination of physical and chemical methods (physicochemical) consists of ammonia fiber explosion, ultrasonication, autohydrolysis, liquid hot water, wet oxidation, and CO_2 explosion pre-treatment. Table 21.1 demonstrates the comparison of various operational parameters for lignocellulosic pre-treatment methods.

It is difficult to identify and apply a universal pre-treatment process across lignocellulosic biomass due to a range of factors that impact their characteristics and mechanisms. A method to improve the breakdown performance is to combine different lignocellulosic pre-treatment techniques operating in conjunction with each other to improve yield of sugars and the overall process efficiency. Combination of these pre-treatments merges the advantages of individual techniques and applies a synergistic impact on the lignocellulosic biomass for a superior conversion (Ana et al., 2010; Bhatia et al., 2020).

Yet, overall conversion efficiency is influenced by biomass structure and complexity, thereby making it problematic to oversee a universally applicable combination of pre-treatment methods effective in treating lignocellulosic biomass. In addition, a combination of various techniques would entail further requirements such as added chemicals, equipment, thereby increasing the overall cost of the process.

21.4 BIO-VALORIZATION OF LIGNOCELLULOSIC MATERIALS

21.4.1 Biofuels/Bioenergy

Biofuels exist in liquid forms such as bio-oil, bioethanol, biodiesel, biobutanol and in gaseous forms, for example, syngas, biogas, biomethane and biohydrogen. These biofuels/bioenergy can be produced using product-specific biochemical (enzymatic hydrolysis, fermentation and anaerobic digestion) and thermochemical pathways (pyrolysis, liquefaction, gasification, transesterification, etc.) (Okolie et al., 2021).

Bio-oils from biomass can be obtained using pyrolysis and liquefaction thermochemical conversion methods. On comparing the oxygen content in the bio-oil produced from these two methods, bio-oil after liquefaction has lower oxygen content compared to pyrolysis thus, showing a higher heating value and better fuel properties. Bio-oils from lignocellulosic biomass contain different proportions of cellulose, hemicellulose and lignin degradation products. They contain mixtures of aldehydes, ketones, acids, phenols, alcohols, carbonyls, esters, ethers, vanillin, etc. due to the decomposition of cellulose, hemicellulose and lignin (Mohan et al., 2006).

TABLE 21.1
Comparison of Operational Parameters for Lignocellulosic Pre-Treatment Methods

	Pre-treatment Methods		
	Chemical	**Physicochemical**	**Biological**
Methods	• Dilute sulphuric acid • Alkaline hydrolysis • Ionic liquids (ILs) • Organosolv process • Deep Eutectic Solvents (DES) • γ-Valerolactone (GVL)	• Liquid hot water • AFEX • Pyrolysis • Steam explosion	• Microbes (bacteria, fungi, actinomycetes) • Enzymes (laccases, peroxidases, etc.)
Breakdown mechanism	Fractionation of lignin releasing the polysaccharides for further hydrolysis and generation of monomeric sugars	The delamination of cell wall microfibrils enhances the digestibility of lignocellulosic biomass	Decomposes the cellulose and hemicellulose to yield intermediate compounds that can be further hydrolyzed by enzymes
Advantages	• Higher product/sugar yields on hydrolysis • Very high delignification • High conversion rate • Moderate reaction conditions	• Non-toxic • Less corrosive • Highly effective on lignocellulosic biomass	• No chemical requirement • Lower energy consumption • Mild reaction conditions • Economic
Disadvantages	• High water usage • Loss of hemicellulose and lignin • The chemicals are not sustainable as harmful	• Expensive setup due to high pressure and temperature needs • Special design for the reactor	• Huge setups • Slow process • Lower hydrolysis sugar outputs • Continuous monitoring
Energy input	High	High	Very low
Process efficiency	High	Medium	Low
Environmental impact	High	Low	Very low
Inhibitor generation	High	–	Very low
Odour generation	High	Low	High
Applicability to lignocellulosic biomass	High	–	High
Biomass solubilization	High	Medium	Medium

The quality and quantity of bio-oil can depend on operating conditions like temperature, heating rate, residence time, pressure, catalysts, concentration of feedstock, type of biomass and reactor type (Okolie et al., 2021).

Bioethanol is produced from biomass via enzymatic hydrolysis and microbial fermentation and can be used as a transportation fuel that could reduce the greenhouse gases (GHGs) emission from fossil fuels. Bioethanol contains oxygen fraction of 35% that improves combustion and decreases the emissions of CO, NO, NO_x, etc. as compared to the conventional fuels. Bioethanol from sugar/ starchy feedstocks can impact the economics, food security and environment because of using the food crops for fuel generation therefore, lignocellulosic feedstock can be a potential alternative from both environmental and economic perspective for the production of bioethanol. It can also increase the economy in rural areas, reduce fossil fuel dependency, and increase energy security (Nanda et al., 2015). Biobutanol has superior fuel properties and high-energy content than bioethanol. Moreover, it can be used in pure form or blended with gasoline in various ratios without any mechanical modification. Both these biofuels i.e., bioethanol and biobutanol are promising alternatives to the existing conventional fuels, the overall process feasibility depends on the separation process used as it can be energy-intensive which can further affect the economics of the process. Bioethanol can be separated using fractional distillation whereas biobutanol separation can be done using adsorption, liquid-liquid extraction, gas stripping, supercritical-fluid extraction, etc. which can be more challenging, time-consuming, costly and energy-intensive (Nanda et al., 2017).

Biomethane is another promising biofuel that can be produced from lignocelluloses via anaerobic digestion or biomethanation. It is used for heat and power generation or for energy production in cooking, heating and electricity generation. Biomethane has low density, high calorific value and can be used in vehicles. When used in heavy vehicles, there can be 63% reduction in GHGs emissions compared to compressed natural gas (CNG). However, the emissions of NOx, hydrocarbons and carbon monoxide are higher than CNG (Khan et al., 2017).

Biodiesel is a renewable, environmentally friendly and non-toxic biofuel that can be used in diesel engines. Biodiesel from lignocellulosic material can be produced using transesterification which involves a chemical reaction of vegetable oil with alcohol to obtain fatty acid alkyl esters i.e., biodiesel and glycerol. Researchers have shown the production of biodiesel from sugarcane bagasse hydrolysates using *Yarrowia lipolytica* (Vasaki et al., 2022). Pre-treatment using 4% v/v H_2SO_4 at 25 min ultrasonication resulted in highest depolymerization and yielded 16.39 g/L biomass concentration. *In-situ* transesterification with K_2CO_3 as a catalyst resulted in 80% biodiesel.

21.4.2 Enzymes and Chemicals

Various enzymes have been produced via submerged and solid culture processes. Some of the enzymes are ligninolytic enzymes like lignin peroxidase (LiP), manganese peroxidase (MnP), versatile peroxidise (VP), laccases, endoglucanases, β-glucosidases and cellobiohydrolases. Because of the lower capital investment and operating costs of solid-state fermentation, this method has been extensively favoured for the production of fungal microbial enzymes using different lignocellulosic materials (Iqbal et al., 2013). Fungi such as *Trichoderma, Aspergillus, Penicillium, Fusarium,* etc. have been used for cellulases production (Batista Meneses et al., 2020). Ligninolytic, cellulases and hemicellulases have various applications in industries such as chemicals, food, fuel, brewery, animal feed, pulp-paper, agriculture, textile and many more. Researchers have studied several lignocellulolytic organisms mainly, white rot fungi that have been identified as most efficient and extensive lignin degraders. Enzymes named LiP, E.C. 1.11.1.14, MnP, E.C. 1.11.1.13, and laccase E.C. 1.10.3.2 are produced by white rot fungi during lignin degradation from lignocelluloses.

Various lignocellulosic materials that have been studied for the production of different microbial enzymes having industrial importance are given in Table 21.2.

Lignocellulosic materials have been extensively studied for the production of useful chemicals. Using renewable raw materials to produce fine chemicals can cover the high demand for these products as well as reduce dependency on fossil resources. Products such as glucose from cellulose and hemicellulose, xylose, arabinose, mannose, glucose, galactose and acetate from hemicellulose and phenolic compounds from lignin are produced during the hydrolysis process of lignocellulosic biomass. Various precursor compounds of different carbon number have been produced during the lignocellulosic biomass degradation such as citric acid, 5-hydroxymethylfurfural, lysine, sorbitol (6-carbon); itaconic acid, furfural, levulinic acid, glutamine acid, xylonic acid (5-carbon); succinic acid, fumaric acid, malic acid, aspartic acid, acetoin, threonine (4-carbon) and glycerol, 3-hydroxypropionate, propionic acid and malonic acid (3-carbon) (Batista Meneses et al., 2020). The precursor compounds can generate a wide range of organic compounds such as acetic acid, oxalic acid, lactic acid, itaconic acid, butyric acid, succinic acid, propionic acid, etc. It has also been reported that various agricultural residues such as rice straw, corn stover, bagasse, spent grain and corncob have been successfully used for the production of xylitol, phenols, catechols, vanillin, vanillic acid, benzene, biphenyls, cyclohexane, syringaldehyde, etc. (Iqbal et al., 2013).

TABLE 21.2
Recent Studies for Production of Microbial Enzymes from Lignocellulosic Material

Lignocellulosic Material	Pre-Treatment	Microbe	Enzyme	References
Orange peel waste	Chemical (acidic)	*Trichoderma viride*	Endoglucanase (655±5.5 U/mL), exoglucanase (412±4.3 U/mL), β-glucosidase (515±3.7 U/mL)	Irshad et al. (2013)
Rice straw and orange peel	n.r.	*Lactobacillus paracasei* MK852178	β-glucosidase	Wahab et al. (2021)
Sugarcane bagasse	Chemical (alkali)	*Pycnoporus sanguineus*	Cellulase	Yoon et al. (2012)
	–	*Aspergillus flavus* KUB2	Cellulase and xylanase	Namnuch et al. (2021)
	–	*Aspergillus niger*	Xylanase and protease	Valladares-Diestra et al. (2021)
Corncobs	–	*Trametes versicolor* IBL-04	Lignin peroxidase	Asgher et al. (2012)
Wheat straw	Chemical (acidic)	*Trichoderma viride*	Cellulase	Hafiz Muhammad Nasir et al. (2011)
Banana stalk Corn stover Rice straw	–	*Phanerochaete chrysosporium* IBL-03	Lignin peroxidase, manganese peroxidase	Asgher et al. (2011)
Rice bran	–	*Aspergillus niger*	Protease	Ahmed et al. (2011)
	–	*Rhizopus* sp.	Protease	Sumantha et al. (2006)
Wheat bran, oats straw, beetroot press	–	*Trametes versicolor*	Laccase	Ivanka et al. (2010)
Wheat bran	–	*Aspergillus niger*	Protease	Ahmed et al. (2011)
Wheat bran, wheat bran+corn starch, rolled oat	–	*Morchella esculenta*	Laccase	Papinutti and Lechner (2008)
Apple pomace	Chemical (acidic)	*Trichoderma viride*	Endoglucanase, exoglucanase, β-glucosidase	Irshad et al. (2013)
Oil palm empty fruit bunch fibre	Chemical and thermal	*Thermobifida fusca*	Cellulase	Harun et al. (2013)

One of the major drawbacks of using lignocellulosic materials is the lack of efficient conversion technology of biomass into value-added chemicals. Other limitations include high pre-treatment costs, high costs of corrosion-resistant equipment, partial degradation of various components of lignocelluloses, generation of toxic components and low product yield.

21.4.3 Other Value-Added Products

21.4.3.1 Bioplastics/Biocomposites

Due to the complex and diverse nature and content, lignocellulosic raw materials are in demand for their utilization as a source for the synthesis of biocomposites and bioplastics. Based on the type of plant (softwoods, hardwoods, or gramineous species), lignocellulosic materials have different chemical compositions (Govil et al., 2020). These carbohydrates can be molded into various shapes, such as films (Al-Battashi et al., 2019). Cellulose films provide better surface gloss, greater durability, sturdiness, and transparency. Films made from cellulose xanthate, commonly known as cellophane, are frequently used during packaged foods (Isikgor and Becer, 2015). Similarly, hemicellulose films are fragile, however, adding plasticizer compounds with high durability, reduced oxygen permeability, and elasticity makes them suitable for packing purposes. To produce polyhydroxyalkanoate (PHA) and polyhydroxybutyrate (PHB), lignocellulosic components, preferably wastes from food and agricultural waste (due to their great amount and zero value), have been used as feedstock (Li and Wilkins, 2020).

Just a few microbes can effectively synthesize PHAs from cellulose, whereas most microorganisms can only utilize simple sugars. Cellulose is mainly broken down via the acetyl-CoA pathway (Kosseva and Rusbandi, 2018). A marine microbe known as *Saccharophagus degradans* has demonstrated the ability to utilize cellulose and other polysaccharides to accumulate PHA in the cells (Kourmentza et al., 2017). Riley and Alva Munoz investigated the fermentation kinetics of *S. degradans* utilizing unprocessed tequila bagasse for PHA synthesis (Alva Munoz and Riley, 2008). On the other hand, microbes use two alternative paths to metabolize hemicelluloses, like celluloses, to accumulate PHA. Hemicelluloses go through autohydrolysis or thermo-hydrolysis (Brunner, 2014). Given the variety of its primary sugars, hemicellulose autohydrolysis results in a more prominent generation of inhibitors. Additionally, hemicelluloses differ in various feedstocks, this implies both the purification and separation procedure, which significantly raises the whole process cost (Pan et al., 2012). Through the action of microorganisms, bio-valorized lignin has been transformed into substances including lipids, adipic acids, and PHA (Tomizawa et al., 2014). *Pseudomonas* species have been shown to synthesize PHA by decomposing lignin. *Cupriavidus basilensis* increased the original lignin content of 5 g/L to 128 mg/L (Wang et al., 2018). Furthermore, following four rounds of fed-batch fermentation and optimization, 319.4 mg/L of PHA was generated from 6 g/L of lignin. There are many ways to enhance the synthesis of PHA, including promoting cell proliferation, microbial genetic manipulation, increasing the source of carbon, and even pre-treating or hydrolyzing lignin (Shi et al., 2017). In a related investigation, *Pandoraea* sp. B-6 was used by Liu and colleagues to break down lignin and produce PHA. In 4 days, this bacterium decomposed 40% of lignin and produced PHA that was equivalent to conventional bioplastics in terms of biocompatibility and biodegradability. Notably, *Pandoraea* sp. B-6 synthesized PHA from lignin without any prior processing (Liu et al., 2019).

To scale up PHA conversion for imminent industrial consumption of PHA-using material, an adjusted eco-competent mixed microbial culture-dependent process that helps to lower PHA conversion's financial and operating costs relative to pure culture must be used. To tailor the cost and use of PHA-based materials to food utilization requirements as well as transport, mechanical, and economic qualities, the combination of synthesized PHAs and lignocellulosic fibers as bio-composites should continue to be studied (Morya et al., 2021).

21.4.3.2 Biosurfactants

Lignocellulose is an abundant organic carbon source that is highly prevalent. The cellulose is predominantly obtained from the plants, potentially grown for their cellulosic content (Mohanty et al., 2021). The tendency of a microorganism to produce biosurfactant using lignocellulosic media as a substrate has been studied on *Lactobacillus pentosus*, utilizing hydrolyzed distilled grape marc consisting of 10.8% cellulose, 11.2% hemicellulose, and about 51% lignin. Growth media supplemented by yeast extract and corn steep liquor was used. An intercellular biosurfactant production of about 4.8 mg/L was stated for this experimental setup. Likewise, the bacterial strain of *Bacillus tequilensis*, isolated from Mexican brines was used to generate both intracellular and extracellular biosurfactant (Bezerra et al., 2019).

The extensive selection of studies in context to biosurfactant production utilizing lignocellulosic substrates has emphasized the possibility of production of various bio-surfactants using different

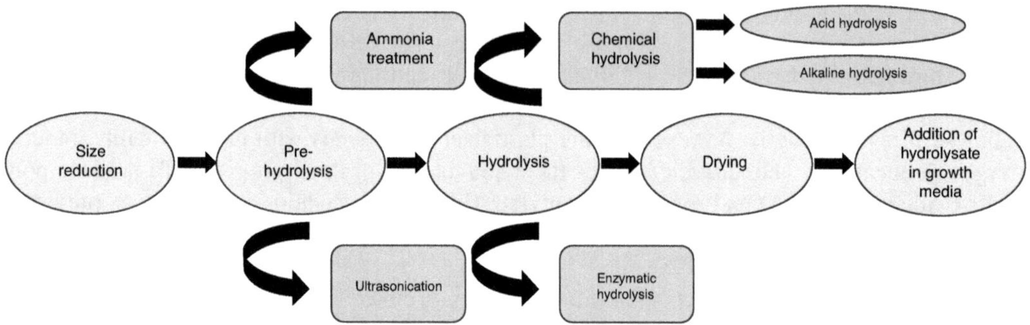

FIGURE 21.2 Pre-treatment technology used during the production of biosurfactants using lignocellulosic biomass.

carbon sources, although some strains might intensify the industrial application potential of the microbial strains and biosurfactants.

Hence, the lignocellulosic substrates are exceptional cost-efficient carbon source for biosurfactant production. Yet, the cost of biosurfactant production has been observed to increase due to the pre-treatment processes required to ensure that the lignocellulosic residues are available for the microbes to act on.

Pre-treatment of lignocellulosic residue includes particle size reduction, pre-hydrolysis, chemical/enzymatic hydrolysis, and drying. The microbe strains that use lignocellulosic residues are *Lactobacillus paracasei*, *Starmerella bombicola*, *C. bombicola*, *Cutaneotrichos poronucleootides* (Joy et al., 2019; Konishi et al., 2015). Pre-treatment of the substrates such as lignin helps in the production of biosurfactants by de-crystallizing the cellulosic structure, reducing the content in the substrate, and increasing the surface area to enhance the enzymatic activity of the enzymes produced by the microbes using the substrate. Figure 21.2 represents the various pre-treatment methods for biosurfactant production from lignocellulosic biomass. The pre-treatment of the substrate biomass makes higher amount of sugar available for the microorganism to act upon.

21.4.3.3 Bioactive Compounds

Pigments are molecules capable of absorbing light in different wavelengths, manifesting a given color. Although synthetic pigments still predominate over their natural counterparts, the demand for natural bio-pigments has been constantly growing. In addition to being used as natural colorants, these compounds, report interesting biological properties, such as antioxidant, antiproliferative, antimicrobial, among other possible activities. Bio-pigments are basically biotechnologically produced coloring molecules.

Combining other carbon sources and lignocellulosic substrates is one strategy to increase yield and pigment production by fermentation (Silva et al., 2021). Glucose, glycerol, and starch, for example, have been added to several agro-industrial residues when formulating the media culture to produce γ-linolenic acid and β-carotene by the strain *Mucor wosnessenskii*.

Bostrycin, an antimicrobial red pigment from *Nigrospora* sp. no. 47 could be produced by submerged fermentation using cane molasses and solid-state fermentation using sugarcane bagasse, rice bran, corn flour and soy meal. Another method is to produce bioactive compounds using hydrolysates. Lignocellulosic hydrolysates from wood are rich in organic carbon sources and, therefore, they can also be used for cultivating microorganisms. In this case, the tree *Fagus sylvatica* was treated by diluted acid (3% H_2SO_4, 100°C, 1 hour), and its isolated components were used as a substrate for growing the microalgae *Chlorella sorokiniana* in mixotrophic conditions to produce carotenoids and fatty acids.

Researchers have investigated decoction extraction (DE) and microwave-assisted extraction of bioactive compounds from lignocellulosic halophyte *Salicornia ramosissima*. The biomass was milled to open the lignocellulosic structure, and particles of 1 mm were obtained. DE was conducted by boiling 300 mg biomass in 10 mL distilled water for 5 minutes and leaving it to cool for 25 minutes. The extract was filtered and freeze-dried into a powder (Silva et al., 2020).

21.5 BIOREFINERY APPROACHES

Lignocellulosic biomass is a prevalent and sustainable feedstock available worldwide. As a source of polysaccharides, cellulose, and hemicellulose, it has the ability to be transformed into biofuels and other platform chemicals (Usmani et al., 2021). It is derived from natural processes or sources being constantly replenished. Sustainable processes of utilizing lignocellulosic biomass to produce bio-based products that obtain "zero concept" waste are to be set up based on recent studies. Hence, to achieve this, the model of biorefineries; lignocellulosic biorefineries has been propositioned (Ana et al., 2010).

The main aim of biorefineries is to transition to sustainable economic systems showing efficient usage of resources, reduction of overall waste generation, in addition to enabling recycling of unavoidable waste as a source to produce new products. Nevertheless, discovering technologies both efficient and sustainable is challenging (Ali et al., 2020; Tišma et al., 2021). There are different pathway techniques to be taken by biorefineries, from feedstock to product, based on the composition and availability of the feedstock, the conversion technologies used, and the production of the desired products.

Although lignocellulosic biomass offers clear value in terms of its green advantages and sustainability, there has been very low commercial success at industrial production levels. This can be accredited to inefficient or complex pre-treatment techniques, complication saccharification techniques, irregular biomass supply chain, and elevated operational and capital expenses caused by scale-up challenges (Adewuyi, 2022; Hassan et al., 2019).

Figure 21.3 shows the common routes to derive value-added bioproducts from lignocellulosic biomass. First-generation biorefineries used feedstocks such as sweet sorghum, beet sugar, sugarcane,

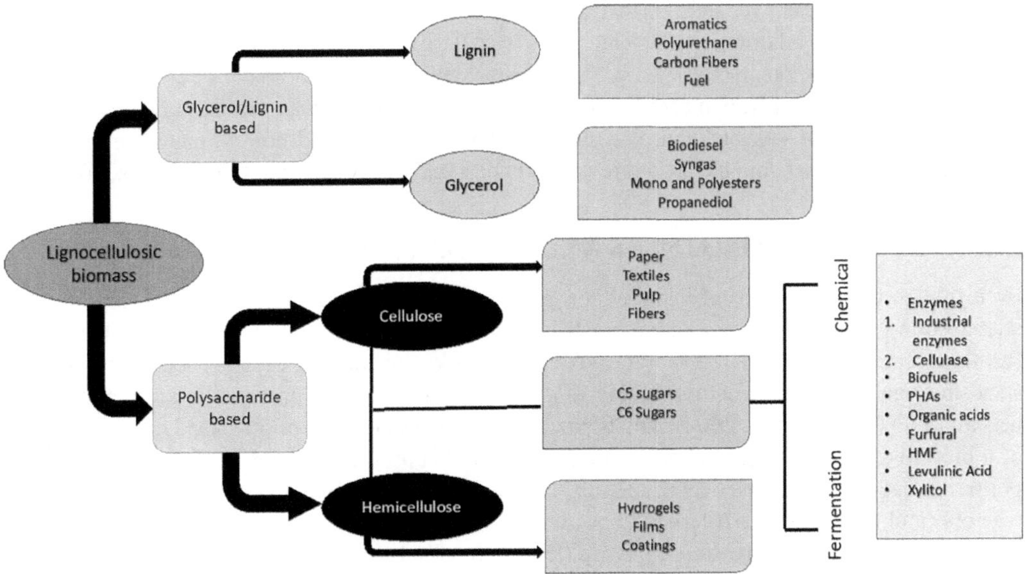

FIGURE 21.3 Valorization of lignocellulosic biomass into value-added bio-products and chemicals.

soybean, rye, barley, wheat, or corn. Sugar extraction from these feedstocks was done by squeezing, steam jet cooking and water-based extraction, etc. Catalytic transformation, in addition to biological methods, was used to treat these extracted sugars into platform chemicals (lactic acid, propionic acid) and biofuels (ethanol, butanol, etc.). In addition, there were several other by-products formed during the process-feed material and food such as corn liquor, oil, corn syrup, dry distillery grains and soluble processed cake, etc. (Usmani et al., 2021).

Applying treatment methods on these products resulted in the production of surfactants, paints and dyes, detergents, adhesives, paper, and biopolymers. However, growing concerns on using food-grade feedstock into biorefineries led to the development of second-generation biorefineries focusing on non-edible portions of the food production value chain. Materials such as agro-industrial waste, municipal solid waste, crop residues, and forestry waste constituted the feedstocks for second-generation biorefineries. These feedstocks could be exposed to an array of thermo-chemical and biochemical processes such as gasification, pyrolysis, torrefaction, and enzymatic hydrolysis to generate biofuels and sugars. Yet these processes brought with them several challenges such as usage of harmful chemicals, extensive energy requirements, making them unsustainable in the long run. There are several other challenges in addition to these such as the viability of the supply chain (from the point of generation of the feedstock until it gets into a processing facility), scale-up of laboratory-scale experiments into commercial-scale processes and the technical maturity of the technologies used in biorefineries.

An integrated biorefinery using waste or residual lignocellulosic biomass together with by-products to produce co-products such as heat, energy, fertilizer, bio-chemicals, etc. might add to the overall economic value of the biorefinery. The cost-benefit of this setup is realized when high-value low-volume co-products are generated in addition to the primary product. This would permit a wide range of feedstocks to be utilized with higher efficiency than their current utilization.

Most obvious and valuable co-products consist of heat, fuel, and electricity. Considering the concept of circular economy, green focus, optimization and bioconversion of lignocellulosic biomass should consist of a systematic enhancement to the performance of microorganisms and enzymes, in addition to being focused on cost and its environmental impact.

Lignocellulosic biorefineries have the potential to become sustainable sources of value-added chemicals and products such as organic acids, PHA, biofuels, biomaterials, etc. at competitive prices. Yet, the inherent recalcitrant nature of lignocellulose adds to the complexity and cost in achieving a full-fledged commercial scale.

Some of the key factors to be considered while establishing techno-economic assessments for industrial scale-up/commercialization should comprise pre-processing, biomass availability, transportation and handling, variable costs (chemical costs, labour, maintenance, and power), plant runtime, waste processing and disposal, and fixed costs (equipment cost and infrastructure) (Usmani et al., 2021).

21.6 TECHNICAL CHALLENGES AND FUTURE PERSPECTIVES

Even though the production of value-added products from lignocellulosic waste has significantly increased in the past few years, there are still some concerns with modern technology that require further research. To develop a comprehensive framework, considerable effort is necessary to understand the lignocellulose structure, laying the groundwork for major progress in sustainable products and fuel (Werpy et al., 2004). The lignocellulosic substrate constraints based on the definition of a biorefinery are abundant choice and a year-round supply of appropriate feedstock, effective pre-treatment procedure, the expense of the procedure itself including the cost of the infrastructure, reactors, and reagents, etc. Another important factor is toxic substances released during various treatments, and post-pre-treatment procedures, such as substrate washing, process-related waste produced, and related ecological dangers (Tan et al., 2021).

Consequently, while selecting various pre-treatment methods, it should adhere to the following conditions, i.e., ensure the effective breakdown of the three-dimensional lignocellulose structures with reduced cellulose crystallinity. This will contribute to increasing the porosity and surface area of lignocellulosic feedstocks to increase the accessibility of the enzymes which will result in better hydrolysis (Everard, 2020). Pre-treatment generates high sugar yields with reduced formation of fermentation-inhibitory chemicals such as furfural, acetic acid, hydroxymethyl furfurals, and phenolic substances. The expense of operations after pre-treatment techniques including washing, pulp separation, and neutralizing must be reduced or deleted if possible. The reactor design should be straightforward, modest in size, affordable, and high capacity for loading solids requiring low chemical and water inputs (Mahmood et al., 2019). This will subsequently increase the process's overall sustainability. To enhance the yield and effectiveness of pre-treatments, advancements in fermenting strategies and media optimization must be made in addition to genetic engineering methods. It is necessary to use synthetic biology and metabolic engineering techniques to create reliable and effective microorganisms for various fermentation procedures. Merging technologies for lowering energy requirements, reusing process streams, and minimizing the number of processing stages to make the conversion process more cost-effective (Yoo et al., 2020).

Due to the absence of integrated process designs, improvements in the cost-effective conversion of lignocellulosic materials are frequently challenging for precise quantification. Despite recent predictions, there is still a great deal of uncertainty about how well second-generation biofuel will work on a large scale given many sustainability and ecological considerations. Massive R&D efforts are enhancing the conversion of lignocellulose to value-added products, but political risks still exist over the accessibility of substrate and the reliability of revenue. Various elements still need to be correctly put together and optimized before an effective industrial configuration is attained by maintaining a clear perspective (Kumar et al., 2020).

On the other hand, even though globally, biorefineries are being established for environmentally sustainable synthesis of various high-end products, maintaining socio-economic-environmental viability is one of the biggest issues facing biorefineries in the future. However, few bio-based enterprises are now active or in the trial stage, and the idea of biorefineries is already gaining appeal. Presently, the biorefinery industries encounter several difficulties that can be largely divided into biomass-related and process-related difficulties (Dietrich et al., 2019). The major difficulties for each category are displayed in Figure 21.4. The economic sustainability of biorefineries is usually a challenging task. At all phases, including lab-, pilot-, and industrialization stages, cutting-edge research, and state-of-the-art technologies must be created and deployed, which may call for huge investments from the public sector,

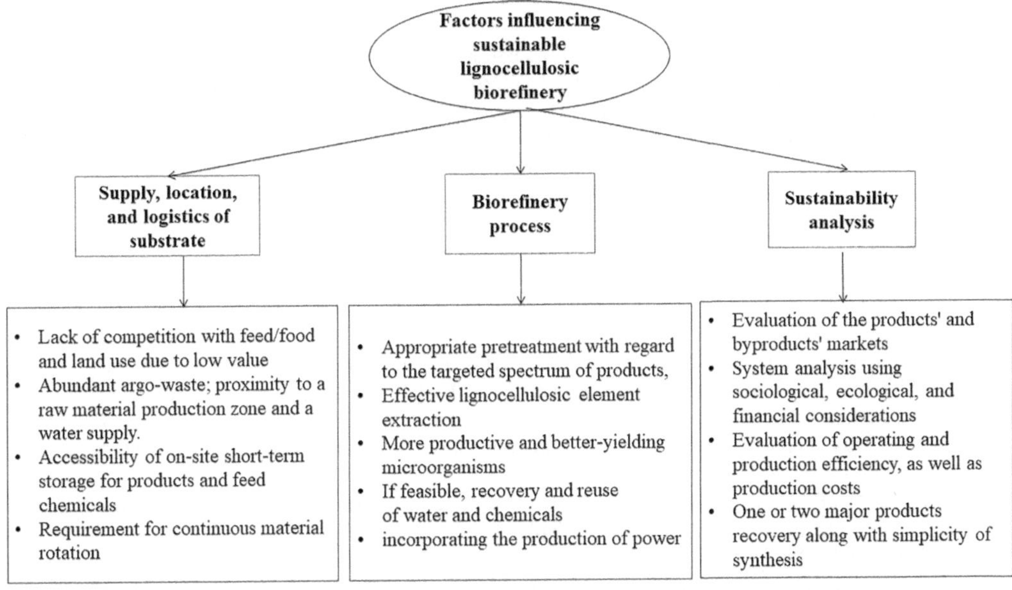

FIGURE 21.4 Several crucial factors to consider while developing a sustainable lignocellulosic biorefinery.

the corporate industry, and academics. To evaluate and validate emerging technologies, continuous and committed efforts in research and development are required. In a lifecycle-based study, there is an obvious requirement for effective modeling, assessing, and evaluating sustainability consequences.

21.7 CONCLUSIONS

The increase in energy and environment problems have led the researchers to focus on utilization of the alternate resources such as lignocellulosic biomass which has an immense potential to meet the energy demands of today's world. It will also help in reducing the excessive dependence on non-renewable fossil resources for liquid fuels. Lignocellulosic biomass has been studied for the production of biofuels, biochemicals, organic acids, enzymes, biomaterials, biosurfactants and bioactive compounds. However, advanced technologies are needed for the efficient and effective conversion of lignocellulosic materials, especially in terms of existing pre-treatment, reduction in associated costs and decrease the toxic/inhibitory products formation. In addition, various studies have been conducted at lab-scale and preliminary levels, and there is a large gap between the laboratory results and the industrial scale application that needs to be closed to obtain the environmental and economic feasibility of the conversion process.

REFERENCES

Adapa, P., Tabil, L., Schoenau, G. (2009). Compaction characteristics of barley, canola, oat and wheat straw. *Biosystems Engineering*, 104(3), 335–344.

Adewuyi, A. (2022). Underutilized lignocellulosic waste as sources of feedstock for biofuel production in developing countries. *Frontiers in Energy Research*, 10, 741570.

Ahmed, I., Zia, M.A., Iftikhar, T., Iqbal, H.M. (2011). Characterization and detergent compatibility of purified protease produced from Aspergillus niger by utilizing agro wastes. *BioResources*, 6(4), 4505–4522.

Al-Battashi, H.S., Annamalai, N., Sivakumar, N., Al-Bahry, S., Tripathi, B.N., Nguyen, Q.D., Gupta, V.K. (2019). Lignocellulosic biomass (LCB): a potential alternative biorefinery feedstock for polyhydroxyalkanoates production. *Reviews in Environmental Science and Bio/Technology*, 18(1), 183–205.

Ali, N., Zhang, Q., Liu, Z.Y., Li, F.L., Lu, M., Fang, X. C. (2020). Emerging technologies for the pretreatment of lignocellulosic materials for bio-based products. *Applied Microbiology and Biotechnology*, 104(2), 455–473.

Alva Munoz, L.E., Riley, M.R. (2008). Utilization of cellulosic waste from tequila bagasse and production of polyhydroxyalkanoate (PHA) bioplastics by Saccharophagus degradans. *Biotechnology and Bioengineering*, 100(5), 882–888.

Anwar, Z., Gulfraz, M., Irshad, M. (2014). Agro-industrial lignocellulosic biomass a key to unlock the future bio-energy: a brief review. *Journal of Radiation Research and Applied Sciences*, 7(2), 163–173.

Asgher, M., Ahmed, N., Iqbal, H.M.N. (2011). Hyperproductivity of extracellular enzymes from indigenous white rot fungi (Phanerochaete chrysosporium) by utilizing agro-wastes. *BioResources*, 6(4), 4454–4467.

Asgher, M., Iqbal, H.M.N., Irshad, M. (2012). Characterization of purified and xerogel immobilized novel lignin peroxidase produced from Trametes versicolor IBL-04 using solid state medium of corncobs. *BMC Biotechnology*, 12(1), 1–8.

Batista Meneses, D., Montes de Oca-Vásquez, G., Vega-Baudrit, J.R., Rojas-Álvarez, M., Corrales-Castillo, J., Murillo-Araya, L.C. (2020). Pretreatment methods of lignocellulosic wastes into value-added products: recent advances and possibilities. *Biomass Conversion and Biorefinery*, 12, 1–18.

Bezerra, K.G., Gomes, U.V., Silva, R.O., Sarubbo, L.A., & Ribeiro, E. (2019). The potential application of biosurfactant produced by Pseudomonas aeruginosa TGC01 using crude glycerol on the enzymatic hydrolysis of lignocellulosic material. *Biodegradation*, 30, 351–361.

Bhatia, S.K., Jagtap, S.S., Bedekar, A.A., Bhatia, R.K., Patel, A.K., Pant, D., ... Yang, Y.H. (2020). Recent developments in pretreatment technologies on lignocellulosic biomass: effect of key parameters, technological improvements, and challenges. *Bioresource Technology*, 300, 122724.

Brunner, G. (2014). Processing of biomass with hydrothermal and supercritical water. In: Erdogan Kiran (ed) *Supercritical fluid science and technology*, Vol. 5. Elsevier, pp. 395–509.

Chavan, S., Yadav, B., Atmakuri, A., Tyagi, R., Wong, J.W., Drogui, P. (2022). Bioconversion of organic wastes into value-added products: A review. *Bioresource Technology*, 344, 126398.

Dietrich, K., Dumont, M.-J., Del Rio, L.F., Orsat, V. (2019). Sustainable PHA production in integrated lignocellulose biorefineries. *New Biotechnology*, 49, 161–168.

Everard, M. (2020). Twenty years of the polyvinyl chloride sustainability challenges. *Journal of Vinyl and Additive Technology*, 26(3), 390–402.

Govil, T., Wang, J., Samanta, D., David, A., Tripathi, A., Rauniyar, S., Salem, D.R., Sani, R.K. (2020). Lignocellulosic feedstock: A review of a sustainable platform for cleaner production of nature's plastics. *Journal of Cleaner Production*, 270, 122521.

Hafiz Muhammad Nasir, I., Ishtiaq, A., Muhammad Anjum, Z., Muhammad, I. (2011). Purification and characterization of the kinetic parameters of cellulase produced from wheat straw by Trichoderma viride under SSF and its detergent compatibility. *Advances in Bioscience and Biotechnology*, 2(03), 149–156.

Harun, N.A.F., Baharuddin, A.S., Zainudin, M.H.M., Bahrin, E.K., Naim, M.N., Zakaria, R. (2013). Cellulase production from treated oil palm empty fruit bunch degradation by locally isolated Thermobifida fusca. *BioResources*, 8(1), 676–687.

Hassan, S.S., Williams, G.A., Jaiswal, A.K. (2019). Lignocellulosic biorefineries in Europe: current state and prospects. *Trends in Biotechnology*, 37(3), 231–234.

Howard, R., Abotsi, E., Van Rensburg, E.J., Howard, S. (2003). Lignocellulose biotechnology: issues of bioconversion and enzyme production. *African Journal of Biotechnology*, 2(12), 602–619.

Iqbal, H.M.N., Kyazze, G., Keshavarz, T. (2013). Advances in the valorization of lignocellulosic materials by biotechnology: an overview. *BioResources*, 8(2), 3157–3176.

Irshad, M.N., Anwar, Z., But, H.I., Afroz, A., Ikram, N., Rashid, U. (2013). The industrial applicability of purified cellulase complex indigenously produced by Trichoderma viride through solid-state bio-processing of agro-industrial and municipal paper wastes. *BioResources*, 8(1), 145–157.

Isikgor, F.H., Becer, C.R. (2015). Lignocellulosic biomass: a sustainable platform for the production of bio-based chemicals and polymers. *Polymer Chemistry*, 6(25), 4497–4559.

Ivanka, S., Albert, K., Veselin, S. (2010). Properties of crude laccase from Trametes versicolor produced by solid-substrate fermentation. *Advances in Bioscience and Biotechnology*, 2010.

Joy, S., Rahman, P.K., Khare, S.K., Sharma, S. (2019). Production and characterization of glycolipid biosurfactant from Achromobacter sp.(PS1) isolate using one-factor-at-a-time (OFAT) approach with feasible utilization of ammonia-soaked lignocellulosic pretreated residues. *Bioprocess and Biosystems Engineering*, 42, 1301–1315.

Khan, I.U., Othman, M.H.D., Hashim, H., Matsuura, T., Ismail, A., Rezaei-DashtArzhandi, M., Azelee, I.W. (2017). Biogas as a renewable energy fuel-A review of biogas upgrading, utilisation and storage. *Energy Conversion and Management*, 150, 277–294.

Konishi, M., Yoshida, Y., Horiuchi, J. I. (2015). Efficient production of sophorolipids by Starmerella bombicola using a corncob hydrolysate medium. *Journal of Bioscience and Bioengineering*, 119(3), 317–322.

Kosseva, M.R., Rusbandi, E. (2018). Trends in the biomanufacture of polyhydroxyalkanoates with focus on downstream processing. *International Journal of Biological Macromolecules*, 107, 762–778.

Kourmentza, C., Plácido, J., Venetsaneas, N., Burniol-Figols, A., Varrone, C., Gavala, H.N., Reis, M.A. (2017). Recent advances and challenges towards sustainable polyhydroxyalkanoate (PHA) production. *Bioengineering*, 4(2), 55.

Kumar, M., Rathour, R., Singh, R., Sun, Y., Pandey, A., Gnansounou, E., Lin, K.-Y.A., Tsang, D.C., Thakur, I.S. (2020). Bacterial polyhydroxyalkanoates: Opportunities, challenges, and prospects. *Journal of Cleaner Production*, 263, 121500.

Li, M., Wilkins, M.R. (2020). Recent advances in polyhydroxyalkanoate production: feedstocks, strains and process developments. *International Journal of Biological Macromolecules*, 156, 691–703.

Liu, D., Yan, X., Si, M., Deng, X., Min, X., Shi, Y., Chai, L. (2019). Bioconversion of lignin into bioplastics by Pandoraea sp. B-6: molecular mechanism. *Environmental Science and Pollution Research*, 26(3), 2761–2770.

Mahmood, H., Moniruzzaman, M., Iqbal, T., Khan, M.J. (2019). Recent advances in the pretreatment of lignocellulosic biomass for biofuels and value-added products. *Current Opinion in Green and Sustainable Chemistry*, 20, 18–24.

Mohan, D., Pittman Jr, C.U., Steele, P.H. (2006). Pyrolysis of wood/biomass for bio-oil: a critical review. *Energy & Fuels*, 20(3), 848–889.

Mohanty, S.S., Koul, Y., Varjani, S., Pandey, A., Ngo, H.H., Chang, J.S., ... Bui, X.T. (2021). A critical review on various feedstocks as sustainable substrates for biosurfactants production: a way towards cleaner production. *Microbial Cell Factories*, 20(1), 120.

Morya, R., Sharma, A., Kumar, M., Tyagi, B., Singh, S.S., Thakur, I.S. (2021). Polyhydroxyalkanoate synthesis and characterization: A proteogenomic and process optimization study for biovalorization of industrial lignin. *Bioresource Technology*, 320, 124439.

Namnuch, N., Thammasittirong, A., Thammasittirong, S.N.-R. (2021). Lignocellulose hydrolytic enzymes production by Aspergillus flavus KUB2 using submerged fermentation of sugarcane bagasse waste. *Mycology*, 12(2), 119–127.

Nanda, S., Azargohar, R., Dalai, A.K., Kozinski, J.A. (2015). An assessment on the sustainability of lignocellulosic biomass for biorefining. *Renewable and Sustainable Energy Reviews*, 50, 925–941.

Nanda, S., Dalai, A.K., Kozinski, J.A. (2017). Butanol from renewable biomass: highlights of downstream processing and recovery techniques. In: Prasenjit Mondal, Ajay K. Dalai (eds) *Sustainable utilization of natural resources*. CRC Press, pp. 187–211.

Nanda, S., Mohanty, P., Pant, K.K., Naik, S., Kozinski, J.A., Dalai, A.K. (2013). Characterization of North American lignocellulosic biomass and biochars in terms of their candidacy for alternate renewable fuels. *Bioenergy Research*, 6(2), 663–677.

Nanda, S., Reddy, S.N., Mitra, S.K., Kozinski, J.A. (2016). The progressive routes for carbon capture and sequestration. *Energy Science & Engineering*, 4(2), 99–122.

Okolie, J.A., Nanda, S., Dalai, A.K., Kozinski, J.A. (2021). Chemistry and specialty industrial applications of lignocellulosic biomass. *Waste and Biomass Valorization*, 12(5), 2145–2169.

Pan, W., Perrotta, J.A., Stipanovic, A.J., Nomura, C.T., Nakas, J.P. (2012). Production of polyhydroxyalkanoates by Burkholderia cepacia ATCC 17759 using a detoxified sugar maple hemicellulosic hydrolysate. *Journal of Industrial Microbiology and Biotechnology*, 39(3), 459–469.

Papinutti, L., Lechner, B. (2008). Influence of the carbon source on the growth and lignocellulolytic enzyme production by Morchella esculenta strains. *Journal of Industrial Microbiology and Biotechnology*, 35(12), 1715–1721.

Sánchez, C. (2009). Lignocellulosic residues: biodegradation and bioconversion by fungi. *Biotechnology Advances*, 27(2), 185–194.

Shi, Y., Yan, X., Li, Q., Wang, X., Xie, S., Chai, L., Yuan, J. (2017). Directed bioconversion of Kraft lignin to polyhydroxyalkanoate by Cupriavidus basilensis B-8 without any pretreatment. *Process Biochemistry*, 52, 238–242.

Silva, A.M., Lago, J.P., Pinto, D., Moreira, M.M., Grosso, C., Cruz Fernandes, V., ... Rodrigues, F. (2021). Salicornia ramosissima bioactive composition and safety: Eco-friendly extractions approach (microwave-assisted extraction vs. conventional maceration). *Applied Sciences*, 11(11), 4744.

Sumantha, A., Deepa, P., Sandhya, C., Szakacs, G., Soccol, C.R., Pandey, A. (2006). Rice bran as a substrate for proteolytic enzyme production. *Brazilian Archives of Biology and Technology*, 49, 843–851.

Tan, D., Wang, Y., Tong, Y., Chen, G.-Q. (2021). Grand challenges for industrializing polyhydroxyalkanoates (PHAs). *Trends in Biotechnology*, 39(9), 953–963.

Tišma, M., Bucić-Kojić, M., & Planinić, M. (2021). Bio-based products from lignocellulosic waste biomass: A state of the art. *Chemical and Biochemical Engineering Quarterly*, 35(2), 139–156.

Tomizawa, S., Chuah, J.-A., Matsumoto, K., Doi, Y., Numata, K. (2014). Understanding the limitations in the biosynthesis of polyhydroxyalkanoate (PHA) from lignin derivatives. *ACS Sustainable Chemistry & Engineering*, 2(5), 1106–1113.

Usmani, Z., Sharma, M., Awasthi, A.K., Lukk, T., Tuohy, M.G., Gong, L., ... Gupta, V.K. (2021). Lignocellulosic biorefineries: the current state of challenges and strategies for efficient commercialization. *Renewable and Sustainable Energy Reviews*, 148, 111258.

Valladares-Diestra, K.K., de Souza Vandenberghe, L.P., Soccol, C.R. (2021). A biorefinery approach for enzymatic complex production for the synthesis of xylooligosaccharides from sugarcane bagasse. *Bioresource Technology*, 333, 125174.

Vasaki, M., Sithan, M., Ravindran, G., Paramasivan, B., Ekambaram, G., Karri, R.R. (2022). Biodiesel production from lignocellulosic biomass using Yarrowia lipolytica. *Energy Conversion and Management*, 13, 100167.

Wahab, W.A.A., Ahmed, S.A., Kholif, A., Abd El Ghani, S., Wehaidy, H.R. (2021). Rice straw and orange peel wastes as cheap and eco-friendly substrates: A new approach in β-galactosidase (lactase) enzyme production by the new isolate L. paracasei MK852178 to produce low-lactose yogurt for lactose-intolerant people. *Waste Management*, 131, 403–411.

Wang, X., Lin, L., Dong, J., Ling, J., Wang, W., Wang, H., Zhang, Z., Yu, X. (2018). Simultaneous improvements of Pseudomonas cell growth and polyhydroxyalkanoate production from a lignin derivative for lignin-consolidated bioprocessing. *Applied and Environmental Microbiology*, 84(18), e01469–18.

Werpy, T., Petersen, G., Aden, A., Bozell, J., Holladay, J., White, J., Manheim, A., Eliot, D., Lasure, L., Jones, S. (2004). *Top value added chemicals from biomass, Volume 1: Results of screening for potential candidates from sugars and synthesis gas*. US Department of Energy. Pacific Northwest National Laboratory/US Department of Energy: Oak Ridge, TN.

Yoo, C.G., Meng, X., Pu, Y., Ragauskas, A.J. (2020). The critical role of lignin in lignocellulosic biomass conversion and recent pretreatment strategies: A comprehensive review. *Bioresource Technology*, 301, 122784.

Yoon, L.W., Ngoh, G.C., Chua, A.S.M. (2012). Simultaneous production of cellulase and reducing sugar from alkali-pretreated sugarcane bagasse via solid state fermentation. *BioResources*, 7(4), 5319–5332.

22 Agro-residue Wastes as a Key to Unlocking Future Bio-energy Potentials

Aditi Roy, Priya Dubey, Ekta Gupta, Anju Patel, Suchi Srivastava, and Pankaj Kumar Srivastava

22.1 INTRODUCTION

The advancement in the past two decades has elevated primary energy consumption up to 38% (Singh et al., 2022). Majority of developing countries generate huge amounts of energy through non-renewable fossil fuels. This has led to an upsurge in GHGs and CO_2 emissions. Also, the limitations in fossil fuel reserves have shown a paradigm shift towards the renewable energy sector. The harmful environmental consequences of using conventional energy resources have raised international concern for utilizing renewable energy resources. There has been constant pressure to find new energy supply choices, particularly those based on renewable sources like bioenergy due to an overall increase in energy costs, a forecast future shortage of fossil fuels and concerns about climate change. Bioenergy derived from biomass provides a handy means of energy storage because of its carbon-neutral property. By boosting domestic agricultural development, lowering import costs and enhancing energy self-sufficiency, it promotes regional growth. According to the geographic region or origin from which it is gathered, agro-industrial residue is widely regarded as a rich source of energy. They are a good choice for bioenergy because they are inexpensive and readily available in huge quantities.

Agro-residue forms the left-over materials after crop harvest like straw, stalks, leaves, fibrous materials, husk and bagasse, and other components that are varied sizes, shapes, forms, and densities (Lee et al., 2019) (Figure 22.1). Agricultural biomass residues like corn stalks, rice husks, corn stover, wheat straw, cropping residues, unmarketable products, vegetable processing leftovers, sugar beet bagasse, tomato waste, cardoon waste, and vegetable waste provide better alternatives for bioenergy production (Gupta and Verma, 2015). In current times, the energy obtained from these biomass sources can be exploited to power vehicles and other machinery. This marks a better eco-friendly alternative as compared to conventional fossil fuels by reducing CO_2 emissions (Saleem et al., 2017). The generation of biogas and liquid fuels for transportation, power generation, and cooking are among the most effective uses of biomass that are anticipated to become more prevalent in the near future (Chanakya and Malayil, 2012). Utilizing biomass wastes, especially for the creation of non-food-centred biofuels, is seen to be an effective way to diminish harmful effects of climate change. As a result, studies are being conducted to determine how to turn lignocellulose, which is easily present in non-edible plant parts, especially in agricultural wastes into bioenergy resulting from biofuels (Trivedi, 2020). However, in order to realise the potential for bioenergy generation it is necessary to maximise mineral fertilisation in accordance with agronomic, environmental, and commercially viable standards while preserving sustainable crop production. Apart from this, it is important to determine the ideal withdrawal level for crop residues which guarantees a balance between the needs for both continuing crop production and bioenergy (Anwar et al., 2014). Moreover, agro-residual biomass made of lignocellulosic waste is a cheap,

DOI: 10.1201/9781003441144-22

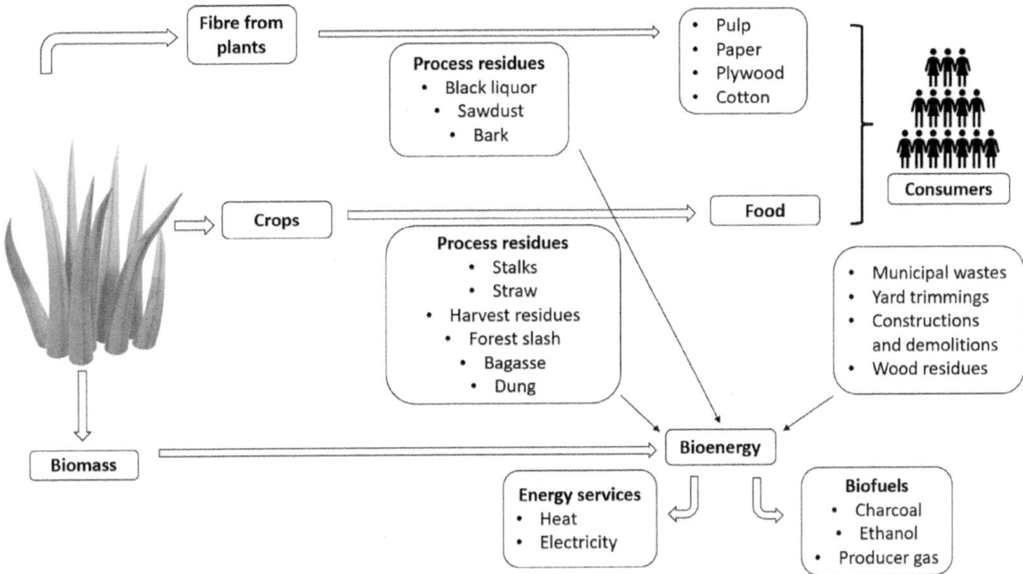

FIGURE 22.1 A comprehensive pathway for obtaining bioenergy from biomass.

Source: Modified from Zabaniotou, 2018).

plentiful, renewable natural resource that offers a special natural resource for efficient and successful large-scale bio-energy collection. However, most of this lignocellulosic waste is disposed through burning, thus putting harmful effects on environment. Currently, it is being researched as a powerful biofuel, value-added fine chemicals, and inexpensive energy sources for microbial fermentation and enzyme production.

22.1.1 Overview of Bioenergy

A renewable energy source that can take the place of fossil fuels is referred to as bioenergy. Energy crops, biomass leftovers from forestry and agriculture, trash and by-products from the agro-industry, pulp and paper industry, moist organic wastes, and the organic portion of municipal solid wastes are just a few examples of the many biological resources that can be used to create bioenergy. Future biological resources are also viewed as being very attractive and include macroalgae, microalgae, seaweeds, and aquatic plants (OECD/IEA, 2017). Regardless of resources, bioenergy can be used as a reliable power resource to provide heat, gas, and fuel for cogeneration, transportation fuels, and electricity. Recently, the manufacturing of biomaterials and bioproducts has also been possible using biomass (Reid et al., 2020). If alternative and renewable energy fuels are employed for essential energy demands, the current energy crisis can be solved. Many nations have started moving in this direction after coming to this realisation. Because the transportation industry is a significant user of primary energy and a significant contributor to greenhouse gas emissions, industrialised and developing nations have devised biofuels policies for large-scale production of biofuels to address this issue (Ullah et al., 2015).

22.2 BIOMASS

Humanity has experimented with a variety of energy sources during the course of its evolution, including wood, coal, oil, and nuclear energy. Recent years have seen the promotion and advancement in renewable energy resources due to various emerging environmental and energy security

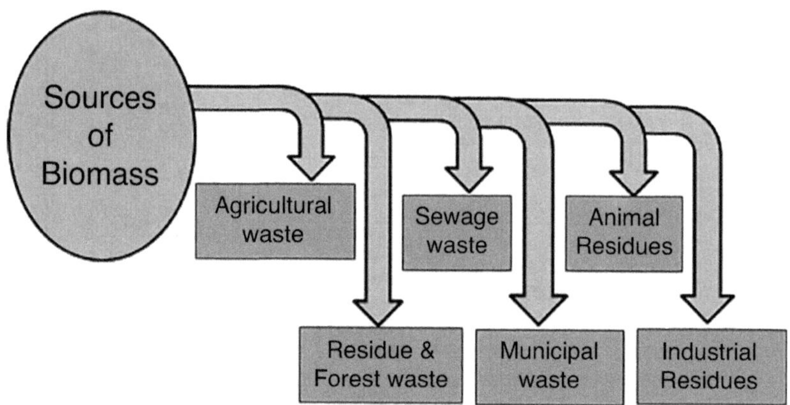

FIGURE 22.2 Sources of biomass.

Source: Modified from Kumar et al. (2015).

issues. One such resource that might contribute significantly to a more varied and sustainable energy mix is biomass. In contrast to using fossil fuels, biomass forms renewable type of energy source. To be more specific, biomass is the organic material derived from plants and animals. Biomass has been utilised as a fuel ever since mankind first used fire to warm themselves or make food thousands of years ago. Approximately 220 billion dry tonnes of biomass are produced annually through photosynthesis with a 1% conversion efficiency worldwide (Umar et al., 2015).

22.2.1 Source of Biomass

Sources of biomass require any organic substance derived from living things (Mafakheri and Nasiri, 2014). As shown in Figure 22.2, it involves plant and animal components as well as leftovers from agricultural and forestry activities, such as wood from forests, crops, and seaweed, along with organic industrial, human, municipal, and animal wastes (Saidur et al., 2011).

Biomass energy is not a form of erratic energy like solar and wind energy. The biomass energy stores the carbon through photosynthesis for as long as the sun is visible. In actuality, solar energy may be captured biologically via biomass technologies. As a replacement for fossil fuels, producing industrial crops for energy is becoming more and more popular nowadays. However, the overall amount of energy from industrial crops is insufficient when compared to the volume of fossil fuel production. Municipal Solid Waste (MSW) is a biomass resource, and the quantity of energy from waste (EfW) relies on the population of people and the composition of the waste materials. EfW plants can produce power from MSW as part of a waste management system, and they provide trash disposal with little impact on the environment (Saleem, 2022). Direct combustion, pyrolysis, fermentation, gasification, and anaerobic digestion are five most popular processes for converting biomass into energy. However, the choice of process is subjected to a variety of elements, including the type and quantity of biomass, environmental regulations, and economic considerations (Saidur et al., 2011).

22.2.2 Biomass Energy Conversion Technologies

The prospective of biomass in India makes it evident that there are a variety of feedstocks available for use in both power generating and biofuel conversion processes. There are many different procedures for converting biomass, and they vary depending on factors like type and quantity of biomass feedstock, surrounding environment, and economy. Two primary process technologies

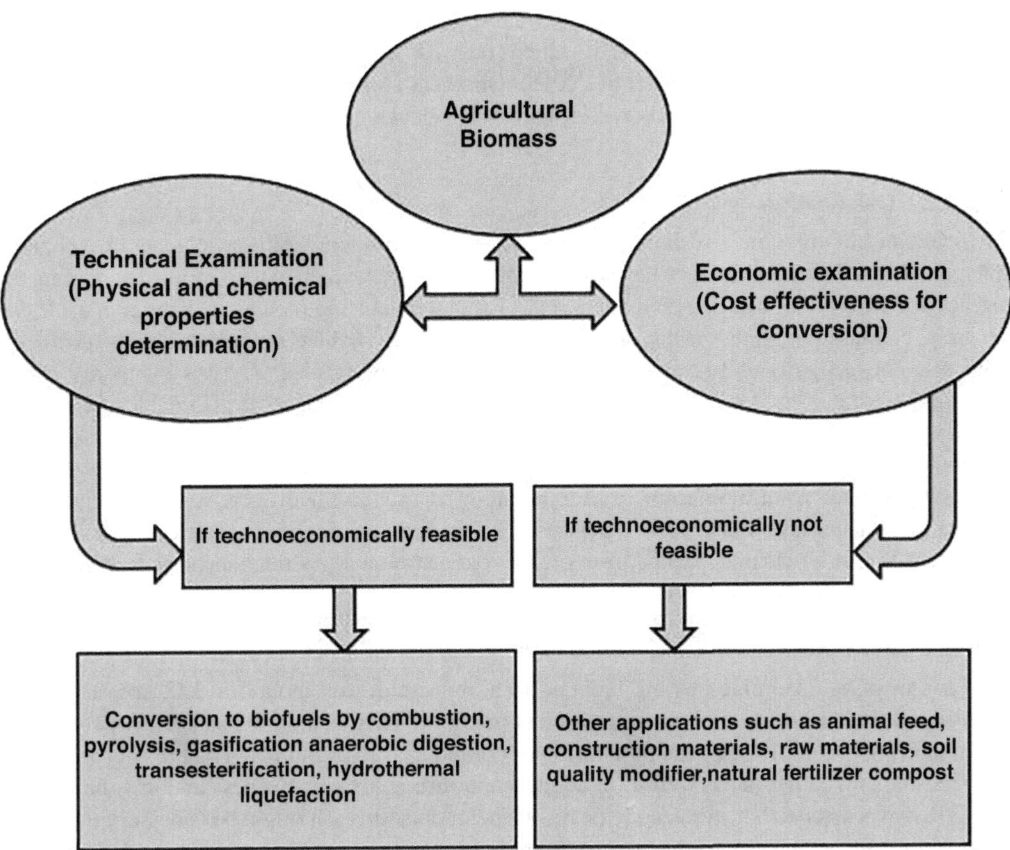

FIGURE 22.3 Overview of the biofuel production process using agricultural biomass through techno-economic analysis.

Source: Modified from Saleem et al. (2017).

like thermo-chemical and bio-chemical or biological are utilised to transform biomass into energy (Figure 22.3). The third approach for generating energy from biomass includes mechanical extraction (with esterification), such as rapeseed methyl ester (RME) biodiesel. The thermal conversion procedures include combustion, liquefaction, biomass gasification, and pyrolysis (Mohan et al., 2006).

22.2.3 Thermo-chemical Conversion Process

Combustion, gasification, and pyrolysis form the three basic thermochemical conversion methods utilised for biomass (Katyal, 2007).

22.2.3.1 Combustion

Combustion of biomass involves aerobic burning that is applied in various equipment like furnaces, stoves, steam turbines, boilers, etc., in order to transform chemical energy stored into heat energy, mechanical power, and electricity. For any combustion process, pre-dried biomass with a maximum moisture content of around 50% is used (Sharma et al., 2014). The size of a combustion plant can be extremely tiny (such as for home heating) or large-scale (between 100 and 3,000 MW). For biomass combustion power plants, net bioenergy conversion efficacies should be in the range from 20% to 40% (Devi et al., 2017). The high conversion efficiency of coal-fired power plants makes the co-combustion of biomass in these facilities a particularly appealing alternative. Maximum efficiencies can be observed in systems over 100 MWe or when biomass is co-combusted in coal-fired power plants (Kumar et al., 2015). One heat engine Stirling cycle directly generates shaft power through combustion; however, the cycle's development is currently restricted to low power outputs.

22.2.3.2 Gasification

Gasification is the fractional oxidation of biomass at maximum temperatures to generate a combustible gas fusion. This resultant gas has low calorific value (CV) which can be utilised as fuel for gas engines and turbines or can be applied as syngas for manufacturing methanol (Kumar et al., 2015). Recently, biomass integrated gasification/combination cycle (BIG/CC), which uses gas turbines to efficiently convert gaseous fuel into electricity, is one intriguing idea. These systems are able to clean gas before it is burned in turbine, thus allowing the adoption for more affordable and compact gas-cleaning equipment. High conversion efficiency is ensured by the combination of gasification and combustion, which results in net efficiencies of 40%–50% for a plant with a capacity of 30–60 MW. Syngas made from biomass is used to create methanol and hydrogen, two fuels that can be utilised for transportation and other purposes. The production of higher value CV gases (usually 9–11 MJ14N m3) by various procedures, such as oxygen blown gasification or hydrogen indirect gasification, is preferable for the synthesis of methanol.

22.2.3.3 Pyrolysis

Pyrolysis involves anaerobic heating biomass to a temperature of about 500 LC, converting it to solid, liquid-like bio-oil or biocrude, or gaseous fractions. This technique has been widely used in manufacturing bio-oil through the transformation of biomass to bio-crude. The bio-oil can be used as a feedstock for refineries as well as in engines and turbines. Corrosiveness and low thermal stability are two problems that still need to be resolved, among others. It might be necessary to upgrade bio-oils for some applications by lowering their oxygen content and removing their alkalis through hydrogenation and catalytic cracking (Mohan et al., 2006).

22.2.4 Bio-chemical Conversion

In addition to two primary processes of fermentation and anaerobic digestion, bio-chemical conversion process forms the third, less common process centred on mechanical extraction and chemical conversion.

22.2.4.1 Fermentation

Commercially, ethanol is generated through the fermentation of sugar crops such as sugar cane and starch crops e.g. maize, wheat. The crop biomass is fragmented which helps in the conversion of starch to sugars by enzymes. Finally, this sugar product is converted to ethanol using yeast additives. With around 450L of ethanol produced from 1,000kg of dry maize, the distillation process is an energy-intensive phase in its purification. On the safer side, the solid waste product generated can be fed to cattle, and obtained bagasse can be utilised for subsequent gasification or as boiler fuel (Kumar et al., 2015). Because lignocellulosic biomass (such as wood and grasses) has longer-chain polysaccharide molecules, transformation is more difficult and requires acid or enzymatic hydrolysis afore the resulting sugars can be fermented to ethanol.

22.2.4.2 Anaerobic Digestion

Anaerobic digestion (AD) process rapidly converts organic material to biogas. The main gases in the combination are methane and carbon dioxide, with traces of other gases like hydrogen sulphide. Bacteria break down biomass in an anaerobic environment, generating a gas with energy equivalent to 20%–40% of lower heating value of material (Dar et al., 2021). For the treatment of organic wastes with high moisture contents, such as those with 80%–90% moisture, AD is an extensively used and commercially tested method. Direct use of biogas in spark ignition gas engines (s.i.g. e.) and gas turbines is possible. By removing CO_2, biogas can be converted to a better quality, similar to natural gas quality. The potential conversion efficiency is 21% overall (Zhou et al., 2017). Any power generation system using an internal combustion engine as the primary mover could utilise a combined heat and power system to collect waste heat from the engine's oil, water, and exhaust cooling systems and from the exhaust.

22.2.4.3 Mechanical Extraction

Oil is created through the mechanical conversion process of extraction from the seeds of different biomass crops including cotton, groundnuts, and others. Along with oil, the process leaves behind a solid residue or "cake" that can be fed to animals. Around three tonnes of rapeseed are required for every tonne of rapeseed oil produced. Rapeseed oil and alcohol can be used in the process of esterification to further process oil and create (Kumar et al., 2015).

22.3 BIOFUELS

Liquid or gaseous fuels obtained from agricultural biomass are usually referred to as "biofuels." According to the biomass resources that were used to convert them into biofuels, they are separated into first-generation and second-generation biofuels. Given that the world's oil production is about to peak, billions of tonnes of carbon emissions have been released into the atmosphere, and there are threats from climatic change, it should go without saying that the scientific community should pay special attention to clean energy. This is especially true in the context of developing countries. At the moment, depleting fuels like coal and nuclear energy, as well as carbon-rich fossil fuels including oil (35%), natural gas (21%), and nuclear energy account for roughly 87% of the world's energy mix (International Energy Agency, 2007). At the current rate of consumption, the economically recoverable proven reserves of oil, natural gas, and coal, respectively, represented 41.6, 60.3, and 133 years of supply at the end of 2007, according to the OECD/IEA. A simple calculation suggests that these verified reserves will be completely consumed after 75 years at the current rate of fossil fuel usage due to rising global energy demand. As a result, conventional oil production 1 could reach its peak in 20 years, followed by natural gas and coal (Ullah et al., 2015). Growing greenhouse gas emissions from fossil fuels combustion increases the risk of global warming and ocean acidification (Ullah et al., 2015). The economics of climate change have recently been examined by Stern et al. (2010). It was suggested to capture and store carbon dioxide. It was suggested that alternatives to the usage of fossil fuels be explored. In the aforementioned framework, bioenergy has been acknowledged as an important element in several future energy scenarios. For the majority of developing nations, biomass is the main source of energy. A verbatim impending future energy problem and a reduction in carbon emissions from fossil fuels are both addressed by the substitution of fossil fuels with biofuels.

22.3.1 First-generation Biofuels

First-generation biofuels can be generated using traditional technologies and are composed of feedstock that is gathered for its sugar, starch, and oil content. The most well-known first-generation biofuel is ethanol, which is created by fermenting starch from maize kernels or other starchy crops and sugar derived from agricultural plants. Straight vegetable oils (SVO) from oleaginous plants are trans-esterified to produce biodiesel. The last three decades have seen the use of first-generation fuels; however, they have shown to be drastically insufficient to meet growing worldwide demands. As a result, a gradual change into then second-generation on biofuel resources, which offer larger potentials, is now necessary because their continuous use has exacerbated the worldwide food for fuel dilemma. However, land availability and the preservation of the world's ecosystems are the key justifications for opposing second-generation fuels. Even though lignocellulose biomass appears to have the ability to significantly increase global demand for the bioenergy sector, it is true that these fuels have a great deal of promise. Cellulosic lingo output surged in Brazil (up 46% to 2.3 billion litres) and Argentina (up 57% over 2009 to 2.1 billion litres), both of which continued their strong expansion (Chakma et al., 2016). Out of this production, 75% was exported. Numerous evaluations conducted by numerous researchers indicate that lingo cellulosic has a lot of potential.

22.3.1.1 First Generation Biofuel Status in Developing Countries

A number of nations have established large-scale targets, regulations, standards, and action plans in order to greatly boost biofuel production and usage in the near future. With the exception of Brazil and Argentina, biofuel production fell short of the goals established. However, it was anticipated that 2010 would see a 17% rise in bioethanol output over the previous year, reaching 86 billion litres. The two largest producers of the year were the US and Brazil, each accounting for 86% of global output, with the US producing 57% of it.

22.3.2 Second Generation Biofuels

As a result, compared to first-generation biofuels, second-generation biofuels will further boost and utilise efficiency. Second-generation biofuels are superior to first-generation biofuels in the following ways, allowing for the use of a variety of agricultural and wood-related waste products as feedstock without laying any direct claim to land. These feedstocks can employ a wider range of land because they use less land than first-generation biofuels. Farmers may therefore make better profit as a result. Future and emerging biofuel technologies have the capacity to convert biomass leftovers and energy crops into feedstocks for the production of very low carbon biofuels (Table 22.1). Second-generation biofuels have the potential to benefit everyone on the planet, but they can notably benefit a sizable portion of the population in developing countries by offering energy security, business opportunities, job creation, and improved environmental conditions.

TABLE 22.1
Types of Biofuels Obtained from Biomass

Types of Biofuels	Substrate	Process	References
Bioethanol	• Sugarcane • Sweet sorghum • Corn • Jatropha • Jojoba	Fermentation	Kang et al. (2014)
Cellulosic Bioethanol	Lignocellulosic biomass	Hydrolysis Fermentation	Zabed et al. (2016)
Biogas	Animal wastes Agricultural wastes	Anaerobic digestion	Lee et al. (2001); Paul and Dutta (2018)
Biohydrogen	Lignocellulosic biomass	Gasification Bio-photolysis Dark fermentation	Cheng et al. (2011); Rittmann (2008)
Biodiesel	• Jojoba • Jatropha	Extraction Trans-estrification	Koh and Ghazi (2011); Borugadda and Goud (2012)

Source: Modified from Dar et al. (2021).

22.4 PRODUCTION TO AGRO-RESIDUE WASTES RATIO GLOBALLY

Data based on crop production can be easily gathered on a global scale. However, the total generation of crop residues remains restricted till date. Some reports have tried to combine and compute the total yield and its residue production. Various datasets from FAOSTAT have been used to represent an overview on the ratio of residue to crop production. Although various reports are available analysing the total generation of crop residues from different crops, but on the regional scale. A very comprehensive data has been reported by Ullah et al. (2015) stating about 2,900 million tonnes per year of total agro-residue production across 25 developing countries. China produces the most residues overall, with more than 800 million tonnes, followed by Brazil (600 million tonnes), India (550 million tonnes), Indonesia (165 million tonnes) and lastly Argentina (129 million tonnes) (Gabhane et al., 2016). Countries like Malaysia, Zambia, Mozambique, Czech Republic, Sri Lanka, Kenya and Bulgaria produce around 5–10 million tonnes of agro-residues annually (Jain et al., 2022). Also, countries like Rwanda and Niger generate around less than 5 million tonnes of residues per year.

Developing countries majorly produce crops like rice, sugarcane, maize, wheat, soyabean, groundnut, cotton, cassava, etc. According to the study reported by Ullah et al. (2015), among these crops, residual contribution is maximum in rice (31%), followed by sugarcane (30%), maize (15%), wheat (11%), and soyabean (5%). One of the most important crops in world, rice, has large residue potential. The largest rice residual producing countries consist of China (37%) followed by India (23%), Indonesia (13%), Vietnam (9%), Thailand (6%), and Myanmar (6%) (Figure 22.4) (Ullah et al., 2015). Furthermore, sugarcane generates second highest residue of around 800 million tonnes. The countries producing maximum sugarcane residues are Brazil (52%) followed by India (20%), China (8%), and Thailand (5%) (Anwar et al., 2014). While the rest of the countries all-together generate around 15% of sugarcane residue only. Maize is a key crop in developing nations and can produce waste products like maize straw and maize cob that is estimated to be around 450 million tonnes. The leading producers of maize residues involve countries like China (46%), Brazil (14.6%), Argentina (6%), Indonesia (5%) and India (4%) (Rogers et al., 2017).

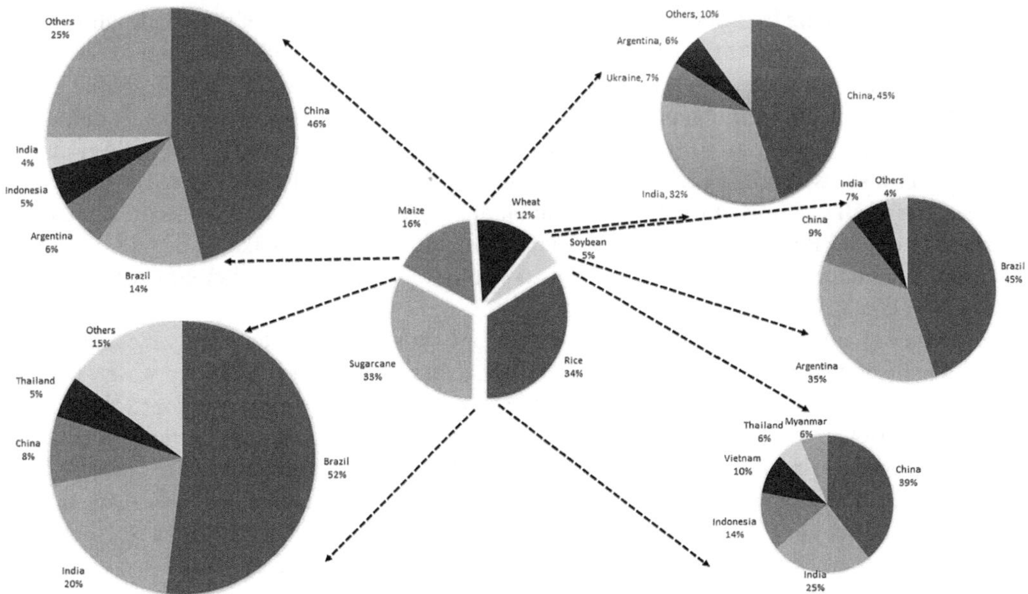

FIGURE 22.4 Country wise contribution on production of different crop residues (values in parenthesis).

Source: Ullah et al. (2015).

Soybean, wheat, and sorghum are further enlisted crops with high residue potential of around 152 million tonnes, 326 million tonnes, and 44 million tonnes respectively (Ullah et al., 2015). Brazil remarks the world's leading soya bean residues producer of around 45%. However, crops like cassava, coconut, cotton, and peanuts, have a high potential for production confined to a particular area. Nigeria is the world's top producer of cassava, with an annual residual capacity of 7.5 million tonnes (Mirza et al., 2008). Groundnut being the leading production in China marks around 36 million tonnes of groundnut straw. This represents around 56% of total production among developing nations. Similarly, residues from cotton mark around 44% and 42% from India and China, respectively forming 86% of total production (Sharma et al., 2014).

22.5 AGRO-RESIDUE WASTES GENERATION IN INDIA

Most commonly grown crops in India consists of sugarcane, rice, wheat, maize, and cotton. These contributed significantly to India's overall agro-residue output of nearly 80%. India's production of agricultural residue counts around 560 million tonnes currently. However, analysis from past years has shown a compound annual growth rate of 2.53% (Trivedi, 2020). Out of the major crops produced rice contributes around 33% of gross residual production followed by wheat (31%) and sugarcane (20%) (Singh et al., 2022). Although 41% of theoretical surplus residue is marked by sugarcane along with rice (25%), wheat (23%), maize (7%), and cotton (4%). This theoretical surplus residue consists of only one-third portion of gross residual production. As per state-wise data, Uttar Pradesh has the highest estimate of crop residue of around 60 million tonnes trailed by Punjab (51 million tonnes) and Maharashtra (46 million tonnes).

22.6 ADVANCES ON APPLICATION OF AGRO-RESIDUES

Agricultural residues are an important source of energy and material production. Globally they account for a significant portion of total annual biomass production. In relation to operational handling, agricultural residues vary in bulk density, moisture content, particle size, and distribution (Mohammed et al., 2018). They are typically fibrous, low in nitrogen, and vary according to location. Agro-residues are traditionally used for a variety of purposes, many of which are site-specific. The ultimate use of agricultural residues varies depending on how and where the crop is harvested. Chemical composition depends on a number of factors, including species types, residue age or harvesting period, physical composition, storage length, and harvesting practices (Ullah et al., 2015). They are generally used as fodder, fertiliser, fibre, feedstock, and construction. Agriculture residues are also used in the production of biofuels (ethanol or biodiesel). Cellulosic-rich crops, such as corn stalks, are used to produce ethanol, and lipid-rich crops like oil seeds, are used to produce biodiesel. Bagasse, a sugarcane residue is highly utilised to make biodegradable bowls, plates, and containers. They are an excellent substitute for styro-foam and other plastic products (Hiloidhari et al., 2014).

Examples of agricultural residues that are frequently left on fields after harvesting and utilised for feed and landfill material include rice straw, wheat straw, rice husk, and maize stover. In many countries, most of the rice straw is burned into ash to be used as organic fertiliser (Mohammed et al., 2018). However, small amounts are used for fodder and raw material, for the production of paper and boards. Also, countries like China, Bangladesh, Vietnam, and India use rice straw as a domestic fuel. Similarly, rice husk is widely used as a source of insulation and electricity in large rice mills (Ullah et al., 2015). About 50%–70% of rice husk is utilised for generating electricity and the remaining is used by the brick industry in Thailand (Hiloidhari et al., 2014). In Malaysia and the Philippines, it works as an insulator for the cement industry. Additionally, rice husk ash is used by the steel industry as a carbon source as well as an insulator (feedstock). Rice straw is a convenient, practical, and inexpensive source of fodder for ruminants such as buffaloes, cattle, goats, and sheep.

Wheat and maize straws have high nutritive values and are used as fodder in almost every country (Ullah et al., 2015). Wheat straw, along with horse manure and poultry litter, is used as a substrate for mushroom cultivation.

Agricultural residues can also increase the fertility and stability of soil to environmental factors (Scarlat et al., 2010). For instance, straw mulching is used to control soil erosion in a number of locations. Coconut residues too have wide applications. Coconut coir dust is applied as a fertiliser and soil conditioner, and straw is used as a mushroom-growing medium. Coconut husk is used as an orchid-growing medium and a packing material. The pure cellulosic component of agro residues has the potential to be used in the textile industry as natural or regenerated cellulosic fiber. Corn is the most distributed food crop in the world and 40% of global corn production is done by the United States. Corn husk is a byproduct rich in natural cellulosic fibers. About 45 million tonnes of corn husk is available annually and suitable for textile applications. Due to a lack of research and technology, the textile industry pays little attention to the cellulosic present in coconut and corn. Agro residues may provide environmental as well as socio-economic benefits to the farmer by adding value to their products (Ullah et al., 2015).

22.7 CHALLENGES ASSOCIATED WITH AGRO-RESIDUES

Due to fast depletion of non-renewable fossil fuels, agro-residues provide a better alternative for generating biofuels (Gontard et al., 2018). However, this area still needs to be more investigated and explored. Even currently, the production process of biofuels from these agricultural residue biomasses requires some non-renewable resources for generating outputs. It would be better if the whole process becomes more environment and energy-friendly. Apart from this the overall generation and consumption of agro-residues in different fields is dispersed and requires to be penned down for general perspective. This may be due to discretion in residue generation, seasonal fluctuations, and disparities in confined conditions (Mohammed et al., 2018). Also, it should be kept in mind that these residues play an important role in soil fertility. Therefore, a total removal from the ground area may lead to soil infertility or even soil erosion and thus, involving proper management. This is certainly not an easy process owing to the numerous obstacles to overcome. Proper crop residue management necessitates knowledge, skills, and effective measures (Scarlat et al., 2010). The major issue to be addressed for the management of agro residues includes the absence of proper and early prediction tools capable of providing clear supervision to policymakers and end-users (Hiloidhari et al., 2014). Another shortcoming of agricultural residue is conversion technologies like anaerobic digestion processes, the most widely used technology for agricultural residue energetic valorisation, which provide a combined route for converting many agricultural residues into biofuels like biogas and fertiliser (Gontard et al., 2018). Furthermore, the challenge is to overcome limitations in the development of inventive building blocks and materials derived from agro-residues (Ullah et al., 2015).

22.8 CONCLUSION

As a clean energy source, bioenergy resources are a crucial part of attempts to address the threat of environmental problems and social security issues brought on by the use of fossil fuels, which are now the main energy source. As a sustainable, low-carbon substitute for fossil fuels, bioenergy and biofuels based on biomass, such as biorefinery, plant materials, manure, and waste resources, can be used to generate electricity and for transportation. A reasonably abundant resource, agricultural biomasses can be used to create energy while simultaneously preserving the environment. Additionally, the amount of residue produced by the food processing industry is often considerable, and using these leftovers to generate energy can result in a sizable amount of renewable energy. To reduce the threat of climatic, economic, environmental, and political concerns related to the burning of fossil fuels, renewable energy resources for transportation and electric power generation are essential. Researchers are attempting to boost biomass capacity to dramatically increase the amount of biofuels and other chemicals and products that can be produced from it. Biofuels and bioelectricity play a vital part in sustainability. Agricultural biomass resources must be used wisely when creating technical practices and regulations so that communities can gain economically and environmentally as the country minimises its usage of coal and oil and its emissions that contribute to global warming. To attain this sustainability aim, wise public investment and policy are needed. The use of biomass as a feedstock for the manufacture of biofuel and bioenergy promises to be a key strategy, and the produced biofuel products are known to have characteristics that are similar to those of petroleum products. One of the comprehensive climate strategies that phase out coal-based electricity generation and incorporates carbon capture is bioenergy.

ACKNOWLEDGEMENT

The authors are thankful to ICMR and University Grant Commission for providing financial support and CSIR-NBRI for institutional support.

REFERENCES

Anwar, Z., Gulfraz, M., & Irshad, M. (2014). Agro-industrial lignocellulosic biomass a key to unlock the future bio-energy: A brief review. *Journal of Radiation Research and Applied Sciences*, 7(2), 163–173.

Borugadda, V. B., & Goud, V. V. (2012). Biodiesel production from renewable feedstocks: Status and opportunities. *Renewable and Sustainable Energy Reviews*, 16, 4763–4784. https://doi.org/10.1016/J.RSER.2012.04.010.

Chakma, S., Ranjan, A., Choudhury, H. A., Dikshit, P. K., & Moholkar, V. S. (2016). Bioenergy from rice crop residues: Role in developing economies. *Clean Technologies and Environmental Policy*, 18(2), 373–394.

Chanakya, H. N., & Malayil, S. (2012). Anaerobic digestion for bioenergy from agro-residues and other solid wastes-An overview of science, technology and sustainability. *Journal of the Indian Institute of Science*, 92(1), 111–144.

Cheng, C. L., Lo, Y. C., Lee, K. S., & Lin, C. Y. (2011). Biohydrogen production from lignocellulosic feedstock. *Bioresource Technology*, 102, 8514–8523. https://doi.org/10.1016/J.BIORTECH.2011.04.059.

Dar, R. A., Parmar, M., Dar, E. A., Sani, R. K., & Phutela, U. G. (2021). Biomethanation of agricultural residues: Potential, limitations and possible solutions. *Renewable and Sustainable Energy Reviews*, 135, 110217.

Devi, S., Gupta, C., Jat, S. L., & Parmar, M. S. (2017). Crop residue recycling for economic and environmental sustainability: The case of India. *Open Agriculture*, 2(1), 486–494.

Gabhane, J., Tripathi, A., Athar, S., William, S. P., Vaidya, A. N., & Wate, S. R. (2016). Assessment of bioenergy potential of agricultural wastes: A case study cum template. *Journal of Biofuels and Bioenergy*, 2(2), 122–131.

Gontard, N., Sonesson, U., Birkved, M., Majone, M., Bolzonella, D., Celli, A., ... Sebok, A. (2018). A research challenge vision regarding management of agricultural waste in a circular bio-based economy. *Critical Reviews in Environmental Science and Technology*, 48(6), 614–654.

Gupta, A., & Verma, J. P. (2015). Sustainable bio-ethanol production from agro-residues: A review. *Renewable and Sustainable Energy Reviews*, 41, 550–567.

Hiloidhari, M., Das, D., & Baruah, D. C. (2014). Bioenergy potential from crop residue biomass in India. *Renewable and Sustainable Energy Reviews*, 32, 504–512.

International Energy Agency. (2007). *Key world energy statistics* (p. 6). Paris: International Energy Agency.

Jain, A., Sarsaiya, S., Awasthi, M. K., Singh, R., Rajput, R., Mishra, U. C., ... Shi, J. (2022). Bioenergy and bio-products from bio-waste and its associated modern circular economy: Current research trends, challenges, and future outlooks. *Fuel*, 307, 121859.

Kang, Q., Appels, L., Tan, T., & Dewil, R. (2014). Bioethanol from lignocellulosic biomass: Current findings determine research priorities. *Scientific World Journal*, 2014(1), 298153. https://doi.org/10.1155/2014/298153.

Katyal, S. (2007). Effect of carbonization temperature on combustion reactivity of bagasse char. *Energy Sources, Part A: Recovery, Utilization, and Environmental Effects*, 29(16), 1477–1485.

Koh, M. Y., & Ghazi, T. I. M. (2011). A review of biodiesel production from Jatropha curcas L. oil. *Renewable and Sustainable Energy Reviews*, 15, 2240–2251. https://doi.org/10.1016/J.RSER.2011.02.013

Kumar, A., Kumar, N., Baredar, P., & Shukla, A. (2015). A review on biomass energy resources, potential, conversion and policy in India. *Renewable and Sustainable Energy Reviews*, 45, 530–539.

Lee, S. Y., Sankaran, R., Chew, K. W., Tan, C. H., Krishnamoorthy, R., Chu, D. T., & Show, P. L. (2019). Waste to bioenergy: A review on the recent conversion technologies. *Bmc Energy*, 1(1), 1–22.

Lee, S. S., Ha, J. K., & Cheng, K. J. (2001). The effects of sequential inoculation of mixed rumen protozoa on the degradation of orchard grass cell walls by anaerobic fungus *Anaeromyces mucronatus* 543. *Canadian Journal of Microbiology*, 47, 754–760. https://doi.org/10.1139/cjm-47-8-754.

Mafakheri, F., & Nasiri, F. (2014). Modeling of biomass-to-energy supply chain operations: Applications, challenges and research directions. *Energy Policy*, 67, 116–126.

Mirza, U. K., Ahmad, N., & Majeed, T. (2008). An overview of biomass energy utilization in Pakistan. *Renewable and Sustainable Energy Reviews*, 12(7), 1988–1996.

Mohammed, N. I., Kabbashi, N., & Alade, A. (2018). Significance of agricultural residues in sustainable biofuel development. *Agricultural Waste and Residues*, 71–88.

Mohan, D., Pittman Jr, C. U., & Steele, P. H. (2006). Pyrolysis of wood/biomass for bio-oil: A critical review. *Energy & Fuels*, 20(3), 848–889.

OECD/IEA. (2017). Technology roadmap: Delivering technology roadmap. Delivering Sustainable Bioenergy. International Energy Agency, 94.

Paul, S., & Dutta, A. (2018). Challenges and opportunities of lignocellulosic biomass for anaerobic digestion. *Resources, Conservation and Recycling*, 130, 164–174. https://doi.org/10.1016/j.resconrec.2017.12.005.

Rittmann, B. E. (2008). Opportunities for renewable bioenergy using microorganisms. *Biotechnology and Bioengineering*, 100, 203–212. https://doi.org/10.1002/bit.21875.

Reid, W. V., Ali, M. K., & Field, C. B. (2020). The future of bioenergy. *Global Change Biology*, 26(1), 274–286.

Rogers, J. N., Stokes, B., Dunn, J., Cai, H., Wu, M., Haq, Z., & Baumes, H. (2017). An assessment of the potential products and economic and environmental impacts resulting from a billion ton bioeconomy. *Biofuels, Bioproducts and Biorefining*, 11(1), 110–128.

Saidur, R., Abdelaziz, E. A., Demirbas, A., Hossain, M. S., & Mekhilef, S. (2011). A review on biomass as a fuel for boilers. *Renewable and Sustainable Energy Reviews*, 15(5), 2262–2289.

Saleem, A., Irshad, M., Hassan, A., Mahmood, Q., & Eneji, A. E. (2017). Extractability and bioavailability of phosphorus in soils amended with poultry manure co-composted with crop wastes. *Journal of Soil Science and Plant Nutrition*, 17(3), 609–623.

Saleem, M. (2022). Possibility of utilizing agriculture biomass as a renewable and sustainable future energy source. *Heliyon*, 8(2).

Scarlat, N., Martinov, M., & Dallemand, J. F. (2010). Assessment of the availability of agricultural crop residues in the European Union: Potential and limitations for bioenergy use. *Waste Management*, 30(10), 1889–1897.

Sharma, S., Meena, R., Sharma, A., & Goyal, P. (2014). Biomass conversion technologies for renewable energy and fuels: A review note. *IOSR Journal of Mechanical and Civil Engineering*, 11(2), 28–35.

Singh, A. D., Gajera, B., & Sarma, A. K. (2022). Appraising the availability of biomass residues in India and their bioenergy potential. *Waste Management*, 152, 38–47.

Trivedi, V. (2020). *Agro-residue for power: Win-win for farmers and the environment*. New Delhi: Centre for Science and Environment (CSE).

Ullah, K., Sharma, V. K., Dhingra, S., Braccio, G., Ahmad, M., & Sofia, S. (2015). Assessing the lignocellulosic biomass resources potential in developing countries: A critical review. *Renewable and Sustainable Energy Reviews*, 51, 682–698.

Zabaniotou, A. (2018). Redesigning a bioenergy sector in EU in the transition to circular waste-based bioeconomy – A multidisciplinary review. *Journal of Cleaner Production*, 177, 197–206.

Zabed, H., Sahu, J. N., Boyce, A. N., & Faruq, G. (2016). Fuel ethanol production from lignocellulosic biomass: An overview on feedstocks and technological approaches. *Renewable and Sustainable Energy Reviews*, 66, 751–774. https://doi.org/10.1016/J.RSER.2016.08.038.

Zhou, K., Chaemchuen, S., & Verpoort, F. (2017). Alternative materials in technologies for Biogas upgrading via CO_2 capture. *Renewable and sustainable energy reviews*, *79*, 1414-1441.

23 Circular Economy and Sustainability
The Role of Technology, Producers, Consumers and Policies to Achieve Sustainable Development Goals

Nabarupa Dhar, Suchismita Das, and Parag Shil

23.1 INTRODUCTION

It is one of the most important matters of discussion across the globe to find out various ways of protecting our Mother Earth and of leading a more sustainable means of life. Circular economy is one such means which aims at zero waste by the industries. It aims to increase the longevity of the product through the reuse of the waste again and again, which means maximum use of natural resources. The waste that has been produced today would become raw material or input for the production of tomorrow. Therefore, the natural resources are utilized in a circular manner and it is quite different from the linear economic system. More precisely, the previous system was centered on the goal of minimum waste whereas the circular economy is based on the pillars of "no waste". The producers of our economy play a pervasive role in achieving the Circularity and sustainability goals as it is the producers only who will bring necessary changes in their production process in commensuration with the principles of circular economy. This will be possible with the help of modified and improvised technology that isdesigned to reuse resources repeatedly. Also, the government should also play an important function in enforcing industries through various policies to implement the practices of CE so that no one can avoid it. Nonetheless, consumers being the king of the market have great influence on achieving the CE model, as it is the consumers only for whom various products are produced to meet their needs. Consumers should try to choose such products which are sustainable. They should demand more of those goods thatare subject to circular economy. Not only this, consumers should try to use more and more of a good and repair them by taking them to the recycling stores. This paper mainly discusses the literature that are available regarding the role of technology, producers, consumers and policies in achieving sustainable and circular economy.

23.2 CONCEPTUAL FRAMEWORK

We all are well acquainted with linear economy where the nonreusable wastes are discarded. Whereas, as defined by World Economic Forum, Circular economic model is a regenerative system designed to replace the end of life concept of the materials with the help of restoration, increasing the use of more of renewable resources, total elimination of the use of toxin chemicals in the production process, and more focus and stress should be given on the eliminating the waste generation so far as possible through improvised designing of the materials or inputs used, process used in production, systems and the business models.In other words, CE is a system where the raw materials used and the final products lose the very least of their value and are used in the correct sense of an

DOI: 10.1201/9781003441144-23

optimized way. CE model is in fact a complete change of people's attitude towards the production and the consumption of goods.For instance, there are some companies thatoffer rented cars. With this fewer people will be buying new vehicles, reducing the use of raw materials. This is indeed a part of CE.

23.3 IMPORTANT ELEMENTS OF CIRCULAR ECONOMY

1. Close Cycles: In the CE model, a very closed, the materials used are closed, that is, the waste of one production will become the input for other production processes. So there should be zero waste generation.
2. Renewable Energy: Apart from the raw materials some forms of energy also have a limited life. Those should be avoided and more stress should be given on the use of renewable sources of energies.
3. Re-thinking of the system: Proper designing should be done and necessary modifications in the economic system so that all the players of this economy become capable and willing to adapt the model (Figures 23.1 and 23.2).

23.4 EVOLUTION OF CIRCULAR ECONOMY GLOBALLY

- The law for promoting the CE model was published for the first time in China in the year 2008. It was the first national law on CE in the entire globe.
- In 2012, CE was first supported by the European Union (EU) for making a resource-efficient Europe, in the European Commission, 2012.
- The CE concept was promoted indirectly in various Environmental Action Programs and the 7th EAP introduced it for the very first time through its well-known program viz. "Living well, within the limits of our planet".
- The first action plan on CE was placed on 2nd Dec, 2015 by the European Commission (EC) which mainly stresses on steps to be promoted for transforming the so-called linear economic model. It was implemented in the year 2016.
- After the outbreak of COVID-19 pandemic, which originated in Asia, created havoc throughout the world, posing various challenges for the governments of various countries, the CE model has to revise its principles in order to adapt to the new terms and conditions posed by the crisis to continue its development process.

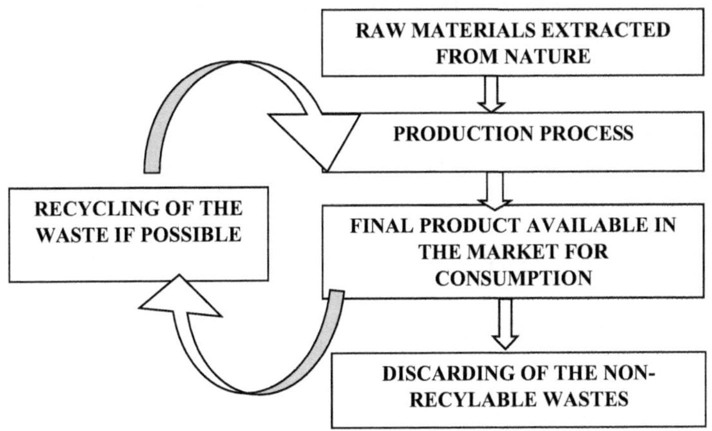

FIGURE 23.1 Flow chart for linear economy.

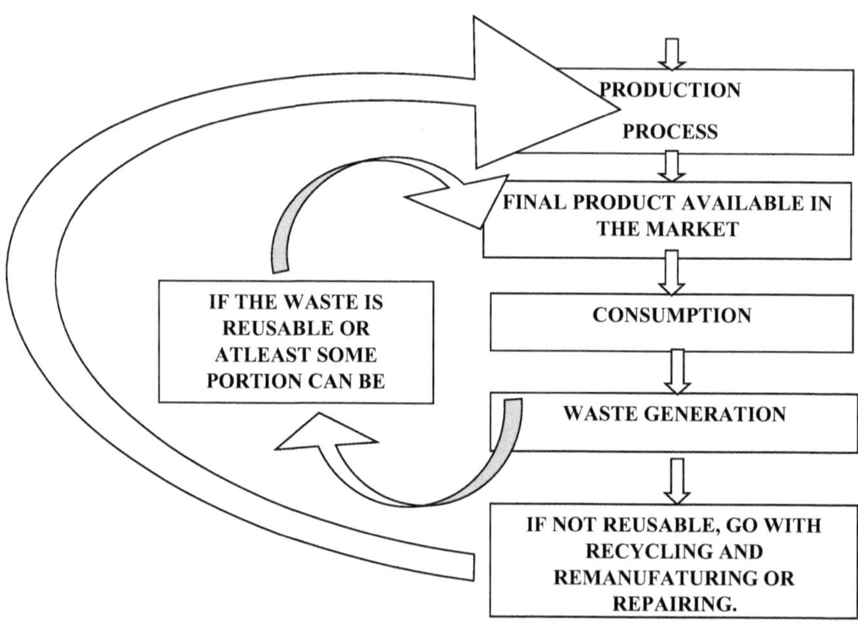

FIGURE 23.2 Flow chart for circular economy.

Source: Compiled.

23.5 FAVOURED ARGUMENTS OF CIRCULAR ECONOMY

With every single increase in the population across the globe, there is every possible increase in the demand forraw materials, energies, etc. where they are available in a very limited amount. Therefore, it has become very essential to have a change in the overall perception of people worldwide to think judiciously while consuming any good for any purpose so nothing can go waste, nothing can lead to natural degradation and environment harming, and nothing can cause extinction of any of our non-renewable resources. In light of this, the following are some of the worth-mentioning facts about circular economy:

- The main agenda of CE is to think and implement every possible way for regenerating the natural resources of our planet, especially the non-renewable ones. It aims at improving the overall health of our local ecosystem.
- It mainly helps in reducing the negative impacts of climate change by decreasing the greenhouse gas emissions.
- It also focuses on economic developments of the entire world as a whole through various sustainable innovations in the production process and also creates a lot of job opportunities. It is reflected in one of the reports of ILO.
- CE model influences business towards recycled products, thus reducing the exposure of the business towards volatile fluctuations in market prices and disruptions in the supply chain.

23.6 CIRCULAR ECONOMY IN INDIA

India has always been known for its natural treasures and various important raw materials as well as for some energy sources too. One of the basic principles of Atmanirbhar Bharat is achieving sustainable growth and the optimum utilization of its resources. However, these days India is also a victim of various environmental problems, and as such it is on the verge of complete transformation from linear economy to Circular Economic model. The Government has been very keen on formulating various policies and projects that would promote CE. Various rules and regulations have already been published with this regard, such as Plastic Waste Management (PWM) Rules, metal recycling policy, e-waste management rules, etc. in order to expedite the process of transformation, Indian Government have infact made 11 committees which are to be led by the NITI AAYOG and Ministry of Environment, Forest And Climatic Change of India, industrial representations, experts and academicians too. They will be responsible to create necessary modalities for ensuring effective implementations of the principles of the CE model in India.

23.7 OBJECTIVE

To discuss the role of technology producers, consumers and policies in implementing circular economy through reviewing various existing literatures.

23.8 SIGNIFICANCE OF THE STUDY

With the environmental problems growing more and more day by day, it has become very important to find out new and innovative ways to deal with such global crisis. One such solution is adopting the principles of circular economy. This study mainly highlights the research already done in this area and makes aware of how technology, producers, consumers and policies play their role in achieving CE.

23.9 RESEARCH METHODOLOGY

It is a descriptive paper based on qualitative information collected from secondary sources available in various journals, articles, websites, newspapers, etc.

23.10 REVIEW OF LITERATURES

Deutsch (2020) made a review of literature related to the connection between circular economyand technology-based theories.This paper is based on two objectives: one, to find out the differences and similarities between circular economy and the classical technological concepts of sustainability; second, to discuss the technological innovations that have been conceptualized under circular economy model. The results of the study state that though the concept of circular economy is new in terms of its nature and scope, it has many similarities with the traditional technology-based theories for attaining sustainability. Circular economy is in fact much influenced by the principles of blue economy as well as natural capitalism. It makes a deeper analysis of how the strategies and techniques of the corporate entities can be moulded to adapt reverse cycles.

Rizvi et al. (2021)made a review research on whether the IT tools have impacted the concept of circular economy. When a product is used rarely or else it is disposed of out of being obsolete, the certain proportion of material or energy that was used while producing it definitely goes on waste (Hannon et al., 2016). The basic aim behind CE is to manufacture such goods thatare long lasting, profitable of course, but at the same time, they should be reusable or recycled or can be re-manufactured.

Laskurain-Iturbe et al. (2021) made a study on the extent of influence made by the technologies used by Industry 4.0 on some main areas under the circular economy. The technologies selected for this study include Artificial Intelligence, Cyber security, Artificial Vision, Robotics, etc. Similarly, the areas of CE on which impact have been measured consists of reuse, reduction of consumption of inputs, wastes and emissions recycling and reduction. The objectives and research questions are defined on the basis of reviews done. Both qualitative and quantitative approaches have been adopted by the researchers in the analysis of the data so collected with the help of a questionnaire. The study has targeted industry 4.0 technology-based projects that were conducted in Europe, Asia,

Africa and America. Likert scale was used as the answering tool of the questionnaire. 120 numbers of responses were received from the group of 168 numbers of I40T projects. As per the study, both business and academic houses have paid much attention towards the impact of implementing I40T on achieving competitiveness in the market, whereas the least focus has been given on fulfilling the goals of circular economy.

Sukiennik et al. (2021) have made a study on the conditions of circular economy in Poland with special reference to the Mining sector and the role of universities in achieving it. As per the study, Poland is following a very slow process in implementing circular economy. One of the main reasons behind this situation is the diversity prevailing in the organization of all those groups which are involved in the implementation of changes to achieve sustainable development and circular economy in Polish enterprises. The main objective of this study is to discuss how universities can play their role in throwing light for the achievement of sustainable development along with circular economy. In other words, the paper highlights how the universities can frame their syllabus and curriculum to educate the students who are actually the future of any nation, to attain the principles of sustainable development and circular economy. The study finally concluded with the suggestion that "Environmental Education" should be included in academics since pre-school class and it should continue university level. This will be possible only when the implementation process is undertaken in confidence with the schools, colleges, universities and also the NGOs and this process should be carried out by well-prepared team of educators. The focus or target of environmental education will be different for school level and higher education level. Education on sustainable development and circular economy should have to be carried out on a global basis to actually prevent our nature from degradation. The universities can shape the minds of the students towards environmental awareness through various trainings which will mean that such behavior will be permanently induced into human life. AGH-UST of Poland has proved it through such environmental programs which showed a clear increase in the awareness level of their students.

Calzolari et al. (2021) conductedan investigation to assess how European MNCs adopt the practices of circular economy in the supply chain. As MNCs dominate by lion share in the European economies, they also the supply networks worldwide and actively participate in the determination of what should be produced and consumed. But, on the other hand, they also extend a major role in the use of natural resources in an unsustainable way, leading to degradation of the global environment. As per the study, since 1988, just 100 of multinational enterprises are responsible for contributing 71% of GHG emission. This study mainly reviews what the CE practices are being adopted by 50 numbers of MNCs of European countries to combat their harmful contribution to the environment, which is based on the secondary source of information collected from Corporate Sustainable reports. The results of the study show that the adaptation of CE practices hasbeen increasing day by day. Use of renewable resources, increasing the efficiency of the resource, recycling of materials, resource reduction, etc. are some of the practices being focused more on achieving circular economy, whereas concern for reusing the products weighed less attention of the producers.

Christensen (2021) conducteda study to find out the local modes of governance towards the transition to circular economy. The study mainly analyses various forms of governance where municipalities play a role in facilitating and supporting the transition to CE at local level. In other words, the paper mainly discusses how the cities or the municipalities manage to achieve CE through different governance modes which include using own assets and materials to support the activities of CE, using the companies involved in waste management to support CE, or through use of enforcement of certain rules for achieving the goals of sustainability and CE. This study is based on action research where a case study has been done for two such projects of Denmark that seek CE. The study mainly tried to figure out whether the collaboration of municipalities, knowledge institutions and local waste management companies can play a major role in transition towards CE, and the result was positive. They concluded that cities and municipalities with the help of other stakeholders can facilitate a lot in the transformation towards sustainable and circular economy goals.

Capetillo et al. (2022) in their research article discussed the support extended by emerging technologies in achieving circular economy in the area of plastic material value chain. As per the article, four different sets of technologies have been identified to study their role and impact in achieving the goals of CE, these are Industry 4.0, distributed economies, bio-based systems and chemical recycling. The technologies under Industry 4.0 have immense influence in increasing circularity in the manufacturing phase of the plastic material value chain. These technologies which are a blend of different subject areas are capable of re-processing the polymer, where polymer is one of the most important materials of plastic industry. The paper concludes that as it is human beings who actually interact with all the emerging technologies, therefore it is the responsibility of the humans to ensure that the technologies are developed in the way that the potential of these new technologies to contribute in circular economy can be effectively and efficiently materialized.

Arauzo-Carod et al. (2022) conducteda study on policies thatsupports achieving sustainability and circular economy at regional level. The techniques and processes used under circular economy are designed in such a way that they demand minimum from our environment and thus result in least generation of wastes. Thus, it will lead to control the harmful impacts of the climatic changes taking place these days. Circular economy and sustainability are two most important topics of concern for the public administration before making any policies as guidelines have to make for firms, consumers, producers and institutions, which indicate that broader approaches, strategies and measures have to be taken while implementing the principles of CE. This might become little easier with the involvement of the regional governments which will facilitate interactions in a positive way. Moreover, in order to achieve the goal of sustainability it is very important to reduce the transportation of waste which again refers to the involvement of local as well as regional stakeholders to function in implementing such policies. It was a review paper based on the research already done regarding policies for implementing sustainability and circularity and this issue has been written in a very descriptive manner.

Gomes and Ometto (2022) had made a review on the challenges faced in changing the mindset and behavior of the consumers to implement circularity in the consumption system. The researchers reviewed 107 articles, out of which 53 were found to good for analysis. Consumers are animportant group of people in the entire consumption system whose mindset actually guides the entire product flow in a particular industry. The factors that influence their behavior can be categorized into categories such as political and legal factors, economic factors, environmental factors, demographic factors, etc. Their mindsets are indeed reflected in their behavior. The paper is based on a systematic literature review (SLR). It is descriptive research based on data collected from various secondary sources.

23.11 FINDINGS OF THE STUDY

- Unlike linear economy which is unidirectional, circular economic model doesn't stops with the discarding of nonreusable wastes; rather, it extends to covert these wastes into some other valuable products through repairing or recycling.
- Also, its scope is much more than 3R's and includes renting, renewing, repairing and also sharing of any product and the materials used in it.
- The main objective of the CE model is to make the world a zero waste state.
- This economy is mainly based on the use of materials and inputs that are renewable in nature, and it tries maximum utilization of the digital technologies.
- It is a package that influences total in the consumption pattern of the consumers in the economy and their way of thinking. The consumers are very important players so they should be aware of the advantages of the CE model. Because once they start demanding more CE goods and services, producers of the economy will automatically design their manufacturing accordingly. Thus, consumer awareness regarding CE is a very important matter of discussion.

- However, there are yet many loopholes and lack of introduction of various legal frameworks to make it mandatory to implement CE model across the globe.
- Technical up-gradation is also yet a barrier tothe success of the CE model.

23.12 CONCLUSION

Circular economy is the most sustainable way of dealing with reducing the exploitation of natural resources in an inequitable manner leading to environmental crisis and thus leading to a positive impact on the ecosystem of the earth. CE model helps to decrease the burden on non-renewable resources by increasing more use of renewable natural resources that have long-term life and are less polluting as compared to the fossil fuels. Technology has a great influence on achieving this model as the entire production process needs a lot of changes to adapt the innovations proposed by circular economy. Moreover, the producers should also be prepared enough to use CE supportive technologies in the production process. Not only this, the customers who are the king of the market should also be aware enough about the products that are produced under CE model and demand more of such products. Also, government policies should be made such that can induce the introduction of CE-supportive technologies by various industries. It is such a model that aims at zero wastage indicatingthat the system has fewerside effects and more of positive impact on our Mother Earth and its residents.

REFERENCES

Arauzo-Carod, J. M., Liviano Solis, D., & Manresa-Prades, E. (2022). Policies supporting sustainability and circular economy at the regional level. *Journal of Environmental Policy and Planning*, 24(1), 15-29.

Aristi Capetillo, A., Bauer, F., & Chaminade, C. Emerging technologies supporting the transition to a circular economy in the plastic materials value chain. *Circular Economy and Sustainability*, 3, 953–982 (2023). https://doi.org/10.1007/s43615-022-00209-2

Calzolari, T., Genovese, A., & Brint, A. (2021). The adoption of circular economy practices in supply chains – An assessment of European Multi-National Enterprises. *Journal of Cleaner Production*, 312, 127616.

Capetillo, A., Bauer, F., & Chaminade, C. (2022). Emerging technologies supporting the transition to a circular economy in the plastic materials value chain. *Circular Economy and Sustainability*. https://doi.org/10.1007/s43615-022-00209-2

Christensen, T. B. (2021). Towards a circular economy in cities: Exploring local modes of governance in the transition towards a circular economy in construction and textile recycling. *Journal of Cleaner Production*, 305(127058), 127058. https://doi.org/10.1016/j.jclepro.2021.127058.

Deutsch, N. (2020). Note on the link between Circular Economy and technology-oriented theories of sustainable development: A literature review. *Közgazdász Fórum Forum on Economics and Business,* 22(139), 3–24.

Gomes, C. M., & Ometto, A. R. (2022). Challenges in changing consumer mindset and behavior to implement circularity in the consumption system: A systematic literature review. *Sustainability*, 14(4), 2341. https://doi.org/10.3390/su14042341

Hannon, M. J., Foxon, T. J., & Gale, W. F. (2016). The co-evolutionary relationship between energy efficiency and energy demand: A critical review. *Energy Research & Social Science*, 12, 1–14. https://doi.org/10.1016/j.erss.2015.10.004

Laskurain-Iturbe, I., Arana-Landín, G., Landeta-Manzano, B., & Uriarte-Gallastegi, N. (2021). Exploring the influence of Industry 4.0 Technologies on the circular economy. *Journal of Cleaner Production*, 321, 128944. https://doi.org/10.1016/j.jclepro.2021.128944

Rizvi, S. W., Agrawal, S., & Murtaza, Q. (2020). Circular economy under the impact of IT tools: A content-based review. *International Journal of Sustainable Engineering*, 14(2), 87–97. https://doi.org/10.1080/19397038.2020.1773567

Sukiennik, M., Zybała, K., Fuksa, D., & Kęsek, M. (2021a). The role of universities in Sustainable Development and Circular Economy Strategies. *Energies*, 14(17), 5365. https://doi.org/10.3390/en14175365

24 Constructed Wetlands Coupled with Microbial Fuel Cells for Bioelectricity Generation

Possible Applications and Challenges

Manoj Kumar, Naveen Chand, and Sanjeev Kumar Prajapati

24.1 INTRODUCTION

Constructed wetlands (CWs) have evolved over the years, basically imitating distinct biological, physical, and chemical mechanisms for the treatment process (Nuamah et al., 2020; Bakhshoodeh et al., 2017). Constructed wetland (CW) systems have been confirmed to be one of the solutions to the water crisis that can contribute to environmental protection and human health security (David et al., 2022). The treatment mechanism includes uptake via plant, biodegradation, and so on (i.e., biological treatment), and adding to that volatilization, chemical degradation, adsorption, oxidation, hydrolysis, photodegradation, etc. (i.e., physical-chemical treatment). All these factors are highly influential in the performance of the CWs (Hijosa-Valsero et al., 2016). However, severe risks to freshwater supply and the exponential rise in energy consumption are anticipated over the next decade. According to the literature, there will be 40% shortage of fresh water and a 36% increase in energy consumption, which calls for an urgent sustainable solution to address these issues (Jingyu et al., 2020; Reddy et al., 2019). In contrast, the imbalance in the work-energy ratio of the current existing wastewater treatment techniques to reach up-to-the-discharge standards (Yuan and He, 2015; Yang et al., 2018) is also a concern. Even the previously reported literature has pointed out the various wastewater treatment techniques developed over the years. For instance, considering a few of them, such as ultrasonication, advanced oxidation, adsorption, biological process, and precipitation (Chaturvedi et al., 2021), which have their own drawbacks (Crini and Lichtfouse, 2019). For electrochemical treatment and precipitation techniques, a considerable amount of chemicals, energy, and costly instruments are involved. Even the biological process needs additional waste disposal steps. Although these techniques are quite effective in treating wastewater, they include high energy consumption or high cost (such as installation, operation, maintenance, etc.), or both. Hence, as a solution, the rise of MFC for wastewater treatment over the years could be seen with many refinements for wastewater treatment and for the generation of electrical energy (Cai et al., 2020), which makes these MFC treatment facilities a unique and sustainable method.

It is now understood that wastewater is a renewable energy source, storing chemical energy at levels many times greater than those required for its treatment (Pandey et al., 2016). As a result, the ideas of converting waste into energy and creating less energy-intensive wastewater treatment systems have emerged and have been extensively researched globally. The most desired strategy at the moment is the development of technologies that are both affordable and energy-free

DOI: 10.1201/9781003441144-24

(Reddy et al., 2019). One of the most promising wastewater treatment solutions in the water-energy context is microbial fuel cell (MFC) technology (Santoro et al., 2019). Hence over the last decade, research on constructed wetlands (CWs) as an integrated system has been going on with bio-electrochemical systems (BESs) extensively under the names constructed wetland-microbial fuel cell (CW-MFC), electroactive wetlands, electro-wetlands and constructed wetland based on microbial electrochemical technologies. Incorporating the BES in CW makes it easier to adjust redox activities and the balance of electron flow in aerobic and anaerobic zones of the wetland's matrix, relieving the restriction caused by the scarcity of electron acceptors, which finally improves the operational controllability.

The CW-MFC technology from its inception to the present is covered in this chapter, along with current research status, bio-electrocatalysts' use, influencing variables for microbial interactions, and operational parameters that regulate various processes. The CW-MFC systems' main difficulties and prospective applications are also covered.

24.2 CONSTRUCTED WETLANDS

Constructed wetlands came into the picture in the early 1950s; since then CWs, have gained the attention of the scientific community and researchers for wastewater treatment. The clean water source has become a severe matter of concern throughout the globe because of an exponential increase in the human population, which is directly proportional to industrial growth and social advancement. Owing to low construction, operational, and maintenance costs, CWs are now used extensively for onsite treatments of domestic wastewater (Tan et al., 2017). These constructed wetlands as treatment facilities have a quite simple design, i.e., an excavated basin generally filled with substrate material commonly known as a conventional substrate which includes gravels, sand, pebbles, and soil (See Figure 24.2). Along with that, CW systems also consist of vegetation that is tolerant to saturated conditions. The design and operation of CWs imitate the principle of natural wetlands to treat wastewater (sewage water, industrial effluents, agriculture, and urban runoff, and landfill leachates) (Lu et al., 2020; Engida et al., 2020; Wang et al., 2021). CWs with different operational and design configuration has been used for wastewater treatment, as represented in the flow chart (Figure 24.1).

The mechanistic track for reducing pollutants in CWs is mainly an exhaustive act of its biotic and abiotic elements, including macrophytes, bacteria, algae, filtration, fungi, media type, adsorption, and sedimentation. Further, these organisms payout for eliminating organic and inorganic pollutants and pathogens via different methods (Shang et al., 2018). Abiotic processes consist of filtration, sedimentation, adsorption, and chemical precipitation. Organic substance is reduced by the aerobic bacteria and plant roots. The plant root offers the necessary breeding surface for microbes and regulates oxygen in the support matrix. Nitrogen elimination is attained through microbial interaction, plant uptake, and adsorption. Further, phosphorus elimination in CWs is caused by bacteria elimination, plant uptake, adsorption through the absorbent media and precipitation, in which phosphorus ions act in response to the porous material. The key attributes thataccount for the elimination effectiveness of CWs are vegetation kind, hydraulic residence time, heat, porous media, and microbial activities (Kasak et al., 2018; Feng et al., 2020a, b) (Figure 24.2).

24.3 MICROBIAL FUEL CELL

MFC isa well-known sustainable technology that uses the energy from the breakdown of organic materials. The mechanism involves anaerobic microorganisms, sometimes called electroactive bacteria (EAB), which are responsible for oxidizing organic materials inside the anode chamber of the MFC to generate protons and electronsFigure 24.3. An internal pathway is followed by the protons to travel from the anode to the cathode chamber, often through a proton exchange membrane, while an external path for the transportation of electrons from the anode chamber (anaerobic zone) to the

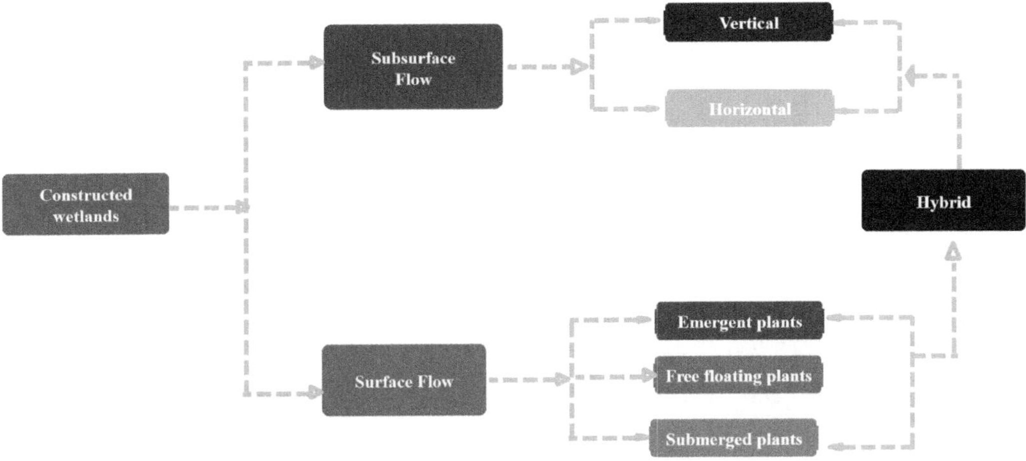

FIGURE 24.1 Classifications of constructed wetlands.

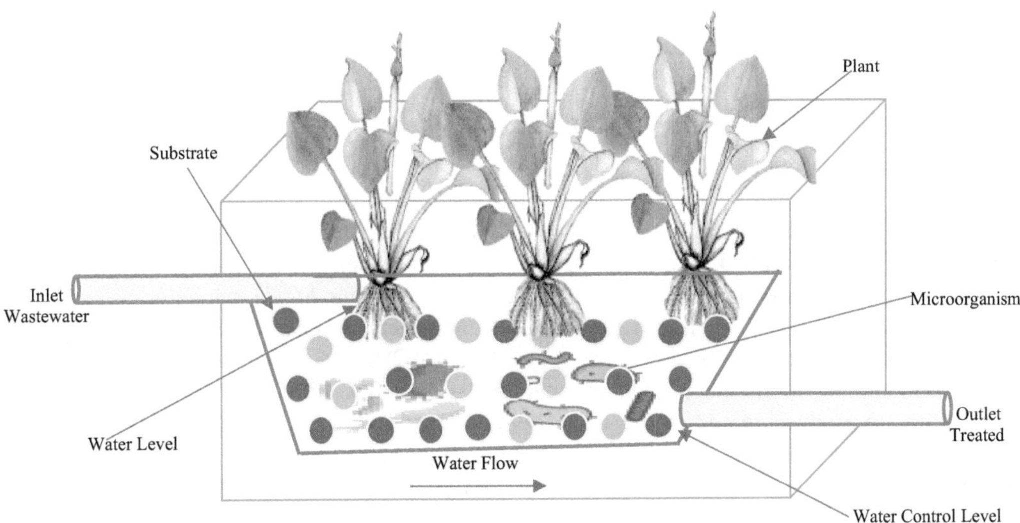

FIGURE 24.2 Schematic view of constructed wetlands.

cathode chamber (aerobic zone) is taken. Microbial fuel cells are regarded as a low-residual-biomass wastewater treatment method that may effectively remove all pollutants with an energy-efficient process. These technologies, however, have flaws such as high cost, weak power production, and limited commercial availability. These flaws severely restrict the use of MFCs as a stand-alone technology, preventing widespread deployment of these devices at practical "field" scales (Pandit et al., 2020).

The concept of a microbial fuel cell was initially introduced by Doo Hyun and J. Gregory (Park and Zeikus, 2003), and it was just one chamber having both the cathode and the anode. In which proton exchange membrane (PEM) was used for the anode and cathode separation, either far apart or close to one another. This MFC is simpler to construct than the double-chamber MFC and yields a sizable amount of fuel. However, single chamber (SC) MFCs have significant issues, such as microbial adulteration and inversion into the cathode and anodes. These MFCs often just include an anodic chamber because the cathodic section does not require air (Hubenova and Mitov, 2015; Morath et al., 2012).

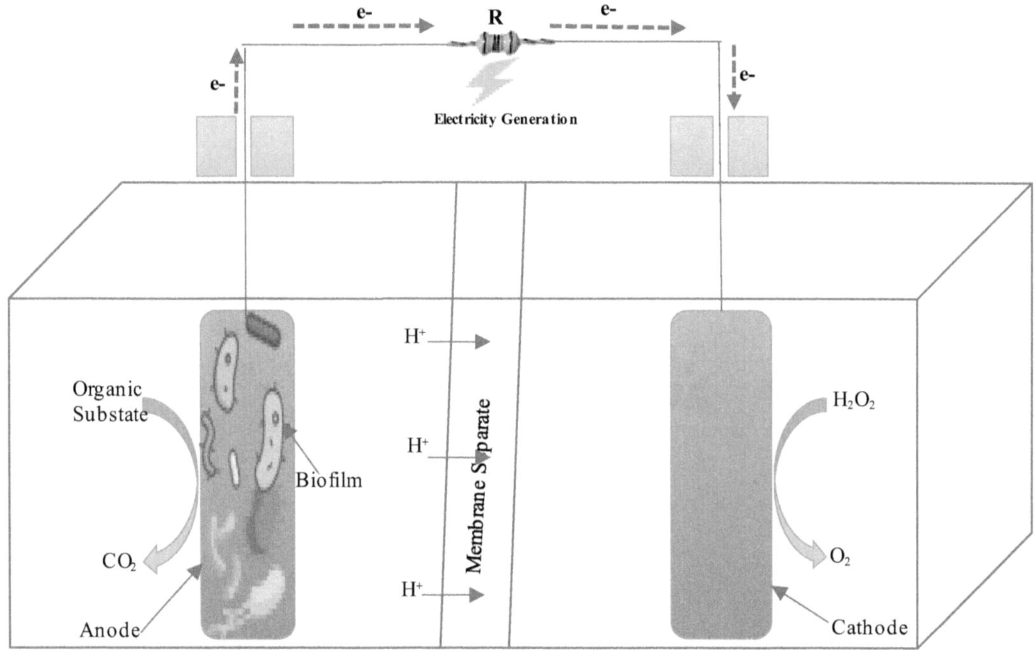

FIGURE 24.3 Schematic view of microbial fuel cell.

Double chamber MFC design is the most straightforward MFC arrangement. A two-chamber MFC might be shaped like a bottle or a cube. The design nomenclature could be affected by the catholyte employed in the MFC. Designs serve as the cathode and anode, which proton exchange membrane(PEM) separates. The function of the PEM generally prevents the transfer of oxygen from the cathode chamber to the anode chamber (Li et al., 2021). A specific substrate (medium) is used in anode in a two-chamber MFC to generate energy, along with a particular cathode catholyte solution to produce energy (Marco-Urrea et al., 2009; Kim et al., 2008; Liu et al., 2005. By using electrogenic microorganisms to oxidize organic materials, the MFC device's basic operating principle entails the conversion of chemical energy into electrical energy. The electrodes (anode and cathode), substrate, PEM, electrogenic microorganism's molecular oxygen, fuel, and electron transport system are the main elements of an MFC device (Peera et al., 2021; Suresh et al., 2022). However, the current situation demands an MFC-CW system for bioelectricity generation and wastewater treatment as an integrated system which could be a sustainable method for electrical energy production and reduce the pollutants from the wastewater. Further, in remote sensing applications, these MFCs may help to generate electricity (Ragauskas et al., 2006; Song, 2002, Ramya and Kumar, 2022).

24.4 INTEGRATION OF CONSTRUCTED WETLANDS WITH MICROBIAL FUEL CELL

Life is based on the component that includes food, energy, and water, for the overall development and for the growth of a sustainable economy. Even though nutrients, particularly phosphorus and nitrogen, are necessary for plant growth,when their limit exceeds in water, it causes various adverse effects. Hence the excessive nutrients are nonetheless regarded as an environmental burden when present in wastewater. Even though microbial fuel cells (MFCs) are gaining popularity, treating wastewater with an MFC has emerged as a low-cost wastewater treatment solution. The combination of a constructed wetland (CW) with an MFC (CW-MFC) has the potential to offer wastewater treatment with a low carbon footprint, low energy consumption, and the recovery of nutrients and energy from the waste.

The set-up mainly consists of an anaerobic and aerobic zone in a standard MFC. These zones offer a redox gradient for the passage of electrons and protons from anaerobic to aerobic zones. Similar to this, redox gradients are found naturally in CWs, allowing integration of an MFC into a CW. Recently, attempts to combine MFC and CWs to create an integrated CW-MFC technology have been made (Corbella and Puigagut, 2016; Zhao et al., 2013; Yadav, 2010). Researchers are becoming more interested in MFC technology integration with CW to enhance treatment outcomes and other advantages (Doherty et al., 2015; Yadav et al., 2012; Fang et al., 2013). This is mainly because of its unique combination of MFC without an ion exchange membrane integrated with constructed wetland, i.e., CW-MFC (See Figure 24.4). In this instance, the idea is to add, a non-consumable anode (electron acceptor) to the anaerobic zone of the wetland to speed up the breakdown of pollutants. This integrated system has gained interest since it offers the benefit of producing electricity while treating wastewater sustainably (Yadav et al., 2018), and enabling a sizable decrease in construction and operational costs. Due to their compatible traits, such as the presence of naturally occurring electroactive bacteria in CWs and (ii) the presence of redox gradients in CWs that an MFC may utilize to produce a bioelectric current flow, which has been successfully integrated.

Various researchers have evaluated the integration of the MFC technology into a CW; according to reports, adding MFC technology to a CW can boost treatment effectiveness by 27%–49% over a conventional CW (Srivastava et al., 2015). As per the documentation of Corbella and Puigagut (2018), CW-MFC operated in closed circuit mode had a treatment efficacy that was enhanced by 18% for COD, 15% for TOC, 31% for PO_4^{3-}, and 25% for NH_4^+ reduction when compared with CW-MFC operated in open circuit mode. When an electric circuit is cut off, the MFC operates in open circuit mode, which ultimately impedes the passage of electrons from the anode (anaerobic zone) to the cathode (aerobic zone). Such stated improvements result from the anaerobic CW-MFC having electrodes that serve as a transient electron acceptor during oxidation, transferring the received electron to the cathode through an external wire or another pathway (Srivastava et al., 2019).

If we talk about the electrical performance when compared to the minimum values published for conventional MFCs (2,110 and 2,850 mW/m^2), the electrical performance of integrated systems

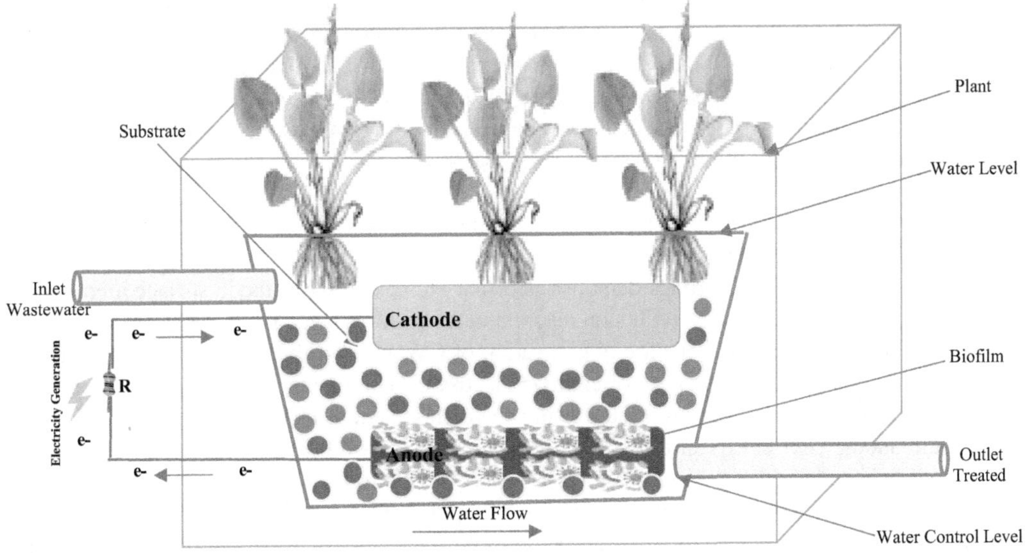

FIGURE 24.4 Integration of microbial fuel cell with constructed wetland.

has remained low (0.093–226.3 mW/m^2) (Pandey et al., 2016). The system experiences significant energy losses due to several reasons. These include biological elements (biocatalysts, plants biofilm), operational elements (pH, cell configuration, salinity, organic loading rate), and structural elements (electrode material, electrode separation, surface area) that affect power generation (Butti et al., 2016). Many of these elements have been researched to enhance electric performance (Srivastava et al., 2015; Hartl et al., 2019).

24.5 PERFORMANCE EVALUATION OF THE WASTEWATER TREATMENT AND BIOELECTRICITY GENERATION

It has been discovered and widely investigated that the combination of CWs and MFCs can be used as a renewable energy source for bioelectricity generation and for the concurrent treatment of wastewater, as given in Table 24.1. The literature has documented the successful implementation of the CW-MFCs for a variety of wastewater, including swine wastewater (Liu et al., 2019, 2020), domestic wastewater (Wen et al., 2020; Gonzálezet al., 2021), pharmaceutical wastewater (Wen et al., 2020; Wang et al., 2016b; Zhang et al., 2017), petroleum refinery wastewater Hussain and Ismail (2020), and azo dye (Fang et al., 2017; Oon et al., 2020). For instance, the literature reports an effective and efficient treatment of municipal wastewater, recording the efficacy of100% for COD and TN (Oon et al., 2015), 97% for total phosphorus (TP) (Saz et al., 2018), 99% for NH_4 (Wu et al., 2015) and apart from that, a power generation of 1300 mW/m^3 (Xu et al., 2018c).

However, under a high initial load of COD in CW-MFC, the performance in terms of treatment efficacy, power generation, and Coulombic efficiency (CE) declines. Moreover, the previous documentation has also supported this point of view, the documentation by Wang et al. (2017b) revealed a decline in pollutants removal with rising COD because of the predominately acidic circumstances. The performance of the CW-MFC was also evaluated by Liu et al. (2014) under a range of initial COD concentrations (i.e., 50–1000 mg/L), and it was discovered that as the influent COD increased, COD removal, power production, and CE gradually decreased. Similar results were reported by Bolton and Randall (2019), who noted a 44% drop in the power output when COD rose from 432 – 807 mg/L. These discoveries lead to the emergence of two distinct causes. The first study to show that methanogenic bacteria predominate over electrogenic bacteria when influent COD is high (Wang et al., 2019). Even this has been highlighted by prior research, which showed a very low CE under high initial COD, demonstrating that a greater portion of the substrate was utilized in anaerobic digestion or methanogenic processes rather than being consumed for the production of electricity (Song et al., 2018; Wang et al., 2016a; Zhao et al., 2013). Apart from that, Un-degraded substrate flows to the cathode layer and creates an anaerobic environment there, impeding the cathodic oxidation-reduction processes since it is impossible to degrade all substrate in the anodic layer at high initial COD (Liu et al., 2014; Villasenor et al., 2013; Wang et al., 2019).

Additionally, several parameters, including plants, substrate, electrode size, etc., have been demonstrated to increase electricity generation and removal efficacy. For instance, Fang et al. (2018) investigated the performance of CW-MFC using cathodes of varied diameters between 20 and 30 cm (at intervals of 2.5 cm). The findings demonstrated that increasing the cathode surface area boosted energy production. The enhanced reduction reaction in the cathode layer and lower internal resistance were credited for the increased electricity generation of CW-MFC with the larger cathode. Xu et al. (2018b) also investigated the quantity of cathode electrodes as a practical technique to improve the active surface area. They discovered that raising the cathode count from one to three (while maintaining the anodic site) led to an 82% increase in nitrification and denitrification rates. According to research by Cheng and Logan (2011) on the impact of electrode size on CW-MFC performance, doubling the cathode surface area increased power output by 62%, whereas doubling the anode size only increased it by 12%. Therefore, it can be deduced that increasing the relative size of the electrodes, particularly cathode electrodes can increase the rate of half-reactions, mainly cathodic reduction reactions.

TABLE 24.1
Current Status of Constructed Wetlands Microbial Fuel Cell (CWs-MFC)

S. No.	CWs- MFC	Electrode Materials	Substrate Materials	Planted Vegetation	Bioelectricity Generation(mV)	Pollutant Removal (%)			References
						Ammonia	Total Nitrogen	COD	
1.	CW-MFCs	Titanium-Embedded Graphite	Pebbles	*Bruguiera gymnorrhiza*	389.15	51.84	77.58	97.19	Liu et al. (2022)
2.	CW-MFC	Graphite felt Stainless steel wire mesh	Gravel	*Leersia hexandra*	426.00	–	–	95.70	Wang et al. (2023)
3.	CW-MFCs	Titanium Mesh Box	Ceramsite and Sand	*Iris tectorum*	293.16	84.6	61.3	87.01	Dai et al. (2022)
4.	CW-MFCs	Carbon Cloth	Sand and Rice husk	Canna indica	1260.00	–	–	98.00	Sonu et al. (2022)
5.	CW-MFCs	Graphite rod	Graphite Gravels	Unplanted	450	–	–	95.80	Saket et al. (2022)
6.	CW-MFCs	CW-MFCs	Activated carbon Stainless steel mesh	*Reed*	85.00	86.11	–	91.01	Han et al. (2021)
7.	CW-MFCs	graphite rod	Sand	Algae	–	85.14	–	96.37	Gupta et al. (2021)
8.	CW-MFCs	Graphite Rod	Gravels	*Eichhornia crassipes*	373.01	96.54	87.34	85.14	Kumar and Singh (2020)
9.	CW-MFC	circular graphite rod	Graphite gravel/stone gravel	*Canna indica*	650	–	–	98.90	Srivastava et al. (2020)
10.	CW-MFC	–	Gravel	*Canna indica*	–	–	–	75.00	Yadav et al. (2012)

Moreover, the main reactions that take place in a planted CW-MFC. It demonstrates that plant roots' exudates serve as a portion of the carbon source and oxygen source in the CW-MFC (Fang et al., 2018; Lu et al., 2015). Therefore, the CW-MFC relies on the coordinated actions of plants and microbes for its energy generation and treatment efficiency (Feng et al., 2020a). Microorganisms in the rhizosphere have access to additional nutrients from organic substances generated by a plant's root (Oon et al., 2017). Plants also raise the oxygen level in the area around the root, which enhances cathodic potential. It is advised that the plant's root be situated in the cathode zone and a suitable distance from the anode area because the oxygen released from the plant root may interfere with the anaerobic bacteria's ability to function in the anode layer. In this situation, the plant's oxygen and root exudates will benefit the rhizosphere microorganisms.

Additionally, plants increase the redox gradient within the CW-MFC column, which enhances electron transmission across the external circuit (Das et al., 2019; Liu et al., 2019; Oodally et al., 2019). In planted CW-MFC compared to the unplanted system, these features were reported to enhance the bacterial density by ten times (Gagnon et al., 2007) and DO by nine times (Shen et al., 2018). As an illustration, Zhu et al. (2019) examined the treatment effectiveness of the planted (Ipomoea aquatica) and unplanted CW-MFCs and discovered that the planted system could reduce more pollutants, which enhanced COD removal from 92.1% to 94.8% and TN removal from 54.4% to 90.8%. Because the plant utilized NH_4^+-N and NO_3-N as the nitrogen sources needed for growth, hence TN reduction significantly increased. In another documentation, Shen et al. (2018) also observed that the voltage generation and NH_4^+-N reduction were much higher in the planted system than in the unplanted system.

For an appropriate filter, media selection is quite important, as it's vital to consider the filter medium's porosity, specific surface area, cost, and adsorption capacity. Additionally, despite using thinner filter media to improve treatment efficacy (Wang et al., 2017a), the use of very fine media could lead to system clogging that lowers anode and cathode potential (Corbella et al., 2016; Hartl et al., 2019), this is crucial for a CW-MFC packed with fine media to operate well over the long term. It is unclear what elements may cause system clogging, and nothing is known about the ideal filter media size and type for CW-MFC systems.

24.6 CONCLUSION: CHALLENGES AND FUTURE PROSPECTS

The use of integrated systems CW-MFCs is a sophisticated sewage treatment and green energy recovery technique that offsets the requirement to employ fossil fuels in the traditional treatment procedure. The low practical power output for their direct use is the main issue that CW-MFCs are facing. For typical CW-MFCs, the PDs (9–72mW/m^2) and coulombic efficiencies 0.05%–10.48% which are not that significant (Liu et al., 2017). To drastically enhance the power output and attain financial sustainability, more study is required to understand the workings of CW-MFC systems better. Things that are needed to be addressed are:

- More detailed attention should be directed towards the genetic isolation and engineering of microorganism strains based on actual operating conditions.
- Design configuration is affected by DO and redox gradient; hence attention to design is required.
- Another factor is electrode size and materials, which significantly impact power output and treatment efficacy.
- Interaction and relationship among the plant rhizosphere, and rhizosphere microorganisms.
- Relationship of rhizosphere microorganisms with the strength of wastewater.
- Compatibility between microorganisms and electrodes dictates how biofilms grow on electrode surfaces, and the electron transfer mechanism affects internal resistance and energy loss when harvesting renewable energy.

- Importance of the critical variables that affect electrode design, configuration, and change to prevent electrode materials' deterioration over time. Factors that are limiting the current, such as transfer resistance.

To enhance electrical performance and get around current limitations in CW-MFCs, it is crucial to take into account and thoroughly investigate factors like over-potential during activation (Nitisoravut and Regmi, 2017), transfer resistance (Doherty et al., 2015), competition among microorganisms (Liu et al., 2017), and organic load (Wang et al., 2017b). As mentioned above, there has been substantial research on COD and N removal by CW-MFCs and practical ways to increase efficiency.

There are a lot of opportunities for CW-MFCs to target heavy metals, and emerging contaminants (i.e., pharmaceuticals, flame retardants, industrial chemicals, etc.). Most coupling CW-MFC set-up research and development is now being done in laboratories. The unavoidable process of scaling CW-MFCs to field scale is a potential future course to satisfy commercial applications. It is necessary to enhance energy-harvesting parts at the current capacity (Xu et al., 2018c). Internal resistance should be decreased for engineering applications to ensure power generation (Cai et al., 2020).

Additionally, CW-MFCs can be used as biosensors to monitor the DO level, pollutant concentration, and anaerobic digester performances because electricity signal fluctuations inside the system can reflect the change in parameters. Several investigations on pure MFC-based sensors for monitoring water quality and DO have been done (Di Lorenzo et al., 2014, Zhang and Angelidaki, 2012). The future applications of CW-MFC-based biosensors merit investigation due to the feature, which includes the transducer converting the parameter signals to electric signals (Xu et al., 2017).

ACKNOWLEDGEMENTS

The authors are thankful for the support ofUttarakhand State Council for Science & Technology (UCOST) under the sponsored Project Grant (Grant No. UCS&T/ R&D-10/2021/19072/02) and IIT Roorkee for encouragingand providing research facilities.

REFERENCES

Bakhshoodeh, R., Alavi, N., Majlesi, M. and Paydary, P. (2017). Compost leachate treatment by a pilot-scale subsurface horizontal flow constructed wetland. *Ecological Engineering*, 105, pp. 7–14.

Bolton, C.R. and Randall, D.G. (2019). Development of an integrated wetland microbial fuel cell and sand filtration system for greywater treatment. *Journal of Environmental Chemical Engineering*, 7(4), p. 103249.

Butti, S.K., Velvizhi, G., Sulonen, M.L., Haavisto, J.M., Koroglu, E.O., Cetinkaya, A.Y., Singh, S., Arya, D., Modestra, J.A., Krishna, K.V. and Verma, A. (2016). Microbial electrochemical technologies with the perspective of harnessing bioenergy: Maneuvering towards upscaling. *Renewable and Sustainable Energy Reviews*, 53, pp. 462–447.

Cai, T., Meng, L., Chen, G., Xi, Y., Jiang, N., Song, J., Zheng, S., Liu, Y., Zhen, G. and Huang, M. (2020). Application of advanced anodes in microbial fuel cells for power generation: A review. *Chemosphere*, 248, p.125985.

Chaturvedi, P., Giri, B.S., Shukla, P. and Gupta, P. (2021). Recent advancement in remediation of synthetic organic antibiotics from environmental matrices: Challenges and perspective. *Bioresource Technology*, 319, p. 124161.

Cheng, S. and Logan, B.E. (2011). Increasing power generation for scaling up single-chamber air cathode microbial fuel cells. *Bioresource Technology*, 102(6), pp. 4468–4473.

Corbella, C., García, J. and Puigagut, J. (2016). Microbial fuel cells for clogging assessment in constructed wetlands. *Science of the Total Environment*, 569, pp. 1060–1063.

Corbella, C. and Puigagut, J. (2016). Microbial fuel cells implemented in constructed wetlands: Fundamentals, current research and future perspectives. *Contributions to Science*, 11(1), pp. 113–120.

Corbella, C., and Puigagut, J. (2018). Improving domestic wastewater treatment efficiency with constructed wetland microbial fuel cells: Influence of anode material and external resistance. *Science of the Total Environment*, 631, pp. 1406–1414.

Crini, G. and Lichtfouse, E. (2019). Advantages and disadvantages of techniques used for wastewater treatment. *Environmental Chemistry Letters*, 17(1), pp. 145–155.

Dai, M., Wu, Y., Wang, J., Lv, Z., Li, F., Zhang, Y. and Kong, Q. (2022). Constructed wetland-microbial fuel cells enhanced with iron carbon fillers for ciprofloxacin wastewater treatment and power generation. *Chemosphere*, 305, p. 135377.

Das, B., Thakur, S., Chaithanya, M.S. and Biswas, P. (2019). Batch investigation of constructed wetland microbial fuel cell with reverse osmosis (RO) concentrate and wastewater mix as substrate. *Biomass and Bioenergy*, 122, pp. 231–237.

David, G., Rana, M.S., Saxena, S., Sharma, S., Pant, D. and Prajapati, S.K. (2022). A review on design, operation, and maintenance of constructed wetlands for removal of nutrients and emerging contaminants. *International Journal of Environmental Science and Technology*, 20, pp. 9249–9270. https://doi.org/10.1007/s13762-022-04442-y

Di Lorenzo, M., Thomson, A.R., Schneider, K., Cameron, P.J. and Ieropoulos, I. (2014). A small-scale air-cathode microbial fuel cell for on-line monitoring of water quality. *Biosensors and Bioelectronics*, 62, pp. 182–188.

Doherty, L., Zhao, Y., Zhao, X., Hu, Y., Hao, X., Xu, L. and Liu, R. (2015). A review of a recently emerged technology: Constructed wetland-microbial fuel cells. *Water Research*, 85, pp. 38–45.

Engida, T., Alemu, T., Wu, J., Xu, D., Zhou, Q. and Wu, Z. (2020). Analysis of constructed wetlands technology performance efficiency for the treatment of floriculture industry wastewater, in Ethiopia. *Journal of Water Process Engineering*, 38, p. 101586.

Fang, Z., Cao, X., Li, X., Wang, H. and Li, X. (2018). Biorefractory wastewater degradation in the cathode of constructed wetland-microbial fuel cell and the study of the electrode performance. *International Biodeterioration & Biodegradation*, 129, pp. 1–9.

Fang, Z., Cheng, S., Cao, X., Wang, H. and Li, X. (2017). Effects of electrode gap and wastewater condition on the performance of microbial fuel cell coupled constructed wetland. *Environmental Technology*, 38(8), pp. 1051–1060.

Fang, Z., Song, H.L., Cang, N. and Li, X.N. (2013). Performance of microbial fuel cell coupled constructed wetland system for decolorization of azo dye and bioelectricity generation. *Bioresource Technology*, 144, pp. 165–171.

Feng, L., Liu, Y., Zhang, J., Li, C. and Wu, H. (2020a). Dynamic variation in nitrogen removal of constructed wetlands modified by biochar for treating secondary livestock effluent under varying oxygen supplying conditions. *Journal of Environmental Management*, 260, p. 110152.

Feng, L., Wang, R., Jia, L. and Wu, H. (2020b). Can biochar application improve nitrogen removal in constructed wetlands for treating anaerobically-digested swine wastewater?. *Chemical Engineering Journal*, 379, p. 122273.

Gagnon, V., Chazarenc, F., Comeau, Y. and Brisson, J.(2007). Influence of macrophyte species on microbial density and activity in constructed wetlands. *Water Science and Technology*, 56(3), pp. 249–254.

González, T., Puigagut, J. and Vidal, G. (2021). Organic matter removal and nitrogen transformation by a constructed wetland-microbial fuel cell system with simultaneous bioelectricity generation. *Science of the Total Environment*, 753, p. 142075.

Gupta, S., Nayak, A., Roy, C. and Yadav, A.K. (2021). An algal assisted constructed wetland-microbial fuel cell integrated with sand filter for efficient wastewater treatment and electricity production. *Chemosphere*, 263, p. 128132.

Han, J., Yang, Z., Wang, H., Zhong, H., Xu, D., Yu, S. and Gao, L. (2021). Decomposition of pollutants from domestic sewage with the combination systems of hydrolytic acidification coupling with constructed wetland microbial fuel cell. *Journal of Cleaner Production*, 319, p. 128650.

Hartl, M., Bedoya-Ríos, D.F., Fernández-Gatell, M., Rousseau, D.P., Du Laing, G., Garfí, M. and Puigagut, J. (2019). Contaminants removal and bacterial activity enhancement along the flow path of constructed wetland microbial fuel cells. *Science of the Total Environment*, 652, pp. 1195–1208.

Hijosa-Valsero, M., Reyes-Contreras, C., Domínguez, C., Bécares, E. and Bayona, J.M. (2016). Behaviour of pharmaceuticals and personal care products in constructed wetland compartments: Influent, effluent, pore water, substrate and plant roots. *Chemosphere*, 145, pp. 508–517.

Hubenova, Y. and Mitov, M. (2015). Extracellular electron transfer in yeast-based biofuel cells: A review. *Bioelectrochemistry*, 106, pp. 177–185.

Hussain, T.A. and Ismail, Z.Z. (2020). Effect of petrolume refinery wastewater on plant growth in integrated microbial fuel cell-constructed wetlands systems. *The Iraqi Journal of Agricultural Science*, 51(4), pp. 1239–1248.

Jingyu, H., Miwornunyuie, N., Ewusi-Mensah, D. and Koomson, D.A. (2020). Assessing the factors influencing the performance of constructed wetland-microbial fuel cell integration. *Water Science and Technology*, 81(4), pp. 631–643.

Kasak, K., Truu, J., Ostonen, I., Sarjas, J., Oopkaup, K., Paiste, P., Kõiv-Vainik, M., Mander, Ü. and Truu, M. (2018). Biochar enhances plant growth and nutrient removal in horizontal subsurface flow constructed wetlands. *Science of the Total Environment*, 639, pp. 67–74.

Kim, I.S., Chae, K.J., Choi, M.J. and Verstraete, W. (2008). Microbial fuel cells: Recent advances, bacterial communities and application beyond electricity generation. *Environmental Engineering Research*, 13(2), pp. 51–65.

Kumar, M. and Singh, R. (2020). Sewage water treatment with energy recovery using constructed wetlands integrated with a bioelectrochemical system. *Environmental Science: Water Research & Technology*, 6(3), pp. 795–808.

Li, C., Luo, M., Zhou, S., He, H., Cao, J. and Luo, J. (2021). Comparison analysis on simultaneous decolorization of Congo red and electricity generation in microbial fuel cell (MFC) with l-threonine-/conductive polymer-modified anodes. *Environmental Science and Pollution Research*, 28(4), pp. 4262–4275.

Liu, F., Sun, L., Wan, J., Shen, L., Yu, Y., Hu, L. and Zhou, Y. (2020). Performance of different macrophytes in the decontamination of and electricity generation from swine wastewater via an integrated constructed wetland-microbial fuel cell process. *Journal of Environmental Sciences*, 89, pp. 252–263.

Liu, F., Sun, L., Wan, J., Tang, A., Deng, M. and Wu, R. (2019). Organic matter and ammonia removal by a novel integrated process of constructed wetland and microbial fuel cells. *RSC Advances*, 9(10), pp. 5384–5393.

Liu, F.F., Zhang, Y.X. and Lu, T. (2022). Performance and mechanism of constructed wetland-microbial fuel cell systems in treating mariculture wastewater contaminated with antibiotics. *Process Safety and Environmental Protection*,169, pp. 293–303

Liu, H., Cheng, S. and Logan, B.E. (2005). Production of electricity from acetate or butyrate using a single-chamber microbial fuel cell. *Environmental Science & Technology*, 39(2), pp. 658–662.

Liu, S., Feng, X. and Li, X. (2017). Bioelectrochemical approach for control of methane emission from wetlands. *Bioresource Technology*, 241, pp. 812–820.

Liu, S., Song, H., Wei, S., Yang, F. and Li, X. (2014). Bio-cathode materials evaluation and configuration optimization for power output of vertical subsurface flow constructed wetland-Microbial fuel cell systems. *Bioresource Technology*, 166, pp. 575–583.

Lu, J., Guo, Z., Kang, Y., Fan, J. and Zhang, J. (2020). Recent advances in the enhanced nitrogen removal by oxygen-increasing technology in constructed wetlands. *Ecotoxicology and Environmental Safety*, 205, p. 111330.

Lu, L., Xing, D. and Ren, Z.J. (2015). Microbial community structure accompanied with electricity production in a constructed wetland plant microbial fuel cell. *Bioresource Technology*, 195, pp. 115–121.

Marco-Urrea, E., Pérez-Trujillo, M., Vicent, T. and Caminal, G. (2009). Ability of white-rot fungi to remove selected pharmaceuticals and identification of degradation products of ibuprofen by Trametes versicolor. *Chemosphere*, 74(6), pp. 765–772.

Morath, S.U., Hung, R. and Bennett, J.W. (2012). Fungal volatile organic compounds: A review with emphasis on their biotechnological potential. *Fungal Biology Reviews*, 26(2–3), pp. 73–83.

Nitisoravut, R. and Regmi, R. (2017). Plant microbial fuel cells: A promising biosystems engineering. *Renewable and Sustainable Energy Reviews*, 76, pp. 81–89.

Nuamah, L.A., Li, Y., Pu, Y., Nwankwegu, A.S., Haikuo, Z., Norgbey, E., Banahene, P. and Bofah-Buoh, R. (2020). Constructed wetlands, status, progress, and challenges. The need for critical operational reassessment for a cleaner productive ecosystem. *Journal of Cleaner Production*, 269, p. 122340.

Oodally, A., Gulamhussein, M. and Randall, D.G. (2019). Investigating the performance of constructed wetland microbial fuel cells using three indigenous South African wetland plants. *Journal of Water Process Engineering*, 32, p. 100930.

Oon, Y.L., Ong, S.A., Ho, L.N., Wong, Y.S., Dahalan, F.A., Oon, Y.S., Lehl, H.K., Thung, W.E. and Nordin, N. (2017). Role of macrophyte and effect of supplementary aeration in up-flow constructed wetland-microbial fuel cell for simultaneous wastewater treatment and energy recovery. *Bioresource Technology*, 224, pp. 265–275.

Oon, Y.L., Ong, S.A., Ho, L.N., Wong, Y.S., Dahalan, F.A., Oon, Y.S., Teoh, T.P., Lehl, H.K. and Thung, W.E. (2020). Constructed wetland-microbial fuel cell for azo dyes degradation and energy recovery: Influence of molecular structure, kinetics, mechanisms and degradation pathways. *Science of the Total Environment*, 720, p. 137370.

Oon, Y.L., Ong, S.A., Ho, L.N., Wong, Y.S., Oon, Y.S., Lehl, H.K. and Thung, W.E. (2015). Hybrid system up-flow constructed wetland integrated with microbial fuel cell for simultaneous wastewater treatment and electricity generation. *Bioresource Technology*, 186, pp. 270–275.

Pandey, P., Shinde, V.N., Deopurkar, R.L., Kale, S.P., Patil, S.A. and Pant, D. (2016). Recent advances in the use of different substrates in microbial fuel cells toward wastewater treatment and simultaneous energy recovery. *Applied Energy*, 168, pp. 706–723.

Pandit, S., Savla, N. and Jung, S.P. (2020). Recent advancements in scaling up microbial fuel cells. In Abbassi, R.,Yadav A.K., Khan F., Garaniya V. (eds) *Integrated microbial fuel cells for wastewater treatment* (pp. 349–368). Butterworth-Heinemann. https://doi.org/10.1016/C2017-0-03157-9

Park, D.H. and Zeikus, J.G. (2003). Improved fuel cell and electrode designs for producing electricity from microbial degradation. *Biotechnology and Bioengineering*, 81(3), pp. 348–355.

Peera, S.G., Maiyalagan, T., Liu, C., Ashmath, S., Lee, T.G., Jiang, Z. and Mao, S. (2021). A review on carbon and non-precious metal-based cathode catalysts in microbial fuel cells. *International Journal of Hydrogen Energy*, 46(4), pp. 3056–3089.

Ragauskas, A.J., Williams, C.K., Davison, B.H., Britovsek, G., Cairney, J., Eckert, C.A., Frederick, W.J., Hallett, J.P., Leak, D.J. and Liotta, C.L. (2006). The path forward for biofuels and biomaterials." *Science*, 311(5760), pp. 484–489.

Ramya, M. and Kumar, P.S. (2022). A review on recent advancements in bioenergy production using microbial fuel cells. *Chemosphere*, 288, p. 132512.

Reddy, C.N., Nguyen, H.T., Noori, M.T. and Min, B. (2019). Potential applications of algae in the cathode of microbial fuel cells for enhanced electricity generation with simultaneous nutrient removal and algae biorefinery: Current status and future perspectives. *Bioresource Technology*, 292, p. 122010.

Saket, P., Mittal, Y., Bala, K., Joshi, A. and Yadav, A.K. (2022). Innovative constructed wetland coupled with microbial fuel cell for enhancing diazo dye degradation with simultaneous electricity generation. *Bioresource Technology*, 345, p. 126490.

Santoro, C., Walter, X.A., Soavi, F., Greenman, J. and Ieropoulos, I. (2019). Self-stratified and self-powered micro-supercapacitor integrated into a microbial fuel cell operating in human urine. *Electrochimica Acta*, 307, pp. 241–252.

Saz, Ç., Türe, C., Türker, O.C. and Yakar, A. (2018). Effect of vegetation type on treatment performance and bioelectric production of constructed wetland modules combined with microbial fuel cell (CW-MFC) treating synthetic wastewater. *Environmental Science and Pollution Research*, 25(9), pp. 8777–8792.

Shang, K., Zhang, G. and Gilles, V. (2018). Effect of low-polluted water treatment on plant growth by subsurface flow constructed wetland. *Water Purification Technology*, 37(9), pp. 120–125.

Shen, X., Zhang, J., Liu, D., Hu, Z. and Liu, H. (2018). Enhance performance of microbial fuel cell coupled surface flow constructed wetland by using submerged plants and enclosed anodes. *Chemical Engineering Journal*, 351, pp. 312–318.

Song, C. (2002). Fuel processing for low-temperature and high-temperature fuel cells: Challenges, and opportunities for sustainable development in the 21st century. *Catalysis Today*, 77(1–2), pp. 17–49.

Song, H.L., Li, H., Zhang, S., Yang, Y.L., Zhang, L.M., Xu, H. and Yang, X.L. (2018). Fate of sulfadiazine and its corresponding resistance genes in up-flow microbial fuel cell coupled constructed wetlands: Effects of circuit operation mode and hydraulic retention time. *Chemical Engineering Journal*, 350, pp. 920–929.

Sonu, K., Sogani, M., Syed, Z., Rajvanshi, J. and Sengupta, N. (2022). Effectiveness of rice husk in the removal of methyl orange dye in Constructed Wetland-Microbial Fuel Cell. *Bioresource Technology Reports*, 20, p. 101223.

Srivastava, P., Abbassi, R., Garaniya, V., Lewis, T. and Yadav, A.K. (2020). Performance of pilot-scale horizontal subsurface flow constructed wetland coupled with a microbial fuel cell for treating wastewater. *Journal of Water Process Engineering*, 33, p. 100994.

Srivastava, P., Yadav, A.K., Garaniya, V. and Abbassi, R. (2019). Constructed wetland coupled microbial fuel cell technology: Development and potential applications. In S. Venkata Mohan, Sunita Varjani and Ashok Pandey (eds.). *Microbial Electrochemical Technology* (pp. 1021–1036). https://doi.org/10.1016/B978-0-444-64052-9.00042-X.

Srivastava, P., Yadav, A.K. and Mishra, B.K. (2015). The effects of microbial fuel cell integration into constructed wetland on the performance of constructed wetland. *Bioresource Technology*, 195, pp. 223–230.

Suresh, R., Rajendran, S., Kumar, P.S., Dutta, K. and Vo, D.V.N. (2022). Current advances in microbial fuel cell technology toward removal of organic contaminants-A review. *Chemosphere*, 287, p. 132186.

Tan, E., Hsu, T.C., Huang, X., Lin, H.J. and Kao, S.J. (2017). Nitrogen transformations and removal efficiency enhancement of a constructed wetland in subtropical Taiwan. *Science of the Total Environment*, 601, pp. 1378–1388.

Villasenor, J., Capilla, P., Rodrigo, M.A., Canizares, P. and Fernandez, F.J. (2013). Operation of a horizontal subsurface flow constructed wetland-microbial fuel cell treating wastewater under different organic loading rates. *Water Research*, 47(17), pp. 6731–6738.

Wang, J., Song, X., Wang, Y., Abayneh, B., Li, Y., Yan, D. and Bai, J. (2016a). Nitrate removal and bioenergy production in constructed wetland coupled with microbial fuel cell: Establishment of electrochemically active bacteria community on anode. *Bioresource Technology*, 221, pp. 358–365.

Wang, J., Song, X., Wang, Y., Bai, J., Bai, H., Yan, D., Cao, Y., Li, Y., Yu, Z. and Dong, G. (2017a). Bioelectricity generation, contaminant removal and bacterial community distribution as affected by substrate material size and aquatic macrophyte in constructed wetland-microbial fuel cell. *Bioresource Technology*, 245, pp. 372–378.

Wang, J., Song, X., Wang, Y., Zhao, Z., Wang, B. and Yan, D. (2017b). Effects of electrode material and substrate concentration on the bioenergy output and wastewater treatment in air-cathode microbial fuel cell integrating with constructed wetland. *Ecological Engineering*, 99, pp. 191–198.

Wang, L., Liu, Y., Ma, J. and Zhao, F. (2016b). Rapid degradation of sulphamethoxazole and the further transformation of 3-amino-5-methylisoxazole in a microbial fuel cell. *Water Research*, 88, pp. 322–328.

Wang, X., Tian, Y., Liu, H., Zhao, X. and Wu, Q. (2019). Effects of influent COD/TN ratio on nitrogen removal in integrated constructed wetland-microbial fuel cell systems. *Bioresource Technology*, 271, pp. 492–495.

Wang, Y., Zhang, X., Xiao, L. and Lin, H. (2023). The in-depth revelation of the mechanism by which a downflow Leersia hexandra Swartz constructed wetland-microbial fuel cell synchronously removes Cr (VI) and p-chlorophenol and generates electricity. *Environmental Research*, 216, p. 114451.

Wang, Y., Zhou, J., Shi, S., Zhou, J., He, X. and He, L. (2021). Hydraulic flow direction alters nutrients removal performance and microbial mechanisms in electrolysis-assisted constructed wetlands. *Bioresource Technology*, 325, p. 124692.

Wen, H., Zhu, H., Yan, B., Xu, Y. and Shutes, B. (2020). Treatment of typical antibiotics in constructed wetlands integrated with microbial fuel cells: Roles of plant and circuit operation mode. *Chemosphere*, 250, p. 126252.

Wu, D., Yang, L., Gan, L., Chen, Q., Li, L., Chen, X., Wang, X., Guo, L. and Miao, A. (2015). Potential of novel wastewater treatment system featuring microbial fuel cell to generate electricity and remove pollutants. *Ecological Engineering*, 84, pp. 624–631.

Xu, L., Wang, B., Liu, X., Yu, W. and Zhao, Y. (2018b). Maximizing the energy harvest from a microbial fuel cell embedded in a constructed wetland. *Applied Energy*, 214, pp. 83–91.

Xu, L., Zhao, Y., Fan, C., Fan, Z. and Zhao, F. (2017). First study to explore the feasibility of applying microbial fuel cells into constructed wetlands for COD monitoring. *Bioresource Technology*, 243, pp. 846–854.

Xu, L., Zhao, Y., Tang, C. and Doherty, L. (2018c). Influence of glass wool as separator on bioelectricity generation in a constructed wetland-microbial fuel cell. *Journal of Environmental Management*, 207, pp. 116–123.

Yadav, A.K. (2010). October. Design and development of novel constructed wetland cum microbial fuel cell for electricity production and wastewater treatment. In P. F. Cooper, B. C. Findlater (eds.), *Proceedings of 12th international conference on wetland systems for water pollution control (IWA)* (pp. 4–10).

Yadav, A.K., Dash, P., Mohanty, A., Abbassi, R. and Mishra, B.K. (2012). Performance assessment of innovative constructed wetland-microbial fuel cell for electricity production and dye removal. *Ecological Engineering*, 47, pp. 126–131.

Yadav, A.K., Srivastava, P., Kumar, N., Abbassi, R. and Mishra, B.K. (2018). Constructed wetland microbial fuel cell: an emerging integrated technology for potential industrial wastewater treatment and Bioelectricity generation. In Stefanakis Alexandros (ed.), *Constructed wetlands for industrial wastewater treatment* (pp. 493–510).

Yang, Z., Pei, H., Hou, Q., Jiang, L., Zhang, L. and Nie, C. (2018). Algal biofilm-assisted microbial fuel cell to enhance domestic wastewater treatment: nutrient, organics removal and bioenergy production. *Chemical Engineering Journal*, 332, pp. 277–285.

Yuan, H. and He, Z. (2015). Integrating membrane filtration into bioelectrochemical systems as next generation energy-efficient wastewater treatment technologies for water reclamation: A review. *Bioresource Technology*, 195, pp. 202–209.

Zhang, S., Song, H.L., Yang, X.L., Huang, S., Dai, Z.Q., Li, H. and Zhang, Y.Y. (2017). Dynamics of antibiotic resistance genes in microbial fuel cell-coupled constructed wetlands treating antibiotic-polluted water. *Chemosphere*, 178, pp. 548–555.

Zhang, Y. and Angelidaki, I. (2012). A simple and rapid method for monitoring dissolved oxygen in water with a submersible microbial fuel cell (SBMFC). *Biosensors and Bioelectronics*, 38(1), pp. 189–194.

Zhao, Y., Collum, S., Phelan, M., Goodbody, T., Doherty, L. and Hu, Y. (2013). Preliminary investigation of constructed wetland incorporating microbial fuel cell: batch and continuous flow trials. *Chemical Engineering Journal*, 229, pp. 364–370.

Zhu, J., Zhang, T., Zhu, N., Feng, C., Zhou, S. and Dahlgren, R.A. (2019). Bioelectricity generation by wetland plant-sediment microbial fuel cells (P-SMFC) and effects on the transformation and mobility of arsenic and heavy metals in sediment. *Environmental Geochemistry and Health*, 41(5), pp. 2157–2168.

25 Nature-based Solutions
Circularity Assessment through LCA Approach

Hema Diwan, Binilkumar Amarayil Sreeraman, and Gautam Yadav

25.1 INTRODUCTION

The human dependence on the ecosystems, particularly water ecosystems for their livelihoods has been established through the historical assessment on the origin and evolution of human societies. It has been observed that most of the world civilizations have been developed in close proximity to water (e.g. ancient civilizations have been developed along major rivers like the Nile, Indus, Tigris, and Euphrates). The waterbodies like oceans, rivers, lakes or all such wetland ecosystems have, thus, historically remained part of human civilizations. But unfortunately, due to various reasons, many waterbodies like lakes and other wetlands have been reclaimed for other purposes like industrial, residential, etc. So it becomes imperative to protect these waterbodies with appropriate management as they can deliver different ecosystem services like water cleansing, groundwater recharging, and overall biodiversity conservation.

Nature-based solutions (NBS) as an appropriate management approach emerged recently to efficiently conserve and manage these valuable resources. NBS are the solutions to environmental problems, which are largely motivated and adopted from natural ecosystems. It mimics the natural processes in the preservation of water, which may involve not only conserving natural ecosystems but also the creation of processes that are inspired by nature but modified with artificial processes (WWAP, 2018). The definition of NBS as given by Cohen-Shacham et al. (2016) states "Nature-based solutions are actions to protect, sustainably manage and restore natural and modified ecosystems in ways that address societal challenges effectively and adaptively, to provide both human well-being and biodiversity benefits.". Hence, NBS largely consists of ecosystem-based approaches or conservation frameworks, which are in tandem with the nature. NBS emerged as a popular as well as useful tool in ecosystem management by IUCN in 2009, but was later adopted as an umbrella term to underline the common traits of numerous conservation methods, while in the case of water-related ecosystem services and the conservation programmed targeting these resources, the adoption of NBS is still in the initial stages (WWAP, 2018).

In recent times, NBS has become a common framework of ecosystem-based approaches, which are adopted largely to tackle numerous challenges faced by human societies (Cohen-Shacham et al., 2016). It has been highlighted the importance of NBS to face environmental challenges confronted by the societies, through different programmes and projects of various international organizations, like IUCN and the World Bank. The importance of depending on natural ecosystems or processes to generate solutions instead of depending on conventional engineering solutions (such as seawalls), to address and alleviate adverse climate change impacts. The foremost importance of adopting NBS is that, while achieving the desired solutions, it will also ensure the improvement of sustainable livelihoods and the protection of natural ecosystems and biodiversity (Dudley et al., 2010; Cohen-Shacham et al., 2016). The development of new approaches in the ecosystem conservation has led to the emergence of two broad classes of conservation initiatives. One with a primary aim is to protect biodiversity due to its inherent social and environmental values, and the other with an objective to safeguard societal welfare, which is termed as 'Nature-based Solutions' (NBS).

DOI: 10.1201/9781003441144-25

As per the NBS framework developed by IUCN (2016), NBS aim to find sustainable solutions to major socio-environmental challenges like mitigation and adaptation of adverse climate change impacts, reduce the risks of natural calamities, attain the social and economic development, secure human health, food and water security, slowing the rate of environmental degradation as well as biodiversity loss, while maintaining the biodiversity or the ecosystem intact (Figures 25.1 and 25.2). The NBS is evolving and approaches environmental issues and climate change resilience with more socioecological reliance (Dumitru and Wendling, 2021). But there is still ambiguity in the definitions and application fields of NBS, which has resulted in applying the same in different disciplines (Sowińska-Świerkosz and García, 2022).

Ecological systems and hydrological processes in the ecosystem significantly influence the origin, movement, and quality of water in a system. NBS can assist in addressing water supply by managing

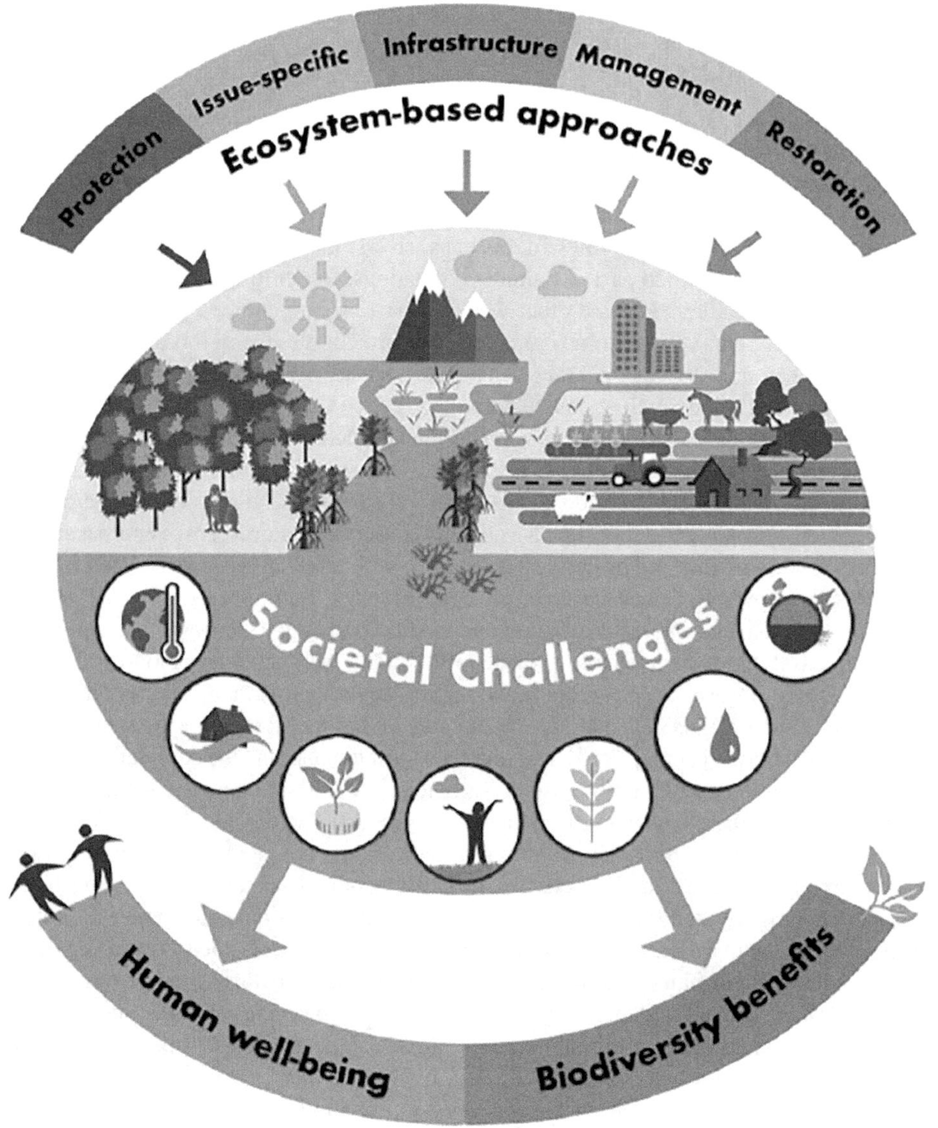

FIGURE 25.1 NBS framework.

Source: IUCN (2020).

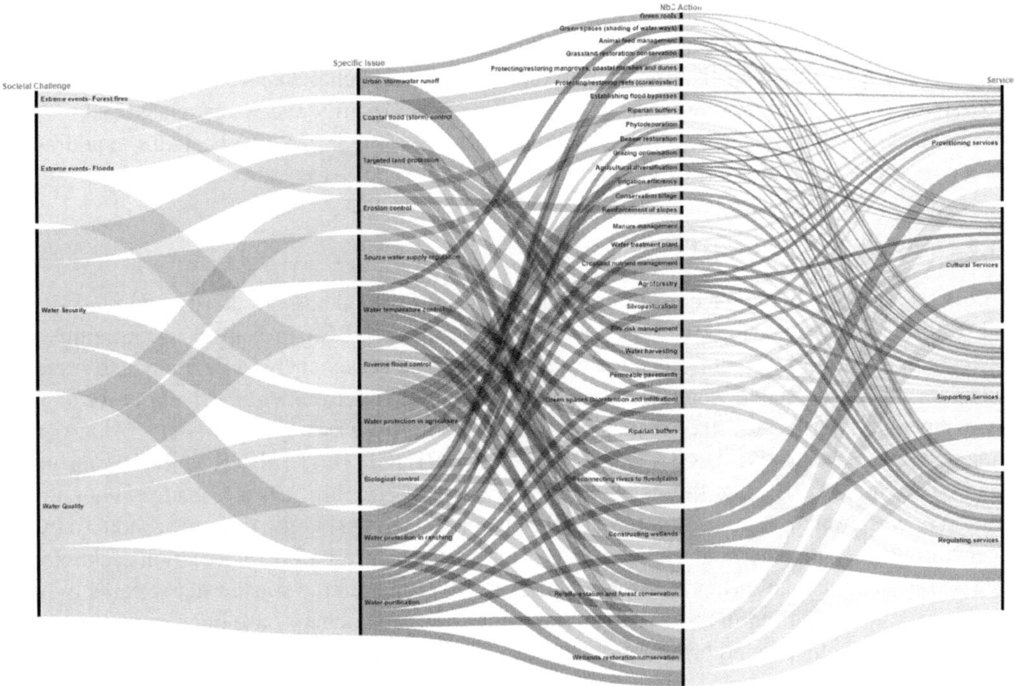

FIGURE 25.2 NBS approaches to different water-related issues.

the hydrological processes like precipitation, humidity, water storage, infiltration and transmission. NBS can develop more sustainable as well as nature-friendly options of storing water such as natural wetlands and preserving and recharging the groundwater, which are more cost-effective than traditional capital intensive as well as unsustainable artificial Dams. NBS assist in ensuring the supply and quality of the water, which will in turn, may lower costs for treating the water, which have to be incurred by the urban suppliers. NBS may also increase access to safe drinking water in rural areas, which will have a significant positive impact on the living standard and morbidity level of the rural communities. Further, the ecosystem-based approaches emphasize on protecting natural resources such as forests, wetlands and grasslands, as well as soils. If these natural resources managed properly, there will be significant improvement in the water quality, as they will reduce the sediment loadings, cleanse water by capturing and retaining pollutants, and recycling nutrients. The required condition of an NBS is not only to ensure the ecosystem used is 'natural' but also to ensure the natural processes are managed to achieve an efficient solution to water-related challenges. Hence an NBS largely depends on ecosystem services to bring desirable outcomes related to a water management issue. But the applicability of one type of NBS solution may not be universal. There exist considerable differences in the impacts of various ecosystems on water systems or hydrological systems, depending on their location, climate and management scenarios. Hence, it is imperative to develop site-specific knowledge on the implementation of NBS on field (WWAP, 2018; Magni et al., 2020). For example, it is assumed that wetlands act as sponge area and preclude floods as well as reduce the severity of draught, while certain wetlands can lead to downstream flooding.

NBS also helps in addressing water-related risks such as floods and draught. It has been estimated that around 30% global population are residing in areas and regions, which are vulnerable to such extreme climatic events like floods and draughts. It has been found that ecosystem degradation is the major driver of such extreme risks. Hence adopting ecosystem-based solutions will address such issues in tandem with nature. Therefore, NBS may realize water security by ensuring the availability and quality of water while reducing the occurrence of water-related adverse events like floods and draughts. Further, it will also ensure more social, economic, and environmental benefits in the process. The water-related ecosystem services provide different essential services to people in supporting their livelihood. But, there are different challenges that affect water-related ecosystem services, that is water security/availability, water quality and water-related risks (floods and draughts). NBS can be used to address these challenges (Figure 25.2). NBS helps in addressing these challenges and thereby contributing the water-related ecosystem services.

25.2 NBS & CIRCULARITY

NBS is increasingly being used as an important concept for inducing circularity and regenerating and restoring ecosystem and associated ecosystem services. NBS can help in minimizing waste production, restoring nutrient cycles, material recovery and reuse, energy efficiency and recovery, and restoring contaminated water/sites. It is applied in urban environment particularly for negative environmental externalities like waste water generation, solid waste generation etc. Countries are implementing projects like URBANGEENUP, NATURE4CITIES, THINKNATURE on understanding the role of NBS like these, endorse the concept of circularity as it leads to eventual reutilization of wastewater/waste land as a secondary resource, increasing the resource productivity and regenerating/restoring the contaminated site. Employing nature-based remediation technologies, such as *remediation*, can bring in resource efficiency by remediating wastewater/streams to better quality parameters, aiding in utilization as secondary use, lesser energy use in comparison to technical treatment methods, and bringing in material efficiency (Liang and Wang, 2017).

25.3 REMEDIATION APPROACH: USING NBS

NBS are supported by nature especially using plants to manage water quality aiding in wastewater management. NBS are being implemented for contaminated land remediation and land development/regeneration. Remediation technologies could be one of the ways of leading to the management of polluted/grey waters. This holds a special reference to industry sector whereby natural/plants can be utilized to remediate or reclaim a contaminated site. Phytoremediation is the plant-based decontamination strategy for ecological restoration of pollutant water/waste sites. The technology uses plants to remove contaminants from wastewater either by accumulation, adsorption or immobilization (Shanker et al., 2005; Diwan et al., 2008, 2009, 2012). Terrestrial plants, aquatic plants are good accumulators e.g.. potential of *Brassicaceae, Pistia stratiotes*, *Hydrilla verticillata*, *Ceratophyllum demersum*, *Bacopa monnieri*, etc. has been demonstrated for remediation purposes. Metal accumulation ability of algal species like *Oscillatoria* spp., *Phormedium* spp., *Spirogyra* spp., etc. has also been reported. Thus, aquatic/terrestrial plants showcasing higher removal rates can be considered for the restoration of such sites/waste streams.

25.4 NBS AND LIFE CYCLE ASSESSMENT

As NBS strategies are getting implemented, the environmental, social and environment assessment of various technologies, substrates, can be carried out to identify the best possible intervention in various remediation technologies. Within a restorative thinking framework, circularity concept can be implemented for sustainable management of wastewater or wasteland. Life Cycle Assessment (LCA) approach can be applied on various systems of study eg. Water reuse, grey water recycling, urban land/water restoration, built environment water mapping, or performance evaluation of technologies like bioremediation, biodegradation, rhizofiltration for treating wastewater for reuse, or mineral recovery, or restoration of habitats. Mapping of the various technologies supported by NBS, e.g., remediation, biodeterioration, and biodegradation can be done with special reference to resource use, energy use, waste generated, and wastewater remediated. An LCA study on performance assessment of nature-based interventions has been conducted by Vigil et al. (2015) and O'Connor et al. (2018).

The environmental performance evaluation can be carried out in terms of impacts like Climate Change, Global warming, Energy & Material Consumption, Resource Consumption, etc. LCA is being increasingly considered for comparing soil and water remediation processes.

Various risk scenarios can be hypothesized and the comparative performance assessment can be conducteO' Connor et al., 2019), carried out a LCA study to understand the impacts of sea level rise and remediation performance with respect to environmental footprints. Another study carried LCA for a technological assessment of three treatment strategies considered namely soil washing, electrokinetic treatment and enhanced landfarming. The paper used LCA as an approach to study the recovery strategies of contaminated marine sites. A similar study identified bioretention systems, green roofs and wetlands as suitable NBS instruments used in urban environment for storm water hydrology management (Biswal et al., 2022). Similar studies have been conducted for managing urban ecosystems by Zheng et al. (2021), Barwise and Kumar (2020), Prigioniero et al. (2021). Song et al. (2019), have reviewed the NBS for remediation of land in the context of brown field development.

25.5 LCA APPLICATION FOR NBS

Dominguez et al. 2018 studied the LCA application for water circularity using restorative thinking. A comparative study was undertaken to understand the performance of photocatalysis and biological membrane reactor for grey water treatment. Energy was the indicator of assessment as energy is instrumental to both technologies. Similarly, understanding the application of water sustainability through LCA studies is gaining momentum. Wastewater recycling systems, with reference to resource efficiency as a unit of assessment, have been implemented (Wilcox et al., 2016). Treatment of grey water particularly onsite is a powerful application of NBS; however, LCA can help in understanding the positive and negative impacts for optimum utilization of these solutions (Fountoulakis et al., 2016). Remediation approaches can be benchmarked using LCA for easy design making at planning stage.

Comprehensive assessment of environmental impacts generated by NBS is required to undertake the environmental performance evaluation of NBS solutions to implemented template. LCA is a tool that can help in mapping the impact of NBS on environment like impact on ecosystem, resources (depletion). This can entail in gap analysis and better decision-making ability to the stakeholders of concern. LCA provides a comprehensive assessment methodology to understand the impacts of water across the value chain. NBS today play a differentiating role in mitigating water-related impacts. Environmental assessment of NBS in wastewater treatment can be used and literature supports the application of LCA in the context of NBS (Suer and Andersson-Sköld, 2011; Nie, 2010; Nika et al., 2020).

25.6 LCA METHODOLOGY: GOAL STETTING

The context of the study is set out the Goal and scope setting phase. This involves understanding the system of study, life cycle of the process being studied, boundaries and a functional unit (Figure 25.3).

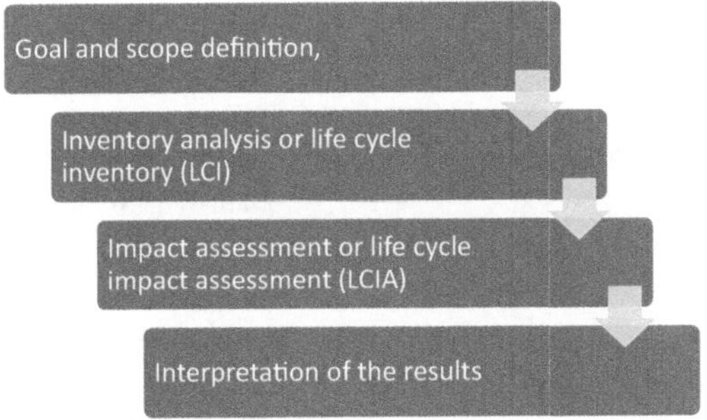

FIGURE 25.3 LCA: methodology.

25.6.1 Life Cycle Inventory

The LC Inventory phase identifies and earmarks the quantifies, and flow of materials in and out of the system of study. It quantifies all the flows in the technosphere, listing down all technical and biological materials, including the flows involving interaction with the external environment. Some of the flows can be identified as resource use as inflows, and outflows like emission and discharges to air, water or land. With reference to NBS, involves carbon capturing/quenching, carbon balance, and emissions in the environment through fertilizer/ pesticide use.

25.6.2 Life Cycle Impact Assessment

This phase of LCA characterizes the impacts caused due to consumption of resources, release of discharges/emissions and final assessment. The impacts are characterized by using characterizing factors (CF) and coming up with impact. The impacts are characterized as mid point and end point indicators. Midpoint indicators involve an assessment of impacts on the environment, and end point indicators consider the impacts on human health, ecosystem and resources. The various methods considered in LCIA methods are CML 2001, Traci, RECIPE, IMPACT 2002+, USECTOX, ILCD method (Table 25.1).

25.6.3 Impact Interpretation

Interpretation phase of LCA helps in evaluating the impacts by giving out the areas of maximum and least impact across midpoint/end point category. The study can be further extended to conducting a sensitivity analysis and benchmarking of various scenarios of assessment (Figure 25.4).

25.7 IMPACT ASSESSMENT METHOD COMMONLY USED FOR NBS

Various methodologies like Environmental Sustainability metrics, IPCC 2007, Recipe, USETOX are being used to identify the mid-point and end point indicators with reference to NBS (Table 25.2). Some of the common ones are Global warming Potential Climate Change, Cumulative Energy Demand, Land Use Change, Human toxicity, and aquatic toxicity.

TABLE 25.1
Relevant Indicators in NBS

Impact Categories	
Environment	Climate change
	GWP
	Photochemical ozone creation
Resource depletion	Abiotic
	biotic
Human toxicity	Toxicity potential
Ecosystem (Toxicity)	Terrestrial ecotoxicity
	Marine ecotoxicity
	Marine sediment ecotoxicity
Energy	Cumulative energy demand
Land use	Land use change
Biodiversity impacts	Land competition
Eutrophication	PO_4- eq, NO_3-
Water stress index	Water withdrawal/availability

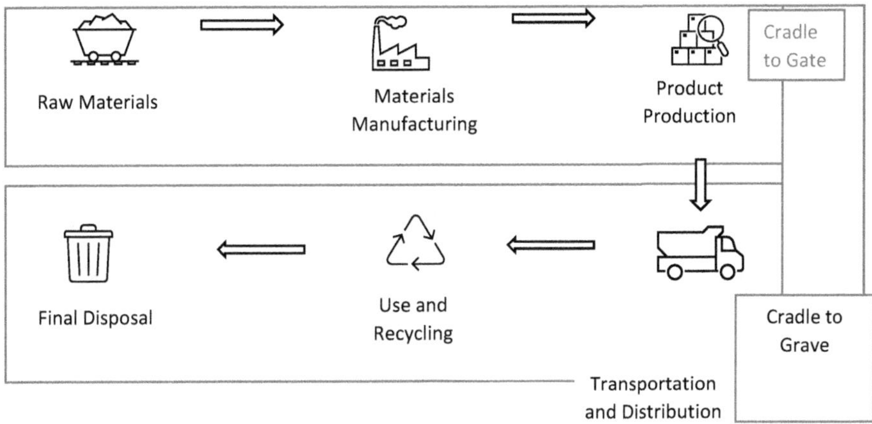

FIGURE 25.4 Life cycle assessment.

TABLE 25.2
Impact Assessment Method Commonly used for NBS

Methods	Indicators
Environmental Sustainability Metrics	Atmospheric acidification
	Global warming
	Human health effects
	Photochemical ozone formation
	Stratospheric ozone depletion
IPCC 2007	Climate change
	Acidification
	Eutrophication
ReCiPe	Climate change
	Climate change ecosystems
	Acidification
	Eutrophication
	Agricultural land occupation
	Fossil depletion
Water scarcity footprint	Water scarcity
Impact world +	Aware model

25.7.1 Natural Managed System

The natural water cycle works re-optimize, reuse, and replenish water (Figure 25.5).

25.7.2 Human Managed System

On the human-managed (Figure 25.5) system, there is human activity that highly affects water circularity when we change the natural water cycle, such as by:

- Taking the freshwater faster than it replenishes.
- Polluting the water, which reduces its usefulness for other users.

With Close loop Economy for Water, there is a possibility to synchronize the human water cycle with the natural water cycle by doing the following actions:

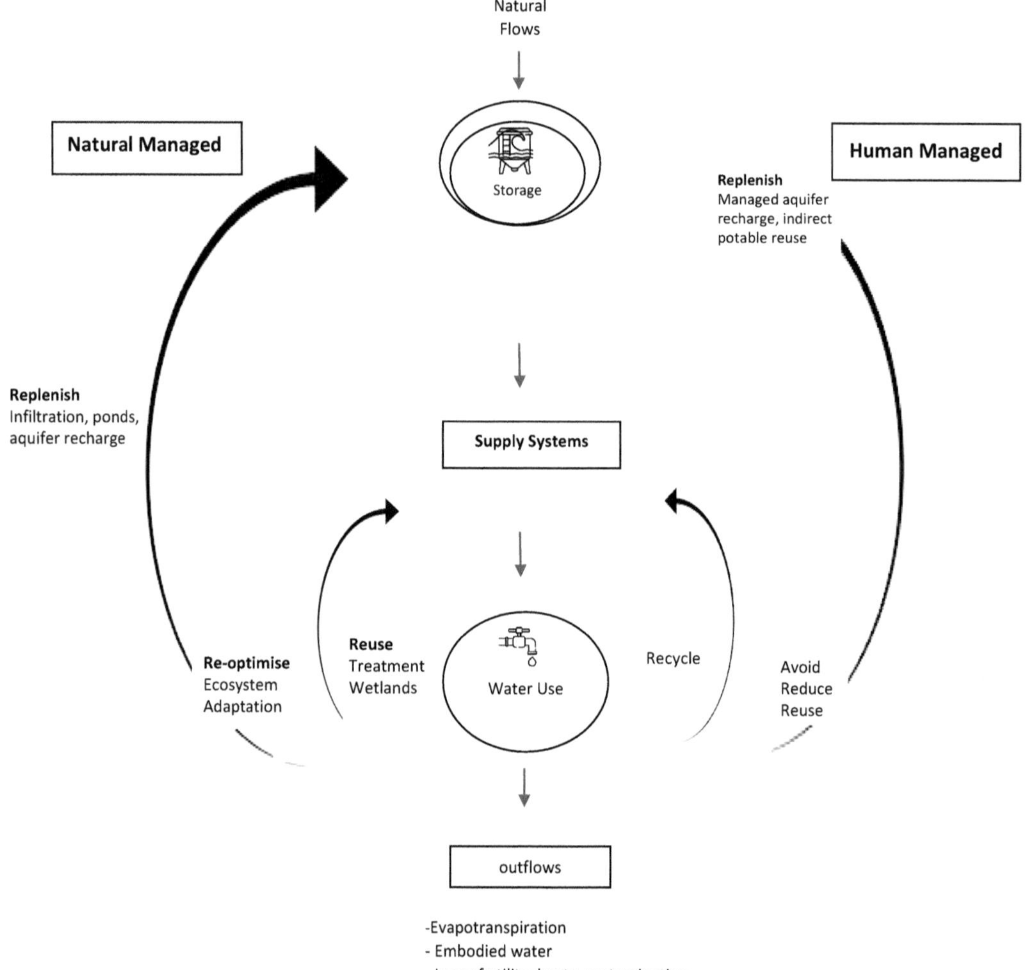

FIGURE 25.5 Natural-managed and human-managed system.

25.7.2.1 Decreasing the Amount of Water Use

- Avoid Use
- Reduce Usage
- Replace

25.7.2.2 Optimising the Use of Water

- **Reuse:** Using water again for the same or a different purpose without any treatment is known as reuse.
- **Recycle:** Recycling entails reusing treated water for the same or different applications. Various technologies and treatment methods exist, and they vary based on the intended recycling use.
- **Cascading:** Water is used in cascading in a series of sequential operations for various purposes. Multiple solutions incorporating various tactics can be merged and combined to accomplish this goal. Water can be utilized repeatedly in many phases of both home and industrial operations (whether it has been treated or not) (Figure 25.5).

25.7.2.3 Retaining Water

- Store
- Recover

25.8 CONCLUSION

With the growing concept of circularity, it is becoming imperative to transition from a linear to a circular model of development. Resources, particularly water, are valuable and have competing users ranging from industry, commercial, and household, along with water required for nature-based ecosystem services. Circularity concepts specifically hold potential for water and allied sectors and NBS can be the tools in aiding this transition. Methodologies like LCA become instrumental in mapping the water flows, and impact assessment of water-linked interventions. Technological assessment of water-linked systems can be done using methodologies like LCA, aiding in identifying the impacts/hotspots across the life cycle. There is a need to understand the processes and entails in benchmarking optimum solutions, with reference to water reuse, recycling and regeneration.

REFERENCES

Barwise, Y., & Kumar, P. (2020). Designing vegetation barriers for urban air pollution abatement: A practical review for appropriate plant species selection. *Climate and Atmospheric Science*, 3, 12. https://doi.org/10.1038/s41612-020-0115-3.

Biswal, B.K., Bolan, N., Zhu, Y.G., & Balasubramanian, R. (2022). Nature-based Systems (NbS) for mitigation of stormwater and air pollution in urban areas: A review. *Resources, Conservation and Recycling*, 186, 106578. https://doi.org/10.1016/j.resconrec.2022.106578

Cohen-Shacham, E., Walters, G., Janzen, C., & Maginnis, S. (Eds.) (2016). *Nature-based Solutions to address global societal challenges*. Gland: IUCN.

Diwan, H., Ahmad, A., & Iqbal, M. (2008). Genotypic variation in the phytoremediation potential of Indian mustard for chromium. Environmental *Management*, 41(5), 734–741.

Diwan, H., Ahmad, A., & Iqbal, M. (2009). Chromium-induced modulation in the antioxidant defense system during phenological growth stages of Indian mustard. *International Journal of Phytoremediation*, 12(2), 142–158.

Diwan, H., Ahmad, A., & Iqbal, M. (2012). Chromium-induced alterations in photosynthesis and associated attributes in Indian mustard. *Journal of Environmental Biology*, 33(2), 239.

Dudley, N., Stolton, S., Belokurov, A., Krueger, L., Lopoukhine, N., MacKinnon, K., Sandwith, T., & Sekhran, N. (Eds.) (2010). *Natural Solutions: Protected areas helping people cope with climate change*. Gland, Washington, DC and New York: IUCNWCPA, TNC, UNDP, WCS, The World Bank and WWF.

Dumitru, A., & Wendling, L. (2021). *Evaluating the impact of nature-based solutions: A handbook for practitioners*. Luxembourg: European Commission.

Fountoulakis, M.S., Markakis, N., Petousi, I., & Manios, T. (2016). Single house on-site grey water treatment using a submerged membrane bioreactor for toilet flushing. *Science of the Total Environment*, 551–552, 706–711.

González-Domínguez, J., Sánchez-Barroso, G., Zamora-Polo, F., & García-Sanz-Calcedo, J. (2020). Application of circular economy techniques for design and development of products through collaborative project-based learning for industrial engineer teaching. *Sustainability*, 12(11), 4368. https://doi.org/10.3390/su12114368

IUCN. (2020). *Guidance for using the IUCN global standard for nature-based solutions. A user-friendly framework for the verification, design and scaling up of nature-based solutions*. First edition. Gland: IUCN.

Liang, D., & Wang, S. (2017). Development and characterization of an anaerobic microcosm for reductive dechlorination of PCBs. *Frontiers in Environmental Sciences & Engineering*, 11, 1–10.

Magni, F., Musco, F., Litt, G., & Carraretto, G. (2020). The mainstreaming of NBS in the SECAP of San Donà di Piave: The LIFE master adapt methodology. *Sustainability*, 12(23), 10080. https://doi.org/10.3390/su122310080

Nika, C.E., Gusmaroli, L., Ghafourian, M., Atanasova, N., Buttiglieri, G., & Katsou, E. (2020). Nature-based solutions as enablers of circularity in water systems: A review on assessment methodologies, tools and indicators. *Water Research*, 183, 115988. https://doi.org/10.1016/j.watres.2020.115988.

O'Connor, D., Hou, D., Ok, Y.S., Song, Y., Sarmah, A.K., & Li, X. (2018). Sustainable in situ remediation of recalcitrant organic pollutants in groundwater with controlled release materials: A review. *Journal of Controlled Release*, 283, 200–213.

O'Connor, D., Zheng, X., Hou, D., Shen, Z., Li, G., Miao, G., & Guo, M. (2019). Phytoremediation: Climate change resilience and sustainability assessment at a coastal brownfield redevelopment. *Environment International*, 130, 104945.

Prigioniero, A., Zuzolo, D., Niinemets, Ü., & Guarino, C. (2021). Nature-based solutions as tools for air phytoremediation: A review of the current knowledge and gaps. *Environmental Pollution*, 277, 116817. https://doi.org/10.1016/j.envpol.2021.116817

Shanker, A. K., Cervantes, C., Loza-Tavera, H., & Avudainayagam, S. (2005). Chromium toxicity in plants. *Environment International*, 31(5), 739–753.

Sheng-Wei Nie, Wang-Sheng Gao, Yuan-Quan Chen, Peng Sui, & A. Egrinya Eneji (2010). Use of life cycle assessment methodology for determining phytoremediation potentials of maize-based cropping systems in fields with nitrogen fertilizer over-dose. *Journal of Cleaner Production*, 18(15), 1530–1534.

Song, Y., Kirkwood, N., Maksimović, Č., Zheng, X., O'Connor, D., Jin, Y., & Hou, D. (2019). Nature based solutions for contaminated land remediation and brownfield redevelopment in cities: A review. *Science of the Total Environment*, 1(663), 568–579. https://doi.org/10.1016/j.scitotenv.2019.01.347.

Sowińska-Świerkosz, B., & García, J. (2022). What are Nature-based solutions (NBS)? Setting core ideas for concept clarification, *Nature-Based Solutions*, 2, 100009. https://doi.org/10.1016/j.nbsj.2022.100009

Suer, P., & Andersson-Sköld, Y. (2011). Biofuel or excavation? – Life cycle assessment (LCA) of soil remediation options. *Biomass Bioenergy*, 35, 969–981.

Vigil, M., Marey-Pérez, M.F., Martinez Huerta, G., & Álvarez Cabal, V. (2015). Is phytoremediation without biomass valorization sustainable? - Comparative LCA of landfilling vs. anaerobic co-digestion. *Science of the Total Environment*, 505, 844–850.

Wilcox, J., Nasiri, F., Bell, S., & Rahaman, M.S. (2016). Urban water reuse: A triple bottom line assessment framework and review. *Sustainable Cities and Society*, 27, 448–456.

WWAP (United Nations World Water Assessment Programme) (2018). *The United Nations World Water Development Report 2018: Nature-based solutions for water*. Paris: UNESCO.

Zheng, T., Zhang, S., Li, X.B., Wu, Y., Jia, Y.P., Wu, C.L., He, H.D., & Peng, Z.R. (2021). Impacts of vegetation on particle concentrations in roadside environments. *Environmental Pollution*, 282, 117067. https://doi.org/10.1016/j.envpol.2021.117067.

26 Biorefinery

Algal-based Valorisation of Wastewater for Nutrients Recovery and Energy Production

Supratim Ghosh

26.1 INTRODUCTION

It has been widely proven (Bogan et al., 1960; Davis et al., 1990; Doran and Boyle, 1979; Gates and Borchardt, 1964; Oswald and Gotaas, 1957) that microalgae have the ability to accumulate nutrients from wastewater, and this process is seen to be an environmentally friendly way to clean nutrients (Shi et al., 2007; Whitton et al., 2015). Aside from allowing the remediation of wastewater, algal cultivation has other added benefits including: (i) CO_2 sequestration which is enabled by the inherent property of photosynthesis (Chandrasekhar et al., 2022), (ii) oxidising the effluent thereby treating it in case of secondary wastewater treatment (Das, 2016), (iii) because the algae are photosynthetic organisms, a continuous supply of external carbon is not required which is not the case for other biological treatment methods (Susmita Ghosh et al., 2022; Yang et al., 2021), and (iv) trace pollutant removal by accumulation in biomass. Furthermore, the biomass obtained after wastewater treatment can be utilised for various purposes which include biofuels (biodiesel, biomethane, bioethanol), cosmeceuticals, animal feed, etc. in a biorefinery concept (Amulya et al., 2015; Kalia, 2016).

Microalgae are prevalent in wastewater settings, notwithstanding at diluted absorptions, suggesting that the nutritional properties of such habitats are favourable for growth, with microalgae demonstrating the potential to repair effluents at concentrations routinely found after secondary treatment (Christenson and Sims, 2011; Hemalatha and Venkata Mohan, 2016). Therefore, microalgae are being considered for tertiary wastewater treatment as a sustainable and cost-effective alternative (Gómez-Serrano et al., 2015; Selvaratnam et al., 2015; Sukačová et al., 2015) The constraints for algae-mediated tertiary treatment of wastewater include a reasonable hydraulic retention time (HRT). Current methods for tertiary treatment which include wetlands generally have a HRT of one day (Butterworth et al., 2013). In the present scenario, microalgal wastewater treatment has a HRT of 4–10 days which can only be feasible when land is abundant in countries such as the USA (Cai et al., 2013). This is one of the challenges of algae-mediated wastewater treatment which has to be overcome in order to provide a sustainable solution.

Microalgae have evolved three metabolic pathways for development and subsistence: autotrophic, heterotrophic, and mixotrophic. As the environment changes, they have the ability to adjust their metabolism (Ghosh et al., 2015). Autotrophy involves the fixation of CO_2 in the presence of sunlight through the process of photosynthesis (Kumar et al., 2016). The autotrophic process has its flaws which include lower biomass productivity, penetration of light in the depths of the bioreactor (which is very shallow due to the shading effect) and requirement of a photobioreactor which has a very high surface area and shallow depth in order to increase the light penetrability. When there is no light, the process of photosynthetic energy production is hindered, and instead, algae obtain their energy from alternate organic processes through heterotrophic mode of nutrition that channelises the excess energy to storage products such as starch and lipids (Kumar et al., 2013).

DOI: 10.1201/9781003441144-26

Increased lipid yields are made possible by this route, which results in biomass that is noticeably denser. A biorefinery approach that includes biodiesel production along with CO_2 sequestration is envisaged as a cost-effective and sustainable method that offers extra initiative of waste remediation. For recovering lipids and different value-added products from microalgae, a variety of bioprocesses and downstream processing methods are widely employed. Physical, thermochemical, biochemical, and biological treatments are also used to manufacture energy-rich products from the original biomass (Ahmed et al., 2022). Utilising algal biomass effectively and reducing the overall residual waste component of biomass will be made possible by combining the biorefinery idea with wastewater treatment, which will support sustainable economics. By examining current literature in conjunction with new advances, an effort has been made in this chapter to outline the fundamental and practical features of the various types of wastewaters and their characteristics, the varied methods of conversion for algal biomass and its integration into an algal biorefinery concept keeping in mind the technoeconomic and life cycle aspects of the process.

26.2 WASTEWATER CHARACTERISTICS AND USAGE

Urbanisation, irrigation, and industrialisation all result in significant daily wastewater production, and untreated discharge of this wastewater has a substantial negative environmental impact (Ilyas et al., 2019). This negative side stream is caused by the accumulation of high amounts of evolving pollutants, inorganic nutrients, and heavy metals (Barat et al., 2013; Hena et al., 2021). The spectrum of wastewater is different under different conditions due to the source from which it is obtained (agricultural, industrial, domestic, municipal). Moreover, the local environment and socio-economic parameters also determine the characteristics and quality of wastewater (Das, 2016; Von Sperling, 2007). Wastewater has a high concentration of organic and inorganic substances and chemicals, total solids (TS), and microorganisms (Silva-Bedoya et al., 2016). Numerous micronutrients (vitamins and heavy metals) and macronutrients (phosphorus, nitrogen and carbon) are also present (Razzak et al., 2017). However, additional toxins present in wastewater, such as heavy metals and evolving pollutants, can be hazardous to its development and survival. These nutrients are necessary for microalgal growth (del Mar Morales-Amaral et al., 2015; Maryjoseph and Ketheesan, 2020; Razzak et al., 2017). Due to the difficulty of executing the necessary laboratory studies, insufficient attention has been paid to identifying the many components that comprise wastewater in the development of a wastewater treatment plant (WWTP) in the past. But wastewater with suspended particles has a measurable impact on the effectiveness of the system (Marcilhac et al., 2014). Enhancing water's ability to self-cleanse and reducing pollutant concentrations to lower specified levels that meet sustainability goals are the main goals of wastewater treatment plants (Mainardis et al., 2020). Wastewater treatment procedures can be classified as primary, secondary, or tertiary. Wastewater treatment can be broadly divided into three categories. The primary treatment generally separates the solids by sedimentation and floc formation. Secondary treatment utilises the effluent from the primary treatment and treats it aerobically or anaerobically to reduce the organic load. Tertiary treatment includes the removal of nitrogen and phosphorus and practically all suspended particles from wastewater. (Hena et al., 2021). Table 26.1 shows the main characteristics of wastewater from different sources.

26.3 ALGAL BIOREFINERY CONCEPT IN THE WASTEWATER PERSPECTIVE

A biorefinery is a facility that integrates various biomass conversion processes to create energy and value-added chemicals. According to a wide definition, it transforms all types of biomasses (organic residues, energy crops, and aquatic biomass) into a variety of goods (fuels, chemicals, electricity and heat, materials, food and feed, etc.) (Figure 26.1). Due to the low environmental impact of the

TABLE 26.1
Main Characteristics of Wastewater of Different Sources

Source	BOD (ppm)	COD (mg/L)	pH	TS (mg/L)	TSS (mg/L)	VS (mg/L)	TN (mg/L)	TP (mg/L)	EC
Cheese industry	2.42	2.42	12.08	–	5.07	–	–	–	–
	27.36	50–70	6.0–6.5	55–65	10–15	–	–	–	–
Dairy industry	442	8960	7.10	797.2	253.6	–	120.1	–	1082.2
	170	1007.3	4.53	–	299.67	–	–	–	1091.67
Ice cream industry	2.45	5.2	5.2	3.9	3.1	2.6	–	–	–
Domestic	250	500	–	700	220	150	40	12	–

Source: Aderibigbe et al. (2017).

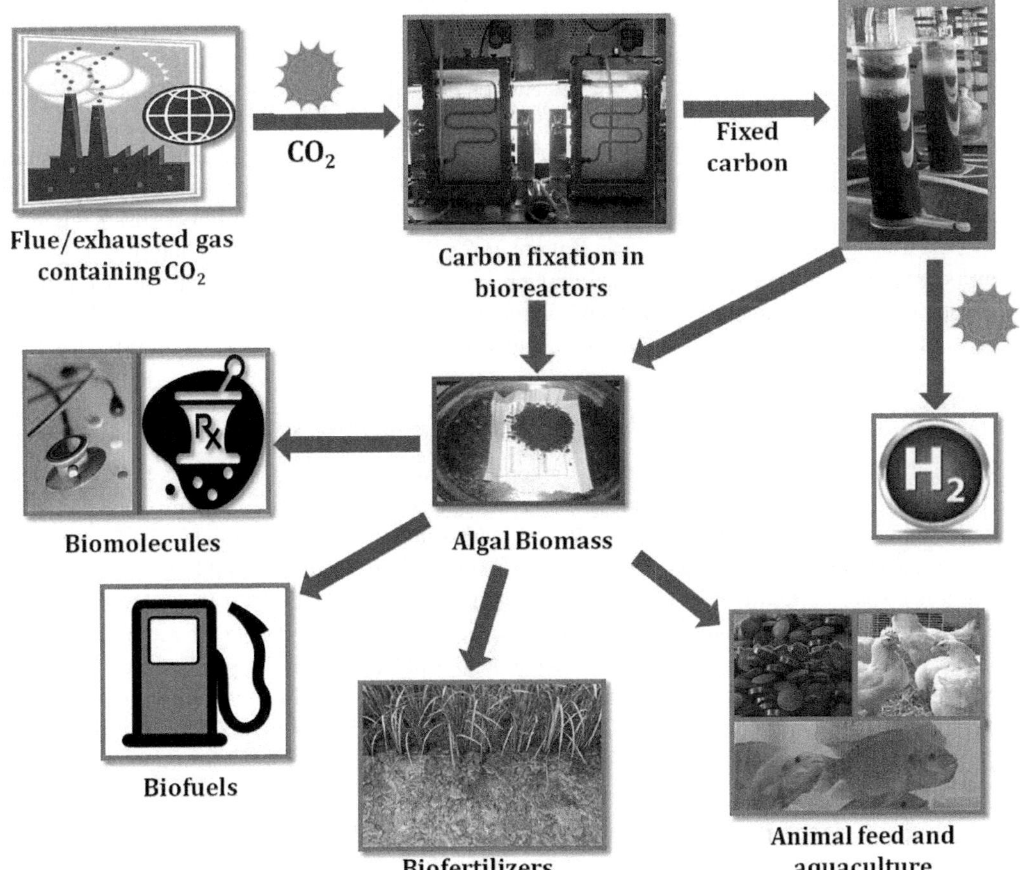

FIGURE 26.1 A pictorial interpretation of the algal biorefinery concept.

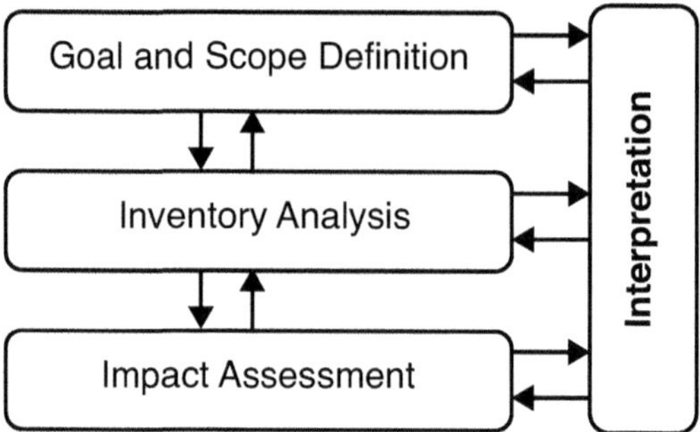

FIGURE 26.2 Schematic representation of Life Cycle Analysis (LCA).

conversion processes, biomass may be used to make a variety of goods. Similar to how different products are generated at various stages of petroleum refining in crude oil refineries, so is this idea. A conceptual approach for future biofuel generation and high value-added product manufacture is provided by the biorefinery idea. As a result, the cost of producing liquid fuels using the most biomass is decreased. Future biorefineries must be more effective, utilising both the heat produced by the process to the fullest and the biomass to the greatest possible degree. The heat created during the process may be circulated to meet the biorefinery's heat needs (Ghosh and Das, 2016).

Biomass is utilised as a raw resource for the manufacturing of a variety of goods, much like a petroleum refinery. To produce goods for commercial use, many conversion processes (physical, chemical, biological, and thermal) are utilised either singly or in combination. After conversion, the products are either segregated into different independent products or may go through additional processing to produce goods with value-added features. The waste products generated after each phase of treatment are either reused or recycled back into the manufacturing chain to be utilised as raw materials in the process. Transportation fuels, medicines, food additives, and biofertilisers are just a few of the uses for items made from algae (Figure 26.2). When the primary raw materials utilised in the process are waste products, the process may be made more cost-effective. This in turn fulfils a dual function of both energy production and bioremediation. To reduce production costs and carbon impact, the biorefineries may potentially be linked with the current power generation infrastructure.

26.4 CONVERSION PATHWAYS IN ALGAL BIOREFINERY

Microalgal biomass is composed of various macronutrients which include proteins in a major proportion followed by carbohydrates and lipids. As they are photosynthetic in nature, they also contain chlorophyll and accessory pigments. Various cultivation conditions can vary the biochemical composition accordingly. Algal biomass is grown using wastewater in the algal biorefinery process, where it is converted to different products using varied alteration platforms. There are three main algal conversion platforms that are noteworthy: (i) biochemical conversion which produces biofuels such as biomethane, bioethanol and biohydrogen and biopolymers such as polyhydroxyalkanoates (PHAs); (ii) chemical conversion produces biodiesel and (iii) thermochemical conversion produces syngas, charcoal, biocrude or bio-oil. The most affordable and environmentally beneficial method of conversion is biochemical.

26.4.1 Biochemical Conversion

The biochemical conversion of organic material to biomethane via anaerobic digestion (Gao et al., 2018; Schenk et al., 2008) and to biohydrogen (Nayak et al., 2014; Roy et al., 2014; Singh et al., 2022), bioethanol (Banerjee et al., 2022, 2021; H. Singh et al., 2019), and bioplastics (Ghosh et al., 2021, 2019; Susmita Ghosh et al., 2022) via fermentation, plays a crucial role in the usage of enzymes

TABLE 26.2
Algal Biorefinery using Biochemical Conversion Route

Wastewater	Algal Species	Cultivation Conditions	Conversion Platform	Product	Reference
Primary treated wastewater	Mixed species	Pilot scale	Anaerobic digestion	Biomethane	Passos et al. (2013)
Domestic wastewater	Mixed species	Laboratory scale	Anaerobic digestion	Biomethane	Kinnunen et al. (2014)
Municipal wastewater	Mixed species	Pilot scale	Anaerobic digestion	Biomethane	Banu et al. (2020a)
Diluted swine manure	Mixed species	Laboratory scale	Dark fermentation	Biohydrogen	Kumar et al. (2018)
Urban wastewater	*Chlorella vulgaris*	Laboratory scale	Dark fermentation	Biohydrogen	Batista et al. (2015)
Urban wastewater	*Scenedesmus obliquus*	Laboratory scale	Dark fermentation	Biohydrogen	Batista et al. (2015)
Synthetic wastewater	*Chlamydomonas reinhardtii*	Laboratory scale	Yeast fermentation	Bioethanol	Banerjee et al. (2022)
Domestic wastewater	*Microcystis* sp.	Pilot scale	Fermentation	Bioplastics	Abdo and Ali (2019)
Domestic wastewater	*Haematococcus pluvalis*	Pilot scale	Fermentation	Bioplastics	Abdo and Ali (2019)
Domestic wastewater	*Chlorococcus turgidus*	Pilot scale	Fermentation	Bioplastics	Abdo and Ali (2019)
Domestic wastewater	*Microcystis aeruginosa*	Pilot scale	Fermentation	Bioplastics	Abdo and Ali (2019)

produced by bacteria or other microorganisms. Algal biomass has a complex cell wall that prevents hydrolysis by enzymes and anaerobic substrate biodegradability, which leads to a low product yield. Numerous studies have used various pre-treatments or cell fragmentation procedures preceding the fermentation step in attempt to get around this problem. Many researchers have employed various biochemical conversion pathways to utilise algal biomass for biofuel generation. A mixed microalgal consortium was utilised for the production of biomethane through anaerobic digestion using municipal wastewater as a substrate (Banu et al., 2020a). Mixed microalgae was also utilised for biohydrogen production using swine manure where they studied the impact of pretreatment and organic load on the production process (Kumar et al., 2018). To demonstrate the scalability, pilot-scale reactors were studied for biofuel production. Studies were performed for biomethane production using wastewater as a substrate where the feasibility and longevity of growing mixed microalgae were observed (Passos et al., 2014). Depending on the source, substrate, growing conditions, and biomass concentration, different microalgal species may have different potential for producing biofuel. Ho et al. successfully produced 3.55 g/L of bioethanol when growing the microalgal species *Chlorella vulgaris* in a lab-scale reactor with a 1 L capacity using synthetic growth medium (Ho et al., 2013). Bioethanol production was also studied using *Scenedesmus dimorphus* where they studied the synergistic effect of pretreatment and fermentation (Chng et al., 2017). The various pathways for conversion of microalgal biomass to various biochemical platforms are shown in Table 26.2.

26.4.1.1 Biomethane

Biomethane-rich biogas will be produced by the anaerobic digestion (AD) of organic feedstock. The AD process typically consists of four main steps: hydrolysis, acidogenesis, acetogenesis, and methanogenesis. The rate-limiting step in AD is hydrolysis, which is one of these processes. Many studies advise pretreatment or cell disintegration as a solution to this problem. The synthesis of biomethane using an algal biorefinery was covered in this section, although growing algae in fresh water has significant start-up and ongoing expenses. As a result, Mahdy et al. cultivated microalgae utilising just municipal wastewater as a source of nutrients, and the biomass was subsequently utilised to produce biomethane. According to their findings, *Chlorella vulgaris* is useful for removing excess nutrients from wastewater and accumulating carbohydrates. Anaerobic digestion is used to produce biomethane from the collected biomass. As a consequence, at an ideal organic loading rate of 1.5 kg COD/m^{3}d, the greatest biomethane output of 137 mL of CH_4/g of COD is attained (Mahdy et al., 2016). According to another study, mixed macroalgae effectively use the nutrients present in wastewater to boost biomass productivity. Open cultivation in raceway ponds was established with significant biomass productivity with simultaneous reduction in COD of wastewater (Kannah et al., 2021a). The collected microalgae's tough cell wall and extracellular matrix protection result in a very low methane production. To enhance methane production, it can be eliminated by using any form of pretreatment as a precursor to anaerobic digestion. Studies have focussed on the collection of native strains of microalgae for biomethane production. Anaerobic digestion of the microalgae produced biomethane in a significant quantity (Passos et al., 2013). In a different research, the researchers mainly employed treated wastewater to effectively produce microalgae which increased the biomethane productivity thereby emphasising the importance of pretreatment of wastewater (Passos et al., 2014). Direct wastewater usage limits biomass productivity; however, treated wastewater allows algal biomass to more efficiently utilise dissolved nutrients for biomass productivity.

26.4.1.2 Biohydrogen

Clean energy is said to include hydrogen. Its energy conversion efficiency is great (142 kJ/g), and it has a high energy density of 120 MJ/kg (Das and Veziroglu, 2008). In general, hydrogen may be produced using the following processes: gasification, pyrolysis, natural gas steam reforming, fractional oxidation of CH_4, electrolysis of water, and fermentation of organic biomass (Mahata et al., 2020). Among these, anaerobic fermentation-based biological hydrogen generation is more affordable, environmentally benign, and energy-efficient than other methods (Béligon et al., 2018). Biological hydrogen production is supplementarily separated into two groups: photo (fermentation that occurs in the presence of light source) and dark (fermentation that occurs in the absence of light source). The method most frequently chosen for producing biological hydrogen is dark fermentation. A few researchers do, however, strongly advise pre-treating algae before dark fermentation in order to facilitate simple hydrolysis of carbohydrate polymers. The effects of various colour light energies employed for the development of immobilised algal biomass and the generation of biohydrogen were investigated by various researchers. Microalgae were grown in urban wastewater to boost biomass output. Blue and purple lights were used by the authors to evaluate the effects of algae biomass productivity and biohydrogen output. As a result, blue light-grown algae exhibit a greater biomass concentration than purple light-grown algae (Ruiz-Marin et al., 2020).

Similar to this, Kumar et al. produced biohydrogen by cultivating a variety of algae species in various dilutions of swine dung. Increased dilution ratios decreased biomass productivity. The gathered biomass was then put through a dark fermentation process to create biohydrogen. At similar loading rates of swine dung, the highest biohydrogen production was observed (Kumar et al., 2018). To produce biohydrogen, Batista et al. cultivated *Scenedesmus obliquus* and *Chlorella vulgaris* in a tubular photobioreactor. *Scenedesmus obliquus* and *Chlorella vulgaris* were found to produce the largest amounts of biohydrogen according to the researchers (Batista et al., 2015).

26.4.1.3 Bioethanol

Because of its physiochemical similarities to gasoline, bioethanol does have the potential to replace gasoline consumption. The bioethanol production is more influenced by the biochemical makeup of the feedstock. The most popular substrate for yeast fermentation to make bioethanol is one that is high in carbohydrates. Algal species can accumulate starch and various other polysaccharides to a large extent when grown under stress conditions. Algal biomass is therefore regarded as a promising fuel for the manufacture of bioethanol (Banerjee et al., 2022, 2021). Researchers studied the various saccharification techniques utilising cultivated microalgae biomass (*Hindakia tetrachotoma*) and municipal effluent. For the purpose of growing microalgae, a flat airlift photo-bioreactor with a volume of 1 L is designed. The highest biomass output is seen at the lowest (25%) growth medium concentration, which is at 0.78 g/L. Higher dilutions lead to further reduction in biomass productivity. The microalgae were then put through three separate hydrolysis processes: enzymatic, alkaline, and acidic. Enzymatic hydrolysis stands out among them because it produces bioethanol at a greater yield (11.2 g/L) and converts substrates at a faster pace (94%) (Onay, 2019). Similar to this, another study reported on the outdoor pond scale growth of algal biomass for the generation of bioethanol in a household wastewater treatment facility. *Microcystis* sp., the algal biomass that was extracted, has a high glucose content (45%). Prior to yeast fermentation, they advised acid hydrolysis, which increases the yield of bioethanol (El-Mekkawi et al., 2019). In order to produce bioethanol, the impact of several acids and their concentrations on microalgae were investigated. In their research, sulfuric and acetic acids were chosen with a range of concentrations, including 1%, 3%, 5%, 7%, and 9%. At 5% of acetic and sulfuric acid, respectively, the maximum bioethanol yields of 0.231 and 0.281 g/g were observed. They found that using sulfuric acid, which has a higher bioethanol output than acetic acid at all doses, is more successful than using acetic acid, which is a weaker acid, in increasing the synthesis of bioethanol from microalgae. This shows that using a strong acid pretreatment increases bioethanol output while requiring less energy and chemical input (Phwan et al., 2019).

26.4.1.4 Bioplastic

Petrochemical-derived plastic has poor natural biodegradability and stays in the environment for lengthy periods of time, which contributes to a number of environmental problems. However, the creation of bioplastics from organic biomass offers two advantages, including the replacement of conventional plastics and the amelioration of a number of environmental problems. Polyhydroxyalkanoates (PHA), often known as bioplastics, decompose entirely into CO_2 and water (Supratim Ghosh et al., 2022). A feedstock high in proteins is needed to produce bioplastic. Abdo and Ali investigated the impact of household wastewater-cultivated algae and the biomass used to make bioplastics. In their investigation, the scientists chose three microalgal strains for the generation of bioplastics. Among the three strains, *Microcystis aeruginosa* had the highest PHA yield (Abdo and Ali, 2019). Pilot scale studies on bioplastic production using microalgae were also performed where the effect of various substrates was observed. Protein-rich microalgae were utilised for bioplastic production (Rocha et al., 2020). Compared to other translation processes, the biochemical conversion process uses less energy and produces a significant quantity of CO_2 as a by-product, which may be used as a carbon source for algae development.

26.4.2 Chemical Conversion

Because microalgae-derived bio-oils have a higher viscosity than diesel oils by nature, transesterification must be done on the microalgae oils before using them in engines to reduce their overall viscosity (Akubude et al., 2019). The direct transesterification approach is thought to be more advantageous for increasing the effectiveness of biodiesel made from microalgae. Notably, in a transesterification process, alcohol serves as both the reactant and the solvent. Methanol is frequently employed in place of other sorts of alcohol in this procedure due to its accessibility and price. The inclusion of a catalyst is also significant since it increases alcohol's solubility, which is vital given that alcohol is not very soluble in various types of oils. Direct transesterification of algal biomass removes the problematic steps of drying and moisture removal thereby reducing the costs involved in the process and increasing the final yields of biodiesel. As a result, scientists have looked into the creation of sophisticated wet lipid extraction procedures (Lakshmikandan et al., 2020). Microalgal pretreatment for lipid extraction has been explored mainly by two ways: physical (which includes ultrasonication, microwave, etc.) and chemical (which includes ionic detergents, etc.) (Howlader and French, 2020; Ramola et al., 2019). Microwave treatment has been used for lipid extraction from *Spirulina* sp. (R. Singh et al., 2019). Furthermore, it has been demonstrated that large-scale operations may be successful when hexane and methanol are combined in an equal proportion (Shin et al., 2018). Ionic liquid use can boost the effectiveness of lipid extraction from microalgae. Hexane or the supercritical CO_2 extraction process can produce biodiesel with a relatively high conversion efficiency. Higher conversion rates for biodiesel synthesis can be observed with improved catalysts (Umdu et al., 2009). Additionally, it was discovered that the use of metal-based catalysts resulted in a 90.2% conversion efficiency when operating at 350–400°C and 2500 pressure for the transesterification of green microalgae (Cheng et al., 2009; McNeff et al., 2008). As much as 41% and 65% less biodiesel might be produced from lipids extracted using chloroform and n-hexane, respectively (Cheng et al., 2020). Conversely, the transesterification of lipids in the presence of TEPDA resulted in an up to 9% increase in FAME output, according to the same investigators. The use of organic solvents has been observed to increase the biodiesel yield as compared to other physical methods. To (Ghosh et al., 2017; Goh et al., 2019; Mandik et al., 2020).

The production of biodiesel from microalgae is now of great interest to the scientific research community for a number of reasons, including the microalgae's quick growth and improved lipid build up. Additionally, compared to traditional ethanol, microalgae-based biodiesel displays a greater energy content (up to 34% increase) (Moshood et al., 2021). Notwithstanding this promise, additional study is required to determine whether expanding biodiesel production using the wet lipid extraction process is feasible. One of the oldest common and advanced processes is the alkaline-catalysed procedure. However, this approach has drawbacks due to the significant energy required, the difficulty in extracting glycerine, and the post-process removal of catalyst. Scientists have suggested alternate techniques of producing biodiesel that do not require a catalyst called as the vapor phase approach, taking into account the adverse effects from the usage of alkaline-based catalysts (Ortiz-Martínez et al., 2019). With this innovative method, microalgae biomass is transformed into biodiesel at temperatures ranging from 250°C to 350°C in a single reactor. The supercritical approach uses a great deal less energy than conventional co-solvent-supported transesterification and lipid extraction routes (Dickinson et al., 2017).

26.4.3 Thermochemical Conversion

Liquid algal feedstock may be converted to bio-oil using the hydrothermal liquefaction method at low temperature and high pressure. The employment of a catalyst in this process is optional (Arvindnarayan et al., 2017). The energy-intensive drying process is eliminated in such a method, which has advantages. Biomolecules found in algal biomass are converted into bio-oil by the process of liquefaction. Another often used method is pyrolysis, which involves burning biomass at temperatures between 350°C and 700°C without the use of oxygen to produce bio-oil, syngas, and charcoal (Yang et al., 2019). With a fairly brief hot vapour residence period of about one second, rapid pyrolysis takes place in a medium temperature range of around 500°C. (Venderbosch, 2019). Fast pyrolysis produces more bio-oil outputs while using less total energy because of its quick response. Furthermore, compared to its equivalent acquired through slow pyrolysis, bio-oil produced by the quick pyrolysis procedure has a lower viscosity (Tan et al., 2015). The dark brown liquid that results from the pyrolysis or hydrothermal liquefaction (HTL) of microalgae biomass is known as bio-oil. The primary distinction between these two techniques is that the former can only be used in dry circumstances (i.e., when the biomass feedstock has less than 5% moisture), whereas HTL may be used with any kind of microalgal biomass (Toro-Trochez et al., 2019). These techniques include the thermal decomposition of organic molecules under anaerobic conditions (Sun et al., 2020). Prior to further processing, the algal biomass material undergoes a number of thermophysical processes, including dehydration, dehydrogenation, deoxygenation, and decarboxylation. Microalgae biomass components are among the main determinants of variation in biocrude output and quality. When *Chlorella* species of microalgae are pyrolysed, there is a great possibility for extracting high energy content (Arvindnarayan et al., 2017). Additionally, shared mobility fuels may be simply combined with bio-oil produced from biomass microalgae without requiring significant engine adjustments. In another experiment, *Spirulina platensis* was pyrolysed at 350°C to produce biochar, bio-oil, and gases in proportions of 39.7, 23.8, and 19.2 wt%, respectively (Jena and Das, 2011). Higher yields of bio-crude were observed at temperatures of hydrothermal liquefaction (Shakya et al., 2017). Additionally, the biomass of two different species of microalgae, *Scenedesmus* and *Spirulina*, was utilised to create bio-oil. The scientists compared the outcomes of applying pyrolysis (heating to 450°C at a rate of 50°C/min) and hydrothermal liquefaction (300°C and 10–12 MPa) to the aforementioned microalgae strains. As a consequence, the output of bio-oil from both strategies was equivalent. A noteworthy result of the tests was bio-oil with a high heating value of 24%–45% (35–37 MJ/kg). As a result, the species and pyrolysis circumstances affect the outputs of the pyrolysis of microalgae biomass. In addition, the study confirmed that hydrothermal liquefaction has a lower energy consumption ratio than pyrolysis (Xia et al., 2020). To turn microalgae into biofuels, researchers have recently achieved substantial advancements in the development of microalgae biomass pyrolysis. Higher yields were observed when mixed microalgal cultures from natural sources were used for liquefaction. One study successfully achieved about 54.4% (wt/wt) yield of bio-oil from microwave-enhanced pyrolysis treated to a CO_2 atmosphere (Zhang et al., 2016). Further example showed that, when the temperature range was 200°C, the pyrolysis of the native microalgae consortium produced up to 31% of bio-oil (Choudhary et al., 2017). Heterogeneous catalysis has been used to increase the yield of bio-oil through pyrolysis (Kohansal et al., 2019; Xu et al., 2019). *Nannochloropsis* was used as the main biomass feedstock for low-temperature hydrothermal liquefaction. Nano-Ni/SiO_2 produced the most bio-oil (30 wt%) among these catalysts, followed by Na_2CO_3 (24.2 wt%) and zeolite (24 wt%) (Saber et al., 2016). In a similar experiment, *Nannochlopsis* hydrothermal liquefaction at 300°C with Ni-supported TiO_2 produced a high yield of biocrude (Wang et al., 2018). Gasification is the process of heating biomass at relatively high temperatures (800°C–1,000°C) in low-oxygen environments to produce flammable gas mixtures (Clark and Deswarte, 2014). Syngas is well suited for residential heating and cooking by burning in gas engines or gas turbines despite having a low energy density (4–6 MJ/m^3) (McKendry, 2002). The co-gasification of microalgae and wood produced greater H_2, CO, and CH_4 by 3%–20%, 6%–31%, and 9%–20%, respectively, than utilising only wood biomass (Zhu et al., 2016). Similarly, Raheem et al. gasified *Chlorella* biomass at 950°C, producing H_2, CO, and CH_4 that accounted for up to 2.9, 22.8, and 10.1 wt% of the biomass, respectively (Raheem et al., 2015). The feedstocks' steam and moisture contents, as well as to a lesser extent the combination of the microalgae substrate, were other elements the scientists validated as affecting the cold gas efficiency. More crucially, the primary element influencing the composition of the output syngas is moisture content in the biomass and gasification feedstock (Azadi et al., 2014).

26.5 INTEGRATED ALGAL BIOREFINERY WITH VALUE-ADDED PRODUCT RECOVERY

It has been extensively found by several studies over the past 20 years that algal biomass may be used as a substrate for biorefinery. The integrated algal biorefinery idea is depicted in Figure 26.1. It is not economically possible to grow algal biomass using wastewater and harvest it to make just one type of bioproduct, such as value-added goods or biofuels. Economically viable biomass usage will only result in multiproduct recovery when upstream and downstream processes are integrated. Through total biomass utilisation and reduced waste disposal costs, the integrated algal biorefinery (IAB) establishes a foundation for a sustainable algae circular bioeconomy or the commercialisation of algal-based bioproducts. Additionally, IAB significantly decreases freshwater and energy requirements for algae production in comparison to traditional algal biorefineries. The feedstock supply chain may be balanced and optimised by using different conversion platforms, which can also help to lessen the negative effects on the economy and the environment (Kannah et al., 2021b).

26.6 TECHNO-ECONOMIC ASSESSMENT OF ALGAL BIOREFINERY

Before the procedure is implemented on a small scale as a pilot, the techno-economic assessment of IAB is seen to be a crucial effort (Banu et al., 2020b; Kannah et al., 2021b). Assessing the economic viability of the pilot scale method is helpful. The suggested biorefinery pathway, on the other hand, will be more valuable for forecasting bottleneck technical concerns, and the technoeconomic evaluation also provides cost-effective alternatives to address issues that are found. For instance, Davis et al. found that the downstream process in a marine biorefinery requires between 40% and 50% of the project's overall cost (Davis et al., 2016). The above-mentioned problem may be resolved by utilising downstream and upstream process integration, which establishes a foundation for a circular bioeconomy (Gifuni et al., 2019). The techno-economic evaluation methodology involves three main steps: Utilising economic models, (i) this same recommended IAB route's expenditure are calculated which include capital and operational expenditures. (ii) The total costs incurred for capital as well as expenditure are diligently calculated for the IAB. (iii) Finally, the suggested IAB route's profit level is predicted using an economic model (Wu et al., 2018). Further detailed analysis can be performed by adding the environmental parameters calculated by LCA analysis and presenting a detailed techno-economic analysis of the process (Wu and Chang, 2019). The selling price of lipids extracted from microalgae biomass was determined to be 1.8 USD/L (Nezammahalleh et al., 2018). The ABF selling price for thermochemical and biochemical platforms, respectively, would be roughly 10.41 USD/GGE and 12.85 USD/GGE (DeRose et al., 2019). The cost to produce bioethanol from algal biomass was estimated to be 0.93 USD/L, and it uses 0.07 USD/kWh of power (Soleymani and Rosentrater, 2017). Algal biomass is considered to be the most suitable substrate for the production of biofuels in a biorefinery concept but commercialisation seems to be the Achilles heel. The price of production of biofuels as well as consequent product recovery poses a challenge for researchers. Biomass production expenses might be reduced by 90% by diverting the waste stream toward the production of microalgae and potential design advancements. In addition, compared to integrated algal biorefineries, solo biomass mono-processing demands greater production costs (Wijffels et al., 2010). A suitable and cost-effective alternative could be integrated algal biorefineries which can be implemented at an industrial scale since single processing units are more production intensive as compared to integrated biorefineries.

26.7 LIFE CYCLE ASSESSMENT OF ALGAL BIOREFINERY

A vital technique for assessing the entire environmental effect of the life cycle of the products produced by a process is life cycle assessment (LCA). The major goals of LCA are to forecast energy use and GHG emissions. Research has focussed on thermochemical and biochemical conversion platforms to compare the LCA of IAB. The findings of the LCA study reveal that higher emissions were observed for biochemical conversion as compared to thermochemical conversion, based on cradle-to-grave system boundary conditions (DeRose et al., 2019). This could be because total energy required and gaseous emissions were higher for the biochemical process as compared to the thermochemical process. The process for LCA is broken down into four parts, according to the International Organization for Standardization (ISO): (i) aim and scope definition, (ii) life cycle inventory assessment, (iii) life cycle impact assessment, and (iv) life cycle interpretation (Finkbeiner et al., 2006). The LCA analysis for IAB comprises evaluations of the effects on the four phases of product recovery, energy consumption, and feedstock usage. The following section discusses the LCA's several stages.

i. The goal and scope definition process are the first steps in the LCA analysis process. It is done to gather crucial data for the study, including details about the need for the LCA analysis, potential applications of the process, system boundaries, and functional units that will be used to calculate the life cycle of the varied products.
ii. The second step in an LCA study is a life cycle inventory evaluation. The amount of raw material utilised, the influence of pretreatment, product recovery, the degree of GHG emission, and the energy consumption during the processing of algae are all taken into account at this step. Algal yield from wastewater treatment is also evaluated (Brentner et al., 2011). The life cycle inventory can sometimes be used as an indication to give specific details about the causes and scope of environmental impacts caused by the biorefinery process.
iii. The third stage of an LCA analysis is the life cycle effect analysis. It is regarded as a crucial step in the LCA analysis since the IAB's important environmental consequences are being measured at this stage (Chandra et al., 2019). Numerous academics have recommended various software programmes, such as open LCA, Simapro, or Gabi, for calculating the environmental effects of processes (Silva et al., 2019; Wu et al., 2020). Fossil depletion, mineral resource depletion, natural land transformation, urban land occupation, agricultural land occupation, marine ecotoxicity, freshwater ecotoxicity, terrestrial ecotoxicity, freshwater eutrophication, terrestrial acidification, climate change, ionising radiation, particulate matter formation, photochemical oxidant formation, human toxicity, ozone depletion, and human toxicity are just a few of the 18 types of environmental impacts (Wu et al., 2020).
iv. The last step of an LCA investigation is life cycle interpretation. It interprets the findings of 18 different environmental effects that were discovered during the LCA analysis's preliminary stage. This stage involves comparing similar processes in order to understand the overall effect of the process under study. The conversion platform or technology was created with an eye on evaluating the product's environmental effects over the course of its lifespan. According to Tao et al., reliable indicators including life cycle carbon emission (i.e., CO_2 eq), life cycle energy needs, and ecoindicator are often used to validate the life cycle environmental evaluation of IAB (Tao et al., 2018). A schematic representation of the LCA process is shown in Figure 26.2.

26.8 CHALLENGES AND FUTURE PERSPECTIVES

Algal biorefineries have attracted a lot of attention since they provide cost-effective wastewater treatment and a yield of algae for the manufacture of biofuels, as well as being utilised as a raw material in a variety of sectors including the cosmetics and pharmaceutical industries. However, there are still many difficult difficulties that need to be resolved, such as methods to boost biomass yield and the high energy requirements for algae culture and harvesting. The considerations mentioned above have an impact on how much more expensive ABF is to produce than PBF. Many academics have proposed a few fundamental concepts to address these problems, like utilisation of wastewater and syngas to lower the cost of substrates involved for algae culture. Algal biorefinery will cut the cost of cultivation in half, although there are still certain restrictions on the practical implementation of ABF. It is still necessary to find a solution to the production and harvesting issues related to biomass. Modern genetic manipulation technologies such as rDNA technology and other omics technologies have paved a path for modification of existing algal strains which can produce robust strains with higher productivity. Moreover, wastewater with higher inorganic load may be utilised as a growth medium to increase the rate of biomass output. By creating and implementing this sort of strategy, algal biomass productivity for the generation of biofuels and the extraction of bioactive compounds for various commercial uses will be increased. However, choosing and identifying a certain species of algae for culture can lessen the adverse effects of the desired component extraction. Integration of the upstream and downstream processes will lead to the establishment of a sustainable system for low-cost, value-added product manufacturing. Therefore, using an IAB system will be the best option and a financially sound way to scale up a process or commercialise an ABF.

26.9 CONCLUSION

By functioning as a justifiable substratum for the manufacture of biofuels, the mining of value-added products, and the treatment of wastewater, algae biotechnology offers a green solution for a variety of environmental issues. In the process, it offers beneficial compounds including carotenoids and polyunsaturated fatty acids. Additionally, to improve its biomass production, algal biomass ingests atmospheric CO_2, which indirectly helps to lessen the effects of global warming. Therefore, creating an integrated algal biorefinery idea will increase the value chain and lessen the obstacles associated with the economy.

REFERENCES

Abdo, S.M., Ali, G.H. (2019). Analysis of polyhydroxybutrate and bioplastic production from microalgae. *Bull. Natl. Res. Cent.* 43, 1–4.

Aderibigbe, D.O., Giwa, A.R.A., & Bello, I.A. (2017). Characterization and treatment of wastewater from food processing industry: A review. *Imam Journal Appl. Sci.*, 2(2), 27–36.

Ahmed, S.F., Mofijur, M., Parisa, T.A., Islam, N., Kusumo, F., Inayat, A., Badruddin, I.A., Khan, T.M.Y., Ong, H.C. (2022). Progress and challenges of contaminate removal from wastewater using microalgae biomass. *Chemosphere* 286, 131656.

Akubude, V.C., Nwaigwe, K.N., Dintwa, E. (2019). Production of biodiesel from microalgae via nanocatalyzed transesterification process: a review. *Mater. Sci. Energy Technol.* 2, 216–225.

Amulya, K., Jukuri, S., Venkata Mohan, S. (2015). Sustainable multistage process for enhanced productivity of bioplastics from waste remediation through aerobic dynamic feeding strategy: process integration for up-scaling. *Bioresour. Technol.* 188, 231–239. https://doi.org/10.1016/j.biortech.2015.01.070

Arvindnarayan, S., Prabhu, K.K.S., Shobana, S., Kumar, G., Dharmaraja, J. (2017). Upgrading of micro algal derived bio-fuels in thermochemical liquefaction path and its perspectives: a review. *Int. Biodeterior. Biodegrad.* 119, 260–272.

Azadi, P., Brownbridge, G.P.E., Mosbach, S., Inderwildi, O.R., Kraft, M. (2014). Production of biorenewable hydrogen and syngas via algae gasification: a sensitivity analysis. *Energy Procedia* 61, 2767–2770.

Banerjee, S., Das, D., Ghosh, A.K. (2022). Production of bioethanol from microalgal feedstock: a circular biorefinery approach. In Agarwal, A.K., Valera, H. (eds.), *Potential and Challenges of Low Carbon Fuels for Sustainable Transport*. Springer, pp. 33–65.

Banerjee, S., Ray, A., Das, D. (2021). Optimization of Chlamydomonas reinhardtii cultivation with simultaneous CO2 sequestration and biofuels production in a biorefinery framework. *Sci. Total Environ.* 762, 143080.

Banu, J.R., Kannah, R.Y., Kavitha, S., Ashikvivek, A., Bhosale, R.R., Kumar, G. (2020a). Cost effective biomethanation via surfactant coupled ultrasonic liquefaction of mixed microalgal biomass harvested from open raceway pond. *Bioresour. Technol.* 304, 123021.

Banu, J.R., Kavitha, S., Kannah, R.Y., Kumar, M.D., Atabani, A.E., Kumar, G. (2020b). Biorefinery of spent coffee grounds waste: viable pathway towards circular bioeconomy. *Bioresour. Technol.* 302, 122821.

Barat, R., Serralta, J., Ruano, M. V, Jiménez, E., Ribes, J., Seco, A., Ferrer, J. (2013). Biological Nutrient Removal Model No. 2 (BNRM2): a general model for wastewater treatment plants. *Water Sci. Technol.* 67, 1481–1489.

Batista, A.P., Ambrosano, L., Graça, S., Sousa, C., Marques, P.A.S.S., Ribeiro, B., Botrel, E.P., Castro Neto, P., Gouveia, L. (2015). Combining urban wastewater treatment with biohydrogen production – an integrated microalgae-based approach. *Bioresour. Technol.* 184, 230–235. https://doi.org/10.1016/j.biortech.2014.10.064

Béligon, V., Noblecourt, A., Christophe, G., Lebert, A., Larroche, C., Fontanille, P. (2018). Proof of concept for biorefinery approach aiming at two bioenergy production compartments, hydrogen and biodiesel, coupled by an external membrane. *Biofuels* 9, 163–174. https://doi.org/10.1080/17597269.2016.1259142

Bogan, R.H., Albertson, O.E., Pluntze, J.C. (1960). Use of algae in removing phosphorus from sewage. *J. Sanit. Eng. Div.* 86, 1–20.

Brentner, L.B., Eckelman, M.J., Zimmerman, J.B. (2011). Combinatorial life cycle assessment to inform process design of industrial production of algal biodiesel. *Environ. Sci. Technol.* 45, 7060–7067.

Butterworth, E., Dotro, G., Jones, M., Richards, A., Onunkwo, P., Narroway, Y., Jefferson, B. (2013). Effect of artificial aeration on tertiary nitrification in a full-scale subsurface horizontal flow constructed wetland. *Ecol. Eng.* 54, 236–244.

Cai, T., Park, S.Y., Li, Y. (2013). Nutrient recovery from wastewater streams by microalgae: status and prospects. Renew. *Sustain. Energy Rev.* 19, 360–369.

Chandra, R., Iqbal, H.M.N., Vishal, G., Lee, H.-S., Nagra, S. (2019). Algal biorefinery: a sustainable approach to valorize algal-based biomass towards multiple product recovery. *Bioresour. Technol.* 278, 346–359.

Chandrasekhar, K., Raj, T., Ramanaiah, S. V, Kumar, G., Banu, J.R., Varjani, S., Sharma, P., Pandey, A., Kumar, S., Kim, S.-H. (2022). Algae biorefinery: a promising approach to promote microalgae industry and waste utilization. *J. Biotechnol.* 345, 1–16.

Cheng, J., Guo, H., Qiu, Y., Zhang, Z., Mao, Y., Qian, L., Yang, W., Park, J.-Y. (2020). Switchable solvent N, N, N′, N′-tetraethyl-1, 3-propanediamine was dissociated into cationic surfactant to promote cell disruption and lipid extraction from wet microalgae for biodiesel production. *Bioresour. Technol.* 312, 123607.

Cheng, Y., Zhou, W., Gao, C., Lan, K., Gao, Y., Wu, Q. (2009). Biodiesel production from Jerusalem artichoke (Helianthus Tuberosus L.) tuber by heterotrophic microalgae Chlorella protothecoides. *J. Chem. Technol. Biotechnol. Int. Res. Process. Environ. Clean Technol.* 84, 777–781.

Chng, L.M., Lee, K.T., Chan, D.J.C. (2017). Synergistic effect of pretreatment and fermentation process on carbohydrate-rich Scenedesmus dimorphus for bioethanol production. *Energy Convers. Manag.* 141, 410–419.

Choudhary, P., Malik, A., Pant, K.K. (2017). Mass-scale algal biomass production using algal biofilm reactor and conversion to energy and chemical precursors by hydropyrolysis. *ACS Sustain. Chem. Eng.* 5, 4234–4242.

Christenson, L., Sims, R. (2011). Production and harvesting of microalgae for wastewater treatment, biofuels, and bioproducts. *Biotechnol. Adv.* 29, 686–702. https://doi.org/10.1016/j.biotechadv.2011.05.015

Clark, J.H., Deswarte, F. (2014). *Introduction to Chemicals from Biomass.* John Wiley & Sons.

Das, D. (2016). *Algal Biorefinery: An Integrated Approach.* Springer.. https://doi.org/10.1007/978-3-319-22813-6

Das, D., Veziroglu, T.N. (2008). Advances in biological hydrogen production processes. *Int. J. Hydrogen Energy* 33, 6046–6057. https://doi.org/10.1016/j.ijhydene.2008.07.098

Davis, L.S., Hoffmann, J.P., Cook, P.W. (1990). Production and nutrient accumulation by periphyton in a wastewater treatment facility. *J. Phycol.* 26, 617–623.

Davis, R., Markham, J., Kinchin, C., Grundl, N., Tan, E.C.D., Humbird, D. (2016). Process design and economics for the production of algal biomass: algal biomass production in open pond systems and processing through dewatering for downstream conversion. National Renewable Energy Lab. Golden, CO: NREL.

del Mar Morales-Amaral, M., Gómez-Serrano, C., Acién, F.G., Fernández-Sevilla, J.M., Molina-Grima, E. (2015). Outdoor production of Scenedesmus sp. in thin-layer and raceway reactors using centrate from anaerobic digestion as the sole nutrient source. *Algal Res.* 12, 99–108.

DeRose, K., DeMill, C., Davis, R.W., Quinn, J.C. (2019). Integrated techno economic and life cycle assessment of the conversion of high productivity, low lipid algae to renewable fuels. *Algal Res.* 38, 101412.

Dickinson, S., Mientus, M., Frey, D., Amini-Hajibashi, A., Ozturk, S., Shaikh, F., Sengupta, D., El-Halwagi, M.M. (2017). A review of biodiesel production from microalgae. *Clean Technol. Environ. Policy* 19, 637–668.

Doran, M.D., Boyle, W.C. (1979). Phosphorus removal by activated algae. *Water Res.* 13, 805–812.

El-Mekkawi, S.A., Abdo, S.M., Samhan, F.A., Ali, G.H. (2019). Optimization of some fermentation conditions for bioethanol production from microalgae using response surface method. *Bull. Natl. Res. Cent.* 43, 1–8.

Finkbeiner, M., Inaba, A., Tan, R., Christiansen, K., Klüppel, H.-J. (2006). The new international standards for life cycle assessment: ISO 14040 and ISO 14044. *Int. J. Life Cycle Assess.* 11, 80–85.

Gao, G., Clare, A.S., Rose, C., Caldwell, G.S. (2018). Ulva rigida in the future ocean: potential for carbon capture, bioremediation and biomethane production. *GCB Bioenergy* 10, 39–51. https://doi.org/10.1111/gcbb.12465

Gates, W.E., Borchardt, J.A. (1964). Nitrogen and phosphorus extraction from domestic wastewater treatment plant effluents by controlled algal culture. *J. Water Pollut. Control Fed. 36,* 443–462.

Ghosh, S., Banerjee, S., Das, D. (2017). Process intensification of biodiesel production from Chlorella sp. MJ 11/11 by single step transesterification. *Algal Res.* 27, 12–20. https://doi.org/10.1016/j.algal.2017.08.021

Ghosh, Supratim, Coons, J., Yeager, C., Halley, P., Chemodanov, A., Belgorodsky, B., Gozin, M., Chen, G.Q., Golberg, A. (2022). Halophyte biorefinery for polyhydroxyalkanoates production from Ulva sp. Hydrolysate with Haloferax mediterranei in pneumatically agitated bioreactors and ultrasound harvesting. *Bioresour. Technol.* 344, 125964. https://doi.org/10.1016/j.biortech.2021.125964

Ghosh, S., Das, D. (2016). Improvement of harvesting technology for algal biomass production, algal biorefinery: an integrated approach. https://doi.org/10.1007/978-3-319-22813-6_8

Ghosh, S., Gnaim, R., Greiserman, S., Fadeev, L., Gozin, M., Golberg, A. (2019). Macroalgal biomass subcritical hydrolysates for the production of polyhydroxyalkanoate (PHA) by Haloferax mediterranei. *Bioresour. Technol.* 271, 166–173. https://doi.org/10.1016/j.biortech.2018.09.108

Ghosh, S., Greiserman, S., Chemodanov, A., Slegers, P.M., Belgorodsky, B., Epstein, M., Kribus, A., Gozin, M., Chen, G.-Q., Golberg, A. (2021). Polyhydroxyalkanoates and biochar from green macroalgal Ulva sp. biomass subcritical hydrolysates: process optimization and a priori economic and greenhouse emissions break-even analysis. *Sci. Total Environ.* 770, 145281. https://doi.org/10.1016/j.scitotenv.2021.145281

Ghosh, S., Roy, S., Das, D. (2015). Improvement of biomass production by Chlorella sp. MJ 11/11 for use as a feedstock for biodiesel. *Appl. Biochem. Biotechnol.* 175, 3322–35. https://doi.org/10.1007/s12010-015-1503-8

Ghosh, Susmita, Sarkar, T., Pati, S., Kari, Z.A., Edinur, H.A., Chakraborty, R. (2022). Novel bioactive compounds from marine sources as a tool for functional food development. *Front. Mar. Sci.* 9, 1–28. https://doi.org/10.3389/fmars.2022.832957

Gifuni, I., Pollio, A., Safi, C., Marzocchella, A., Olivieri, G. (2019). Current bottlenecks and challenges of the microalgal biorefinery. *Trends Biotechnol.* 37, 242–252.

Goh, B.H.H., Ong, H.C., Cheah, M.Y., Chen, W.-H., Yu, K.L., Mahlia, T.M.I., (2019). Sustainability of direct biodiesel synthesis from microalgae biomass: a critical review. Renew. *Sustain. Energy Rev.* 107, 59–74.

Gómez-Serrano, C., Morales-Amaral, M. del M., Acién, F.G., Escudero, R., Fernández-Sevilla, J.M., Molina-Grima, E. (2015). Utilization of secondary-treated wastewater for the production of freshwater microalgae. *Appl. Microbiol. Biotechnol.* 99, 6931–6944.

Hemalatha, M., Venkata Mohan, S. (2016). Microalgae cultivation as tertiary unit operation for treatment of pharmaceutical wastewater associated with lipid production. Bioresour. Technol. https://doi.org/10.1016/j.biortech.2016.04.101

Hena, S., Gutierrez, L., Croué, J.-P. (2021). Removal of pharmaceutical and personal care products (PPCPs) from wastewater using microalgae: a review. *J. Hazard. Mater.* 403, 124041.

Ho, S.-H., Huang, S.-W., Chen, C.-Y., Hasunuma, T., Kondo, A., Chang, J.-S. (2013). Bioethanol production using carbohydrate-rich microalgae biomass as feedstock. *Bioresour. Technol.* 135, 191–198.

Howlader, M.S., French, W.T. (2020). Pretreatment and lipid extraction from wet microalgae: challenges, potential, and application for industrial-scale application. In *Microalgae Biotechnology for Food, Health and High Value Products*. Springer, pp. 469-483.

Ilyas, M., Ahmad, W., Khan, H., Yousaf, S., Yasir, M., Khan, A. (2019). Environmental and health impacts of industrial wastewater effluents in Pakistan: a review. *Rev. Environ. Health* 34, 171–186.

Jena, U., Das, K.C. (2011). Comparative evaluation of thermochemical liquefaction and pyrolysis for bio-oil production from microalgae. *Energy & fuels* 25, 5472–5482.

Kalia, V.C. (2016). *Microbial Factories: Biofuels, Waste Treatment.* Springer. (Vol. 1, pp. 1–353). https://doi.org/10.1007/978-81-322-2598-0

Kannah, R.Y., Kavitha, S., Banu, J.R., Sivashanmugam, P., Gunasekaran, M., Kumar, G. (2021a). A mini review of biochemical conversion of algal biorefinery. *Energy and Fuels* 35, 16995–17007. https://doi.org/10.1021/acs.energyfuels.1c02294

Kannah, R.Y., Kavitha, S., Karthikeyan, O.P., Kumar, G., Dai-Viet, N.V., Banu, J.R. (2021b). Techno-economic assessment of various hydrogen production methods–A review. *Bioresour. Technol.* 319, 124175.

Kinnunen, V., Craggs, R., & Rintala, J. (2014). Influence of temperature and pretreatments on the anaerobic digestion of wastewater grown microalgae in a laboratory-scale accumulating-volume reactor. *Water Res.*, 57, 247–257.

Kohansal, K., Tavasoli, A., Bozorg, A. (2019). Using a hybrid-like supported catalyst to improve green fuel production through hydrothermal liquefaction of Scenedesmus obliquus microalgae. *Bioresour. Technol.* 277, 136–147.

Kumar, G., Nguyen, D.D., Sivagurunathan, P., Kobayashi, T., Xu, K., Chang, S.W. (2018). Cultivation of microalgal biomass using swine manure for biohydrogen production: impact of dilution ratio and pretreatment. *Bioresour. Technol.* 260, 16–22.

Kumar, K., Ghosh, S., Angelidaki, I., Holdt, S.L., Karakashev, D.B., Morales, M.A., Das, D. (2016). Recent developments on biofuels production from microalgae and macroalgae. *Renew. Sustain. Energy Rev.* 65, 235–249. https://doi.org/10.1016/j.rser.2016.06.055

Kumar, K., Sirasale, A., Das, D. (2013). Use of image analysis tool for the development of light distribution pattern inside the photobioreactor for the algal cultivation. *Bioresour. Technol.* 143, 88–95. https://doi.org/10.1016/j.biortech.2013.05.117

Lakshmikandan, M., Murugesan, A.G., Wang, S., Abomohra, A.E.-F., Jovita, P.A., Kiruthiga, S. (2020). Sustainable biomass production under CO_2 conditions and effective wet microalgae lipid extraction for biodiesel production. *J. Clean. Prod.* 247, 119398.

Mahata, C., Ray, S., Das, D. (2020). Optimization of dark fermentative hydrogen production from organic wastes using acidogenic mixed consortia Optimization of dark fermentative hydrogen production from organic wastes using acidogenic mixed consortia. *Energy Convers. Manag.* 219, 113047. https://doi.org/10.1016/j.enconman.2020.113047

Mahdy, A., Ballesteros, M., González-Fernández, C. (2016). Enzymatic pretreatment of Chlorella vulgaris for biogas production: influence of urban wastewater as a sole nutrient source on macromolecular profile and biocatalyst efficiency. *Bioresour. Technol.* 199, 319–325.

Mainardis, M., Buttazzoni, M., De Bortoli, N., Mion, M., Goi, D. (2020). Evaluation of ozonation applicability to pulp and paper streams for a sustainable wastewater treatment. *J. Clean. Prod.* 258, 120781.

Mandik, Y.I., Cheirsilp, B., Srinuanpan, S., Maneechote, W., Boonsawang, P., Prasertsan, P., Sirisansaneeyakul, S. (2020). Zero-waste biorefinery of oleaginous microalgae as promising sources of biofuels and biochemicals through direct transesterification and acid hydrolysis. *Process Biochem.* 95, 214–222.

Marcilhac, C., Sialve, B., Pourcher, A.-M., Ziebal, C., Bernet, N., Béline, F. (2014). Digestate color and light intensity affect nutrient removal and competition phenomena in a microalgal-bacterial ecosystem. *Water Res.* 64, 278–287.

Maryjoseph, S., Ketheesan, B. (2020). Microalgae based wastewater treatment for the removal of emerging contaminants: a review of challenges and opportunities. *Case Stud. Chem. Environ. Eng.* 2, 100046.

McKendry, P. (2002). Energy production from biomass (part 3): gasification technologies. *Bioresour. Technol.* 83, 55–63.

McNeff, C. V, McNeff, L.C., Yan, B., Nowlan, D.T., Rasmussen, M., Gyberg, A.E., Krohn, B.J., Fedie, R.L., Hoye, T.R. (2008). A continuous catalytic system for biodiesel production. *Appl. Catal. A Gen.* 343, 39–48.

Moshood, T.D., Nawanir, G., Mahmud, F. (2021). Microalgae biofuels production: a systematic review on socioeconomic prospects of microalgae biofuels and policy implications. *Environ. Chall.* 5, 100207.

Nayak, B.K., Roy, S., Das, D. (2014). Biohydrogen production from algal biomass (Anabaena sp. PCC 7120) cultivated in airlift photobioreactor. *Int. J. Hydrogen Energy* 39, 7553–7560. https://doi.org/10.1016/j.ijhydene.2013.07.120

Nezammahalleh, H., Adams II, T.A., Ghanati, F., Nosrati, M., Shojaosadati, S.A. (2018). Techno-economic and environmental assessment of conceptually designed in situ lipid extraction process from microalgae. *Algal Res.* 35, 547–560.

Onay, M. (2019). Bioethanol production via different saccharification strategies from H. tetrachotoma ME03 grown at various concentrations of municipal wastewater in a flat-photobioreactor. *Fuel* 239, 1315–1323.

Ortiz-Martínez, V.M., Andreo-Martinez, P., Garcia-Martinez, N., de los Ríos, A.P., Hernández-Fernández, F.J., Quesada-Medina, J. (2019). Approach to biodiesel production from microalgae under supercritical conditions by the PRISMA method. *Fuel Process. Technol.* 191, 211–222.

Oswald, W.J., Gotaas, H.B. (1957). Photosynthesis in sewage treatment. *Trans. Am. Soc. Civ. Eng* 122, 73–105.

Passos, F., Hernandez-Marine, M., García, J., Ferrer, I. (2014). Long-term anaerobic digestion of microalgae grown in HRAP for wastewater treatment. Effect of microwave pretreatment. *Water Res.* 49, 351–359.

Passos, F., Solé, M., García, J., Ferrer, I. (2013). Biogas production from microalgae grown in wastewater: effect of microwave pretreatment. *Appl. Energy* 108, 168–175.

Phwan, C.K., Chew, K.W., Sebayang, A.H., Ong, H.C., Ling, T.C., Malek, M.A., Ho, Y.-C., Show, P.L. (2019). Effects of acids pre-treatment on the microbial fermentation process for bioethanol production from microalgae. *Biotechnol. Biofuels* 12, 1–8.

Raheem, A., Azlina, W.W., Yap, Y.H.T., Danquah, M.K., Harun, R. (2015). Thermochemical conversion of microalgal biomass for biofuel production. *Renew. Sustain. Energy Rev.* 49, 990–999.

Ramola, B., Kumar, V., Nanda, M., Mishra, Y., Tyagi, T., Gupta, A., Sharma, N. (2019). Evaluation, comparison of different solvent extraction, cell disruption methods and hydrothermal liquefaction of Oedogonium macroalgae for biofuel production. *Biotechnol. Reports* 22, e00340.

Razzak, S.A., Ali, S.A.M., Hossain, M.M., deLasa, H. (2017). Biological CO_2 fixation with production of microalgae in wastewater – a review. *Renew. Sustain. Energy Rev.* 76, 379–390.

Rocha, C.J.L., Álvarez-Castillo, E., Yáñez, M.R.E., Bengoechea, C., Guerrero, A., Ledesma, M.T.O. (2020). Development of bioplastics from a microalgae consortium from wastewater. *J. Environ. Manage.* 263, 110353.

Roy, S., Kumar, K., Ghosh, S., Das, D. (2014). Thermophilic biohydrogen production using pre-treated algal biomass as substrate. *Biomass Bioenergy* 61, 157–166. https://doi.org/10.1016/j.biombioe.2013.12.006

Ruiz-Marin, A., Canedo-López, Y., Chávez-Fuentes, P. (2020). Biohydrogen production by Chlorella vulgaris and Scenedesmus obliquus immobilized cultivated in artificial wastewater under different light quality. *Amb Express* 10, 1–7.

Saber, M., Golzary, A., Hosseinpour, M., Takahashi, F., Yoshikawa, K. (2016). Catalytic hydrothermal liquefaction of microalgae using nanocatalyst. *Appl. Energy* 183, 566–576. https://doi.org/https://doi.org/10.1016/j.apenergy.2016.09.017

Schenk, P.M., Thomas-Hall, S.R., Stephens, E., Marx, U.C., Mussgnug, J.H., Posten, C., Kruse, O., Hankamer, B. (2008). Second generation biofuels: high-efficiency microalgae for biodiesel production. *BioEnergy Res.* 1, 20–43. https://doi.org/10.1007/s12155-008-9008-8

Selvaratnam, T., Pegallapati, A., Montelya, F., Rodriguez, G., Nirmalakhandan, N., Lammers, P.J., Van Voorhies, W. (2015). Feasibility of algal systems for sustainable wastewater treatment. *Renew. Energy* 82, 71–76.

Shakya, R., Adhikari, S., Mahadevan, R., Shanmugam, S.R., Nam, H., Dempster, T.A. (2017). Influence of biochemical composition during hydrothermal liquefaction of algae on product yields and fuel properties. *Bioresour. Technol.* 243, 1112–1120.

Shi, J., Podola, B., Melkonian, M. (2007). Removal of nitrogen and phosphorus from wastewater using microalgae immobilized on twin layers: an experimental study. *J. Appl. Phycol.* 19, 417–423.

Shin, H.-Y., Shim, S.-H., Ryu, Y.-J., Yang, J.-H., Lim, S.-M., Lee, C.-G. (2018). Lipid extraction from Tetraselmis sp. microalgae for biodiesel production using hexane-based solvent mixtures. *Biotechnol. Bioprocess Eng.* 23, 16–22.

Silva, D.A.L., Nunes, A.O., Piekarski, C.M., da Silva Moris, V.A., de Souza, L.S.M., Rodrigues, T.O. (2019). Why using different Life Cycle Assessment software tools can generate different results for the same product system? A cause-effect analysis of the problem. *Sustain. Prod. Consum.* 20, 304–315.

Silva-Bedoya, L.M., Sánchez-Pinzón, M.S., Cadavid-Restrepo, G.E., Moreno-Herrera, C.X. (2016). Bacterial community analysis of an industrial wastewater treatment plant in Colombia with screening for lipid-degrading microorganisms. *Microbiol. Res.* 192, 313–325.

Singh, H., Rout, S., Das, D. (2022). Dark fermentative biohydrogen production using pretreated Scenedesmus obliquus biomass under an integrated paradigm of biorefinery. *Int. J. Hydrogen Energy* 47, 102–116.

Singh, H., Varanasi, J.L., Banerjee, S., Das, D. (2019). Production of carbohydrate enrich microalgal biomass as a bioenergy feedstock. *Energy* 188, 116039.

Singh, R., Kumar, A., Chandra Sharma, Y. (2019). Biodiesel production from microalgal oil using barium-calcium-zinc mixed oxide base catalyst: optimization and kinetic studies. *Energy & Fuels* 33, 1175–1184.

Soleymani, M., Rosentrater, K.A. (2017). Techno-economic analysis of biofuel production from macroalgae (seaweed). *Bioengineering* 4, 92.

Sukačová, K., Trtílek, M., Rataj, T. (2015). Phosphorus removal using a microalgal biofilm in a new biofilm photobioreactor for tertiary wastewater treatment. *Water Res.* 71, 55–63.

Sun, K., Li, Q., Zhang, L., Shao, Y., Zhang, Z., Zhang, S., Liu, Q., Wang, Y., Hu, X. (2020). Impacts of water-organic solvents on polymerization of the sugars and furans in bio-oil. *Bioresour. Technol. Rep.* 10, 100419.

Tan, C.H., Show, P.L., Chang, J.-S., Ling, T.C., Lan, J.C.-W. (2015). Novel approaches of producing bioenergies from microalgae: a recent review. *Biotechnol. Adv.* 33, 1219–1227.

Tao, J., Li, L., Yu, S. (2018). An innovative eco-design approach based on integration of LCA, CAD\ CAE and optimization tools, and its implementation perspectives. *J. Clean. Prod.* 187, 839–851.

Toro-Trochez, J.L., Carrillo-Pedraza, E.S., Bustos-Martínez, D., García-Mateos, F.J., Ruiz-Rosas, R.R., Rodríguez-Mirasol, J., Cordero, T. (2019). Thermogravimetric characterization and pyrolysis of soybean hulls. *Bioresour. Technol. Rep.* 6, 183–189.

Umdu, E.S., Tuncer, M., Seker, E. (2009). Transesterification of Nannochloropsis oculata microalga's lipid to biodiesel on Al_2O_3 supported CaO and MgO catalysts. *Bioresour. Technol.* 100, 2828–2831.

Venderbosch, R.H. (2019). Fast pyrolysis. In R.C. Brown (Ed.) *Thermochemical Processing of Biomass: Conversion into Fuels, Chemicals and Power* (pp. 175–206). Chichester: John Wiley & Sons.

Von Sperling, M. (2007). *Wastewater Characteristics, Treatment and Disposal.* IWA Publishing.

Wang, W., Xu, Y., Wang, X., Zhang, B., Tian, W., Zhang, J. (2018). Hydrothermal liquefaction of microalgae over transition metal supported TiO_2 catalyst. *Bioresour. Technol.* 250, 474–480.

Whitton, R., Ometto, F., Pidou, M., Jarvis, P., Villa, R., Jefferson, B. (2015). Microalgae for municipal wastewater nutrient remediation: mechanisms, reactors and outlook for tertiary treatment. *Environ. Technol. Rev.* 4, 133–148.

Wijffels, R.H., Barbosa, M.J., Eppink, M.H.M. (2010). Microalgae for the production of bulk chemicals and biofuels. *Biofuels, Bioprod, Biorefin.* 4, 287–295.

Wu, T., Gong, M., Xiao, J. (2020). Preliminary sensitivity study on an life cycle assessment (LCA) tool via assessing a hybrid timber building. *J. Bioresour. Bioprod.* 5, 108–113.

Wu, W., Chang, J.-S. (2019). Integrated algal biorefineries from process systems engineering aspects: a review. *Bioresour. Technol.* 291, 121939.

Wu, W., Lin, K.-H., Chang, J.-S. (2018). Economic and life-cycle greenhouse gas optimization of microalgae-to-biofuels chains. *Bioresour. Technol.* 267, 550–559.

Xia, A., Sun, C., Fu, Q., Liao, Q., Huang, Y., Zhu, X., Li, Q. (2020). Biofuel production from wet microalgae biomass: comparison of physicochemical properties and extraction performance. *Energy* 212, 118581. https://doi.org/https://doi.org/10.1016/j.energy.2020.118581

Xu, D., Guo, S., Liu, L., Wu, Z., Wang, Y., Lin, G. (2019). Water-soluble and-insoluble biocrude production from hydrothermal liquefaction of microalgae with catalyst. *Energy Procedia 1*58, 97–102.

Yang, C., Li, R., Zhang, B., Qiu, Q., Wang, B., Yang, H., Ding, Y., Wang, C. (2019). Pyrolysis of microalgae: a critical review. *Fuel Process. Technol.* 186, 53–72.

Yang, Y.L., Wu, Y., Lu, Y.X., Cai, Y., He, Z., Yang, X.L., Song, H.L. (2021). A comprehensive review of nutrient-energy-water-solute recovery by hybrid osmotic membrane bioreactors. *Bioresour. Technol.* 320, 124300. https://doi.org/10.1016/j.biortech.2020.124300

Zhang, R., Li, L., Tong, D., Hu, C. (2016). Microwave-enhanced pyrolysis of natural algae from water blooms. *Bioresour. Technol.* 212, 311–317.

Zhu, Y., Piotrowska, P., van Eyk, P.J., Boström, D., Wu, X., Boman, C., Brostrom, M., Zhang, J., Kwong, C.W., Wang, D. (2016). Fluidized bed co-gasification of algae and wood pellets: gas yields and bed agglomeration analysis. *Energy Fuels* 30, 1800–1809.

27 Bioconversion of Lignocellulose

Current Trends in Biofuels, Platform Chemicals for Biorefinery Concept

Sampurna Nand, Anju Patel, and Siddharth Shukla

27.1 INTRODUCTION

Since fossil fuels are being depleted and their use in transportation has several disadvantages such as climate change through the emissions of greenhouse gas, environmental contamination, resource reduction, and unstable supply and demand, thus it is becoming more important to find alternative energy sources, such as lignocellulosic biomass (Sindhu., et al., 2016). Waste generation is one of the most pressing environmental issues in recent times. Unsustainable waste management practices contribute to a range of negative impacts, including air and water pollution, soil contamination, and global climate change. The production and disposal of waste pose a great risk to both humans and ecosystems, as toxic materials can leach into surrounding areas and cause long-term damage. Additionally, landfills take up valuable land and resources, while producing greenhouse gases that significantly contribute to changes in climatic conditions. Biowaste has a wide variety of impacts on the environment, depending on how it is managed, that can cause water, air and soil contamination, and may contribute to global temperature rise, ultimately posing serious risks to human and animal health (Prasad et al., 2015). Poorly managed biowaste can release methane, a potent greenhouse gas, into the atmosphere. In addition, biowaste can attract pests and vectors that can spread diseases. On the other hand, biowaste, as shown in Table 27.1, can create odors, promote the growth of noxious plants, and attract scavenging animals that can disturb the local ecosystem. Lignocellulosic waste materials from forestry and agricultural sources include wood chips, bark, sawdust, straw, husks, and other plant-based materials. These materials can be used for a variety of purposes, from creating biofuel to generating electricity, producing building materials, and more (Shirvani et al., 2011). Further, they can also be utilized as soil amendments to enhance soil fertility.

To reduce the ecological influences of waste, countries must invest in sustainable waste management solutions. This includes investing in waste reduction and recycling initiatives, as well as implementing policies that encourage the reuse and repurposing of materials. Governments must also invest in public education and awareness campaigns to promote responsible and appropriate disposal of waste generated from various sources. Waste management is all about the collection, safe transference, processing, proper recycling, and appropriate disposal of waste constituents produced by developmental activities, in order to minimize their effects. Finally, countries must ensure that waste is disposed of properly, either through composting, incineration, or other environmentally-friendly disposal methods.

DOI: 10.1201/9781003441144-27

TABLE 27.1
Biowaste Generated from Different Categories

S. No.	Category	Biowaste
1.	Food waste	Leftover food scraps, spoiled food, and other edible organic matter
2.	Yard waste	Grass clippings, leaves, branches, and other organic matter from gardens and landscaping
3.	Animal waste	Manure, urine, and other organic matter from livestock and companion animals
4.	Sewage sludge	Organic matter from wastewater treatment plants
5.	Human waste	Urine and faecal matter from humans
6.	Medical waste	Disposable medical instruments, syringes, and other organic matter from medical facilities
7.	Pharmaceutical waste	Unused medications, vaccines, and other organic matter from pharmacies and health care facilities.

27.2 LIGNOCELLULOSE: A POTENTIAL CANDIDATE TO GENERATE VALUABLE PRODUCTS

Lignocellulosic materials are plant-based (forestry and agricultural sources include wood chips, bark, sawdust, straw, husks, and other plant-based materials) materials that are composed of hemicellulose, cellulose, and lignin (Okolie et al., 2021). They are abundant, renewable sources of biomass in nature that can be utilized for the development of various valuable products such as bioethanol, bioplastics, biofuels, and biochemicals through biotransformation processes (Usmani et al., 2021). In which, microbes, enzymes, and other biocatalysts convert lignocellulosic waste into valuable bioresources (Bilal et al., 2017). These processes are efficient, cost-effective, and can be applied for commercial production. Not only does the biotransformation of lignocellulosic materials help in the utilization of renewable resources, but also helps reduce the environmental impact of the waste generated from various developmental activities. The process typically begins with pre-treatment of the lignocellulosic material to release fermentable sugars (Mezule et al., 2015). The fermentable sugars are then used by microorganisms to produce desired products. Various microorganisms, including bacteria, fungi, and yeasts, have been used to change lignocellulosic material into beneficial products. The products produced from the bioconversion of lignocellulosic material can be used for a range of applications, comprising the production of fuels, chemicals, materials, and pharmaceuticals (Kabbour & Luque, 2020).

27.2.1 LIGNIN

Lignin is a natural polymer commonly found in plant cell walls. It is a transformational product of the paper and pulp industry, but can also be processed and used as a value-added product (Yuan, 2013). Lignin is used in many industries, including the production of leather, adhesives, and bioplastics. In the energy sector, it can also be used as a fuel source, as it contains energy-rich materials that may be utilized to produce bioelectricity and/or heat. It can also be used as an animal feed additive and as an ingredient in nutraceuticals (Gil-Chávez et al., 2021). Lignin waste from sugar factories can be processed and used as fuel for boilers or other industrial plants (Klein et al., 2019). It can also be converted into activated carbon, and have various industrial and environmental applications such as water and air filtration (Bajwa et al., 2019). There are several kinds of heavy metal ions that can be absorbed by lignin such as cadmium, hexavalent chromium, lead, copper, cobalt, nickel, zinc, and mercury (Arora et al., 2018). Additionally, lignin waste can be used for the production of bioplastics, which can be used for packaging materials and other products. Finally, waste containing lignin can also be utilized as a soil conditioner to enhance soil quality and water retention capacity.

27.2.2 Cellulose and Hemicellulose

Cellulose and hemicellulose are two of the most abundant polysaccharides on the planet and are found in a variety of plant materials. They are the main components of biomass, which is the primary source of bioenergy. Cellulose is composed of glucose molecules linked together in long chains. Hemicellulose is composed of various kinds of different sugar molecules, as well as xylose, mannose, and arabinose (Otieno & Ahring, 2012). Cellulose and hemicellulose can be harvested from plant material, such as wood, grasses, and crop residues, and converted into bioethanol, a renewable fuel. When the polysaccharides are fermented, the sugars are fragmented into ethanol and carbon dioxide. The ethanol can then be distilled and used as fuel. Finally, cellulose and hemicellulose can be used to produce bio-oil, a renewable energy source. During pyrolysis, plant material is heated in the absence of oxygen and breaks down into different organic compounds, including bio-oil. This bio-oil may be used as a fuel directly or as a raw material for the manufacturing of other substances. Cellulose and hemicellulose are biopolymers with a wide range of applications. It has been used as an effective environmental remediator for many years due to its ability to absorb heavy metals, organic compounds, and other contaminants from the environment. It can also be used to clean up contaminated water and soil, reducing pollutant levels and making the environment healthier for humans and animals. Additionally, hemicellulose can be used as a soil conditioner to increase soil health, as it is known to intensify the amount of organics (Chowdhury et al., 2022). Furthermore, hemicellulose can be used to reduce the amount of methane gas produced by landfills, making it an important tool in mitigating the effects of climate change

27.3 BIOREFINERY: POTENTIAL AND PROSPECTS FOR UTILIZATION OF BIOGENIC WASTE

Lignocellulosic material can be used to generate energy through the process of thermochemical conversion, which involves the decomposition and combustion of organic material. Thermochemical conversion of lignocellulosic material encompasses the use of high temperatures and pressure to convert biomass into useful products. The process includes the breaking down of lignocellulosic materials such as grasses, wood, and agricultural remains into simpler molecules such as sugars and volatile fatty acids. These molecules can then be further processed into liquid fuels, heat, power, and chemical feedstocks. Thermochemical conversion technologies include pyrolysis, gasification, and hydrothermal liquefaction. This type of energy such as ethanol and biodiesel production is cleaner and more efficient than traditional fossil fuel sources, and it can produce electricity, heat, and steam. These fuels are renewable and biodegradable and help in the reduction of greenhouse gas, advance air quality, and address climate change making them a valuable resource for the future.

Biorefineries can also be used to produce materials that are more sustainable and environmentally friendly than traditional materials. Lignocellulosic material (wood, grasses, agricultural residues, and other plant-based materials) is used as a feedstock to develop a variety of valuable products, such as biofuels, chemicals, materials, and energy. For example, lignin can be broken down using an acid or a base. This can then be further broken down into monomers, which can be used to create various materials, such as plastics and bioplastics can be made from lignocellulosic materials. Other components, such as hemicellulose and cellulose can be utilized to produce ethanol, while lignin for the generation of heat and electricity. Biorefineries are also used to process non-renewable feedstocks, like petroleum and natural gas into renewable fuels and chemicals for various industrial utilization.

27.4 LIGNOCELLULOSIC BIOMASS AS A POTENTIAL AGENT FOR BIOREFINERY

Biorefineries can be used to produce biochemicals, such as enzymes and proteins (fig. 27.1). For example, Suhara et al. (2012) used selective lignin-biodegrading basidiomycetes, along with biological pre-treatment to make bioethanol from bamboo culms. In the study, 51 fungal isolates were obtained from *Punctularia sp.* and identified as white rot basidiomycete. The unidentified basidiomycete, TUFC20057, was preferred for the removal of lignin (50%) over *Ceriporiopsis subvermispora* and *Phanerochaete sordida* FP90031 and YK624. According to Taha et al. (2015), the co-culture of microorganisms that are capable of degrading lignocellulose to enhance straw saccharification has been found to be twofold greater than those found in bacteria. The results showed that fungal isolates had twice the enzyme activity as compared to bacterial isolates. Using co-culturing gives rise to a seven-fold growth in saccharification rates which leads to a higher commercial potential for the use of microbial consortia. Castoldi et al. (2014) reported that biological pretreatment with white rot fungi resulted in the biodegradation forms and saccharification kinetics of *Eucalyptus grandis* sawdust. Following the pre-treatment, organizational changes occurred in the sawdust fibres and a 20-fold increase in reducing sugars was found. It has been shown that treatment with *Pleurotus pulmonarius* and *Pleurotus ostreatus* caused lignin to be selectively degraded. It is possible to ferment hemicellulosic sugars (xylose and arabinose) into ethanol and butanol using numerous microbes such as *Candida shehatae, Pichia stipitis, Escherichia coli,* and *Zymomonas mobilis*, etc. (Khan & Dwivedi, 2013). In addition to serving as an antiseptic, ethanol can also be utilized as a remedy to ethylene glycol and methanol toxicity, and as fuel. However, butanol may also be used as a biofuel, paint thinner and perfume base material. There has been a growing

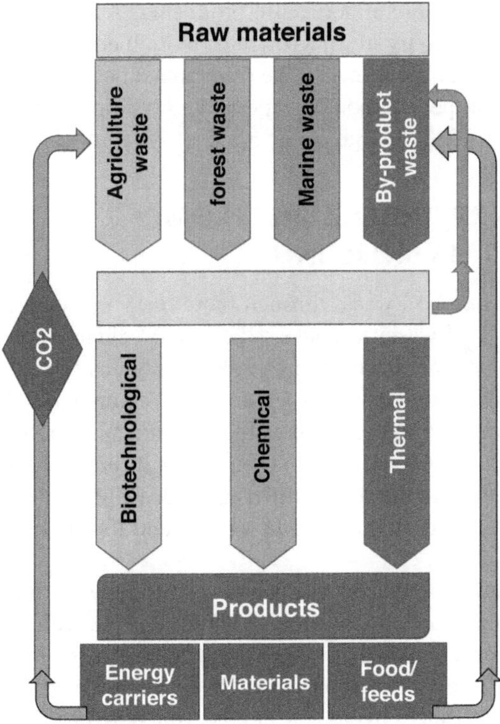

FIGURE 27.1 A bioconversion process to convert waste to energy and other essentials.

Source: Modified A Linares-Pasten et al. (2014).

interest in seaweed biomass as a biogas feedstock recently. Seaweeds (or macroalgae) are considered third-generation raw materials since their usage as energy crops offers crucial benefits over terrestrial crops. Following the extraction of bioproducts, it was discovered that seaweed residues have a high potential for producing methane. During some instances, the stoichiometric methane yield and C: N ratios altered to flavor-enhanced digestibility, with bioconversion rates exceeding 70%, such as in the case of *Laminaria species* and *Fucus serratus* on the West Coast. The two spp., of *Laminaria, were* also studied with the potential to produce the highest CH_4 production, with *Laminaria digitata* achieving 523 mL CH_4 VS^{-1} and *Laminaria saccharina* reaching 523 mL CH_4 VS^{-1} in both non-acclimatized and acclimatized sludge (Tedesco and Daniels, 2018). Consequently, in a biorefinery, seaweeds are considered a viable alternative to biomass feedstock for the development of energy, food, and chemicals.

27.5 BIOREFINERY: WASTE BIOGENESIS POTENTIAL AND PROSPECTS

Biorefineries can take many forms and employ a variety of processes. These processes often involve the alteration of waste biomass into useful products. Waste biorefineries can make use of a variety of organic waste streams, such as municipal solid waste, agricultural waste, and industrial waste, to transform them into valuable products. The potential benefits of waste biorefineries include increased resource efficiency, reduced environmental impacts and improved economic performance (Pagotto, & Halog, 2016). Biorefineries are emerging industrial facilities that convert biomass into a variety of products such as fuels, chemicals, and materials. By utilizing waste streams as a source of energy, biorefineries can reduce emissions, create jobs, and provide a net benefit to the environment. The waste biogenesis potential of biorefineries is vast. For example, waste from dairy farms can be used to produce bioethanol, while municipal solid waste can be converted into biogas. Additionally, a wide range of by-products can be generated from biorefineries such as biochar, bioplastics, and animal feed. The prospects for biorefineries are promising. A number of countries, including the United States, are investing heavily in the growth of biorefineries. The US Department of Energy has set aside over $25 million in funding for the development of biorefineries (Leibensperger et al., 2021). Additionally, the European Union has adopted a binding target to source 14% of its energy from renewable sources by 2020 (Scarlat et al., 2015)

27.6 PRODUCTION OF THE ENZYMES FROM FOOD WASTES VIA BIOREFINERY

Production of enzymes from food waste through biorefinery is a concept that involves the use of enzymes produced from food waste to convert them into valuable products (Uçkun Kiran et al., 2014). This concept involves the use of microbes and enzymes to break down the complex molecules present in food waste into simpler and more useful compounds. These compounds can then be used to develop a variety of products such as fuels, chemicals, and feedstocks. This process can help to reduce the amount of food waste that is sent to landfills and decrease the amount of energy needed to process the waste. Additionally, utilizing food waste in this process can help to reduce the amount of energy needed to produce food products and thus help to reduce the global carbon footprint (Eriksson et. al. 2015).

A lignocellulosic biorefinery concept is a promising strategy for the suitable manufacturing of bio-based products and fuels from lignocellulosic biomass. Plant cell wall-degrading enzymes are key components of this concept as they are necessary for the efficient deconstruction of plant cell walls into the polymeric components from which the biorefinery products and fuels are derived (Silva et al., 2018). Recent advances in enzyme engineering have enabled the production of enzymes with improved properties for biomass conversion. In addition, the use of these enzymes in combination with other technologies, such as fractionation and hydrothermal pre-treatment, can further advance the efficiency and economics of the biorefinery process (Wagle et al., 2022). Thus, the incorporation of plant cell wall-biodegrading enzymes into the lignocellulosic biorefinery offers great potential for the development of bio-based products and fuels (Silva et al., 2018). The green macroalgae *Ulva ohnoi* has been identified as a potentially valuable source of bioproducts. A biorefinery concept using this macroalga could offer a unique opportunity to produce a wide range of economically viable products. *Ulva ohnoi* is a fast-growing species and can be cultivated in large quantities using aquaculture and mariculture techniques (Mata et al., 2016). The biorefinery concept would involve harvesting the macroalgae, followed by processing and extraction of components such as proteins, lipids, carbohydrates, and bioactive compounds. The extracted components could then be used to produce various products such as food, animal feed, fuel, biofertilizers, and pharmaceuticals. The biorefinery concept could also include the use of integrated technologies such as thermochemical conversion to convert *Ulva ohnoi* into biogas or bioethanol (Polikovsky et al., 2020). The biorefinery could serve as an alternative source of revenue for farmers, providing them with an opportunity to diversify their income.

27.7 CONCLUSION

Bioconversion of lignocellulosic waste is important for the efficient and sustainable production of food, pharmaceuticals, fuels, and other useful products. It offers an eco-friendly and cost-effective alternative to conventional chemical syntheses, and it is able to produce high-value products from renewable resources. Bioconversion can also be used to improve the quality of food and reduce waste. Furthermore, bioconversion can help reduce the impact of climate change by reducing the emission of greenhouse gases.

ACKNOWLEDGEMENT

It is with great gratitude that the author extends his thanks to the Environmental Technologies Division of the CSIR-National Botanical Research Institute in Lucknow, India, for providing infrastructure facilities and to the Indian Council of Medical Research for providing a Senior Research Fellowship (file no. 5/3/8/58/ITR-F/2022) to the first author.

REFERENCES

Arora, R., Sharma, N. K., & Kumar, S. (2018). Valorization of by-products following the biorefinery concept: commercial aspects of by-products of lignocellulosic biomass. In *Advances in sugarcane biorefinery* (pp. 163–178). Elsevier.

A Linares-Pasten, J., Andersson, M., & Nordberg Karlsson, E. (2014). Thermostable glycoside hydrolases in biorefinery technologies. *Current Biotechnology*, *3*(1), 26–44.

Bajwa, D. S., Pourhashem, G., Ullah, A. H., & Bajwa, S. G. (2019). A concise review of current lignin production, applications, products and their environmental impact. *Industrial Crops and Products*, *139*, 111526.

Bilal, M., Asgher, M., Iqbal, H. M., Hu, H., & Zhang, X. (2017). Biotransformation of lignocellulosic materials into value-added products-a review. *International Journal of Biological Macromolecules*, *98*, 447–458.

Castoldi, R., Bracht, A., de Morais, G. R., Baesso, M. L., Correa, R. C. G., Peralta, R. A., ... Peralta, R. M. (2014). Biological pretreatment of Eucalyptus grandis sawdust with white-rot fungi: Study of degradation patterns and saccharification kinetics. *Chemical Engineering Journal*, *258*, 240–246.

Chowdhury, S. A., Kaneko, A., Baki, M. Z. I., Takasugi, C., Wada, N., Asiloglu, R., ... Suzuki, K. (2022). Impact of the chemical composition of applied organic materials on bacterial and archaeal community compositions in paddy soil. *Biology and Fertility of Soils*, *58*(2), 135–148.

Eriksson, M., Strid, I., & Hansson, P. A. (2015). Carbon footprint of food waste management options in the waste hierarchy–a Swedish case study. *Journal of Cleaner Production*, *93*, 115–125.

Gil-Chávez, J., Gurikov, P., Hu, X., Meyer, R., Reynolds, W., & Smirnova, I. (2021). Application of novel and technical lignins in food and pharmaceutical industries: Structure-function relationship and current challenges. *Biomass Conversion and Biorefinery*, *11*, 2387–2403.

Hendroko, R., Salafudin, A. W., & Vincevica-Gaile, Z. Study of biorefinery capsule husk from Jatropha Curcas L. waste crude Jatropha Oil as source for biogas. *Journal of Advanced Research in Biofuel and Bioenergy, 1*, 1–7.

Kabbour, M., & Luque, R. (2020). Furfural as a platform chemical: From production to applications. In *Biomass, Biofuels, Biochemicals*, 283–297. Elsevier.

Khan, Z., & Dwivedi, A. K. (2013). Fermentation of biomass for production of ethanol: A review. *Universal Journal of Environmental Research & Technology, 3*(1).

Klein, Bruno Colling, et al. (2019). Beyond ethanol, sugar, and electricity: A critical review of product diversification in Brazilian sugarcane mills. *Biofuels, Bioproducts and Biorefining*, 13(3), 809–821.

Leibensperger, C., Yang, P., Zhao, Q., Wei, S., & Cai, X. (2021). The synergy between stakeholders for cellulosic biofuel development: Perspectives, opportunities, and barriers. *Renewable and Sustainable Energy Reviews, 137*, 110613.

Mata, L., Magnusson, M., Paul, N. A., & de Nys, R. (2016). The intensive land-based production of the green seaweeds Derbesia tenuissima and Ulva ohnoi: biomass and bioproducts. *Journal of Applied Phycology, 28*, 365–375.

Mezule, L., Dalecka, B., & Juhna, T. (2015). Fermentable sugar production from lignocellulosic waste. *Chemical Engineering Transactions, 43*, 619–624.

Okolie, J. A., Nanda, S., Dalai, A. K., & Kozinski, J. A. (2021). Chemistry and specialty industrial applications of lignocellulosic biomass. *Waste and Biomass Valorization, 12*, 2145–2169.

Otieno, D. O., & Ahring, B. K. (2012). The potential for oligosaccharide production from the hemicellulose fraction of biomasses through pretreatment processes: xylooligosaccharides (XOS), arabinooligosaccharides (AOS), and mannooligosaccharides (MOS). *Carbohydrate Research, 360*, 84–92.

Pagotto, M., & Halog, A. (2016). Towards a circular economy in Australian agri-food industry: An application of input-output oriented approaches for analyzing resource efficiency and competitiveness potential. *Journal of Industrial Ecology, 20*(5), 1176–1186.

Polikovsky, M., Gillis, A., Steinbruch, E., Robin, A., Epstein, M., Kribus, A., & Golberg, A. (2020). Biorefinery for the co-production of protein, hydrochar and additional co-products from a green seaweed Ulva sp. with subcritical water hydrolysis. *Energy Conversion and Management, 225*, 113380.

Prasad, R. J., Sourie, S. J., Cherukuri, V. R., Fita, L., & Merera, C. E. (2015). Global warming: Genesis, facts and impacts on livestock farming and mitigation strategies. *International Journal of Agriculture Innovations and Research, 3*, 2319–1473.

Scarlat, N., Dallemand, J. F., Monforti-Ferrario, F., Banja, M., & Motola, V. (2015). Renewable energy policy framework and bioenergy contribution in the European Union–An overview from National Renewable Energy Action Plans and Progress Reports. *Renewable and Sustainable Energy Reviews, 51*, 969–985.

Setyobudi, R. H., Salafudin, A. W., & Vincevica-Gaile, Z. Study of Biorefinery Capsule Husk from Jatropha curcas L. Waste Crude Jatropha Oil as Source for Biogas. In *World Renewable Energy Congress International Conference on Renewable Energy and Energy Efficiency*, Bali, Indonesia

Shirvani, T., Yan, X., Inderwildi, O. R., Edwards, P. P., & King, D. A. (2011). Life cycle energy and greenhouse gas analysis for algae-derived biodiesel. *Energy & Environmental Science, 4*(10), 3773–3778.

Silva, C. O., Vaz, R. P., & Filho, E. X. (2018). Bringing plant cell wall-degrading enzymes into the lignocellulosic biorefinery concept. *Biofuels, Bioproducts and Biorefining, 12*(2), 277–289.

Sindhu, R., Binod, P., & Pandey, A. (2016). Biological pretreatment of lignocellulosic biomass–An overview. *Bioresource Technology*, 199, 76–82.

Suhara, H., Kodama, S., Kamei, I., Maekawa, N., & Meguro, S. (2012). Screening of selective lignin-degrading basidiomycetes and biological pretreatment for enzymatic hydrolysis of bamboo culms. *International Biodeterioration & Biodegradation, 75*, 176–180.

Tedesco, S., & Daniels, S. (2018). Optimisation of biogas generation from brown seaweed residues: Compositional and geographical parameters affecting the viability of a biorefinery concept. *Applied Energy, 228*, 712–723.

Usmani, Z., Sharma, M., Diwan, D., Tripathi, M., Whale, E., Jayakody, L. N., ... Gupta, V. K. (2021). Valorization of sugar beet pulp to value-added products: A review. *Bioresource Technology*, 126580.

Uçkun Kiran, E., Trzcinski, A. P., Ng, W. J., & Liu, Y. (2014). Enzyme production from food wastes using a biorefinery concept. *Waste and Biomass Valorization, 5*, 903–917.

Wagle, A., Angove, M. J., Mahara, A., Wagle, A., Mainali, B., Martins, M., ... Paudel, S. R. (2022). Multi-stage pre-treatment of lignocellulosic biomass for multi-product biorefinery: A review. *Sustainable Energy Technologies and Assessments, 49*, 101702.

Yuan, T. Q., Xu, F., & Sun, R. C. (2013). Role of lignin in a biorefinery: separation characterization and valorization. *Journal of Chemical Technology & Biotechnology, 88*(3), 346–352.

28 Macrophytes as a Potential Source of Food and Fodders for Humans and Animals Due to their High Biochemical Composition

Vartika Gupta, Namita Gupta, Pankaj Kumar Srivastava, and Anju Patel

28.1 INTRODUCTION

The Food and Agricultural Organization (FAO) is alarmed that the world will be unable to produce enough food to meet the demands of the world's ever-increasing population. It is predicted to reach 10 billion by the year 2050, posing a challenge to agricultural scientists to meet growing demand (Parihar et al., 2022). Although numerous edible aquatic plants can be used as food, many people still struggle with hunger and malnutrition. Unconventional, less well-known edile aquatic plants hold promise for supplying food for the population's unchecked increase (Das et al., 2021). Globally, at least 150 of the 3,000 plants that humans have utilized as sustenance throughout history have racially. However, there has been a trend over the years to focus on fewer and fewer sources, and as a result, the majority of the world's population is now fed by roughly 20 different crop species. Macroscopic aquatic plants known as edible macrophytes can be found in ponds, lakes, rivers, streams, and waterlogged soils (Chai et al., 2015).

For more than 50% of the world's population, rice (*Oryza sativa*), an emergent macrophyte species is the world's most important single crop (Bin et al., 2022). Farming aquatic plants is a seriously underutilized field, and it is time to consider these underutilized edible aquatic plants to assess their potential contribution to the expansion of the human food supply. Over 40 different species of comestible aquatic plants were believed to exist, but only about 25% of species are currently cultivated extensively for nutrition or can do so in an economically viable way. Thus, edible aquatic plants can serve as an alternative source of food, contain enough nutrients for daily needs, and serve as possible medications and food supplements (Hernawati et al., 2022).

The intrinsic nutrients included in food, including carbohydrates, proteins, vitamins, fats, antioxidants, and minerals are necessary for the body's system to function normally physiologically. India supports about 17.5% of the world's population and 20% of the world's livestock on just 2.3% of the earth's land area. Contrary to livestock, which is growing at a rate of 0.66% annually, the human population is growing at a rate of 1.6% annually. These expanding human and animal populations are vying ferociously for land resources needed for the production of food and fodder. As a result, only 4% of India's total cultivable area is used for farmed fodders. Currently, there is a net shortage of 44% concentrate feed ingredients, 10.5% dry crop leftovers, and 35.6% green fodder in India. There are very few options for expanding the extent of land used for fodder cultivation (Singh et al., 2022). Therefore, macrophytes offer a viable solution to the food and fodder dilemma because of their high biochemical content and potential to drive the ecosystem.

DOI: 10.1201/9781003441144-28

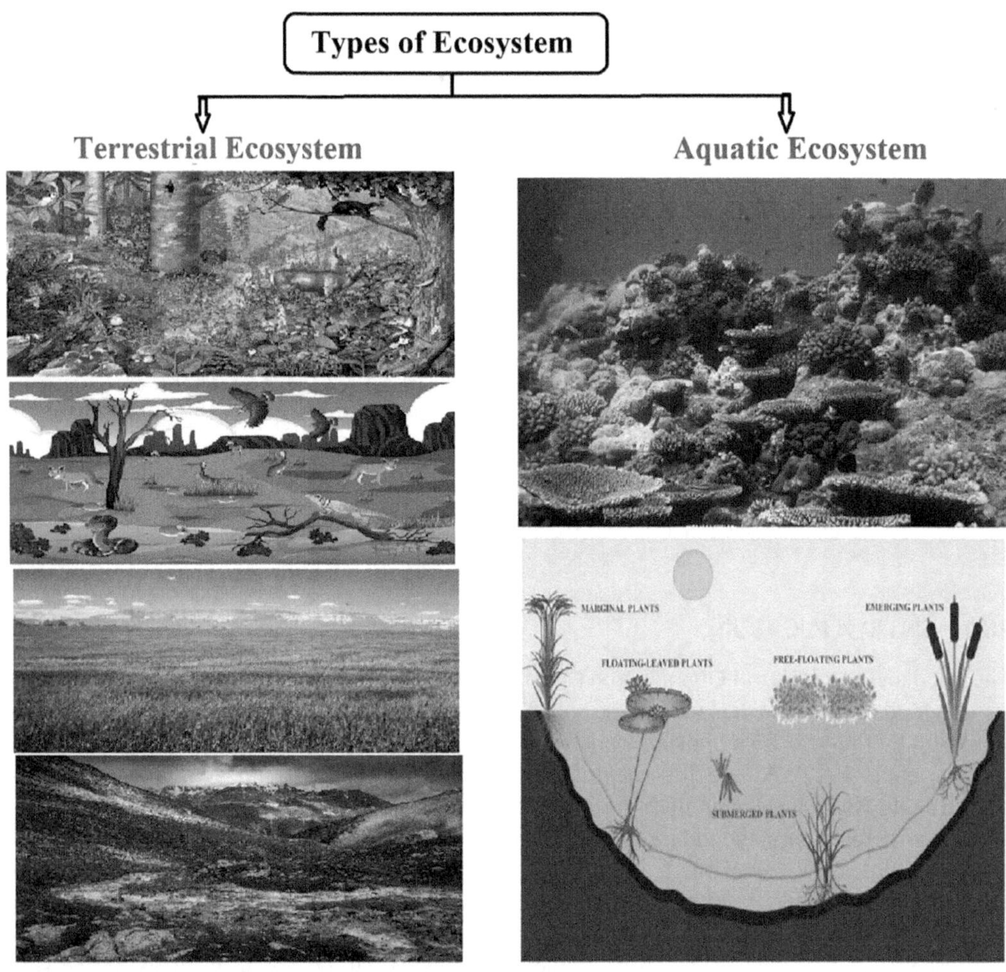

FIGURE 28.1 The aquatic environment of macrophytes.

Macrophytes are found naturally growing worldwide. They are aquatic plants that are large enough to be visible to the naked eye and are typically rooted in the substrate of a body of water (Chambers et al., 2008). They can be found in a variety of aquatic environments, including freshwater, brackish, and marine habitats. They can be emergent, submerged, or floating, that is, have erect parts above the surface of the water (Figures 28.1 and 28.2).

They can include enormous algae plants or flowering plants like ferns, cattails, water hyacinths, duckweeds, hydrillas, etc. Plant size and light intensity are important points found to be of great significance for the growth and reproduction of plants (Yuan et al., 2020). Therefore, macrophytes serve as an important tool for the biomonitoring of such ecosystems, which are communities of shallow lakes and marshes where they are biologically dominating (Wetzel, 2001; Lacoul and Freedman, 2006). According to Murphy et al. (2019), macrophytes species have constricted world ranges within the six ecozones of the world i.e. Afrotropics, Australasia, Nearctic, Neotropics, Orient, Palaearctic (Figure 28.3). They all provide most of the inland waterbody habitat suitable for the colonization of macrophytes. A total of ca. 2,294 species of macrophytes can be considered world-rare, out of which, 1,194 can be classified as "very rare". They usually occur within endorheic basins or catchments of rivers.

They also play an important role in aquatic ecosystems, providing food and habitat for a wide range of aquatic organisms, as well as helping to maintain water quality through the uptake of

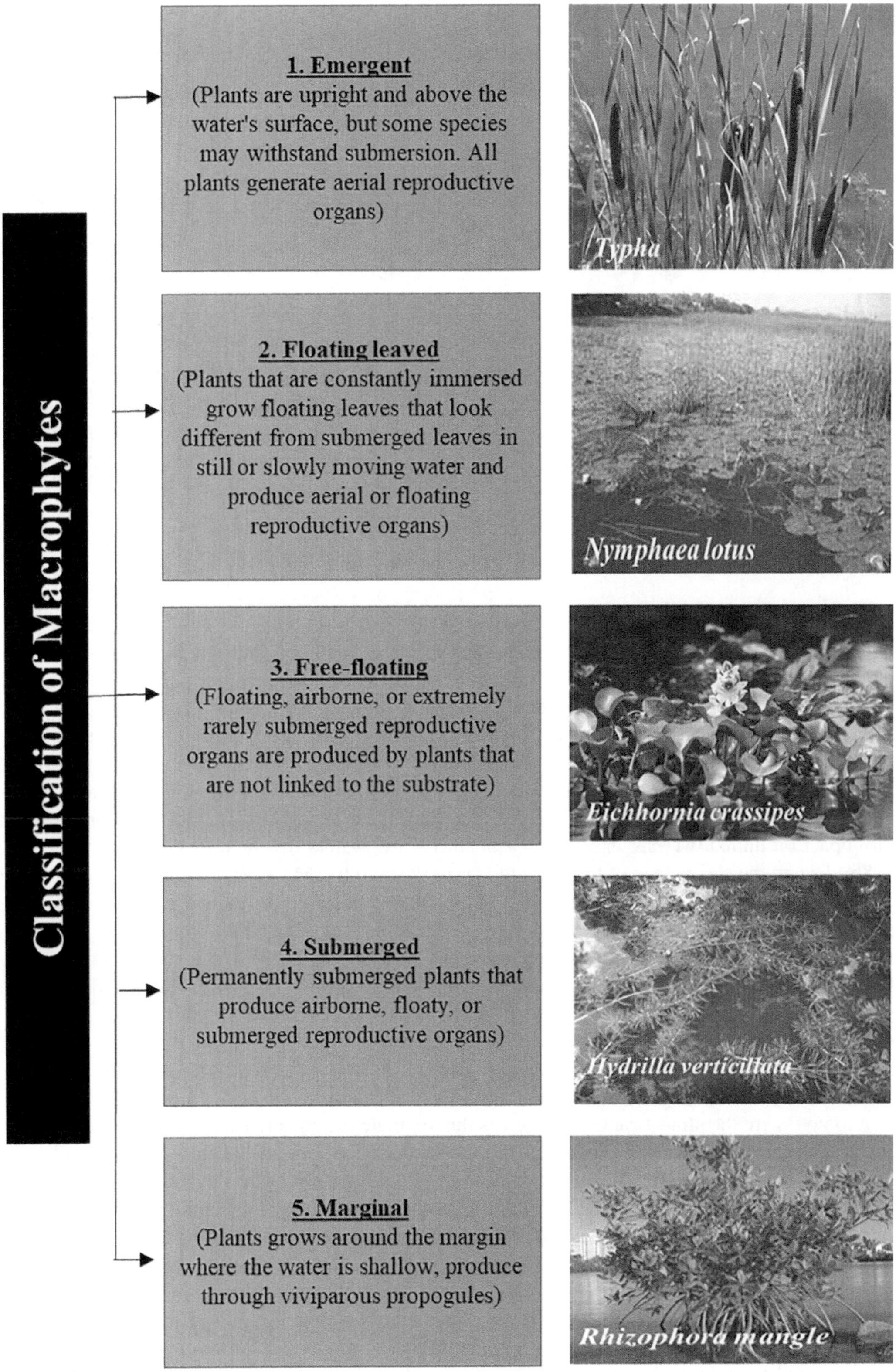

Classification of Macrophytes

1. Emergent
(Plants are upright and above the water's surface, but some species may withstand submersion. All plants generate aerial reproductive organs)

Typha

2. Floating leaved
(Plants that are constantly immersed grow floating leaves that look different from submerged leaves in still or slowly moving water and produce aerial or floating reproductive organs)

Nymphaea lotus

3. Free-floating
(Floating, airborne, or extremely rarely submerged reproductive organs are produced by plants that are not linked to the substrate)

Eichhornia crassipes

4. Submerged
(Permanently submerged plants that produce airborne, floaty, or submerged reproductive organs)

Hydrilla verticillata

5. Marginal
(Plants grows around the margin where the water is shallow, produce through viviparous propogules)

Rhizophora mangle

FIGURE 28.2 Classification of macrophytes with example.

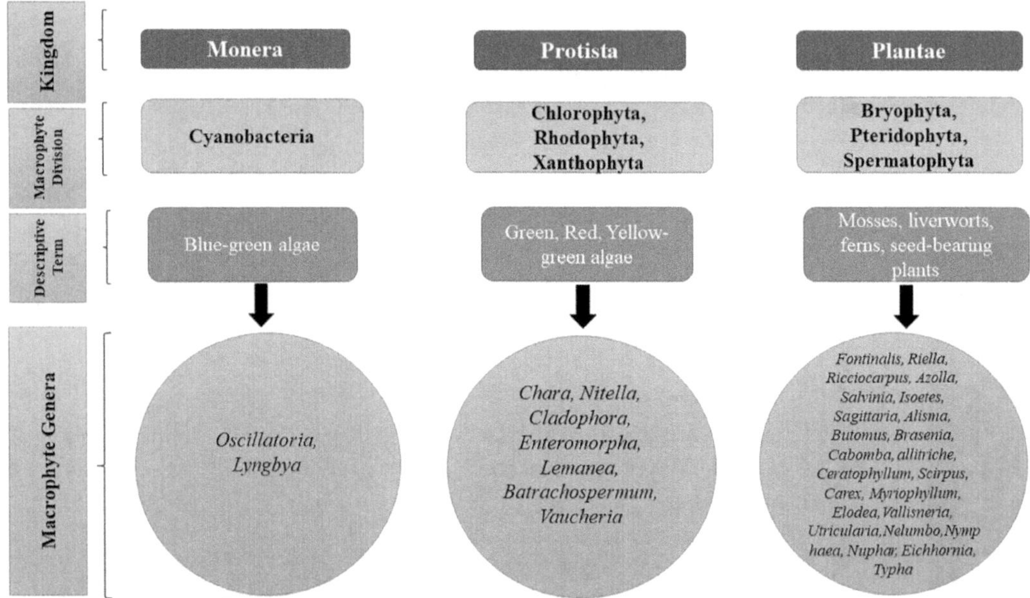

FIGURE 28.3 Division and representative genera of macrophytes.

nutrients and other pollutants. They are also employed as fertilizer, food, fodder, pharmaceutical, etc. However, because of urbanization, industrialization, or anthropogenic activities, they are very vulnerable to any form of nutritional alterations and organic contaminants (Gupta et al., 2018). They also can be used as indicators of water quality, as some species are sensitive to pollutants, while others can tolerate relatively high levels of pollution. They provide a food source for waterfowl and mammals like beavers and moose, as well as forming nesting places for many invertebrate species and certain fish. They also aid in stabilizing shorelines (Hendrey, 2001). They have historically had an impact on human welfare and culture, much like the waters in which they flourish (Mitchell, 1986). Due to the severe lack of food and fodder in the country. Macrophytes can be harvested and used directly as human food and animal feed after being assessed for their nutritional value. The species, seasons, and sites of the plants have a significant impact on their chemical composition (Anon, 1984). So, understanding macrophytes' biochemical composition is essential for assessing their potential as food, fodder sources, and nutritional worth (Hawkins and Hartnoll, 1983; Swapna et al., 2011).

28.2 IMPORTANCE OF MACROPHYTES

Macrophytes are the utmost captivating plants that regulate the aquatic ecosystem structure and the wetland's function. Because of their nutritional worth and health advantages, food technologies, and nutritionists are paying close attention to them. They have phytochemicals that are very useful in eating (Debbarma et al., 2022). Several macrophytes are crucial for human nutrition, thus it's important to look into alternative nutritious nutrients. The importance of aquatic plants is supplying nutrients for dietary requirements, supplements, medication, and health care (Hernawati et al., 2022). Macrophytes that are edible can provide benefits to the environment, as well as food, fodder, nutrition, nutraceuticals, functionality, and medicine (Haroon, 2022). Aside from their dietetic value, macrophytes are utilized for the treatment of various diseases (Nishan, 2020). However, due to social and economic advancements, modernizing lifestyles, ignorance, and other factors, these plants are misused (Butt et al., 2021).

28.3 UTILIZATION OF VARIOUS MACROPHYTES AS FOOD AND FODDER

Macrophytes have organs that may be able to produce food for humans and fodder for animals because of their accumulated food reserves. These food reserves in aquatic plants may be roots, rhizomes, seeds, etc. are discussed in Table 28.1. Several species of macrophytes such as *Eichhornia crassipes, Lemna minor, Ceratophyllum demersum, Nymphoides peltatum, Trapa natans, Ludwigia stolonifera, L. peploids, Nymphaea nouchali, Enhydra fluctuans, Ipomoea aquatic, Hygroryza aristata, and Limnocharis flava* can be used as food and fodder (Abobi et al., 2015; Bhat and Dar, 2015; Gupta et al., 2018; Alfasane et al., 2018).

Macrophytes play a key role in the lifecycle of different organisms and offer several utility services, such as livestock fodder, fish feed, and poultry feed due to their large size. *E. crassipes, Hydrilla verticillata, Nymphoides macrosperma, Pistia stratiotes, Salvinia sp., Utricularia aurea, Mimulus orbicularis, L. adscendens* are some macrophytes that are described as animal feed (Meena and Rout, 2016). Pathways in which macrophytes can be involved in food production processes, directly as human food, livestock fodder, and food for fish, turtles, manatees, and rodents were studied by Edwards (1981). The utilization of different macrophytes as fodder is given in Table 28.2.

28.4 BIOCHEMICAL COMPOSITION (NUTRITIONAL VALUE) OF MACROPHYTES

Macrophytes are well-known for being a good source of food and fodder for agricultural animals, aquatic herbivores, and humans, in addition to serving as the base of the food chain (Rather et al., 2022). Understanding the chemical composition of macrophytes is crucial for assessing their food potential and estimating their nutritional worth (Hawkins and Hartnoll, 1983). Abiotic factors like light, temperature, sediment composition, and water chemistry have a significant impact on metabolic processes like photosynthesis, growth rates, and levels of proximate nutrient constituents (Roslin, 2001; Rajasulochana et al., 2002), necessitating the identification of the biochemical components of the tissues produced under ecological conditions. Light and temperature both have an

TABLE 28.1
Edible Parts of Macrophytes are Utilized as Food Around the Globe

S. No.	Macrophytes Plant Parts	Countries	References
1.	Rhizomes	Bangladesh, Bolgatanga people of	Irvine (1952)
2.	Leaves	Ghana, Guinea, Ethiopia, Finland,	Watt and Breyer-Brandwijk (1962)
3.	Seeds	India, Kenya, Malawi, Nigerian,	Morton (1965)
4.	Stems	Pakistan, Senegal, Southern Florida,	Hujjatullah et al. (1967)
		Sudan, Thailand, West Africa	Kloos (1982)
			Airaksinen et al. (1986)
			Kabuye (1986)
			Chawanje et al. (2001)
			Ibrahim (2007)
			Teklehaymanot and Giday (2010)
			Danhassan et al. (2018)
			Adanse et al. (2019)
			Khan (2019)
			Gueye et al. (2020)
			Merga (2021)
			Bangar et al. (2022)

TABLE 28.2
Edible Parts of Macrophytes are Utilized as Fodder/Feed for Different Animals

S. No.	Macrophytes Species	Plant Parts	Uses	References
1	*Avena sativa*	Seeds	Fish feed	Ghaly et al. (2005)
2	*Avicennia marina*	Leaves	Camel (fodder)	FAO; Ghosh et al. (2015)
3	*Azolla pinnata*	Whole	Duck, Chicken, Cow, Buffalo, Sheep, Rabbit, and Goat feed	Samanta and Tamang (1995); Swain et al. (2018); Joysowal et al. (2018); Kurniawan et al. (2021); El-Naggar and El-Mesery (2022)
4	*Bruguiera gymnorrhiza*	Leaves	Animal feed	Ghosh et al. (2015)
5	*Carex virgata*	Leaves	Animal feed	Kurniawan et al. (2021)
6	*Echinochloa stagnina*	Leaves and stem	Livestock (fodder)	Balogun and Ibeun (1995); Abobi et al. (2015)
7	*Eichhornia crassipes*	Leaves	Cattle, sheep feed	Indulekha et al. (2019); Agarwala (1988); de Vasconcelos et al. (2016)
8	*Elodea canadensis*	Leaves	Fish food	Department of wildlife and fisheries sciences; Abobi et al. (2015)
9	*Ipomea aquatica*	Leaves	Livestock (fodder)	Ita (1994); Abobi et al. (2015)
10	*Lemna minor*	Leaves	Duck feed	Tu (2012); Kurniawan et al. (2021)
11	*Ludwigia adscendens*	Whole	Animal feed	Kurniawan et al. (2021)
12	*L. stolonifera*	Leaves	Livestock (fodder)	Ita (1994); Abobi et al. (2015)
13	*Neptunia oleracea*	Leaves	Livestock (fodder)	Nakamura et al. (1996); Abobi et al. (2015)
14	*Nymphaea lotus*	Fresh stalk & inflorescence	Feed	Ita (1993); Ita (1994); Kwarfo-Apegyah and Ipinjolu (1995); Abobi et al. (2015)
15	*Panicum hemitomon*	Shoots	Livestock (fodder)	U.S. Department of Agriculture; Abobi et al. (2015)
16	*Phragmites karka*	Shoots	Livestock (fodder)	Sushil (2012); Abobi et al. (2015)
17	*Pistia australis*	Leaves	Animal feed	Kurniawan et al. (2021)
18	*P. stratiotes*	Whole plant; leaves	Food, Fish feed, fodder	Ita (1994); Kurniawan et al. (2021)
19	*Rhizophora mangle*	Leaves	Cattle feed	Morton (1965)
20	*Rotala rotundifolia*	Leaves	Animal feed	Kurniawan et al. (2021)
21	*Scripus grossus*	Leaves	Buffalo feed	Singh and Vishwavidyalaya (2017)
22	*Trapa natans*	Seeds	Animal feed	Kurniawan et al. (2021)
23	*Typha angustifolia*	Leaves	Goat feed	Bansal et al. (2019)
24	*Vossia cuspidata*	Shoots	Livestock (fodder)	Burkill (1994); Ita (1994); Abobi et al. (2015)

impact on the growth rate, photosynthesis, morphology, chlorophyll composition, lipids, proteins, and amino acids (Rather et al., 2022). Macrophytes are one of the potential sources of proteins, carbohydrates, vitamins, minerals, fatty acids, dietary fibre, and ash. Due to the richness in biochemical components, they have a high nutritional value for use as food by humans, animals, and aquaculture. The principal biochemical elements, including proteins, fibre, ash, and lipids, serve as the energy-producing nutrients that give human and animal bodies the energy they need to carry out all of their processes of body (Rather and Nazir, 2015).

28.4.1 Protein and Amino Acids

Protein has drawn a lot of attention as a critical ingredient that has been deficient in vast portions of the world's population, leading to serious nutritional and health issues. With the use of single-cell protein cultures and leaf, protein concentrates, efforts have been undertaken to improve the diets of humans and animals. Another source that may have the ability to improve the protein status of both ruminants and non-ruminants is vast and expanding stands of macrophytes populations found in rivers and lakes around the globe. The composition of a protein's amino acid, especially those considered essential, gives a decent indication of its nutritional quality (Muztar et al., 1978). Although leaves of *R. mangle* have a high tannin content, they may be intake as a rich source of protein (Morton, 1965). Protein provides the precise amino acids that the body needs, including nitrogen of the amino acid (Rather et al., 2022). According to Abelti et al. (2023), *Nymphaea petersiana* had a greater protein content (8.1%) than cassava, sweet cassava, potatoes, sweet potatoes, yams, 1.7%, 1.6%, and 5.2% of *N. lotus*. When compared to tubers like cassava, potato, and yam as well as cereals like maize, millet, and rice the rhizome of *N. petersiana* has a higher content of protein. The red and green seeds of *N. lotus* were found to have a protein level that was comparable to that of maize, rice, sorghum, and wheat (Laminu et al., 2021).

For the maintenance and anabolism of muscle protein, water lilies contain both essential and non-essential amino acids (EAAs). According to Danhassan et al. (2018), *N. lotus* seeds showed presence of all nine EAAs *viz.*, Arginine, Histidine, Isoleucine, Lysine, Leucine, Methionine, Phenylalanine, Threonine, and Tryptophan, along with eight non-essential amino acids (Alanine, Aspartate, Cysteine, Glutamate, Glycine, Proline, Serine, and Tyrosine. Thus, *N. lotus* high concentration fulfils the daily intake suggested by the WHO/FAO (0.8 g/kg of body weight). Protein content (%) in different macrophytes is given in Table 28.3.

28.4.2 Dietary Fibre Contents

Dietary fibre (DF) refers to a class of heterogeneous substances that vary greatly in their chemical composition and physical characteristics both within and between plant sources (Graham and Åman, 1991). It is thought to be crucial for maintaining overall human health. According to epidemiological research, lack of DF in the diet has been linked to some chronic diseases, including diabetes, obesity, heart diseases, and several malignancies. DF, whether soluble or insoluble, helps to regulate blood cholesterol levels, maintain weight, lower diabetes menace and colon cancer, and promote the growth of good bacteria (Abelti et al., 2023). The DF content of *N. lotus* seed (4.68% and 8.54%) was studied by Danhassan et al. (2018). The tuber of *N. nouchali* is an affordable nutritive supplement and efficient food with noteworthy macro and micronutrients that can aid in the combat against oxidative stress resulting from metabolic diseases brought on by the contemporary lifestyle (Anand et al., 2019). The *Nymphaea* sp. seeds have dietary fibre levels that are comparable to those of rice and millet, which are a staple food in diets around the globe (Gueye et al., 2020). Some macrophytes that are rich in fibre content are discussed below (Table 28.4).

28.4.3 Minerals Content

Macrophytes are a good source of minerals and nutrients (Tables 28.5 and 28.6). The age, type of the plants, and fertility of the aquatic environment may be responsible factors for the substantial variance in mineral content found in the tissue chemistry of macrophytes (Boyd, 1968).

TABLE 28.3
Protein Content (%) in Different Macrophytes

S. No.	Macrophyte Species	Protein Content (%)	References
1	*Azolla africana*	28.9	Fasakin and Balogun (1998)
2	*A. caroliniana*	18.8	Datta (2011)
3	*A. filiculoides*	19.7	Datta (2011)
4	*A. mexicana*	18.6	Datta (2011)
5	*A. microphylla*	20.2	Datta (2011)
6	*A. pinnata*	15.4–28	Samanta and Tamang (1995); Ahirwar and Leela (2012); Parashuramulu et al. (2013); Bora (2018); El-Fadel et al. (2020); Bhatt et al. (2021)
7	*A. rubra*	19	Datta (2011)
8	*A. sp.*	20–30	Gupta et al. (2018); Rather et al. (2022)
9	*Ceratophyllum demersum*	21.99	Rather et al. (2022)
10	*Colocasia esculenta*	25	Gohl (1981)
11	*Eichornia crassipes*	17.93	Adelakun et al. (2016)
12	*Elodea densa*	20.5	Boyd (1968)
13	*Enhydra fluctuans*	18.2	Alfasane et al. (2018)
14	*Euryale ferox*	15.6	Alfasane et al. (2018)
15	*Hydrilla verticillata*	14.1–21.5	Venkatesh and Shetty (1978); De Silva and Perera (1983); Ray and Das (1994)
16	*Hygroryza aristata*	14.35	Alfasane et al. (2018)
17	*Ipomaea aquatica*	16.8–26.45	Adelakun et al. (2016); Alfasane et al. (2018)
18	*I. reptans*	32.2	Kalita et al. (2007)
19	*Lemna minor*	20.3–36.07	Bora (2018); Kumar et al. (2022)
20	*L. polyrhiza*	18.6	Bairagi et al (2002)
21	*Limnocharis flava*	14.35	Alfasane et al. (2018)
22	*Monochoria hastata*	39.5	Pandey and Srivastava (1991)
23	*Myriophyllum spicatum*	7.5–21.8	Muztar et al. (1978); Pine et al. (1990)
24	*M. verticillatum*	13.01	Rather et al. (2022)
25	*Nelumbo nucifera*	17.5	Rather et al. (2022)
26	*Nymphaea lotus*	7.93–21.66	Adelakun et al. (2016); Stephen et al. (2017); Danhassan etal. (2018); Etse et al. (2018); Aung et al. (2020); Gueye et al. (2020); Keak et al. (2022)
27	*N. nouchali*	17.15	Alfasane et al. (2018)
28	*Nymphoides peltatum*	1.08–1.14	Bhat and Dar (2015)
29	*Phragmites australis*	18.6	Rather et al. (2022)
30	*Pistia* sp.	11.8	Mandal and Ghosh (2019)
31	*P. stratiotes*	23.27	Adelakun et al. (2016)
32	*P. natans*	15.8	Rather et al. (2022)
33	*P. crispus*	10.9–15.2	Boyd (1968); Pine et al. (1990)
34	*Potamogetan lucens*	14	Rather et al. (2022)
35	*Salvinia curculata*	11	Kalita et al. (2007)
36	*Sparganium americanum*	23.8	Tacon (1987)
37	*Spirodela polyrhiza*	23.8–43	Datta (2011); Bora (2018); Gupta et al. (2018); Kumar et al. (2022)
38	*Trapa natans*	11.4–16.82	Kalita et al. (2007); Rather et al. (2022)
39	*Typha angustata*	19.56	Rather et al. (2022)
40	*Wolffia arrhiza*	20.4	Chareontesprasit and Jiwyam (2001)

TABLE 28.4
Fibre Content (%) in Different Macrophytes

S. No.	Macrophyte Species	Fibre Content (%)	References
1	*Alternanthera philoxeroides*	15.1	Tacon (1987)
2	*Azolla pinnata*	12.7–32.17	Ahmed et al. (2016); Anhita et al. (2016); Bora, 2018; Cherryl et al. (2014); El-Fadel et al. (2020)
3	*A. sp.*	15.1	Rather et al. (2022)
4	*Ceratophyllum demersum*	21	Rather et al. (2022)
5	*Cladophora glomerata*	20.7	Muztar et al. (1978)
6	*Colocasia esculenta*	12.1	Gohl (1981)
7	*Eleocharis ochrostachys*	29.2	Klinnavee et al. (1990)
8	*Enhydra fluctuans*	11.5	Alfasane et al. (2018)
9	*Euryale ferox*	7.6	Alfasane et al. (2018)
10	*Hygroryza aristata*	25	Alfasane et al. (2018)
11	*Ipomaea aquatica*	12.3	Alfasane et al. (2018)
12	*Lemna minor*	10	Bora (2018)
13	*Limnocharis flava*	24	Alfasane et al. (2018)
14	*Myriophyllum spicatum*	50.2	Muztar et al. (1978)
15	*M. verticillatum*	11.93	Rather et al. (2022)
16	*Nelumbo nucifera*	2	Rather et al. (2022)
17	*Nymphaea nouchali*	23	Alfasane et al. (2018)
18	*Phragmites australis*	2	Rather et al. (2022)
19	*Potamogetan lucens*	20.95	Rather et al. (2022)
20	*Potamogetan natans*	12.9	Rather et al. (2022)
21	*P.* sp.	21.5	Muztar et al. (1978)
22	*Sagittaria latifolia*	27.6	Boyd (1968)
23	*Sparganium americanum*	20.2	Tacon (1987)
24	*Spirodela polyrhiza*	18.7	Bora (2018)
25	*Trapa natans*	12.3	Rather et al. (2022)
26	*Typha angustata*	1.8	Rather et al. (2022)
27	*Typha latifolia*	33.2	Boyd (1968)
28	*Vallisncria america*	34.6	Muztar et al. (1978)

Minerals have a key role as catalysts in many biological processes. These are vital parts of metabolism, growth, and development that also enable animals to adapt to their always-changing environment. Each mineral has an optimal dosage. The physiology of the organism may be impacted by low or high quantities. The body only needs trace amounts of toxic minerals like arsenic (As), antimony (Sb), cadmium (Cd), mercury (Hg), and others, whereas excessive amounts of beneficial minerals like calcium (Ca), iron (Fe), potassium (K), sodium (Na), and magnesium (Mg) may be hazardous (Skalnaya and Skalny, 2018). Whereas, macrophytes captivate different amounts of energy which enhances the quality of feed supplements (Table 28.7).

When plants are taken into account as a feed source, the Ca and phosphorus (P) ratio is a crucial factor (Alfasane et al., 2018). According to Danhassan et al. (2018), *N. lotus* seed (30 g) can provide the Recommended Daily Intake (RDI) of Fe for children aged 1–3, 9–13, and adults aged 51–70. The recommended daily allowance of Fe for children is 88% and for women is 59% (Abelti et al., 2023). The ratio of sodium to potassium varies from species to species (Figure 28.4). For an adult human, WHO/FAO recommended Na: K ratio is <0.49 (Bailey et al., 2015).

Calcium (Ca) has been the primary mineral for structural purposes and aids in metabolism. It acts as an indication for crucial physiological functions. 350 biological enzymes require Mg,

TABLE 28.5
Mineral Composition of Some Macrophytes

S. No.	Macrophytes Species	Minerals Composition (mg/100 g)								References
		Ca	Cu	Fe	K	Mg	Na	P	Zn	
1	*Enhydra fluctuants*	224.45	–	30.48	–	–	–	175.22	–	Alfasane et al. (2018)
2	*Ipomaea aquatica*	128.26	–	27.7	–	–	–	173.22	–	Alfasane et al. (2018)
3	*Hygroriza aristata*	208.4	–	20.58	–	–	–	228	–	Alfasane et al. (2018)
4	*Nymphaea nouchali*	328.66	–	95.13	–	–	–	250	–	Alfasane et al. (2018)
5	*Limnocharis flava*	493.18	–	17.94	–	–	–	262	–	Alfasane et al. (2018)
6	*Nymphaea lotus*	0.98–188.81	0.42–25.3	1.36–26.1	240–481.1	13–312.71	11.55–210	0.15–81.1	2.19–3.81	Danhassan et al. (2018); Abelti et al. (2023)
7	*N.micrantha*	193–296.52	–	–	454–770.81	518.67	8.99–13.52	–	–	Gueye et al. (2020); Abelti et al. (2023)
8	*N.petersiana*	–	–	100	–	–	–	–	–	Abelti et al. (2023)
9	*N.nouchali*	3–159	–	1.98–14.7	215–746	10.5–91	10.2–441	21.4–98	1.33	Jarapala et al. (2021); Abelti et al. (2023)
10	*N.pubescens*	336	1.36	36.1	968	114	66	66.7	1.33	Tresina et al. (2020)
11	*N.rubra*	336	1.14	32	846	136	48	98	1.78	Tresina et al. (2020)

Ca=calcium; Cu=copper; P=phosphorus; Fe=iron; Na=sodium; Mg=magnesium; K=potassium; Zn=zinc.

TABLE 28.6
Mineral Composition of Some Macrophytes

S.No.	Macrophytes Species	Minerals (mg/kg)											Minerals (%)				References
		B	Co	Cr	Cd	Cu	Fe	Mn	Ni	Pb	Phosphates	Zn	Na	K	P	Ca	
1	*Azolla pinnata*	31	7.1–8.11	2.01–5.06	1.2	4.1–17.15	710.65–1569	207.87–2418	5.33–6.1	8.1	0.34	77.3–325		2.41	0.26–0.9	0.4–2.58	Ahirwar and Leela (2012); Mathur et al. (2013); Cherryl et al. (2014); Ahmed et al. (2016); Anhita et al. (2016); Bora (2018); Bhatt et al. (2021)
2	*Callitriche brutia*	–	–	–	–	25.3	39.05	357.45	–	–	13031	–	0.4	0.25	–	–	Cabrera-Guzmán et al. (2020)
3	*Ceratophyllum demersum*	–	–	–	–	–	–	–	–	–	–	–	–	–	0.32	1.3	Venkatesh and Shetty (1978)
4	*Colocasia esculenta*	–	–	–	–	–	–	–	–	–	–	–	–	–	0.58	1.74	Gohl (1981)
5	*Eichornia crassipes*	–	–	–	–	6.73	197.57	118.3	–	–	–	62.7	–	–	–	–	Adelakun et al. (2016)
6	*Elodea canadensi*	–	–	–	–	–	–	–	–	–	–	–	–	–	0.36	1.75	Pine et al. (1990)
7	*Hydrilla verticillata*	–	–	–	–	–	–	–	–	–	–	–	–	–	0.28	4.4	Venkatesh and Shetty (1978)
8	*Ipomoea aquatica*	–	–	–	–	6.73	197.57	116.5	–	–	–	63.87	–	–	–	–	Adelakun et al. (2016)

(Continued)

TABLE 28.6 (*Continued*)
Mineral Composition of Some Macrophytes

S.No.	Macrophytes Species	Minerals (mg/kg)											Minerals (%)				References
		B	Co	Cr	Cd	Cu	Fe	Mn	Ni	Pb	Phosphates	Zn	Na	K	P	Ca	
9	*Justicia americana*	–	–	–	–	–	–	–	–	–	–	–	–	–	0.12	0.82	Tacon (1987)
10	*Lemna minor*	–	–	–	–	–	–	–	–	–	–	–	–	–	0.7	0.6	Bora (2018)
11	*Myriophyllum alterniflorum*	–	–	–	–	12.65	4423.4	3462.14	–	–	10127	–	0.6	0.53	–	–	Cabrera-Guzmán et al. (2020)
12	*M. spicatum*	–	–	–	–	–	–	–	–	–	–	–	–	–	0.41	2.82	Pine et al. (1990)
13	*Najas guadalupensis*	–	–	–	–	–	–	–	–	–	–	–	–	–	0.15	0.98	Buddington (1979)
14	*Nymphaea lotus*	–	–	–	–	5.7	186.3	113.1	–	–	–	61.37	–	–	–	–	Adelakun et al. (2016)
15	*Potamogeton crispus*	–	–	–	–	–	–	–	–	–	–	–	–	–	0.24	1.68	Boyd (1968)
16	*Ranunculus peltatus*	–	–	–	–	11.95	777.41	291.11	–	-	13331	–	0.39	0.15	–	–	Cabrera-Guzmán et al. (2020)
17	*Spirodela polyrhiza*	–	–	–	–	–	–	–	–	–	–	–	–	–	1.28	2.1	Bora (2018)
18	*Typha latifolia*	–	–	–	–	–	–	–	–	–	–	–	–	–	0.17	0.64	Boyd (1968)

B=Boron; Ca=calcium; Cd=cadmium; Co=cobalt; Cr=chromium; Cu=copper; Fe=iron; K=potassium; Mn=manganese; Na=sodium; Ni=nickel; P=phosphorus; Pb=lead; Zn=zinc

TABLE 28.7
Energy (kcal/100 g) Production Through Some Macrophytes

S. No.	Macrophytes Species	Energy (kcal/100 g)	References
1	*Azolla pinnata*	1759	Parashuramulu et al. (2013)
2	*Cladophora glomera*	2.8	Muztar et al. (1978)
3	*Enhydra fluctuants*	317.28	Alfasane et al. (2018)
4	*Hygroriza aristata*	306.34	Alfasane et al. (2018)
5	*Ipomaea aquatica*	312.2	Alfasane et al. (2018)
6	*Limnocharis flava*	305.39	Alfasane et al. (2018)
7	*Myriophyllum spicatum*	2.3	Muztar et al. (1978)
8	*Nymphaea nouchali*	307.33	Alfasane et al. (2018)
9	*Potamogeton spp.*	3.9	Muztar et al. (1978)
10	*Vallisncria american*	3.2	Muztar et al. (1978)

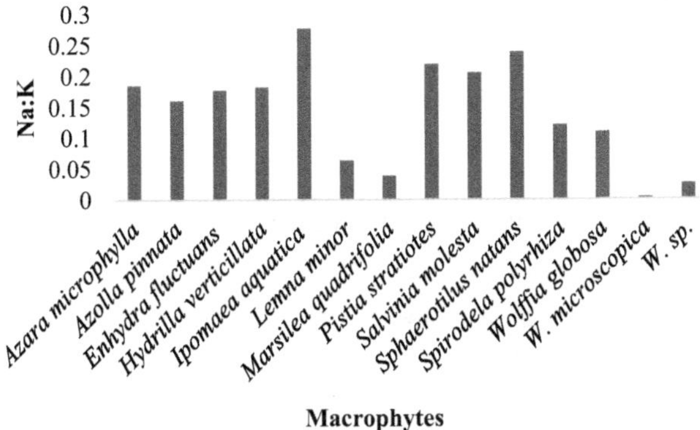

FIGURE 28.4 Na to K ratio of some macrophytes (Kumar et al., 2022).

the body's fourth most abundant cation, as a co-factor, the majority of which are engaged in energy metabolism (World Health Organization, 2009). In aquatic animals, fish with deficiencies in Co, Mn, Fe, Se, and Zn have hydropchromic microcytic anaemia, poor development, cataracts, and skin & caudal fin erosion. Fish with deficiencies in selenium (Se) also develop muscular dystrophy. Cobalt (Co) is a crucial component of fish nutrition. Therefore, adding macrophytes to a diet may assist fish and other animals overcome their mineral deficiencies without having any detrimental effects (Sree et al., 2019). The dietary requirements of various minerals for human and animals are given in Table 28.8.

28.4.4 Fatty Acids Contents

Plants have a unique feature in that their nutritional profile changes significantly as their substrate water changes. Polyunsaturated fatty acids (PUFAs) in the diet have several health benefits for humans and other animals. Thus, the composition of fatty acids in macrophytes must be evaluated (Table 28.9) before they can be used as feed ingredients for fish and other animals (Kumar et al., 2022) (Figure 28.5).

The various substrate sources used in macrophyte production are critical in determining the fatty acids profile. FA plays important biological, functional, and structural roles in the human body. They serve as the primary energy source for cellular membranes. FA also have been used

TABLE 28.8
Dietary Requirement of Different Minerals for Humans and Different Animals

S. No.	Nutrients	Humans (g/day)	Animals (g/kg diet)						
			Cattle	Common Carp	Fishes and Prawns	Grass Carp	Nile Tilapia	Poultry	Other Fishes (mg/kg diet)
1	Na	2.4	0.96	–	–	2	–	0.012–0.2	–
2	Mg	–	–	–	0.4–0.946	–	–	–	–
3	K	3.5	2.4	0.9–12.4	–	4.6	2.1–3.3	0.3	–
4	Ca	1	5.12	0.1	1.9	2	7	8	–
5	Mn	–	–	–	–	–	–	–	12 to 25
6	Fe	–	–	–	–	–	–	–	30–200
7	Zn	–	–	–	–	–	–	–	15–79
8	Co	–	–	–	–	–	–	–	0.01–0.5

Ca = Calcium; Co = Cobalt; Fe = Iron; K = Potassium; Mg = Magnesium;Mn = Manganese; Na = Sodium; Zn = Zinc
Source: Kumar et al. (2022).

TABLE 28.9
Composition of Fatty Acid in Plant Samples (mg/g)

S.No.	Fatty Acids	*A. pinnata*	*L. minor*	*S. polyrhiza*
1	Arachidonic acid (5,8,11,14-Eicosatetraenoic acid)	0.54	–	0.5
2	Behenic acid (Docosanoic acid)	0.23	–	–
3	Lignoceric acid (Tetracosanoic acid)	–	0.36	0.12
4	Linoleic acid (octadeca-9, 12-dienoic acid)	4.8	6.5	4.2
5	Linolenic acid (9,12,15- Octadecatrienoic acid)	15.9	16.2	15.8
6	Palmitic Acid (hexadecanoic acid)	6.89	7.4	6.54
7	Palmitoleic acid (hexadec-9- enoic acid)	0.84	2.1	2.3
8	Stearic acid (octadecanoic acid)	–	0.39	0.8

Source: Bora (2018).

to identify potential biomarkers for a variety of pathologies including polycystic ovary syndrome (PCOS) (Nagy and Tiuca, 2017). It also ensures fluidity, flexibility, and permeability of the membrane as well as passive transport through the membrane. PUFAs (n-3 & n-6) appear to be the most important of the FA due to their multiple biological roles such as inflammatory cascade, reducing oxidative stress and providing neuro and cardiovascular protection. Food with a ratio of 1:1–5:1 ω-6: ω-3 is measured as optimum for human health (Patel et al., 2022).

Animals cannot produce n-3 or n-6 polyunsaturated fatty acids (PUFAs). As a result, these must be brought in relative proportions in the food, either in their long-chain form linoleic acid (18:2n-6) or as their precursors-LA (ALA 18:3n-3). Although the importance of essential fatty acids has long been recognized, little investigation has been piloted on swine essential fatty acid necessity, and dietary consumption values are slightly ambiguous. The National Research Council recommends a maximum concentration of linolenic acid in the diet of 0.1% (Bora, 2018). Some researchers believe that 3.3% linolenic and 0.45% linolenic levels may be optimal for reproduction. (Bhaskaran and Kannapan, 2015). Macrophytes (*A. pinnata*, *L. minor*, *S. polyrhiza*) are high in linolenic and palmitic acid, as well as PUFAs like arachidonic acid, which are essential

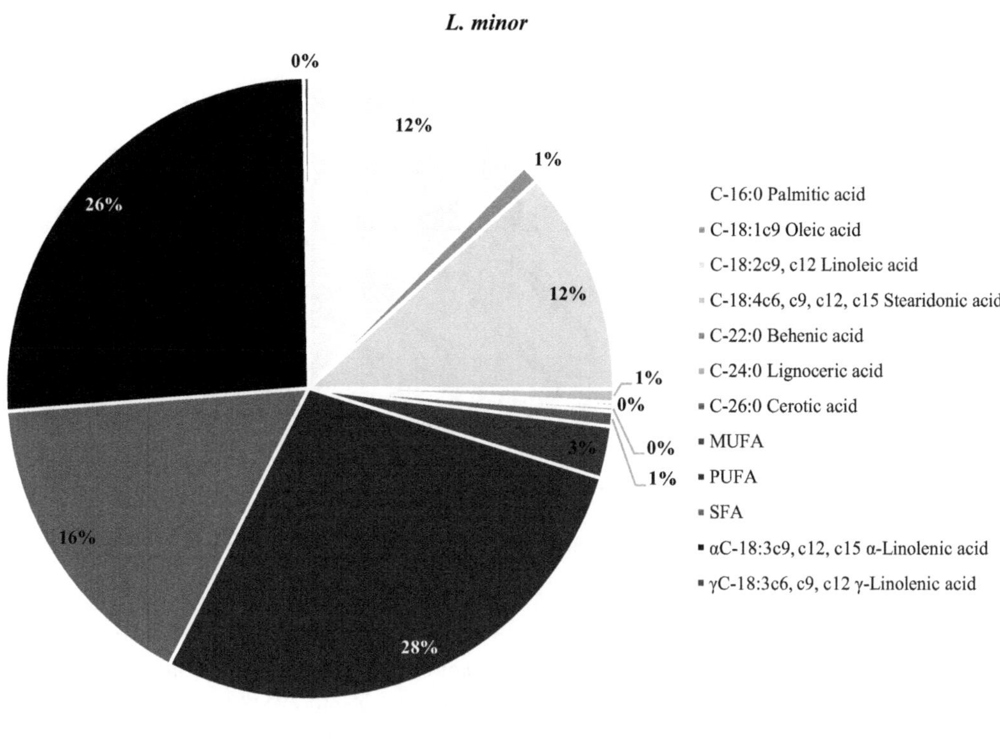

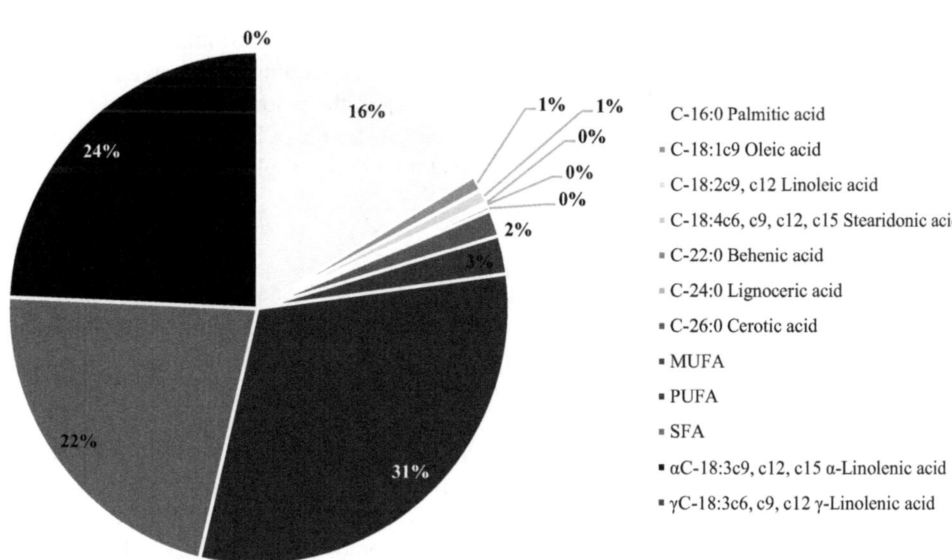

FIGURE 28.5 Fatty acid structure (% total Fatty Acid Methyl Esters) (Bora, 2018).

AAs for pig and poultry development (Bora, 2018). Macrophytes have enormous potential for use as mineral-rich sources of n-6 and n-3 PUFA for fish, poultry, and livestock. According to Kumar et al. (2022), the nutritional value of macrophytes can be increased by cultivating with organic manure (Figure 28.6).

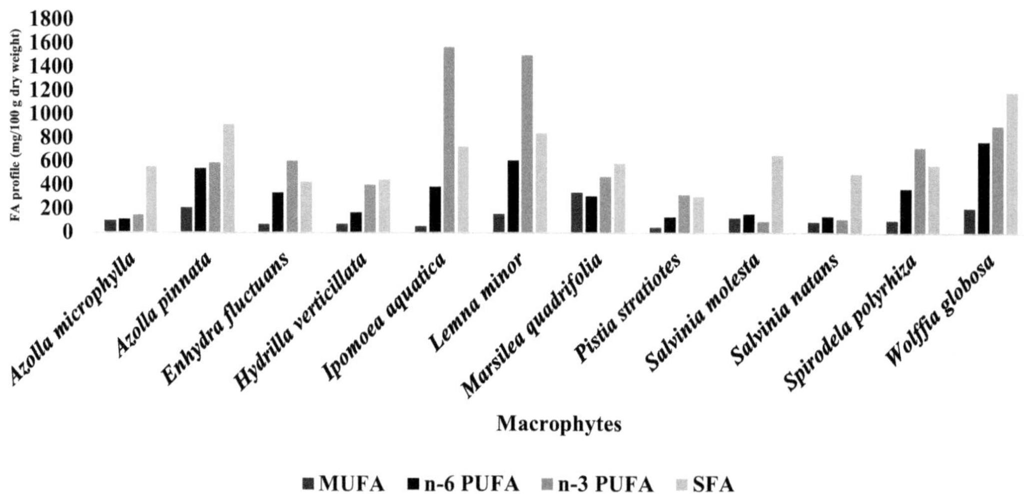

FIGURE 28.6 Fatty acids profiling of different macrophytes cultured with organic manure (Kumar et al., 2022).

Fish typically require omega-3 and omega-6 fatty acids. HUFA (n-3) is essential for marine fish health and growth, and its concentration should range from 0.5% to 2% of a dry intake. This two major groups of EFAs are eicosapentaenoic acid (20:5n-3) and docosahexzenoic acid (22:6n-3). Freshwater requires 18 carbon n-3 FA, linolenic acid (18:3-n-3), in amounts ranging from 0.5% to 1.5% of dry weight of the food. *Tilapia* need n-6 FA, whereas eels or carp entail a mixture of n-3 and n-6 FA.

Macrophytes may effectively serve as alternative fodder if the formulation is prudently fabricated based on the animal age. Some studies have found that feeding macrophytes causes fatty acid deposition in the liver. This necessity more adjustments in the amount of feed given to the animal, but in terms of the macrophyte's nutritional properties, its profile of FA is extremely appropriate and nutritious. Fish also need PUFA for growth that is found in macrophytes.

28.4.5 Vitamins Contents

Macrophytes are a good source of vitamins A, C, D, E, K and the majority of group B substances (Donchenko et al., 2020). Our bodies require trace amounts of vitamins for physiological functions and metabolism to preclude diseases allied with oxidative stress. *N. lotus* rhizome contains a high concentration of carotenoids alike (51.3 mg/100 g) and vitamin C (24.65 mg/100 g) (Stephen et al., 2017). According to Anand et al. (2019), boiled water lily rhizome contains ascorbic acid (3.12 mg/100 g), riboflavin (1.11 mg/100 g), thiamine (0.05 mg/100 g), and niacin (1.45 mg/100 g). Red and green seeds of *N. lotus* were found to have greater riboflavin and thiamine contents than acha, maize, pearl millet, sorghum, and wheat. *Azolla* was found to be high in vitamins (vitamin A, vitamin B12, and beta-carotene) and growth promoter intermediaries (Pillai et al., 2002). *Azolla* recommended feeding animals, poultry, and fish due to the presence of vitamins A and B12 (Gupta et al., 2018). Dietary inclusion of *L. minor* could be an energetic source of vitamin E (26.6 mg/100 g) and vitamin C (3.8 mg/100 g). *Ipomea reptans* can be used as a forage constituent in fish feed including vitamin E (28.5 mg/100 g) and vitamin C (4 mg/100 g) (Kalita et al., 2007). *C. esculentu* (leaves and petioles) and *Rorippu nasturtium* are rich sources of vitamins A, B, and C (Edwards, 1981).

28.4.6 Carbohydrate Contents

Carbohydrates, along with protein and fat, are one of the three macronutrients in the human diet. Carbon, hydrogen, and oxygen atoms are found in these molecules. Carbohydrates provide energy, aid in blood glucose and insulin metabolism, participate in cholesterol and triglyceride metabolism, and aid in fermentation. Upon consumption, the digestive tract begins to break down carbohydrates into glucose, which is used for energy (El-Sayed, 2003). The Food and Drug Administration (FDA) recommends 275 g of carbohydrates per day in a 2,000-calorie diet. Sugars, starches, dextrin, and glycogen are all important carbohydrates in nutrition. It accounts for approximately 75% of an animal's diet. They also provide energy, which powers muscular movements in animals and generates body heat, which aids in keeping the animal warm. It also aids the body's utilization of proteins and fats. It does not accumulate in the body. It must be included in the animal's diet on daily basis (CAERT). According to Adelakun et al. (2016), water hyacinth appears to be very nutritive because the protein concentration is higher than some nutritional constituents used in aquafeed preparation. It contains 31.27% carbohydrates. *I. aquatic* is an emergent macrophyte with a noble biochemical profile that comprises 22.12% content of carbohydrates. It has been suggested that it can be used in fish feed to deliver all diurnal nutrient supplies at a very low cost (El-Sayed, 2003). The plant has a carbohydrate content of 21.8%. It can be used as a fractional spare for fish meals for major carp fingerlings such as *Catla catla, C. mrigala, and Labeo rohita,* of India (Kalita et al., 2007).

The different content of carbohydrates in some macrophytes is discussed in Table 28.10. *Nymphaea* leaf meal can be encompassed in the nutrition of *Cyprinus carpio* up to 40%, resulting in improved fish growth. Beyond this level of inclusion, the growth response and nutrient utilization efficacy were reduced. this can be attributed to the low digestibility of *Nymphaea* leaf incorporation at higher levels owing to the high content of carbohydrates (Sivani et al., 2013). Corms of *Cyrtospermu chamisson* can grow to be 100–300 kg in weight and are high in carbohydrates but low in protein (0.7%–1.4%). They are either cooked as a vegetable or ground into flour. Corms of *Eleocharis duclis* have a high carbohydrate content but a low protein content (1.4%–1.5%) (Edwards, 1981). *Trapa bispinoa*, also known as singhara in India, is cultivated commercially for its edible seasonal fruit, which is a good source of nutrition with a high carbohydrate content (Gupta et al., 2018). *Nymphoides peltatum* contains more starch and carbohydrates. It could be used in the food production process as human food,

TABLE 28.10
Carbohydrate Content (%) of Some Macrophytes

S. No.	Macrophytes	Carbohydrate (%)	References
1	*Azolla africana*	46	Fasakin and Balogun (1998)
2	*A. pinnata*	55.73	Kumari et al. (2018)
3	*Eichornia crassipes*	49.05	Adelakun et al. (2016)
4	*Ipomoea aquatica*	40.5	Adelakun et al. (2016)
5	*I. reptans*	31.8	Kalita et al. (2007)
6	*Lemna minor*	42	Kalita et al. (2007)
7	*L. polyrhiza*	94.9	Bairagi et al. (2002)
8	*Nymphaea lotus*	43.92	Adelakun et al. (2016)
9	*Pistia* sp.	60.7	Mandal and Ghosh (2019)
10	*P. stratiotes*	39.75	Adelakun et al. (2016)
11	*Salvinia cuculata*	50.8	Kalita et al. (2007)
12	*Spirodela polyrhiza*	49.9	Fasakin and Balogun (1998)
13	*Trapa natans*	67.3	Kalita et al. (2007)

livestock fodder, fertilizer (mulch and manure, ash, green manure, compost, biogas slurry), and food for aquatic herbivores (Bhat and Dar, 2015). Water lily seeds and rhizomes contain various carbohydrate fractions. The main component of lotus seed (50% dry weight) is starch (Dhull et al., 2022). Higher intake of foods having starch with a low glycemic index has been shown to improve health, predominantly diabetes mellitus 2 and obesity (Abelti et al., 2023).

28.4.7 Ash Content

Mineral nutrients are critical components of wholesome quality. Excessive ash concentration reduces the number of organic components per unit weight, lowering nutrition value. Though, ash values less than 15% have a diminutive value in determining whether the feed is nutritious because it is an individual element that is vital in metabolic processes (Alfasane et al., 2018). According to Nicholson and Post (1975), macrophyte ash content is required to estimate organic production by harvest methods and is of interest for potential economic uses of macrophytes (Table 28.11).

TABLE 28.11
Ash Content (%) in Nutrient Composition of Macrophytes

S. No.	Macrophytes	Ash (%)	References
1	*Azolla africana*	15	Naseem et al. (2021)
2	*A.caroliniana*	16.7	Naseem et al. (2021)
3	*A. filiculoides*	18.5	Naseem et al. (2021)
4	*A. mexicana*	17.2	Naseem et al. (2021)
5	*A. microphylla*	16.3	Naseem et al. (2021)
6	*A. pinnata*	15.59–27.14	Samanta and Tamang (1995); Cherryl et al. (2014); Ahmed et al. (2016); Anhita et al. (2016); Sihag et al. (2018); El-Fadel et al. (2020); Naseem et al. (2021)
7	*A. rubra*	15.5	Naseem et al. (2021)
8	*A. sp.*	14.6	Rather et al. (2022)
9	*Ceratophyllum demersum*	23.97–25.3	Abu (2017); Rather et al. (2022)
10	*Cladophora glomerata*	42.7	Muztar et al. (1978)
11	*Eichornia crassipes*	8.85	Naseem et al. (2021)
12	*Ipomoea aquatica*	10	Naseem et al. (2021)
13	*I. reptans*	30	Naseem et al. (2021)
14	*Lemna minor*	25	Naseem et al. (2021)
15	*L. polyrhiza*	2.5	Naseem et al. (2021)
16	*Myriophyllum spicatum*	29.31–50.2	Muztar et al. (1978); Rather et al. (2022)
17	*Nelumbo nucifera*	8	Rather et al. (2022)
18	*Nymphaea lotus*	13.6	Naseem et al. (2021)
19	*N.* sp.	15.63	Naseem et al. (2021)
20	*Phragmites australis*	10.2	Rather et al. (2022)
21	*Pistia* sp.	17	Naseem et al. (2021)
22	*P. stratiotes*	22.2	Naseem et al. (2021)
23	*Potamogetan lucens*	20.99	Rather et al. (2022)
24	*P. natans*	11.2	Rather et al. (2022)
25	*P.* sp.	21.5	Muztar et al. (1978)
26	*Salvinia cuculata*	31.2	Naseem et al. (2021)
27	*Spirodela polyrhiza*	15.2	Naseem et al. (2021)
28	*Trapa natans*	10.9–13.3	Naseem et al. (2021); Rather et al. (2022)
29	*Typha angustata*	9.91	Rather et al. (2022)
30	*Vallisncria americana*	34.6	Muztar et al. (1978)
31	*Wolffia arrhiza*	17.6	Naseem et al. (2021)

TABLE 28.12
Moisture Content (%) of Some Macrophytes

S. No.	Macrophytes	Moisture (%)	References
1	*Azolla pinnata*	93.5	Bora (2018)
2	*A. sp.*	94.6	Rather et al. (2022)
3	*Ceratophyllum demersum*	91.3	Rather et al. (2022)
4	*Enhydra fluctuants*	10.06	Alfasane et al. (2018)
5	*Euryale ferox*	12.5	Alfasane et al. (2018)
6	*Hygroriza aristata*	12.7	Alfasane et al. (2018)
7	*Ipomaea aquatica*	10.3	Alfasane et al. (2018)
8	*Lemna minor*	98.3	Bora (2018)
9	*Limnocharis flava*	13.1	Alfasane et al. (2018)
10	*Myriophyllum verticillatum*	92.08	Rather et al. (2022)
11	*Nelumbo nucifera*	11.8–86.04	Alfasane et al. (2018); Rather et al. (2022)
12	*Nymphaea nouchali*	12.3	Alfasane et al. (2018)
13	*Phragmites australis*	90.06	Rather et al. (2022)
14	*Potamogetan lucens*	87	Rather et al. (2022)
15	*P. natans*	89.22	Rather et al. (2022)
16	*Spirodela polyrhiza*	91	Bora (2018)
17	*Trapa bispinosa*	7.3	Alfasane et al. (2018)
18	*T. natans*	91.24	Rather et al. (2022)
19	*Typha angustata*	86.06	Rather et al. (2022)

28.4.8 Moisture Content

The use of macrophytes as sustenance for animals, aquaculture, and humans is based on their high nutritional value, which is due to their high concentration of biochemical constituents such as moisture (Rather et al., 2022). Several macrophyte species are utilized as livestock fodder, but their high content of moisture is a major limitation (Table 28.12). On a dry weight basis, macrophytes compare favorably to conventional forages, but to be used effectively as animal fodder, they must be partially dehydrated, as water weeds contain only about 5%–15% dry matter paralleled to 10%–30% for terrestrial fodders. Animals cannot ingest enough to sustain their weight of body due to the high moisture content. Since cattle and buffalo have been observed eating fresh *E. crassipes*. Its meal produced by dehydrating the entire green plant to <15% moisture content, could provide 10%–20% of the beef cattle diet, however above this quantity, the animals writhed from mineral disproportion due to high intensities of Fe, K, and Mg (Edwards, 1981).

28.5 CONCLUSION

The global use of macrophytes has evolved in tandem with the advancement of human culture (Mitchell, 1986). Besides the human population, the population of livestock is projected to grow. India, one of the world's foremost livestock fabricators, grieves from productivity losses caused primarily by insufficient feed sources and poor health management. This is due to a significant decrease in a forest area that was previously used to produce feed and fodder. High-yielding dwarfs also reduce the availability of fodder from certain crops. This feed shortage is usually encountered by commercially contrived feed, which is quite expensive, making animal husbandry a profitable livelihood. Scientists all over the world are concerned about ensuring a year-round supply of high-quality feed. Food security is frequently questioned in light of unstable feed generation and a swiftly growing populace. Food safety will now be entirely dependent on the effective use of alternative feed sources such as macrophytes (Bora, 2018). Based on overall nutrient composition, macrophytes are found to contain sufficient amounts of nutrients and can be incorporated into human diets and feed. According to Quirino et al. (2022), macrophyte solidity and variety can affect food resource use. Vegetation of macrophytes increases inclusive physical intricacy and mends resource accessibility. Macrophytes have enormous potential for use as rich mineral sources for humans, as well as fish, poultry, and livestock (Kumar et al., 2022). Thus, macrophytes are safe enough to be considered as potential human food and animal fodder. Besides this, cost-benefit scrutiny is also required to determine the commercial viability of the feed ingredients (Naseem et al., 2021).

ACKNOWLEDGMENT

The authors are thankful to the Director, CSIR-National Botanical Research Institute, Lucknow for providing laboratory facilities and basic infrastructure to accomplish the study. We are also thankful to the Head, Department of Botany (Environmental Science), University of Lucknow, Lucknow for providing logistic support.

REFERENCES

Abelti, A. L., Teka, T. A., & Bultosa, G. (2023). Review on edible water lilies and lotus: Future food, nutrition and their health benefits. *Applied Food Research*, 3, 100264. https://doi.org/10.1016/j.afres.2023.100264

Abobi, S. M., Ampofo-Yehoah, A., Kpodonu, T. A., Alhassan, E. H., Abarike, E. D., Atindaana, S. A., Akongyuure, D.. N., Konadu, V., & Twumasi, F. (2015). Socio-ecological importance of aquatic macrophytes to some fishing communities in the Northern region of Ghana.

Abu, T. (2017). A review: Aquatic macrophyte *Ceratophyllum demersum* L.(Ceratophyllaceae): Plant profile, phytochemistry and medicinal properties. *International Journal of Science and Research (IJSR)*, *6*(7), 394–399.

Adanse, J., Ahmed, A., & Apakah, B. (2019). Production and evaluation of muffins from nymphaea lotus and wheat flour. *Think India Journal*, *22*(4), 3892–3900.

Adelakun, K. M., Kehinde, A. S., Amali, R. P., Ogundiwin, D. I., & Omotayo, O. L. (2016). Nutritional and phytochemical quality of some tropical aquatic plants. *Poultry, Fisheries & Wildlife Sciences*, 4, 164.

Agarwala, O. N. (1988). Water hyacinth (*Eichhornia crassipes*) silage as cattle feed. *Biological wastes*, *24*(1), 71–73.

Ahirwar, M. K., & Leela, V. (2012). Nutritive value and in vitro degradability of *Azolla pinnata* for ruminants. *Indian Veterinary Journal*, *89*(4), 101–102.

Ahmed, H. A., Ganai, A. M., Beigh, Y. A., Sheikh, G. G., & Reshi, P. A. (2016). Performance of growing sheep on Azolla based diets. *Indian Journal of Animal Research*, *50*(5), 721–724.

Airaksinen, M. M., Peura, P., Ala-Fossi-Salokangas, L., Antere, S., Lukkarinen, J., Saikkonen, M., & Stenbäck, F. (1986). Toxicity of plant material used as emergency food during famines in Finland. *Journal of Ethnopharmacology*, *18*(3), 273–296.

Alfasane, A., Kauser, S., Shahjadee, U. F., & Khondker, M. (2018). Biochemical composition of some selected aquatic macrophytes under ex-situ conditions. *Journal of the Asiatic Society of Bangladesh, Science*, *44*(1), 53–60.

Anand, A., Priyanka, U., Nayak, V. L., Zehra, A., Babu, K. S., & Tiwari, A. K. (2019). Nutritional composition and antioxidative stress properties in boiled tuberous rhizome of Neel Kamal (Nymphaea nouchali Burm. f.). *Indian Journal of Natural Products and Resources*, *10*(1), 59–67.

Anhita, K. C., Rajeshwari, Y. B., Prabhu, T. M., Vivek Patil, M., Shilpa Shree, J., & Anupkumar, P. K. (2016). Effect of supplementary feeding of azolla on growth performance of broiler rabbits. *ARPN Journal of Agricultural and Biological Science*, *11*, 30–36.

Anon. (1984). *Making aquatic weeds useful: Some perspectives for developing countries*. Washington, DC: National Academy of Sciences, 175.

Aung, T. T., Myat, Y. Y., Mar, M. M., & Kyu, K. K. (2020). Nutritional compositions, elemental compositions and antinutrient factor in different varieties of water lily. *3rd Myanmar Korea Conference Research Journal*, *3*(5), 1917–1922.

Bailey, R. L., Parker, E. A., Rhodes, D. G., Goldman, J. D., Clemens, J. C., Moshfegh, A. J., ... Weaver, C. M. (2015). Estimating sodium and potassium intakes and their ratio in the American diet: Data from the 2011-2012 NHANES. *The Journal of Nutrition*, *146*(4), 745–750.

Bairagi, A., Ghosh, K. S., Sen, S. K., & Ray, A. K. (2002). Duckweed (Lemna polyrhiza) leaf meal as a source of feedstuff in formulated diets for rohu (Labeo rohita Ham.) fingerlings after fermentation with a fish intestinal bacterium. *Bioresource Technology*, *85*(1), 17–24.

Balogun, J. K., & Ibeun, M. O. (1995). Additional information on fish stocks and fisheries of lake Kainji (Nigeria). In: R. C. M. Crul and F. C. Roest (Eds.), *Current status of fisheries and fish stocks of the four largest African reservoirs. CIFA Technical Paper*, No. 30. Rome: FAO

Bangar, S. P., Dunno, K., Kumar, M., Mostafa, H., & Maqsood, S. (2022). A comprehensive review on lotus seeds (Nelumbo nucifera Gaertn.): Nutritional composition, health-related bioactive properties, and industrial applications. *Journal of Functional Foods*, *89*, 104937.

Bansal, S., Lishawa, S. C., Newman, S., Tangen, B. A., Wilcox, D., Albert, D., Anteau, M. J., Chimney, .J., Cressey, R. L., DeKeyser, E., Elgersma, K. J., Finkelstein, S. A., Freeland, J., Grosshans, R., Klug, P. E., Larkin, D. J., Lawrence, B. A., Linz, G., Marburger, J., Noe, G., Otto, C., Reo, N., Richards, J., Richardson, C., Rodgers, L. R., Schrank, A. J., Svedarsky, D., Travis, S., Tuchman, N., & Windham-Myers, L. (2019). Typha (Cattail) invasion in north american wetlands: Biology, regional problems, impacts, ecosystem services, and management. *Wetlands*, *39*, 645–684.

Bhaskaran, S. K., & Kannapan, P. (2015). Nutritional composition of four different species of *Azolla. European Journal of Experimental Biology*, *5*(3), 6–12.

Bhat, J. I., & Dar, Z. A. (2015). Study on exploring the potential of Nymphoides peltatum as animal feed growing in Dal lake of Kashmir Himalaya. *IJAR*, *1*(9), 514–517.

Bhatt, N., Tyagi, N., Chandra, R., Meena, D. C., & Prasad, C. K. (2021). Growth performance and nutrient digestibility of azolla pinnata feeding in sahiwal calves (Bos indiens) by replacing protein content of concentrate with Azolla pinnata during winter season. *Indian Journal of Animal Research*, *55*(6), 663–668.

Bin Rahman, A. R., & Zhang, J. (2022). Trends in rice research: 2030 and beyond. *Food and Energy Security*, e390.

Bora, P. (2018). Biochemical analysis of common macrophytes and their potential uses (Doctoral dissertation).

Boyd, C. E. (1968). Fresh-water plants: A potential source of protein. *Economic Botany*, *22*(4), 359–368.

Buddington, R. K. (1979). Digestion of an aquatic macrophyte by Tilapia zillii (Gervais). *Journal of Fish Biology*, *15*(4), 449–455.

Burkill, H. M. (1994). The useful plants of west tropical Africa. *Volume 2: Families EI* (No. Edn 2). Royal Botanic Gardens.

Butt, M. A., Zafar, M., Ahmed, M., Shaheen, S., Sultana, S., Butt, M. A., Shaheen, S., & Sultana, S. (2021). Nutritive value of wetland flora. In Wetland Plants: A Source of Nutrition and Ethno-medicines (pp. 75–90).

Cabrera-Guzmán, E., Díaz-Paniagua, C., & Gomez-Mestre, I. (2020). Differential effect of natural and pigment-supplemented diets on larval development and phenotype of anurans. *Journal of Zoology*, *312*(4), 248–258.

CAERT Inc. E-unit: The importance of water, carbohydrates, and fats. AgEdLibrary.com, pp. 1–4.

Chai, T. T., Ooh, K. F., Quah, Y., & Wong, F. C. (2015). Edible freshwater macrophytes: A source of anticancer and antioxidative natural products-a mini-review. *Phytochemistry Reviews*, *14*(3), 443–457.

Chambers, P. A., Lacoul, P., Murphy, K. J., & Thomaz, S. M. (2008). Global diversity of aquatic macrophytes in freshwater. In Freshwater animal diversity assessment (pp. 9–26).

Chareontesprasit, N., & Jiwyam, W. (2001). An evaluation of Wolffia meal (*Wolffia arrhiza*) in replacing soybean meal in some formulated rations of Nile tilapia (*Oreochromis niloticus* L.). *Pakistan Journal of Biological Sciences*, *4*, 618–620.

Chawanje, C. M., Barbeau, W. E., & Grün, I. (2001). Nutrient and antinutrient content of an underexploited malawian water tuber Nymphaea Petersiana (Nyika). *Ecology of Food and Nutrition*, *40*(4), 347–366.

Cherryl, D. M., Prasad, R. M. V., JagadeeswaraRao, S., Jayalaxmi, P., & Srinivas Kumar, D. (2014). A study on the nutritive value of Azolla pinnata. *Livestock research international*, *2*(1), 13–15.

Danhassan, M. S., Salihu, A., & Inuwa, H. M. (2018). Effect of boiling on protein, mineral, dietary fibre and antinutrient compositions of Nymphaea lotus (Linn) seeds. *Journal of Food Composition and Analysis*, *67*, 184–190.

Das, J. K., Saikia, K., Handique, G. K., & Handique, A. K. (2021). Nutritive values and dietary antioxidant of some under-utilized seeds and nuts from ethnic sources of North East India. *Crop Research (0970-4884)*, *56*.

Datta, S. N. (2011). Culture of Azolla and its efficacy in diet of *Labeo rohita*. *Aquaculture*, *310*(3-4), 376–379.

De Silva, S. S., & Perera, M. K. (1983). Digestibility of an aquatic macrophyte by the cichlid Etroplus suratensis (Bloch) with observations on the relative merits of three indigenous components as markers and daily changes in protein digestibility. *Journal of Fish Biology*, *23*(6), 675–684.

de Vasconcelos, G. A., Véras, R. M. L., de Lima Silva, J., Cardoso, D. B., de Castro Soares, P., de Morais, N. N. G., & Souza, A. C. (2016). Effect of water hyacinth (Eichhornia crassipes) hay inclusion in the diets of sheep. *Tropical Animal Health and Production*, *48*(3), 539–544.

Debbarma, J., Viji, P., Rao, B. M., & Ravishankar, C. N. (2022). Seaweeds: Potential applications of the aquatic vegetables to augment nutritional composition, texture, and health benefits of food and food products. *Sustainable Global Resources of Seaweeds*, *2*, 3–54.

Dhull, S. B., Chandak, A., Collins, M. N., Bangar, S. P., Chawla, P., & Singh, A. (2022). Lotus seed starch: A novel functional ingredient with promising properties and applications in food-A review. *Starch-Stärke*, *74*(9-10), 2200064.

Donchenko, L., Bitutskaya, O., Vlaschik, L., & Limareva, N. (2020). Biologically active complex with high antioxidant properties based on macrophytes of the Azov-Black Sea Basin. *KnE Life Sciences*, 592–603.

Edwards, P. (1981). Food potential of aquatic macrophytes [as human food, livestock feed and fertilizer; study conducted in Thailand]. *ICLARM [International Center for Living Aquatic Resources Management] Studies and Reviews (Philippines).*

El Naggar, S., & El-Mesery, H. S. (2022). Azolla pinnata as unconventional feeds for ruminant feeding. *Bulletin of the National Research Centre*, *46*(1), 1–5.

El-Fadel, A., Hassanein, H. A., & El-Sanafawy, H. A. (2020). Effect of partial replacement of protein sun flower meal by Azolla meal as source of protein on productive performance of growing lambs. *Journal of Animal and Poultry Production*, *11*(4), 149–153.

El-Sayed, A. F. M. (2003). Effects of fermentation methods on the nutritive value of water hyacinth for Nile tilapia *Oreochromis niloticus* (L.) fingerlings. *Aquaculture*, *218*(1-4), 471–478.

Etse, W. J., Annang, T. Y., & Ayivor, J. S. (2018). Nutritional composition of aquatic plants and their potential for use as animal feed: A case study of the lower volta basin. *Biofarmasi Journal of Natural Product Biochemistry*, *16*(2), 99–112.

Fasakin, A. E., & Balogun, A. M. (1998). Evaluation of dried water fern (*Azolla pinnata*) as a replacer for soybean dietary components for Clarias gariepinus fingerlings. *Journal of Aquaculture in the Tropics*, *11*(4):83–92. DOI:10.1300/J028v11n04_09.

Ghaly, A. E., Kamal, M., & Mahmoud, N. S. (2005). Phytoremediation of aquaculture wastewater for water recycling and production of fish feed. *Environment International*, *31*(1), 1–13.

Ghosh, S., Chattoraj, S., & Nandi, A. (2015). Proximate composition of some mangrove leaves used as alternative fodders in Indian Sunderban region. *International Journal of Livestock Research*, *3*(1), 1.

Gohl, B. (1981). *Tropical feeds*. Rome. 529 p. https://searchworks.stanford.edu/view/1479772

Graham, H., & Åman, P. (1991). Nutritional aspects of dietary fibres. *Animal Feed Science and Technology*, *32*(1-3), 143–158.

Gueye, F. K., Ayessou, N. C., Mbaye, M. S., Diop, M. B., Cissé, M., & Noba, K. (2020). *Nymphaea lotus* L. and *Nymphaea micrantha* Guill. et perr seeds as cereal substitute in the Delta of Senegal river. *Food and Nutrition Sciences*, *11*(5), 375.

Gupta, P., Tamot, S., & Shrivastava, V. K (2018). Role of macrophytes for the upliftment of Socio-Economic weekened section of Upper Lake. *International Journal of Research in Advent Technology*, *6*(8), 2239–2244.

Haroon, A. M. (2022). Review on aquatic macrophytes in Lake Manzala, Egypt. *The Egyptian Journal of Aquatic Research*, *48*(1), 1–12. https://doi.org/10.1016/j.ejar.2022.02.002.

Hawkins, S. J., & Hartnoll, R. G. (1983). Grazing of intertidal algae by marine invertebrates. *Oceanography and Marine Biology*, *21*, 195–282.

Hendrey, George R. (2001). Acid rain and deposition IV.G.2.iii. In *Macrophytes. Encyclopedia of Biodiversity.*

Hernawati, D., Meylani, V., Putra, R. R., & Agustian, D. (2022). Assistance in sustainable food programs through the introduction of edible plants as potential foodstuffs. *Engagement: Journal Pengabdian Kepada Masyarakat*, *6*(1), 150–162.

Hujjatullah, S., Bloch, A. K., & Jabbar, A. (1967). Chemical composition and utilisation of the roots of Nymphaea lotus L. *Journal of the Science of Food and Agriculture*, *18*(10), 470–473.

Ibrahim, K. I. B. (2007). Properties of Soutab (Nymphaea lotus) tubers flour and it's utilization with wheat flour for bread making. Universityof Khartoum.

Indulekha, V. P., Thomas, C. G., & Anil, K. S. (2019). Utilization of water hyacinth as livestock feed by ensiling with additives. *Indian Society of Weed Science.* doi:10.5958/0974-8164.2019.00014.5.

Irvine, F. R. (1952). Supplementary and emergency food plants of West Africa. *Economic Botany*, *6*(1), 23–40.

Ita, E. O. (1993). Aquatic and wildlife resource of Nigeria. CIFA Occasional paper No. 21. Rome: FAO, pp. 10–12.

Ita, E. O. (1994). Aquatic plants and wetland wildlife resources of Nigeria. CIFA Occasional Paper. No. 21. Rome: FAO, p. 52.

Jarapala, S. R., Shivudu, G., Mangathya, K., Rathod, A., Panda, H., & Reddy, P. K. (2021). A New approach for identifying potentially effective indigenous plants consumed by Chenchu Tribes and their nutritional composition-India. *American Journal of Plant Sciences*, *12*(8), 1180–1196.

Joysowal, M., Aziz, A., Mondal, A., Singh, S. M., Boda, S. S., Chirwatkar, B., & Chhaba, B. (2018). Effect of *Azolla (Azolla pinnata)* feed on the growth of broiler chicken. *Journal of Entomology and Zoology Studies*, *6*(3), 391–393.

Kabuye, C. H. S. (1986). Edible roots from wild plants in arid and semi-arid Kenya. *Journal of Arid Environments*, *11*(1), 65–74. https://doi.org/10.1016/s0140-1963(18)31310-7

Kalita, P., Mukhopadhyay, P. K., & Mukherjee, A. K. (2007). Evaluation of the nutritional quality of four unexplored aquatic weeds from northeast India for the formulation of cost-effective fish feeds. *Food Chemistry*, *103*(1), 204–209.

Keak, D., Nurfeta, A., Banerjee, S., & Ali, S. (2022). Growth performance and carcass quality characteristics of Cobb 500 broiler chicken fed rations with different levels of water lily (*Nymphae lotus*) seed meal. *Ethiopian Journal of Agricultural Sciences*, *32*(2), 1–12.

Khan, M. A. H. (2019). *Nutritional Composition, Phytochemical And Antioxidant Activity Of Stem of (Nymphaea nouchali) and (Nymphaea rubra)* (Doctoral dissertation, Chattogram Veterinary & Animal Sciences University).

Klinnavee, S., Tansakul, R., & Promkuntong, W. (1990). Growth of Nile tilapia (Oreochromis niloticus) fed with aquatic plant mixtures. 2nd Asian Fish. Forum, Asian. Fish. Soc., Manila, Philippines (pp. 283–286).

Kloos, H. (1982). Development, drought, and famine in the Awash Valley of Ethiopia. *African Studies Review*, *25*(4), 21–48. https://doi.org/10.2307/524399.

Kumar, G., Sharma, J., Goswami, R. K., Shrivastav, A. K., Tocher, D. R., Kumar, N., & Chakrabarti, R. (2022). Freshwater macrophytes: A potential source of minerals and fatty acids for fish, poultry, and livestock. *Frontiers in Nutrition*, *9*, 869425. doi.org/10.3389/fnut.2022.869425

Kumari, R., Dhuria, R. K., Patil, N. V., Sawal, R. K., & Singh, S. (2018). Chemical composition and pellet quality of Azolla pinnata grown in semi-arid zone of India. *International Journal of Chemical Studies*, *6*(3), 2031–2033.

Kurniawan, S. B., Ahmad, A., Said, N. S. M., Imron, M. F., Abdullah, S. R. S., Othman, A. R., ... Hasan, H. A. (2021). Macrophytes as wastewater treatment agents: Nutrient uptake and potential of produced biomass utilization toward circular economy initiatives. *Science of the Total Environment*, *790*, 148219.

Kwarfo-Apegyah, K., & Ipinjolu, J. K. (1995). Some aspects of the ecology and utilization of macrophytes of Kware Lake Sokoto State. Nigeria, pp. 1–2.

Lacoul, P., & Freedman, B. (2006). Environmental influences on aquatic plants in freshwater ecosystems. *Environmental Reviews*, *14*(2), 89–136.

Laminu, H. H., Sa'ad, R. S., Damasak, A. A., Abubakar, M. I., Bala, A. M., Madu, D. K., & Theresa, C. I. (2021). Nutritional value of green and red *Nymphaea lotus* seeds and their glycemic index. *IOSR Journal of Biotechnology and Biochemistry*, *7*(4), 25–33.

Mandal, S., & Ghosh, K. (2019). Utilization of fermented Pistia leaves in the diet of rohu, *Labeo rohita* (Hamilton): Effects on growth, digestibility and whole body composition. *Waste and Biomass Valorization*, *10*(11), 3331–3342.

Mathur, G. N., Sharma, R., & Choudhary, P. C. (2013). Use of azolla (Azolla pinnata) as cattle feed. Scientists joined as life member of Society of Krishi Vigyan, p. 73.

Meena, T., & Rout, J. (2016). Macrophytes and their ecosystem services from natural ponds in Cachar district, Assam. *Indian Journal of Traditional Knowledge*, *15*(4), 553–560.

Merga, L. B. (2021). *Impacts of anthropogenic activities on the ecology and ecosystem service delivery of Lake Ziway, Ethiopia* (Doctoral dissertation, Wageningen University and Research).

Mitchell, D. S. (1986). Aquatic macrophytes and man. In Patrick Deckker, W. D. Williams (eds) *Limnology in Australia* (pp. 587–598). Dordrecht: Springer.

Morton, J. F. (1965). Can the red mangrove provide food, feed and fertilizer?. *Economic Botany*, *19*(2), 113–123.

Murphy, K., Efremov, A., Davidson, T., Molina-Navarro, E., Fidanza, K., Crivelari Betiol, T. C., Chambers, P. A., Grimaldo, J. T., Varandas Martins, S., Springuel, I., Kennedy, M., Mormul, R., Dibble, E., Hofstra, D., Lukács, B. A., Gebler, D., Båstrup-Spohr, L., & Urrutia Estrada, J. (2019). World distribution, diversity and endemism of aquatic macrophytes. *Aquatic Botany*, *158*, 103127.

Muztar, A. J., Slinger, S. J., & Burton, J. H. (1978). Chemical composition of aquatic macrophytes I. Investigation of organic constituents and nutritional potential. *Canadian Journal of Plant Science*, *58*(3), 829–841.

Nagy, K., & Tiuca, I. D. (2017). Importance of fatty acids in physiopathology of human body. In *Fatty acids*. IntechOpen. https://doi.org/10.5772/67407

Nakamura, Y., Murakami, A., Koshimizu, K., & Ohigashi, H. (1996). Identification of pheophorbide and its related compounds as possible anti-tumor promoters in the leaves of Neptunia oleracea. *Bioscience, Biotechnology, and Biochemistry*, *60*(6), 1028–1030.

Naseem, S., Bhat, S. U., Gani, A., & Bhat, F. A. (2021). Perspectives on utilization of macrophytes as feed ingredient for fish in future aquaculture. *Reviews in Aquaculture*, *13*(1), 282–300.

Nicholson, S. A., & Post, L. W. (1975). Ash content of macrophytes from Chautauqua Lake. *The Ohio Journal of Science*, *75*(1), 29–32.

Nishan, K. F. (2020). *Nutraceutical Study And Potential Antidiabetic Activity (In-Vitro) of Two Species Of Tropical Water Lily* (Doctoral dissertation, Chattogram Veterinary and Animal Sciences University Chattogram-4225, Bangladesh).

Pandey, V. N., & Srivastava, A. K. (1991). Yield and quality of leaf protein concentrates from Monochoria hastata (L.) Solms. *Aquatic Botany*, *40*(3), 295–299.

Parashuramulu, S., Swain, P. S., & Nagalakshmi, D. (2013). Protein fractionation and in vitro digestibility of Azolla in ruminants. *Online Journal of Animal and Feed Research*, *3*(3), 129–132.

Parihar, A., Mondal, S., Singh, P. K., & Singh, R. L. (2022). Introduction, scope, and applications of biotechnology and genomics for sustainable agricultural production. InRam Lakhan Singh, Sukanta Mondal, Akarsh Parihar, Pradeep Kumar Singh (eds) *Proceedings of the plant genomics for sustainable agriculture* (pp. 1–14). Springer.

Patel, A., Desai, S. S., Mane, V. K., Enman, J., Rova, U., Christakopoulos, P., & Matsakas, L. (2022). Futuristic food fortification with a balanced ratio of dietary ω-3/ω-6 omega fatty acids for the prevention of lifestyle diseases. *Trends in Food Science & Technology*, 140–153. https://doi.org/10.1016/j.tifs.2022.01.006.

Pillai, P. K., Premalatha, S., & Rajamony, S. (2002). *Azolla*-A sustainable feed substitute for livestock. *Leisa India*, *4*(1), 15–17.

Pine, R. T., Anderson, L. W. J., & Hung, S. S. O. (1990). Control of aquatic plants in static and flowing water by yearling triploid grass carp. *Journal of Aquatic Plant Management*, *28*(1), 36–40.

Quirino, B. A., Thomaz, S. M., Jeppesen, E., Søndergaard, M., Dainez-Filho, M. S., & Fugi, R. (2022). Aquatic macrophytes shape the foraging efficiency, trophic niche breadth, and overlap among small fish in a neotropical river. *Water*, *14*(21), 3543.

Rajasulochana, N., Baluswami, M., Parthasarathy, M. D. V., & Krishnamurthy, V. (2002). Chemical analysis of Grateloupia lithophila Boergesen. *Seaweed Research and Utilization*, *24*, 79–82.

Rather, Z. A., & Nazir, R. (2015). Biochemical composition of selected macrophytes of Dal Lake, Kashmir Himalaya. *Journal of Ecosystem Ecography*, *5*, 1–5.

Rather, Z. A., Sharma, P., & Dar, N. A. (2022). Macrophytes and their nutrient content analysis-a study of Dal Lake, Kashmir. *Sustainability and Biodiversity Conservation*, *1*(1), 12–24.

Ray, A. K., & Das, I. S. I. T. A. (1994). Apparent digestibility of some aquatic macrophytes in rohu, Labeo rohita (Ham.), fingerlings. *Journal of Aquaculture in the Tropics*, *9*(4), 335–342.

Roslin, A. S. (2001). Seasonal variations in the lipid content of some marine algae in relation to environmental parameters in the Arockiapuram coast. *Seaweed Research Utilness*, *23*, 119127.

Samanta, G., & Tamang, Y. (1995). Feeding value of azolla (Azolla pinnata) in goats. *Annales de zootechnie*, *44*(Suppl. 1), 62–62.

Sihag, S., Sihag, Z. S., Kumar, S., & Singh, N. (2018). Effect of feeding *Azolla* (*Azolla pinnata*) based total mixed ration on growth performance and nutrients utilization in goats. *Forage Research*, *43*(4), 314–318.

Singh, D. N., Bohra, J. S., Tyagi, V., Singh, T., Banjara, T. R., & Gupta, G. (2022). A review of India's fodder production status and opportunities. *Grass and Forage Science*, *77*(1), 1–10.

Singh, L. R., & Vishwavidyalaya, D. S. (2017). Medicinal and ecological potential of Kaseru (Scirpus grossus): A review. *Journal of Harmonized Research*, *6*(3), 51–53.

Sivani, G., Reddy, D. C., & Bhaskar, M. (2013). Effect of *Nymphaea* meal incorporated diets on growth, feed efficiency and body composition in fingerlings of *Cyprinus carpio* L. *Journal of Applied and Natural Science*, *5*(1), 5–9.

Skalnaya, M. G., & Skalny, A. V. (2018). *Essential trace elements in human health: A physician's view*. Tomsk: Publishing House of Tomsk State University.

Sree, K. S., Dahse, H. M., Chandran, J. N., Schneider, B., Jahreis, G., & Appenroth, K. J. (2019). Duckweed for human nutrition: No cytotoxic and no anti-proliferative effects on human cell lines. *Plant Foods for Human Nutrition*, *74*(2), 223–224.

Stephen, E. C., Adebisi, A. K., Chinedu, I., & Samuel, A. A. (2017). Chemical composition of Water lily (*Nymphaea lotus*) bulbs. *American Journal of Food Sciences and Nutrition*, *4*(2), 7–12.

Sushil, M. (2012). Performance Evaluation of Reed Grass (*Phragmites karka*) in Constructed Reed Bed System (CRBs) on Domestic sludge, Ujjain city, India. *Research Journal of Recent Sciences, 1,* 41–46

Swain, B. K., Naik, P. K., Sahoo, S. K., Mishra, S. K., & Kumar, D. (2018). Effect of Feeding *Azolla (Azolla pinnata)* on the performance of White Pekin Laying Ducks. *International Journal of Livestock Research*, *8*, 248.

Swapna, M. M., Prakashkumar, R., Anoop, K. P., Manju, C. N., & Rajith, N. P. (2011). A review on the medicinal and edible aspects of aquatic and wetland plants of India. *Journal of Medicinal Plants Research*, *5*(33), 7163–7176.

Tacon, A. G. J. (1987). The nutrition and feeding of farmed fish and shrimp – A training manual. 2. Nutrient sources and composition. FAO Field Document, Project GCP/RLA/075/ITA Field Document No. 5, Brasilia, Brazil, 129pp.

Teklehaymanot, T., & Giday, M. (2010). Ethnobotanical study of wild edible plants of Kara and Kwego semi-pastoralist people in Lower Omo River Valley, Debub Omo Zone, SNNPR, Ethiopia. *Journal of Ethnobiology and Ethnomedicine*, *6*(1), 1–8.

Tresina, P.S., Doss, A. & Mohan, V.R. (2020). Fatty acid composition of underutilized corms, rhizomes and tubers. *Food Research*, *4*(5), 1569–1572. https://doi.org/10.26656/fr.2017.4(5).123

Tu, D. T. M. (2012). *Manipulation of the nutritive value of duckweed (Lemna minor) as a feed resource for local Muscovy ducks* (Doctoral dissertation, MSc. Thesis in Agricultural Sciences Animal Husbandry, Cantho University).

USDA, NRCS, Louisiana State Office, National Plant Data Center, & the Grazing Land Conservation Initiative-South Central Region FAO. 2023. Mangroves trees and shrubs. https://www.fao.org/4/ai387e/ai387e06.htm

Venkatesh, B., & Shetty, H. P. C. (1978). Nutritive value of two aquatic weeds and a terrestrial grass as feed for the grass carp, Ctenopharyngodon idella (Valenciennes). *Mysore Journal of Agricultural Sciences*, *12*(4), 597–604.

Watt, J. M., & Breyer-Brandwijk, M. G. (1962). *The medicinal and poisonous plants of southern and eastern africa being an account of their medicinal and other uses, chemical composition, pharmacological effects and toxicology in man and animal* (2nd edn). E. & S. Livingstone.

Wetzel, R. G. (2001). *Limnology: Lake and river ecosystems* (3rd edn, p. 1006). Philadelphia: Academic Press.

World Health Organization. (2009). *Calcium and magnesium in drinking water: Public health significance.* World Health Organization.

Yuan, G., Fu, H., Zhang, M., Lou, Q., Dai, T., & Jeppesen, E. (2020). Effects of plant size on the growth of the submersed macrophyte *Vallisneria spinulosa* SZ Yan at different light intensities: Implications for lake restoration. *Hydrobiologia*, *847*(17), 3609–3619.

29 Valorization of Bio-based Waste through Research and Community Service in University towards Circular Economy Transition and Future Sustainability

Case Study in Indonesia

Ratna Dewi Kusumaningtyas, Dwi Widjanarko, Haniif Prasetiawan, Sucihatiningsih Dian Wisika Prajanti, Amin Retnoningsih, and M. Margunani

29.1 INTRODUCTION

The 2030 development agenda for sustainable development was launched by the United Nations in September 2015. It comprises 17 sustainable development goals (SDGs), 169 sub-targets, and five pillars (5P) (Khajuria et al., 2022). The 17 SDGs are: (i) no poverty, (ii) zero hunger, (iii) good health and well-being, (iv) quality education, (v) gender equality, (vi) clean water and sanitation, (vii) affordable and clean energy, (viii) decent work and economic growth, (ix) industry, innovation and infrastructure, (x) reduced inequality, (xi) make cities and human settlement inclusive, safe, resilient, and sustainable, (xii) responsible consumption and production, (xiii) climate action, (xiv) life below water, (xv) life on land, (xvi) peace, justice, and strong institution, (xvii) partnership for the goal. SDGs are created to stop poverty, guard the planet, and assure wealth for people by 2030. The five pillars of the sustainability are people (Goal 1, 2, 3, 4, 5, and 6), planet (Goal 13, 14, and 15), prosperity (Goals 7, 8, 9, 10, 11, and 12), peace and partnership (Goals 16 and 17) (Morton et al., 2017). However, the diminishing of natural resources, climate change, energy resource shortage, the rapid growing population, the rising consumption, and the increasing of the waste become a big challenge to accomplish sustainability (Knäble et al., 2022; Kara et al., 2022). This problem arises as a result of the current linear economy applied by the industry and business which comprises these order: extraction, production, consumption and disposal or "Take-Make-Use-Dispose" (Kara et al., 2022).

To overcome this obstacle, the circular economy principle should be implemented. Circular economy can be described as an economic model targeting on the effective utilization of resources via waste minimization and close loop of material which results in environmental and social economic advantage (Morseletto, 2020). Circular economy applies a closed loop of material and resources which can minimize waste disposal. Circular economy concept involves eco-friendly design, repair, reuse, refurbishment, remanufacture, product sharing, waste avoidance, and waste recycle (Schroeder et al., 2019). It is also often known as 10R: Refuse, Rethink, Reduce, Reuse, Repair, Refurbish, Remanufacture,

DOI: 10.1201/9781003441144-29

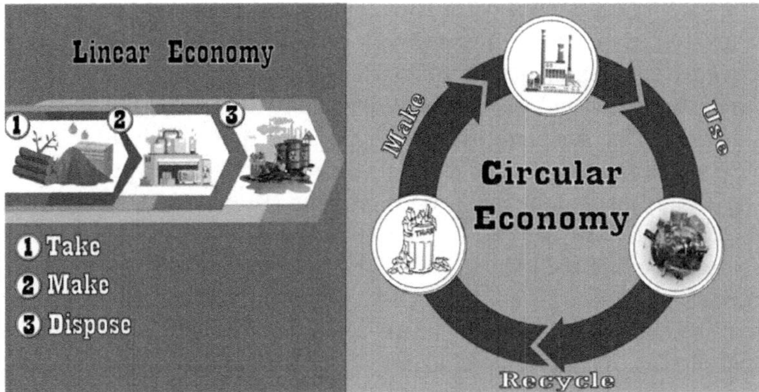

FIGURE 29.1 Illustration of linear and circular economy comparison.

Repurpose, Recycle, and Recovery (Morseletto, 2020). Furthermore, Velenturf and Purnell (2021), emphasized that by employing circular economy principle, the resources are used optimally waste disposal can be minimized. It is revealed that the circular economy in the industrial process is a closed loop consisting of the sequence: "Take-Make-Use-Recycle". The illustration of linear and circular economy comparison is displayed in Figure 29.1.

Waste recycling and disposal minimization is an important issue in circular economy to reach sustainable development. Waste recycle can prolong the lifecycle of substance, create value-added to the waste, minimize the cost for purchasing fresh raw materials, minimize the over-exploitation of resources and nature, reduce energy consumption, protect the environment, and regenerate ecosystem. Waste recycle also reduces the carbon footprint which contributes to climate change (Hagelüken and Goldmann, 2022). It can be stated that the circular economy is an essential factor for the transition towards sustainable consumption and production. Thus, it is evident that circular economy and zero waste process are potential to assist the realization of the sustainable development (Valverde and Avilés-Palacios, 2021).

However, to successfully implement circular economy towards sustainable development, the active role and collaboration among parties, such as industry, government, business, university, scholar, practitioners, NGO, and community, are essential. University as the agents of change is expected to create innovation which concerned with the circular economy and enhance sustainable development. This work describes the role of the university in promoting circular economy and sustainable development through research and community service with a case study in Indonesia.

29.2 THE ROLE OF UNIVERSITY TO PROMOTE CIRCULAR ECONOMY AND SUSTAINABLE DEVELOPMENT: CASE STUDY IN INDONESIA

University as higher education is an agent of change and sustainability. The university is an institution that contributes to human resources development and provokes the next leader through education. Education in the university is expected to create qualified human resources who are ready to cope with the environmental, societal, and technical change, and actively participate in the thriving UN SDGs implementation. University also have a central responsibility in the development of knowledge, science, technology, and innovation. Capability to transform science and technology into new products and process is a vital aspect to nurture economic growth, social welfare, and sustainable development (Mormina, 2019).

In Indonesia, based on the Regulation of Republic Indonesia No. 12/2012, university are obligated to implement the Three Pillars of Higher Education (Tri Dharma Perguruan Tinggi) to conduct education and teaching, research, and community service (Nur et al., 2018; Siregar et al., 2016).

It is named Tri Dharma because one pillar is interconnected to the others and affects each other. The improved quality and quantity of research will provide innovative technology and knowledge and thenceforth enhance the excellence of teaching and community service. Community service is a connection between education and research in the university with the real problem-solving practices in the society. The faculties, students, and educational staff should actively fulfil the Tri Dharma. The lecturers apply community service through the community service program, and the students participate in field study and community service program (Kuliah Kerja Nyata/ KKN). KKN program for the students is usually handled by the Centre of KKN, which is part of the Research and Community Service Institute (LPPM). In the KKN program, students engage with a community for certain duration to conduct a well-designed problem-solving program and empower the society (Syamsi and Heriyanti, 2022). Among the main program of KKN is socialization of the waste management and carrying out the zero waste principle in daily life. Implementation of Tri Dharma fosters the university to develop civilization and education and to produce new findings and innovations in sciences, technology and social humanity that are useful for the society and nation. The interconnected pillars in Tri Dharma are excellent foundation for the university to play an important role to promote circular economy and sustainable development.

Higher education is among the public service institutions and it has academic and non-academic autonomy. Based on the Government Regulation No. 4/ 2014 Article 27 about the Higher Education Management, it is stated that the state universities (PTNBH) are categorized into three different types of financial management: (i) State university with general state financial management (PTN Satker), (ii) state university with the financial management system of public service board (PTN BLU), and (iii) state university as autonomous legal entity (PTNBH) (Nur et al., 2018; Triatmoko et al., 2018). The legal entity university (PTNBH) is given full autonomy to handle their resources and conversely the government interference reduced. Thus, the university should have the ability to achieve its mission based on its own preference.

Universitas Negeri Semarang (UNNES) is a university in Indonesia that has been granted a decree as a legal entity university since October 2022. This status grants a bigger opportunity and wider autonomy for UNNES to realize its vision to become a world-reputable university and a pioneer of excellence in education with a conservation perspective (UNNES, 2022b). Conservation perspective comprises conservation in values and human character, arts and culture, and natural resources and environment. Monitoring of the conservation principle in the university is managed by the Technical Implementation Unit of Conservation Development (Amin et al., 2022). Renewable and non-renewable natural resources are sources of raw materials that are always utilized by humans for giving benefit to humans. The national conservation strategy refers to three main things, namely: protecting and saving, studying, and using. Conservation of natural resources and environment which involves the aspects of green and smart building, biodiversity, waste management, clean water, clean energy, and green transportation is strongly associated with the circular economy and sustainable development, especially SDG 2 (specifically sustainable agriculture), SDG 6 (sustainable water management), SDG 7 (clean and sustainable energy), SDG 8 (green economy), SDG 9 (especially foster innovation), SDG 12 (sustainable consumption and production), SDG 13 (prevent climate change and the impact), SDG 14 (conservation of ocean, seas, and marine), and SDG 15 (promote sustainable ecosystem, forest, land, and biodiversity).

The successfulness of the conservation and sustainable principle implementation in UNNES is supported by the work of the university's Research and Community Service Institute to the highest degree. The Institute of Research and Community Service is the leading unit in UNNES in the development of research, innovation, and community service. This unit has to play an active role in developing science and technology and enhancing societies' welfare as well as the national competitiveness in research and community service. Ultimately, LPPM is expected to contribute to resolve and overwhelm the problems in the society at the national level and the world community, improve the quality of life, and accelerate sustainable development (UNNES, 2022a). To achieve this goal, LPPM concerns on the funding facilitation for the faculty members and educational staff to conduct research and community service. The resources of the research funding are among the

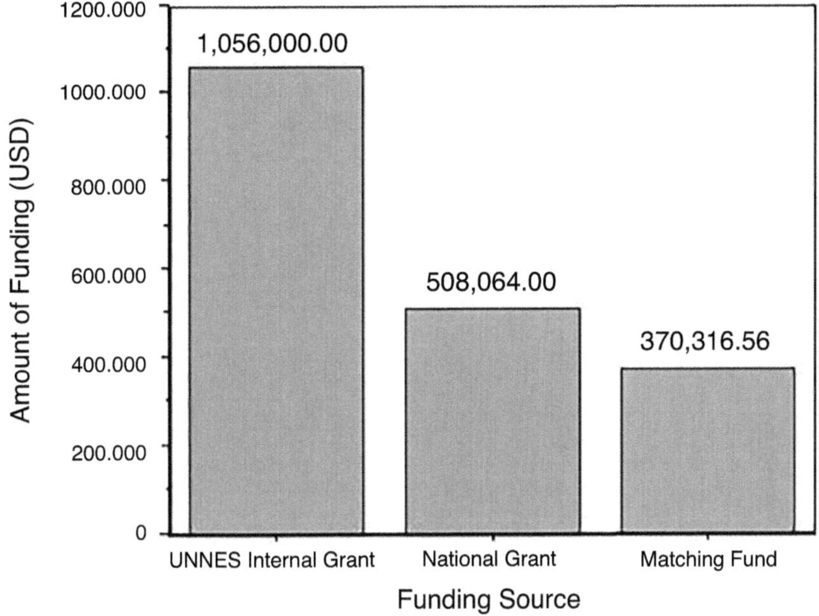

FIGURE 29.2 The amount of research funding provided in 2022 from main funding resources.

other from the university internal funding, national grant provided by the Ministry of Education, Culture, Research, and Technology for basic and applied research, matching fund grant for the research focusing on commercialization and development. The amount of each grant is exhibited in Figure 29.2. Besides these main grant resources, there are also several other grants from international institutions, Corporate Social Responsibility funding from the companies, local government, National Research And Innovation Agency, etc.

To effectively nurture the research which relevant to the conservation perspective, the sustainable research should be executed. The Technical Implementation Unit of Conservation Development has mapped the amount of the research funding provided by LPPM of UNNES which supported the research concerning sustainability during 2019–2021. The result is displayed in Figure 29.3.

The research conducted in the university generally creates innovation which will act as the key driver of economic development and has high potential for commercialization. Innovation is activities that create new or improve a product, process, or service. Innovation can be protected by means of intellectual property rights (IPR). The World Intellectual Property Organization (WIPO) defines the terminology of IPR as "all the rights generated from intellectual activity in industrial, scientific, literary or artistic areas" (Sharif et al., 2018). There are five major types of IPR, namely patents, industrial design, trademarks, copyrights, and geographical indicators which are usually used to guard intangible resources. IPR can boost technological progress by protecting innovation from piracy and illegal use, and by providing information related to the existence of innovation to the society. In terms of innovation commercialization, patent is the most important since it is simply granted for an invention that meets a significant novelty, involves invention steps, and shows a prospective industrial and commercial implementation. It can be awarded for processes or products (Nath Saha and Bhattacharya, 2011). IPR currently becomes one among the measuring instruments of the university innovation. IPR can also act as a bridge for technology transfer and commercialization, which potentially results in income generating for the university and has positive impact on the local economy (Katzman and Azziz, 2021).

Concerning the urgent role of IPR in the university research, UNNES has established a Centre for Technology Dissemination and Intellectual Property Rights as a sub-unit of LPPM since 2011. This centre supports the implementation of Tri Dharma by disseminating IPR-oriented technology

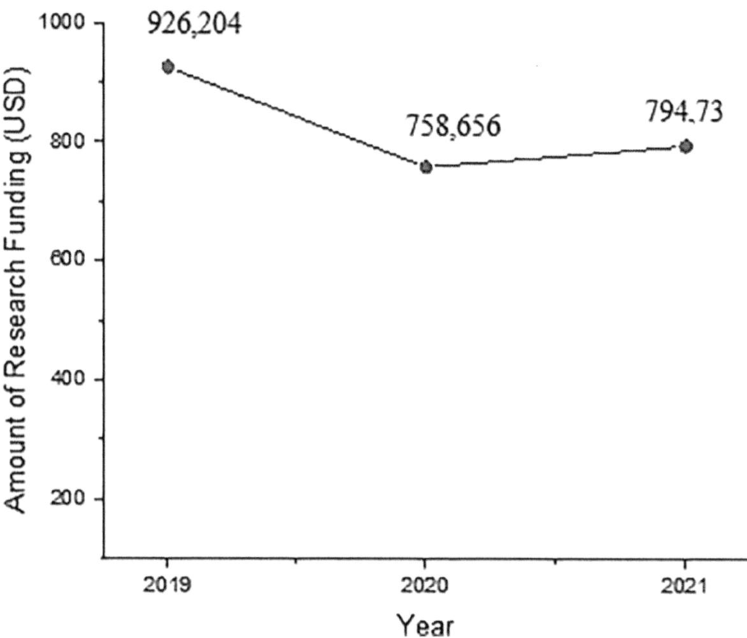

FIGURE 29.3 The amount of research funding to support the research concerning sustainability.

and facilitating IPR management, including drafting assisting, registration, examination, and obtaining the IPR certificate. This centre provides the service for the internal academic community, general public, and industry. The number of the patents obtained in UNNES since 2010 is depicted in Figure 29.4.

To acquire technological innovation and intellectual property commercialization, the Centre for Innovation and Commercialization is founded under LPPM. It has a main task to accelerate the university innovation and facilitate the downstreaming of the innovation, specifically with the purpose of income generating. This centre also assists the researcher to win the Matching Fund for product commercialization. Matching fund is a research scheme provided by the Ministry of Education, Culture, Research and Technology design for creating synergic collaboration between academicians in higher education with industries to commercialize the result of the research which also supports the achievement of eight main performance indicators of the higher education in Indonesia. Additionally, the Centre for Innovation and Commercialization is expected to ally the IPR commercialization with the sustainable business model which creates positive impacts on the society and environment. Thus, the IPR downstreaming holds sustainable benefits for the public, environment, and business (Hernández-Chea et al., 2020).

To aid the launch of the start-up in UNNES, a Business Unit Incubator is also founded as a subunit of LPPM UNNES. Business incubation is an activity of coaching, mentoring, and developing beginner entrepreneurs to survive and independently build up their businesses. Business incubation is conducted by a third party called Business Unit Incubator (BUI). BUI acts as an intermediate party that conduct activities to facilitate the problem solving, marketing, and accessing the capital for the tenant. BUI of LPPM UNNES assists to incubate and launch the start-up specifically in the areas of food products and engineering innovation for sustainable natural resources utilization. It provides service for UNNES' students, lecturers, alumni, and assisted local entrepreneurs who are pioneering start-up companies (Margunani et al., 2020). BUI affiliated with the university demonstrates a vast benefit to the entrepreneurs since it confers a network with the government, banking institutions, companies, and community (Hassan, 2020). In recent times, university

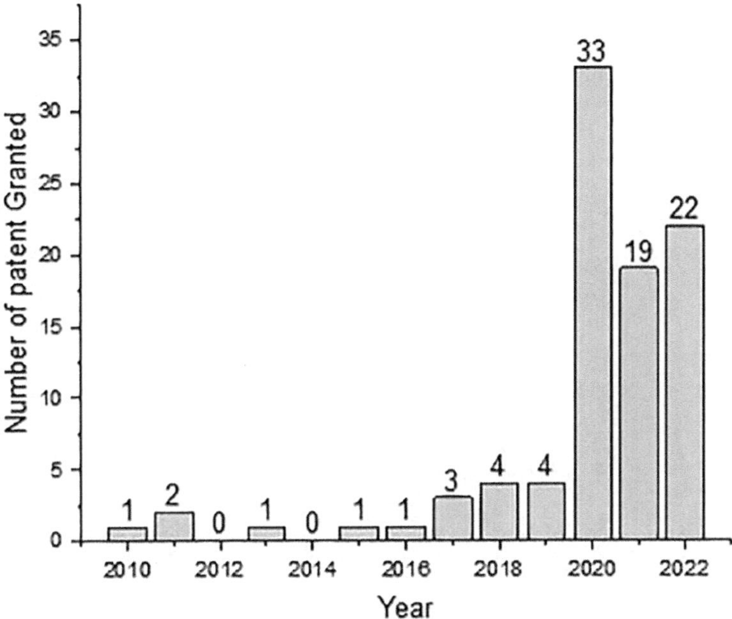

FIGURE 29.4 Number of the patents granted.

business incubators have shown an increasing function as a hub for entrepreneurship (Pellegrini and Johnson-Sheehan, 2021). The university has altered its paradigm from merely teaching activity to an economic growth driven through research, innovation, and entrepreneurship (Hassan, 2020). It therefore contributes to the success of SDG 8, i.e. promoting sustainable economic growth, higher productivity, and technological innovation.

Product commercialization and business development can simply be developed when the innovative products are recognized by society, industry, government, and other stakeholders. Therefore, to introduce the innovative products and technology produced by the faculty members, UNNES built a "House of Innovation" (Rumah Inovasi) inside the LPPM building. House of Innovation displays various sustainable products created by the researchers in UNNES in various categories such as crafts made from waste recycling, textile and fashion, functional food such as tempeh fortified by moringa leaves (National Patent IDS 000004375), bioethanol in gel form (energel, bioflame gel), water hyacinth biopellet for domestic stove fuel (National Patent IDS 000004494), equipment for clean water production from air, digital radiograph from recycling material (National Patent No. IDS 000002866), metal smelting crucibles made from evaporation boats waste (National Patent No. IDS 000004495), soft thorns milkfish machine (National Patent No. IDP 000058557), STEM learning robot, solar cell car, electric vehicles, and other innovative machineries. Examples of unique and sustainable art products are crafts made from leaves skeleton. The raw material is waste tree leaf which has been processed and leaves the skeleton. The leaves skeleton is used among the others for artistic painting media (Copyright registration No. 000137351) and for making decoupage bags (Copyright registration No. 000379163). The brand for the leaves skeleton craft has also been nationally registered with the name Rasendriya (Trademark No. IDM000842375). Another art product exposed is sustainable textile and fashion produced using the natural color of the leaves waste by applying the eco-print technique. Batik products which are colored using natural dyes are also exhibited. The products exposed in House Innovation have attracted public, company, government, schools, partner universities, and the minister of state-owned enterprises to visit and promote. It becomes an effective media to endorse the innovative products and technology in UNNES.

It is also a hub to introduce products which is developed based on conservation and sustainability principles, which in the future will open the opportunity for collaboration with the company towards commercialization. The situation inside and outside of the House of Innovation is shown in Figures 29.5 and 29.6, respectively.

FIGURE 29.5 The innovative products display inside house of innovation of UNNES.

FIGURE 29.6 The display of innovation on machinery in house of innovation of UNNES.

29.3 WASTE TO ORGANIC FERTILIZER

Based on the analysis conducted by the Technical Implementation Unit of Conservation Development, the most urgent problem is waste. The continual generation of waste does not equalize with the waste processing. Hence, waste accumulation occurs and results in various negative impacts. Therefore, it is essential to handle and manage the waste properly. Waste minimalization in UNNES is carried out through the Independent Green Campus Milestone program initiated by the Technical Implementation Unit of Conservation Development. This program primarily focuses on the waste minimalization and clean and renewable energy utilization. Waste minimalization is executed by converting waste into valuable and economical products. This program has led UNNES to rank 6th in the UI Green Metric Ranking at the national level and 42nd world rank. UI green metric is a ranking on green campus and environmental sustainability pioneered by Universitas Indonesia since 2010 (Suwartha and Berawi, 2019). The zero waste programs in UNNES are in line with the current trend to shift the linear economy to the circular economy, as well.

There are many works related to the valorization of waste and among the popular ones is converting waste into organic fertilizer. The work involving the conversion of organic waste from the domestic, Small and Medium Enterprises (SME) production, and large industry processing activities into eco-friendly fertilizers are reported by several UNNES researchers. Sulistyaningsih et al. (2022) transferred technology to society related to the method of making eco-enzymes from household organic waste. New and fresh domestic waste was mixed with sugar and water. The mixtures were then kept for 90 days to produce eco-enzyme. Eco-enzyme can function as organic fertilizer, biopesticides, and cleaning agent. On the other hand, old waste can be converted into solid fertilizer or compost. Kusumaningtyas et al. (2020) treated the liquid waste of traditional tofu industry, which had bad odor, high COD and BOD, and pollute the rivers, into organic fertilizer by using a fermentation process assisted by EM4. Organic fertilizer yielded was used to cultivate plants in the alley garden.

It is also possible to utilize industrial waste as raw material for fertilizer production. Sugarcane-based ethanol industry is among the dominant producers of liquid waste. The liquid waste is vinasse, which is abundantly produced and very harmful to the water source, rivers, and land. However, vinasse contains high nutrients which make it prospective to be used for fertilizer. Vinasse can be treated to produce several types of fertilizer. The classical form of vinasse-based fertilizer is liquid fertilizer. Meanwhile, Qudus et al. (2021) introduced a novel method of producing slow release fertilizer (SRF) from vinasse. SRF is superior to the liquid fertilizer. Liquid fertilizer releases nutrients very rapidly and excessively. Thus, it requires frequent reapplication on the plants. On the other hand, SRF discharges nutrients slowly and balances with the plant's need. It accordingly can long last, limit the fertilizer residue in the land, and show a sustainable characteristic. Valorization of waste into fertilizer significantly contributes to reaching SDG 2 sustainable agriculture, SDG 3 good health and well-being, and SDG 12 responsible consumption and production.

29.4 CASE STUDY: VALORIZATION OF WASTE COOKING OIL TO VALUABLE PRODUCTS

Frying and deep frying are among the major cooking methods in Indonesia. It is particularly due to the fact that Indonesia is the world's biggest producer of sustainable palm oil producer, which is the most productive and efficient oil crop (Nomanbhay et al., 2017; Murphy et al., 2021; Limaho et al., 2022; PASPI-Monitor, 2022). There was a huge consumption of cooking oil utilization in culinary sectors, including the usage at domestic, restaurant, and industrial levels. The Ministry of Energy and Mineral Resources of the Republic of Indonesia reported that the national use of palm oil cooking oil was 16.2 million kL in 2019 (Perdana, 2021). Cooking oil can be reused but there is a limitation since it will deteriorate and worsen due to the cooking process (Warsiki et al., 2020). Repetitive heating of cooking oil will also produce polycyclic aromatic hydrocarbons

(PAH), which are categorized as carcinogenic compounds and cancer-contributing factors to human beings (Ganesan et al., 2019). Thus, the massive use of cooking oil will generate a large amount of waste cooking oil (WCO) after the cooking process. It was stated that the WCO resulting from the cooking activity was about 40%–60% of the cooking oil or equal to 6.46 – 9.97 million kL (Perdana, 2021). WCO is regarded as a precarious waste since inappropriate dumping of WCO brings about an environmental problem. However, it is apprised that the common disposal of WCO is directly thrown out to the drains (66%), to the land (8.6%), and to the trash 15.8% (Hartini et al., 2020). The direct disposal of WCO to the water drainage can pollute the surface water, groundwater, drinking source water, and sea water and trigger undesired physical process and biological process as well as unforeseen chemical reactions. It will also be harmful to the aquatic life (Foo et al., 2021). Discarding untreated WCO in the landfill also causes a negative impact. Thode Filho et al. (2017) reported that WCO was hazardous to the earthworms. On the contrary, the presence of earthworms in soil is essential to decompose the dead organic matter. Placing WCO over the soil will also deteriorate the sprouting and growth of the plant. It potentially altered the plant morphology, as well. Hence, improper disposal of WCO is not sustainable and risky to the environment.

The unwell managed dumping of WCO is disadvantageous not only from the environmental point of view but also from the economic aspect. A good management and collection of WCO will extend the produces' life cycle and promote its economic potential. WCO is a low cost promising raw material for manufacturing valuable products such as biofuel, soap, lubricants, biopolymers, and other bio-based chemicals (Hanisah et al., 2013; Tsai, 2019; De Feo et al., 2020; Foo et al., 2021). Moreover, WCO is abundantly available. Therefore, transforming WCO into a value-added product will provide benefits towards circular economy development and future sustainability (Perdana, 2021; Hidalgo-Crespo et al., 2022). It nurtures to accomplish the SDGs, specifically SDGs 12 (responsible consumption and production) by recycling waste, SDGs 7 (affordable and clean energy) by producing low-cost renewable energy, SDGs 3 (good health and well-being) by producing hand soap for human-hygiene, SDGs 6 (clean water and sanitation) by reducing wastewater, SDGs 13 (climate action), and SDGs 4 (life below water) by water pollutant.

Realizing this situation, Universitas Negeri Semarang greatly supports the technology development and dissemination of WCO conversion to value-added products through the funding and facilities for research and community service activities. Some projects concerning WCO valorization were the utilization of WCO as feedstock for producing biodiesel, household cooking fuel, and hand soap, just to name a few. The technology prototype and product results are disseminated to society and encouraged for commercialization to bring advantages at a wider scale.

29.4.1 Design and Construction of Reactor Prototype for WCO Biodiesel Production

Biodiesel is a promising renewable and sustainable energy and it has been broadly applied as a substitute for diesel machine fuel. Currently, the mixture of biodiesel-petro diesel for fuel has been implemented in many countries such as the United States (B20), Indonesia (B30), Malaysia (B15), and Thailand (B10) (Kusumaningtyas et al., 2022a). Biodiesel is generally yielded via the transesterification reaction of vegetable oil with short-chain alcohol over an alkaline catalyst such as KOH (Perea et al., 2016). For the high acidic vegetable oil feedstock, an esterification reaction pre-treatment to reduce the free fatty acid (FFA) content should be conducted to avoid the undesired saponification reaction during the main transesterification process (Kusumaningtyas et al., 2017). Various vegetable oils have been studied for the raw material of biodiesel production such as rapeseed oil, soybean oil, palm and coconut oil, jatropha curcas oil, and waste cooking oil (Kusumaningtyas et al., 2022a). WCO commonly has FFA content above the limit, thus, the esterification process as a pre-treatment step is necessary (Kusumaningtyas et al., 2018).

As a university with high concern for conservation and sustainability, UNNES supports the research on biodiesel production and process intensification. The support includes the research funding and intellectual properties application facility. Various innovations have been resulted and granted national patents such as reactive distillation for biodiesel production (P00201507855), biodiesel production using multi-capacity reactors (S00202100468), and double reactors system for biodiesel production (S00202106437). In this work, an improvement of the previous double-system reactor was designed and constructed to produce biodiesel from WCO. The production apparatus was equipped with a filter to clarify the WCO before entering the reactor as presented in Figure 29.7. The technical scheme of the apparatus is revealed in Figure 29.8.

The apparatus was tested to produce biodiesel via a two-step process, i.e. esterification pre-treatment and transesterification reactions. Biodiesel production process can be described as follows. Initially, WCO raw material was inputted into the filter to remove the solid impurities. Filter comprised three sieves which have different sizes, i.e. large, medium, and small. After the screening process, 1,200 mL WCO was introduced into the first reactor for undergoing esterification reaction as a pre-treatment to decrease the FFA content. Methanol was mixed with the sulphuric acid catalyst and was introduced to the reactor at the same time. The esterification process was carried out using methanol with a molar ratio of 1:6 and sulfuric acid catalyst with a concentration of 0.5% at a constant temperature of 60°C and stirring speed of 800 rpm for 60 minutes. The reaction products were ester and water. Water was then separated from the oil.

WCO and ester layer entered the transesterification reactor. Methanol was mixed with the KOH catalyst to form a methoxide solution in the methoxide tank. The methoxide solution was then flowed to the reactor. WCO subsequently underwent the main transesterification reaction to produce

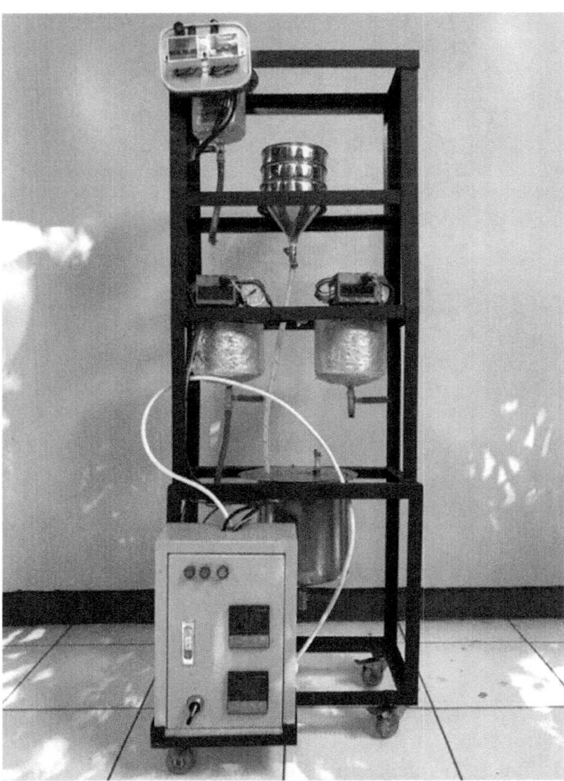

FIGURE 29.7 Double reactors system equipped with filter for WCO biodiesel production.

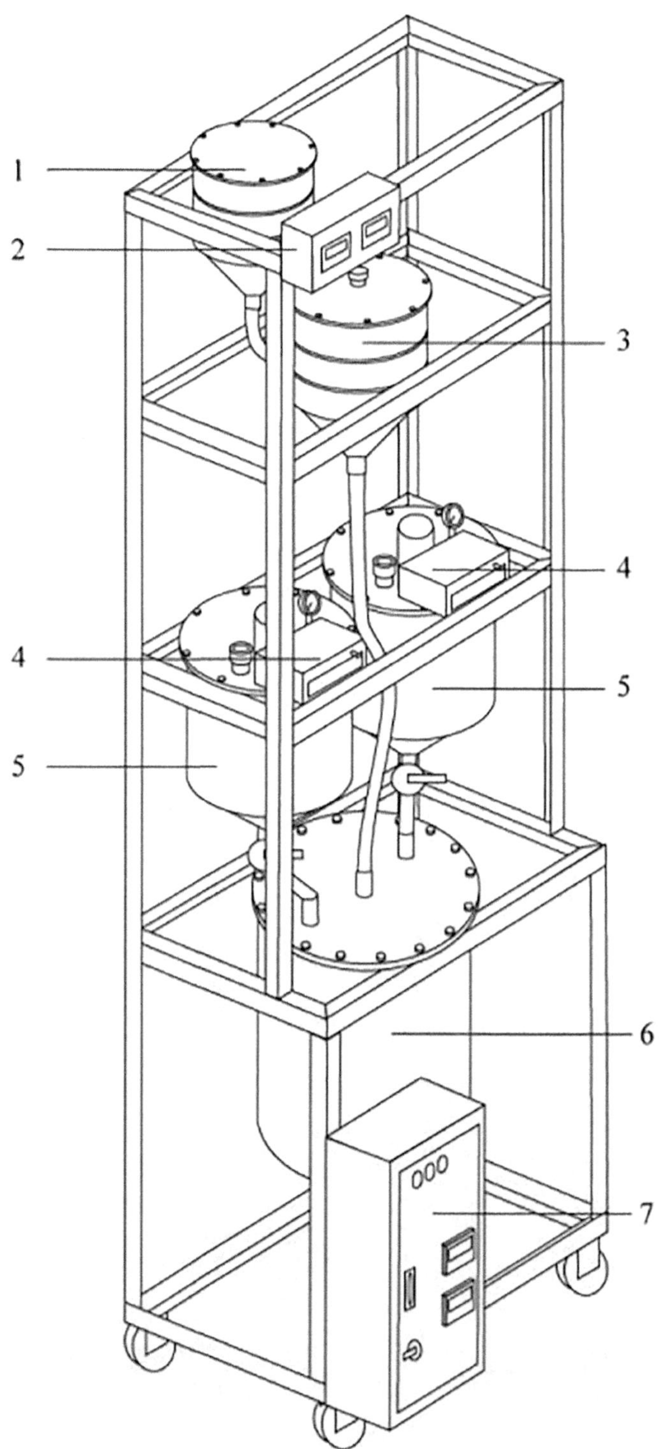

FIGURE 29.8 Technical scheme of the double reactors system equipped with filter for WCO biodiesel production: (1) Methoxide Tank, (2) Temperature Controller, (3) WCO Feedstock Filter, (4) Mixing Speed Controller, (5) Reactors, (6) Product Tank, (7) Electrical Control Panel.

biodiesel. Transesterification was conducted using methanol with a molar ratio of 1:6 and 1% KOH catalyst at 60°C. The reaction ran for 60 minutes. The reaction produced biodiesel and glycerol as by-product. Glycerol was separated by washing with water which was repeated three times. The washing was followed by decantation for 8–24 hours until it reached equilibrium. The water and glycerol layer was removed, and the biodiesel product was collected. The water residual water contained in the biodiesel product was removed by heating. This process can produce high-purity biodiesel with a yield of 80% v/v (960 mL biodiesel from 1,200 mL WCO feedstock). The product of WCO biodiesel yielded in this work and the commercial petroleum diesel "Dexlite" from PT Pertamina Indonesia as a comparison are exhibited in Figure 29.9.

The WCO biodiesel was planned to be produced at a large scale in the future in order to fulfil the low-cost and renewable energy necessity for society. Thus, the biodiesel production apparatus needs to be scaled up. Besides, WCO biodiesel should be examined to determine its performance as a fuel in the diesel machine. Figures 29.10, 29.11, and 29.12 show the testing car, the dynamometer and the computer for analysis, and the biodiesel fuel performance testing process. This evaluation is essential for the recommendation of future production at a large scale. UNNES policy is driving

FIGURE 29.9 WCO biodiesel and the commercial petroleum diesel "Dexlite".

FIGURE 29.10 The testing car for WCO biodiesel fuel performance evaluation.

FIGURE 29.11 Dynamometer and the computer for analysis.

FIGURE 29.12 Testing process of the WCO biodiesel's performance on the diesel machine.

the downstream of the research result for product commercialization, fulfilling the society demand, promoting the community welfare, and enhancing sustainable development. This research strengthens the achievement of SDGs 7 (affordable and clean energy). It also supports SDGs 6 (clean water and sanitation) by reducing wastewater and SDGs 12 (responsible consumption and production).

29.4.2 WCO for Household Cooking Fuel

Liquefied Petroleum Gas (LPG) is the main fuel for stove and household cooking in Indonesia. It has superior combustion performance. However, it holds some drawbacks such as the possibility of gas leakage which leads to the fire, expensive, and non-renewable. As an alternative, in this work, WCO was prepared and utilized as fuel for the stove or household cooking. WCO stove is simple, portable, and similar to the traditional kerosene stove (Susilawati and Buchori, 2020). The preparation of WCO was easy. It just needed a filtering process to eliminate the solid particle content, then it was ready to use in the stove. The problem associated with the application of the WCO fuel was

FIGURE 29.13 WCO as alternative fuel for household cooking.

the delay on the ignition. It was accelerated by adding a small amount of alcohol at the beginning. Subsequently, the flame can be produced on the stove and the combustion was stable. The project on the application of WCO for household cooking fuel was funded by UNNES and disseminated to the society through the community empowerment project. The society used this technology for home cooking and for small-scale culinary business. Thus, the WCO for stove fuel provided contribution in enhancing the waste utilization, providing low-cost energy source for domestic cooking, and promote circular economy in the society. Bio-based fuel is also low carbon compared to the fossil-based fuel. Implementation of WCO for cooking using simple stove in the society is shown in Figure 29.13. This program promotes the realization of SDGs 7, i.e. affordable and clean energy, and SDG 12 (ensure sustainable consumption and production pattern).

29.4.3 Conversion of Waste Cooking Oil to Liquid Hand Soap

Soap is made via saponification reaction of vegetable oil. Saponification is the process in which triglycerides contained in vegetable oil are reacted with strong alkaline such as KOH and NaOH to result in fatty acid metal salt (soap), glycerol, and free fatty acid. The type of alkaline affects the form of the soap produced. Utilization of NaOH or lye will result in a solid and harder bar soap. On the other hand, the presence of KOH or potash will make a transparent liquid soap (Prieto Vidal et al., 2018; Rahayu et al., 2021). Soap can be made from various vegetable oil as feedstock.

WCO is a cheap, abundant, and continuously available, and renewable raw material for soap making (Li et al., 2020; Mustakim et al., 2020). WCO derived from palm oil has advantage since it contains shorter chain length saturated fatty acid (C8:0 – C14:0) which will improve the soap solubility in water and lead to a better foaming characteristic of the soap (Prieto Vidal et al., 2018). The soap manufacturing process is an eco-friendly and zero waste process. It yields no undesired side product and consumes low energy (Antonić et al., 2020).

The project on the liquid hand soap making from WCO was funded by UNNES via a community empowerment program. The process of soap making was simple and strategic to be applied by the society to nurture the hygiene, especially in the pandemic era. The process consisted of three main processes, namely purification, soap-base making, and liquid hand soap making. Purification step include filtration to separate solid impurities, despicing process to remove spice, neutralization, and decolorization. Despicing was carried out by mixing WCO with water and heating them. The spice was then separated from the solution and filtered. Netralization was performed using 15% KOH solution at 40°C... The decolorization process was carried out using active carbon. Active carbon for separated from the WCO after the decolorization process by using filtration. Subsequently, the soap base making was conducted through the saponification process of the purified WCO using 36% KOH. The reaction was performed at 70°C for 45 minutes in a stirred tank to produce the soft soap base. The last step was the liquid hand soap making. The soap base was mixed with water with a volume ratio of 1:1. The mixture was then heated at 60°C for 1 hour to completely dissolve the soap base in the water. Citric acid solution was then added to neutralize the solution as well as serve as a preservative. Additives such as coloring matter and perfume were also added. Glycerine for also introduced into the liquid hand soap product to act as hand moisturizer (Kusumaningtyas et al., 2022b). The product was then packed as seen in Figure 29.14.

The homemade soap business is promising to date. Since the emergence of the coronavirus disease 2019 pandemic, the public habit of washing hands using soap has dramatically increased. As a consequence, it has drastically enhanced the demand for the hand soap (Chirani et al., 2021). The liquid hand soap is more practical in use than the bar type. Choi et al. (2021) reported that since the occurrence of Covid-19, the number of people who used bar soap reduced from 71.8% to 51.4%. Conversely, the number of consumers who used liquid hand soap rose from 23.5% to 41%. The consumers also showed a higher perception of the liquid hand soap in preventing COVID-19 transmission. It denoted that the market opportunity on liquid hand soap is prospective. The community empowerment project on liquid hand soap making presented significant impacts in enhancing the

FIGURE 29.14 Hand soap from WCO.

zero waste process, developing circular economy of WCO, increasing public hygiene, and initiating a golden opportunity of small business related to the homemade soap. This project contributes towards the achievement of SDGs 3, namely good health and well-being and SDG 12 (ensure sustainable consumption and production pattern).

29.5 CONCLUSION

University holds promising contributions to the circular economy transition and sustainable development realization through research, innovation, and community service activities, in particular, the work concerning waste management and recycle. In Indonesia, this function is strengthened by the government regulation related to the Tri Dharma as the main duties of higher education, i.e. teaching and education, research, and community service. Several factors that will enhance the university's role in fostering circular economy and sustainability are among others the university policy support, good institutional management of research and community service, adequate funding, IPRs for the products and technologies created, and excellent dissemination and commercialization of the research output. Universitas Negeri Semarang as the university with high concern on conservation has applied the best practice in Universitas Negeri Semarang through the synergic work under the coordination of the Research and Community Service Institute and the Technical Implementation Unit of Conservation Development. The existence of House of Innovation also promotes the sustainable products and technology created by the researchers to the public, company, and government. Thus, it enlarges the opportunity for product commercialization and downstreaming with regard to the circular economy implementation.

ACKNOWLEDGEMENT

The authors thank the Head of the Research and Community Service of UNNES (Prof. Dr. H. R. Benny Riyanto, S.H., M. Hum., CN and the staffs, the Management and Staffs of Technical Implementation Unit of Conservation Development (especially Elli Dwi Astuti, S.Si.) for providing the supporting information, and Sciment for the product photography and graphical illustration.

REFERENCES

Amin, R., Utomo, A.P.Y., Fathoni, K., Prihanto, T., Ekiyardi, N.Y.P., Astuti, E.D., Rahmanudin, Therawati, C.A., Abdimmuniib, A., Pujiyono. (2022). Konsevasi Berkelanjutan Kampus UNNES 2022. Semarang: LPPM UNNES.

Antonić, B., Dordević, D., Jančíková, S., Tremlova, B., Kushkevych, I. (2020). Physicochemical characterization of home-made soap from waste-used frying oils. *Processes*. 8(10):1219. doi:10.3390/pr8101219

Chirani, M., Kowsari, E., Teymourian, T., Ramakrishna, S. (2021). Since January 2020 Elsevier has created a COVID-19 resource centre with free information in English and Mandarin on the novel coronavirus COVID-19. The COVID-19 resource centre is hosted on Elsevier Connect, the company's public news and information. *Sci Total Environ*. 796:149013.

Choi, K.O., Sim, S., Choi, J., Park, C., Uhm, Y., Lim, E., Kim, A.Y., Yoo, S.J., Lee, Y.J. (2021). Changes in handwashing and hygiene product usage patterns in Korea before and after the outbreak of COVID-19. *Environ Sci Eur*. 33:79. doi:10.1186/s12302-021-00517-8

De Feo, G., Di Domenico, A., Ferrara, C., Abate, S., Osseo, L.S. (2020). Evolution of waste cooking oil collection in an area with long-standing waste management problems. *Sustainability*. 12(20):1–16. doi:10.3390/su12208578

Foo, W.H., Chia, W.Y., Tang, D.Y.Y., Koay, S.S.N., Lim, S.S., Chew, K.W. (2021). The conundrum of waste cooking oil: Transforming hazard into energy. *J Hazard Mater*. 417(March):126129. doi:10.1016/j.jhazmat.2021.126129

Ganesan, K., Sukalingam, K., Xu, B. (2019). Impact of consumption of repeatedly heated cooking oils on the incidence of various cancers – A critical review. *Crit Rev Food Sci Nutr*. 59(3):488–505. doi:10.1080/10408398.2017.1379470

Hagelüken, C., Goldmann, D. (2022). Recycling and circular economy-towards a closed loop for metals in emerging clean technologies. *Miner Econ.* 35(3-4):539–562. doi:10.1007/s13563-022-00319-1

Hanisah, K., Kumar, S., Tajul, A. (2013). The management of waste cooking oil: a preliminary survey. *Heal Environ J.* 4(1):76–81.

Hartini, S., Puspitasari, D., Roudhatul, Aisy, N., Widharto, Y. (2020). Eco-efficiency level of production process of waste cooking oil to be biodiesel with life cycle assessment. *E3S Web Conf.* 202:10004. DOI: 10.1051/e3sconf/202020210004

Hassan, N.A. (2020). University business incubators as a tool for accelerating entrepreneurship: theoretical perspective. Rev Econ Polit Sci (ahead-of-print). doi:10.1108/REPS-10-2019-0142

Hernández-Chea, R., Vimalnath, P., Bocken, N., Tietze, F., Eppinger, E. (2020). Integrating intellectual property and sustainable business models: *The SBM-IP canvas. Sustainability.* 12(21):8871. doi:10.3390/su12218871

Hidalgo-Crespo, J., Alvarez-Mendoza, C.I., Soto, M., Amaya-Rivas, J.L. (2022). Towards a Circular economy development for household used cooking oil in Guayaquil: Quantification, Characterization, modeling, and geographical mapping. *Sustainability.* 14(15):9565. doi:10.3390/su14159565

Kara, S., Hauschild, M., Sutherland, J., McAloone, T. (2022). Closed-loop systems to circular economy: A pathway to environmental sustainability? *CIRP Ann.* 71(2):505–528. doi:10.1016/j.cirp.2022.05.008

Katzman, R.S., Azziz, R. (2021). Technology transfer and commercialization as a source for new revenue generation for higher education institutions and for local economies Richard. In: AI-Youbi, A.O., Zahed, A.H.M., Atalar, A. (Eds.) *International Experience in Developing the Financial Resources of Universities*. Springer International Publishing, pp. 89–111. doi:10.1007/978-3-030-78893-3.

Khajuria, A., Atienza, V.A., Chavanich, S., Henning, W., Islam, I., Kral, U., Liu, M., Liu, X., Murthy, I.K., Oyedotun, T.D.T., et al. (2022). Accelerating circular economy solutions to achieve the 2030 agenda for sustainable development goals. *Circ Econ.* 1(1):100001. doi:10.1016/j.cec.2022.100001

Knäble, D., de Quevedo Puente, E., Pérez-Cornejo, C., Baumgärtler, T. (2022). The impact of the circular economy on sustainable development: A European panel data approach. *Sustain Prod Consum.* 34:233–243. doi:10.1016/j.spc.2022.09.016

Kusumaningtyas, R., Putut, M., Cahya, W., Qoni, A., Muhammad, A., Waliyuddin, S. (2020). Utilization of tofu industrial liquid waste as organic fertilizer to support the alley garden project development. *Abdimas.* 24(3):200–204. doi:10.15294/abdimas.v24i3.21859

Kusumaningtyas, R.D., Kusuma, A.D.H., Budiono, Y.W.P., Prasetiawan, H., Nanggala, P.L.A., Istadi, I. (2022). Biodiesel production from used cooking oil using integrated double column reactive distillation: simulation study. *J Adv Res Fluid Mech Therm Sci.* 94(1):152–162. doi:10.37934/arfmts.94.1.152162

Kusumaningtyas, R.D., Prasetiawan, H., Pratama, B.R.B.R., Prasetya, D., Hisyam, A. (2018). Esterification of non-edible oil mixture in reactive distillation column over solid acid catalyst: Experimental and simulation study. *J Phys Sci.* 29(Suppl 2):212–226. doi:10.21315/jps2018.29.s2.17

Kusumaningtyas, R.D., Purnamasari, I., Mahmudati, R., Prasetiawan, H. (2022a). Interesterification reaction of vegetable oil and alkyl acetate as alternative route for glycerol-free biodiesel synthesis. In: Gurunathan, B., Sahadevan, R. (Eds.) *Biofuels and Bioenergy*. Elsevier, pp. 435–452. doi:10.1016/B978-0-323-90040-9.00020-5

Kusumaningtyas, R.D., Ratrianti, N., Purnamasari, I., Budiman, A. (2017). Kinetics study of Jatropha oil esterification with ethanol in the presence of tin (II) chloride catalyst for biodiesel production. *AIP Conf Proc.* 1788(January):030086. doi:10.1063/1.4968339

Kusumaningtyas, R.D., Widjanarko, D., Cahyati, W.H., Wulansarie, R., Maksiola, M., Meysanti, D., Salsabilla, M.T., Nugraha, D.D., Najuda, M.D., Rachmadi, M.F. (2022b). Pengolahan Limbah Minyak Jelantah Menjadi Sabun Cuci Tangan sebagai Upaya Konservasi Lingkungan dan Pencegahan Penularan Virus Covid-19. *Abdimas.* 26(2):110–121.

Li, W., Guan, R., Yuan, X., Wang, H., Zheng, S., Liu, L., Chen, X. (2020). Product soap from waste cooking oil. *IOP Conf Ser Earth Environ Sci.* 510:042038. doi:10.1088/1755-1315/510/4/042038

Limaho, H., Sugiarto, Pramono, R., Christiawan, R. (2022). The need for global green marketing for the palm oil industry in Indonesia. *Sustainability.* 14(14):8621. doi:10.3390/su14148621

Margunani, M., Ardiansari, A., Mutiatari, D.P. (2020). *Pedoman Umum Inkubator Bisnis Lembaga Penelitian dan Pengabdian Kepada Masyarakat Universitas Negeri Semarang*. Pramono, S.E. (Ed.). Semarang: LPPM UNNES.

Mormina, M. (2019). Science, technology and innovation as social goods for development: rethinking research capacity building from Sen's capabilities approach. *Sci Eng Ethics.* 25:671–692. doi:10.1007/s11948-018-0037-1

Morseletto, P. (2020). Targets for a circular economy. *Resour Conserv Recycl.* 153:104553. doi:10.1016/j.resconrec.2019.104553
Morton, S., Pencheon, D., Squires, N. (2017). Sustainable Development Goals (SDGs), and their implementation. *Br Med Bull.* 124(1):81–90. doi:10.1093/bmb/ldx031
Murphy, D.J., Goggin, K., Paterson, R.R.M. (2021). Oil palm in the 2020s and beyond: challenges and solutions. *CABI Agric Biosci.* 2:39. doi:10.1186/s43170-021-00058-3
Mustakim, M., Taufik, R., Trismawati, T. (2020). The utilization of waste cooking oil as a material of soap. *J Dev Res.* 4(2):86–91. doi:10.28926/jdr.v4i2.114
Nath Saha, C., Bhattacharya, S. (2011). Intellectual property rights: An overview and implications in pharmaceutical industry. *J Adv Pharm Technol Res.* 2(2):88–93. doi:10.4103/2231-4040.82952.
Nomanbhay, S., Salman, B., Hussain, R., Ong, M.Y. (2017). Microwave pyrolysis of lignocellulosic biomass – a contribution to power Africa. *Energy Sustain Soc.* 7:23. doi:10.1186/s13705-017-0126-z
Nur, T.F., Widodo, A., Mutiara, R. (2018). State university with legal entity and the impact to the rights and obligations of the tax transaction: case of the University of Indonesia. *KnE Soc Sci.* 3(11):1388–1398. doi:10.18502/kss.v3i11.2857
PASPI-Monitor. (2022). Indonesia is the largest producer of certified sustainable palm oil in the world. *Palm Oil J.* II:467–472.
Pellegrini, M., Johnson-Sheehan, R. (2021). The evolution of university business incubators: transnational hubs for entrepreneurship. *J Bus Tech Commun.* 35(2):185–218. doi:10.1177/1050651920979983
Perdana, B.E.G. (2021). Circular economy of used cooking oil in Indonesia: current practices and development in special region of Yogyakarta. *J World Trade Stud.* 6(1):28–39.
Perea, A., Kelly, T., Hangun-Balkir, Y. (2016). Utilization of waste seashells and Camelina sativa oil for biodiesel synthesis. *Green Chem Lett Rev.* 9(1):27–32. doi:10.1080/17518253.2016.1142004
Prieto Vidal, N., Adeseun Adigun, O., Huong Pham, T., Mumtaz, A., Manful, C., Callahan, G., Stewart, P., Keough, D., Horatio Thomas, R. (2018). The effects of cold saponification on the unsaponified fatty acid composition and sensory perception of commercial natural herbal soaps. *Molecules.* 23(9):2356. doi:10.3390/molecules23092356
Qudus, N., Kusumaningtyas, R.D., Syamrizal, Z., Hartanto, D., Zakaria, Z.A. (2021). Jurnal Bahan Alam Terbarukan Vinasse-based slow-release organo-mineral fertilizer with chitosan-bentonite. *J Bahan Alam Terbarukan.* 10(1):1–8.
Rahayu, S., Pambudi, K.A., Afifah, A., Fitriani, S.R., Tasyari, S., Zaki, M., Djamahar, R. (2021). Environmentally safe technology with the conversion of used cooking oil into soap. *J Phys Conf Ser.* 1869:012044. doi: 10.1088/1742-6596/1869/1/012044
Schroeder, P., Anggraeni, K., Weber, U. (2019). The relevance of circular economy practices to the sustainable development goals. *J Ind Ecol.* 23(1):77–95. doi:10.1111/jiec.12732
Sharif, S.M., Ahamat, A., Abdullah, M.M., Jabar, J., Bakri, M.H. (2018). University intellectual property commercialization: a critical review of literature. *Turkish Online J Des Art Commun.* 8(4):874–886. doi:10.7456/1080sse/124.
Siregar, Z., Lumbanraja, P., Salim, S.R.A. (2016). The implementation of Indonesia's three principles of higher education standard towards increasing competitiveness of local universities for ASEAN economic community. *Pertanika J Soc Sci Humanit.* 24(S):1–12.
Sulistyaningsih, T., Susanti, R., Rosanti, Y.M., Mulyani, G. (2022). Community empowerment in processing household organic waste through eco enzymes. *Glob Community Serv.* 1(1):1–5.
Susilawati, Zamzami, R., Buchori, A.S. (2020). The utilization of waste cooking oil (wco) in simple stove as an alternative fuel for household scale. *J Phys Conf Ser.* 1700:012052. doi:10.1088/1742-6596/1700/1/012052
Suwartha, N., Berawi, M.A. (2019). The role of Ui greenmetric as a global sustainable ranking for higher education institutions. *Int J Technol.* 10(5):862–865. doi:10.14716/ijtech.v10i5.3670
Syamsi, N., Heriyanti, H. (2022). Implementation of community service-based Indonesian learning at UINSI Samarinda. *Southeast Asian J Islam Educ.* 4(2):157–170. doi:10.21093/sajie.v4i2.4226
Thode Filho, S., de Paiva, J.L., Franco, H.A., Perez, D.V., Marques, M.R. da C. (2017). Environmental impacts caused by residual vegetable oil in the soil-plant system. *Ciência e Nat.* 39(3):748. doi:10.5902/2179460x27645
Triatmoko, H., Kurniasih, L., Muhtar, Bandi. (2018). Financial management of government (state) universities in Indonesia. *Rev Integr Bus Econ Res.* 7(Supp. 2):253–260.
Tsai, W.T. (2019). Mandatory recycling of waste cooking oil from residential and commercial sectors in Taiwan. *Resources.* 8(1):38. doi:10.3390/resources8010038
UNNES. (2022a). *Profil LPPM.* https://lppm.unnes.ac.id/profil-lppm [accessed 2022 Dec 20].

UNNES IC. (2022b). *Vision Mission*. https://webu.unnes.ac.id/about-unnes/vision-mission/ [accessed December 20, 2022].

Valverde, J.M., Avilés-Palacios, C. (2021). Circular economy as a catalyst for progress towards the sustainable development goals: a positive relationship between two self-sufficient variables. *Sustainability.* 13(22):12652. doi:10.3390/su132212652

Velenturf, A.P.M., Purnell, P. (2021). Principles for a sustainable circular economy. *Sustain Prod Consum.* 27:1437–1457. doi:10.1016/j.spc.2021.02.018

Warsiki, E., Iskandar, A., Hidayati, M. (2020). Degradation quality of reused palm cooking oil during storage: Case study in fried shallot industry. *IOP Conf Ser Earth Environ Sci.* 460:012010. doi:10.1088/1755-1315/460/1/012010

Index

E

F

G

H

I

K

L

M

N

O

P

R

S

T

U

V

W

X

Z

For Product Safety Concerns and Information please contact our EU representative GPSR@taylorandfrancis.com Taylor & Francis Verlag GmbH, Kaufingerstraße 24, 80331 München, Germany

Batch number: 10392095

Printed by Printforce, the Netherlands